Pulse Width Modulation For Power Converters

IEEE Press
445 Hoes Lane
Piscataway, NJ 08854

Books of Related Interest from the IEEE Press

Electric Power Systems: Analysis and Control
Fabio Saccomanno
2003 Hardcover 728pp 0-471-23439-7

Power System Protection
P. M. Anderson
1999 Hardcover 1344pp 0-7803-3472-2

Understanding Power Quality Problems: Voltage Sags and Interruptions
Math H. J. Bollen
2000 Hardcover 576pp 0-7803-4713-7

Electric Power Applications of Fuzzy Systems
Edited by M. E. El-Hawary
1998 Hardcover 384pp 0-7803-1197-3

Principles of Electric Machines with Power Electronic Applications, Second Edition
M. E. El-Hawary
2002 Hardcover 496pp 0-471-20812-4

Analysis of Electric Machinery and Drive Systems, Second Edition
Paul C. Krause, Oleg Wasynczuk, and Scott D. Sudhoff
2002 Hardcover 624pp 0-471-14326-X

Pulse Width Modulation For Power Converters

Principles and Practice

D. Grahame Holmes
Monash University
Melbourne, Australia

Thomas A. Lipo
University of Wisconsin
Madison, Wisconsin

IEEE PRESS SERIES ON POWER ENGINEERING

IEEE Series on Power Engineering,
Mohamed E. El-Hawary, *Series Editor*

IEEE

IEEE PRESS

WILEY-INTERSCIENCE

A JOHN WILEY & SONS, INC., PUBLICATION

Library of Congress Cataloging-in-Publication Data is available.

Printed in the United States of America.

ISBN 0-471-20814-0

10 9 8 7 6 5 4 3

Contents

Preface xiii

Acknowledgments xiv

Nomenclature xv

Chapter 1 Introduction to Power Electronic Converters 1

1.1 Basic Converter Topologies 2

1.1.1 Switch Constraints 2

1.1.2 Bidirectional Chopper 4

1.1.3 Single-Phase Full-Bridge (H–Bridge) Inverter 5

1.2 Voltage Source/Stiff Inverters 7

1.2.1 Two-Phase Inverter Structure 7

1.2.2 Three-Phase Inverter Structure 8

1.2.3 Voltage and Current Waveforms in Square-Wave Mode .. 9

1.3 Switching Function Representation of Three-Phase Converters 14

1.4 Output Voltage Control 17

1.4.1 Volts/Hertz Criterion 17

1.4.2 Phase Shift Modulation for Single-Phase Inverter 17

1.4.3 Voltage Control with a Double Bridge 19

1.5 Current Source/Stiff Inverters 21

1.6 Concept of a Space Vector 24

1.6.1 *d–q–*0 Components for Three-Phase Sine Wave Source/Load 27

1.6.2 *d–q–*0 Components for Voltage Source Inverter Operated in Square-Wave Mode 30

1.6.3 Synchronously Rotating Reference Frame 35

1.7 Three-Level Inverters 38

1.8 Multilevel Inverter Topologies 42

1.8.1 Diode-Clamped Multilevel Inverter 42

1.8.2 Capacitor-Clamped Multilevel Inverter 49

1.8.3 Cascaded Voltage Source Multilevel Inverter 51

1.8.4 Hybrid Voltage Source Inverter 54
1.9 Summary 55

Chapter 2 Harmonic Distortion 57
2.1 Harmonic Voltage Distortion Factor 57
2.2 Harmonic Current Distortion Factor 61
2.3 Harmonic Distortion Factors for Three-Phase Inverters 64
2.4 Choice of Performance Indicator 67
2.5 WTHD of Three-Level Inverter 70
2.6 The Induction Motor Load 73
2.6.1 Rectangular Squirrel Cage Bars 73
2.6.2 Nonrectangular Rotor Bars 78
2.6.3 Per-Phase Equivalent Circuit 79
2.7 Harmonic Distortion Weighting Factors for Induction Motor Load 82
2.7.1 WTHD for Frequency-Dependent Rotor Resistance 82
2.7.2 WTHD Also Including Effect of Frequency-Dependent Rotor Leakage Inductance 84
2.7.3 WTHD for Stator Copper Losses 88
2.8 Example Calculation of Harmonic Losses 90
2.9 WTHD Normalization for PWM Inverter Supply 91
2.10 Summary 93

Chapter 3 Modulation of One Inverter Phase Leg 95
3.1 Fundamental Concepts of PWM 96
3.2 Evaluation of PWM Schemes 97
3.3 Double Fourier Integral Analysis of a Two-Level Pulse Width–Modulated Waveform 99
3.4 Naturally Sampled Pulse Width Modulation 105
3.4.1 Sine–Sawtooth Modulation 105
3.4.2 Sine–Triangle Modulation 114
3.5 PWM Analysis by Duty Cycle Variation 120
3.5.1 Sine–Sawtooth Modulation 120
3.5.2 Sine–Triangle Modulation 123

3.6 Regular Sampled Pulse Width Modulation 125
3.6.1 Sawtooth Carrier Regular Sampled PWM 130
3.6.2 Symmetrical Regular Sampled PWM 134
3.6.3 Asymmetrical Regular Sampled PWM 139
3.7 "Direct" Modulation 146
3.8 Integer versus Non-Integer Frequency Ratios 148
3.9 Review of PWM Variations 150
3.10 Summary 152

Chapter 4 Modulation of Single-Phase Voltage Source Inverters 155
4.1 Topology of a Single-Phase Inverter 156
4.2 Three-Level Modulation of a Single-Phase Inverter 157
4.3 Analytic Calculation of Harmonic Losses 169
4.4 Sideband Modulation 177
4.5 Switched Pulse Position 183
4.5.1 Continuous Modulation 184
4.5.2 Discontinuous Modulation 186
4.6 Switched Pulse Sequence 200
4.6.1 Discontinuous PWM — Single-Phase Leg Switched ... 200
4.6.2 Two-Level Single-Phase PWM 207
4.7 Summary 211

Chapter 5 Modulation of Three-Phase Voltage Source Inverters 215
5.1 Topology of a Three-Phase Inverter (VSI) 215
5.2 Three-Phase Modulation with Sinusoidal References 216
5.3 Third-Harmonic Reference Injection 226
5.3.1 Optimum Injection Level 226
5.3.2 Analytical Solution for Third-Harmonic Injection 230
5.4 Analytic Calculation of Harmonic Losses 241
5.5 Discontinuous Modulation Strategies 250
5.6 Triplen Carrier Ratios and Subharmonics 251
5.6.1 Triplen Carrier Ratios 251
5.6.2 Subharmonics 253

5.7 Summary 257

Chapter 6 Zero Space Vector Placement Modulation Strategies 259

6.1 Space Vector Modulation 259
6.1.1 Principles of Space Vector Modulation 259
6.1.2 SVM Compared to Regular Sampled PWM 265
6.2 Phase Leg References for Space Vector Modulation 267
6.3 Naturally Sampled SVM 270
6.4 Analytical Solution for SVM 272
6.5 Harmonic Losses for SVM 291
6.6 Placement of the Zero Space Vector 294
6.7 Discontinuous Modulation 299
6.7.1 120° Discontinuous Modulation 299
6.7.2 60° and 30° Discontinuous Modulation 302
6.8 Phase Leg References for Discontinuous PWM 307
6.9 Analytical Solutions for Discontinuous PWM 311
6.10 Comparison of Harmonic Performance 322
6.11 Harmonic Losses for Discontinuous PWM 326
6.12 Single-Edge SVM 330
6.13 Switched Pulse Sequence 331
6.14 Summary 333

Chapter 7 Modulation of Current Source Inverters 337

7.1 Three-Phase Modulators as State Machines 338
7.2 Naturally Sampled CSI Space Vector Modulator 343
7.3 Experimental Confirmation 343
7.4 Summary 345

Chapter 8 Overmodulation of an Inverter 349

8.1 The Overmodulation Region 350
8.2 Naturally Sampled Overmodulation of One Phase Leg of an Inverter 351

8.3 Regular Sampled Overmodulation of One Phase Leg of an Inverter 356

8.4 Naturally Sampled Overmodulation of Single- and Three-Phase Inverters 360

8.5 PWM Controller Gain during Overmodulation 364

8.5.1 Gain with Sinusoidal Reference 364

8.5.2 Gain with Space Vector Reference 367

8.5.3 Gain with 60° Discontinuous Reference 371

8.5.4 Compensated Modulation 373

8.6 Space Vector Approach to Overmodulation 376

8.7 Summary 382

Chapter 9 Programmed Modulation Strategies 383

9.1 Optimized Space Vector Modulation 384

9.2 Harmonic Elimination PWM 396

9.3 Performance Index for Optimality 411

9.4 Optimum PWM 416

9.5 Minimum-Loss PWM 421

9.6 Summary 430

Chapter 10 Programmed Modulation of Multilevel Converters 433

10.1 Multilevel Converter Alternatives 433

10.2 Block Switching Approaches to Voltage Control 436

10.3 Harmonic Elimination Applied to Multilevel Inverters 440

10.3.1 Switching Angles for Harmonic Elimination Assuming Equal Voltage Levels 440

10.3.2 Equalization of Voltage and Current Stresses 441

10.3.3 Switching Angles for Harmonic Elimination Assuming Unequal Voltage Levels 443

10.4 Minimum Harmonic Distortion 447

10.5 Summary 449

Chapter 11 Carrier-Based PWM of Multilevel Inverters 453

11.1 PWM of Cascaded Single-Phase H-Bridges 453

11.2 Overmodulation of Cascaded H-Bridges 465
11.3 PWM Alternatives for Diode-Clamped Multilevel Inverters ... 467
11.4 Three-Level Naturally Sampled PD PWM 469
11.4.1 Contour Plot for Three-Level PD PWM 469
11.4.2 Double Fourier Series Harmonic Coefficients 473
11.4.3 Evaluation of the Harmonic Coefficients 475
11.4.4 Spectral Performance of Three-Level PD PWM 479
11.5 Three-Level Naturally Sampled APOD or POD PWM 481
11.6 Overmodulation of Three-Level Inverters 484
11.7 Five-Level PWM for Diode-Clamped Inverters 489
11.7.1 Five-level Naturally Sampled PD PWM 489
11.7.2 Five-Level Naturally Sampled APOD PWM 492
11.7.3 Five-Level POD PWM 497
11.8 PWM of Higher Level Inverters 499
11.9 Equivalent PD PWM for Cascaded Inverters 504
11.10 Hybrid Multilevel Inverter 507
11.11 Equivalent PD PWM for a Hybrid Inverter 517
11.12 Third-Harmonic Injection for Multilevel Inverters 519
11.13 Operation of a Multilevel Inverter with a Variable Modulation Index 526
11.14 Summary 528

Chapter 12 Space Vector PWM for Multilevel Converters 531
12.1 Optimized Space Vector Sequences 531
12.2 Modulator for Selecting Switching States 534
12.3 Decomposition Method 535
12.4 Hexagonal Coordinate System 538
12.5 Optimal Space Vector Position within a Switching Period 543
12.6 Comparison of Space Vector PWM to Carrier-Based PWM 545
12.7 Discontinuous Modulation in Multilevel Inverters 548
12.8 Summary 550

Chapter 13 Implementation of a Modulation Controller 555

13.1 Overview of a Power Electronic Conversion System 556

13.2 Elements of a PWM Converter System 557

13.2.1 VSI Power Conversion Stage..563

13.2.2 Gate Driver Interface 565

13.2.3 Controller Power Supply 567

13.2.4 I/O Conditioning Circuitry...568

13.2.5 PWM Controller 569

13.3 Hardware Implementation of the PWM Process 572

13.3.1 Analog versus Digital Implementation........................572

13.3.2 Digital Timer Logic Structures 574

13.4 PWM Software Implementation 579

13.4.1 Background Software 580

13.4.2 Calculation of the PWM Timing Intervals....................581

13.5 Summary 584

Chapter 14 Continuing Developments in Modulation 585

14.1 Random Pulse Width Modulation..586

14.2 PWM Rectifier with Voltage Unbalance..................................590

14.3 Common Mode Elimination 598

14.4 Four Phase Leg Inverter Modulation 603

14.5 Effect of Minimum Pulse Width 607

14.6 PWM Dead-Time Compensation...612

14.7 Summary 619

Appendix 1 Fourier Series Representation of a Double Variable Controlled Waveform 623

Appendix 2 Jacobi–Anger and Bessel Function Relationships 629

A2.1 Jacobi–Anger Expansions...629

A2.2 Bessel Function Integral Relationships...................................631

Appendix 3 Three-Phase and Half-Cycle Symmetry Relationships 635

Appendix 4 Overmodulation of a Single-Phase Leg ... 637

A4.1 Naturally Sampled Double-Edge PWM ... 637

A4.1.1 Evaluation of Double Fourier Integral for Overmodulated Naturally Sampled PWM ... 638

A4.1.2 Harmonic Solution for Overmodulated Single-Phase Leg under Naturally Sampled PWM ... 646

A4.1.3 Linear Modulation Solution Obtained from Overmodulation Solution ... 647

A4.1.4 Square-Wave Solution Obtained from Overmodulation Solution ... 647

A4.2 Symmetric Regular Sampled Double-Edge PWM ... 649

A4.2.1 Evaluation of Double Fourier Integral for Overmodulated Symmetric Regular Sampled PWM ... 650

A4.2.2 Harmonic Solution for Overmodulated Single-Phase Leg under Symmetric Regular Sampled PWM ... 652

A4.2.3 Linear Modulation Solution Obtained from Overmodulation Solution ... 653

A4.3 Asymmetric Regular Sampled Double-Edge PWM ... 654

A4.3.1 Evaluation of Double Fourier Integral for Overmodulated Asymmetric Regular Sampled PWM ... 655

A4.3.2 Harmonic Solution for Overmodulated Single-Phase Leg under Asymmetric Regular Sampled PWM ... 660

A4.3.3 Linear Modulation Solution Obtained from Overmodulation Solution ... 661

Appendix 5 Numeric Integration of a Double Fourier Series Representation of a Switched Waveform ... 663

A5.1 Formulation of the Double Fourier Integral ... 663

A5.2 Analytical Solution of the Inner Integral ... 666

A5.3 Numeric Integration of the Outer Integral ... 668

Bibliography ... 671

Index ... 715

Preface

The work presented in this book offers a general approach to the development of fixed switching frequency pulse width–modulated (PWM) strategies to suit hard-switched converters. It is shown that modulation of, and resulting spectrum for, the half-bridge single-phase inverter forms the basic building block from which the spectral content of modulated single- phase, three-phase, or multiphase, two-level, three-level, or multilevel, voltage link and current link converters can readily be discerned. The concept of harmonic distortion is used as the performance index to compare all commonly encountered modulation algorithms. In particular, total harmonic distortion (THD), weighted total harmonic distortion (WTHD), and harmonic distortion criterion specifically designed to access motor copper losses are used as performance indices.

The concept of minimum harmonic distortion, which forms the underlying basis of comparison of the work presented in this book, leads to the identification of the fundamentals of PWM as

- Active switch pulse width determination.
- Active switch pulse placement within a switching period.
- Active switch pulse sequence across switching periods.

The benefit of this generalized approach is that once the common threads of PWM are identified, the selection of a PWM strategy for any converter topology becomes immediately obvious, and the only choices remaining are to trade-off the "best possible" performance against cost and difficulty of implementation, and secondary considerations. Furthermore, the performance to be expected from a particular converter topology and modulation strategy can be quickly and easily identified without complex analysis, so that informed trade-offs can be made regarding the *implementation* of a PWM algorithm for any particular application. All theoretical developments have been confirmed either by simulation or experiment. Inverter implementation details have been included at the end of the text to address practical considerations.

Readers will probably note the absence of any closed loop issues in this text. While initially such material was intended to be included, it soon became apparent that the inclusion of this material would require an additional volume. A further book treating this subject is in preparation.

Acknowledgments

The authors are indebted to their graduate students, who have contributed greatly to the production of this book via their Ph.D. theses. In particular the important work of Daniel Zmood (Chapter 7), Ahmet Hava (Chapter 8) and Brendan McGrath (Chapter 11) are specifically acknowledged. In addition, numerous other graduate students have also assisted with the production of this book both through their technical contributions as well as through detailed proof-reading of this text. The second author (Lipo) also wishes to thank the David Grainger Foundation and Saint John's College of Cambridge University for funding and facilities provided respectively. Finally, we wish to thank our wonderful and loving wives, Sophie Holmes and Chris Lipo, for nuturing and supporting us over the past five years as we have written this book.

Nomenclature

Generic Variable Usage Conventions

Variable Format	Meaning
F	CAPITALS: peak AC or average DC value
f	LOWER CASE: instantaneous value
$\langle f \rangle$	BRACKETED: low-frequency average value
$\bar{f}$	OVERBAR: space vector (complex variable)
$\bar{f}^{\dagger}$	DAGGER: conjugate of space vector
$\boldsymbol{f}$	BOLD LOWER CASE: column vector
$\boldsymbol{F}$	BOLD CAPITAL: matrix
$\boldsymbol{f}^T$	TRANSPOSED VECTOR: row vector

Specific Variable Usage Definitions

Variable	Meaning	Page First Used
a, b, c	Phase leg identifiers for three phase inverter	9
$\bar{a}$	Complex vector $e^{j2\pi/3}$	34
γ	Third-harmonic component magnitude M_3/M	227
A_{mn}, B_{mn}	Coefficients of Fourier expansion	102
$\overline{C_{mn}}$	Complex Fourier coefficient $\overline{C_{mn}} = A_{mn} + jB_{mn}$	102
D_k, k=1,2..	Diode section of inverter switch	7
e_{az}	Motor EMF w.r.t. DC bus midpoint	169
f_a, f_b, f_c	Generic variables in a–b–c reference frame	26
f_{as}, f_{bs}, f_{cs}	Generic variables in a–b–c reference frame referenced to load neutral (star) point	29
f_c	Frequency of carrier waveform	112
f_o	Frequency of fundamental component	112

Variable	Meaning	Page First Used
$\bar{f}_s$	Stationary space vector $f_{qs}^s - jf_{ds}^s$	34
$\boldsymbol{f}_{qd0}$	Vector $[f_{qs}, f_{ds}, f_{0s}]'$	36
f_q^s, f_d^s, f_0	Generic variables in d–q–0 stationary reference frame	26
$f_{qs}^s, f_{ds}^s, f_{0s}$	Stationary reference frame (d–q–0) variables referenced to load neutral (star) point	29
$f(x, y)$	Unit cell variable	100
HDF	Harmonic distortion factor	248
i_a, i_b, i_c	Three phase line currents	13
I_{dc}	DC link current	13
I_h	RMS value of the overall harmonic currents	172
$\overline{i_{h,k}}$	Instantaneous harmonic current over internal k	385
Δi_a	Ripple component of current in phase a	170
j	$\sqrt{-1}$	34
$J_n(x)$	Bessel function of order n and argument x	110
L	Number of multilevel inverter voltage levels	434
L_1'	Thevenin equivalent stator leakage inductance of induction motor	81
L_σ	Effective motor inductance of one phase	170
m_k , k=1,2.. or a,b,c	Inverter switching functions	14
m,n	Harmonic index variables	102
M	Modulation index (modulation depth)	92
M_3	Modulation index for third harmonic	227
n	Negative inverter DC rail	9
n	Harmonic component number	18
p	Positive inverter DC rail	9
p	$p = d/dt$, time derivative operator	16

Variable	Meaning	Page First Used
p	p^{th} carrier interval	131
p	Pulse ratio	250
p	Pulse number	384
$P_{h,cu}$	Harmonic copper loss	173
q	Charge	26
q	$m + n(\omega_o / \omega_c)$	137
$\boldsymbol{R}$	Rotating transformation matrix	36
r_1'	Thevenin Equivalent stator resistance of induction motor	81
R_e	Equivalent load resistance	172
RMS	Root mean square	10
$\overline{SV_x}, x = 1, \ldots, 7$	Voltage space vector corresponding to three-phase inverter states	31
$\overline{SC_x}, x = 1, \ldots, 7$	Current space vector corresponding to three-phase inverter states	338
$\mathrm{S_k}$, k=1,2..	Inverter switch	31
T_c	Carrier interval	99
$\mathrm{T_k}$, k=1,2..	Transistor section of inverter switch	7
$\boldsymbol{T}$	Transformation matrix	37
THD	Total harmonic distortion	58
T_o	Period of fundamental waveform	100
T_i	Switching time of inverter switch "i"	218
ΔT	Carrier period — $1/f_c$	158
u	per unit EMF — e_{az}/V_{dc}	170
U	Unbalance factor	597
v_{as}, v_{bs}, v_{cs}	Phase voltages with respect to load neutral	11

Variable	Meaning	Page First Used
v_{ab}, v_{bc}, v_{ca}	Line-to-line (l–l) voltages for a three phase inverter	11
v_{az}, v_{bz}, v_{cz}	Phase voltages with respect to DC link midpoint	14
$v_{qs}^{s}, v_{ds}^{s}, v_{0s}^{s}$	Stationary reference frame (d–q–0) voltages	28
v_{sn}	Voltage between load neutral and negative DC bus	23
V_1	Peak magnitude of fundamental voltage component	13
V_{bus}	DC link voltage	7
V_{dc}	One-half the DC link voltage	5
V_m	Space vector magnitude or phase voltage amplitude	35
V_{mp}, V_{mn}	Amplitude of positive and negative phase voltages	595
$\overline{V}_o^*$	Target output space vector	260
V_p	Peak input l–l voltage	226
V_{rms}	RMS voltage	57
WTHD	Weighted total harmonic distortion	63
WTHD2	Weighted THD for rotor bar losses	85
WTHD1	Weighted THD for stator losses	89
WTHD0	Weighted THD normalized to base frequency	92
W_x	Pulse width	146
$x(t)$	Time variable corresponding to modulation angular frequency $\omega_c t = 2\pi f_c t$	99
x_r, x_f	Rising and falling switching instants for phase leg	128
$y(t)$	Time variable corresponding to fundamental angular frequency $\omega_o t = 2\pi f_o t$	99
y'	Variable for regular sampling: $y - \frac{\omega_o}{\omega_c}(x - 2\pi p)$	131
z	DC bus midpoint (virtual)	9
$Z(p)$	Load impedance	16

Variable	Meaning	Page First Used
α	Phase shift delay	17
α	Skin depth	76
α_1	Amplitude of modulating function	178
$\alpha_1, \alpha_1, \ldots, \alpha_{2N}$	Switching angles for harmonic elimination	397
$\delta_{advance}$	Advance compensation for PWM sampling delay	581
θ_c	Phase offset angle of carrier waveform	99
θ_o	Phase offset angle of fundamental component	99
$\theta_o(k)$	Phase offset angle of fundamental component at sampling time k	581
λ	Flux linkage	17
ϕ_{mp}, ϕ_{mn}	Phase angle of positive and negative sequence phase voltages respectively	595
ψ	Overmodulation angle	353
ω_c	Angular frequency of carrier waveform	99
ω_o	Angular frequency of fundamental component	7
ω_o/ω_c	Fundamental to carrier frequency ratio	106

1

Introduction to Power Electronic Converters

Power electronic converters are a family of electrical circuits which convert electrical energy from one level of voltage/current/frequency to another using semiconductor-based electronic switches. The essential characteristic of these types of circuits is that the switches are operated only in one of two states – either fully ON or fully OFF – unlike other types of electrical circuits where the control elements are operated in a (near) linear active region. As the power electronics industry has developed, various families of power electronic converters have evolved, often linked by power level, switching devices, and topological origins. The process of switching the electronic devices in a power electronic converter from one state to another is called *modulation,* and the development of optimum strategies to implement this process has been the subject of intensive international research efforts for at least 30 years. Each family of power converters has preferred modulation strategies associated with it that aim to optimize the circuit operation for the target criteria most appropriate for that family. Parameters such as switching frequency, distortion, losses, harmonic generation, and speed of response are typical of the issues which must be considered when developing modulation strategies for a particular family of converters.

Figure 1.1 presents a categorization of power electronic converters into families according to their type of electrical conversion. Of these families, converters that change energy to or from alternating current (AC) form involve much more complex processes than those that solely involve direct current (DC). The purpose of this book is to explore the converter modulation issue in detail as it relates to high power DC/AC (inverting) and AC/DC (rectifying) converters, with particular emphasis on the process of open-loop pulse width modulation (PWM) applied to these types of converters. This chapter presents the fundamentals of inverter structures, block-switching voltage control, and space vector concepts, as a foundation for the material to follow.

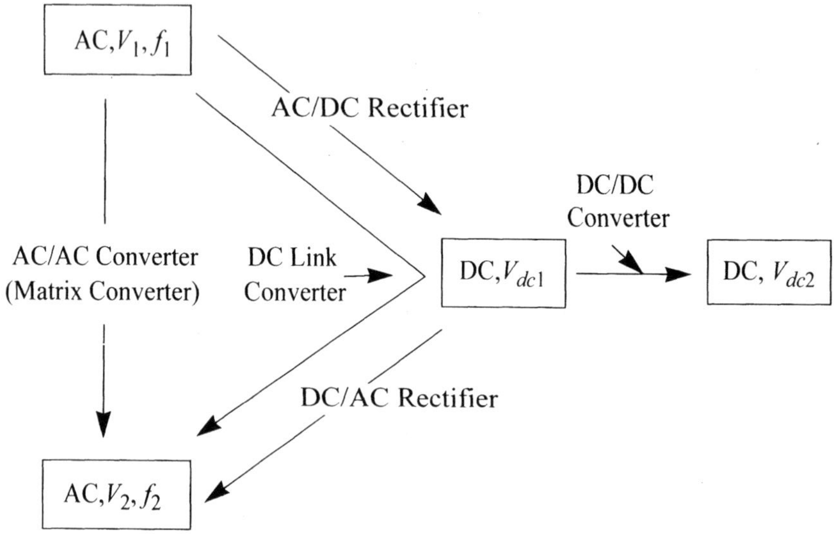

Figure 1.1 Families of solid state power converters categorized according to their conversion function.

1.1 Basic Converter Topologies

1.1.1 Switch Constraints

The transistor switch used for solid state power conversion is very nearly approximated by a resistance which either approaches zero or infinity depending upon whether the switch is closed or opened. However, regardless of where the switch is placed in the circuit, Kirchoff's voltage and current laws must, of course, always be obeyed. Translated to practical terms, these laws give rise to the two basic tenets of switch behavior:

- The switch cannot be placed in the same branch with a current source (i.e., an inductance) or else the voltage across the inductor (and consequently across the switch) will become infinite when the switch turns off. As a corollary to this statement it can be argued that *at least one* of the elements in branches connected via a node to the branch containing the switch must be non-inductive for the same reason.

- The switch cannot be placed in parallel with a voltage source (i.e., a true source or a capacitance) or else the current in the switch will become infinite when the switch turns on. As a corollary it can be stated that if more than one branch forms a loop containing the switch branch then *at least one* of these branch elements must not be a voltage source.

If the purpose of the switch is to aid in the process of transferring energy from the source to the load, then the switch must be connected in some manner so as to select between two input energy sources or sinks (including the possibility of a zero energy source). This requirement results in the presence of two branches delivering energy to one output (through a third branch). The presence of three branches in the interposing circuit implies a connecting node between these branches.

One of the three branches can contain an inductance (an equivalent current source frequently resulting from an inductive load or source), but the other branches connected to the same node must not be inductive or else the first basic tenet will be violated. The only other alternatives for the two remaining branches are a capacitance or a resistance. However, when the capacitor is connected between the output or input voltage source and the load, it violates the second tenet. The only choice left is a resistance.

The possibility of a finite resistance can be discarded as a practical matter since the circuit to be developed must be as highly efficient as possible, so that the only possibility is a resistor having either zero or infinite resistance, i.e., a second switch. This switch can only be turned on when the first switch is turned off, or vice versa, in order to not violate Kirchoff's current law. For the most common case of unidirectional current flow, a unidirectional switch which inhibits current flow in one direction can be used, and this necessary complementary action is conveniently achieved by a simple diode, since the demand of the inductance placed in the other branch will assure the required behavior. Alternatively, of course, the necessary complementary switching action can be achieved by a second unidirectional switch. The resulting circuits, shown in Figure 1.2, can be considered to be the *basic switching cells* of power electronics. The switches having arrows in (b) and (c) denote unidirectional current flow devices.

When the circuit is connected such that the current source (inductance) is connected to the load and the diode to the source, one realizes what is termed a *step-down chopper*. If the terminals associated with input and output are

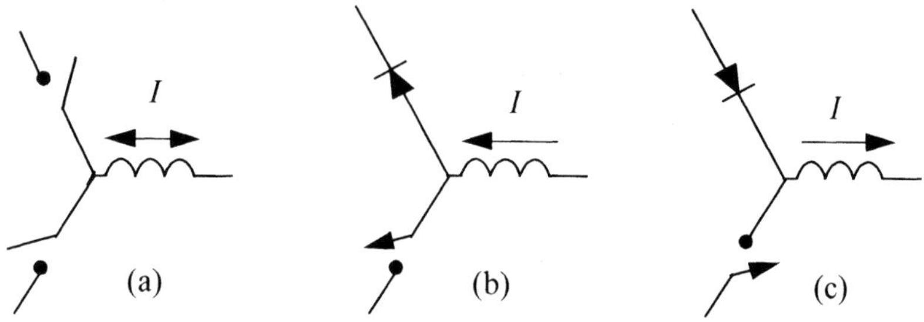

Figure 1.2 Basic commutation cells of power electronic converters using (a) bidirectional switches and (b) and (c) unidirectional switches.

reversed, a *step-up chopper* is produced. Energy is passed from the voltage source to the current "source" (i.e., the load) in the case of the step-down converter, and from the current source to the voltage "source" (load) in the case of the step-up converter.

Since the source voltage sums to the voltage across the switch plus the diode and since the load is connected across the diode only, the voltage is the quantity that is stepped down in the case of the step-down chopper. Because of the circulating current path provided by the diode, the current is consequently stepped up. On the other hand the sum of the switch plus diode voltage is equal to the output voltage in the case of the step-up chopper so that the voltage is increased in this instance. The input current is diverted from the output by the switch in this arrangement so that the current is stepped down.

Connecting the current source to both the input and output produces the *up-down chopper* configuration. In this case the switch must be connected to the input to control the flow of energy into/out of the current source. Since the average value of voltage across the inductor must equal zero, the average voltage across the switch must equal the input voltage while the average voltage across the diode equals the output voltage. Ratios of input to output voltages greater than or less than unity (and consequently current ratios less or greater than unity) can be arranged by spending more or less than half the available time over a switching cycle with the switch closed. These three basic DC/DC converter configurations are shown in Figure 1.3.

1.1.2 Bidirectional Chopper

In cases where power flow must occur in either direction a combination of a step-down and a step-up chopper with reversed polarity can be used as shown

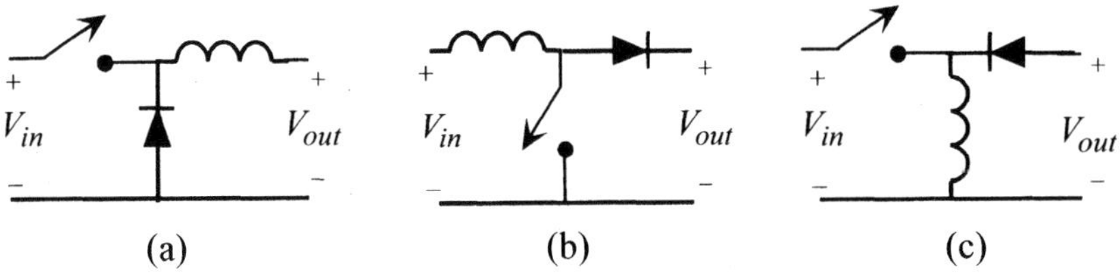

Figure 1.3 The three basic DC/DC converters implemented with a basic switching cell (a) step-down chopper, (b) step-up chopper, and (c) up-down chopper.

in Figure 1.4. The combination of the two functions effectively places the diodes in inverse parallel with switches, a combination which is pervasive in power electronic circuits. When passing power from left to right, the step-down chopper transistor is operated to control power flow while the step-up chopper transistor operates for power flow from right to left in Figure 1.4. The two switches need never be (and obviously should never be) closed at the same instant.

1.1.3 Single-Phase Full-Bridge (H–Bridge) Inverter

Consider now the basic switching cell used for DC/AC power conversion. In Figure 1.4 it is clear that current can flow bidirectionally in the current source/sink of the up-down chopper. If this component of the circuit is now considered as an AC current source load and the circuit is simply tipped on its side, the half-bridge DC/AC inverter is realized as shown in Figure 1.5. Note that in this case the input voltage is normally center-tapped into two equal DC voltages, $V_{dc1} = V_{dc2} = V_{dc}$, in order to produce a symmetrical AC voltage waveform. The total voltage across the DC input bus is then $2V_{dc}$. The parallel combination of the unidirectional switch and inverse conducting diode forms

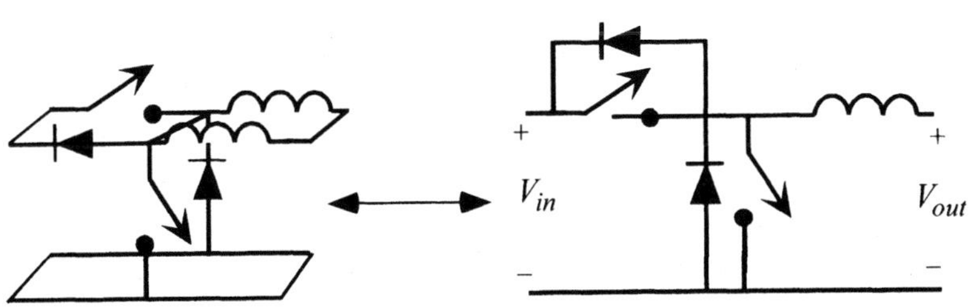

Figure 1.4 Bidirectional chopper using one up-chopper and one down-chopper.

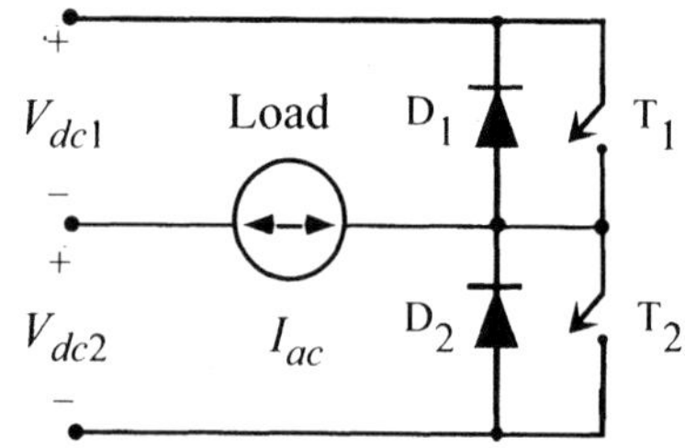

Figure 1.5 Half-bridge single-phase inverter.

the first type of practical inverter switch. The switch combination permits unidirectional current flow but requires only one polarity of voltage blocking ability and hence is suitable, in this case, for operating from a DC voltage source.

It is important to note that in many inverter circuits the center-tap point of the DC voltage shown in Figure 1.5 will not be provided. However, this point is still commonly used either as an actual ground point or else, in more elaborate inverters, as the reference point for the definition of multiple DC link voltages. Hence in this book, the total DC link voltage is considered as always consisting of a number of DC levels, and with conventional inverters that can only switch between two levels it will always be defined as $2V_{dc}$.

The structure of a single-phase full-bridge inverter (also known as a H–bridge inverter) is shown in Figure 1.6. This inverter consists of two single-phase leg inverters of the same type as Figure 1.5 and is generally preferred over other arrangements in higher power ratings. Note that as discussed above, the DC link voltage is again defined as $2V_{dc}$. With this arrangement, the maximum output voltage for this inverter is now twice that of the half-bridge inverter since the entire DC voltage can be impressed across the load, rather than only one-half as is the case for the half-bridge. This implies that for the same power rating the output current and the switch currents are one-half of those for a half-bridge inverter. At higher power levels this is a distinct advantage since it requires less paralleling of devices. Also, higher voltage is preferred since the cost of wiring is typically reduced as well as the losses in many types of loads because of the reduced current flow.

In general, the converter configurations of Figures 1.5 and 1.6 are capable of bidirectional power flow. In the case where power is exclusively or primarily intended to flow from DC to AC the circuits are designated as *inverters,* while the same circuits are designated *rectifiers* if the reverse is true. In cases

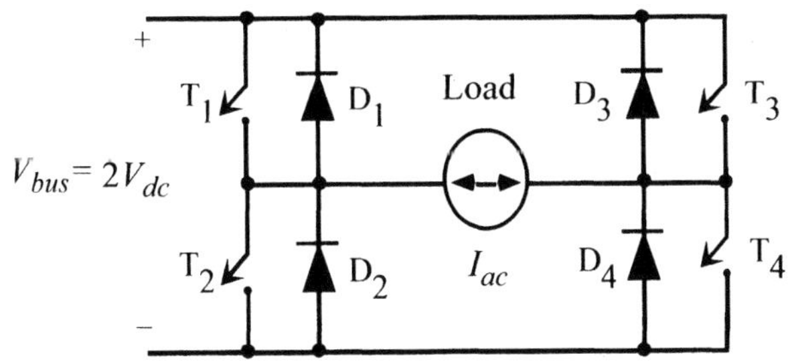

Figure 1.6 Single-phase full-bridge (H–bridge) inverter

where the DC supplies are derived from a source such as a battery, the inverter is designated as a *voltage source inverter* (VSI). If the DC is formed by a temporary DC supply such as a capacitor (being recharged ultimately, perhaps, from a separate source of energy), the inverter is designated as a *voltage stiff inverter* to indicate that the link voltage tends to resist sudden changes but can alter its value substantially under heavy load changes. The same distinction can also be made for the rectifier designations.

1.2 Voltage Source/Stiff Inverters

1.2.1 Two-Phase Inverter Structure

Inverters having additional phases can be readily realized by simply adding multiple numbers of half-bridge (Figure 1.5) and full-bridge inverter legs (Figure 1.6). A simplified diagram of a two-phase half-bridge inverter is shown in Figure 1.7(a). While the currents in the two phases can be controlled at will, the most desirable approach would be to control the two currents so that they are phase shifted by 90° with respect to each other (two-phase set) thereby producing a constant amplitude rotating field for an AC machine. However, note that the sum of the two currents must flow in the line connected to the center point of the DC supplies. If the currents in the two phases can be approximated by equal amplitude sine waves, then

$$\begin{aligned} i_{\text{neutral}} &= I\sin\omega_o t + I\sin\left(\omega_o t + \frac{\pi}{2}\right) \\ &= \sqrt{2}I\sin\left(\omega_o t + \frac{\pi}{4}\right) \end{aligned} \qquad (1.1)$$

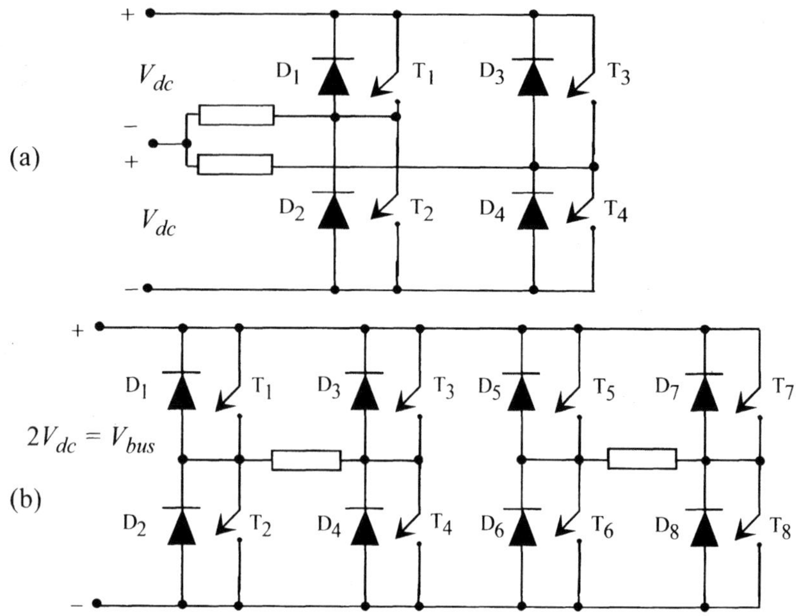

Figure 1.7 Two-phase (a) half-bridge and (b) full-bridge inverters.

Since a relatively large AC current must flow in the midpoint connection, this inverter configuration is not commonly used. As an alternative, the midpoint current could be set to zero if the currents in the two phases were made equal and opposite. However, this type of operation differs little from the single-phase bridge of Figure 1.6 except that the neutral point of the load can be considered as being grounded (i.e., referred to the DC supply midpoint). As a result this inverter topology is also not frequently used.

The full-bridge inverter of Figure 1.7(b) does not require the DC midpoint connection. However, eight switches must be used which, in most cases, makes this possibility economically unattractive.

1.2.2 Three-Phase Inverter Structure

The half-bridge arrangement can clearly be extended to any number of phases. Figure 1.8 shows the three-phase arrangement. In this case, operation of an AC motor requires that the three currents are a balanced three-phase set, i.e., equal amplitude currents with equal 120° phase displacement between them. However it is easily shown that the sum of the three currents is zero, so that the connection back to the midpoint of the DC supply is not required. The

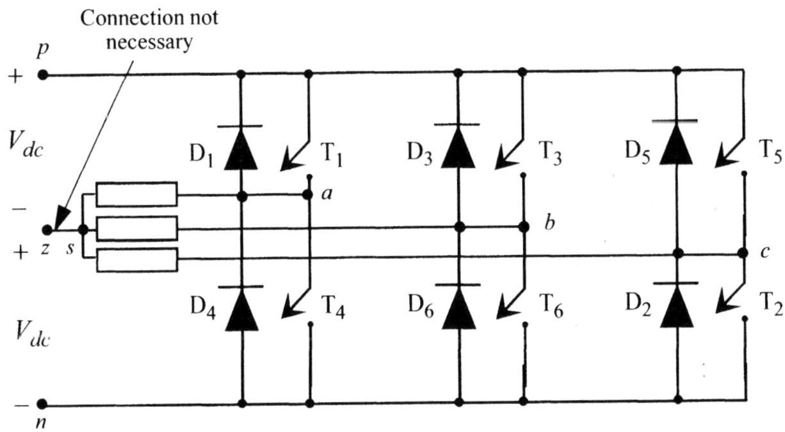

Figure 1.8 Three-phase bridge-type voltage source inverter.

simplification afforded by this property of three-phase currents makes the three-phase bridge-type inverter the de facto standard for power conversion. However while the connection from point s (neutral of the *star-connected stationary* load to the midpoint z (*zero* or *reference* point of the DC supply) need not be physically present, it remains useful to retain the midpoint z as the reference (ground) for all voltages. Also note that p and n are used in this text to denote the positive and negative bus voltages respectively, with respect to the midpoint z.

1.2.3 Voltage and Current Waveforms in Square-Wave Mode

The basic operation of the three-phase voltage inverter in its simplest form can be understood by considering the inverter as being made up of six mechanical switches. While it is possible to energize the load by having only two switches closed in sequence at one time (resulting in the possibility of one phase current being zero at instances in a switching cycle), it is now accepted that it is preferable to have one switch in each phase leg closed at any instant. This ensures that all phases will conduct current under any power factor condition. If two switches of each phase leg are turned on for a half cycle each in nonoverlapping fashion, this produces the voltage waveforms of Figure 1.9 at the output terminals *a, b,* and *c*, referred to the negative DC bus n. The numbers on the top part of the figure indicate which switches of Figure 1.8 are closed. The sequence is in the order 123, 234, 345, 456, 561, 612, and back to 123.

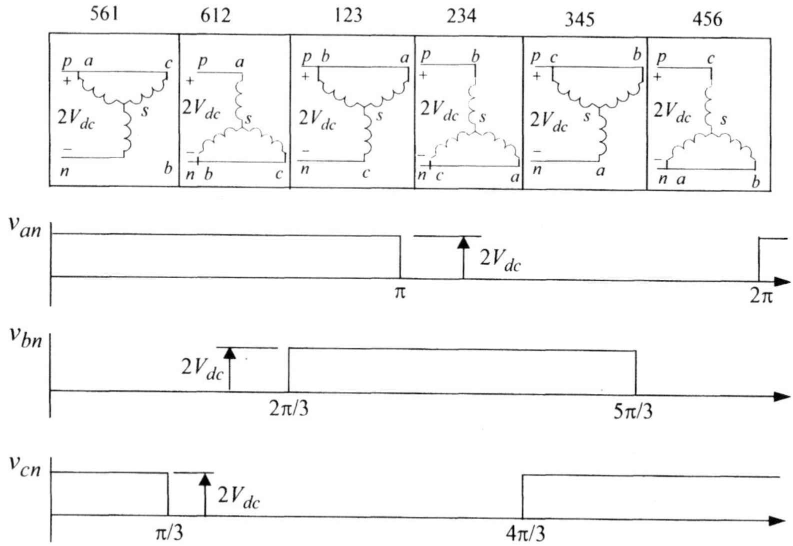

Figure 1.9 The six possible connections of a simple three-phase voltage stiff inverter. The three waveforms show voltages from the three-phase leg outputs to the negative DC bus voltage.

The line-to-line (*l–l*) voltage v_{ab} then has the *quasi-square* waveform shown in Figure 1.10. As will be shown shortly, the line-to-line voltage contains a root-mean-square (RMS) fundamental component of

$$V_{1,ll,rms} = \frac{2\sqrt{6}V_{dc}}{\pi} \cong 1.56V_{dc} \tag{1.2}$$

Thus, a standard 460 V, 60 Hz induction motor would require 590 V at the DC terminals of the motor to operate the motor at its rated voltage and speed. For this reason a 600 V DC bus (i.e., $V_{dc} = 300\text{ V}$) is quite standard in the United States for inverter drives.

Although motors function as an active rather than a passive load, the effective impedances of each phase are still *balanced*. That is, insofar as voltage drops are concerned, active as well as passive three-phase loads may be represented by the three equivalent impedances [and electromotive forces (EMFs)] shown in Figure 1.10 for the six possible connections. Note that each individual phase leg is alternately switched from the positive DC rail to the negative DC rail and that it is alternately in series with the remaining two phases con-

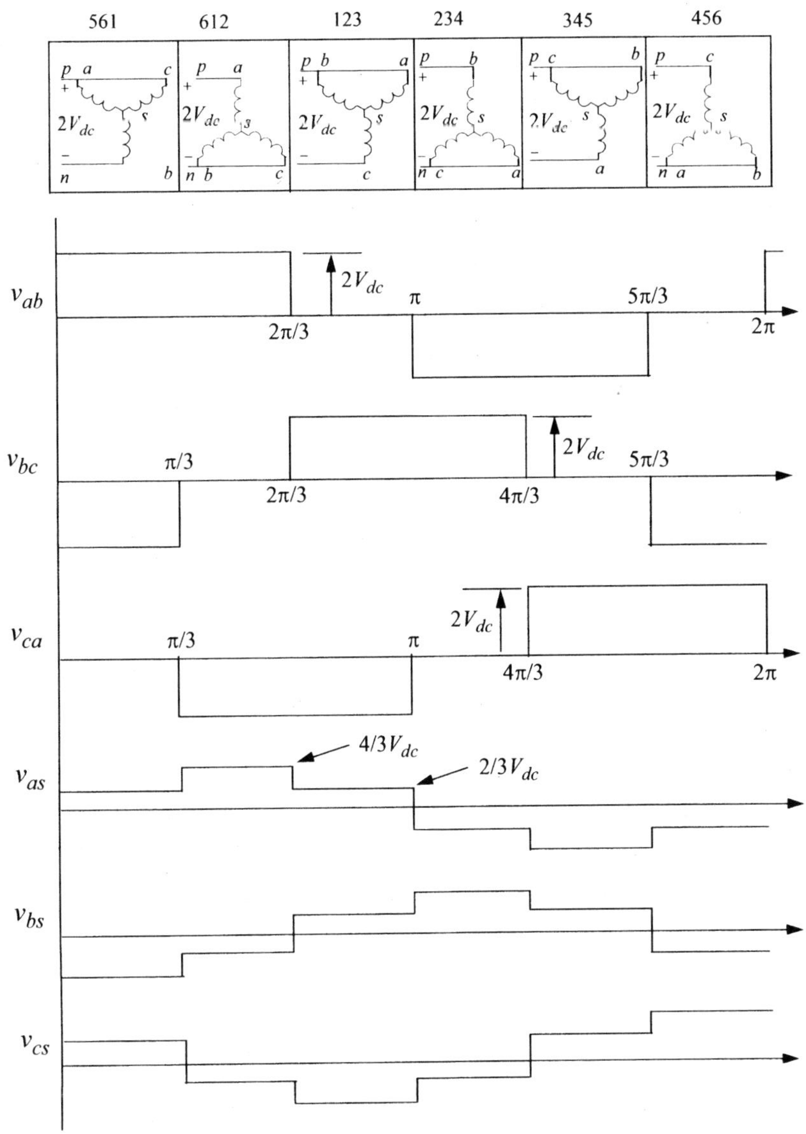

Figure 1.10 The three line-to-line and line-to-neutral load voltages created by the six possible switch connection arrangements of a six-step voltage stiff inverter.

nected in parallel, or it is in parallel with one of the other two phases and in series with the third. Hence the voltage drop across each phase load is always one-third or two-thirds of the DC bus voltage, with the polarity of the voltage drop across the phase being determined by whether it is connected to the positive or negative DC rail.

A plot of the line and phase voltages for a typical motor load is included in Figure 1.10. The presence of six "steps" in the load line-to-neutral voltage waveforms v_{as}, v_{bs}, and v_{cs}, is one reason this type of inverter is called a *six-step* inverter, although the term *six-step* in reality pertains to the method of voltage/frequency control rather than the inverter configuration itself.

A Fourier analysis of these waveforms indicates a simple square-wave type of geometric progression of the harmonics. When written as an explicit time function, the Fourier expansion for the time-varying *a* phase to negative DC bus voltage *n* can be readily determined to be

$$v_{an}(t) = V_{dc}\frac{4}{\pi}\left[\frac{\pi}{4} + \sin\omega_o t + \frac{1}{3}\sin 3\omega_o t + \frac{1}{5}\sin 5\omega_o t + \frac{1}{7}\sin 7\omega_o t + \cdots\right] \quad (1.3)$$

The *b* and *c* phase to negative DC bus voltages can be found by replacing $\omega_o t$ with $(\omega_o t - 2\pi/3)$ and $(\omega_o t + 2\pi/3)$, respectively, in Eq. (1.3).

The v_{ab} line-to-line voltage is found by subtracting v_{bn} from v_{an} to give

$$v_{ab}(t) = V_{dc}\frac{4\sqrt{3}}{\pi}\left[\sin\left(\omega_o t + \frac{\pi}{6}\right) + \frac{1}{5}\sin\left(5\omega_o t - \frac{\pi}{6}\right) + \frac{1}{7}\sin\left(7\omega_o t + \frac{\pi}{6}\right) + \cdots\right] \quad (1.4)$$

Similar relationships can be readily found for the v_{bc} and v_{ca} voltages, phase shifted by $-2\pi/3$ and $+2\pi/3$, respectively. Note that harmonics of the order of multiples of three are absent from the line-to-line voltage, since these *triplen* harmonics cancel between the phase legs.

In terms of RMS values, each harmonic of the line-to-neutral voltages has the value of

$$V_{n,\,ln,\,rms} = V_{dc}\frac{2\sqrt{2}}{\pi}\frac{1}{n} \quad (1.5)$$

and, for the line-to-line voltages,

$$V_{n,\,ll,\,rms} = V_{dc}\frac{2\sqrt{6}}{\pi}\frac{1}{n} \qquad \text{where } n = 6k \pm 1,\ k = 1,2,3,\ldots \quad (1.6)$$

Because of its utility as a reference value for pulse width modulation in later chapters, it is useful to write the fundamental component of the line-to-neutral voltage in terms of its peak value referred to half the DC link voltage, in which case

$$V_{1,\,ln,\,pk} = \frac{4}{\pi} V_{dc} = V_1 \tag{1.7}$$

This value is, of course, the fundamental component of a square wave of amplitude V_{dc}. It should be noted also that since the use of peak rather than RMS quantities will predominate in this book, quantities in capital letters will denote only DC or peak AC quantities. Hence, for simplicity V_1 in Eq. (1.7) has the same meaning as $V_{1,\,ln,\,pk}$. When the quantity is intended to be root-mean-square, the subscript *rms* will always be appended. For example, the term $V_{1,\,ln,\,rms}$ designates the RMS fundamental value of the line-to-neutral voltage.

Assuming an *R–L–*EMF load, the current as well as voltage waveforms are sketched for both wye and delta connections in Figure 1.11(a). Note that when the inverter current flows in opposite polarity to the voltage, the current is carried by the feedback diode (in a step-up chopper mechanism) in much the same manner as for the single-phase inverter. The transfer of current from main to auxiliary switches is illustrated by the conduction pattern of Figure 1.11 and can be used to determine the DC side inverter current waveform I_{dc}. For example, from the moment that T_3 is turned off to the instant that D_2 turns on, the input current is equal to the current in T_1, that is, i_a. This interval lasts one-sixth of a period or 60°. During the next 60°, switch T_6 returns current to the

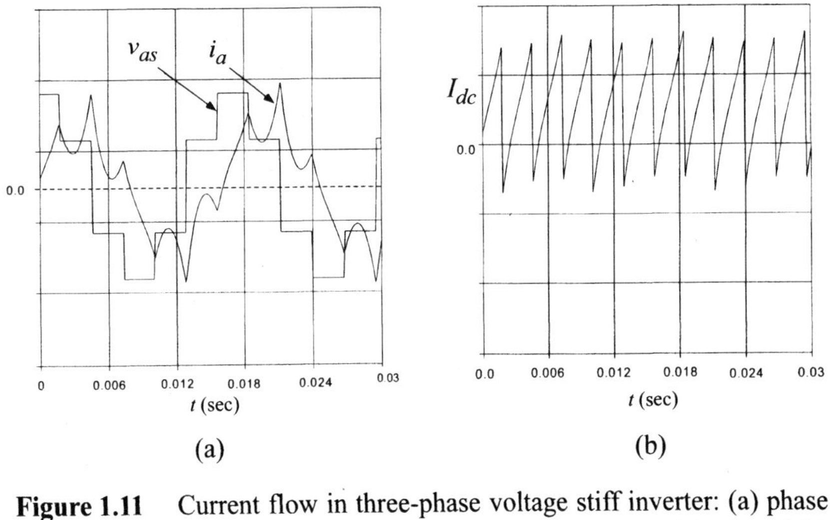

Figure 1.11 Current flow in three-phase voltage stiff inverter: (a) phase voltage and current waveform, wye-connected load, and (b) DC link current.

DC link. In effect, the link current is equal to $-i_c$. Continuing through all six 60° intervals generates the DC link current shown in Figure 1.11(b). For the case shown, I_{dc} is both positive and negative so that a certain amount of energy transfers out of and into the DC supplies. If the load current is considered to be sinusoidal, it can be shown that I_{dc} is always positive only when the fundamental power factor is greater than 0.55. However, in any case, the source supplies the average component of the link current while a current with frequency six times the fundamental frequency component circulates in and out of the DC capacitor. The sizing of the capacitor to accommodate these harmonics, regardless of the modulation algorithm, is a major consideration in inverter design.

1.3 Switching Function Representation of Three-Phase Converters

The basic three-phase inverter circuit operation shown in Figures 1.9 and 1.10 can be condensed to equation form by defining logic-type *switching functions* which express the closure of the switches [1, 2][1]. For example, let $m_1, m_2, \ldots, m_6$ take on the value "+1" when switches T_1, T_2, ...,T_6 are closed and the value "zero" when opened. The voltages from the three-phase legs to the DC center point can then be expressed as

$$\begin{aligned} v_{az} &= V_{dc}(m_1 - m_4) \\ v_{bz} &= V_{dc}(m_3 - m_6) \\ v_{cz} &= V_{dc}(m_5 - m_2) \end{aligned} \tag{1.8}$$

Considering now the constraints imposed by the circuit it is apparent that both the top and bottom switches of a given phase cannot be closed at the same time. Furthermore, from current continuity considerations in each phase leg

$$\begin{aligned} m_1 + m_4 &= 1 \\ m_3 + m_6 &= 1 \\ m_5 + m_2 &= 1 \end{aligned} \tag{1.9}$$

1 References referred to throughout this text are given at the end of each chapter. A more exhaustive set of references are located in the Bibliography.

Substituting Eq. (1.9) into Eq. (1.8) gives

$$
\begin{aligned}
v_{az} &= V_{dc}(2m_1 - 1) \\
v_{bz} &= V_{dc}(2m_3 - 1) \\
v_{cz} &= V_{dc}(2m_5 - 1)
\end{aligned}
\tag{1.10}
$$

Since the quantities in the parentheses of Eq. (1.10) take on the values ±1, it is useful to define new variables m_a, m_b, m_c, such that $m_a = 2m_1 - 1$, etc. Hence, more compactly,

$$
\begin{aligned}
v_{az} &= V_{dc}m_a \\
v_{bz} &= V_{dc}m_b \\
v_{cz} &= V_{dc}m_c
\end{aligned}
\tag{1.11}
$$

The current in the DC link can be expressed as

$$
I_{dc} = i_a\frac{m_a + 1}{2} + i_b\frac{m_b + 1}{2} + i_c\frac{m_c + 1}{2} \tag{1.12}
$$

However, since

$$
i_a + i_b + i_c = 0
$$

Equation (1.12) reduces to

$$
I_{dc} = \frac{1}{2}(i_a m_a + i_b m_b + i_c m_c) \tag{1.13}
$$

The line-to-line AC voltages are

$$
\begin{aligned}
v_{ab} &= v_{az} - v_{bz} = V_{dc}(m_a - m_b) \\
v_{bc} &= v_{bz} - v_{cz} = V_{dc}(m_b - m_c) \\
v_{ca} &= v_{cz} - v_{az} = V_{dc}(m_c - m_a)
\end{aligned}
\tag{1.14}
$$

If the load is star connected, the load line-to-neutral (phase) voltages can be expressed as

$$
\begin{aligned}
v_{as} &= v_{az} - v_{sz} \\
v_{bs} &= v_{bz} - v_{sz} \\
v_{cs} &= v_{cz} - v_{sz}
\end{aligned}
\tag{1.15}
$$

For most practical cases, the phase impedances in all three legs of the star load are the same. Hence, in general,

$$\begin{aligned} v_{as} &= Z(p)i_a \\ v_{bs} &= Z(p)i_b \\ v_{cs} &= Z(p)i_c \end{aligned} \tag{1.16}$$

where the operator $p = d/dt$ and the impedance $Z(p)$ is an arbitrary function of p (which is the same in each phase). The phase voltages can now be solved by adding together the three parts of Eq. (1.16), to produce

$$\begin{aligned} v_{as}+v_{bs}+v_{cs} &= v_{az}+v_{bz}+v_{cz}-3v_{sz} \\ &= Z(p)(i_a+i_b+i_c) = 0 \end{aligned} \tag{1.17}$$

Thus

$$\begin{aligned} v_{sz} &= \frac{1}{3}(v_{az}+v_{bz}+v_{cz}) \\ &= \frac{1}{3}V_{dc}(m_a+m_b+m_c) \end{aligned} \tag{1.18}$$

The phase voltages can now be expressed as

$$v_{as} = \frac{2}{3}v_{az}-\frac{1}{3}v_{bz}-\frac{1}{3}v_{cz} \tag{1.19}$$

so that, from Eq. (1.8),

$$v_{as} = V_{dc}\left(\frac{2}{3}m_a-\frac{1}{3}m_b-\frac{1}{3}m_c\right) \tag{1.20}$$

Similarly

$$v_{bs} = V_{dc}\left(\frac{2}{3}m_b-\frac{1}{3}m_a-\frac{1}{3}m_c\right) \tag{1.21}$$

$$v_{cs} = V_{dc}\left(\frac{2}{3}m_c-\frac{1}{3}m_a-\frac{1}{3}m_b\right) \tag{1.22}$$

Finally, the power flow through the inverter is given by

$$P_{dc} = 2V_{dc}I_{dc} = V_{dc}(i_am_a+i_bm_b+i_cm_c) \tag{1.23}$$

Equations (1.20) to (1.22) are convenient for use in defining switching functions representing the converter's behavior in different frames of reference [2].

1.4 Output Voltage Control

A power electronic inverter is essentially a device for creating a variable AC frequency output from a DC input. The frequency of the output voltage or current is readily established by simply switching for equal time periods to the positive and the negative DC bus and appropriately adjusting the half-cycle period. However, the variable frequency ability is nearly always accompanied by a corresponding need to adjust the amplitude of the fundamental component of the output waveform as the frequency changes, i.e., *voltage control.* This section introduces the concept of voltage control, a central theme of this book.

1.4.1 Volts/Hertz Criterion

In applications involving AC motors, the load can be characterized as being essentially inductive. Since the time rate of change of flux linkages λ in an inductive load is equal to the applied voltage, then

$$\lambda = \int v \, dt \tag{1.24}$$

If one is only concerned with the fundamental component, then, if a phase voltage is of the form $v = V_1 \cos \omega_o t$, the corresponding flux linkage is

$$\lambda_1 = \frac{V_1}{\omega_o} \sin \omega_o t \tag{1.25}$$

suggesting that the fundamental component of voltage must be varied in proportion to the frequency if the amplitude of the flux in the inductive load is to remain sensibly constant.

1.4.2 Phase Shift Modulation for Single-Phase Inverter

The method by which voltage adjustment is accomplished in a solid state power converter is the heart of the issue of *modulation.* Much more detail will be developed concerning modulation techniques in later chapters. However, a very simple introductory example of modulation can be obtained by taking a single-phase inverter as shown in Figure 1.12(a) and operating each phase leg with a 50% duty cycle but with a phase delay of $\pi - \alpha$ between the two phase legs. Typical waveforms for this inverting operation (DC-to-AC power conversion) in what can be termed *phase shift voltage control* or *phase shift modula-*

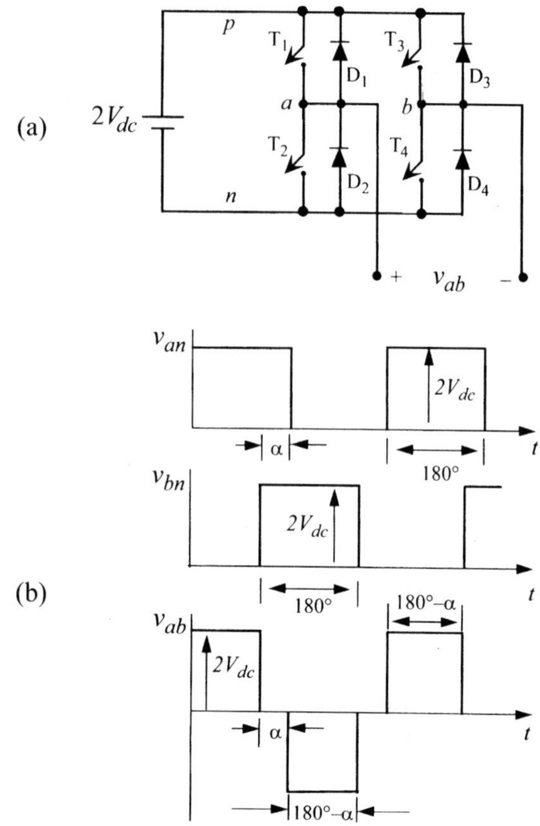

Figure 1.12 Full-bridge, single-phase inverter control by phase shift cancellation: (a) power circuit and (b) voltage waveforms.

tion are shown in Figure 1.12(b). Clearly, as the phase delay angle α changes, the RMS magnitude of the line-to-line output voltage changes.

The switched output voltage of this inverter can be represented as the sum of a series of harmonic components (a Fourier series in fact). The magnitude of each harmonic can be conveniently evaluated using the quantity $\beta = 90° - \alpha/2$ where α is as shown in Figure 1.12. Conventional Fourier analysis gives, for each harmonic n, a peak harmonic magnitude of

$$V_{ab(n)} = \frac{2}{\pi}\int_{-\pi/2}^{\pi/2} 2V_{dc}\cos n\theta \, d\theta \tag{1.26}$$

$$= V_{dc}\frac{4}{\pi}\int_{-\beta}^{\beta}\cos n\theta d\theta$$

$$= V_{dc}\frac{8}{\pi n}\sin n\beta$$

$$= V_{dc}\frac{8}{\pi n}\cos\frac{n\alpha}{2} \qquad \text{where } n \text{ is odd} \tag{1.27}$$

Figure 1.13 shows the variation of the fundamental frequency and harmonic components as a function of the overlap angle α. The components are normalized with respect to $2V_{dc}$.

1.4.3 Voltage Control with a Double Bridge

While voltage control is not possible with a conventional six-step inverter without adjusting the DC link voltage, some measure of voltage control is pos-

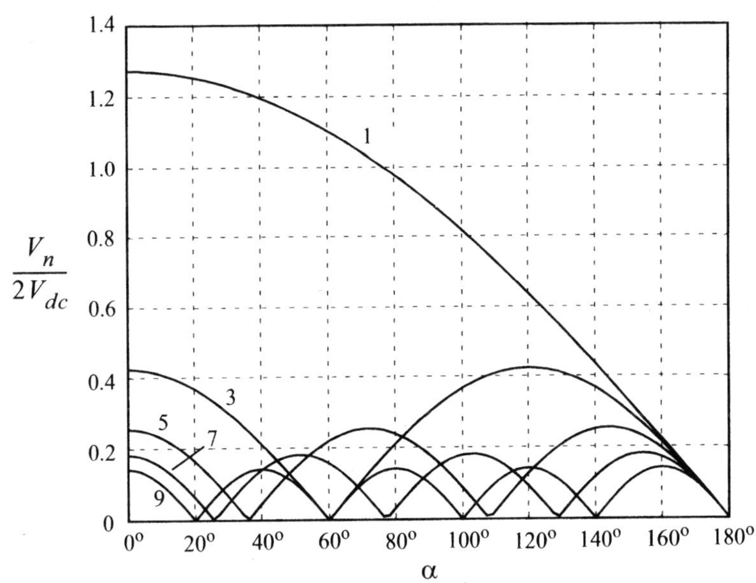

Figure 1.13 First five odd (nonzero) harmonic components of single-phase inverter with phase shift control as a function of phase shift angle α normalized with respect to $2V_{dc}$.

sible with a double bridge as shown in Figure 1.14. Note that this type of bridge is essentially three single-phase bridges so that voltage control can again be accomplished by phase shifting in much the same manner as the overlap method described by Figure 1.12. To avoid short circuits the three-phase load must either be separated into three electrically isolated single-phase loads or a transformer must be used to provide electrical isolation. Figure 1.14 shows the output phase voltages of this inverter.

Recall also that when the phase output voltages are coupled through a transformer into a three-phase voltage set with a common neutral, harmonics of multiples of three are eliminated in the line-to-line output voltages by virtue of the 120° phase shift between the quasi square waves of each phase.

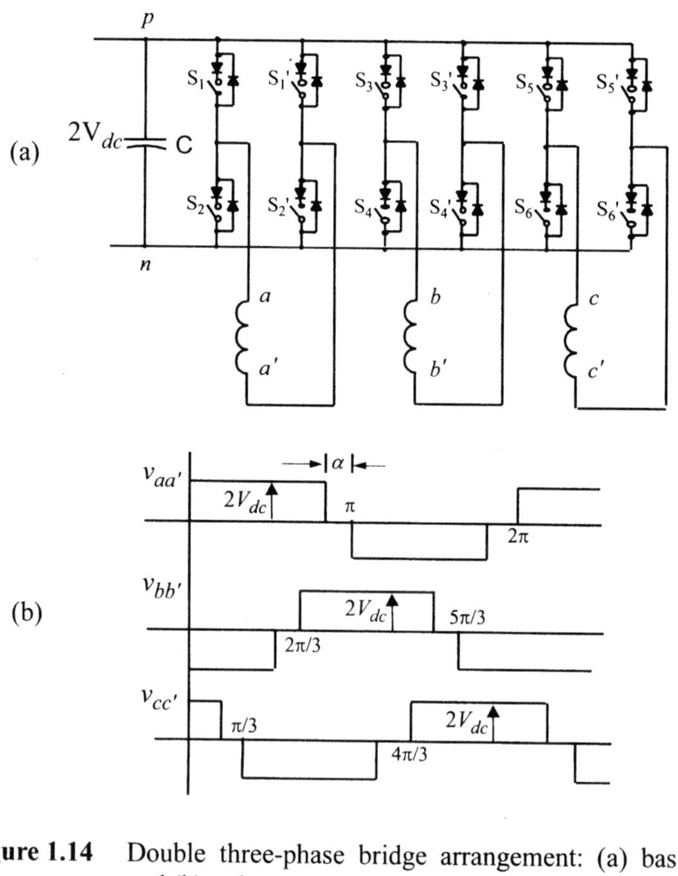

Figure 1.14 Double three-phase bridge arrangement: (a) basic circuit and (b) voltage waveforms.

1.5 Current Source/Stiff Inverters

Up to this point the focus has been on the most popular class of power converters, i.e., those operating with a voltage source or with a stiff capacitor on the DC side of the converter. However, another class of inverters evolve from the dual concept of a current source or stiff inductor on the DC side. These converters can be developed from essentially the same starting point using the basic commutation cells of Figure 1.2, except that the diode is replaced with a second switch in order to have complete control over the direction of the inductor current. Figure 1.15 briefly depicts the evolution of the three-phase current source/stiff inverter. In Figure 1.15(a) the current source commutation cell is shown, and in Figure 1.15(b) the inductor is chosen as the source so that the switch branches become loads. Since the switch branches are connected in series with the load, these loads must clearly be noninductive so as to not produced infinite voltage spikes across the switches. In order to create AC currents in the load two such commutation cells are used – one to produce positive current and the other to produce negative current in the load as shown in Figure 1.15(c). A single-phase bridge is produced by recognizing that no current need flow in the center point connection between the two current sources if they produce the same amplitude of current, as shown in Figure 1.15(d). Finally, a third phase is added in the same manner to produce a three-phase current source inverter, Figure 1.15(e). This evolution realizes the second practical switch combination suitable for DC current sources, a bidirectional voltage blocking, unidirectional current conducting switch. At the present time, such a switch is typically realized by a series-connected transistor and diode arrangement, as shown in Figure 1.15(f).

The basic switching strategy for this converter can again be summarized using switching functions. If $m_1, m_2, \ldots, m_6$ are defined as +1 when switches $T_1, T_2, \ldots, T_6$ are closed and zero when they are open, then to ensure current continuity in the DC side inductor, it is evident from current continuity considerations and Figure 1.15 that

$$m_1 + m_3 + m_5 = 1 \tag{1.28}$$

and

$$m_2 + m_4 + m_6 = 1 \tag{1.29}$$

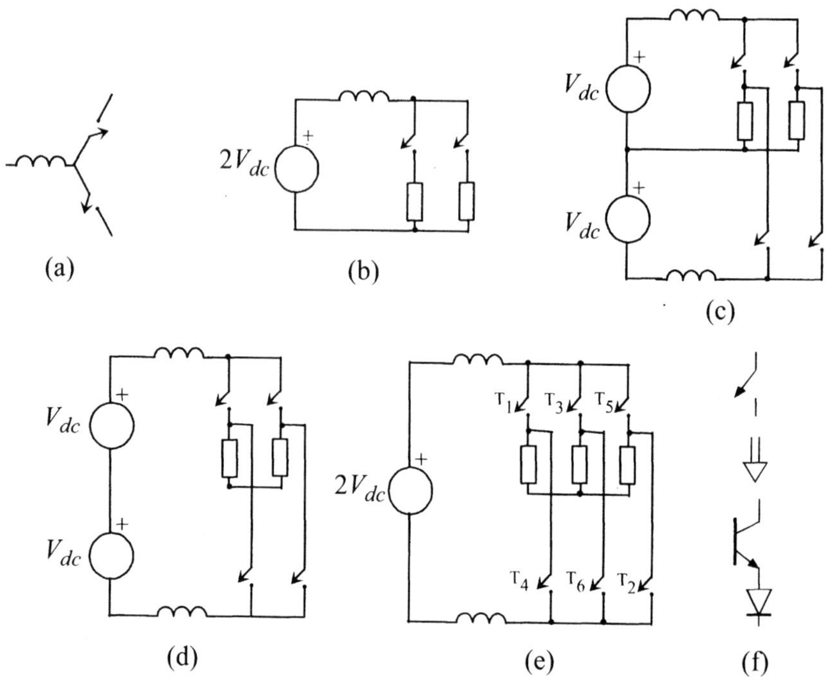

Figure 1.15 Evolution of three-phase current source/stiff inverter from basic commutation cell.

The load currents can also be defined as

$$
\begin{aligned}
i_a &= I_{dc}(m_1 - m_4) \\
i_b &= I_{dc}(m_3 - m_6) \\
i_c &= I_{dc}(m_5 - m_2)
\end{aligned}
\tag{1.30}
$$

The line voltages can then be expressed in terms of the switching functions as

$$
\begin{aligned}
v_{ab} &= 2V_{dc}(m_1 m_6 - m_4 m_3) \\
v_{bc} &= 2V_{dc}(m_3 m_2 - m_6 m_5) \\
v_{ca} &= 2V_{dc}(m_5 m_4 - m_2 m_1)
\end{aligned}
\tag{1.31}
$$

where it is assumed that the voltage drop across the link inductor is negligible for any reasonable size of inductor, since the current will then be very nearly constant.

The phase voltages can be determined in much the same manner as for the voltage link converter, i.e.,

$$\begin{aligned} v_{an} &= v_{as} + v_{sn} \\ v_{bn} &= v_{bs} + v_{sn} \\ v_{cn} &= v_{cs} + v_{sn} \end{aligned} \tag{1.32}$$

where n again represents the voltage at the negative bus of the DC link voltage and s denotes the center point of the load. Adding together the voltages of Eq. (1.32) gives

$$\begin{aligned} v_{an} + v_{bn} + v_{cn} &= v_{as} + v_{bs} + v_{cs} + 3v_{sn} \\ &= 0 + 3v_{sn} \end{aligned} \tag{1.33}$$

from which

$$\begin{aligned} v_{sn} &= \frac{1}{3}(v_{an} + v_{bn} + v_{cn}) \\ &= \frac{2}{3}(m_1 + m_3 + m_5)V_{dc} \\ &= \frac{2}{3}V_{dc} \end{aligned} \tag{1.34}$$

Thus

$$\begin{aligned} v_{as} &= \left(m_1 - \frac{1}{3}\right)2V_{dc} \\ v_{bs} &= \left(m_3 - \frac{1}{3}\right)2V_{dc} \\ v_{cs} &= \left(m_5 - \frac{1}{3}\right)2V_{dc} \end{aligned} \tag{1.35}$$

A plot of the load current assuming a star- and wye-connected load is given in Figure 1.16. If the load is inductive, it is apparent that the idealized current waveforms of Figure 1.16 would produce infinite spikes of voltage. Hence, strictly speaking, the harmonic content for this converter is infinite. In reality, the slopes corresponding to the rapidly changing *di/dt* would not be infinite but would change at a rate dominated by the capacitance of a commutating circuit. For example, the autosequentially commutated inverter (ASCI) of Figure 1.17 is widely used for implementing a current source/stiff converter. Alternatively, capacitive filters can be placed on AC output terminals to absorb the rapid changes in current.

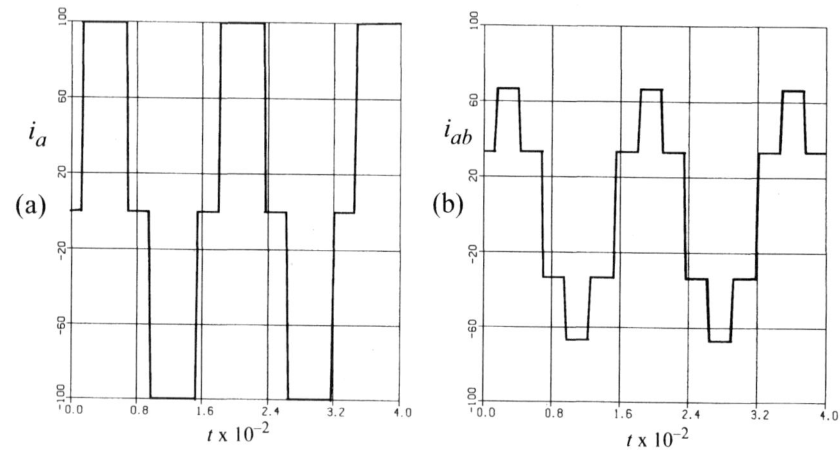

Figure 1.16 Current source inverter waveforms: (a) line current for a star-connected load and (b) phase current for a delta-connected load assuming a DC link current of 100 A.

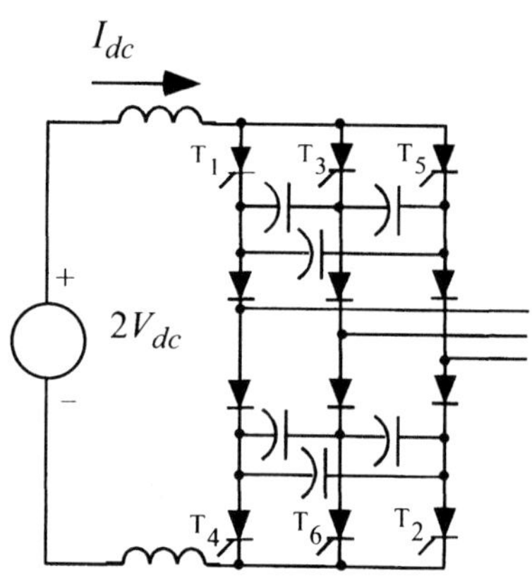

Figure 1.17 Autosequentially commutated current source/stiff converter.

1.6 Concept of a Space Vector

The highly coupled nature of inverter loads such as induction and synchronous machines has led to the use of artificial variables rather than actual (phase) variables for the purpose of simulation as well as for visualization. The essence of the nature of the transformation of variables that is utilized can be understood by reference to Figure 1.18, which shows three-dimensional orthogonal axes labeled *a*, *b*, and *c* [3]. Consider, for instance, the stator currents of a three-phase induction machine load which is, in general, made up of three independent variables. These currents (phase variables) can be visualized as being components of a single three-dimensional vector (*space vector*) existing in a three-dimensional orthogonal space, i.e., the space defined by Figure 1.18. The projection of this vector on the three axes of Figure 1.18 produces the instantaneous values of the three stator currents.

In most practical cases as has been noted already, the sum of these three currents adds up to zero since most three-phase loads do not have a neutral return path. In this case, the stator current vector is constrained to exist only on a plane defined by

$$i_a + i_b + i_c = 0 \tag{1.36}$$

The fact that Eq. (1.36) defines a particular plane is evident if it is recalled from analytic geometry that the general definition of a plane is $ax + by + cz = d$. This plane, the so-called *d–q plane*, is also illustrated in Figure 1.18. Components of the current and voltage vector in the plane are called the *d–q components* while the component in the axis normal to the plane (in the event that the currents do not sum to zero) is called the *zero component.*

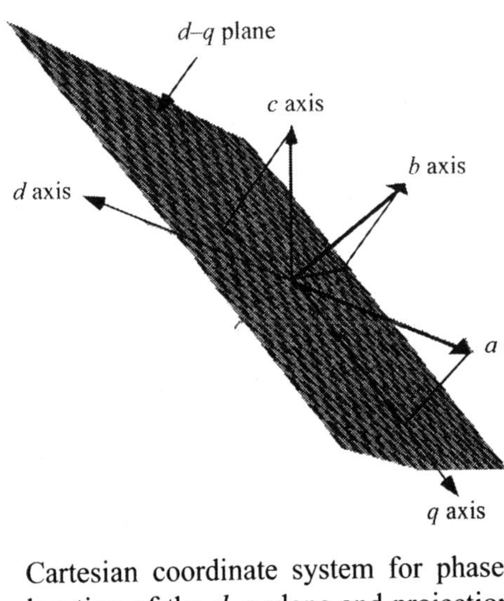

Figure 1.18 Cartesian coordinate system for phase variables showing location of the *d–q* plane and projection of phase variables onto the plane.

When the phase voltages and phase flux linkages also sum to zero, as is the case with most balanced three-phase loads (including even a salient pole synchronous machine), this same perspective can be applied to these variables as well. By convention it is assumed that the projection of the phase a axis on the d–q plane forms the reference q axis for the case where the d–q axes are not rotating. A second axis on the plane is defined as being orthogonal to the q axis such that the cross product $d \times q$ yields a third axis, by necessity normal to the d–q plane, that produces a third component of the vector having the conventional definition of the zero sequence quantity. The components of the phase current, phase voltage, or phase flux linkage vectors in the d–q–0 stationary coordinate system in terms of the corresponding physical variables are

$$\begin{bmatrix} f_q^s \\ f_d^s \\ f_0 \end{bmatrix} = \sqrt{\frac{2}{3}} \begin{bmatrix} 1 & -\frac{1}{2} & -\frac{1}{2} \\ 0 & -\frac{\sqrt{3}}{2} & \frac{\sqrt{3}}{2} \\ \frac{1}{\sqrt{2}} & \frac{1}{\sqrt{2}} & \frac{1}{\sqrt{2}} \end{bmatrix} \begin{bmatrix} f_a \\ f_b \\ f_c \end{bmatrix} \tag{1.37}$$

where f is a general variable used to denote the current variable i, voltage v, flux linkage λ or charge q. The superscript s on the d–q variable is used to denote the case where the d–q axes are *stationary* and fixed in the d–q plane.

In the dominant case where the three-phase variables sum to zero (i.e., the corresponding current, voltage, and flux linkage vectors are located on the d–q plane and have no zero sequence component) this transformation reduces to

$$\begin{bmatrix} f_q^s \\ f_d^s \\ f_0 \end{bmatrix} = \begin{bmatrix} \sqrt{\frac{3}{2}} & 0 & 0 \\ 0 & -\frac{1}{\sqrt{2}} & \frac{1}{\sqrt{2}} \\ 0 & 0 & 0 \end{bmatrix} \begin{bmatrix} f_a \\ f_b \\ f_c \end{bmatrix} \tag{1.38}$$

where the last row is now clearly not necessary and often can be discarded.

Figure 1.19 shows the location of the various axes when projected onto the d–q plane. Note that the projection of the a phase axis on the d–q plane is considered to be lined up with the q axis (the a phase axis corresponds to the magnetic axis of phase a in the case of an electrical machine). The other axis on the

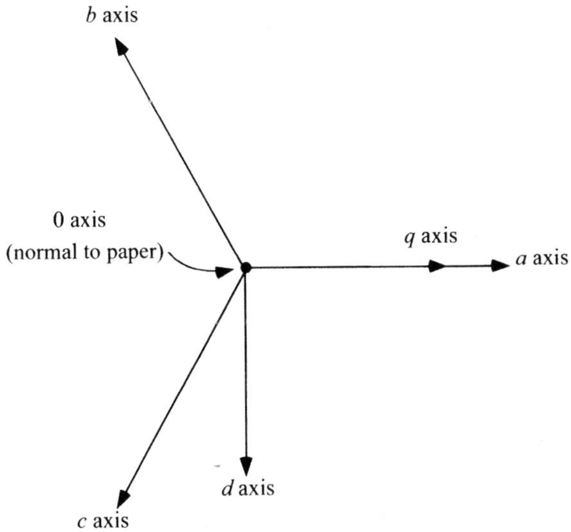

Figure 1.19 Physical a–b–c and conceptual stationary frame d–q–0 axes when viewed from an axis normal to the d–q plane.

plane is, by convention, located 90° clockwise with respect to the q axis. The third axis (necessarily normal to the d–q plane) is chosen such that the sequence d–q–0 forms a right-hand set.

Sometimes another notation, using symbols α,β (Clarke's components), is used to denote these same variables. However, the third component, Fortesque's zero sequence component, is normally not scaled by the same factor as the two Clarke components, and this can cause some confusion. With the transformation shown, when viewed from the zero sequence axis, the d axis is located 90° clockwise with respect to the q axis. Unfortunately, these two axes are sometimes interchanged so that the reader should exercise caution when referring to the literature. When the d–q axes are fixed in predefined positions in the d–q plane, they are said to define the *stationary reference frame* [2].

1.6.1 *d*–*q*–0 Components for Three-Phase Sine Wave Source/Load

When balanced sinusoidal three-phase AC voltages are applied to a three-phase load, typically, with respect to the supply midpoint z

$$
\begin{aligned}
v_{az} &= V_1 \sin\omega_o t \\
v_{bz} &= V_1 \sin\left(\omega_o t - \frac{2\pi}{3}\right) \\
v_{cz} &= V_1 \sin\left(\omega_o t + \frac{2\pi}{3}\right)
\end{aligned}
\tag{1.39}
$$

It can be recalled from Eqs. (1.15) and (1.16) that for a three-wire wye-connected load with balanced impedances the load voltages can be expressed in terms of the supply voltages as [2]

$$
\begin{aligned}
v_{az} &= v_{as} + v_{sz} = Z(p)i_a + v_{sz} \\
v_{bz} &= v_{bs} + v_{sz} = Z(p)i_b + v_{sz} \\
v_{cz} &= v_{cs} + v_{sz} = Z(p)i_c + v_{sz}
\end{aligned}
\tag{1.40}
$$

where, again, s is the load neutral point, p represents the time derivative operator $p = d/(dt)$, and $Z(p)$ denotes the impedance operator made up of an arbitrary circuit configuration of resistors, inductors, and capacitors. If the circuit is at rest at $t = 0$, then summing the rows of Eq. (1.40) gives

$$
v_{az} + v_{bz} + v_{cz} = Z(p)(i_a + i_b + i_c) + 3v_{sz} \tag{1.41}
$$

Since the three currents sum to zero and $Z(p)$ is common to all three-phases, the voltage between load neutral and inverter zero voltage points, for balanced loads but arbitrary source voltages, is

$$
v_{sz} = \frac{1}{3}(v_{az} + v_{bz} + v_{cz}) \tag{1.42}
$$

In the special case of balanced source voltages [Eq. (1.39)] the right-hand side of Eq. (1.42) is zero and the corresponding phase and source voltages are identical. From this result it can readily be determined that, in the d–q–0 coordinate system,

$$
\begin{aligned}
v_{qs}^{s} &= \sqrt{\frac{3}{2}}V_1 \sin\omega_o t \\
v_{ds}^{s} &= -\sqrt{\frac{3}{2}}V_1 \cos\omega_o t \\
v_{0s}^{s} &= 0
\end{aligned}
\tag{1.43}
$$

The use of the subscript s used here to denote the load neutral point can be remembered as the *star point*, *c(s)enter point*, or *neutral point* of the stationary circuit. It should be apparent from the orthogonality of the d–q axes and the sine/cosine relationships that the phase voltage vector traces out a circle on the d–q plane with radius $\sqrt{\frac{3}{2}}V_1$ where V_1 is the amplitude of the phase voltage. The vector rotates with an angular velocity equal to the angular frequency of the source voltage (377 rad/s in the case of 60 Hz). The current and flux linkage vectors, being a consequence of applying the voltage to a linear, balanced load will also trace out circles on the d–q plane in the steady state.

The fact that the length of the rotating vector differs from the amplitude of the sinusoidal variable has prompted researchers to introduce methods to "correct" this supposed deficiency. The difference in length essentially comes about because the a–b–c axes are not in the plane of the d–q axes but have a component in the third direction (0 axis) as evidenced by the third row of Eq. (1.37). However, if the transformation of Eq. (1.37) is multiplied by $\sqrt{2/3}$ a scale change is made in moving from a–b–c to d–q–0 variables which eliminates this difference. The transformation becomes

$$\begin{bmatrix} f_{qs}^s \\ f_{ds}^s \\ f_{0s} \end{bmatrix} = \begin{bmatrix} \frac{2}{3} & -\frac{1}{3} & -\frac{1}{3} \\ 0 & -\frac{1}{\sqrt{3}} & \frac{1}{\sqrt{3}} \\ \frac{\sqrt{2}}{3} & \frac{\sqrt{2}}{3} & \frac{\sqrt{2}}{3} \end{bmatrix} \begin{bmatrix} f_{as} \\ f_{bs} \\ f_{cs} \end{bmatrix} \tag{1.44}$$

with the inverse relationship of

$$\begin{bmatrix} f_{as} \\ f_{bs} \\ f_{cs} \end{bmatrix} = \begin{bmatrix} 1 & 0 & \frac{1}{\sqrt{2}} \\ -\frac{1}{2} & -\frac{\sqrt{3}}{2} & \frac{1}{\sqrt{2}} \\ -\frac{1}{2} & \frac{\sqrt{3}}{2} & \frac{1}{\sqrt{2}} \end{bmatrix} \begin{bmatrix} f_{qs}^s \\ f_{ds}^s \\ f_{0s} \end{bmatrix} \tag{1.45}$$

Since the same scale change has been made for all three components, the zero component uses somewhat unconventional scaling. More conventionally, Fortesque's scaling for this component is

$$f_{0s} = \frac{1}{3}(f_{as} + f_{bs} + f_{cs}) \tag{1.46}$$

and is also widely used. When the projection of the vector on the zero sequence axis is zero, Eq. (1.44) reduces to

$$\begin{bmatrix} f_{qs}^{s} \\ f_{ds}^{s} \\ f_{0s} \end{bmatrix} = \begin{bmatrix} 1 & 0 & 0 \\ 0 & -\frac{1}{\sqrt{3}} & \frac{1}{\sqrt{3}} \\ 0 & 0 & 0 \end{bmatrix} \begin{bmatrix} f_{as} \\ f_{bs} \\ f_{cs} \end{bmatrix} \tag{1.47}$$

or inversely as

$$\begin{bmatrix} f_{as} \\ f_{bs} \\ f_{cs} \end{bmatrix} = \begin{bmatrix} 1 & 0 & 0 \\ -\frac{1}{2} & -\frac{\sqrt{3}}{2} & 0 \\ -\frac{1}{2} & \frac{\sqrt{3}}{2} & 0 \end{bmatrix} \begin{bmatrix} f_{qs}^{s} \\ f_{ds}^{s} \\ f_{0s} \end{bmatrix} \tag{1.48}$$

Note that Eq. (1.47) does not formally have an inverse, but a suitable equivalent can be obtained by first inverting Eq. (1.44) and then setting the zero sequence component to zero. This modified definition of the stationary frame *d–q–0* components will be used in the remainder of this book.

1.6.2 *d–q–0* Components for Voltage Source Inverter Operated in Square-Wave Mode

In general it is readily determined that there are only eight possible switch combinations for a three-phase inverter, as shown in Figure 1.20. Two of these states ($\overline{SV_0}$ and $\overline{SV_7}$) are a short circuit of the output while the remaining six produce active voltages, as previously indicated in Figure 1.10. Figure 1.21 shows how the eight switch combinations can be represented as stationary vectors in the *d–q* plane simply by repetitive application of Eq. (1.47). It is relatively easy to determine the values of the six vectors in the *d–q* plane by investigating each nonzero switch connection. For example, for vector $\overline{SV_3}$, switch S_3 is closed connecting phase *b* to the positive terminal, while switches S_4 and S_2 are closed connecting phases *a* and *c* to the negative terminal. Assuming a wye-connected load, the phase *b* voltage receives two-thirds of the DC pole-to-pole voltage while the parallel connected phases *a* and *c* receive one-third of this voltage. With due regard for polarity, the *d–q* voltages are, from Eq. (1.44),

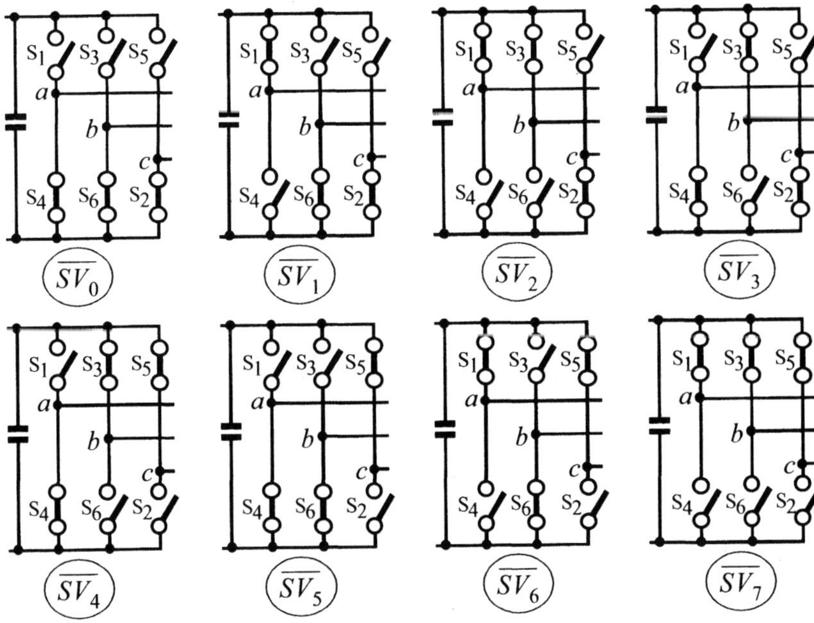

Figure 1.20 The eight possible phase leg switch combinations for a VSI.

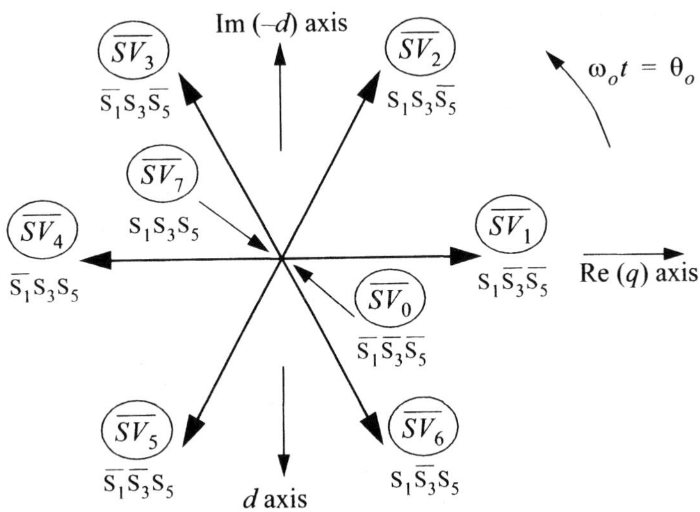

Figure 1.21 Eight possible stationary vectors on the *d–q* plane for a VSI.

$$v_{qs} = -\frac{2}{3}V_{dc}$$
$$v_{ds} = \frac{1}{\sqrt{3}}\left(-\frac{2}{3}V_{dc} - \frac{4}{3}V_{dc}\right) = -\frac{2}{\sqrt{3}}V_{dc} \quad (1.49)$$

The magnitude of this vector is

$$\left|\overline{SV}_3\right| = \sqrt{v_{qs}^2 + v_{ds}^2} = \sqrt{\left(\frac{2}{3}V_{dc}\right)^2 + \left(\frac{2}{\sqrt{3}}V_{dc}\right)^2} = \frac{4}{3}V_{dc} \quad (1.50)$$

as is the case for all six of the nonzero vector locations. For convenience, the projection of axes of the three-phase voltages can also be located on the *d–q* plane since vectors $\overline{SV}_1$, $\overline{SV}_3$, and $\overline{SV}_5$ also result in a positive maximum voltage on phases *as*, *bs*, and *cs*, respectively.

Note that the lower phase leg switches (S_4, S_6, S_2) are represented as "NOT" the upper phase leg switches ($\overline{S}_1$, $\overline{S}_3$, $\overline{S}_5$) in Figure 1.21, reflecting the fact that the upper or lower switch in each phase leg must always be turned on to maintain current continuity through each phase leg for a voltage source inverter.

In general, the inverter attempts to follow the circle defined by the balanced set of voltages, Eqs. (1.39) and (1.43). However, since only nonzero inverter states are possible, as illustrated by $\overline{SV}_1$, $\overline{SV}_2$, $\overline{SV}_3$, ..., $\overline{SV}_6$ in Figure 1.20, the vector representing the voltage applied to the load jumps abruptly by 60 electrical degrees in a continuous counterclockwise fashion, approximating the circle by the points on a hexagon. While only crudely approximated in this case, more accurate tracking of the target circle on the *d–q* plane can be accomplished by more sophisticated pulse width modulation techniques, as will be considered later in this book. If a simple inductive load is assumed, Figure 1.22 shows a typical plot of the transient progression of the voltage vector and current vector for a typical *r–L* load initially at rest (zero initial conditions) without PWM. The voltage vector remains confined to the six points of a hexagon while the current vector traces out a hexagon rotated by roughly 90 electrical degrees.

Since any vector has spatial content (length and direction), it is frequently convenient to abandon the matrix notation and to assign directional unit vectors to identify the components of the vector in the three-dimensional space defined by the *d–q–0* coordinates. In this case, however, one is concerned

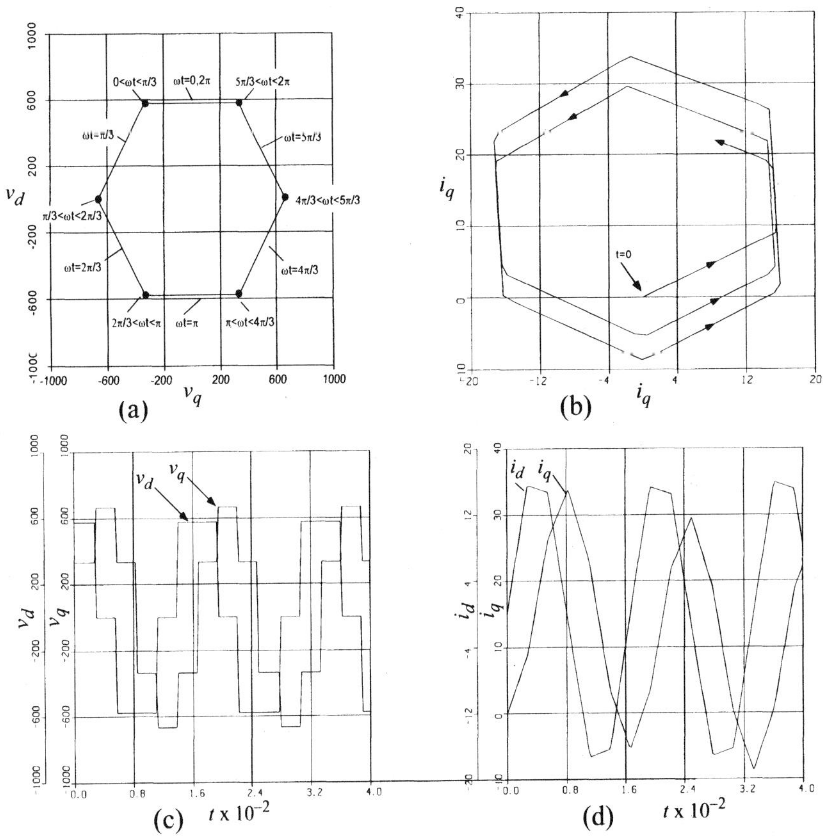

Figure 1.22 Locus of the (a) voltage and (b) current space vectors for square-wave voltage source inverter assuming a balanced r–L load and (c), (d) corresponding time domain waveforms. Parameters: $V_{dc} = 500$ V, $r = 2.0\ \Omega$, $L = 0.1$ mH.

almost entirely with rotation in the d–q plane as opposed to linear translation in the three dimensional d–q–0 space. In such cases, rotation is most readily represented in complex polar form so the most convenient method of representing the rotating vector is simply to convert the d–q plane to one which is complex, whereupon, by definition,

$$\begin{aligned} \mathrm{Re}(\bar{f}_s) &= f_{qs}^s \\ \mathrm{Im}(\bar{f}_s) &= -f_{ds}^s \end{aligned} \tag{1.51}$$

The "jumps" in the vector representing the inverter voltage can now be conveniently represented by defining the operator

$$\bar{a} = e^{j(2\pi/3)} = -\frac{1}{2}+j\frac{\sqrt{3}}{2} \tag{1.52}$$

where $j = \sqrt{-1}$. In terms of the vector $\bar{a}$ the axes of the three-phase voltages can be located by the vectors 1, $\bar{a}$, and $\bar{a}^2$, respectively.

Since the phase voltages have been assigned spatial attributes, it is now possible to visualize the phase variables as *space vectors* in the complex plane. In general, this vector is given by

$$\bar{f}_s = f_{qs}^s - j f_{ds}^s \tag{1.53}$$

and, from Eq. (1.47),

$$\bar{f}_s = f_{as} - j\left(\frac{f_{cs}-f_{bs}}{\sqrt{3}}\right) \tag{1.54}$$

The quantity $-j$ can be written in terms of the operator $\bar{a}$ as

$$-j = \frac{1}{\sqrt{3}}(\bar{a}^2 - \bar{a}) \tag{1.55}$$

so that Eq. (1.54) becomes

$$\begin{aligned}\bar{f}_s &= f_{as} + \frac{1}{3}(\bar{a}^2 - \bar{a})(f_{cs}-f_{bs}) \\ &= f_{as} + \frac{1}{3}(\bar{a}^2 f_{cs} + \bar{a} f_{bs}) - \frac{1}{3}(\bar{a}^2 f_{bs} + \bar{a} f_{cs})\end{aligned} \tag{1.56}$$

Since the sum of the three components f_{as}, f_{bs}, f_{cs} equals zero, this equation can be written as

$$\bar{f}_s = f_{as} + \frac{1}{3}(\bar{a}^2 f_{cs} + \bar{a} f_{bs}) - \frac{1}{3}[\bar{a}^2(-f_{as}-f_{cs}) + \bar{a}(-f_{as}-f_{bs})]$$

which simplifies to

$$\bar{f}_s = \frac{2}{3}(f_{as} + \bar{a} f_{bs} + \bar{a}^2 f_{cs}) \tag{1.57}$$

Note the presence of the 2/3 factor, which is necessary to preserve the correct amplitude when represented in phase variable coordinates.

The six nonzero switch combinations can also be considered to be stationary snapshots of a three-phase set of time-varying sinusoids with a phase voltage magnitude V_m as shown in Figure 1.23. The magnitude of each of the six active vectors is determined by recognizing that the line voltage at each snap-

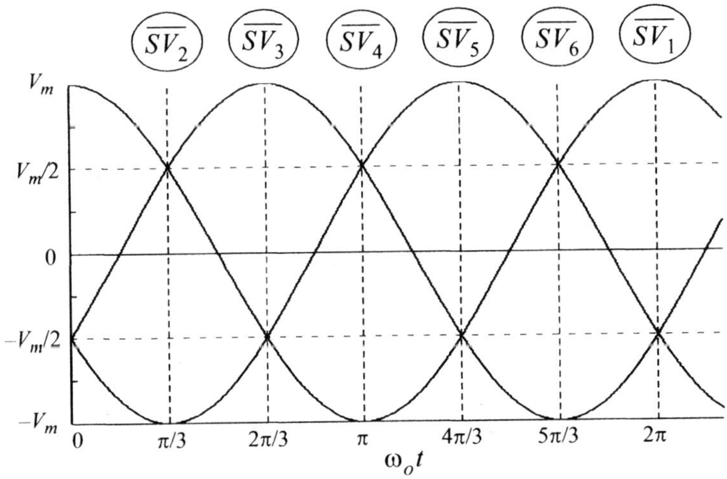

Figure 1.23 VSI phasor angular positions in fundamental cycle for space vector.

shot of voltage in Figure 1.23 is $1.5V_m$ where V_m is the peak phase voltage, and that this voltage is equal to the DC bus voltage $2V_{dc}$. That is,

$$V_m = \frac{4}{3}V_{dc} \tag{1.58}$$

1.6.3 Synchronously Rotating Reference Frame

The visualization of vector rotation on the d–q plane has also led to transformations which rotate with these vectors. For example, if axes are defined which rotate with the stator voltage, one realizes the *synchronous voltage reference frame*. In general, it is not necessary to define rotating axes to rotate synchronously with one particular vector but to simply define a general rotating transformation which transforms the phase variables to rotating axes on the d–q plane, viz:

$$\begin{bmatrix} f_{qs}(\theta) \\ f_{ds}(\theta) \\ f_{0s} \end{bmatrix} = \frac{2}{3}\begin{bmatrix} \cos\theta & \cos\left(\theta - \frac{2\pi}{3}\right) & \cos\left(\theta + \frac{2\pi}{3}\right) \\ \sin\theta & \sin\left(\theta - \frac{2\pi}{3}\right) & \sin\left(\theta + \frac{2\pi}{3}\right) \\ \frac{1}{\sqrt{2}} & \frac{1}{\sqrt{2}} & \frac{1}{\sqrt{2}} \end{bmatrix}\begin{bmatrix} f_{as} \\ f_{bs} \\ f_{cs} \end{bmatrix} \tag{1.59}$$

For completeness, the zero component is also again included, together with the scale change of $\sqrt{2/3}$. The angle θ is the angular displacement of the vector on the d–q plane measured with respect to the projection of the as axis onto this plane. Since the same scale change has been made for all three components, the zero component again uses somewhat unconventional scaling. Alternatively, Fortesque's scaling for this component can also be selected, namely

$$f_{0s} = \frac{1}{3}(f_{as} + f_{bs} + f_{cs}) \tag{1.60}$$

and this is also widely used.

Note that the zero axis does not enter into the rotational transformation. Hence, the zero axis can be considered as the axis about which the rotation takes place, i.e., the axis of rotation. Because of the scaling, the power (and subsequently the torque if the load is a motor) is different in d–q–0 components than in a–b–c variables and a 3/2 multiplier must be added to the power as calculated in the transformed system of equations since both current and voltage variables have been scaled by $\sqrt{2/3}$.

In vector notation, Eq. (1.59) can be written as

$$\boldsymbol{f}_{qd0} = \boldsymbol{T}_{qd0}(\theta)\boldsymbol{f}_{abc} \tag{1.61}$$

where

$$\boldsymbol{T}_{qd0}(\theta) = \frac{2}{3}\begin{bmatrix} \cos\theta & \cos\left(\theta - \frac{2\pi}{3}\right) & \cos\left(\theta + \frac{2\pi}{3}\right) \\ \sin\theta & \sin\left(\theta - \frac{2\pi}{3}\right) & \sin\left(\theta + \frac{2\pi}{3}\right) \\ \frac{1}{\sqrt{2}} & \frac{1}{\sqrt{2}} & \frac{1}{\sqrt{2}} \end{bmatrix} \tag{1.62}$$

The transformation $\boldsymbol{T}_{qd0}(\theta)$ can, for convenience and for computational advantage, be broken into two portions, one of which takes variables from physical phase quantities to nonrotating d–q–0 variables (stationary reference frame) while the other goes from nonrotating to rotating d–q–0 variables (rotating reference frame). For this case one can write,

$$\boldsymbol{f}_{qd0} = \boldsymbol{T}_{qd0}(\theta)\boldsymbol{f}_{abc} = \boldsymbol{R}(\theta)\boldsymbol{T}_{qd0}(0)\boldsymbol{f}_{abc} \tag{1.63}$$

where

$$\boldsymbol{T}_{qd0}(0) = \begin{bmatrix} \frac{2}{3} & -\frac{1}{3} & -\frac{1}{3} \\ 0 & -\frac{1}{\sqrt{3}} & \frac{1}{\sqrt{3}} \\ \frac{\sqrt{2}}{3} & \frac{\sqrt{2}}{3} & \frac{\sqrt{2}}{3} \end{bmatrix} \tag{1.64}$$

and

$$\boldsymbol{R}(\theta) = \begin{bmatrix} \cos\theta & -\sin\theta & 0 \\ \sin\theta & \cos\theta & 0 \\ 0 & 0 & 1 \end{bmatrix} \tag{1.65}$$

Note that $\boldsymbol{T}_{qd0}(0)$ is obtained by simply setting $\theta = 0$ in Eq. (1.62). The inverse transformation is

$$\boldsymbol{f}_{abc} = \boldsymbol{T}_{qd0}(\theta)^{-1}\boldsymbol{f}_{qd0} = \boldsymbol{T}_{qd0}(0)^{-1}\boldsymbol{R}(\theta)^{-1}\boldsymbol{f}_{qd0} \tag{1.66}$$

where

$$\boldsymbol{T}_{qd0}(\theta)^{-1} = \frac{3}{2}\boldsymbol{T}_{qd0}(\theta)^{T} = \begin{bmatrix} \cos\theta & \sin\theta & \frac{1}{\sqrt{2}} \\ \cos\left(\theta - \frac{2\pi}{3}\right) & \sin\left(\theta - \frac{2\pi}{3}\right) & \frac{1}{\sqrt{2}} \\ \cos\left(\theta + \frac{2\pi}{3}\right) & \sin\left(\theta + \frac{2\pi}{3}\right) & \frac{1}{\sqrt{2}} \end{bmatrix} \tag{1.67}$$

$$\boldsymbol{T}_{qd0}(0)^{-1} = \frac{3}{2}\boldsymbol{T}_{qd0}(0)^{T} = \begin{bmatrix} 1 & 0 & \frac{1}{\sqrt{2}} \\ -\frac{1}{2} & -\frac{\sqrt{3}}{2} & \frac{1}{\sqrt{2}} \\ -\frac{1}{2} & \frac{\sqrt{3}}{2} & \frac{1}{\sqrt{2}} \end{bmatrix} \tag{1.68}$$

$$\boldsymbol{R}(\theta)^{-1} = \boldsymbol{R}(\theta)^{T} = \begin{bmatrix} \cos\theta & \sin\theta & 0 \\ -\sin\theta & \cos\theta & 0 \\ 0 & 0 & 1 \end{bmatrix} \tag{1.69}$$

Figure 1.24 shows the locus of the same voltage and current vector as Figure 1.22, but viewed in a synchronously rotating reference frame. Here, the superscript e is used on the d–q variables to denote that they are rotating with the *electrical* frequency. In this case the locus of the vector continually jumps ahead by 60 electrical degrees at each switching instant. However, since the vector subsequently remains stationary while the synchronous frame continues to rotate, the locus of the vector slowly retreats backward and then jumps forward, repeating the locus shown six times each cycle.

1.7 Three-Level Inverters

Another voltage source/stiff inverter configuration which is becoming increasingly important for high power applications is the so-called *neutral-point clamped* (NPC) inverter [4]. This inverter, shown in Figure 1.25, has a zero DC voltage center point, which is switchable to the phase outputs, thereby creating the possibility of switching each inverter phase leg to one of three voltage levels. The major benefit of this configuration is that, while there are twice as many switches as in the two-level inverter, each of the switches must block only one-half of the DC link voltage (as is also the case for the six center-tapped diodes). However, one problem, that does not occur with a two-level inverter, is the need to ensure voltage balance across the two series-connected capacitors making up the DC link. One solution is to simply connect each of

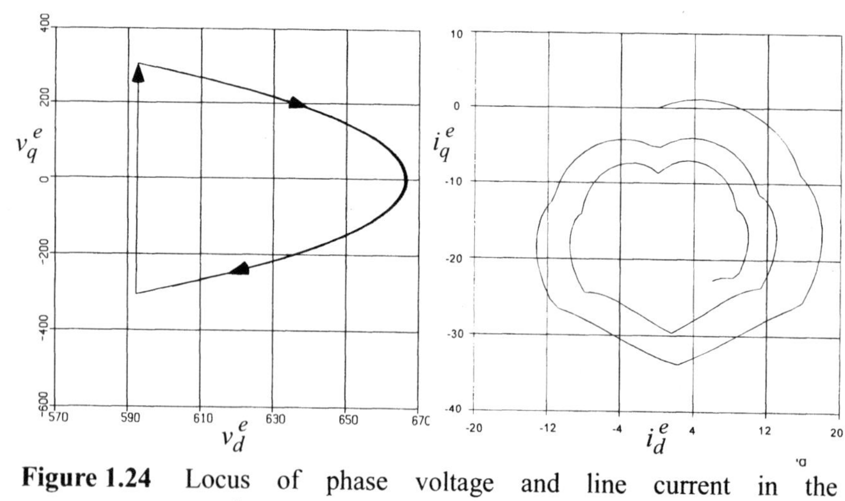

Figure 1.24 Locus of phase voltage and line current in the synchronously rotating reference frame for the same parameters as Figure 1.22.

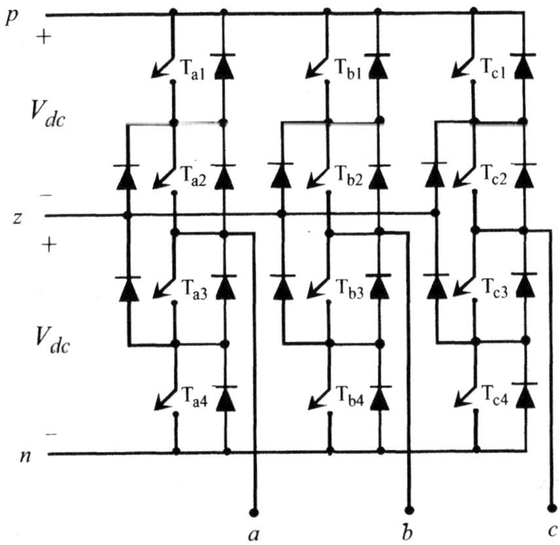

Figure 1.25 Three-level voltage source/stiff inverter.

the capacitors to its own isolated DC source (for example the output of a diode bridge fed from a transformer secondary). The other method is to balance the two capacitor voltages by feedback control. In this case the time that each inverter leg dwells on the center point can be adjusted so as to regulate the average current into the center point to be zero [5, 6].

In order to produce three levels, the switches are controlled such that only two of the four switches in each phase leg are turned on at any time. For example, when switches T_{a1} and T_{a2} are turned on, the positive DC link voltage is applied to the phase a terminal, when T_{a2} and T_{a3} are turned on, the center point voltage (zero volts) is applied, while when T_{a3} and T_{a4} are turned on, the negative DC link voltage appears at the terminal of phase a. Defining variables m_{x1}, m_{x2}, and m_{x3} ($x = a, b,$ or c) as a logical one when switch combinations (T_{x1}–T_{x2}), (T_{x2}–T_{x3}), (T_{x3}–T_{x4}) are true, respectively, and zero otherwise, the three terminal voltages can be written as

$$\begin{aligned} v_{az} &= V_{dc}(m_{a1} - m_{a3}) \\ v_{bz} &= V_{dc}(m_{b1} - m_{b3}) \\ v_{cz} &= V_{dc}(m_{c1} - m_{c3}) \end{aligned} \tag{1.70}$$

In a similar manner to previous work, the line-to-line and line-to-neutral voltages (in the case of a star-connected load) are given by

$$\begin{aligned} v_{ab} &= V_{dc}(m_{a1} - m_{a3} - m_{b1} + m_{b3}) \\ v_{bc} &= V_{dc}(m_{b1} - m_{b3} - m_{c1} + m_{c3}) \\ v_{ca} &= V_{dc}(m_{c1} - m_{c3} - m_{a1} + m_{a3}) \end{aligned} \tag{1.71}$$

and

$$\begin{aligned} v_{as} &= \frac{2}{3}V_{dc}\left[m_{a1} - m_{a3} - \frac{1}{2}(m_{b1} - m_{b3} + m_{c1} - m_{c3})\right] \\ v_{bs} &= \frac{2}{3}V_{dc}\left[m_{b1} - m_{b3} - \frac{1}{2}(m_{a1} - m_{a3} + m_{c1} - m_{c3})\right] \\ v_{cs} &= \frac{2}{3}V_{dc}\left[m_{c1} - m_{c3} - \frac{1}{2}(m_{a1} - m_{a3} + m_{b1} - m_{b3})\right] \end{aligned} \tag{1.72}$$

The load neutral-to-inverter-midpoint voltage is, for this case,

$$v_{sz} = \frac{1}{3}V_{dc}(m_{a1} - m_{a3} + m_{b1} - m_{b3} + m_{c1} - m_{c3}) \tag{1.73}$$

The three DC link currents are

$$\begin{aligned} I_{dc}^{+} &= m_{a1}i_a + m_{b1}i_b + m_{c1}i_c \\ I_{dc}^{-} &= m_{a3}i_a + m_{b3}i_b + m_{c3}i_c \\ I_{dc}^{z} &= m_{a2}i_a + m_{b2}i_b + m_{c2}i_c \end{aligned} \tag{1.74}$$

It is clear from the constraints of the circuit that

$$m_{a1} + m_{a2} + m_{a3} = 1 \tag{1.75}$$

and so forth, so that I_{dc}^{z} from Eq. (1.74) can also be written as

$$I_{dc}^{z} = -(m_{a1} + m_{a3})i_a - (m_{b1} + m_{b3})i_b - (m_{c1} + m_{c3})i_c \tag{1.76}$$

Examination of Eq. (1.72) identifies that the line-to-neutral voltage can take on seven distinct values, namely $\pm 4V_{dc}/3$, $\pm V_{dc}$, $\pm 2V_{dc}/3$, and zero.

Figure 1.26 shows the switched phase leg, line-to-line and line-to-neutral phase voltages for a three level NPC inverter, where the phase legs are switched with a 30° zero voltage plateau between the $+V_{dc}$ and the $-V_{dc}$ steps to achieve minimum total harmonic distortion (THD). It is obvious that the switched waveforms of Figure 1.26 are significantly improved compared to those of a square-wave, two-level converter shown in Figure 1.10.

The fundamental component of the load voltage can now be determined by calculating the fundamental component of the voltage from one phase leg to

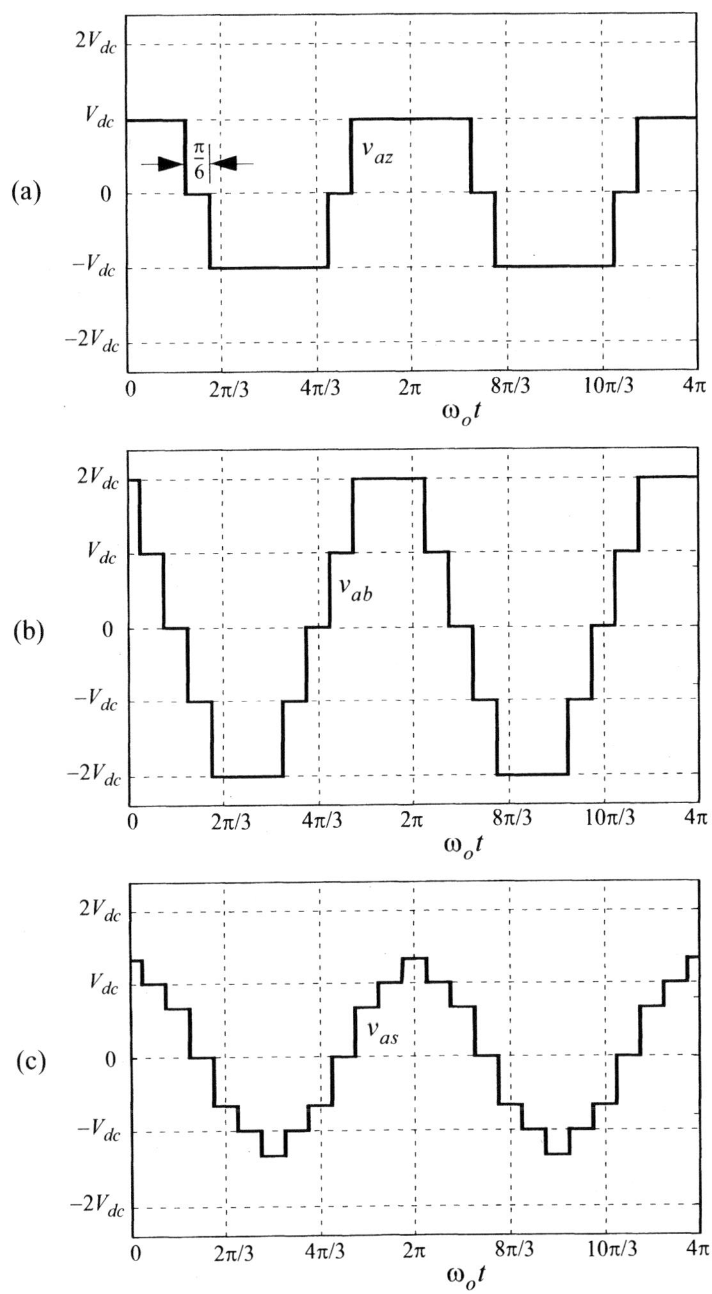

Figure 1.26 Switched (a) line-to-midpoint, (b) line-to-line and (b) line-to-neutral voltages produced by three-level NPC inverter. Phase leg zero voltage step width arranged for minimum THD.

the DC center point, which will be the same as the fundamental component measured with respect to the load neutral. Hence

$$V_{1,rms} = \frac{1}{\sqrt{2}}\frac{4}{\pi}\int_{0}^{5\pi/12} V_{dc}\cos\theta\, d\theta$$

$$= \frac{2\sqrt{2}}{\pi}\sin\frac{5\pi}{12}V_{dc} = 0.87V_{dc} \tag{1.77}$$

This is not quite as large as the $0.90V_{dc}$ RMS voltage that is obtained from a two-level inverter with the same overall DC link voltage of $2V_{dc}$ by dividing the result of Eq. (1.7) by $\sqrt{2}$ to convert it to RMS. Of course, the same RMS voltage could still be obtained if the midpoint voltage was never selected, but this results in a higher harmonic content and more or less defeats the purpose of having a three-level inverter.

The space vector approach, discussed in Section 1.6, can be utilized to analyze these types of converters as well. For example, when the possible phase voltage switching states of the three-level inverter of Figure 1.25 are expressed as space vectors, a double hexagon is obtained as shown in Figure 1.27.

1.8 Multilevel Inverter Topologies

1.8.1 Diode-Clamped Multilevel Inverter

It has been shown that the principle of diode-clamping to DC link voltages can be extended to any number of voltage levels [7]. Figure 1.28 shows the circuit implementation for four- and five-level inverters. Since the voltages across the semiconductor switches are limited by conduction of the diodes connected to the various DC levels, this class of multilevel inverters is termed *diode-clamped multilevel inverters*. For the case of the four-level inverter, switches T_{a1}, T_{a2}, and T_{a3} are turned on simultaneously, then T_{a2}, T_{a3}, and T_{a4} and so forth to produce the desired level. A possible total of 12 voltage vectors are produced, being $\pm kV_{dc}/3$, $k = 1,\ldots,6$ where the $k = 2$ state does not appear in the line-to-neutral voltage output. In the same manner, four switches are always triggered to select a desired level in the five-level inverter, producing 19 voltage levels, $\pm kV_{dc}/3$, $k = 1,\ldots,9$ plus zero where the $k = 1$ and $k = 3$ levels do not appear in the output line-to-neutral voltage. Since several intermediate levels are now created, the problem of ensuring voltage balance across each

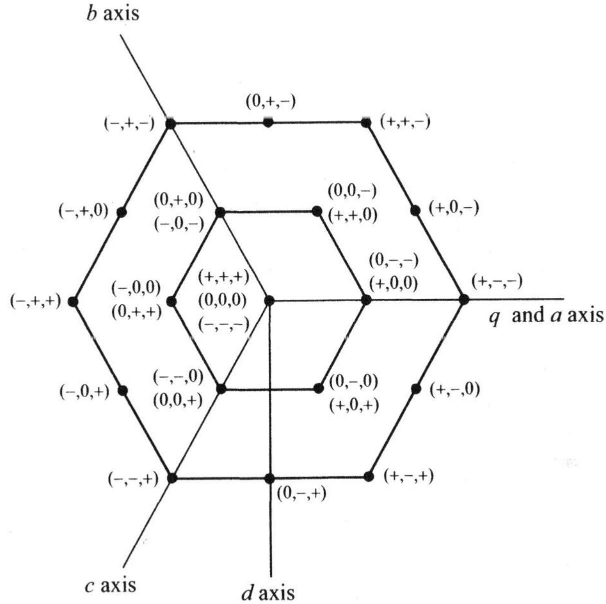

Figure 1.27 Hexagons of possible switching states for three-level DC voltage source/stiff inverter.

of the series-connected capacitors becomes a challenging problem. The problem is most easily solved by feeding each DC link capacitor with an independent DC supply but this sometimes adds undesirable cost to the system. When supplied by a single DC supply, the issue has been shown to be solvable by the use of appropriate control algorithms [8]. However, complete control of the DC link voltages is lost when the output phase voltage reaches roughly 65% of its maximum value, somewhat restricting the use of this circuit [9].

The switching operation of these converters and the current path for positive and negative current can further identified by considering the seven-level diode-clamped inverter of Figure 1.29. The three DC voltages are labeled as V_{dc1}, V_{dc2}, V_{dc3} to distinguish them in the inverter output, although in most cases $V_{dc1} = V_{dc2} = V_{dc3}$. The phase leg switch states required to achieve the seven output levels can be determined by connecting say phase leg b to the negative DC bus by triggering all switches in the lower portion of its phase leg. Then the phase leg a output voltage with respect to the negative DC rail, v_{an}, can be identified for various switch combinations, as summarized in Table 1.1.

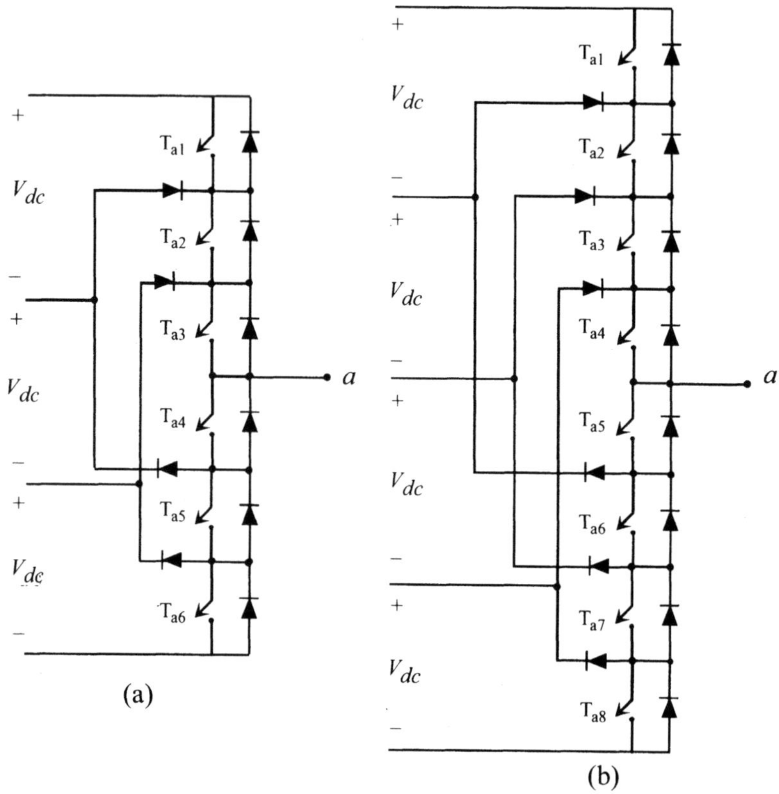

Figure 1.28 (a) Four-level and (b) five-level diode-clamped inverter implementations (one phase only shown).

Figures 1.30, 1.31, and 1.32, show the switched phase leg, line-to-line and line-to-neutral phase voltages for a four-level, five-level, and a seven-level, diode-clamped inverter, respectively. In this case the phase legs have been switched between the voltage levels at the appropriate times to eliminate low-order harmonics, as discussed in Section 10.3.1 and summarized in Table 10.1. The progressive improvement in the quality of the switched waveform is obvious as the number of inverter voltage levels increases.

Regardless of the number of levels, the blocking voltage of the switches in this type of topology is limited to V_{dc}, so that inverters operating at the medium AC voltage range (2 to 13.2 kV) can be implemented with low cost, high-performance Insulated Gate Bipolar Transistor (IGBT) switches. Unfortunately the same is not true of the diodes connecting the various DC levels to

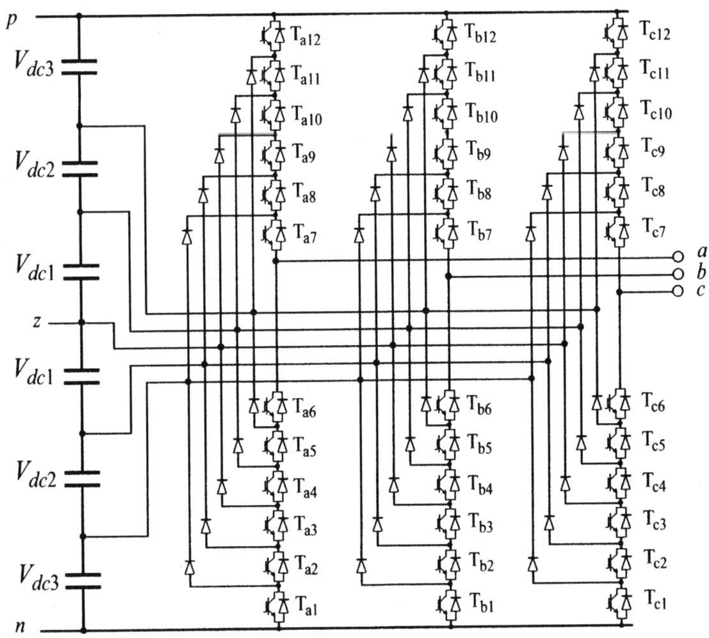

Figure 1.29 A seven-level diode-clamped inverter.

Table 1.1 Switch States and Corresponding Current Path for Diode-Clamped Converter Illustrating Seven Positive Levels for Phase Leg *a* Voltage v_{an}, Phase *b* Assumed Connected to the Negative DC Bus

Switch State	Phase Leg *a* Voltage v_{an}	Devices Turned on	Current Path with I_a Positive	Current Path with I_a Negative
1	$v_{an} = 0$	T_{a1} to T_{a6}	T_{a6}>T_{a5}>T_{a4}> T_{a3}>T_{a2}>T_{a1}	T_{a6}>T_{a5}>T_{a4}> T_{a3}>T_{a2}>T_{a1}
2	$v_{an} = V_{dc3}$	T_{a2} to T_{a7}	D_{a7}>T_{a7}	T_{a6}>T_{a5}>T_{a4}> T_{a3}>T_{a2}>D_{a2}
3	$v_{an} = V_{dc3}+V_{dc2}$	T_{a3} to T_{a8}	D_{a8}>T_{a8}>T_{a7}	T_{a6}>T_{a5}>T_{a4}> T_{a3}>D_{a3}
4	$v_{an} = V_{dc3}+V_{dc2} +V_{dc1}$	T_{a4} to T_{a9}	D_{a9}>T_{a9}>T_{a8}> T_{a7}	T_{a6}>T_{a5}>T_{a4}> D_{a4}
5	$v_{an} = V_{dc3}+V_{dc2} +2V_{dc1}$	T_{a5} to T_{a10}	D_{a10}>T_{a10}>T_{a9}> T_{a8}>T_{a7}	T_{a6}>T_{a5}>D_{a5}
6	$v_{an} = V_{dc3} + 2(V_{dc2}+V_{dc1})$	T_{a6} to T_{a11}	D_{a11}>T_{a11}>T_{a10}> T_{a9}>T_{a8}>T_{a7}	T_{a6}>D_{a6}
7	$v_{an} = 2(V_{dc3} + V_{dc2} + V_{dc1})$	T_{a7} to T_{a12}	T_{a12}>T_{a11}>T_{a10}> T_{a9}>T_{a8}>T_{a7}	T_{a7}>T_{a8}>T_{a9}> T_{a10}>T_{a11}>T_{a12}

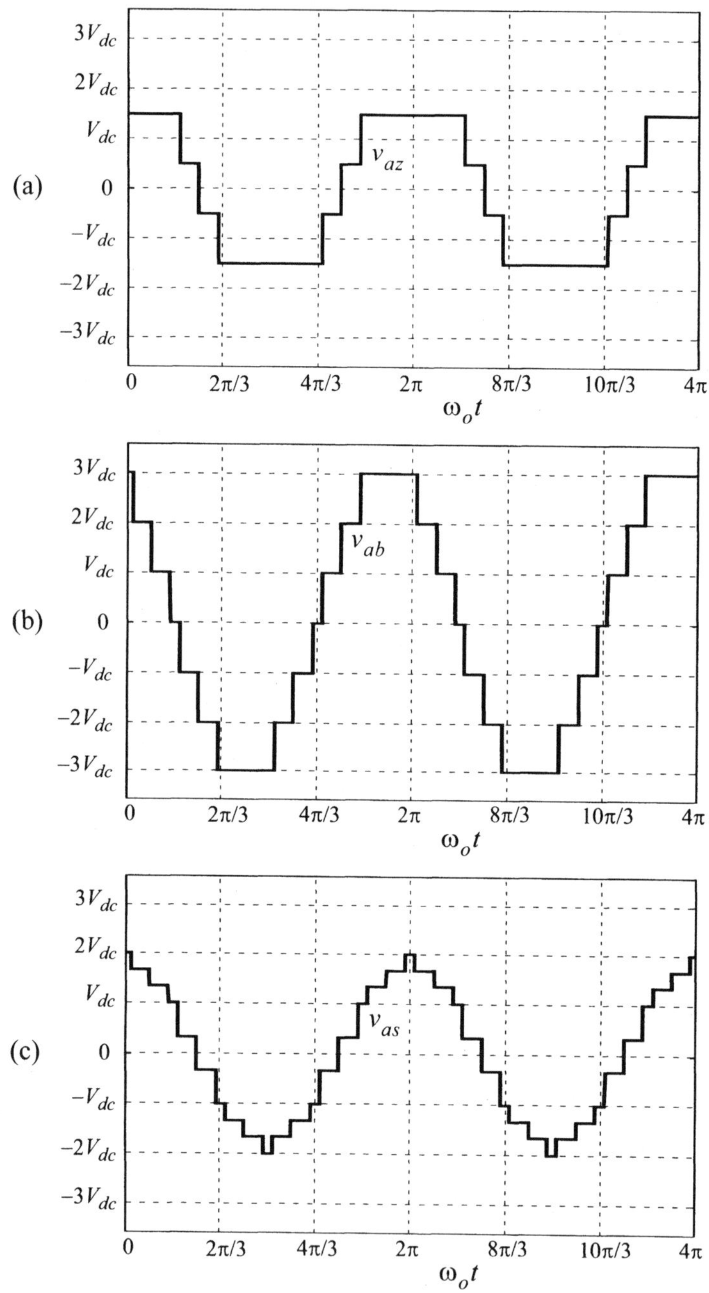

Figure 1.30 Switched (a) line-to-midpoint, (b) line-to-line and (b) line-to-neutral voltages produced by four-level diode-clamped inverter. Switching times defined in Table 10.1.

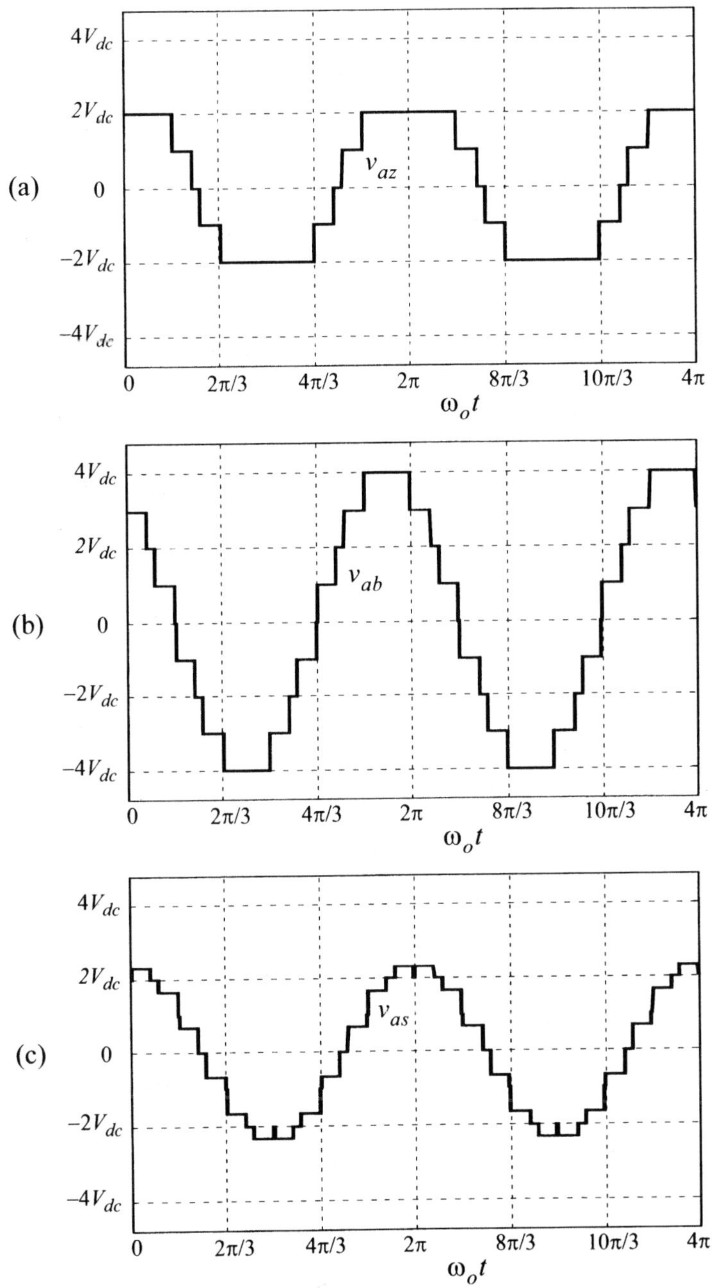

Figure 1.31 Switched (a) line-to-midpoint, (b) line-to-line and (b) line-to-neutral voltages produced by five-level diode-clamped inverter. Switching times defined in Table 10.1.

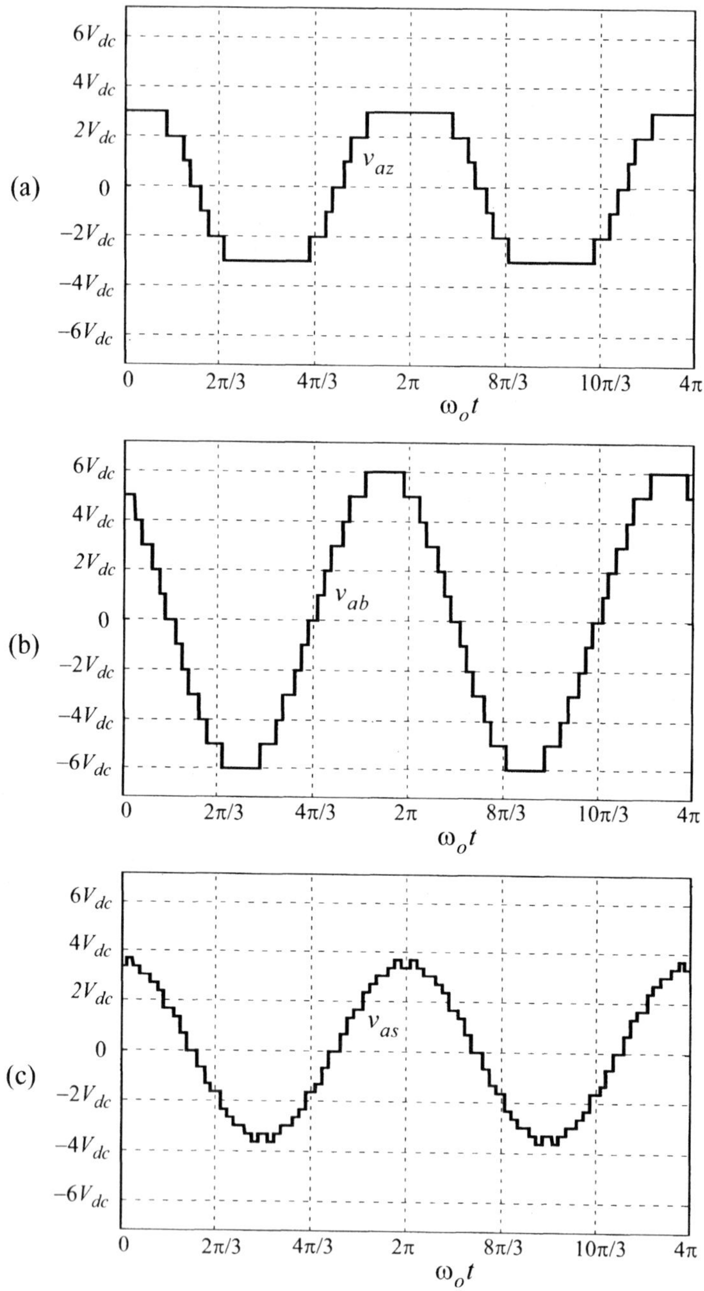

Figure 1.32 Switched (a) line-to-midpoint, (b) line-to-line and (b) line-to-neutral voltages produced by seven-level diode-clamped inverter. Switching times defined in Table 10.1.

the switches, some of which must be rated at $(k-2)V_{dc}$ where k is the number of levels ($k \geq 3$). The voltage rating of the diodes therefore quickly becomes a problem and levels greater than five are not considered as practical at the present time. This problem can be overcome by simply connecting several diodes in series, but the stress across the series-connected devices must then be carefully managed. Also since the number of series-connected switches increases with the number of levels, the switch conduction losses clearly increase in the same proportion. Fortunately, the power rating also increases at the same rate so the efficiency of the inverter remains roughly unaffected by the number of series-connected switches.

1.8.2 Capacitor-Clamped Multilevel Inverter

The capacitor-clamped VSI proposed by Meynard and Foch is an alternative to the diode-clamped VSI and shares many of the advantages of this topology [10]. Figure 1.33 illustrates the capacitor-clamped VSI topology for a seven-level system.

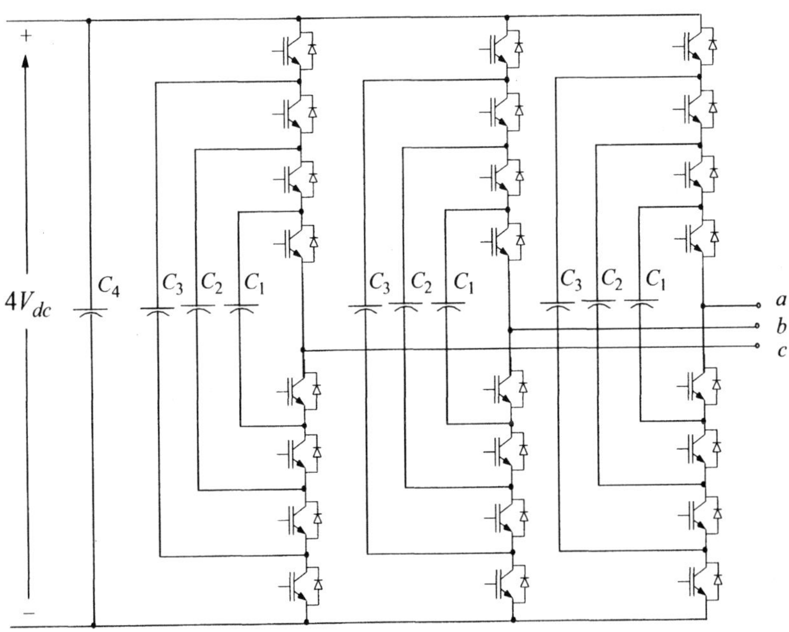

Figure 1.33 Topology of a five-level capacitor-clamped inverter.

The circuit operation of the capacitor-clamped topology is best explained by considering the simple three-level single-phase version shown in Figure 1.34. To obtain an output voltage level of $2V_{dc}$, switches S_1, S_2, S_7, and S_8 are turned on. To obtain $-2V_{dc}$, switches S_3, S_4, S_5, and S_6 are turned on. To obtain V_{dc}, S_1–S_3–S_7–S_8 are turned on, and to obtain $-V_{dc}$, S_3–S_4–S_5–S_7 are turned on. If clamping capacitors C_1 are maintained at voltage V_{dc}, the voltage on these capacitors in series with the load will partially cancel the source voltage. To obtain the zero voltage level, the combinations of either S_1–S_3–S_5–S_7 or S_2–S_4–S_6–S_8 can be turned on. The charge of C_1 can be balanced by proper management of the switch combinations.

This topology shares the advantages of all multilevel voltage source converters (VSCs) listed above, as well as those of a diode-clamped system, in that the required number of voltage levels can be obtained without the use of a transformer. This assists in reducing the cost of the converter and again reduces power loss. Unlike the diode-clamped structure where the series string of capacitors share the same voltage, in the capacitor-clamped VSC the capacitors within a phase leg are charged to different voltage levels. To synthesize the phase voltage waveform the various switches within the phase leg are switched ON to combine the various capacitor voltage levels with the constraint that no capacitor is short circuited and current continuity with the DC link is maintained for each capacitor. This leads to a significant amount of redundancy in

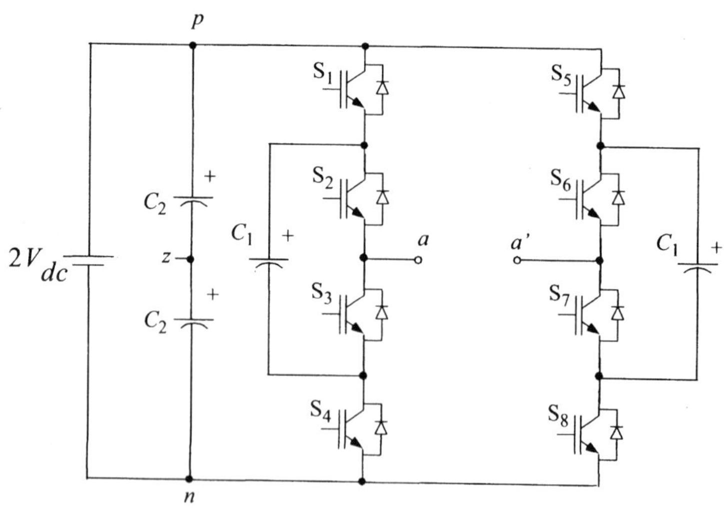

Figure 1.34 Three-level capacitor-clamped inverter, voltage on capacitors C_1 regulated to V_{dc}.

the switching states that lead to the same phase voltage levels. A similar table to Table 1.1 can be readily prepared for the capacitor-clamped circuit of Figure 1.33.

The capacitor-clamped (alternatively known as flying-capacitor) topology has several disadvantages that have limited its use. The first of these is the converter initialization. Before the capacitor-clamped VSC can be modulated, the clamping capacitors must be set up with the required voltage levels. This complicates the modulation process and hinders the performance of the converter under ride-through conditions. The capacitor voltages must also be regulated under normal operation in a similar fashion to the capacitor neutral points for a diode-clamped VSC. However, due to the large degree of redundancy in the phase voltage states, this regulation problem is generally combined with the modulation strategy. Another problem concerns the rating of the capacitors that form the clamping network. Since these have large fractions of the DC bus voltage across them, the voltage rating of the capacitors must be large when compared to the diode-clamped topology. It is this requirement and the initialization problems of the capacitor-clamped VSC that have limited its continued development.

1.8.3 Cascaded Voltage Source Multilevel Inverter

The third topological alternative for a multilevel inverter is to series-cascade single-phase full-bridges to make up each phase leg of the main inverter, as shown in Figure 1.35 for a seven-level system. Each full-bridge can switch between $+V_{dc}$, 0, and $-V_{dc}$, so that, for the seven-level example shown in Figure 1.35, each phase leg as a whole can switch in steps of V_{dc} over the range $+3V_{dc}$ to $-3V_{dc}$. Note also that the DC link voltage of each full-bridge has been redefined to V_{dc} rather than $2V_{dc}$ as was used in Section 1.1.3. This is the usual convention for cascaded voltage source multilevel inverter systems. In general, N series connected full-bridges can generate $(2N+1)$ switching levels per phase leg.

For this topology, the DC link supply for each full-bridge converter element must be provided separately, and this is typically achieved using diode rectifiers fed from isolated secondary windings of a three-phase transformer. A very simple approach is shown in Figure 1.36 using single-phase diode bridges. Other types of input side converters can also be used such as three-phase diode bridge rectifiers fed from a secondary star or delta connection. It

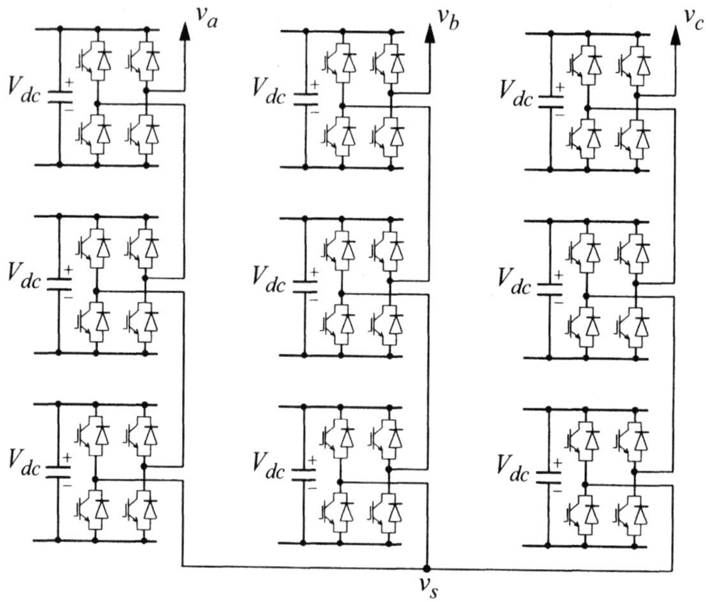

Figure 1.35 Seven-level series-bridge cascaded inverter.

can be shown that with progressive phase shifting of the three-phase secondaries by 60/(k–1) degrees where k is the number of levels, significant harmonic cancellation can be achieved in the transformer input current.

The cascaded inverter topology has several advantages that have made it attractive in power conditioning systems and medium to high power drive applications [11, 12, 13]. The first advantage concerns the ease of regulation of the DC buses described above, while the second advantage concerns the modularity of control that can be achieved. Unlike the diode-clamped inverter and the capacitor-clamped inverter where individual phase legs must be modulated by a central controller, the full-bridge inverters of a cascaded structure can be modulated independently of each other. Communication between the full-bridge inverters is required in order to achieve the synchronization of reference and carrier waveforms, but the controller can be distributed. This makes for a simpler controller structure than for either of the two previously discussed topologies. However, the cascaded inverter topology has not been applied particularly successfully at low power levels to date because of the need to provide separate isolated DC supplies for each full-bridge converter element.

Clearly, while the control logic to realize the switched state varies, the same switched phase leg voltages can be accomplished by any of the three

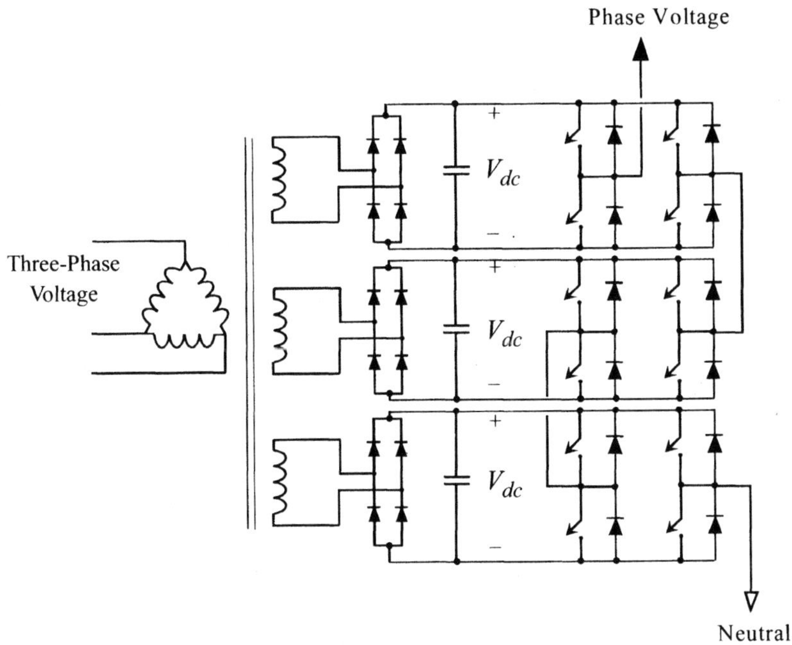

Figure 1.36 Isolated DC voltage supplies obtained from separate transformer secondary windings for 7-level cascaded multilevel inverter (one output phase only shown).

multilevel inverter topologies described here, provided, of course, they have the same number of available voltage levels. Hence, one would anticipate that the harmonic signature produced by the three circuits would be essentially the same. This property will be shown to be the case later in Chapter 11. However, in general the various switches of the phase leg of a multilevel converter are not equally loaded. This presents some difficulties with a practical multilevel inverter implementation, but it is possible at least for a cascaded inverter to rotate its switching between the series-connected full-bridges and hence balance the average power flow through each bridge. This approach is presented in Chapters 10 and 11.

One important difference between the three multilevel circuits, which will not be addressed in this book, concerns the issue of capacitor voltage balancing. Since the capacitors of each individual single-phase inverter derive from a different DC source, balancing is inherent. However, this is not the case for the diode-clamped or capacitor-clamped inverters. In this case the inner capacitors, such as those producing voltages V_{dc2} and V_{dc1} in Figure 1.29 will become over or undercharged if the average current fed to the inner taps on the DC link is not zero. Regulation of the link capacitor voltages has been resolved for these inverters [14, 15], but the details will not be pursued in this text.

1.8.4 Hybrid Voltage Source Inverter

The hybrid VSC proposed by Manjrekar et al. is a cascaded structure that has been modified such that the full-bridge inverters that comprise the phase leg of a hybrid inverter have different magnitude DC buses [16]. Figure 1.37 shows the structure of a seven-level hybrid inverter, in which it can be seen that each phase leg is constructed from a high-voltage (HV) stage and a low-voltage (LV) stage. In terms of operation, the hybrid converter uses the HV stage to achieve the bulk power transfer and uses the LV stage as a means to improve the spectral performance of the overall converter. Also note that the HV stage is shown to be constructed using devices that have high-voltage blocking characteristics but not necessarily fast switching characteristics [e.g., integrated gate controlled thyristors (IGCTs)], while the LV stage is constructed using devices that have fast switching characteristics but not necessarily high-voltage blocking characteristics (e.g., IGBTs). The major advantages of the hybrid structure include the fact that it marries the best performance characteristics of two different power electronic devices, and that it can achieve similar perfor-

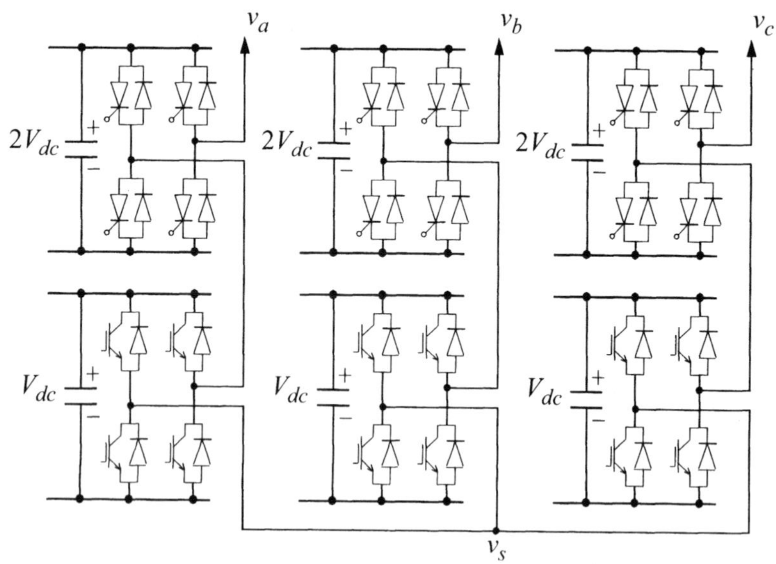

Figure 1.37 Topology of a seven-level hybrid voltage source inverter.

mance to other multilevel VSCs with a reduced switch count (e.g., 24 switches for a seven-level hybrid VSC as opposed to 36 for diode-clamped, cascaded, and capacitor-clamped VSCs).

The hybrid system again requires the use of a transformer to produce the isolated DC supplies for each full-bridge inverter, and the control of the converter is more complex than the standard cascaded structure. However, the control is still modular in that the LV stage and HV stage have their own reference waveforms, but the LV stage reference must be created from the HV reference. Another problem that must be addressed for the hybrid converter is that the HV stage will supply more power than the load requires in the middle ranges of the modulation index. Under these operating conditions the LV stage will be required to operate in a rectification mode, which means that the DC link must be capable of bidirectional power flow [16]. This necessitates the use of a PWM rectifier on the front end of the LV stage and further complicates the control of the hybrid converter system. However, the reduced switch count and more effective use of the power electronic devices that comprise the hybrid system make it a particularly attractive system at medium to high power levels.

1.9 Summary

This introductory chapter has presented the *switching circuits* and *block modulation* concepts involved in the application of power electronic converters. While block modulation is still used to produce a variable frequency AC supply in some applications, it has largely been supplanted by more sophisticated modulation strategies such as pulse width modulation, which have the major advantage of allowing simultaneous phase and amplitude control of the output voltage. Some time has also been spent here introducing the concept of *space vectors*, which have great utility in the analysis of the more complex inverter switching processes that are presented later in this book.

References

[1] P.C. Krause and T.A. Lipo, "Analysis and simplified representation of a rectifier-inverter induction motor drive," *IEEE Trans. on Power Apparatus and Systems*, vol. PAS–88, no. 5, May 1969, pp. 55–66.

[2] D.W. Novotny and T.A. Lipo, *Vector Control and Dynamics of AC Drives*, Oxford University Press, London, 1996.

[3] T.A. Lipo, "A Cartesian vector approach to reference frame theory of AC machines," in *Conf. Rec. International Conference on Electrical Machines (ICEM)*, Lausanne, 1994, pp. 239–242.

[4] A. Nabae, I. Takahashi, and H. Akagi, "A new neutral-point-clamped PWM inverter," *IEEE Trans. on Industry Applications*, vol. IA–17, no. 5, Sept./Oct. 1981, pp. 518–523.

[5] M.C. Klabunde, Y. Zhao, and T.A. Lipo, "Current Control of a 3-Level Rectifier/Inverter Drive System," in *Conf. Rec. IEEE Industry Applications Society Annual Mtg.*, Denver, 1994, pp. 859–866.

[6] S. Ogasawara and H. Akagi, "Analysis of variation of neutral point potential in neutral-point-clamped voltage source PWM inverters," in *Conf. Rec. IEEE Industry Applications Society Annual Mtg.*, Toronto, 1993, pp. 965–970.

[7] P. Bhagwat and V.R. Stefanovic, "Generalized structure of a multilevel PWM inverter," *IEEE Trans. on Industry Applications*, vol. 19, no. 6, Nov./Dec. 1983, pp. 1057–1069.

[8] G. Sinha and T.A. Lipo, "A four-level inverter based drive with a passive front end," *IEEE Trans. on Power Electronics*, vol. 15, no. 2, March 2000, pp. 285–294.

[9] M. Marchesoni and P. Tenca, "Theoretical and practical limits in multilevel npc inverters with passive front ends," in *Conf. Record European Power Electronics Conf. (EPE)*, Graz, 2001, in CD–ROM, 12pp.

[10] T.A. Meynard and H. Foch, "Multi-level conversion: high voltage choppers and voltage-source inverters," in *Conf. Rec. IEEE Power Electronics Specialists Conf. (PESC)*, Toledo Spain, 1992, pp. 397–403.

[11] Y. Liang and C.O. Nwankpa, "A new type of STATCOM based on cascading voltage source inverters with phase-shifted unipolar SPWM," in *Conf. Rec. Industry Applications Society Annual Mtg.*, St. Louis, 1998, pp. 978–985.

[12] N.P. Schibli, T. Nguyen, and A.C. Rufer, "A three-phase multilevel converter for high-power induction motors," *IEEE Trans. on Power Electronics*, vol. 13, no. 5, Sept. 1998, pp. 987–985.

[13] R.H. Osman, "A medium voltage drive utilizing series-cell multilevel topology for outstanding power quality," in *Conf. Rec. IEEE Industry Applications Society Annual Mtg.*, Phoenix, 1999, pp. 2662–2669.

[14] X. Yuan, H. Stemmler, and I. Barbi, "Investigation on the clamping voltage self-balancing of the three-level capacitor clamping inverter," in *Conf. Rec. Power Electronics Specialists Conf. (PESC)*, Charleston, 1999, pp. 1059–1064.

[15] C. Keller, R. Jakob, and S. Salama, "Topology and balance control of medium voltage multilevel drives," in *Conf. Rec. European Power Electronics Conf. (EPE)*, Graz, 2001, in CD–ROM.

[16] M.D. Manjrekar, R. Lund, P. Steimer, and T.A. Lipo, "Hybrid multilevel power conversion system: A competitive solution for high power applications," *IEEE Trans. on Industry Applications*, vol. 36, no. 3, May/June 2000, pp. 834–841.

2

Harmonic Distortion

In Chapter 1 the need to produce voltage control for adjustable frequency loads, and the concept of regulating the switched state of power electronic devices to achieve this control, has been introduced. Unfortunately, voltage control can only be accompanied by undesirable harmonics as a result of the inherent switched nature of modern power electronic equipment. The need to consider the degree to which undesirable harmonics are created during the process of voltage control leads to the development of basic performance indices which quantify *harmonic distortion*. These indices are introduced in this chapter and will be used as a means of comparing various modulation algorithms throughout this book.

2.1 Harmonic Voltage Distortion Factor

The proliferation of converter modulation algorithms that have appeared over the years has created confusion as to the effectiveness of one method over another for comparing various algorithms. One method of comparing the effectiveness of modulation processes is by comparing the unwanted components, i.e., the *distortion* in the output voltage or current waveform relative to that of an ideal sine wave. Given that the output voltage of a solid state converter $v(t)$ is a periodic function with period T, the root-mean-square (RMS) value of the function is, by definition,

$$V_{rms} = \sqrt{\frac{1}{T}\int_0^T v(t)^2 dt} \tag{2.1}$$

Since $v(t)$ is periodic, then it can be represented by the Fourier series

$$v(t) = V_0 + V_1 \cos\omega_1 t + V_2 \cos 2\omega_1 t + V_3 \cos 3\omega_1 t + \cdots \tag{2.2}$$

whereupon

$$V_{rms} = \sqrt{\frac{1}{T}\int_0^T \sum_{n=0}^{\infty} \sum_{k=0}^{\infty} V_n V_k \cos n\omega_1 t \cos k\omega_1 t \, dt} \tag{2.3}$$

Upon expanding, the integration of terms in which $n \neq k$ become zero, so that,

$$V_{rms} = \sqrt{\frac{1}{T}\int_0^T \sum_{n=0}^{\infty} V_n^2 \cos^2 n\omega_1 t \, dt}$$

$$= \frac{1}{T}\int_0^T \left[V_0^2 + \sum_{n=1}^{\infty} \frac{V_n^2}{2}(1 + \cos 2n\omega_1 t) \right] dt \tag{2.4}$$

The double-frequency term integrates to zero over a complete period, so that, finally,

$$V_{rms} = \sqrt{V_0^2 + \sum_{n=1}^{\infty} \frac{V_n^2}{2}} \tag{2.5}$$

or, in terms of RMS values of the individual harmonics,

$$V_{rms} = \sqrt{V_0^2 + \sum_{n=1}^{\infty} V_{n,rms}^2} \tag{2.6}$$

In most practical cases the fundamental component can be considered as the desired output. The remainder of this expression is then considered as the "distortion". Factoring out this desired component gives

$$V_{rms} = V_{1,rms} \sqrt{1 + \left(\frac{V_0}{V_{1,rms}}\right)^2 + \sum_{n=2}^{\infty} \left(\frac{V_{n,rms}}{V_{1,rms}}\right)^2} \tag{2.7}$$

The *total harmonic distortion* (THD) of the voltage can now be defined as

$$\text{THD} = \sqrt{\left(\frac{V_0}{V_{1,rms}}\right)^2 + \sum_{n=2,3,\ldots} \left(\frac{V_{n,rms}}{V_{1,rms}}\right)^2} \tag{2.8}$$

and the RMS voltage becomes

$$V_{rms} = V_{1,rms}\sqrt{1+\text{THD}^2} \tag{2.9}$$

Note that the RMS value of the waveform is equal to the fundamental RMS voltage if the THD is zero. Solving Eq. (2.9) for the THD provides an alternative means for calculating this quantity, namely

$$\text{THD} = \sqrt{\left(\frac{V_{rms}}{V_{1,rms}}\right)^2 - 1} \tag{2.10}$$

It is clear from the intent of the definition of THD that, if the desired output voltage is not the fundamental component, then the definition of THD would have to be modified accordingly.

Alternatively it is possible to eliminate the dependence of the distortion expression on RMS values by writing Eq. (2.6) in terms of peak values as

$$\frac{V_{rms}}{V_{1,rms}} = \sqrt{1 + \left(\frac{2V_0}{V_1}\right)^2 + \sum_{n=2,3,\ldots}^{\infty}\left(\frac{V_n}{V_1}\right)^2} \tag{2.11}$$

Hence another expression for total harmonic distortion is

$$\text{THD} = \sqrt{\left(\frac{2V_0}{V_1}\right)^2 + \sum_{n=2,3,\ldots}^{\infty}\left(\frac{V_n}{V_1}\right)^2} \tag{2.12}$$

If this equation is applied to the inverter circuit of Figure 1.12, it can be assumed that by proper control, the positive and negative portions of the output wave are symmetrical (no DC or even harmonics). The voltage harmonic distortion factor reduces to,

$$\text{THD} = \sqrt{\sum_{n=3,5,7,\ldots}^{\infty}\left(\frac{V_n}{V_1}\right)^2} \tag{2.13}$$

As an example of the use of this equation, consider the case where phase overlap is used to control the voltage. From Eq. (1.27), the THD becomes

$$\text{THD} = \sqrt{\sum_{n=3,5,7,\ldots}^{\infty}\frac{1}{n^2}\left(\frac{\cos n\frac{\alpha}{2}}{\cos\frac{\alpha}{2}}\right)^2} \tag{2.14}$$

A plot of the total harmonic distortion for the single-phase inverter with phase shift voltage control is shown in Figure 2.1. Note that the harmonic distortion approaches infinity as the fundamental component of the voltage becomes zero.

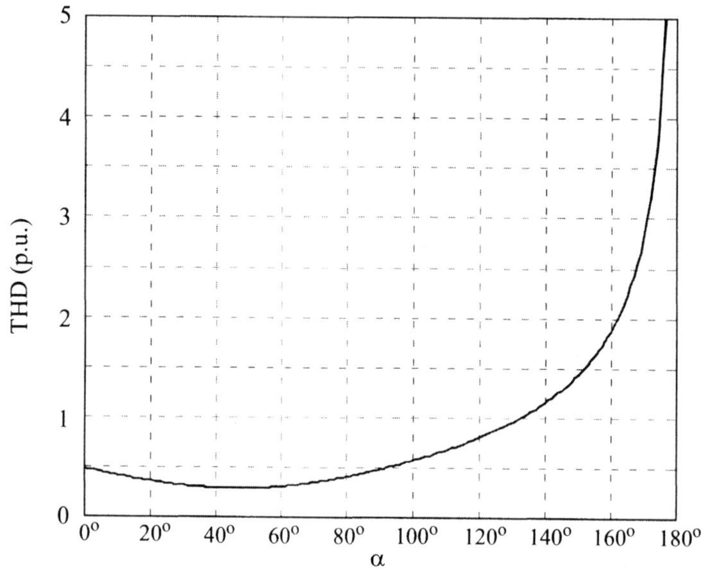

Figure 2.1 Total harmonic distortion (THD) factor of single-phase inverter with phase shift voltage control versus phase shift angle α.

Another means of expressing waveform distortion of an inverter is to normalize or reference the voltage distortion produced, to that of a simple square-wave (*sqw*) inverter without voltage control. Referring again to Eq. (2.14), when $\alpha = 1$, then the voltage harmonic distortion for this waveform is[1]

$$\mathrm{THD}_{sqw} = \sqrt{\sum_{n=3,5,7,\ldots}^{\infty} \frac{1}{n^2}} \qquad (2.15)$$

1 It should be mentioned that another definition of THD is also in use. In particular, Europeans frequently use the form

$$\mathrm{THD} = \sqrt{\sum_{n=3,5,7,\ldots}^{\infty} \left(\frac{V_{n,rms}}{V_{rms}}\right)^2}$$

where V_{rms} denotes the RMS value of the overall waveform. This expression has the advantage of always being between zero and one but is rarely used in the U.S. literature.

From Dwight [1] entry number 48.12, it can be determined that

$$1 + \frac{1}{3^2} + \frac{1}{5^2} + \frac{1}{7^2} + \cdots = \frac{\pi^2}{8} \tag{2.16}$$

so that

$$\mathrm{THD}_{sqw} = \sqrt{\left(\frac{\pi^2}{8} - 1\right)} = 0.4834 \tag{2.17}$$

The effectiveness of the phase shift method of voltage control can be quantified if one expresses the voltage harmonic distortion as a per unit of the distortion obtained with the pure square-wave inverter:

$$D_v = \frac{\mathrm{THD}}{\mathrm{THD}_{sqw}}$$

An alternative and perhaps more realistic per unitization for motor drive applications is to compare the harmonics produced by the phase shift method of voltage control to the harmonics produced by a square-wave inverter in which the DC link voltage is controlled so as to produce constant volts per hertz. That is, the DC voltage is continuously varied so as to obtain the desired fundamental component. In this case it can be readily shown that

$$D_v = \frac{1}{\left(\frac{V_1}{V_{1,max}}\right)} \frac{\mathrm{THD}}{\mathrm{THD}_{sqw}} \tag{2.18}$$

where $V_{1,max}$ denotes the fundamental component when the DC link voltage is at its maximum value. A plot of the normalized total harmonic distortion as a function of the overlap angle α is shown in Figure 2.2. Note that when compared with DC link voltage adjustment, the harmonic distortion produced by phase shift type control is only effective near rated voltage, i.e., in the range from 100 to 80% voltage below which point the distortion increases rapidly.

2.2 Harmonic Current Distortion Factor

While it is the output voltage which is the quantity that is controlled in a voltage source/stiff inverter-type system, it is the current which is frequently of most interest, since losses, output power, etc. typically involve this quantity rather than voltage directly.

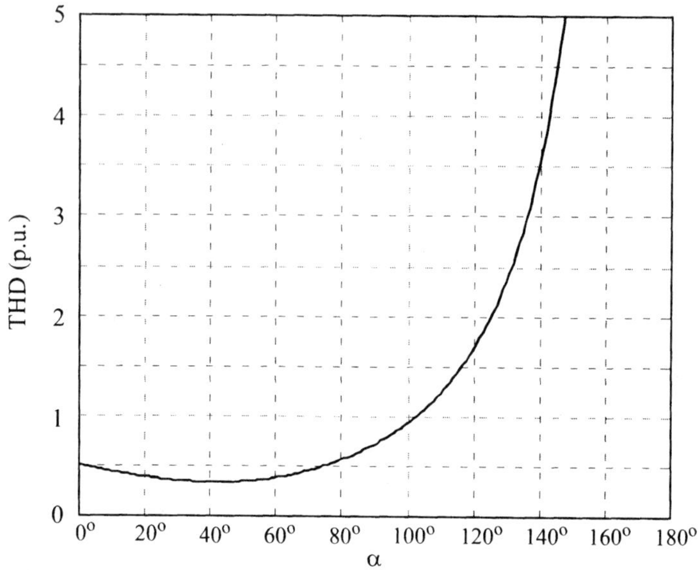

Figure 2.2 Total harmonic distortion of phase-shift method of voltage control normalized to distortion produced by a variable DC link type of voltage control.

Defining the *current harmonic distortion* in the same manner as for voltage, one can state that

$$\mathrm{THD}_i = \sqrt{I_0^2 + \sum_{n=2,3,\ldots}^{\infty} I_n^2} \tag{2.19}$$

The current waveform is, of course, dependent upon the load impedance and, as such, cannot be predicted or characterized in advance. However, in many applications, the load can be characterized by a lossy inductance, that is, by an inductance with a relatively small but nontrivial resistance. In this case, the harmonic current amplitudes can be approximated by the expression,

$$I_n \cong \frac{V_n}{n\omega_1 L} \quad n = 2,3,4,\ldots \tag{2.20}$$

where ω_1 is the angular frequency of the fundamental component of the current waveform. If DC components do not exist,

$$\text{THD}_i = \frac{1}{\omega_1 L}\sqrt{\sum_{n=2}^{\infty}\left(\frac{V_n}{n}\right)^2} \tag{2.21}$$

Normalizing this expression with respect to the quantity $(V_1/\omega_1 L)$ the weighted total harmonic distortion (WTHD) becomes defined as

$$\text{WTHD} = \frac{\sqrt{\sum_{n=2}^{\infty}\left(\frac{V_n}{n}\right)^2}}{V_1} \tag{2.22}$$

It should be noted that the inductance L in the normalizing expression is taken to be essentially the total motor (stator plus rotor) leakage inductance in the case of an electrical machine load. In more exact form the proper inductance to be used for an induction machine is

$$L_\sigma = L_1 + \frac{L_m L_2}{L_m + L_2} \tag{2.23}$$

where L_1, L_2, L_m correspond to stator leakage, rotor leakage, and magnetizing inductances, respectively. In the case of an induction motor load the normalizing ratio $V_1/\omega_1 L_\sigma$ thus corresponds to the maximum motor inrush current [2].

Again, specifically considering the phase-shifted single-phase inverter as an example, its WTHD becomes

$$\text{WTHD} = \sqrt{\sum_{n=3,5,\ldots}^{\infty} \frac{1}{n^4}\left(\frac{\cos n\frac{\alpha}{2}}{\cos\frac{\alpha}{2}}\right)^2} \tag{2.24}$$

The weighted total harmonic distortion factor for phase shift voltage control is plotted in Figure 2.3.

For purposes of comparing various types of switching algorithms the WTHD can be normalized by comparing it to the WTHD obtained with a simple square-wave voltage inverter, in which case

$$\text{WTHD}_{sqw} = \sqrt{\frac{1}{3^4} + \frac{1}{5^4} + \frac{1}{7^4} + \cdots} \tag{2.25}$$

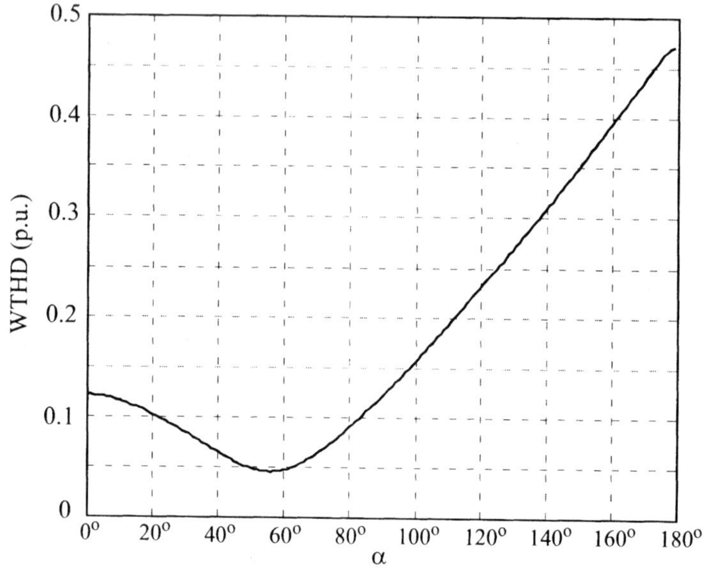

Figure 2.3 Weighted harmonic distortion factor (WTHD) as a function of the switching angle α for the single-phase inverter with phase shift voltage control.

Referring again to Dwight [1], entry 48.14 can be found in which

$$1 + \frac{1}{3^4} + \frac{1}{5^4} + \frac{1}{7^4} + \cdots = \frac{\pi^4}{96} \tag{2.26}$$

so that

$$WTHD_{sqw} = \sqrt{\frac{\pi^4}{96} - 1} = 0.1212 \tag{2.27}$$

2.3 Harmonic Distortion Factors for Three-Phase Inverters

The voltage distortion factor for the three-phase quasi-square-wave ($qsqw$) VSI can be found in much the same way as for the single-phase inverter. In this case

$$THD_{qsqw} = \sqrt{\frac{1}{5^2} + \frac{1}{7^2} + \frac{1}{11^2} + \frac{1}{13^2} + \frac{1}{17^2} + \cdots} \tag{2.28}$$

The infinite series inside the radical can be written as

$$\frac{1}{5^2}+\frac{1}{7^2}+\cdots = \left(1+\frac{1}{3^2}+\frac{1}{5^2}+\frac{1}{7^2}+\frac{1}{9^2}+\cdots\right)-1-\left(\frac{1}{3^2}+\frac{1}{9^2}+\frac{1}{15^2}+\cdots\right)$$

$$= \frac{\pi^2}{8}-1-\frac{1}{3^2}\left(1+\frac{1}{3^2}+\frac{1}{5^2}+\cdots\right)$$

$$= \frac{\pi^2}{8}-1-\frac{1}{9}\frac{\pi^2}{8} = \frac{\pi^2}{9}-1 \tag{2.29}$$

Thus

$$\text{THD}_{qsqw} = \sqrt{\frac{\pi^2}{9}-1} = 0.3108 \tag{2.30}$$

Note that this result is a substantial reduction from the single-phase case where the equivalent factor was 0.4834.

A weighted total harmonic distortion factor can also be derived for this converter. It is not difficult to show that if one again assumes a highly inductive load

$$\text{WTHD}_{qsqw} = \sqrt{\frac{\pi^4}{96}-1-\frac{1}{3^4}\frac{\pi^4}{96}}$$

$$= \sqrt{\frac{5\pi^4}{486}-1}$$

$$= 0.0464 \tag{2.31}$$

This figure is also a vast improvement over the 0.1212 factor obtained for the single-phase voltage source inverter.

The THD for a three-level inverter can be computed in a similar fashion. The RMS value of the line-to-neutral voltage for the three-level case with a 30° zero phase leg voltage step (minimum THD) can be computed as

$$v_{rms}^2 = \frac{2}{\pi}\int_0^{\pi/2} v_{as}^2\, d\theta$$

$$= \frac{2}{\pi}\left[\frac{\pi}{12}\left(\frac{4}{3}\right)^2+\frac{\pi}{6}(1)^2+\frac{\pi}{6}\left(\frac{2}{3}\right)^2+\frac{\pi}{12}0\right]V_{dc}^2 \tag{2.32}$$

whereupon

$$v_{rms} = \frac{\sqrt{7}}{3}V_{dc} \tag{2.33}$$

Utilizing Eq. (1.77) the THD of the line-to-neutral voltage for a three-level inverter is therefore, from Eq. (2.10),

$$\text{THD} = \sqrt{\left(\frac{\frac{\sqrt{7}}{3}V_{dc}}{\frac{2\sqrt{2}}{\pi}\sin\frac{5\pi}{12}V_{dc}}\right)^2 - 1} = 0.1686 \tag{2.34}$$

One can note that this result is a substantial improvement over the value of 0.3108 obtained for the two-level case. It can be shown in the same manner that the minimum THD for four- and five-level inverters is 0.1186 and 0.0943, respectively.

Since additional capacitance enters into the picture as a parameter, generic figures of merit such as THD and WTHD in the case of a current source inverter become difficult to define. However, if the finite slopes of the current waveform are ignored, the weighted THD can be defined in the same manner as for the VSI, as

$$\text{WTHD} = \frac{\sqrt{\sum_{n = 5, 7, 11, \ldots}^{\infty} I_n^2}}{I_1} \tag{2.35}$$

From previous results, however,

$$\left(\frac{1}{5}\right)^2 + \left(\frac{1}{7}\right)^2 + \left(\frac{1}{11}\right)^2 + \left(\frac{1}{13}\right)^2 + \left(\frac{1}{17}\right)^2 + \cdots = \frac{\pi^2}{9} - 1 \tag{2.36}$$

from which

$$\text{WTHD}_{qsqwc} = \sqrt{\frac{\pi^2}{9} - 1} = 0.3108 \tag{2.37}$$

which is, of course, the same value as the THD for the quasi-square-wave voltage inverter. This result is roughly six times greater than the WTHD obtained for the quasi-square-wave voltage source inverter. However, this result is misleading since the WTHD in the case of the VSI was normalized by the quantity

$V_{1,rms}/\omega_1 L$ which is essentially equal to the inrush current of the machine while the WTHD for the CSI was normalized by the rated current. Hence, assuming a motor load, if the per unit inrush current is about six times the rated current (a very typical value), then the two inverters produce the same amount of harmonic current under rated current conditions. Moreover, assuming that the DC link voltage for the VSI remains constant independent of load, the current distortion for the CSI, for the same ratio of inrush to rated current, will in reality be less than the VSI for all loads less than rated load.

2.4 Choice of Performance Indicator

The single-phase voltage source inverter redrawn as Figure 2.4 is one of the simplest power electronic converter topologies and is used to create a variable magnitude, variable frequency AC voltage output from a fixed magnitude DC supply voltage. However, since the PWM techniques used to modulate this converter form the basis of almost all other fixed frequency converter modulation strategies, it is the most appropriate place to begin. Note that the DC bus voltage is again shown as a split supply arranged around a zero voltage midpoint, for convenience of representation of the modulating reference waveform. In practice, this supply midpoint usually does not physically exist.

As an example of how the THD performance indicator fails to provide a useful comparison, consider the results of two modulation algorithms shown in Figure 2.5 in which the switching points have been adjusted to provide the same fundamental component. The first method is derived from what is normally termed naturally sampled modulation in which the pulses vary nearly

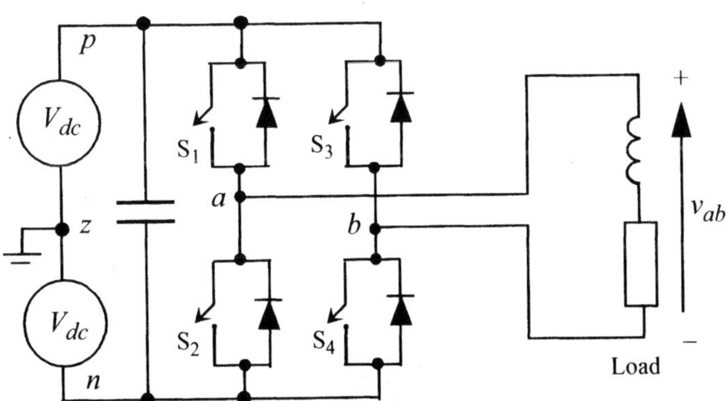

Figure 2.4 Topology of single-phase voltage source inverter.

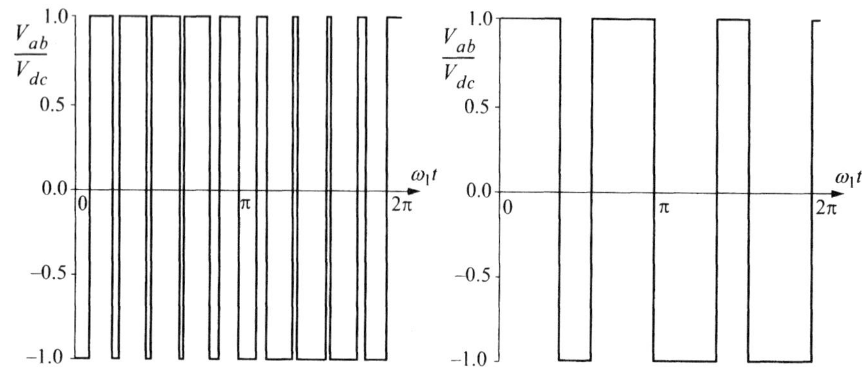

Figure 2.5 Two switched voltage output waveforms of a single-phase inverter having the same fundamental component.

sinusoidally across a period in order that its time integral approximates a sine wave. The second waveform is a simple center-notching method with little regard toward harmonic suppression but with a center "notch" removed to produce a desired fundamental component. By definition, from Eq. (2.10), the THD can be expressed as

$$\text{THD} = \sqrt{\left(\frac{V_{rms}}{V_{1,rms}}\right)^2 - 1} \tag{2.38}$$

where

$$V_{rms} = \sqrt{\frac{1}{T}\int_0^T v(t)^2 dt} \tag{2.39}$$

Upon squaring either of the two waveforms it is evident that all of the negative portions have the same squared value as the positive portions. The RMS value of either waveform is

$$V_{rms} = V_{dc} \tag{2.40}$$

Hence, the RMS value of either waveform is identical and the THD is, in fact, nearly the same, differing only because the normalizing fundamental components differ. It can be recalled that the energy stored in a capacitor is proportional to the square of its voltage. In like manner the square of the RMS value of any waveform can be equated to an energylike average per cycle associated with the waveform. Since a spectral plot is simply another means for portraying this energy, this result says, in essence, that while the location of the frequency domain spectral lines can be shifted by modulation, the total energy content in a given waveform cannot be changed no matter what PWM algorithm is used.

While this simple example uses a single-phase inverter, similar examples can easily be formulated with the three-phase half- or full-bridge versions. Clearly, very misleading results can be produced when THD is used as a figure of merit. Nonetheless, the THD remains a useful figure of merit for sizing filter requirements at the input side converter since voltage distortion is an important criterion for specifying a clean voltage waveform.

Another possibility for a comparison base is the weighted value of THD. In this case the lower portion of the frequency spectra is weighted more heavily (i.e., inversely with frequency), accurately portraying the expected harmonic current of an inductive load. While other types of weighting could be introduced to account for nonlinear effects such as hysteresis or eddy current effect, this performance index is clearly the most widely applicable choice and will be adopted in this text.

Although the WTHD uses a spectral weighting factor, it is useful to note that this quantity need not be computed by spectral decomposition, i.e., calculation of the harmonic spectrum of the waveform. In general, because of its inherent symmetry, the AC inverter output phase voltage can be expressed as an even function in terms of cosinusoidal harmonics by

$$v(\theta) = \sum_{n=1}^{\infty} V_n \cos n\theta \quad \text{where } \theta = \omega t \tag{2.41}$$

If one integrates both sides and squares the result,

$$\left(\int_0^{\theta} v(\theta) d\theta\right)^2 = \left(\sum_{n=1}^{\infty} \frac{V_n}{n} \sin n\theta\right)^2 \tag{2.42}$$

Integrating both sides of the equation again, this time over a period of the fundamental component, yields

$$\int_0^{2\pi} \left(\int_0^{\theta} v(\theta)\, d\theta\right)^2 d\theta = \pi \sum_{n=1}^{\infty} \left(\frac{V_n}{n}\right)^2 \tag{2.43}$$

Note that all of the terms on the right-hand side of this expression containing the product of trigonometric functions of two different frequencies equal zero when integrated over one period.

Recall from Eq. (2.22) that

$$\text{WTHD} = \frac{\sqrt{\sum_{n=2}^{\infty}\left(\frac{V_n}{n}\right)^2}}{V_1} \tag{2.44}$$

In terms of the results of Eq. (2.43),

$$\text{WTHD} = \sqrt{\frac{\frac{1}{\pi}\int_0^{2\pi}\left(\int_0^{\theta} v(\theta)d\theta\right)^2 d\theta}{V_1^2} - 1} \tag{2.45}$$

Equation (2.45) can be used to numerically calculate the WTHD in closed form rather than from the summation of harmonic components which is the usual method. This method removes the uncertainty in terminating the harmonic series at some point to provide a given accuracy.

2.5 WTHD of Three-Level Inverter

As an example of computing the weighted THD using Eq. (2.45), consider again the three-level inverter of Figure 1.25, and its voltage waveform plotted in Figure 1.26. Expressing the waveform as an even function, then over the first 90° the line-to-load neutral voltage is

$$v_{l-n}(\theta) = \frac{4}{3}V_{dc} \qquad \text{for } 0 < \theta < \frac{\pi}{12}$$

$$v_{l-n}(\theta) = V_{dc} \qquad \text{for } \frac{\pi}{12} \le \theta < \frac{\pi}{4}$$

$$v_{l-n}(\theta) = \frac{2}{3}V_{dc} \qquad \text{for } \frac{\pi}{4} \le \theta < \frac{5\pi}{12}$$

$$v_{l-n}(\theta) = 0 \qquad \text{for } \frac{5\pi}{12} \le \theta \le \frac{\pi}{2}$$

Because of symmetry the waveform needs only be defined and WTHD computed over one-quarter cycle, taking care to define an even (cosine) function so as to not produce an average voltage component as a result of the initial condition. In this case the WTHD becomes

$$\text{WTHD} = \sqrt{\frac{\dfrac{4}{\pi}\displaystyle\int_0^{\pi/2}\left(\int_0^{\theta} v(\theta)d\theta\right)^2 d\theta}{V_1^2} - 1} \tag{2.46}$$

The integral inside the square brackets can be obtained piecemeal by integrating in four parts, as follows:

$$\int_0^{\theta} v_{l-n}(\theta)d\theta = V_{dc}\frac{4}{3}\theta \qquad \text{for} \qquad 0 \le \theta \le \frac{\pi}{12}$$

$$\int_0^{\theta} v_{l-n}(\theta)d\theta = V_{dc}\left\{\frac{4}{3}\frac{\pi}{12} + \left(\theta - \frac{\pi}{12}\right)\right\} \qquad \text{for} \qquad \frac{\pi}{12} \le \theta \le \frac{\pi}{4}$$

$$\int_0^{\theta} v_{l-n}(\theta)d\theta = V_{dc}\left\{\frac{4}{3}\frac{\pi}{12} + \frac{\pi}{6} + \frac{2}{3}\left(\theta - \frac{\pi}{4}\right)\right\} \qquad \text{for} \qquad \frac{\pi}{4} \le \theta \le \frac{5\pi}{12}$$

$$\int_0^{\theta} v_{l-n}(\theta)d\theta = V_{dc}\left\{\frac{4}{3}\frac{\pi}{12} + \frac{\pi}{6} + \frac{2}{3}\frac{\pi}{6}\right\} \qquad \text{for} \qquad \frac{5\pi}{12} \le \theta \le \frac{\pi}{2}$$

so that

$$\int_0^{\frac{\pi}{12}}\left(\int_0^{\theta} v_{l-n}(\theta)d\theta\right)^2 d\theta = V_{dc}^2\frac{\pi^3}{2916} \qquad \text{over} \qquad 0 \le \theta < \frac{\pi}{12}$$

$$\int_{\frac{\pi}{12}}^{\frac{\pi}{4}}\left(\int_0^{\theta} v_{l-n}(\theta)d\theta\right)^2 d\theta = V_{dc}^2\frac{13\pi^3}{1944} \qquad \text{over} \qquad \frac{\pi}{12} \le \theta < \frac{\pi}{4}$$

$$\int_{\frac{\pi}{4}}^{\frac{5\pi}{12}}\left(\int_0^{\theta} v_{l-n}(\theta)d\theta\right)^2 d\theta = V_{dc}^2\frac{109\pi^3}{5832} \qquad \text{over} \qquad \frac{\pi}{4} \le \theta < \frac{5\pi}{12}$$

$$\int_{\frac{5\pi}{12}}^{\frac{\pi}{2}}\left(\int_0^{\theta} v_{l-n}(\theta)d\theta\right)^2 d\theta = V_{dc}^2\frac{49\pi^3}{3888} \qquad \text{over} \qquad \frac{5\pi}{12} \le \theta \le \frac{\pi}{2}$$

Summing these terms gives

$$\int_0^{\frac{\pi}{2}}\left(\int_0^{\theta} v_{l-n}(\theta)\, d\theta\right)^2 d\theta = \pi^3\left(\frac{149}{3888}\right)V_{dc}^2 = 1.188\,V_{dc}^2 \qquad (2.47)$$

Substituting Eqs. (2.47) and (1.77) (multiplied by $\sqrt{2}$ to return to peak fundamental magnitude) into Eq. (2.46), one obtains finally

$$\text{WTHD} = \sqrt{\frac{\frac{4}{\pi}\pi^3\frac{149}{3888}V_{dc}^2}{\left(\frac{4}{\pi}\sin\frac{5\pi}{12}V_{dc}\right)^2} - 1} = 0.0161 \qquad (2.48)$$

Comparison with the similar expression for a conventional two-level nonmodulated ("square-wave") inverter having two-levels, 0.0464 in Eq. (2.31), indicates a reduction of 65% when adopting three-level inverter operation. Table 2.1 summarizes the performance of nonmodulated square-wave inverters for up to five levels. In this case, because of the increasing complexity, the WTHD of the four- and five-level inverters have been simply computed numerically by use of Eq. (2.46). These values of WTHD can serve as good values for comparison with the various pulse width-modulated algorithms to be developed.

Table 2.1 Figures of Merit for Five Types of Nonmodulated (Square-Wave) Inverter Topologies

Inverter Type	RMS Fundamental Phase Voltage	RMS Phase Voltage	THD (%)	WTHD (%)
Single-phase bridge	$0.4502V_{dc}$	$0.5V_{dc}$	48.34	12.12
Three-phase two-level bridge	$0.4502V_{dc}$	$0.4714V_{dc}$	31.08	4.64
Three-phase three-level bridge	$0.4348V_{dc}$	$0.441V_{dc}$	16.86	1.61
Three-phase four-level bridge	$0.4321V_{dc}$	$0.4351V_{dc}$	11.86	1.12
Three-phase five-level bridge	$0.4311V_{dc}$	$0.4330V_{dc}$	9.43	0.987

2.6 The Induction Motor Load

It has already been demonstrated that WTHD is superior to THD as a figure of merit for a nonsinusoidal converter waveform since WTHD predicts the distortion in the current and subsequent additional losses which is typically the major issue in the application of such converters. However, it should be recalled that Eq. (2.22) was derived assuming that the (small) load resistance and inductance were constant. When the load is passive, this assumption is generally quite valid. However, when used with a motor load, special precautions must be taken because of the nonlinear nature of the load impedance. The assumption of constant inductance and resistance remains valid for the stator circuit of any random wound AC machine. However, the shorted bars of the rotor of an induction motor produce special problems due to *deep bar effect* which causes the rotor current to rise to the air gap side of the rotor bars. The overall result is that the resistance and inductance of the rotor circuit become frequency dependent.

2.6.1 Rectangular Squirrel Cage Bars

Consider the simple rectangular bar placed in an iron slot as shown in Figure 2.6. The total IR drop across the length of the bar at any height y can be obtained by integrating the electric field over the length of the bar, i.e.,

$$\tilde{V}_R = \int_0^l \tilde{E} \cdot dl \tag{2.49}$$

where l denotes the length of the rotor bar in the laminations. It is shown in [2] that the voltage drop reduces to

$$\tilde{V}_R = \frac{\gamma \rho I_m l}{b}\left(\frac{\cosh \gamma y}{\sinh \gamma d}\right) \tag{2.50}$$

where

$$\gamma = \sqrt{\frac{j\omega_b \mu_o}{\rho}} \tag{2.51}$$

and ω_b is the angular frequency of the EMF impressed on the rotor bar, μ_o is the permeability of air, and ρ is the resistivity of the conducting bar. Note that the resistive drop is a maximum at the top of the slot where the parameter $y = d$.

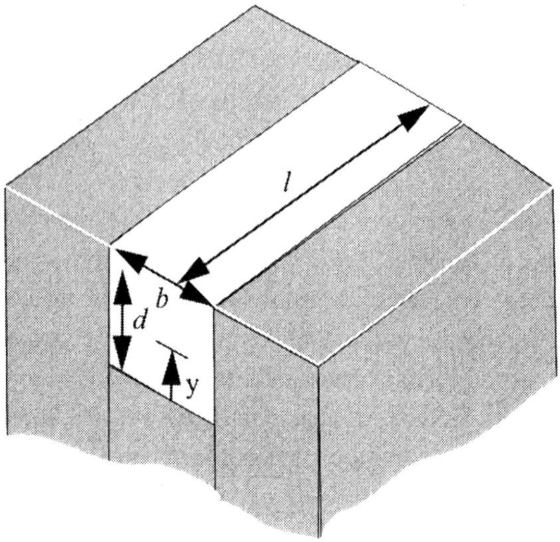

Figure 2.6 Rotor bar embedded in laminations.

The total flux crossing the slot above the height y which therefore links the current below y is

$$\tilde{\phi}_m(y) = \int_0^l \int_y^d \tilde{B}_m \, dy \, dz$$

$$= \frac{\mu_o I_m l}{\gamma b_s}\left(\frac{\cosh\gamma d - \cosh\gamma y}{\sinh\gamma d}\right) \quad (2.52)$$

The total IX drop across the length of the bar at any height y is therefore

$$\tilde{V}_L = \frac{\partial}{\partial t}\int_0^l \int_y^d \tilde{B}_m \, dy \, dz \quad (2.53)$$

or in complex form

$$\tilde{V}_L = j\omega_b \tilde{\phi}_m$$

whereupon

$$\tilde{V}_L = \frac{j\omega_b \mu_o I_m l}{\gamma \; b_s}\left(\frac{\cosh\gamma d - \cosh\gamma y}{\sinh\gamma d}\right) \quad (2.54)$$

Note that the reactive drop is zero at the top of the slot and a maximum at the bottom of the slot, just the reverse of the resistive drop.

The total voltage drop along the bar at an arbitrary height y is the sum of the inductive and resistive drop or

$$\tilde{V}_{bar} = \tilde{V}_R + \tilde{V}_L$$

$$= \frac{\gamma\rho I_m l}{b}\left(\frac{\cosh\gamma y}{\sinh\gamma d}\right) + \frac{j\omega_b\mu_o I_m l}{\gamma \quad b_s}\left(\frac{\cosh\gamma d - \cosh\gamma y}{\sinh\gamma d}\right) \quad (2.55)$$

which can also be written as [2]

$$\tilde{V}_{bar} = \frac{\gamma\rho I_m l}{b}\frac{\cosh\gamma d}{\sinh\gamma d} \quad (2.56)$$

Note that the total drop down the length of the bar is constant, which is to be expected since all of the current filaments are "connected" in parallel. The current distribution in the bar may be thought of as the superposition of a uniform (average) current and a circulating current flowing additively at the top of the bar and negatively at the bottom, directed in such a way as to oppose the time rate of change of slot leakage flux. The deep bar effect forces current to flow in the part of the conductor area which lies at the top of the slot and the effective resistance is many times (usually three to four times) as large as the round bar or shallow bar rotor. In the case of the double bar rotor, the resistance of the top bar usually has a higher resistance than the bottom bar, making the effective resistance at starting even larger if desired.

Equation (2.56) can be written in terms of the DC resistance of the bar if it is recalled that for a rectangular bar

$$R_{dc} = \frac{\rho l}{bd} \quad (2.57)$$

Hence, the effective impedance of the bar is

$$\tilde{Z}_{bar} = \frac{\tilde{V}_{bar}}{I_m}$$

$$= R_{dc}\left(\frac{\gamma d\cosh\gamma d}{\sinh\gamma d}\right) \quad (2.58)$$

The real portion of the impedance representing the AC resistance of the bar can be readily evaluated as

$$R_{ac} = \alpha d R_{dc}\left(\frac{\sinh 2\alpha d + \sin 2\alpha d}{\cosh 2\alpha d - \cos 2\alpha d}\right) \tag{2.59}$$

where

$$\alpha = \sqrt{\frac{\omega_b \mu_o}{2\rho}} \tag{2.60}$$

The imaginary component of the impedance represents the reactance of the bar and is

$$X_{ac} = \alpha d R_{dc}\left(\frac{\sinh 2\alpha d - \sin 2\alpha d}{\cosh 2\alpha d - \cos 2\alpha d}\right) \tag{2.61}$$

The inductance of a rectangular bar in a slot is [2]

$$L_{dc} = \mu_o \frac{dl}{3b} \tag{2.62}$$

so that Eq. (2.61) can also be written as

$$X_{ac} = \frac{3}{2}\frac{\omega_b L_{dc}}{\alpha d}\left(\frac{\sinh 2\alpha d - \sin 2\alpha d}{\cosh 2\alpha d - \cos 2\alpha d}\right) \tag{2.63}$$

The AC inductance is therefore

$$L_{ac} = \frac{3}{2}\frac{L_{dc}}{\alpha d}\left(\frac{\sinh 2\alpha d - \sin 2\alpha d}{\cosh 2\alpha d - \cos 2\alpha d}\right) \tag{2.64}$$

If αd is large, Eqs. (2.59) and (2.61) can be approximated by

$$R_{ac} = \alpha d R_{dc} \tag{2.65}$$

and

$$L_{ac} = \frac{3}{2}\frac{L_{dc}}{\alpha d} \tag{2.66}$$

Equation (2.65) can, by use of Eq. (2.57), be expressed in the form

$$\begin{aligned} R_{ac} &= \frac{\rho l}{bd}(\alpha d) \\ &= \rho\left[\frac{l}{b/\alpha}\right] \end{aligned} \tag{2.67}$$

Hence, for sufficiently high frequencies, the AC resistance can be calculated in the same manner as for the DC resistance if one replaces the actual depth of the bar d by an equivalent depth $1/\alpha$. The quantity $1/\alpha$ is called the *skin depth* and the fact that the current distributes itself unevenly over a conductor due to sinusoidal excitation is called the *skin effect*. A plot of R_{ac} and L_{ac} normalized with respect to their DC values is given in Figure 2.7.

One can write Eqs. (2.65) and (2.66) directly in terms of frequency if it is noted that

$$\alpha d = \sqrt{\frac{\omega_b \mu_o}{2\rho}} d = \sqrt{\frac{\omega_b \mu_o d^2}{2\rho}}$$

Defining a "bar" or "break" frequency

$$\omega_{b0} = \frac{2\rho}{\mu_o d^2} \tag{2.68}$$

one can also write R_{ac} and L_{ac} for large αd (high frequency) as

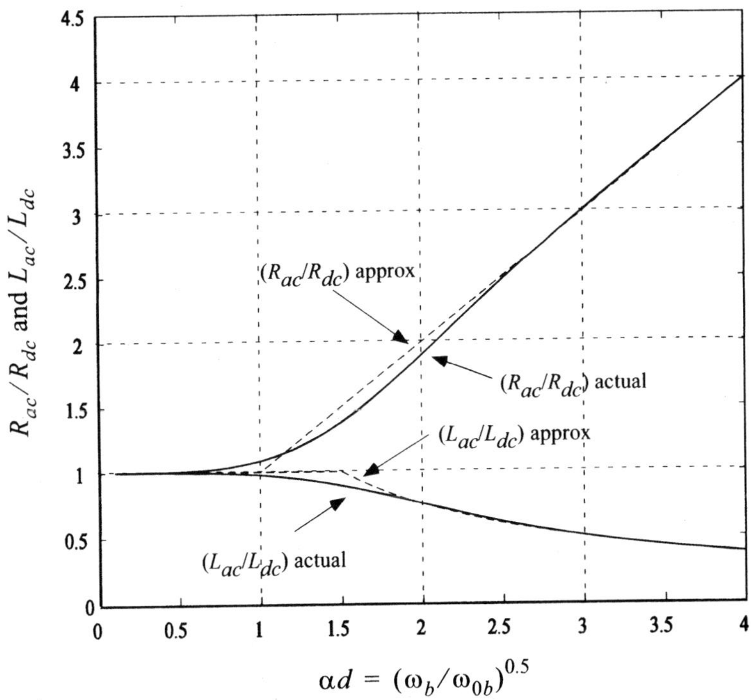

Figure 2.7 Sketch of normalized resistance and inductance of solid bar in a slot as a function of αd and their approximations.

$$R_{ac} = \sqrt{\frac{\omega_b}{\omega_{0b}}} R_{dc} \tag{2.69}$$

$$L_{ac} = \frac{3}{2} \frac{L_{dc}}{\sqrt{\omega_b / \omega_{0b}}} \tag{2.70}$$

If the DC value of R_{ac} and L_{ac} are assumed below $\omega_b / \omega_{0b} = 1$ and 9/4, respectively, and high-frequency approximations are used above these values, the dashed lines of Figure 2.7 are obtained. A good approximation is obtained which can be used to estimate the harmonic distortion caused by the frequency-dependant rotor bar parameters.

2.6.2 Nonrectangular Rotor Bars

It is well known that the bar shape of a practical induction motor design is rarely a simple rectangular structure as considered in Section 2.6.1. Indeed bar shapes have been devised so as to minimize the losses due to inverter harmonics. Since the analysis of the previous section was limited to this case, it is appropriate to question the validity of the results for nonrectangular bars. While a variety of shapes are possible, it appears that essentially all bar configurations possess a bar break frequency of the type identified as Eq. (2.68). The frequency behavior of the rotor resistance and leakage inductance can typically be approximated as a constant below this break point and as proportional (i.e., R_{ac}) or inversely proportional (L_{ac}) to $\sqrt{\omega_b / \omega_{b0}}$ beyond this point. The bar shapes can be shown to mainly affect the slope of the curves beyond the break point. However, the analysis that has been performed can always be modified to accommodate a different slope beyond the break point. For example, let K_b denote the multiplier which changes the slope from a unity value described in Eqs. (2.69) and (2.70) obtained, for example, by finite element analysis. Above the break frequency the AC resistance for the modified case can then be expressed as

$$\begin{aligned} R'_{ac} &= [K_b(\sqrt{\omega_b / \omega_{b0}} - 1) + 1] R_{dc} \\ &= K_b R_{ac} + (1 - K_b) R_{dc} \end{aligned} \tag{2.71}$$

Similarly

$$L'_{ac} = \frac{3}{2} \frac{L_{dc}}{K_b(\sqrt{\omega_b / \omega_{b0}} - 1) + 1} \tag{2.72}$$

which becomes

$$L'_{ac} = \frac{\sqrt{\omega_b/\omega_{b0}}L_{ac}}{K_b(\sqrt{\omega_b/\omega_{b0}} - 1) + 1} \tag{2.73}$$

Hence, when the AC resistance and inductance for a particular bar harmonic has been calculated for a simple rectangular bar, the equivalent result can be readily translated to a nonrectangular bar by Eqs. (2.71), (2.72), and (2.73).

2.6.3 Per-Phase Equivalent Circuit

Although thc impedance has been determined for a single bar, in the balanced steady state all of the bars of an induction motor experience the same impressed flux (only with different phase relationships) so that the motor equivalent circuit rotor impedance maintains the same frequency dependence. Although a portion of the rotor circuit is comprised of the end ring portion, again the same frequency dependence exists. Since the rotor can, for practical purposes, be considered as rotating near synchronous speed, differing only by a small slip, it can be assumed that the bar frequency is related to an arbitrary harmonic impressed on the stator by

$$\omega_b = \omega_n \pm \omega_1 \tag{2.74}$$

where the plus sign is taken when the stator harmonic rotates in the negative direction, (i.e., 2,5,8,11,...) and the minus sign is used when the harmonic rotates in the positive direction (i.e., 1,4,7,10,13,...). The frequency-dependent rotor resistance and inductance can then be well approximated by

$$R_{ac} = \sqrt{\frac{\omega_n \pm \omega_1}{\omega_{0b}}}R_{dc} \tag{2.75}$$

and

$$L_{ac} = \frac{3}{2}\frac{L_{dc}}{\sqrt{(\omega_n \pm \omega_1)/\omega_{0b}}} \tag{2.76}$$

The resulting per-phase equivalent circuit for an induction motor with frequency-dependent parameters is shown in Figure 2.8(a).

When high-frequency harmonics are superimposed on the fundamental, superposition principles can be applied assuming that saturation is not too severe. When the frequency is high, the slip frequency corresponding to an arbitrary nontriplen harmonic is

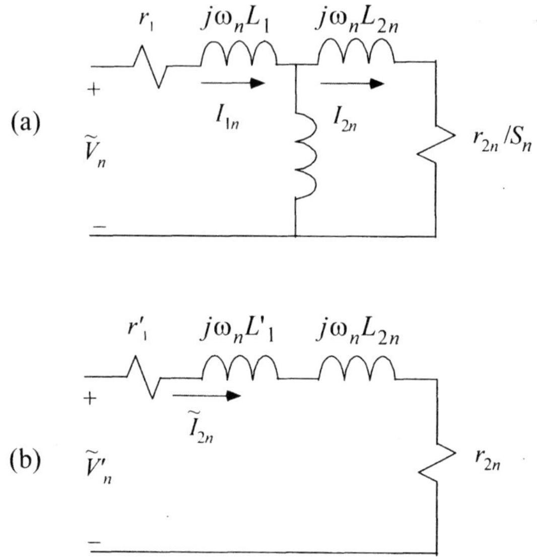

$$r_{2n} = r_2\sqrt{(\omega_n \pm \omega_1)/\omega_{0b}} \qquad L_{2n} = \frac{3}{2}\frac{L_2}{\sqrt{[(\omega_n \pm \omega_1)/\omega_{0b}]}}$$

$$r_1' = r_1\frac{L_m}{L_1 + L_m} \qquad L_1' = \frac{L_1 L_m}{L_1 + L_m}$$

$$\tilde{V}_n' = \frac{L_m}{L_1 + L_m}\tilde{V}_n$$

Figure 2.8 (a) Frequency-dependent induction motor equivalent circuit for nontriplen harmonic component and (b) approximation with stator side referred to the rotor by Thevenin's theorem.

$$S_n = \frac{\pm\omega_n - \omega_r}{\pm\omega_n} \rightarrow 1 \tag{2.77}$$

where ω_r is the rotor speed in electrical radians per second, and either the plus or minus sign applies depending upon whether the harmonic is a positive or negative sequence, respectively. If ω_n is sufficiently high, this expression approaches unity regardless of the polarity of ω_n.

Because the rotor parameters are frequency dependent, it is useful to refer the stator side of the circuit to the rotor rather than vice versa. Using Thevenin's theorem, the voltage observed at the air gap from the rotor side is

$$V'_n = \frac{j\omega_n L_m}{r_1 + j\omega_n(L_1 + L_m)} V_n \tag{2.78}$$

For sufficiently high frequencies, Eq. (2.78) can be conveniently approximated as

$$V'_n \cong \frac{L_m}{L_1 + L_m} V_n \tag{2.79}$$

The Thevenin driving point impedance is obtained by shorting the voltage source in which case

$$Z'_{1n} = \frac{(r_1 + j\omega_n L_1) j\omega_n L_m}{r_1 + j\omega_n(L_1 + L_m)} \tag{2.80}$$

Again, if the angular frequency ω_n is sufficiently high,

$$\begin{aligned} Z'_{1n} &\cong r_1 \frac{L_m}{L_1 + L_m} + j\omega_n\left(\frac{L_1 L_m}{L_1 + L_m}\right) \\ &= r'_1 + j\omega_n L'_1 \end{aligned} \tag{2.81}$$

where

$$r'_1 = r_1 \frac{L_m}{L_1 + L_m} \tag{2.82}$$

$$L'_1 = \frac{L_1 L_m}{L_1 + L_m} \tag{2.83}$$

It is clear that the frequency dependence of the rotor parameters must reflect the fact that the rotor is rotating, so that the fifth-harmonic stator voltage, for example, impresses essentially a fourth- or a sixth-harmonic voltage on the rotor depending upon its direction of rotation, due to the fact that it is positive or negatively rotating with respect to the rotor. In general, for harmonics related to the fundamental component, the three-phase voltage waveforms remain "balanced". That is, the three voltages have identical wave shapes but are only phase shifted in time by one-third of a complete cycle. The rotation sequence can be determined by dividing the harmonic number n by 3 and examining the numerator of the resulting fraction (if any). If the numerator is 1, the sequence is positive and the minus sign applies in Eq. (2.77), while if the numerator of the fraction is 2, the sequence is negative (the plus sign applies). If the value of n is divisible by 3, the component corresponds to a zero sequence term and should be omitted from the summation since this component does not link the rotor.

2.7 Harmonic Distortion Weighting Factors for Induction Motor Load

2.7.1 WTHD for Frequency-Dependent Rotor Resistance

A very simple figure of merit for the effect of impressing a nonsinusoidal voltage waveform on the rotor branch can be developed if one considers only the frequency dependence of the rotor resistance and assumes that the rotor inductance remains constant. In this case, referring to Eq. (2.69) and Figure 2.7, one can approximate the variation of r_2 with frequency as

$$r_{2n} = r_2 \quad \text{when } \omega_b \leq \omega_{0b} \tag{2.84}$$

and

$$r_{2n} = r_2\sqrt{\frac{\omega_b}{\omega_{0b}}} \quad \text{when } \omega_b > \omega_{0b} \tag{2.85}$$

In the first case the loss in the rotor resistance due to a single harmonic is simply

$$P_{2n} = 3I_{2n}^2 r_2 \cong 3\left[\frac{V_n'}{\omega_n(L_1' + L_2)}\right]^2 r_2 \tag{2.86}$$

if stator resistance can be neglected. When normalized to the loss occurring during the starting period (due to inrush current), this becomes

$$\frac{P_{2n}}{P_{2,\,inrush}} = \frac{3\left[\dfrac{V_n'}{\omega_n(L_1' + L_2)}\right]^2 r_2}{3\left[\dfrac{V_1'}{\omega_1(L_1' + L_2)}\right]^2 r_2} = \left(\frac{\omega_1}{\omega_n}\right)^2\left(\frac{V_n'}{V_1'}\right)^2 \tag{2.87}$$

which reduces finally to

$$\frac{P_{2n}}{P_{2,\,inrush}} = \frac{1}{n^2}\left(\frac{V_n}{V_1}\right)^2 \tag{2.88}$$

if it is assumed that the stator resistance drop is negligible compared to the drop across the stator leakage inductance at the frequencies of interest.

When $\omega_b > \omega_{0b}$,

$$P_{2n} = 3\left[\frac{V_n'}{\omega_n(L_1' + L_2)}\right]^2 r_2\sqrt{\frac{\omega_n \pm \omega_1}{\omega_{0b}}} \tag{2.89}$$

or

$$\frac{P_{2n}}{P_{2,\,inrush}} = \frac{3\left[\frac{V_n'}{\omega_n(L_1' + L_2)}\right]^2 r_2\sqrt{\frac{\omega_n \pm \omega_1}{\omega_{0b}}}}{3\left[\frac{V_1'}{\omega_1(L_1' + L_2)}\right]^2 r_2} \tag{2.90}$$

$$= \left(\frac{\omega_1}{\omega_n}\right)^2\sqrt{\frac{\omega_n \pm \omega_1}{\omega_{0b}}}\left(\frac{V_n'}{V_1'}\right)^2$$

$$= \left(\frac{\omega_1}{\omega_n}\right)^2\sqrt{\frac{\omega_n}{\omega_1}\frac{\omega_1}{\omega_{0b}} \pm \frac{\omega_1}{\omega_{0b}}}\left(\frac{V_n'}{V_1'}\right)^2$$

$$\cong \frac{1}{n^2}\sqrt{(n \pm 1)\frac{\omega_1}{\omega_{0b}}}\left(\frac{V_n}{V_1}\right)^2 \tag{2.91}$$

One can now define a weighted harmonic distortion factor which includes two sets of terms such that

$$\text{WTHD}_{bar} = \sqrt{\sum_{n=2}^{n_0}\frac{1}{n^2}\left(\frac{V_n}{V_1}\right)^2 + \sum_{n_0+1}^{\infty}\frac{1}{n^2}\sqrt{(n \pm 1)\frac{\omega_1}{\omega_{0b}}}\left(\frac{V_n}{V_1}\right)^2} \tag{2.92}$$

where n_0 denotes the highest harmonic for which $\omega_b < \omega_{0b}$.

It is apparent that the geometry-dependent parameter ω_1/ω_{0b} has an important effect on WTHD since the second group of terms are weighted more heavily with respect to frequency than the first group. In most practical cases where pulse width modulation is used, the lower frequency harmonics are suppressed and all of the harmonics are greater than ω_{0b}, so that one can define

$$\text{WTHD}_{bar} = \left(\frac{\omega_1}{\omega_{0b}}\right)^{1/4}\sqrt{\sum_{n=2}^{\infty}\frac{\sqrt{n \pm 1}}{n^2}\left(\frac{V_n}{V_1}\right)^2} \tag{2.93}$$

The quantity ω_1/ω_{0b} can be considered as a “bar factor” related to the design of the machine over which the inverter specialist has little control, whereas the remaining portion is most relevant for the purpose of selecting a PWM algorithm. For 60 Hz operation the value of ω_1/ω_{0b} takes on values ranging from 0.25 for small machines (5 HP) to 4.0 for large machines (500 HP).

2.7.2 WTHD Also Including Effect of Frequency-Dependent Rotor Leakage Inductance

Thus far, only the effects of frequency-dependent rotor resistance has been considered. When the variation in rotor leakage inductance is also taken into account, the procedure is similar. In this case three regions can be identified:

$$r_{2n} = r_2 \qquad L_{2n} = L_2 \qquad \text{when } \omega_b \le \omega_{0b} \tag{2.94}$$

$$r_{2n} = r_2\left(\sqrt{\frac{\omega_n \pm \omega_1}{\omega_{0b}}}\right) \qquad L_{2n} = L_2 \qquad \text{when } \omega_{0b} < \omega_b \le \frac{9}{4}\omega_{0b} \tag{2.95}$$

$$r_{2n} = r_2\left(\sqrt{\frac{\omega_n \pm \omega_1}{\omega_{0b}}}\right) \qquad L_{2n} = \frac{3}{2}\frac{L_2}{\sqrt{(\omega_n \pm \omega_1)/\omega_{0b}}} \qquad \text{when } \omega_b > \frac{9}{4}\omega_{0b} \tag{2.96}$$

The solution for the contributions to the WTHD from the first two ranges of ω_n is the same previous work as in Section 2.7.1. A typical loss term for a harmonic belonging to the third region is

$$P_{2n} = 3\left[\frac{V_n'}{\omega_n\left(L_1' + 1.5\frac{L_2}{\sqrt{(\omega_n \pm \omega_1)/\omega_{0b}}}\right)}\right]^2 r_2\sqrt{\frac{\omega_n \pm \omega_1}{\omega_{0b}}} \tag{2.97}$$

or

$$P_{2n} = 3\left[\frac{V_n'}{\omega_n L_1'\left(\sqrt{(\omega_n \pm \omega_1)/\omega_{0b}} + 1.5\frac{L_2}{L_1'}\right)}\right]^2 r_2\left(\frac{\omega_n \pm \omega_1}{\omega_{0b}}\right)^{3/2} \tag{2.98}$$

which, when normalized to the power dissipated during the current inrush interval (assuming constant parameters under this condition), becomes

$$\frac{P_{2n}}{P_{2,\,inrush}} = \frac{3\left[\frac{V_n'}{\omega_n L_1'\left(\sqrt{(\omega_n \pm \omega_1)/\omega_{0b}} + 1.5\frac{L_2}{L_1'}\right)}\right]^2 r_2\left(\frac{\omega_n \pm \omega_1}{\omega_{0b}}\right)^{3/2}}{3\left[\frac{V_1'}{\omega_1(L_1' + L_2)}\right]^2 r_2} \tag{2.99}$$

This rearranges to

$$\frac{P_{2n}}{P_{2,\,inrush}} = \frac{1}{n^2}\frac{\left(1+\frac{L_2}{L_1'}\right)^2}{\left[\sqrt{(\omega_n \pm \omega_1)/\omega_{0b}} + 1.5\frac{L_2}{L_1'}\right]^2}\left(\frac{\omega_n \pm \omega_1}{\omega_{0b}}\right)^{3/2}\left(\frac{V_n'}{V_1'}\right)^2 \tag{2.100}$$

Some simplification is possible if it is assumed that the primed stator inductance L_1' and the rotor leakage inductance L_2 are equal, a standard assumption in induction machine analysis. Then, neglecting the effects of stator resistance on the voltage ratio V_n'/V_1', Eq. (2.100) becomes

$$\frac{P_{2n}}{P_{2,\,inrush}} = \frac{1}{n^2}\frac{4}{\left[\sqrt{(n \pm 1)\frac{\omega_1}{\omega_{0b}}} + 1.5\right]^2}\left[(n \pm 1)\frac{\omega_1}{\omega_{0b}}\right]^{3/2}\left(\frac{V_n}{V_1}\right)^2 \tag{2.101}$$

A weighted THD factor, termed WTHD2, can now be developed as the square root of three sets of terms in which Eq. (2.87) is used in the first region where $\omega_b \le \omega_{0b}$, Eq. (2.91) applies for harmonics in which $\omega_{0b} < \omega_b \le (9/4)\omega_{0b}$, and Eq. (2.101) is applicable for values of harmonics for which $\omega_b > (9/4)\omega_{0b}$. Again, the WTHD must be presented as a function of the machine-dependent parameter ω_1/ω_{0b} rather than as a single number.

The resulting function is plotted in Figure 2.9 for the case of the quasi-square-wave inverter. The complicated nature of WTHD2 prompts one to consider acceptable approximations. In most cases the harmonic content of the lower harmonics are negligible; indeed this result is the goal of most pulse width modulation algorithms. In such cases it can be assumed that the harmonic components of flux rotate in the air gap at a sufficiently high rate of rotation that the rotor appears as essentially stationary. Hence, one can, in effect, replace the bar frequency $b = n \pm 1$ terms simply by n. The final result for WTHD2 then becomes

$$\mathrm{WTHD2} = \sqrt{W1 + W2 + W3} \tag{2.102}$$

where

$$W1 = \sum_{\substack{n = 3k \pm 1 \\ k = 1, 2, \ldots}}^{n_{0b}} \frac{1}{n^2}\left(\frac{V_n}{V_1}\right)^2 \tag{2.103}$$

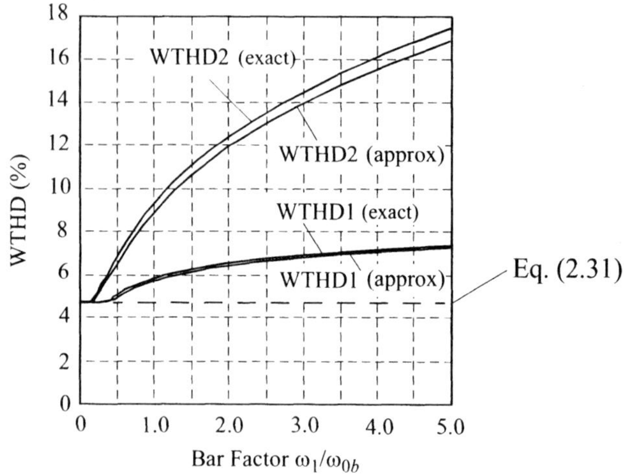

Figure 2.9 Effect of frequency dependant rotor parameters on the stator and rotor weighted total harmonic distortion.

$$W2 = \sum_{n_{0b}+1}^{n_{2b}} \frac{1}{n^{3/2}} \sqrt{\frac{\omega_1}{\omega_{0b}}} \left(\frac{V_n}{V_1}\right)^2 \qquad (2.104)$$

$$W3 = \sum_{n_{2b}+1}^{\infty} \frac{1}{\sqrt{n}} \frac{4}{\left[\sqrt{n\frac{\omega_1}{\omega_{0b}}} + 1.5\right]^2} \left(\frac{\omega_1}{\omega_{0b}}\right)^{3/2} \left(\frac{V_n}{V_1}\right)^2 \qquad (2.105)$$

In these equations n_{0b} represents the highest harmonic for which $\omega_b < \omega_{0b}$ and n_{2b} denotes the highest harmonic for which $\omega_b < 9/4\omega_{0b}$. Included in this result also is the fact that third-harmonic components (zero sequence components) of stator current do not link the rotor if the stator winding is an ideal sinusoidally distributed winding. Hence, only values of $n = 2,4,5,7,8,10,...$ are included in the summation. Since the parameter ω_1/ω_{0b} cannot be isolated from the remainder of the expression, the WTHD2 is best portrayed as a function of ω_1/ω_{0b} or specified for particular values of ω_1/ω_{0b}. Note also that if the stator and rotor leakage inductances differ appreciably, a more general expression for $W3$ based on Eq. (2.100) should be used

The exact and approximate functions are plotted in Figure 2.9 for the case of the quasi-square-wave inverter. Since it can be recalled that the quasi-square-wave inverter output has a full measure of nontriplen odd harmonics, the approximation can be considered sufficiently accurate to be used as a figure of merit.

One should note that at most only two harmonics can exist in the middle region and could easily be approximated by relegating the lower of the two harmonics to the first region and the upper one to region three. Also, if all of the harmonics are sufficiently higher than ω_{0b}, then only values in the third region apply. Furthermore, if

$$\sqrt{(n \pm 1)\frac{\omega_1}{\omega_{0b}}} \gg 1.5$$

Eq. (2.101) simplifies to

$$\frac{P_{2n}}{P_{2,\,inrush}} = \frac{4}{n^2}\sqrt{(n \pm 1)\frac{\omega_1}{\omega_{0b}}}\left(\frac{V_n}{V_1}\right)^2 \tag{2.106}$$

which is simply four times Eq. (2.91). Again making the assumption that the bar frequencies are sufficiently high that $b \cong n$, the frequency dependent WTHD for this condition approximates to

$$\text{WTHD2} = 2\left(\frac{\omega_1}{\omega_{0b}}\right)^{1/4}\sqrt{\sum_{\substack{n = 3k \pm 1 \\ k = 1, 2, \ldots}}^{\infty} \frac{1}{n^{3/2}}\left(\frac{V_n}{V_1}\right)^2} \tag{2.107}$$

This result is to be expected since Eq. (2.106) effectively assumes that the rotor leakage inductance L_{2n} is zero and, since it has been assumed that $L_1' = L_2$, the current for all harmonics will be twice the value obtained when L_{2n} is assumed as a constant equal to L_1'.

It must be mentioned here that the assumption of ideal sinusoidally distributed windings is a convenient mathematical artifice which is never realized exactly in practice. In reality, the stator winding distribution is nonideal, meaning the higher harmonic fields are set up in the air gap, that is, fields with sets of poles which are an odd multiple of the number of poles set up by the fundamental component. In most practical cases a small third-harmonic spatial component will therefore exist in the air gap when third-harmonic stator currents are allowed to flow, so that a field is produced which does link the rotor bars. This field can be shown to be single-phase in nature, i.e., stationary and pulsating in space. Since this effect is generally small, it is neglected here.

2.7.3 WTHD for Stator Copper Losses

The equivalent circuit of Figure 2.8 can also be used to obtain a suitable figure of merit for stator losses. The ratio of stator resistive losses due to harmonic n as a per unit of power dissipated during the inrush period with fundamental frequency voltage applied is

$$\frac{P_{1n}}{P_{1,\,inrush}} = \frac{3\left[\dfrac{V_n'}{\omega_n(L_1'+L_2)}\right]^2 r_1}{3\left[\dfrac{V_1'}{\omega_1(L_1'+L_2)}\right]^2 r_1} \tag{2.108}$$

When $\omega_1 < \omega_{0b}$ and $\omega_{0b} \le \omega_1 \le 9/4\omega_{0b}$, both the stator resistance and the rotor leakage inductance are constant and this expression reduces to

$$\frac{P_{1n}}{P_{1,\,inrush}} = \left(\frac{\omega_1}{\omega_n}\right)^2\left(\frac{V_n}{V_1}\right)^2 \tag{2.109}$$

Hence, for harmonics, $0 \le \omega_b \le 9/4\omega_{b0}$, the stator loss WTHD is the same as previously obtained for the rotor when r_2 and L_2 are constant, namely

$$\text{WTHD1} = \sqrt{\sum_{n=2}^{\infty}\frac{1}{n^2}\left(\frac{V_n}{V_1}\right)^2} \tag{2.110}$$

For operation above $9/4\omega_{0b}$

$$\frac{P_{1n}}{P_{1,\,inrush}} = \frac{3\left[\dfrac{V_n'}{\omega_n\left(L_1' + 1.5\dfrac{L_2}{\sqrt{(\omega_n \pm \omega_1)/\omega_{0b}}}\right)}\right]^2 r_1}{3\left[\dfrac{V_1'}{\omega_1(L_1'+L_2)}\right]^2 r_1} \tag{2.111}$$

which, after some manipulation and using the previous assumptions, becomes

$$\frac{P_{1n}}{P_{1,\,inrush}} = \frac{1}{n^2}\,\frac{4(n\pm1)}{\left[\sqrt{(n\pm1)\dfrac{\omega_1}{\omega_{0b}}}+1.5\right]^2}\left(\frac{\omega_1}{\omega_{0b}}\right)\left(\frac{V_n}{V_1}\right)^2 \tag{2.112}$$

Again if it is assumed that the bar frequency for each harmonic is sufficiently close to the corresponding stator harmonic frequency, then

$$\frac{P_{1n}}{P_{1,\,inrush}} \cong \frac{1}{n}\frac{4}{\left[\sqrt{n\frac{\omega_1}{\omega_{0b}}}+1.5\right]^2}\left(\frac{\omega_1}{\omega_{0b}}\right)\left(\frac{V_n}{V_1}\right)^2 \tag{2.113}$$

A suitable expression for the stator WTHD is consequently

$$\mathrm{WTHD1} = \sqrt{\sum_{n=3k\pm1}^{n_0}\frac{1}{n^2}\left(\frac{V_n}{V_1}\right)^2+\frac{\omega_1}{\omega_{0b}}\sum_{n=n_0+1}^{\infty}\frac{1}{n}\frac{4}{\left[\sqrt{n\frac{\omega_1}{\omega_{0b}}}+1.5\right]^2}\left(\frac{V_n}{V_1}\right)^2} \tag{2.114}$$

where n_0 is the highest harmonic for which $\omega_b < 9/4\omega_{0b}$.

Finally, for sufficiently high harmonic content (above $9/4\omega_{0b}$)

$$\sqrt{(n\pm1)\left(\frac{\omega_1}{\omega_{0b}}\right)} \gg 1.5$$

so that, as a final figure of merit one can reasonably approximate the frequency dependent stator WTHD by means of the simple expression

$$\mathrm{WTHD1} = 2\sqrt{\sum_{n=3k\pm1}^{\infty}\frac{1}{n^2}\left(\frac{V_n}{V_1}\right)^2} \tag{2.115}$$

which is simply twice the WTHD obtained by ignoring the frequency dependence of the motor parameters. The factor of 2 is obtained by neglecting, in effect, the rotor leakage inductance. If the inverter waveform has only high-frequency harmonic content, this result demonstrates that WTHD as defined by Eq. (2.22) remains a reasonable figure of merit for the stator losses of an induction motor but not for the rotor losses.

Figure 2.9 shows how the two WTHD terms vary as a function of ω_1/ω_{0b} for the case of a three-phase quasi-square-wave inverter. Also shown for reference is the WTHD which does not include the effect of frequency-dependent parameters [Eq. (2.31)]. It is apparent that the deep bar effect has an important effect on the losses, particularly when the fundamental component is substantially greater than the characteristic bar frequency. The effect on the loss is particularly dominant in the rotor since the resistance corresponding to the rotor bar harmonics increases with the bar factor while the stator resistance remains independent of frequency.

2.8 Example Calculation of Harmonic Losses

The following induction motor is assumed to be powered by a quasi-square-wave voltage source inverter:

HP = 50 HP	$V_{ll,rms} = 230$ V	$\omega_1 = 377$ rad/s
$r_1 = 0.03\ \Omega$	$r_2 = 0.04\ \Omega$	
$X_1 = 0.1234\ \Omega$	$X_2 = 0.1176\ \Omega$	
$X_m = 2.5\ \Omega$		

For the condition of interest, the inverter operates in the field weakening mode at two per unit frequency (based on the rating of the motor). The bar characteristic frequency is determined to be $\omega_{b0} = 300$ rad/s. The task is to determine (roughly) the motor losses due to the harmonic content of the inverter.

Solution

Since the angular frequency corresponding to two per unit is

$$\omega_1 = 2(377) = 754 \text{ rad/s}$$

the characteristic bar factor is

$$\frac{\omega_1}{\omega_{0b}} = \frac{754}{300} = 2.51$$

The corresponding WTHDs from Figure 2.9 are

$$\text{WTHD1} = 0.067$$

$$\text{WTHD2} = 0.135$$

By definition,

$$\frac{\sum_n P_{1n}}{P_{1,\,inrush}} = (\text{WTHD1})^2 = 0.067^2 = 0.00449$$

Thus

$$\sum_n P_{1n} = 0.00449 P_{1,\,inrush} = \frac{0.00449(V_1')^2}{[\omega_1(L_1' + L_2)]^2} r_1$$

Now

$$V_1' = \frac{X_m}{X_m + X_1} V_1 = \frac{2.5}{2.6234} \frac{230}{\sqrt{3}} = 126.5$$

$$\omega_1 L_1' = 754 \frac{L_m L_1}{L_m + L_1} = \frac{2(0.1234)2.5}{0.1234 + 2.5} = 0.2352$$

Thus

$$\sum_n P_{1n} = \frac{0.00449(V_1')^2}{[\omega_1(L_1' + L_2)]^2} r_1 = \frac{0.00449(126.5)^2}{0.4704^2} 0.03 = 9.74 \text{ W/phase}$$

Similarly

$$\frac{\sum_n P_{2n}}{P_{1,\,inrush}} = \text{WTHD2}^2 = 0.135^2 = 0.0182$$

which gives

$$\sum_n P_{2n} = 0.0182 P_{inrush} = \frac{0.0182(V_1')^2}{[\omega_1(L_1' + L_2)]^2} r_2$$

and eventually

$$\sum_n P_{2n} = \frac{0.0182(126.5)^2}{0.4704^2} 0.04 = 52.64 \text{ W/phase}$$

The total copper loss due to harmonic content in the quasi-square-wave inverter is

$$\sum_n P_n = 3\left(\sum_n P_{1n} + \sum_n P_{2n}\right) = 187 \text{ W}$$

which represents roughly 0.5% of the motor 50 HP rating.

2.9 WTHD Normalization for PWM Inverter Supply

In cases where the fundamental component remains constant or when the DC link voltage is proportional to the fundamental AC component as is the case for constant volts/hertz drives, the normalization of the WTHD by the fundamental AC component is a suitable choice. In particular, note that the THD and WTHD of the five inverter types shown in Table 2.1 do not change with frequency provided that the DC voltage decreases in direct proportion. In this case the fundamental component as well as all of the harmonics change by the same value making the THD and the WTHDs independent of frequency.

In the case of a pulse width-modulated inverter, however, the DC voltage remains constant while the fundamental component varies. On the other hand, the harmonic components change relatively little, assuming the same ratio of switching to output frequency, making the THD and WTHDs vary widely. In fact, it is easy to see that when the fundamental component approaches zero [as the modulation index (M) approaches zero], the distortion factors approach infinity, clearly an unsatisfactory situation. The problem can be avoided by simply choosing a normalization factor which is invariant as frequency changes. For the case of the half-bridge inverter this quantity is conveniently chosen as the value of the fundamental AC voltage existing when the modulation index M equals 1, i.e., V_{dc}. Using the approximation $b = n$, the WTHDs normalized to the inverter DC voltage become

$$\mathrm{WTHD0} = \frac{\sqrt{\sum_{n=2}^{\infty} \frac{1}{n^2} V_n^2}}{V_1\big|_{M=1}} = \sqrt{\sum_{n=2}^{\infty} \frac{1}{n^2}\left(\frac{V_n}{V_{dc}}\right)^2} \tag{2.116}$$

$$\mathrm{WTHD01} = \sqrt{\begin{aligned}&\sum_{\substack{n=3k\pm1\\k=1,2,3,\ldots k_{0b}}}^{n_{0b}} \frac{1}{n^2}\left(\frac{V_n}{V_{dc}}\right)^2\\ &+\frac{\omega_1}{\omega_{0b}} \sum_{\substack{n=3k\pm1\\k=k_{0b}+1,\ldots}}^{\infty} \frac{1}{n} \frac{4}{\left[\sqrt{n\frac{\omega_1}{\omega_{0b}}}+1.5\right]^2}\left(\frac{V_n}{V_{dc}}\right)^2\end{aligned}} \tag{2.117}$$

$$\mathrm{WTHD02} = \sqrt{W1 + W2 + W3} \tag{2.118}$$

where

$$W1 = \sum_{\substack{n = 3k \pm 1 \\ k = 1, 2, 3, \dots}}^{n_{0b}} \frac{1}{n^2} \left(\frac{V_n}{V_{dc}}\right)^2 \tag{2.119}$$

$$W2 = \sum_{n_{0b}+1}^{n_{2b}} \frac{1}{n^{3/2}} \sqrt{\frac{\omega_1}{\omega_{0b}}} \left(\frac{V_n}{V_{dc}}\right)^2 \tag{2.120}$$

$$W3 = \sum_{n_{2b}+1}^{\infty} \frac{1}{\sqrt{n}} \frac{4}{\left[\sqrt{n\frac{\omega_1}{\omega_{0b}}} + 1.5\right]^2} \left(\frac{\omega_1}{\omega_{0b}}\right)^{3/2} \left(\frac{V_n}{V_{dc}}\right)^2 \tag{2.121}$$

In these equations n_{0b} again represents the highest harmonic for which $\omega_b < \omega_{0b}$ while n_{2b} denotes the highest harmonic for which $\omega_b < 9/4\omega_{ob}$.

If desired the WTHD is readily obtained from these expressions by taking

$$\text{WTHD} = \text{WTHD0}\frac{V_{dc}}{V_1} = \frac{\text{WTHD0}}{M} \tag{2.122}$$

$$\text{WTHD1} = \frac{\text{WTHD01}}{M} \tag{2.123}$$

$$\text{WTHD2} = \frac{\text{WTHD02}}{M} \tag{2.124}$$

These figures of merit will now be utilized throughout the remainder of this book.

2.10 Summary

The concept of the harmonic distortion factor has been extensively explored in this chapter. The remainder of this text will treat the analysis of the many PWM algorithms which exist for single-leg, single-phase, three-phase, and multilevel three-phase inverters. A key to interpretation of these algorithms is a suitable performance criterion. This chapter has developed the concept of a weighted total harmonic distortion for passive or active *R–L* loads in which the resistance is either frequency independent or frequency dependent (i.e., induction motor load). It is now appropriate to turn to the major task at hand, the derivation and assessment of inverter PWM algorithms.

References

[1] H.B. Dwight, *Table of Integrals and Other Mathematical Data*, 3rd ed., MacMillan, New York, 1957.

[2] T.A. Lipo, *Introduction to AC Machine Design*, Vol. 1, University of Wisconsin Power Electronics Research Center, University of Wisconsin, Madison, 1996.

3

Modulation of One Inverter Phase Leg

One of the most widely utilized strategies for controlling the AC output of power electronic converters is the technique known as pulse width modulation (PWM), which varies the duty cycle (or mark-space ratio) of the converter switch(es) at a high switching frequency to achieve a target average low-frequency output voltage or current. Modulation theory has been a major research area in power electronics for over three decades and continues to attract considerable attention and interest. This is not surprising, since modulation is at the heart of nearly every modern power electronic converter, but the enormous amount of material published makes it challenging for a user to identify basic modulation principles and apply them to particular implementations. Much reported research has presented "new" or "improved" PWM techniques, which are often only a straightforward variation of a previous approach. But it can at times be quite difficult to see how they are related.

On the other hand, there have been a number of clear trends in the development of PWM concepts and strategies since the 1970s, addressing the main objectives of reduced harmonic distortion and increased output magnitudes for a given switching frequency and the development of modulation strategies to suit different converter topologies.

Since the work presented in this book primarily relates to medium and higher power level hard-switched inverters (i.e., above the 1 kW power level), this chapter considers only modulation strategies which are appropriate for the two major converter topologies in this power range, being the voltage source inverter (VSI) and the current source inverter (CSI). While this chapter concentrates on VSI modulation, the principles are equally applicable to a CSI as will be discussed in Chapter 9.

While there has been a wealth of research investigating the modulation and control of lower power DC/DC converters, the actual PWM process for these converters is usually a simple comparison between a reference waveform and a sawtooth or a triangular carrier waveform. The operation and performance of this strategy is straightforward and needs little elaboration in this text.

A more recent area of research in power electronic converters has been the development of soft-switched resonant converters, and the selection of appropriate modulation strategies for this class of converters is a further area of interest. Generally, closed-loop feedback modulation strategies are used for this class of inverter, and their performance is more dependent on the performance of the closed-loop controller than any intrinsic principles of PWM. Since the focus of this book is more directed toward fixed-frequency hard-switched PWM schemes, the modulation and control of resonant converters and similar topologies will not be considered further in this book.

3.1 Fundamental Concepts of PWM

In principle, all modulation schemes aim to create trains of switched pulses which have the same fundamental volt-second average (i.e., the integral of the voltage waveform over time) as a target reference waveform at any instant. The major difficulty with these trains of switched pulses is that they also contain unwanted harmonic components which should be minimized. Hence for any PWM scheme, a *primary* objective can be identified which is to calculate the converter switch ON times which create the desired (low-frequency) target output voltage or current. Having satisfied this primary objective, the *secondary* objective for a PWM strategy is to determine the most effective way of arranging the switching processes to minimize unwanted harmonic distortion, switching losses, or any other specified performance criterion.

It is interesting to note that despite the wealth of material that has been published relating to PWM, only three significantly different alternatives for determining the converter switch ON times have ever been usefully proposed for fixed-frequency modulation systems. These alternatives are:

1. Switching at the intersection of a target reference waveform and a high-frequency carrier (*naturally sampled PWM*).
2. Switching at the intersection between a regularly sampled reference waveform and a high-frequency carrier (*regular sampled PWM*).
3. Switching so that the integrated area of the target reference waveform over the carrier interval is the same as the integrated area of the converter switched output (*direct PWM*).

Many variations of these three alternatives have been published, and it sometimes can be quite difficult to see their underlying commonality. For example, even the well-known space vector modulation strategy, which is often claimed to be a completely different approach to modulation, is really only a variation of regular sampled PWM which specifies the same switched pulse widths but places them a little differently in each carrier interval.

In this chapter, the principles of PWM are developed in a generalized way which can be used with any converter topology and modulation implementation. This development offers a more integrated perspective than that presented in most technical papers and textbooks.

The fundamental basis of this development is that all fixed-frequency open-loop PWM strategies can be explained in terms of

- Switched pulse width determination.
- Switched pulse position within a carrier interval.
- Switched pulse sequence within and across carrier intervals.

The harmonic performance of a particular PWM implementation then results from the effect of these three factors on

- The harmonics generated by the phase leg switched waveform.
- Harmonic cancellation that may occur between individual phase leg switched outputs.

Note that switching frequency in this context is not regarded as a significant parameter which changes harmonic performance — changes in switching frequency affect the influence of the sideband harmonics because of attenuation through the low-frequency filtering action of loads but have no direct impact on the magnitude of these sideband harmonics.

In this chapter, the means by which naturally sampled and regularly sampled PWM determine the switched pulse *width* of one phase leg of an inverter is developed. The results of this development are then used in later chapters to investigate the performance of multiple phase legs arranged as single- and three-phase inverter systems.

3.2 Evaluation of PWM Schemes

Before proceeding with any investigation into PWM strategies it is necessary to establish a common basis for comparing the performance of various algorithms, and this has been the subject of considerable research work over the years. Some researchers argue the merits of a particular PWM implementation on the basis of diminished harmonics [1], others propose first- or second-order filtered distortion performance factors [2], and others evaluate the RMS harmonic ripple current in a typical load such as an induction motor [3]. Many variations of these approaches have been published, all with particular arguments as to why the performance index proposed is preferable. Unfortunately, one result of all this work is some confusion, since it becomes difficult to precisely compare the performance of PWM strategies presented by different researchers. With recent PWM innovations claiming performance improvements of fractions of a percent over previous approaches, it is important to be sure how much of the improvement claimed is because of more careful physical implementation, rather than any intrinsic advantages of the new scheme.

In this text, particular care has been taken to compare all PWM variations on exactly the same basis. The approach used is threefold:

- First, where they are available, analytical solutions to the PWM strategies that are to be compared are used to determine the magnitude of the various harmonic components. This ensures that harmonics caused by simulation roundoff errors or practical implementation effects such as dead time, switch ON-state voltages, DC bus voltage ripple, etc., are not confused with intrinsic harmonic differences between the PWM strategies being compared.
- Second, PWM strategies are compared at exactly the same phase leg switching frequencies. Where necessary (e.g., with discontinuous modulation strategies) the carrier frequency of a PWM implementation is varied to ensure that harmonic comparisons are done at precisely the same average device switching frequency per fundamental cycle.
- Third, the first-order weighted total harmonic distortion factor WTHD0 (WTHD normalized to the fundamental component at a modulation index of $M = 1$) is used as a single-term performance indicator for more rapid comparison of PWM alternatives. As explained in Section 2.2, WTHD can be interpreted as the normalized current ripple expected into an inductive load when fed from the switched waveform, and is a widely applicable performance indicator which has a useful physical significance.

3.3 Double Fourier Integral Analysis of a Two-Level Pulse Width–Modulated Waveform

The essential concept of a two-level pulse width-modulated converter system is that a low-frequency target reference waveform is compared against a high-frequency carrier waveform, and the comparison output is used to control the state of a switched phase leg. When the reference waveform is greater than the carrier waveform, the phase leg is switched to the upper DC rail. When the reference waveform is less than the carrier waveform, the phase leg is switched to the lower DC rail. The resultant phase leg output is a stream of pulses switching between the upper and lower DC rails which (hopefully) have, as a fundamental component, the target reference waveform, but also incorporate a series of unwanted harmonics generated as a consequence of the switching process.

Determination of the harmonic frequency components of a PWM switched phase leg output is quite complex and is often done by using a Fast Fourier Transform analysis of a simulated time-varying switched waveform. This approach offers the benefits of expediency and reduced mathematical effort but requires considerable computing capacity (especially when analyzing PWM systems with a higher carrier frequency) and always leaves the uncertainty as to whether a subtle simulation roundoff or error may have slightly tarnished the results obtained. In contrast, an analytical solution which exactly identifies the harmonic components of a PWM waveform ensures that precisely the correct harmonics are being considered when various PWM strategies are compared against each other.

The most well-known analytical method of determining the harmonic components of a PWM switched phase leg was first developed by Bowes and Bird [4], who adapted to modulated converter systems an analysis approach originally developed for communication systems by Bennet [5] and Black [6].

The analysis process assumes the existence of two time variables $x(t) = \omega_c t + \theta_c$ and $y(t) = \omega_o t + \theta_o$, where

$$\omega_c = 2\pi / T_c = \text{carrier angular frequency} \tag{3.1}$$

with

T_c = carrier interval

θ_c = arbitrary phase offset angle for carrier waveform

$$\omega_o = 2\pi/T_o = \text{fundamental (sinusoid) angular frequency, } \omega_o < \omega_c \quad (3.2)$$

with

T_o = period of fundamental waveform

θ_o = arbitrary phase offset angle for fundamental waveform

These variables $x(t)$ and $y(t)$ can be thought of as representing the time variation of the high-frequency modulating wave and low-frequency modulated wave, respectively, with each of these variables being considered as independently periodic. The problem of solving for the modulated $f(t)$ waveform can then be addressed by exploring the existence of a unit cell which identifies contours within which $f(t)$ is constant for cyclic variations of $x(t)$ and $y(t)$, as depicted in Figure 3.1.

The value of the function $f(t) = f[x(t), y(t)]$ within each contour region over the unit cell represents the output phase leg voltage and can be represented in topological form as shown by allowing the z axis to take on the values of $f(x,y)$ so that the functional dependence can be considered as three dimensional. The x axis and y axis of the cell are scaled in radians corresponding to the frequency of the modulating carrier and low-frequency fundamental reference waveforms, respectively, and span from $-\pi$ to π. (Note that a span of 0 to 2π is often alternatively used by many texts, but this makes no difference to the solution process which only requires cyclic variation of x and y over 2π intervals. The range of $-\pi$ to π is used in this book to simplify the form of the analytical solutions which are developed during the text.)

In general, the value of the function $f[x(t), y(t)]$ at any point in the unit cell can be represented as a Fourier Series if it is assumed that the function is

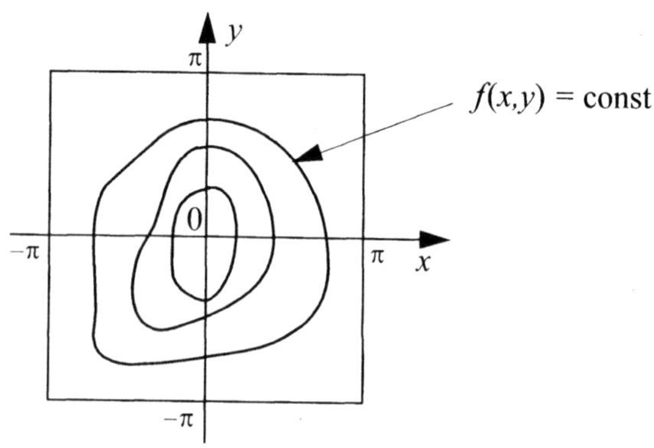

Figure 3.1 The unit cell.

periodic in both x and y. That is, identical unit cells are assumed to exist infinitely in both the x and y directions as shown in Figure 3.2.

From Fourier transform theory [7], any time-varying function $f(t)$ can be expressed as a summation of harmonic components

$$f(t) = \frac{a_0}{2} + \sum_{m=1}^{\infty} [a_m \cos m\omega t + b_m \sin m\omega t] \tag{3.3}$$

where

$$a_m = \frac{1}{\pi} \int_{-\pi}^{\pi} f(t) \cos m\omega t \, d\omega t \qquad m = 0, 1, \ldots, \infty \tag{3.4}$$

$$b_m = \frac{1}{\pi} \int_{-\pi}^{\pi} f(t) \sin m\omega t \, d\omega t \qquad m = 1, 2, \ldots, \infty \tag{3.5}$$

In Appendix 1, it is shown that a Fourier harmonic component form similar to Eq. (3.3) can be developed for a double variable controlled waveform $f(x, y)$, as

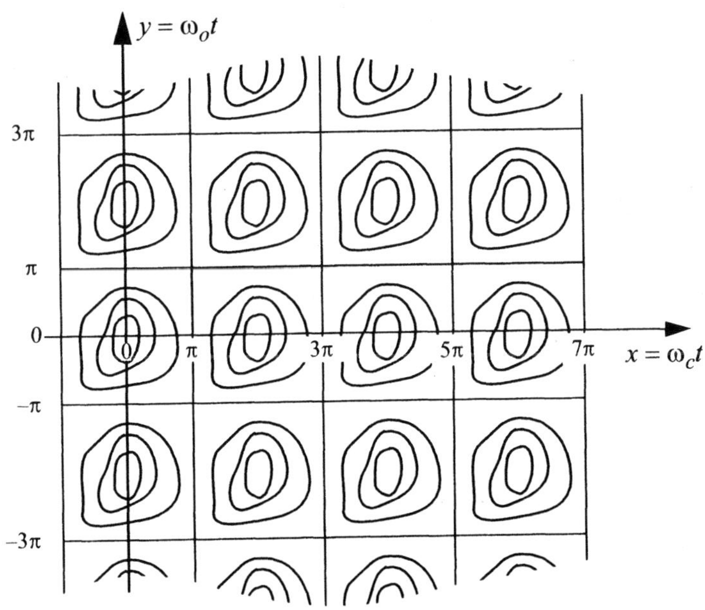

Figure 3.2 Unit cells replicated in two dimensions.

$$f(x,y) = \frac{A_{00}}{2} + \sum_{n=1}^{\infty}[A_{0n}\cos ny + B_{0n}\sin ny] + \sum_{m=1}^{\infty}[A_{m0}\cos mx + B_{m0}\sin mx] \tag{3.6}$$

$$+ \sum_{m=1}^{\infty}\sum_{\substack{n=-\infty \\ (n\neq 0)}}^{\infty}[A_{mn}\cos(mx+ny) + B_{mn}\sin(mx+ny)]$$

where

$$A_{mn} = \frac{1}{2\pi^2}\int_{-\pi}^{\pi}\int_{-\pi}^{\pi} f(x,y)\cos(mx+ny)\,dx\,dy \tag{3.7}$$

$$B_{mn} = \frac{1}{2\pi^2}\int_{-\pi}^{\pi}\int_{-\pi}^{\pi} f(x,y)\sin(mx+ny)\,dx\,dy \tag{3.8}$$

or in complex form

$$\overline{C_{mn}} = A_{mn} + jB_{mn} = \frac{1}{2\pi^2}\int_{-\pi}^{\pi}\int_{-\pi}^{\pi} f(x,y)e^{j(mx+ny)}\,dx\,dy \tag{3.9}$$

Replacing x by $\omega_c t + \theta_c$ and y by $\omega_o t + \theta_o$, Eq. (3.6) can be alternatively expressed in time-varying form as

$$f(t) = \underbrace{\frac{A_{00}}{2}}_{\text{DC Offset}} + \underbrace{\sum_{n=1}^{\infty}[A_{0n}\cos(n[\omega_o t + \theta_o]) + B_{0n}\sin(n[\omega_o t + \theta_o])]}_{\text{Fundamental Component \& Baseband Harmonics}} \tag{3.10}$$

$$+ \underbrace{\sum_{m=1}^{\infty}[A_{m0}\cos(m[\omega_c t + \theta_c]) + B_{m0}\sin(m[\omega_c t + \theta_c])]}_{\text{Carrier Harmonics}}$$

$$+ \underbrace{\sum_{m=1}^{\infty}\sum_{\substack{n=-\infty \\ (n\neq 0)}}^{\infty}\left[\begin{aligned}&A_{mn}\cos(m[\omega_c t + \theta_c] + n[\omega_o t + \theta_o]) \\ &+ B_{mn}\sin(m[\omega_c t + \theta_c] + n[\omega_o t + \theta_o])\end{aligned}\right]}_{\text{Sideband Harmonics}}$$

where m is the carrier index variable and n is the baseband index variable.

It is useful to examine Eq. (3.10) carefully.

The carrier index variable m and the baseband index variable n define the (angular) frequency of each harmonic component of the switched phase leg output voltage, as $(m\omega_c + n\omega_o)$. This means, for example, values of $m = 2$ and $n = 4$ define the fourth sideband harmonic in the group of harmonics that are located around the second carrier harmonic (i.e., within the second carrier sideband group). This sideband harmonic will have an absolute frequency of $(2\omega_c + 4\omega_o)$ rad/s, where ω_c is the carrier frequency and ω_o is the target fundamental frequency.

Special groups of these harmonics occur for $m = 0$, where the harmonic frequencies are defined by n alone, and for $n = 0$, where the harmonic frequencies are defined by m alone. These groups are termed the *baseband* and the *carrier* harmonic components, respectively. Note also that with this definition, the fundamental component of the switched waveform is the first baseband harmonic component, although it is often useful to maintain it as a separate component.

The magnitudes of the harmonic components defined in Eq. (3.10) are the A_{mn} and the B_{mn} coefficients, which must be evaluated for particular values of m and n for each PWM scheme to be considered. Equation (3.9) gives a general form of the Fourier integral that must be solved to determine these coefficient values, once the integrals have been adapted to take account of the particular switching pattern that is to be used. This adaption is discussed in detail in this and succeeding chapters, but for now it is simply assumed that it is possible to evaluate Eq. (3.10) for any particular values of m and n, to generate magnitudes for all the harmonic components that define the switched output waveform of the inverter phase leg.

The first term of Eq. (3.10), $A_{00}/2$ where $m = n = 0$, corresponds to the DC offset component of the pulse width-modulated waveform (if any).

The first summation term, $\left(\sum_{n=1}^{\infty} \ldots\right)$ where $m = 0$, defines the output fundamental low-frequency synthesized waveform and its *baseband* harmonics (if they exist). This term includes low-order undesirable harmonics around the fundamental output which should be minimized or (preferably) eliminated with the modulation process (except for the $n = 1$ element which, of course, defines the desired fundamental output waveform and must be retained).

The second summation term, $\left(\sum_{m=1}^{\infty}\ldots\right)$ where $n = 0$, corresponds to the carrier wave harmonics, which are relatively high-frequency components, since the lowest frequency term is the modulating carrier frequency.

The final double summation term, $\left(\sum_{m=1}^{\infty}\sum_{\substack{n=-\infty \\ (n\neq 0)}}^{\infty}\ldots\right)$ where $m, n \neq 0$, is the ensemble of all possible frequencies formed by taking the sum and difference between the modulating carrier waveform harmonics and the reference waveform and its associated baseband harmonics. These combinations are generally referred to as sideband harmonics, and exist as groups around the carrier harmonic frequencies.

In general the two angular frequencies ω_o and ω_c will not be an integer ratio, and the switched pulse train created by the phase leg will not be periodic. However, if the Fourier coefficient integral Eq. (3.9) is evaluated over very many periodic cycles of the fundamental reference waveform instead of just one cycle, eventually the integration of the carrier frequency cycles will also be periodic for the same overall integration period. Hence, the harmonic solution for $f(t)$ defined by Eq. (3.10) is valid even for the case where the switched waveform is not periodic over one fundamental period T_o.

There does exist a possible source of error if the Fourier coefficients for noninteger carrier ratios are evaluated over less than the very many fundamental cycles which may be required to achieve both periodic carrier and fundamental waveforms. Of course, such an error would affect only the predicted magnitude of each harmonic component not its frequency, based on the harmonic components defined in Eq. (3.10). However, it has been found that if such errors exist in the theoretical solutions (and in most cases they do not), for all reasonable carrier frequency ratios they are much smaller than the computing roundoff errors of simulation studies and can be neglected.

This is in sharp contrast to the more conventional method of determining the harmonic components of a switched waveform by computing a Fast Fourier Transform (FFT) of the waveform. This method is quite sensitive to both the time resolution of the simulation and the periodicity of the overall waveform. In general, harmonic investigations of PWM strategies using time simulation studies are only effective for exact integer carrier ratios, particularly if the small magnitude sideband or baseband harmonics are the focus of study.

3.4 Naturally Sampled Pulse Width Modulation

3.4.1 Sine–Sawtooth Modulation

The earliest and most straightforward modulation strategy is termed naturally sampled PWM, which compares a low-frequency target reference waveform (usually a sinusoid) against a high-frequency carrier waveform. A simple example of this type of modulation is illustrated in Figure 3.3, showing one phase leg of an inverter driven by a sawtooth wave carrier. It can be noted that, as described earlier, the phase leg is switched to the upper DC rail when the reference waveform is greater than the sawtooth wave carrier and to the lower DC rail when the carrier waveform is greater than the reference waveform. To obtain a sinusoidal output using this modulation strategy, the reference waveform has the form

$$v_{az}^{*} = M\cos(\omega_o t + \theta_o) = M\cos y \tag{3.11}$$

where

M = modulation index or modulation depth (i.e., normalized output voltage magnitude) with range $0 < M < 1$

ω_o = target output frequency

θ_o = arbitrary output phase

and the use of the asterisk denotes the "commanded" or "target" value. Note that the phase leg reference waveform can be made symmetrical about zero, because the sawtooth carrier ramps from normalized values of -1 to 1. With this type of sawtooth carrier, only the trailing edge of the pulse varies as M changes value, and hence this type of modulation is termed *trailing edge naturally sampled modulation.*

The unit cell for trailing edge sawtooth modulation is shown in Figure 3.4, where rather than $f(x,y)$ having many areas of constant value in the x, y plane, the function for the case of PWM switching simply has two values of $+2V_{dc}$ and 0 as a result of the circuit switching action. These values correspond to the voltage potential of point a in Figure 3.3 with respect to the negative DC bus point n for the two possible switching states of the phase leg. Note that defining the switched output with respect to the negative DC bus in this way rather than with respect to the midpoint z of the DC bus, considerably simplifies the mathematics of the Fourier solution at the trivial expense of introducing a DC offset of $+V_{dc}$ into the final solution.

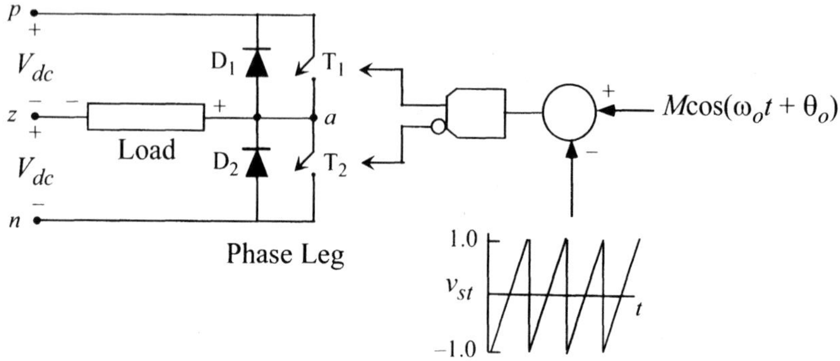

Figure 3.3 Naturally sampled trailing edge PWM with half-bridge (one phase leg) voltage source inverter.

The boundary between the two switched levels represents the locus of all possible solutions to the switching problem defined as the intersection of the sawtooth wave, defined along the x axis, and the sine wave reference command, defined along the y axis.

Over multiple carrier and fundamental cycles the sine wave waveform replicates over the x, y plane as shown in Figure 3.5. As time t unfolds, $x = \omega_c t$ and $y = \omega_o t$ define a straight line in the x, y plane with slope $y/x = \omega_o/\omega_c$ = carrier ratio (note that both the carrier and fundamental angles are assumed to be zero at time $t = 0$, or in other words both θ_c and θ_o are zero for this discussion). Intersections of this line with the boundary locus between the two switching voltage levels correspond to the actual switching instant for particular values of ω_o and ω_c as shown in Figure 3.6. Hence, if p is the number of

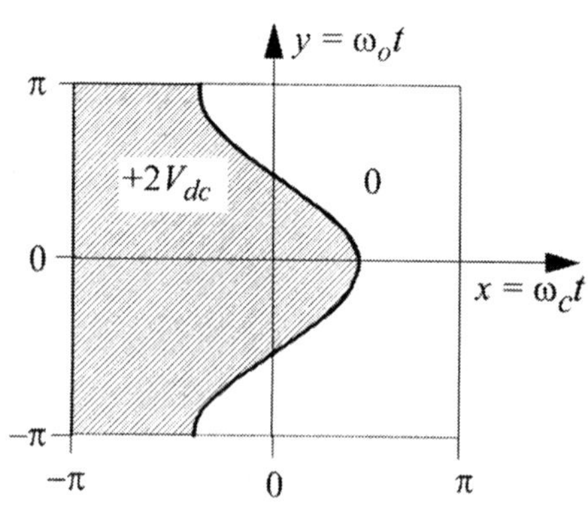

Figure 3.4 Unit cell for trailing edge sawtooth modulation.

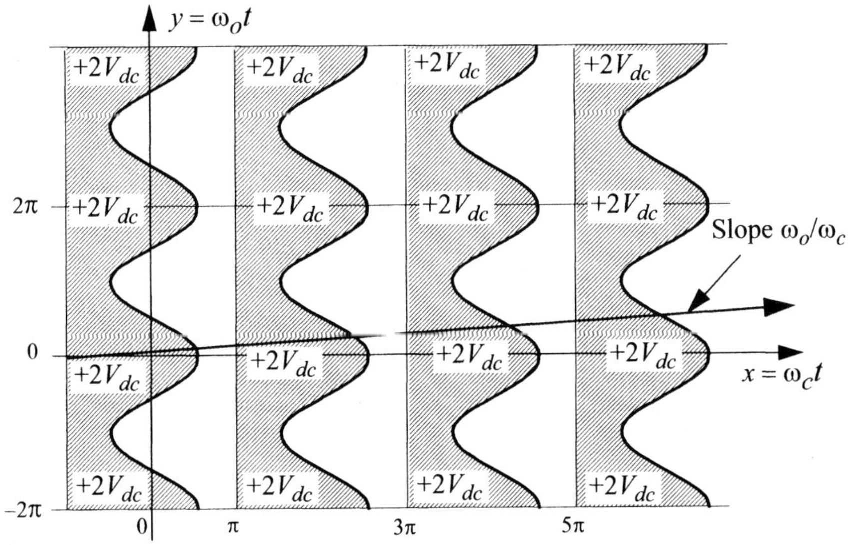

Figure 3.5 The x,y plane for trailing edge sawtooth modulation showing solution for particular values of ω_o and ω_c.

the cell along the x axis counting from the origin, then for naturally sampled trailing edge PWM the switching instants are defined such that

$f(x, y)$ changes from 0 to $2V_{dc}$ when

$$x = \pi(2p-1) \qquad p = 0, 1, 2, \ldots, \infty \tag{3.12}$$

and $f(x, y)$ changes from $2V_{dc}$ to 0 when

$$x = \pi 2p + \pi M \cos\omega_o t \qquad p = 0, 1, 2, \ldots, \infty \tag{3.13}$$

The function $f(x, y)$ plotted versus $\omega_c t$ in Figure 3.6(b) is the switched output of the phase leg. Note that when v_{an}, the voltage from phase leg output to the negative DC bus, is plotted versus $\omega_o t$ (projected onto the y axis instead of the x axis) the waveform is the same as $f(x, y)$, except it is scaled differently with respect to time (in radian measures of the modulated reference frequency rather than the modulating carrier frequency).

Equations (3.12) and (3.13) specify the limits of the integration regions defined in Eq. (3.9) for which $f(x, y)$ is nonzero (and constant). Using these limit definitions, Eq. (3.9) can then be adapted to become

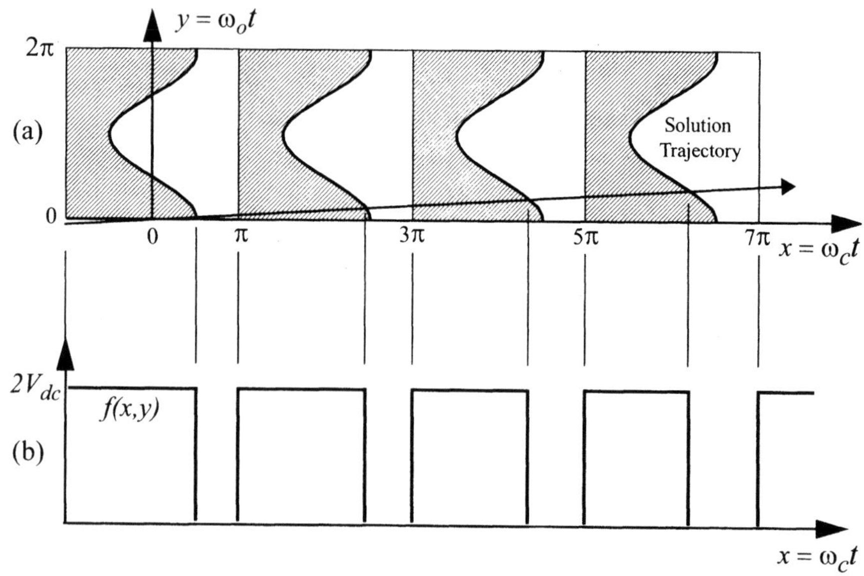

Figure 3.6 Half-bridge switching for *trailing edge* PWM: (a) x, y plane showing intersection of reference with unit cells and (b) x, y plane showing resulting PWM voltage.

$$A_{mn} + jB_{mn} = \frac{1}{2\pi^2} \int_{-\pi}^{\pi} \int_{-\pi}^{\pi M \cos y} 2V_{dc} e^{j(mx+ny)} dx\, dy \tag{3.14}$$

which can be evaluated for particular values of the index variables m and n.

For $m = n = 0$ [the DC offset in Eq. (3.10)], Eq. (3.14) simplifies to

$$A_{00} + jB_{00} = \frac{V_{dc}}{\pi^2} \int_{-\pi}^{\pi} \int_{-\pi}^{\pi M \cos y} dx\, dy$$

$$= \frac{V_{dc}}{\pi^2} \int_{-\pi}^{\pi} (\pi M \cos y + \pi)\, dy = \frac{V_{dc}}{\pi^2}(2\pi^2) = 2V_{dc} \tag{3.15}$$

When substituted back into Eq. (3.10), this gives a DC component of

$$\frac{A_{00}}{2} = V_{dc} \quad \text{with} \quad \frac{B_{00}}{2} = 0 \tag{3.16}$$

For $m = 0, n > 0$ [the baseband harmonics in Eq. (3.10)], Eq. (3.14) simplifies to

$$A_{0n} + jB_{0n} = \frac{V_{dc}}{\pi^2} \int_{-\pi}^{\pi} \int_{-\pi}^{\pi M \cos y} e^{jny}\, dx\, dy$$

$$= \frac{V_{dc}}{\pi^2} \int_{-\pi}^{\pi} (\pi M \cos y + \pi) e^{jny}\, dy$$

$$= \frac{V_{dc}}{\pi^2} \left[\left\{ \int_{-\pi}^{\pi} \pi M \cos y (\cos ny + j \sin ny) dy \right\} + \pi \left(\frac{e^{jn\pi} - e^{-jn\pi}}{jn} \right) \right] \tag{3.17}$$

The integral within the curly brackets is zero except for the case where $n = 1$, while the second term is zero for all nonzero values of n, so that Eq. (3.17) evaluates to

$$A_{01} + jB_{01} = \frac{V_{dc}}{\pi^2}(M\pi^2) = V_{dc}M \tag{3.18}$$

Equation (3.18) defines the magnitude of the fundamental harmonic of the modulated waveform, which equals as anticipated the target reference defined in Eq. (3.11), multiplied by V_{dc}.

For $m > 0, n = 0$ [the carrier harmonics in Eq. (3.10)], Eq. (3.14) simplifies to

$$A_{m0} + jB_{m0} = \frac{V_{dc}}{\pi^2} \int_{-\pi}^{\pi} \int_{-\pi}^{\pi M \cos y} e^{jmx}\, dx\, dy$$

$$= \frac{V_{dc}}{\pi^2} \int_{-\pi}^{\pi} \left(\frac{e^{jm\pi M \cos y} - e^{-jm\pi}}{jm} \right) dy$$

$$= \frac{V_{dc}}{jm\pi^2} \left[\int_{-\pi}^{\pi} e^{jm\pi M \cos y} dy - e^{-jm\pi} \int_{-\pi}^{\pi} dy \right] \tag{3.19}$$

The first integral within the square brackets of Eq. (3.19) can be evaluated using Eq. (A2.16), so that Eq. (3.19) becomes

$$A_{m0} + jB_{m0} = \frac{2V_{dc}}{jm\pi}[J_0(m\pi M) - (-1)^m] \tag{3.20}$$

which can be reexpressed using $(-1)^m = \cos m\pi$ as

$$A_{m0} + jB_{m0} = j\frac{2V_{dc}}{m\pi}[\cos m\pi - J_0(m\pi M)] \tag{3.21}$$

For $m > 0, n \neq 0$ [the sideband harmonics in Eq. (3.10)], the inner integral of Eq. (3.14) evaluates to give

$$A_{mn} + jB_{mn} = \frac{V_{dc}}{jm\pi^2}\int_{-\pi}^{\pi} e^{jny}(e^{jm\pi M\cos y} - e^{-jm\pi})\,dy \tag{3.22}$$

and, using Eq. (A2.15), this quantity becomes

$$A_{mn} + jB_{mn} = -\frac{jV_{dc}}{m\pi^2}\left[2\pi j^n J_n(m\pi M) - e^{-jm\pi}\left(\frac{e^{jn\pi} - e^{-jn\pi}}{jn}\right)\right] \tag{3.23}$$

The second term in the square brackets of this expression is zero for all non-zero values of n, and since

$$j^n = (\sqrt{-1})^n = \left(\sqrt{e^{j\pi}}\right)^n = e^{j\frac{\pi}{2}n}$$

Eq. (3.23) becomes

$$A_{mn} + jB_{mn} = -\frac{j2V_{dc}}{m\pi}e^{jn\frac{\pi}{2}}J_n(m\pi M) \qquad \text{for } n \neq 0 \tag{3.24}$$

which can be reexpressed as

$$A_{mn} + jB_{mn} = \frac{2V_{dc}}{m\pi}J_n(m\pi M)\left(\sin n\frac{\pi}{2} - j\cos n\frac{\pi}{2}\right) \tag{3.25}$$

Equation (3.25) is valid for all nonzero values of m and n, i.e., for $m = 1, 2, \ldots, \infty$ and $n = -\infty, \ldots, -2, -1, 1, 2, \ldots, \infty$.

The complete harmonic solution for *trailing edge naturally sampled modulation* of a half-bridge phase leg can now be formed by substituting the results of Eqs. 3.16, (3.18), (3.21), and (3.25) back into Eq. (3.10), so that the time-varying switched phase leg voltage $v_{an}(t)$ can be expressed in terms of its harmonic components as

$$
\begin{aligned}
v_{an}(t) &= V_{dc} + V_{dc} M \cos(\omega_o t + \theta_o) \\
&+ \frac{2V_{dc}}{\pi} \sum_{m=1}^{\infty} \frac{1}{m} [\cos m\pi - J_0(m\pi M)] \sin(m[\omega_c t + \theta_c]) \\
&+ \frac{2V_{dc}}{\pi} \sum_{m=1}^{\infty} \sum_{\substack{n=-\infty \\ (n \neq 0)}}^{\infty} \frac{1}{m} J_n(m\pi M) \left[\begin{aligned} &\sin n\frac{\pi}{2} \cos(m[\omega_c t + \theta_c] + n[\omega_o t + \theta_o]) \\ &- \cos n\frac{\pi}{2} \sin(m[\omega_c t + \theta_c] + n[\omega_o t + \theta_o]) \end{aligned} \right]
\end{aligned}
\tag{3.26}
$$

From Figure 3.3, it is clear that the phase leg switched output voltage measured with respect to the DC bus midpoint z is the same as the switched output voltage measured with respect to the negative DC rail n, but with a $-V_{dc}$ offset. Hence Eq. (3.26) also defines the harmonic components of $v_{az}(t)$, if the first term which defines the DC offset is discarded.

Figure 3.7 presents the voltage spectrum for trailing edge naturally sampled PWM for the conditions of a carrier ratio of 21 and a modulation index M of 0.9. This plot shows the single fundamental low-frequency component produced by the modulation process, and the groups of sideband harmonics arranged around the carrier and double carrier harmonics. Some sideband harmonics from the triple carrier group can also be seen on the right-hand side of the plot. All harmonic components including the fundamental have magnitudes which have been normalized with respect to V_{dc}, since this is the magnitude of the synthesized output fundamental for a modulation index of 1.0 (this convention is followed throughout this book to allow exact comparisons to be made of the various harmonic plots that are presented).

The WTHD0 for this modulation strategy under these conditions is 4.03%, but this of course varies with both the carrier ratio and the modulation index M. Figure 3.8 shows how the WTHD0 changes with these parameters, where it can be seen how the harmonic significance reduces with increasing modulation index and an increasing carrier ratio. Note that the precise carrier ratios used in this figure have no particular significance, particularly in terms of being odd and/or triplen. The reader is referred to Section 3.8 for further discussion on the issue of carrier pulse ratio.

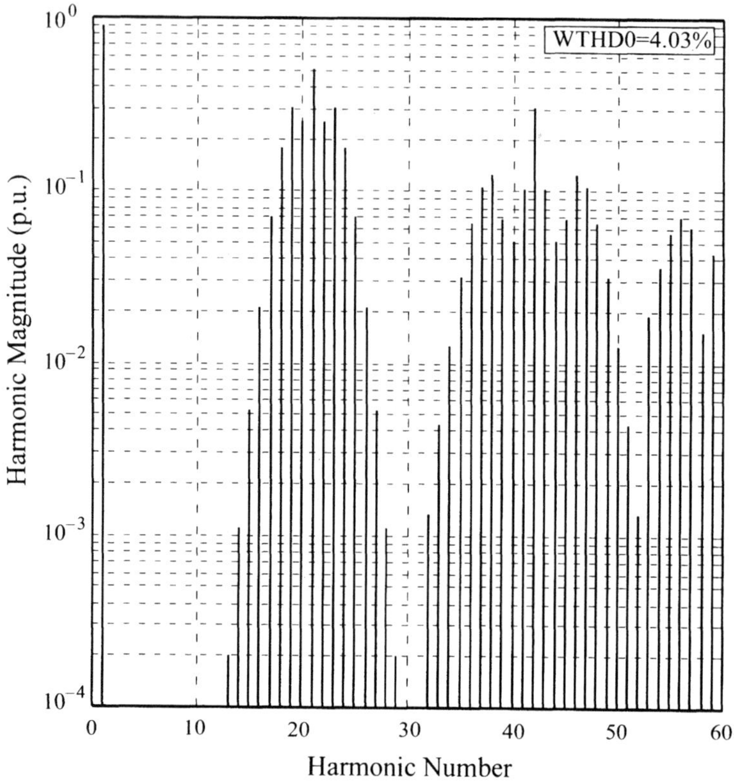

Figure 3.7 Harmonic components for one inverter phase leg with *trailing edge naturally sampled* PWM, $M = 0.9$, $f_c/f_o = 21$.

The reduction of WTHD with modulation index occurs primarily because the magnitude of the first carrier harmonic reduces with increasing modulation, and this harmonic dominates the WTHD calculation. (It will be seen later that the main carrier harmonics usually have no influence when phase legs are paralleled to make complete inverter systems since they become common mode terms and cancel. Consequently the way in which WTHD varies with increasing modulation index for complete inverter systems is quite different to that of one phase leg.) The reduction of WTHD with an increasing carrier ratio is easily identified as the reduced influence of the carrier and sideband harmonics on the WTHD calculation as they move toward higher frequencies.

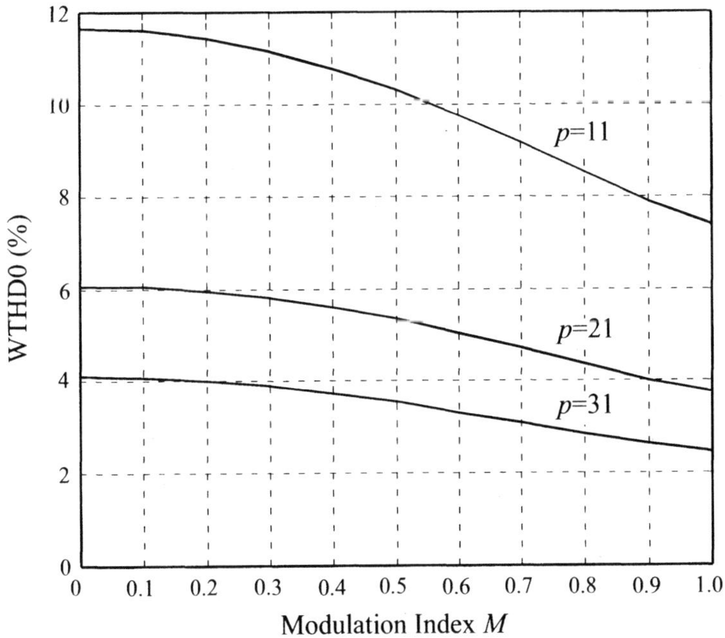

Figure 3.8 WTHD0 for one inverter phase leg with trailing edge naturally sampled PWM for varying modulation indices and carrier ratios.

Note that the WTHD results presented in Figure 3.8 (and in all other similar plots in this book) have been calculated using harmonic components taken directly from Eq. (3.26) and normalized with respect to V_{dc} (i.e., effectively the WTHD0 expression developed in Section 2.9). This means that no adjustment is made for changes to the magnitude of the fundamental component as the modulation index varies. Such an approach provides a convenient comparison of the relative levels of harmonics produced by various modulation strategies and operating conditions, but does not give an indication of the significance of the harmonics compared to the fundamental target component.

If WTHD relative to the fundamental synthesized component is required, the results presented in Figure 3.8 can be scaled by dividing their magnitudes by the modulation index. However, this approach gives a relative WTHD of infinity at a modulation index of $M = 0$, which while mathematically valid, is less useful as a practical measure of the distortion level than an absolute value of WTHD0 based on the normalized harmonic components.

3.4.2 Sine–Triangle Modulation

The more common form of naturally sampled PWM uses a triangular carrier instead of a sawtooth carrier to compare against the reference waveform as shown in Figure 3.9. With this type of carrier, both sides of the switched output pulse from the phase leg are modulated, which considerably improves the harmonic performance of the pulse train. This type of modulation is termed *double-edge naturally sampled modulation.*

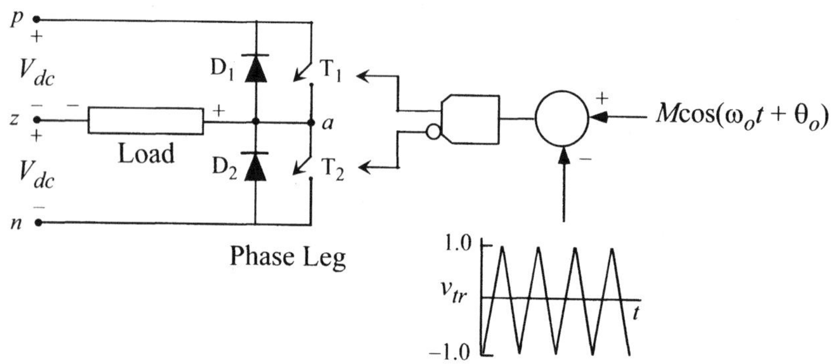

Figure 3.9 Double-edge naturally sampled PWM with half-bridge (one phase leg) voltage source inverter.

A plot of the corresponding unit cell for sine-triangle single-phase modulation is shown in Figure 3.10. Once again, the unit cell is defined over the interval $-\pi < \omega_c t < \pi$ rather than from 0 to 2π for simplicity of result.

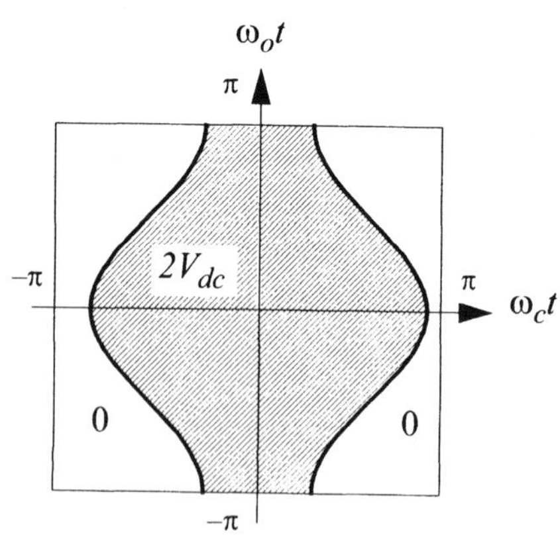

Figure 3.10 Unit cell for double-edge triangular carrier modulation.

Using the same sinusoidal reference waveform as given in Eq. (3.11), the switching instants for double-edge modulation can be expressed such that

$$x = 2\pi p - \frac{\pi}{2}(1 + M\cos\omega_o t) \qquad p = 0, 1, 2, \ldots, \infty \tag{3.27}$$

for $f(x, y)$ changing from 0 to $2V_{dc}$, and

$$x = 2\pi p + \frac{\pi}{2}(1 + M\cos\omega_o t) \qquad p = 0, 1, 2, \ldots, \infty \tag{3.28}$$

for $f(x, y)$ changing from $2V_{dc}$ to 0.

The switched voltage $v_{an}(t)$ measured from the phase leg output to the negative DC bus and plotted versus $\omega_c t$ then becomes the waveform shown in Figure 3.11. Under the integration limits defined by Eqs. (3.27) and (3.28), Eq. (3.9) becomes

$$A_{mn} + jB_{mn} = \frac{1}{2\pi^2} \int_{-\pi}^{\pi} \int_{-\frac{\pi}{2}(1 + M\cos y)}^{\frac{\pi}{2}(1 + M\cos y)} 2V_{dc} e^{j(mx + ny)} \, dx \, dy \tag{3.29}$$

which can now be evaluated for the various possible values of m and n.

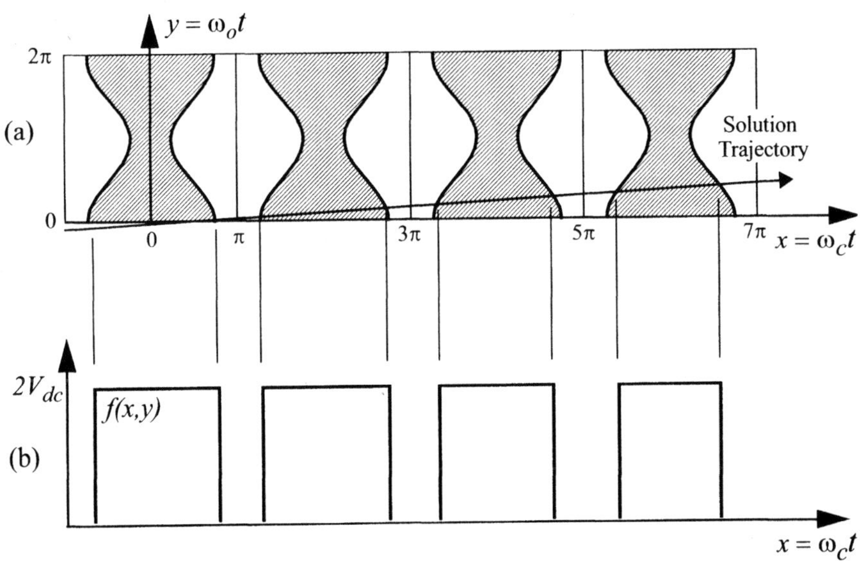

Figure 3.11 Half-bridge switching for double-edge PWM: (a) x, y plane showing intersection of reference with unit cells and (b) x, y plane showing resulting PWM voltage.

For $\underline{m = n = 0}$ [the DC offset in Eq. (3.10)], Eq. (3.29) simplifies to

$$A_{00} + jB_{00} = \frac{V_{dc}}{\pi^2} \int_{-\pi}^{\pi} \int_{-\frac{\pi}{2}(1 + M\cos y)}^{\frac{\pi}{2}(1 + M\cos y)} dx\, dy$$

$$= \frac{V_{dc}}{\pi^2} \int_{-\pi}^{\pi} [\pi(1 + M\cos y)]\, dy = 2V_{dc} \tag{3.30}$$

Thus $A_{00} = 2V_{dc}$, $B_{00} = 0$, which gives the expected V_{dc} offset in the final solution when the negative DC bus is taken as the switched reference voltage.

For $\underline{m = 0, n > 0}$ [the baseband harmonics in Eq.(3.10)], Eq. (3.29) simplifies to

$$A_{0n} + jB_{0n} = \frac{V_{dc}}{\pi^2} \int_{-\pi}^{\pi} \int_{-\frac{\pi}{2}(1 + M\cos y)}^{\frac{\pi}{2}(1 + M\cos y)} e^{jny} dx\, dy$$

$$= \frac{V_{dc}}{\pi^2} \int_{-\pi}^{\pi} [\pi(1 + M\cos y)e^{jny}]\, dy$$

$$= \frac{V_{dc}}{\pi} \int_{-\pi}^{\pi} \left[e^{jny} + \frac{M}{2}(e^{j[n+1]y} + e^{j[n-1]y})\right] dy \tag{3.31}$$

Since $\int_{-\pi}^{\pi} e^{jny}\, dy = 0$ for any nonzero value of n, Eq. (3.31) reduces to

$$A_{01} + jB_{01} = \frac{V_{dc}}{\pi} \int_{-\pi}^{\pi} \frac{M}{2} dy = V_{dc} M \tag{3.32}$$

for the case of $n = 1$ only. For all other $n > 1$, $A_{0n} + jB_{0n} = 0$. Once again this equals the target reference as defined in Eq. (3.11), times V_{dc}.

For $m > 0, n = 0$ [the carrier harmonics in Eq. (3.10)], Eq. (3.29) simplifies to

$$A_{m0} + jB_{m0} = \frac{V_{dc}}{\pi^2} \int_{-\pi}^{\pi} \int_{-\frac{\pi}{2}(1+M\cos y)}^{\frac{\pi}{2}(1+M\cos y)} e^{jmx} dx\, dy$$

$$= \frac{V_{dc}}{jm\pi^2} \int_{-\pi}^{\pi} \left[e^{jm\frac{\pi}{2}(1+M\cos y)} - e^{-jm\frac{\pi}{2}(1+M\cos y)} \right] dy \quad (3.33)$$

Using Eq. (A2.16), this becomes

$$A_{m0} + jB_{m0} = \frac{2V_{dc}}{jm\pi} \left[e^{jm\frac{\pi}{2}} J_0\left(m\frac{\pi}{2}M\right) - e^{-jm\frac{\pi}{2}} J_0\left(-m\frac{\pi}{2}M\right) \right] \quad (3.34)$$

and since $J_0(-\xi) = J_0(\xi)$ from Eq. (A2.22),

$$A_{m0} + jB_{m0} = \frac{4V_{dc}}{m\pi} J_0\left(m\frac{\pi}{2}M\right) \sin m\frac{\pi}{2} \quad (3.35)$$

This term defines the harmonics of the carrier wave itself.

For $m > 0, n \neq 0$ [the sideband harmonics in Eq. (3.10)], the inner integral of Eq. (3.29) evaluates to

$$A_{mn} + jB_{mn} = \frac{V_{dc}}{jm\pi^2} \int_{-\pi}^{\pi} e^{jny} \left[e^{jm\frac{\pi}{2}(1+M\cos y)} - e^{-jm\frac{\pi}{2}(1+M\cos y)} \right] dy$$

$$= \frac{V_{dc}}{jm\pi^2} \int_{-\pi}^{\pi} \left[e^{jm\frac{\pi}{2}} e^{jny} e^{jm\frac{\pi}{2}M\cos y} - e^{-jm\frac{\pi}{2}} e^{jny} e^{-jm\frac{\pi}{2}M\cos y} \right] dy \quad (3.36)$$

This integral expression then becomes, using Eq. (A2.15),

$$A_{mn} + jB_{mn} = \frac{2V_{dc}}{jm\pi} e^{jm\frac{\pi}{2}} j^n J_n\left(m\frac{\pi}{2}M\right) - e^{-jm\frac{\pi}{2}} j^{-n} J_n\left(m\frac{\pi}{2}M\right)$$

$$= \frac{2V_{dc}}{jm\pi} J_n\left(m\frac{\pi}{2}M\right) \left[e^{jm\frac{\pi}{2}} e^{jn\frac{\pi}{2}} - e^{-jm\frac{\pi}{2}} e^{-jn\frac{\pi}{2}} \right] \quad (3.37)$$

which simplifies to

$$A_{mn} + jB_{mn} = \frac{4V_{dc}}{m\pi} J_n\left(m\frac{\pi}{2}M\right) \sin\left([m+n]\frac{\pi}{2}\right) \tag{3.38}$$

These terms define the sidebands of the modulated wave. Note that since n is both positive and negative, the voltage side bands are equally spread above and below the central carrier wave harmonics.

The complete harmonic solution for *double-edge naturally sampled modulation* of a half-bridge phase leg can now be formed by substituting the results of Eqs. (3.30), (3.32), (3.35), and (3.38) back into Eq. (3.10), so that the time-varying switched phase leg voltage $v_{an}(t)$ can be expressed in terms of its harmonic components as

$$\begin{aligned} v_{an}(t) = {} & V_{dc} + V_{dc} M \cos(\omega_o t + \theta_o) \\ & + \frac{4V_{dc}}{\pi} \sum_{m=1}^{\infty} \frac{1}{m} J_0\left(m\frac{\pi}{2}M\right) \sin m\frac{\pi}{2} \cos(m[\omega_c t + \theta_c]) \\ & + \frac{4V_{dc}}{\pi} \sum_{m=1}^{\infty} \sum_{\substack{n=-\infty \\ (n \neq 0)}}^{\infty} \frac{1}{m} J_n\left(m\frac{\pi}{2}M\right) \sin\left([m+n]\frac{\pi}{2}\right) \\ & \qquad \times \cos(m[\omega_c t + \theta_c] + n[\omega_o t + \theta_o]) \end{aligned} \tag{3.39}$$

As before, Eq. (3.39) also defines the harmonic components of $v_{az}(t)$, if the first term which defines the DC offset is discarded.

Figure 3.12 presents the voltage spectrum for double-edge naturally sampled PWM for the conditions of a carrier ratio of 21 and a modulation index M of 0.9. Once again this plot shows the single fundamental low-frequency component produced by the modulation process, and the groups of sideband harmonics arranged around the carrier and double carrier harmonics. Some sideband harmonics from the triple carrier group can also be seen on the right of the plot. However, the significant feature of double-edge naturally sampled PWM is that the *odd* harmonic sideband components around *odd* multiples of the carrier fundamental, and *even* harmonic sideband components around *even* multiples of the carrier fundamental, are completely eliminated by the $\sin[(m+n)\pi/2]$ expression in Eq. (3.39). This is an intrinsic advantage of double-edge naturally sampled modulation using a triangular carrier compared to single-edge modulation using a sawtooth carrier.

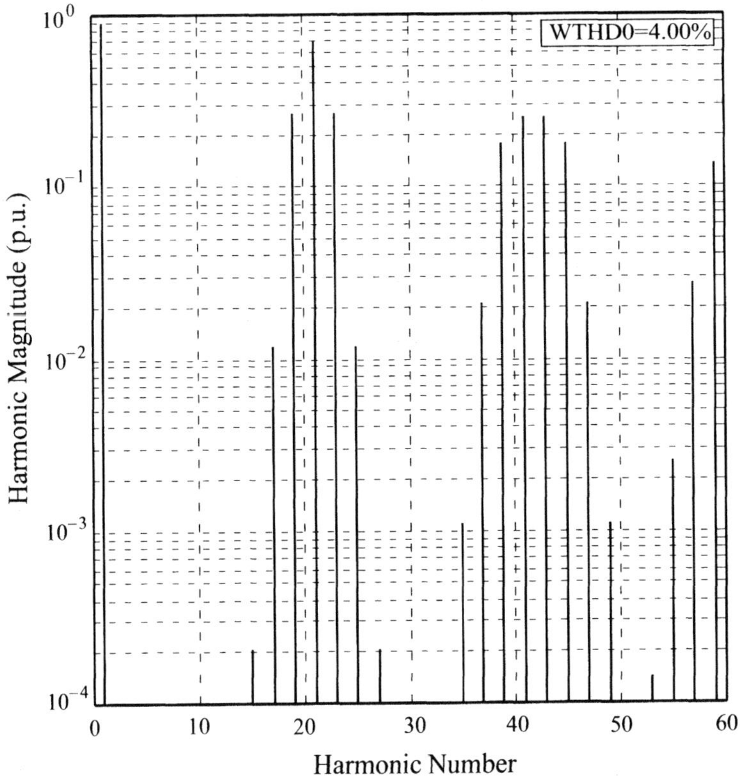

Figure 3.12 Harmonic components for one inverter phase leg with *double-edge naturally sampled* PWM, $M = 0.9$, $f_c/f_o = 21$.

It should be noted that while the WTHD0 for this modulation strategy of 4.00% is almost identical to that of trailing edge naturally sampled PWM, this is because the major harmonic contributor to WTHD in this case is the main carrier fundamental at $21f_o$. As will be seen later this carrier is eliminated for all multiple phase leg converter topologies since it is a common mode component, and the performance of the two PWM strategies then differ markedly.

Because of this carrier dominance, and since the carrier fundamentals are very similar for trailing edge and double-edge naturally sampled PWM, the variation of WTHD0 with modulation index for double-edge naturally sampled PWM is virtually identical to that shown in Figure 3.8 and is not shown again here.

3.5 PWM Analysis by Duty Cycle Variation

3.5.1 Sine–Sawtooth Modulation

An alternative approach to calculate the harmonic spectrum of naturally sampled PWM is to modulate the duty cycle of a simple buck converter with a low-frequency sinusoidal reference [8]. This approach can be developed by considering the switching process of one inverter phase leg during a few arbitrary cycles of the sawtooth carrier waveform, as shown in Figure 3.13.

If the reference waveform is assumed to be constant within each carrier cycle (i.e., $f_c \ll f_o$) with a duty cycle of $d = 1 + M\cos y$, from Eqs. (A1.1), (A1.2), and (A1.3), the Fourier series of the resultant switched waveform is

$$v_{an}(t) = \frac{a_0}{2} + \sum_{m=1}^{\infty}(a_m \cos mx + b_m \sin mx) \tag{3.40}$$

where

$$a_m = \frac{1}{\pi}\int_{-\pi}^{\pi} v_{an}(t)\cos mx\, dx = \frac{2V_{dc}}{\pi}\int_{-\pi}^{\pi M\cos y} \cos mx\, dx$$

$$= \frac{2V_{dc}}{m\pi}[\sin(m\pi M\cos y) + \sin m\pi] \qquad m \neq 0 \tag{3.41}$$

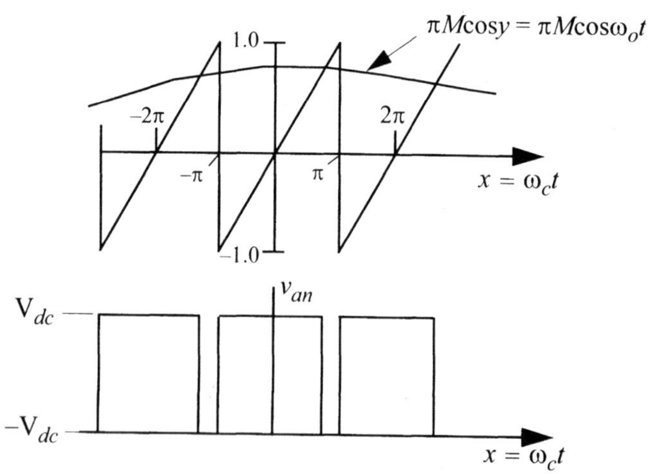

Figure 3.13 Production of PWM waveforms by comparison of sawtooth carrier wave and low-frequency control input wave.

and

$$b_m = \frac{1}{\pi}\int_{-\pi}^{\pi} v_{an}(t)\sin mx\;dx = \frac{2V_{dc}}{\pi}\int_{-\pi}^{\pi M\cos y}\sin mx\;dx$$

$$= \frac{2V_{dc}}{m\pi}[\cos m\pi - \cos(m\pi M\cos y)] \qquad m \neq 0 \tag{3.42}$$

Note that when $m = 0$, $a_0 = 2V_{dc}(1 + M\cos y)$ and $b_0 = 0$.

From Eqs. (A2.2) and (A2.3)

$$\cos(m\pi M\cos y) = J_0(m\pi M) + 2\sum_{n=1}^{\infty}\cos n\frac{\pi}{2}J_n(m\pi M)\cos ny \tag{3.43}$$

$$\sin(m\pi M\cos y) = 2\sum_{n=1}^{\infty}\sin n\frac{\pi}{2}J_n(m\pi M)\cos ny \tag{3.44}$$

Substituting these relationships into Eqs. (3.41) and (3.42) allows Eq. (3.40), the Fourier form of the switched output voltage with respect to the negative bus, to be written as

$$v_{an}(t) = V_{dc}(1 + M\cos y) + \frac{2V_{dc}}{\pi}\sum_{m=1}^{\infty}\frac{1}{m}[\cos m\pi - J_0(m\pi M)]\sin mx \tag{3.45}$$

$$+ \frac{4V_{dc}}{\pi}\sum_{m=1}^{\infty}\frac{1}{m}\sum_{n=1}^{\infty}\sin n\frac{\pi}{2}J_n(m\pi M)\cos ny\cos mx$$

$$- \frac{4V_{dc}}{\pi}\sum_{m=1}^{\infty}\frac{1}{m}\sum_{n=1}^{\infty}\cos n\frac{\pi}{2}J_n(m\pi M)\cos ny\sin mx$$

Now consider next the behavior of the double summation terms.

For the first term, the result would be the same for negative *n*, since only odd *n* terms can exist because $\sin(n\pi/2) = 0$ for even *n*, and from Eq. (A2.18), $J_{-n}(\xi) = -J_n(\xi)$ but also $\sin(n\pi/2) = -\sin(-n\pi/2)$, for odd *n*.

For the second term, the result would also be the same for negative n, since because $\cos(n\pi/2) = 0$ for odd n, only even n terms can exist, and from Eq. (A2.18), $J_{-n}(\xi) = J_n(\xi)$ for even n.

Hence, the summation terms produce the same result if they are summed from $-\infty$ to -1 rather than from 1 to ∞. Alternatively, the summation can be taken over both the ranges $-\infty \le n \le -1$ and $1 \le n \le \infty$ provided the result is divided by two. Thus Eq. (3.45) can be rewritten as

$$v_{an}(t) = V_{dc}(1 + M\cos y) + \frac{2V_{dc}}{\pi}\sum_{m=1}^{\infty}\frac{1}{m}[\cos m\pi - J_0(m\pi M)]\sin mx \tag{3.46}$$

$$+\frac{2V_{dc}}{\pi}\sum_{m=1}^{\infty}\frac{1}{m}\sum_{\substack{n=-\infty \\ (n\neq 0)}}^{\infty}\sin n\frac{\pi}{2}J_n(m\pi M)\cos ny\cos mx$$

$$-\frac{2V_{dc}}{\pi}\sum_{m=1}^{\infty}\frac{1}{m}\sum_{\substack{n=-\infty \\ (n\neq 0)}}^{\infty}\cos n\frac{\pi}{2}J_n(m\pi M)\cos ny\sin mx$$

Now, for each value of m, two terms can be taken from the inner summation expressions with the positive and negative values of n that have the same absolute magnitude, say, for example, $n = \pm N$. Consequently, using the trigonometric identities

$$\cos mx\cos ny = \frac{1}{2}[\cos(mx+ny) + \cos(mx-ny)] \tag{3.47}$$

$$\sin mx\cos ny = \frac{1}{2}[\sin(mx+ny) + \sin(mx-ny)] \tag{3.48}$$

these two terms when added together become

$$\begin{aligned}\cos mx\cos Ny + \cos mx\cos(-Ny) &= \frac{1}{2}[\cos(mx+Ny) + \cos(mx-Ny)] \\ &\quad + \frac{1}{2}[\cos(mx-Ny) + \cos(mx-(-Ny))] \\ &= (\cos(mx+Ny) + \cos(mx-Ny))\end{aligned} \tag{3.49}$$

Similarly

$$\sin mx \cos Ny + \sin mx \cos(-Ny) = \sin(mx + Ny) + \sin(mx - Ny) \quad (3.50)$$

If these substitutions are made for all values of n in Eq. (3.46), it becomes

$$v_{an}(t) = V_{dc}(1 + M\cos y) + \frac{2V_{dc}}{\pi}\sum_{m=1}^{\infty}\frac{1}{m}[\cos m\pi - J_0(m\pi M)]\sin mx \quad (3.51)$$

$$+ \frac{2V_{dc}}{\pi}\sum_{m=1}^{\infty}\sum_{\substack{n=-\infty \\ (n \neq 0)}}^{\infty}\frac{1}{m}J_n(m\pi M)\sin n\frac{\pi}{2}\cos(mx + ny)$$

$$- \frac{2V_{dc}}{\pi}\sum_{m=1}^{\infty}\sum_{\substack{n=-\infty \\ (n \neq 0)}}^{\infty}\frac{1}{m}J_n(m\pi M)\cos n\frac{\pi}{2}\sin(mx + ny)$$

Replacing x by $\omega_c t + \theta_c$ and y by $\omega_o t + \theta_o$, it is evident that this result is the same as Eq. (3.26) obtained in Section 3.4.1 using the double Fourier integral technique.

3.5.2 Sine–Triangle Modulation

A similar approach can be used to develop the harmonic spectrum of a sinusoidal reference modulated by a triangular carrier. For a triangular carrier, the reference waveform has a duty cycle of $d = (1 + M\cos y)/2$, and once again is initially assumed to be constant within the carrier interval. For this case, the Fourier series corresponding to the resultant switched waveform becomes

$$v_{an}(t) = \frac{a_0}{2} + \sum_{m=1}^{\infty}(a_m \cos mx + b_m \sin mx) \quad (3.52)$$

where

$$a_m = \frac{1}{\pi}\int_{-\pi}^{\pi} v_{an}(t)\cos mx\, dx = \frac{2V_{dc}}{\pi}\int_{-\frac{\pi}{2}(1 + M\cos y)}^{\frac{\pi}{2}(1 + M\cos y)} \cos mx\, dx$$

$$= \frac{4V_{dc}}{m\pi}\left[\sin\left(m\frac{\pi}{2}(1 + M\cos y)\right)\right] \qquad m \neq 0 \tag{3.53}$$

and

$$b_m = \frac{1}{\pi}\int_{-\pi}^{\pi} v_{an}(t)\sin mx\, dx = \frac{2V_{dc}}{\pi}\int_{-\frac{\pi}{2}(1 + M\cos y)}^{\frac{\pi}{2}(1 + M\cos y)} \sin mx\, dx = 0 \tag{3.54}$$

Note that when $m = 0$, $a_0 = 2V_{dc}(1 + M\cos y)$.

Equation. (3.53) can be rewritten as

$$a_m = \frac{4V_{dc}}{m\pi}\left[\sin m\frac{\pi}{2}\cos\left(m\frac{\pi}{2}M\cos y\right) + \cos m\frac{\pi}{2}\sin\left(m\frac{\pi}{2}M\cos y\right)\right] \qquad m \neq 0 \tag{3.55}$$

Using Eqs. (A2.2) and (A2.3), Eq. (3.52) then becomes

$$v_{an}(t) = V_{dc}(1 + M\cos y) + \frac{4V_{dc}}{\pi}\sum_{m=1}^{\infty}\frac{1}{m}\sin m\frac{\pi}{2}J_0\left(m\frac{\pi}{2}M\right)\cos mx$$

$$+ \frac{8V_{dc}}{\pi}\sum_{m=1}^{\infty}\frac{1}{m}\sum_{n=1}^{\infty}\left[\begin{array}{l}\sin m\frac{\pi}{2}\cos n\frac{\pi}{2}J_n\left(m\frac{\pi}{2}M\right) \\ + \cos m\frac{\pi}{2}\sin n\frac{\pi}{2}J_n\left(m\frac{\pi}{2}M\right)\end{array}\right]\cos ny\ \cos mx$$

$$= V_{dc}(1 + M\cos y) + \frac{4V_{dc}}{\pi}\sum_{m=1}^{\infty}\frac{1}{m}J_0\left(m\frac{\pi}{2}M\right)\sin m\frac{\pi}{2}\ \cos mx$$

$$+ \frac{8V_{dc}}{\pi}\sum_{m=1}^{\infty}\sum_{n=1}^{\infty}\frac{1}{m}J_n\left(m\frac{\pi}{2}M\right)\sin\left([m+n]\frac{\pi}{2}\right)\cos ny\ \cos mx$$

$$= V_{dc}(1 + M\cos y) + \frac{4V_{dc}}{\pi}\sum_{m=1}^{\infty}\frac{1}{m}J_0\left(m\frac{\pi}{2}M\right)\sin m\frac{\pi}{2}\ \cos mx \tag{3.56}$$

$$+ \frac{4V_{dc}}{\pi}\sum_{m=1}^{\infty}\sum_{n=1}^{\infty}\frac{1}{m}J_n\left(m\frac{\pi}{2}M\right)\sin\left([m+n]\frac{\pi}{2}\right)\left[\begin{array}{l}\cos(mx + ny) \\ + \cos(mx - ny)\end{array}\right]$$

Equation (3.56) can be readily rearranged by changing the inner summation limits to become

$$v_{an}(t) = V_{dc} + V_{dc}M\cos y + \frac{4V_{dc}}{\pi}\sum_{m=1}^{\infty}\frac{1}{m}J_0\left(m\frac{\pi}{2}M\right)\sin m\frac{\pi}{2}\cos mx \quad (3.57)$$

$$+\frac{4V_{dc}}{\pi}\sum_{m=1}^{\infty}\sum_{\substack{n=-\infty\\(n\neq 0)}}^{\infty}\frac{1}{m}J_n\left(m\frac{\pi}{2}M\right)\sin\left([m+n]\frac{\pi}{2}\right)\cos(mx+ny)$$

which is identical to Eq. (3.39) when x and y are replaced by $\omega_c t + \theta_c$ and $\omega_o t + \theta_o$, respectively. The simplicity of this approach compared to the double integral method should be apparent.

3.6 Regular Sampled Pulse Width Modulation

One major limitation with naturally sampled PWM is the difficulty of its implementation in a digital modulation system, because the intersection between the reference sinusoid and the triangular or sawtooth carrier is defined by a transcendental equation and is complex to calculate. To overcome this limitation the modern popular alternative is to implement the modulation system using a "regular sampled" PWM strategy, where the low-frequency reference waveforms are sampled and then held constant during each carrier interval. These sampled values are compared against the triangular carrier waveform to control the switching process of each phase leg, instead of the sinusoidally varying reference.

The sampled reference waveform must change value at either the positive or positive/negative peaks of the carrier waveform, depending on the sampling strategy. This change is required to avoid instantaneously changing the reference during the ramping period of the carrier, which may cause multiple switch transitions if it was allowed to occur.

For a sawtooth carrier, sampling occurs as the carrier waveform falls at the end of the ramping period. For a triangular carrier, sampling can be *symmetrical*, where the sampled reference is taken at either the positive or negative peak of the carrier and held constant for the entire carrier interval, or *asymmetrical*, where the reference is resampled every half carrier interval at both the positive and the negative carrier peak. These alternatives are shown in Figure 3.14. Note that there is no concept of *symmetrical* and *asymmetrical* single-edge PWM, because the phase leg switching transition is calculated every (saw-tooth) carrier cycle.

Figure 3.14 shows that the sampling process produces a stepped reference waveform which is phase delayed with respect to the original reference wave-

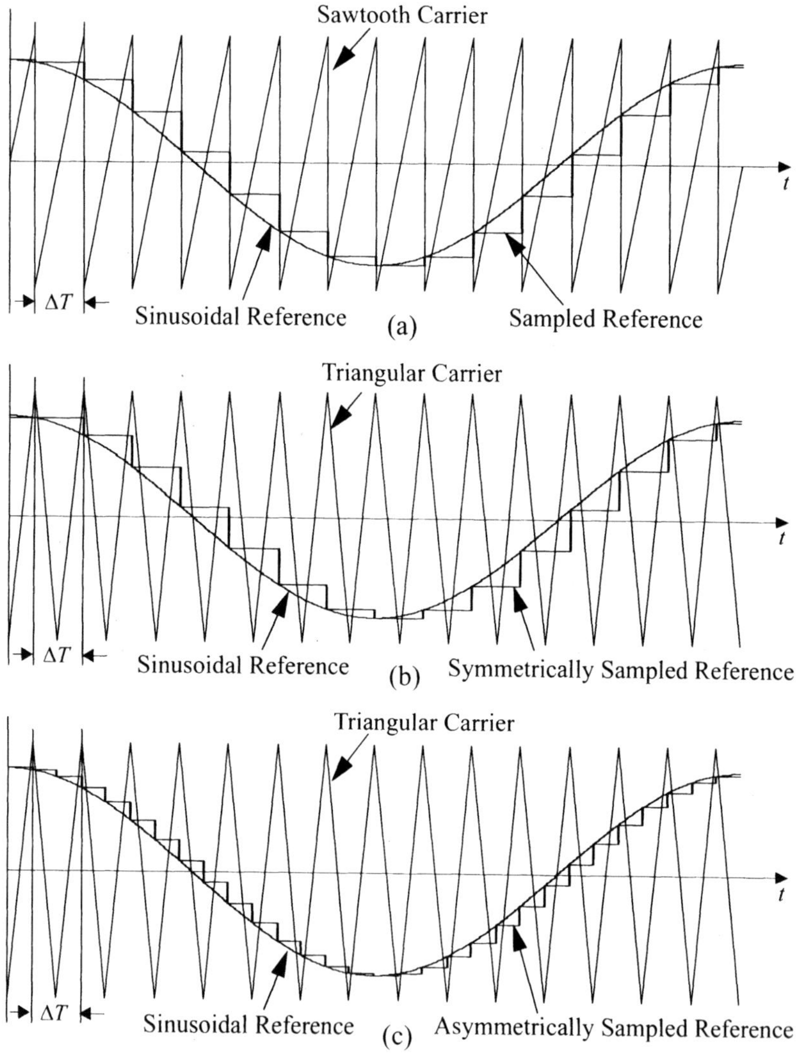

Figure 3.14 Regular sampling for (a) sawtooth carrier, (b) symmetrical sampling with triangular carrier (positive peak sampled), and (c) asymmetrical sampling with triangular carrier.

form. For a sawtooth carrier and symmetrical sampling, this phase delay is one half the carrier interval, while for asymmetrical sampling the phase delay is one quarter the carrier interval. This phase delay can be compensated by phase advancing the reference waveform to produce the result shown in Figure 3.15.

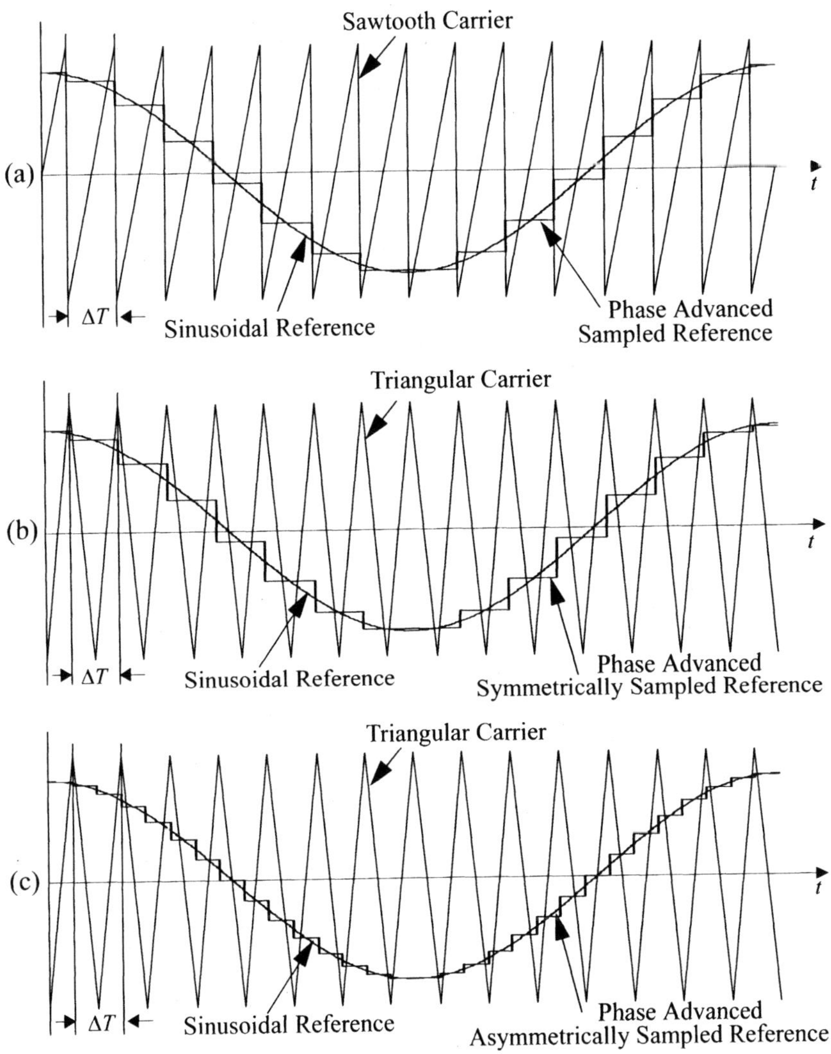

Figure 3.15 Regular sampling with phase advanced reference for (a) sawtooth carrier – 1/2 carrier advance, (b) symmetrically sampled triangular carrier – 1/2 carrier advance, and (c) asymmetrically sampled triangular carrier – 1/4 carrier advance.

Since the same result can be obtained by phase delaying the sampling point on the original reference waveform rather than phase advancing the reference waveform itself, it is clear that this adjustment only affects the phase of the harmonics of the resultant PWM switched waveform, not their amplitudes.

Figure 3.16 shows how the width and the placement of the switching pulse created by a phase leg varies depending on the carrier type and the sampling process used. (Note that the change in the target reference within one carrier interval is grossly exaggerated from normal in Figure 3.16 for purposes of illustration).

As can be seen from this figure, the effect of regular sampling is to change the position of the switching instants within each carrier interval. This effect can be reflected into the double integral harmonic integration of Eq. (3.9) by changing the limits of the inner integral over which the function $f(x, y)$ is non-zero for each type of modulation.

In general terms, the solution to Eq. (3.9) for two-level modulation can be expressed as

$$A_{mn} + jB_{mn} = \frac{1}{2\pi^2} \int_{-\pi}^{\pi} \int_{x_r}^{x_f} 2V_{dc} e^{j(mx + ny)} dx\, dy \tag{3.58}$$

where x_r = instant at which phase leg switches to $2V_{dc}$

x_f = instant at which phase leg switches back to 0 V

For naturally sampled PWM with a *sawtooth* carrier, these inner integral limits have been previously identified as

$$x_r = -\pi \qquad x_f = \pi M \cos \omega_o t \quad (= \pi M \cos y) \tag{3.59}$$

where the 2π rollover included in Eqs. (3.12) and (3.13) has been deleted for convenience since it has no effect on the integration.

For naturally sampled PWM with a *triangular* carrier, the inner integral limits have been previously identified as

$$x_r = -\frac{\pi}{2}(1 + M\cos\omega_o t) \qquad x_f = \frac{\pi}{2}(1 + M\cos\omega_o t) \tag{3.60}$$

where again the 2π rollover included in Eqs. (3.27) and (3.28) has been deleted for convenience.

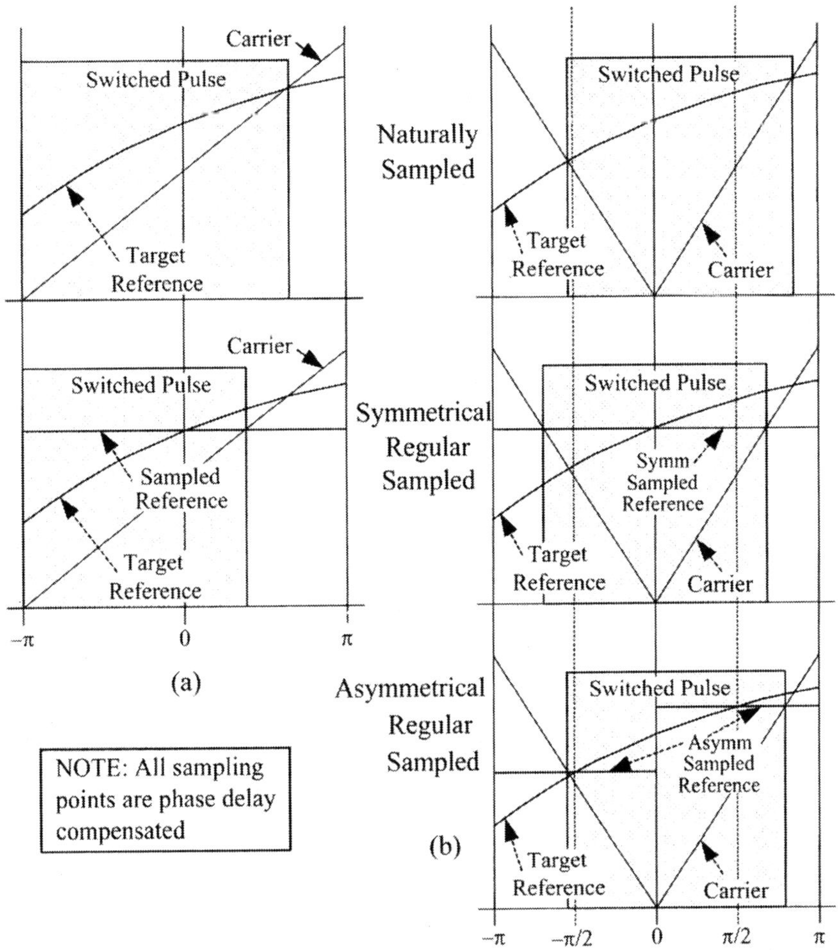

Figure 3.16 Switched pulse from one inverter phase leg for (a) sawtooth carrier and (b) triangular carrier, for naturally and regular sampled PWM alternatives (phase delayed sampling).

Using these limits, the harmonic components of a naturally sampled PWM waveform under a sawtooth or a triangular carrier have been developed previously as Eqs. (3.26) and (3.39), respectively.

In order to similarly calculate the harmonic components of a regular sampled PWM waveform, it is now necessary to modify these inner integral limits to take account of the effects of the sampling process.

3.6.1 Sawtooth Carrier Regular Sampled PWM

Figure 3.17(a) shows how the switching instants for regular sampling can be determined by the intersection between the sampled sinusoidal reference waveform and the solution trajectory line $y = y' + (\omega_o/\omega_c)x$. Figure 3.17(b) shows how the same switching instants can be determined by the intersection between the continuous sinusoidal reference waveform and a sampled solution trajectory where y is held constant within each carrier interval. Both approaches create the same phase leg switched output, but the approach of Figure 3.17(b) is more mathematically tractable to evaluate since the sinusoidal reference waveform remains smooth.

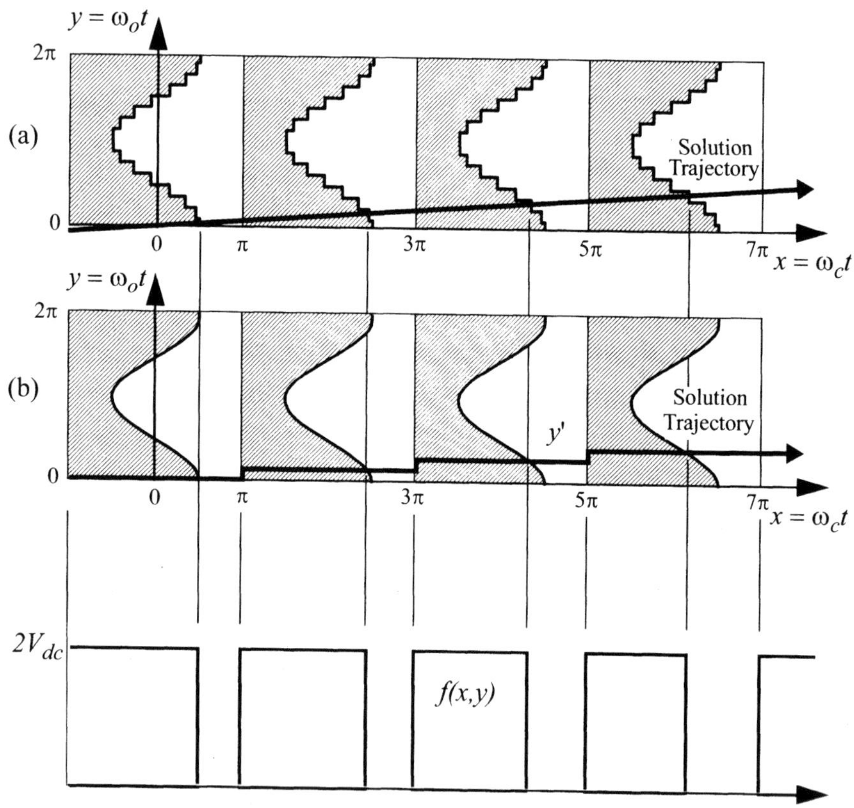

Figure 3.17 Half-bridge switching for trailing edge PWM: (a) sampled sinusoidal reference waveform and (b) solution trajectory held constant within each carrier interval.

The intersection trajectory shown in Figure 3.17(b) is a staircase variable y', which has a constant value frozen at $x = 0$ within each carrier interval. For example, during the first carrier interval, $y' = 0$; during the second carrier interval, $y' = \frac{\omega_o}{\omega_c}2\pi$; and so on. In general the value of y' within each carrier interval can be expressed as

$$y' = \frac{\omega_o}{\omega_c}2\pi p \tag{3.61}$$

where p represents the pth carrier interval in a fundamental cycle.

The staircase variable y' can also be expressed in terms of the continuous variables x and y as

$$y' = y - \frac{\omega_o}{\omega_c}(x - 2\pi p) \tag{3.62}$$

The inner integral limits of Eq. (3.58) for the case of regular sampled PWM with a sawtooth carrier can now be defined using this new staircase variable as

$$x_r = -\pi \qquad x_f = \pi M \cos y' \tag{3.63}$$

so that Eq. (3.58) becomes

$$A_{mn} + jB_{mn} = \frac{V_{dc}}{\pi^2}\int_{-\pi}^{\pi}\int_{-\pi}^{\pi M\cos y'} e^{j(mx+ny)}\,dx\,dy \tag{3.64}$$

Equation (3.64) can be evaluated by changing the integration variable y to $y = y' + (\omega_o/\omega_c)x$ using Eq. (3.62), where the $2\pi p$ rollover term associated with x can once again be discarded since it has no effect on the integration, to become

$$A_{mn} + jB_{mn} = \frac{V_{dc}}{\pi^2}\int_{-\pi}^{\pi}\int_{-\pi}^{\pi M\cos y'} e^{j\left(mx + n\left[y' + \frac{\omega_o}{\omega_c}x\right]\right)}\,dx\,dy'$$

$$= \frac{V_{dc}}{\pi^2}\int_{-\pi}^{\pi}\int_{-\pi}^{\pi M\cos y'} e^{j\left(\left[m + n\frac{\omega_o}{\omega_c}\right]x + ny'\right)}\,dx\,dy' \tag{3.65}$$

Equation (3.65) can be evaluated for particular values of m and n as follows:

For $m = n = 0$, the solution process for Eq. (3.15) remains valid, so that

$$A_{00} = 2V_{dc} \tag{3.66}$$

For $m = 0, n > 0$, the solution process leading to Eq. (3.25) can be used, with m replaced by $n(\omega_o/\omega_c)$. This gives baseband harmonic coefficients of

$$A_{0n} + jB_{0n} = \frac{2V_{dc}}{\left[n\frac{\omega_o}{\omega_c}\right]\pi} J_n\left(n\frac{\omega_o}{\omega_c}\pi M\right)\left(\sin n\frac{\pi}{2} - j\cos n\frac{\pi}{2}\right) \tag{3.67}$$

Equation (3.67) reflects a significant difference for regular sampled PWM compared to naturally sampled PWM, since the baseband harmonic coefficients are no longer zero for $n > 1$. However, in practice the magnitude of the additional harmonics usually roll off fairly rapidly with n for any reasonable carrier ratios.

For $m > 0, n = 0$, the solution process leading to Eq. (3.21) remains valid, so that

$$A_{m0} + jB_{m0} = j\frac{2V_{dc}}{m\pi}[\cos m\pi - J_0(m\pi M)] \tag{3.68}$$

For $m > 0, n \neq 0$, the solution process leading to Eq. (3.25) can be used, with m replaced by $m + n(\omega_o/\omega_c)$. This gives sideband coefficients of

$$A_{mn} + jB_{mn} = \frac{2V_{dc}}{\left[m + n\frac{\omega_o}{\omega_c}\right]\pi} J_n\left(\left[m + n\frac{\omega_o}{\omega_c}\right]\pi M\right)\left(\sin n\frac{\pi}{2} - j\cos n\frac{\pi}{2}\right) \tag{3.69}$$

As before, Eq. (3.69) is valid for all nonzero values of m and n, i.e., for $m = 1, \ldots, \infty$ and $n = -\infty, \ldots, -1, 1, \ldots, \infty$.

The complete harmonic solution for *trailing edge regular sampled modulation* of a half-bridge phase leg can now be formed by substituting the results of Eqs. (3.66), (3.67), (3.68), and (3.69) back into Eq. (3.10), so that the time-varying switched phase leg voltage $v_{an}(t)$ can be expressed in terms of its harmonic components as

$$v_{an}(t) = V_{dc} + \frac{2V_{dc}}{\pi}\sum_{n=1}^{\infty}\frac{1}{\left[n\frac{\omega_o}{\omega_c}\right]}J_n\left(n\frac{\omega_o}{\omega_c}\pi M\right)\begin{bmatrix}\sin n\frac{\pi}{2}\cos(n[\omega_o t + \theta_o]) \\ -\cos n\frac{\pi}{2}\sin(n[\omega_o t + \theta_o])\end{bmatrix}$$

$$+\frac{2V_{dc}}{\pi}\sum_{m=1}^{\infty}\frac{1}{m}[\cos m\pi - J_0(m\pi M)]\sin(m[\omega_c t + \theta_c])$$

$$+\frac{2V_{dc}}{\pi}\sum_{m=1}^{\infty}\sum_{\substack{n=-\infty \\ (n\neq 0)}}^{\infty}\frac{J_n\left(\left[m+n\frac{\omega_o}{\omega_c}\right]\pi M\right)}{\left[m+n\frac{\omega_o}{\omega_c}\right]}\begin{bmatrix}\sin n\frac{\pi}{2}\cos\begin{pmatrix}m[\omega_c t + \theta_c] \\ +n[\omega_o t + \theta_o]\end{pmatrix} \\ -\cos n\frac{\pi}{2}\sin\begin{pmatrix}m[\omega_c t + \theta_c] \\ +n[\omega_o t + \theta_o]\end{pmatrix}\end{bmatrix} \quad (3.70)$$

As before, Eq. (3.70) also defines the harmonic components of $v_{az}(t)$, if the first term which defines the DC offset is discarded.

Figure 3.18 presents the voltage spectrum for trailing edge regularly sampled PWM for the conditions of a carrier ratio of 21 and a modulation index M of 0.9. Comparison of this plot with the spectrum for naturally sampled PWM shown in Figure 3.7 shows the anticipated difference between the two modulation strategies – the presence of low-order baseband harmonics above the desired fundamental for regular sampled PWM. These harmonics are a consequence of the regular sampling process and occur for any regular sampled PWM strategy. However, the roll-off of their magnitude with n is affected by the carrier ratio and the modulation strategy. Higher carrier ratios (and/or sine-triangle sampled PWM as shown in the next section of this chapter) achieve a considerable improvement in the rate of roll-off of the baseband harmonics, and are therefore generally preferable.

The other subtle variation in the harmonics created by regular sampling is a slight shift of carrier sideband energy between the lower and the higher sideband harmonics. This can be seen by careful comparison of Figure 3.18 versus Figure 3.7, where it can be seen how regular sampling has distorted the symmetry of the sidebands compared to the naturally sampled case. This effect is essentially a property of regular sampled PWM, although the exact distortion does vary a little depending on the particular modulation strategy used.

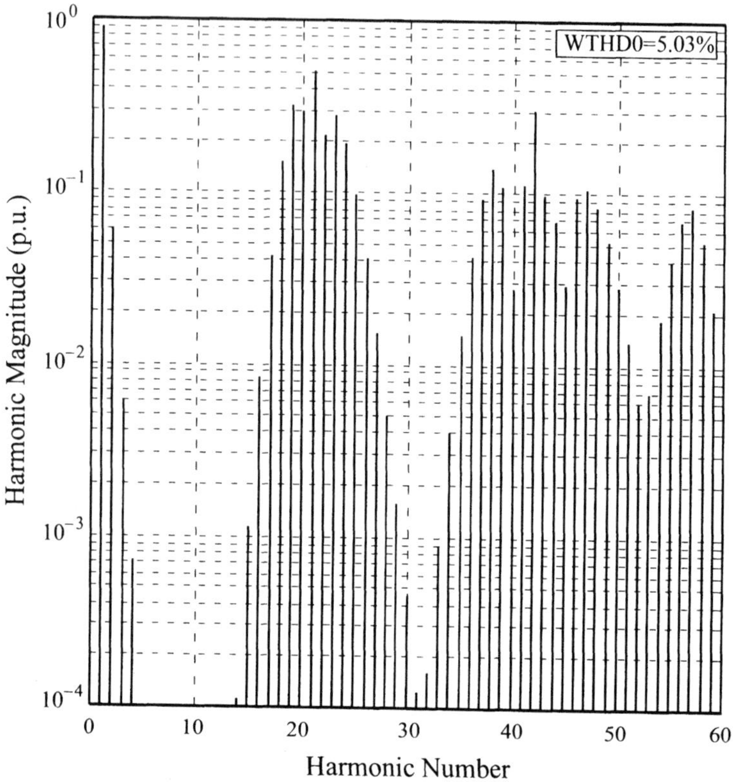

Figure 3.18 Harmonic components for one inverter phase leg with *trailing edge regular sampled* PWM, $M = 0.9$, $f_c/f_o = 21$.

The WTHD0 for regular sampled modulation using a sawtooth carrier under the conditions listed in Figure 3.18 is 5.03%, which is higher than for naturally sampled PWM and primarily reflects the influence of the additional baseband harmonic components.

3.6.2 Symmetrical Regular Sampled PWM

A similar approach can be adopted for symmetrical regular sampled PWM with a triangular carrier, except that the rising edge inner integral limit x_r must now also be defined using the staircase variable y'. Figure 3.19 shows how the intersection occurs between the sinusoidal reference and the staircase solution trajectory for symmetrical regular sampled PWM.

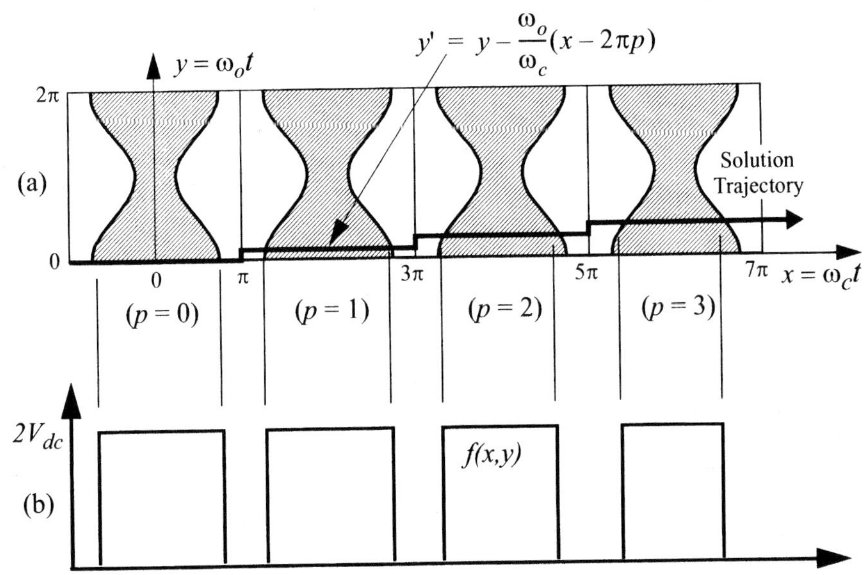

Figure 3.19 Half-bridge switching for symmetrical regular sampled double-edge PWM showing (a) staircase solution trajectory and (b) resultant switched PWM voltage.

The inner integral limits for regular sampled PWM with a triangular carrier can be defined similar to Eq. (3.60) using the same staircase variable y' as

$$x_r = -\frac{\pi}{2}(1 + M\cos y') \qquad x_f = \frac{\pi}{2}(1 + M\cos y') \tag{3.71}$$

so that Eq. (3.58) becomes

$$A_{mn} + jB_{mn} = \frac{V_{dc}}{\pi^2} \int_{-\pi}^{\pi} \int_{-\frac{\pi}{2}(1 + M\cos y')}^{\frac{\pi}{2}(1 + M\cos y')} e^{j(mx + ny)} \, dx \, dy \tag{3.72}$$

Once again this expression can be evaluated by replacing y with $y = y' + (\omega_o/\omega_c)x$ to become

$$A_{mn} + jB_{mn} = \frac{V_{dc}}{\pi^2} \int_{-\pi}^{\pi} \int_{-\frac{\pi}{2}(1 + M\cos y')}^{\frac{\pi}{2}(1 + M\cos y')} e^{j\left(\left[m + n\frac{\omega_o}{\omega_c}\right]x + ny'\right)} \, dx \, dy' \tag{3.73}$$

Equation (3.73) can be evaluated for particular values of m and n as follows:

For $m = n = 0$, the solution process for Eq. (3.30) remains valid, so that

$$A_{00} = 2V_{dc} \tag{3.74}$$

For $m = 0, n > 0$, the solution process leading to Eq. (3.38) can be used, with m replaced by $n(\omega_o/\omega_c)$. This gives baseband harmonic coefficients of

$$A_{0n} + jB_{0n} = \frac{4V_{dc}}{\left[n\frac{\omega_o}{\omega_c}\right]\pi} J_n\left(n\frac{\omega_o}{\omega_c}\frac{\pi}{2}M\right) \sin\left(\left[n\frac{\omega_o}{\omega_c} + n\right]\frac{\pi}{2}\right) \tag{3.75}$$

Equation (3.75) again shows the difference between regular sampled PWM and naturally sampled PWM, with the existence of baseband harmonic coefficients for $n > 1$ for the case of regular sampled PWM.

For $m > 0, n = 0$, the solution process leading to Eq. (3.35) remains valid, so that

$$A_{m0} + jB_{m0} = \frac{4V_{dc}}{m\pi} J_0\left(m\frac{\pi}{2}M\right) \sin m\frac{\pi}{2} \tag{3.76}$$

For $m > 0, n \neq 0$, the solution process leading to Eq. (3.38) can again be used, with m replaced by $m + n(\omega_o/\omega_c)$. This gives sideband coefficients of

$$A_{mn} + jB_{mn} = \frac{4V_{dc}}{\left[m + n\frac{\omega_o}{\omega_c}\right]\pi} J_n\left(\left[m + n\frac{\omega_o}{\omega_c}\right]\frac{\pi}{2}M\right) \sin\left(\left[m + n\frac{\omega_o}{\omega_c} + n\right]\frac{\pi}{2}\right) \tag{3.77}$$

Once again, Eq. (3.77) is valid for all nonzero values of m and n, i.e., for $m = 1, 2, \ldots, \infty$ and $n = -\infty, \ldots, -2, -1, 1, 2, \ldots, \infty$.

The complete harmonic solution for *symmetrical regular sampled modulation* of a half-bridge phase leg can now be formed by substituting the results of Eqs. (3.74), (3.75), (3.76), and (3.77) back into Eq. (3.10), so that the time-varying switched phase leg voltage $v_{an}(t)$ can be expressed in terms of its harmonic components as

$$v_{an}(t) = V_{dc} + \frac{4V_{dc}}{\pi}\sum_{n=1}^{\infty}\frac{J_n\left(n\frac{\omega_o}{\omega_c}\frac{\pi}{2}M\right)}{\left[n\frac{\omega_o}{\omega_c}\right]}\sin\left(n\left[1+\frac{\omega_o}{\omega_c}\right]\frac{\pi}{2}\right)\cos(n[\omega_o t+\theta_o]) \tag{3.78}$$

$$+\frac{4V_{dc}}{\pi}\sum_{m=1}^{\infty}\frac{1}{m}J_0\left(m\frac{\pi}{2}M\right)\sin m\frac{\pi}{2}\cos(m[\omega_c t+\theta_c])$$

$$+\frac{4V_{dc}}{\pi}\sum_{m=1}^{\infty}\sum_{\substack{n=-\infty\\(n\neq 0)}}^{\infty}\frac{J_n\left(\left[m+n\frac{\omega_o}{\omega_c}\right]\frac{\pi}{2}M\right)}{\left[m+n\frac{\omega_o}{\omega_c}\right]}\sin\left(\left[m+n\frac{\omega_o}{\omega_c}+n\right]\frac{\pi}{2}\right) \times\cos(m[\omega_c t+\theta_c]+n[\omega_o t+\theta_o])$$

As before, the DC offset voltage V_{dc} simply reflects the selection of the reference point for the switching waveform and can be discarded without loss of generality.

Equation (3.78) can be expressed more compactly for the switched output voltage measured with respect to the DC bus midpoint, $v_{az}(t)$, as

$$v_{az}(t) = \frac{4V_{dc}}{\pi}\sum_{\substack{m=0\ \leftrightarrow\\ m>0\ \leftrightarrow}}^{\infty}\ \sum_{\substack{n=1\\ n=\infty}}^{\infty}\frac{1}{q}J_n\left(q\frac{\pi}{2}M\right)\sin\left([q+n]\frac{\pi}{2}\right) \times\cos(m[\omega_c t+\theta_c]+n[\omega_o t+\theta_o]) \tag{3.79}$$

where $q = m + n(\omega_o/\omega_c)$ (and is not necessarily integer). Note the change in the lower summation limit for n depending on whether $m = 0$, which is required because baseband harmonics are only defined for positive values of n.

Figure 3.20 shows the voltage harmonic spectrum for symmetrical regular sampled PWM with a triangular carrier for the conditions of a carrier ratio of 21 and a modulation index M of 0.9. This figure shows a number of significant differences for this modulation scheme compared to both double-edge naturally sampled PWM and regular sampled PWM with a sawtooth carrier.

First, the carrier sideband cancellations that were achieved by double-edge naturally sampled PWM have not been fully achieved for the case of symmetric regular sampling. While there is considerable reduction in the magnitude of the odd sideband harmonics around the odd carrier multiples and the even sideband harmonics around the even carrier multiples, the cancellation is nowhere

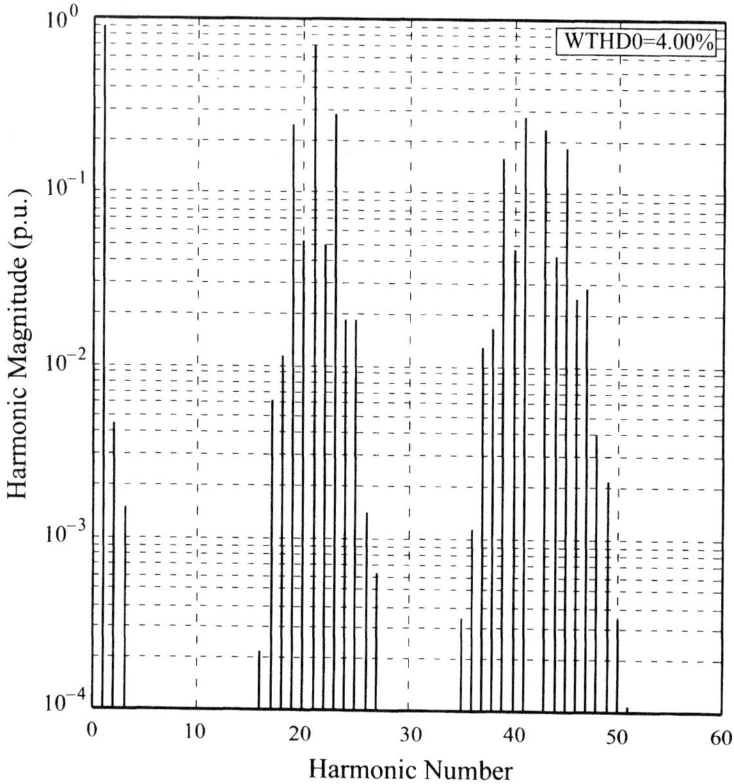

Figure 3.20 Harmonic components for one inverter phase leg with triangular carrier and *symmetrical regular sampled* PWM, $M = 0.9$, $f_c/f_o = 21$.

near complete. This has significant implications for the harmonics produced by phase legs operated together as single- or three-phase bridge inverters, as will be considered in later chapters.

Second, the baseband harmonic components produced by the regular sampling process still exist, but their magnitude has rolled off much more rapidly than for regular sampling with a sawtooth carrier. For the conditions presented in this chapter, the second harmonic in the baseband for sawtooth carrier regular sampled modulation has a magnitude of 6%, while for symmetric regular sampled modulation with a triangular carrier, the magnitude of the second harmonic is below 0.5%. This clearly illustrates the benefits of using a triangular carrier in a regular sampled modulation system.

3.6.3 Asymmetrical Regular Sampled PWM

For asymmetrical regular sampling, the inner integral limits of Eq. (3.58) must now be modified to account for the two sampling points that are established in each carrier interval. This can be achieved by creating two staircase variables, one frozen at $x = -\pi/2$ within each carrier interval to determine the rising edge of the switched pulse, and the other frozen at $x = \pi/2$ within each carrier interval to determine the falling edge of the switched pulse.

These variables can be expressed as

$$y_r' = \frac{\omega_o}{\omega_c}\left(2\pi p - \frac{\pi}{2}\right) \qquad y_f' = \frac{\omega_o}{\omega_c}\left(2\pi p + \frac{\pi}{2}\right) \tag{3.80}$$

or in terms of the continuous variables x and y as

$$y_r' = y - \frac{\omega_o}{\omega_c}\left(x - 2\pi p + \frac{\pi}{2}\right) \qquad y_f' = y - \frac{\omega_o}{\omega_c}\left(x - 2\pi p - \frac{\pi}{2}\right) \tag{3.81}$$

The inner integral limits for asymmetrical regular sampled PWM with a triangular carrier can be defined using these new staircase variables as

$$x_r = -\frac{\pi}{2}(1 + M\cos y_r') \qquad x_f = \frac{\pi}{2}(1 + M\cos y_f') \tag{3.82}$$

so that Eq. (3.58) becomes

$$A_{mn} + jB_{mn} = \frac{V_{dc}}{\pi^2}\int_{-\pi}^{\pi}\int_{-\frac{\pi}{2}(1 + M\cos y_r')}^{\frac{\pi}{2}(1 + M\cos y_f')} e^{j(mx + ny)}\,dx\,dy \tag{3.83}$$

To solve Eq. (3.83), the switched waveform in each carrier interval is split into two sections for analysis, with the results added by superposition. The first section has a modulated rising edge in the first half carrier interval and a falling edge in the center of the carrier interval. The second section has a rising edge at the center of the carrier interval and a modulated falling edge in the second half carrier interval. Each waveform section is evaluated separately, and their harmonics are then combined to determine the harmonics of the overall switched waveform.

Mathematically, this behavior can be expressed as

$$f(x, y) = f_r(x, y) + f_f(x, y) \tag{3.84}$$

where:	$f_r(x, y)$ steps from	0 to $2V_{dc}$ at	$x - 2\pi p = x_r$
	and from	$2V_{dc}$ to 0 at	$x - 2\pi p = 0$
and:	$f_f(x, y)$ steps from	0 to $2V_{dc}$ at	$x - 2\pi p = 0$
	and from	$2V_{dc}$ to 0 at	$x - 2\pi p = x_f$

Equation (3.83) then becomes

$$A_{mn} + jB_{mn} = \frac{V_{dc}}{\pi^2}\left[\int_{-\pi}^{\pi}\int_{-\frac{\pi}{2}(1+M\cos y_r')}^{0} e^{j(mx+ny)}\,dx\,dy + \int_{-\pi}^{\pi}\int_{0}^{\frac{\pi}{2}(1+M\cos y_f')} e^{j(mx+ny)}\,dx\,dy\right] \tag{3.85}$$

The two sections of Eq. (3.85) can be evaluated by replacing y with $y = y_r' + (\omega_o/\omega_c)[x + (\pi/2)]$ and $y = y_f' + (\omega_o/\omega_c)[x - (\pi/2)]$ for each section, respectively, derived from Eq. (3.81), to become

$$A_{mn} + jB_{mn} = \frac{V_{dc}}{\pi^2}\left[\int_{-\pi}^{\pi}\int_{-\frac{\pi}{2}(1+M\cos y_r')}^{0} e^{j\left(mx + n\left[y_r' + \frac{\omega_o}{\omega_c}x + \frac{\omega_o}{\omega_c}\frac{\pi}{2}\right]\right)}\,dx\,dy_r' + \int_{-\pi}^{\pi}\int_{0}^{\frac{\pi}{2}(1+M\cos y_f')} e^{j\left(mx + n\left[y_f' + \frac{\omega_o}{\omega_c}x - \frac{\omega_o}{\omega_c}\frac{\pi}{2}\right]\right)}\,dx\,dy_f'\right] \tag{3.86}$$

Equation (3.86) can be evaluated for particular values of m and n as follows:

For $m = n = 0$, Eq. (3.86) simplifies to

$$A_{00} + jB_{00} = \frac{V_{dc}}{\pi^2}\left[\int_{-\pi}^{\pi}\int_{-\frac{\pi}{2}(1+M\cos y_r')}^{0} dx\,dy_r' + \int_{-\pi}^{\pi}\int_{0}^{\frac{\pi}{2}(1+M\cos y_f')} dx\,dy_f'\right]$$

$$= \frac{V_{dc}}{\pi^2}\left[\int_{-\pi}^{\pi}\frac{\pi}{2}(1 + M\cos y_r')\,dy_r' + \int_{-\pi}^{\pi}\frac{\pi}{2}(1 + M\cos y_f')\,dy_f'\right] = 2V_{dc} \quad (3.87)$$

This result has the expected V_{dc} offset as before.

For $m > 0, n = 0$, Eq. (3.86) simplifies to

$$A_{m0} + jB_{m0} = \frac{V_{dc}}{\pi^2}\left[\int_{-\pi}^{\pi}\int_{-\frac{\pi}{2}(1 + M\cos y_r')}^{0} e^{jmx}\,dx\,dy_r' + \int_{-\pi}^{\pi}\int_{0}^{\frac{\pi}{2}(1 + M\cos y_f')} e^{jmx}\,dx\,dy_f'\right]$$

$$= \frac{V_{dc}}{jm\pi^2}\left[\int_{-\pi}^{\pi}\left[1 - e^{-jm\frac{\pi}{2}(1 + M\cos y_r')}\right]dy_r' + \int_{-\pi}^{\pi}\left[e^{jm\frac{\pi}{2}(1 + M\cos y_f')} - 1\right]dy_f'\right]$$

$$= \frac{V_{dc}}{jm\pi^2}\left[\int_{-\pi}^{\pi}\left[1 - e^{-jm\frac{\pi}{2}}e^{-jm\frac{\pi}{2}M\cos y_r'}\right]dy_r' + \int_{-\pi}^{\pi}\left[e^{jm\frac{\pi}{2}}e^{jm\frac{\pi}{2}M\cos y_f'} - 1\right]dy_f'\right] \quad (3.88)$$

Using Eq. (A2.16), this expression becomes

$$A_{m0} + jB_{m0} = \frac{2V_{dc}}{jm\pi}\left[e^{jm\frac{\pi}{2}}J_0\left(m\frac{\pi}{2}M\right) - e^{-jm\frac{\pi}{2}}J_0\left(-m\frac{\pi}{2}M\right)\right] \quad (3.89)$$

which simplifies using Eq. (A2.22) to

$$A_{m0} + jB_{m0} = \frac{4V_{dc}}{m\pi} J_0\left(m\frac{\pi}{2}M\right) \sin m\frac{\pi}{2} \tag{3.90}$$

Equation (3.90) identifies that the harmonics of the carrier wave for asymmetrical regular sampled PWM are the same as for double-edge naturally sampled PWM and for symmetrical regular sampled PWM, as per Eqs. (3.35) and (3.76).

For $m, n \neq 0$, Eq. (3.86) can be rewritten as

$$A_{mn} + jB_{mn} = \frac{V_{dc}}{\pi^2}\left[\begin{array}{l} \displaystyle\int_{-\pi}^{\pi} \int_{-\frac{\pi}{2}(1 + M\cos y_r')}^{0} e^{j\left(\left[m + n\frac{\omega_o}{\omega_c}\right]x + ny_r' + \left[n\frac{\omega_o}{\omega_c}\right]\frac{\pi}{2}\right)} dx\, dy_r' \\ \displaystyle + \int_{-\pi}^{\pi} \int_{0}^{\frac{\pi}{2}(1 + M\cos y_f')} e^{j\left(\left[m + n\frac{\omega_o}{\omega_c}\right]x + ny_f' - \left[n\frac{\omega_o}{\omega_c}\right]\frac{\pi}{2}\right)} dx\, dy_f' \end{array}\right] \tag{3.91}$$

Substituting $q = m + n(\omega_o/\omega_c)$ and solving the inner integration gives

$$A_{mn} + jB_{mn} = \frac{V_{dc}}{jq\pi^2}\left[\begin{array}{l} \displaystyle\int_{-\pi}^{\pi} e^{j\left(ny_r' + \left[n\frac{\omega_o}{\omega_c}\right]\frac{\pi}{2}\right)} \left[1 - e^{-jq\frac{\pi}{2}(1 + M\cos y_r')}\right] dy_r' \\ \displaystyle + \int_{-\pi}^{\pi} e^{j\left(ny_f' - \left[n\frac{\omega_o}{\omega_c}\right]\frac{\pi}{2}\right)} \left[e^{jq\frac{\pi}{2}(1 + M\cos y_f')} - 1\right] dy_f' \end{array}\right]$$

$$= \frac{V_{dc}}{jq\pi^2}\left[\begin{array}{l} \displaystyle\int_{-\pi}^{\pi} e^{j\left[n\frac{\omega_o}{\omega_c}\right]\frac{\pi}{2}} e^{jny_r} \left[1 - e^{-jq\frac{\pi}{2}} e^{-jq\frac{\pi}{2}M\cos y_r'}\right] dy_r' \\ \displaystyle + \int_{-\pi}^{\pi} e^{-j\left[n\frac{\omega_o}{\omega_c}\right]\frac{\pi}{2}} e^{jny_f'} \left[e^{jq\frac{\pi}{2}} e^{jq\frac{\pi}{2}M\cos y_f'} - 1\right] dy_f' \end{array}\right] \tag{3.92}$$

Only the terms with an $e^{\pm jq\frac{\pi}{2}M\cos y}$ component need be solved for the outer integration, since $\int_{-\pi}^{\pi} Ae^{jny}dy = 0$. Hence Eq. (3.92) can be simplified to

$$A_{mn}+jB_{mn} = \frac{V_{dc}}{jq\pi^2}\left[\begin{array}{l}\int_{-\pi}^{\pi} -e^{jn\frac{\omega_o}{\omega_c}\frac{\pi}{2}} e^{jny_r} e^{-jq\frac{\pi}{2}} e^{-jq\frac{\pi}{2}M\cos y_r'} dy_r' \\ \quad + \int_{-\pi}^{\pi} e^{-jn\frac{\omega_o}{\omega_c}\frac{\pi}{2}} e^{jny_f} e^{jq\frac{\pi}{2}} e^{jq\frac{\pi}{2}M\cos y_f'} dy_f'\end{array}\right]$$

$$= \frac{V_{dc}}{jq\pi^2}\left[\begin{array}{l}\int_{-\pi}^{\pi} -e^{jny_r} e^{-jm\frac{\pi}{2}} e^{-jq\frac{\pi}{2}M\cos y_r'} dy_r' \\ \quad + \int_{-\pi}^{\pi} e^{jny_f} e^{jm\frac{\pi}{2}} e^{jq\frac{\pi}{2}M\cos y_f'} dy_f'\end{array}\right] \tag{3.93}$$

which, using Eq. (A2.15), can be evaluated as

$$A_{mn}+jB_{mn} = \frac{2\pi V_{dc}}{jq\pi^2}j^n\left[-J_n\left(-q\frac{\pi}{2}M\right)e^{-jm\frac{\pi}{2}} + J_n\left(q\frac{\pi}{2}M\right)e^{jm\frac{\pi}{2}}\right] \tag{3.94}$$

and using Eq. (A2.21) to

$$A_{mn}+jB_{mn} = \frac{2V_{dc}}{jq\pi}\left[J_n\left(q\frac{\pi}{2}M\right)\left(j^n e^{jm\frac{\pi}{2}} - j^{-n}e^{-jm\frac{\pi}{2}}\right)\right]$$

$$= \frac{2V_{dc}}{jq\pi}\left[J_n\left(q\frac{\pi}{2}M\right)\left(e^{j[m+n]\frac{\pi}{2}} - e^{-j[m+n]\frac{\pi}{2}}\right)\right] \tag{3.95}$$

and which finally simplifies to

$$A_{mn}+jB_{mn} = \frac{4V_{dc}}{\left[m+\frac{\omega_o}{\omega_c}n\right]\pi}J_n\left(\left[m+\frac{\omega_o}{\omega_c}n\right]\frac{\pi}{2}M\right)\sin\left([m+n]\frac{\pi}{2}\right) \tag{3.96}$$

For $m = 0, n \neq 0$, Eq. (3.96) can be used with m set to 0, to give

$$A_{0n} + jB_{0n} = \frac{4V_{dc}}{\left[\frac{\omega_o}{\omega_c}n\right]\pi} J_n\left(\frac{\omega_o}{\omega_c}n\frac{\pi}{2}M\right)\sin n\frac{\pi}{2} \tag{3.97}$$

which once again contains the baseband harmonic components expected from regular sampled PWM.

The complete harmonic solution for *asymmetrical regular sampled modulation* of a half-bridge phase leg can now be formed by substituting the results of Eqs. (3.87), (3.90), (3.96), and (3.97) back into Eq. (3.10), so that the time-varying switched phase leg voltage $v_{an}(t)$ can be expressed in terms of its harmonic components as

$$\begin{aligned} v_{an}(t) = V_{dc} &+ \frac{4V_{dc}}{\pi}\sum_{n=1}^{\infty}\frac{1}{\left[n\frac{\omega_o}{\omega_c}\right]}J_n\left(n\frac{\omega_o}{\omega_c}\frac{\pi}{2}M\right)\sin n\frac{\pi}{2}\cos(n[\omega_o t + \theta_o]) \\ &+ \frac{4V_{dc}}{\pi}\sum_{m=1}^{\infty}\frac{1}{m}J_0\left(m\frac{\pi}{2}M\right)\sin m\frac{\pi}{2}\cos(m[\omega_c t + \theta_c]) \\ &+ \frac{4V_{dc}}{\pi}\sum_{m=1}^{\infty}\sum_{\substack{n=-\infty \\ (n\neq 0)}}^{\infty}\frac{J_n\left(\left[m + n\frac{\omega_o}{\omega_c}\right]\frac{\pi}{2}M\right)}{\left[m + n\frac{\omega_o}{\omega_c}\right]}\sin\left([m+n]\frac{\pi}{2}\right) \\ &\qquad \times \cos(m[\omega_c t + \theta_c] + n[\omega_o t + \theta_o]) \end{aligned} \tag{3.98}$$

As before the DC offset voltage V_{dc} can be discarded from Eq. (3.98) without loss of generality, and Eq. (3.98) can be expressed more compactly for the switched output voltage with respect to the DC bus midpoint, $v_{az}(t)$, as

$$\begin{aligned} v_{az}(t) = \frac{4V_{dc}}{\pi}\sum_{\substack{m=0 \leftrightarrow n=1 \\ m>0 \leftrightarrow n=\infty}}^{\infty}\sum^{\infty}\frac{1}{q}J_n\left(q\frac{\pi}{2}M\right)\sin\left([m+n]\frac{\pi}{2}\right) \\ \times \cos(m[\omega_c t + \theta_c] + n[\omega_o t + \theta_o]) \end{aligned} \tag{3.99}$$

where again $q = m + n(\omega_o/\omega_c)$ (and is not necessarily integer). Note the change in the lower summation limit for n depending whether $m = 0$, which is once more required because baseband harmonics are only defined for positive values of n.

Figure 3.21 shows the voltage harmonic spectrum for symmetrical regular sampled PWM with a triangular carrier for the conditions of a carrier ratio of 21 and a modulation index M of 0.9. This figure shows a significant harmonic improvement compared to symmetrical regular sampled PWM with a triangular carrier.

In particular, the *odd* harmonic sideband components around *odd* multiples of the carrier fundamental, and *even* harmonic sideband components around *even* multiples of the carrier fundamental, are once again completely eliminated by the $\sin\left([m+n]\frac{\pi}{2}\right)$ expression in Eq. (3.98), despite the regular sampling process. In addition, the *even* low-order baseband harmonics have been eliminated by the $\sin\left(n\frac{\pi}{2}\right)$ expression in the first summation term.

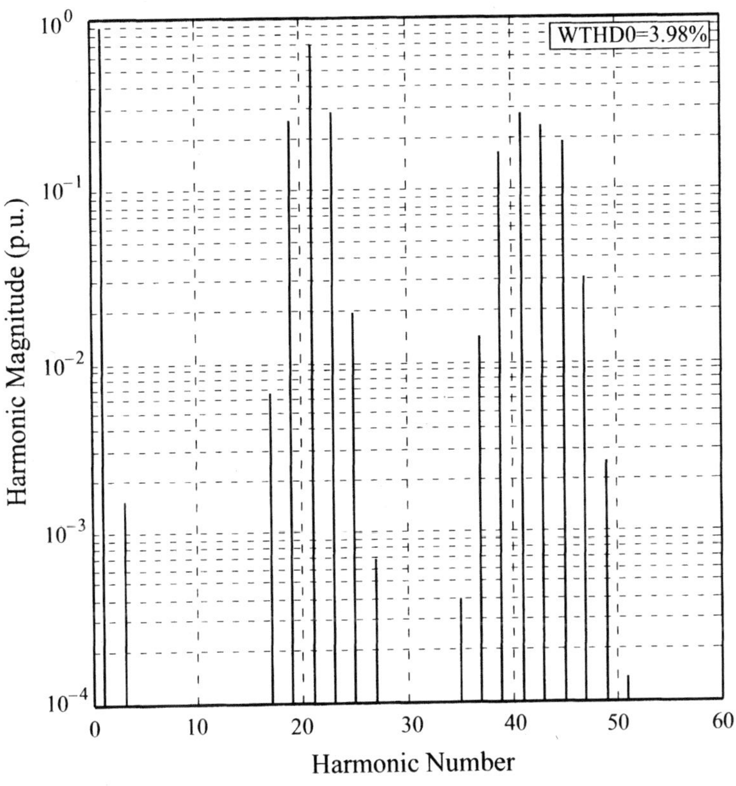

Figure 3.21 Harmonic components for one inverter phase leg with triangular carrier and *asymmetrical regular sampled* PWM, $M = 0.9$, $f_c/f_o = 21$.

These benefits are an intrinsic advantage of asymmetrical sampling compared to symmetrical sampling for triangular carrier (or equivalent) regular sampled modulation systems and have significant consequences when the harmonic canceling that occurs between phase legs is considered for the various converter topologies discussed in later chapters.

3.7 "Direct" Modulation

Another method of determining the width of the switched pulses is occasionally proposed as an alternative to regular sampled PWM. The concept is to switch the inverter to create an active pulse interval for each carrier interval that exactly achieves the same volt-second average as the original target waveform. The method is not usually practical to implement since it requires integration over the carrier interval, but it has been presented in the literature [9] as a more accurate reference against which other strategies can be compared.

In general terms, the required pulse width is given by

$$W_{dutycycle} = \frac{1}{(t_{i+1} - t_i)} \int_{t_i}^{t_{i+1}} M\cos(\omega_o t)\, dt \tag{3.100}$$

This pulse width can be evaluated either on a *symmetrical* basis over a complete carrier interval or on an *asymmetrical* basis at half carrier intervals. Mathematically, these two alternatives can be expressed as

$$W_{symm} = \frac{1}{\Delta T} \int_{t_i}^{t_i + \Delta T} M\cos(\omega_o t)\, dt \tag{3.101}$$

for symmetrical evaluation, and

$$W_{asymm} = \frac{2}{\Delta T} \int_{t_i}^{t_i + \frac{\Delta T}{2}} M\cos(\omega_o t)\, dt \tag{3.102}$$

for asymmetrical evaluation.

For a symmetrically calculated system, the resultant pulse is placed at the center of the carrier interval, to create a rising edge in the first half-carrier interval and a falling edge in the second half-carrier interval. For an asymmetrical system, the pulses in each successive half-carrier intervals are placed alternatively at the start and the end of each interval (i.e., back to back across a complete carrier interval). Since every two pulses therefore run together, no intermediate phase leg switching is required between every second pulse, and there is no increase in switching frequency despite the two pulse width calculations in each carrier interval. The pulses produced by this strategy are very similar to those produced by symmetrical and asymmetrical regular sampling, respectively, as shown in Figure 3.16(b), except that the pulse widths vary slightly because of the alternative technique for pulse width calculation.

Figure 3.22 shows the voltage harmonic spectrum for asymmetrically calculated direct PWM with a triangular carrier for the conditions of a carrier ratio of 21 and a modulation index M of 0.9. As could be anticipated, the performance of this approach is almost exactly the same as for that of asymmetrical regular sampled PWM shown in Figure 3.21. A similar result is obtained when symmetrically evaluated direct modulation is compared against symmetrical regular sampled PWM.

A simple variation of direct modulation is to approximate the reference waveform as a straight line across the evaluation interval. Equation (3.100) is then replaced with

$$W_{dutycycle} = M\frac{\cos[\omega_o t_i] + \cos[\omega_o t_{i+1}]}{2} \tag{3.103}$$

which is very close to the value of $M\cos[\omega_o t_{i+0.5}]$ used in regular sampled systems for all reasonable carrier ratios and consequently produces much the same harmonic performance.

In reality, direction modulation offers virtually no harmonic benefit at any reasonable carrier ratio. For very low carrier ratios, there is a slightly better resolution of the fundamental magnitude compared to the reference (remember that regular sampling does introduce some error in the baseband fundamental because of the $\frac{4V_{dc}}{\pi}\frac{1}{[n(\omega_o/\omega_c)]}J_n\left(n\frac{\omega_o}{\omega_c}\frac{\pi}{2}M\right)$ term in Eqs. (3.78) and (3.98), and this error increases with a decreasing carrier ratio). However, very low carrier ratios also lead to significant carrier sideband harmonics intruding into the spectrum below the fundamental (these harmonics are often referred to as *subharmonics* but are really very low-frequency sidebands from the first carrier

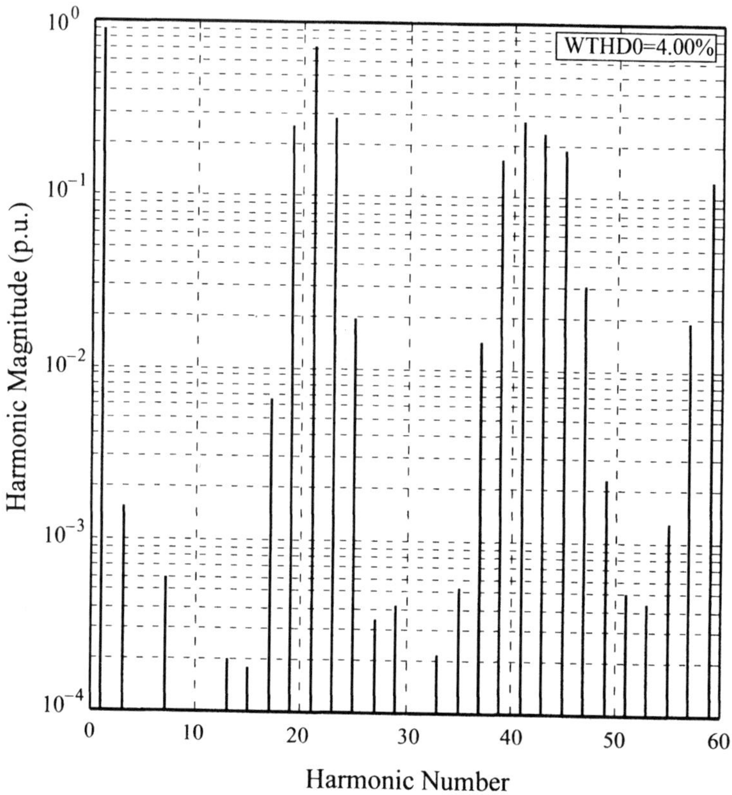

Figure 3.22 Harmonic components for one inverter phase leg with triangular carrier and *direct calculation* of pulse period, $M = 0.9$, $f_c/f_o = 21$.

harmonic group), and hence strategies such as programmed PWM discussed in Chapter 9 are often preferable. As a result of these observations the direct modulation approach will not be considered further in this text.

3.8 Integer versus Non-Integer Frequency Ratios

It should be noted at this point that nothing in the detailed mathematical development presented in this chapter constrains the carrier ratio to integer values, and even irrational values of switching frequency are also permissible. Irrespective of the switching frequency ratio, the PWM process produces a fundamental component, baseband harmonic components when any form of regular sampling is involved, and sideband harmonics grouped around multiples of the carrier frequency, at frequencies of $mf_c \pm nf_o$.

In some literature, non-integer frequency ratios have been identified as undesirable because they are claimed to introduce subharmonics below the fundamental. For example, a carrier frequency ratio of $29/3$ has sometimes been said to create subharmonics of one-third and two-third times the fundamental component. However, the PWM solutions that have been presented in this chapter identify that any harmonics that might exist in this region are really only low-order carrier sideband components. In the case above, for example, the so-called *sub-harmonics* would really correspond to the 9th and 10th lower sidebands grouped around the first carrier frequency harmonic. But as shown in Figures 3.7 and 3.12, the amplitudes of these sideband harmonics are negligible after the first few components.

For applications using very low carrier frequency ratios (say < 11), there is an argument for using an *integer* frequency ratio to ensure that all significant lower sideband harmonics from the first carrier group have a frequency above (or at worst equal to) the fundamental frequency. For natural and asymmetrical regular sampled PWM with a triangular carrier, there is further justification in using an *odd* frequency ratio to ensure that the lowest significant sideband harmonic will be at least three times the fundamental frequency (recall from Sections 3.4.2 and 3.6.2 that these modulation strategies eliminate *odd* sideband components around *odd* carrier multiples, so that for $m = 1$, $f_c \pm nf_o$ must always be *odd* for an integer carrier frequency ratio and the lowest possible sideband component is therefore a third harmonic of the fundamental).

Finally, it should be noted that the selection of an appropriate carrier frequency ratio can become a little more complex for the more sophisticated PWM strategies discussed in the next three chapters of this text. Essentially, the choice depends on how rapidly the sideband harmonics from the first carrier group roll off in magnitude away from the carrier, and this must be considered individually for the various PWM strategies presented. The issue will be revisited as appropriate in later chapters.

3.9 Review of PWM Variations

Table 3.1 summarizes the magnitudes of the significant harmonics for the three approaches to the determination of switched pulse width presented in this chapter, normalized with respect to the naturally sampled PWM fundamental component. From this table it can be seen that there is negligible difference between the fundamental components for the carrier ratio of 21 that was used, and it can therefore be argued that there is virtually no fundamental component differentiation between the techniques at the carrier ratios typically used in modern inverter systems.

In terms of the *secondary* objective of harmonic minimization, it can be seen that the harmonic performance of direct modulation is almost identical to regular sampled modulation. This is only to be expected, since regular sampled

Table 3.1 Harmonic Components for Single-Phase Leg PWM Strategies, $M = 0.9$, $f_c/f_o = 21$, All Harmonics Normalized with respect to V_{dc} (%)

Harmonic Number	Naturally Sampled PWM (%) (Sawtooth Carrier)	Regularly Sampled PWM (%) (Sawtooth Carrier)	Naturally Sampled PWM (%) (Triangular Carrier)	Symmetrical Regular Sampled PWM (%) (Triangular Carrier)	Asymmetrical Regular Sampled PWM (%) (Triangular Carrier)	Direct Modulation Asymmetrical Calculation (%)
1	90.0	89.8	90.0	89.7	89.9	89.9
2	0.0	6.0	0.0	0.5	0.0	0.0
3	0.0	0.61	0.0	0.1	0.2	0.2
4	0.0	0.07	0.0	0.0	0.0	0.0
5	0.0	0.01	0.0	0.0	0.0	0.0
16	2.1	0.8	0.0	0.0	0.0	0.0
17	7.0	4.3	1.2	0.6	0.7	0.7
18	17.7	15.0	0	1.1	0.0	0.0
19	30.5	31.9	26.8	24.8	25.1	25.0
20	25.5	27.9	0.0	5.3	0.0	0.0
21	51.2	51.2	71.2	71.2	71.2	71.3
22	25.5	21.5	0.0	5.0	0.0	0.0
23	30.5	28.3	26.8	28.1	28.4	28.3
24	17.7	19.4	0.0	1.8	0.0	0.0
25	7.0	9.8	1.2	1.9	1.9	2.0
26	2.1	4.1	0.0	0.1	0.0	0.0
WTHD0	3.88	4.91	3.86	3.86	3.84	3.85

PWM simply linearizes the sinusoidal reference about its average for each carrier or half-carrier interval. For carrier ratios above about 15, little difference would be expected between the linearized volt-second average and the exact volt-second average calculated by direct modulation. As a consequence, it would be anticipated that regular sampled PWM and direct PWM would have the same harmonic consequences, since they essentially produce the same switched pulse widths, and these pulses are placed in the center of each half-carrier interval.

In terms of regular sampled PWM compared to naturally sampled PWM, there are a number of specific differences that can be seen from the results presented in Table 3.1. First, as predicted by the analytical results of Section 3.6, regular sampling creates low-order baseband harmonics just above the fundamental component. These low-order baseband harmonic multiples are usually quite small because of the rapid reduction in the Bessel function in the first term of Eqs. (3.78) and (3.98) and so are often overlooked by researchers, especially when carrier ratios are greater than about 15. However, they are an intrinsic consequence of regular sampled modulation, and their rate of attenuation varies with different modulation strategies and carrier ratios. For example, single ended modulation has a much slower decay rate for these terms, so that for this implementation they can make a nontrivial contribution to the WTHD even for the typical pulse ratios of at least 30 which are used in modern inverters.

Second, comparing natural and regular sampled PWM, it is found that the regular sampled modulation process attenuates the lower side sideband harmonics and increases the high-side sideband harmonics around the carrier frequency. This can also be seen by careful comparison of Figures 3.12 and 3.20, and is an intrinsic property of regular sampled PWM [10]. This attenuation preferentially reduces the lower order harmonic components of the switched waveform and can significantly improve the WTHD for modulation implementations with a low pulse ratio. But as a consequence, claims for an improved "new" PWM algorithm compared to naturally sampled PWM may simply reflect these effects of regular sampling rather than any intrinsic advantage of the new scheme, and this may not be appreciated by the proponents.

Third, the major difference between symmetrical and asymmetrical PWM is that symmetrical regular sampled PWM causes additional sideband harmonic components to be present in the output voltage spectra, while asymmetrical regular sampled PWM and naturally sampled PWM do not create these components. This is an intrinsic limitation of symmetrical sampled modulation compared to asymmetrical sampled modulation [10].

3.10 Summary

This chapter has presented the basic concepts of determining switched pulse width by modulation, and has determined from these concepts that:

- Low-order harmonic multiples of the fundamental are produced by the regular sampling process.
- The lower side sideband harmonic components are attenuated slightly, and the high-side sideband harmonic components are increased slightly for regular sampled PWM.
- Symmetrical regular sampled PWM leads to additional carrier sideband harmonics in the phase leg output voltage and is therefore inferior to asymmetrical PWM.
- Direct modulation produces an almost identical harmonic response to regular sampled PWM.

For full inverter systems with multiple phase legs operating together, both the absence of harmonics by virtue of the modulation process and the elimination of harmonics by cancellation between the phase legs of an inverter, play a significant part in determining the harmonics seen in the *l–l* output voltages of the inverter. Hence it is important to have a precise realization of the phase leg switched output when investigating the harmonic performance of various inverter topologies and modulation implementations. The theoretical understanding presented in this chapter, and the exact match between theoretical analysis and numerical simulation that has been achieved (not elaborated on here, see [11]), provide the basis to evaluate the performance of the more complex switched inverter structures investigated in the following chapters of this book.

References

[1] S.R. Bowes and R. Bullough, "PWM switching strategies for current-fed inverter drives," *IEE Proceedings* (*London*), vol. 131 Pt. B, no. 5., Sept. 1984, pp. 195–202.

[2] M.A. Boost and P.D. Ziogas, "State-of-the-art carrier PWM techniques: a critical evaluation," *IEEE Trans. on Industry Applications,* vol. 24, no. 2, March 1988, pp. 271–290.

[3] H.W. Van der Broeck and H.C. Skudelny, "Analytical analysis of the harmonic effects of a PWM ac drive," *IEEE Trans. on Power Electronics,* vol. 3, no. 2, March/April, 1988, pp. 216–223.

[4] S. Bowes and B.M. Bird, "Novel approach to the analysis and synthesis of modulation processes in power converters," *IEE Proceedings* (*London*), vol. 122, no. 5, May 1975, pp. 507–513.

[5] W.R. Bennett, "New results in the calculation of modulation products," *The Bell System Technical Journal*, vol. 12, April 1933, pp. 228–243.

[6] H.S. Black, *Modulation Theory*, Van Nostrand, New York, 1953.

[7] E.O. Brigham, *The Fast Fourier Transform*, Prentice-Hall, Englewood Cliffs, NJ, 1974.

[8] T.H. Barton, "Pulse-width modulation waveform — the bessel approximation," in *Conf. Rec. IEEE Industry Applications Society Annual Mtg.*, Toronto, 1978, pp. 1125–1130.

[9] Y. Kim and M. Ehsani, "An algebraic algorithm for microcomputer-based (direct) inverter pulse width modulation," *IEEE Trans. on Industry Applications*, vol. 23, no. 4, July/Aug. 1987, pp. 654–660.

[10] J.T. Boys and P.G. Handley, "Harmonic analysis of space vector modulated PWM waveforms," *IEE Proceedings* (*London*), Pt. B, vol. 137, no. 4, July 1990, pp. 197–204.

[11] D.G. Holmes, "A generalised approach to the modulation and control of hard switched converters," Ph.D. Thesis, Monash University, Australia, 1997.

4

Modulation of Single-Phase Voltage Source Inverters

A PWM controlled inverter combines both voltage control and frequency control into one system. The inverter typically operates from a fixed voltage DC source, i.e., perhaps an uncontrolled diode rectifier, a battery, or in some cases both. The principle is basically simple. Each of the phase legs of the inverter are switched at a high frequency and operate essentially as choppers. Assuming a motor load, this high carrier frequency is modulated by the desired motor fundamental frequency which consequently determines the speed of the motor. The inverter output voltage amplitude is then controlled by the chopping action. Clearly, the correct control of the imposed switching patterns calls for complex control electronics, but this is no longer a serious problem with the availability of low-cost microprocessors, signal processors, and other custom digital logic chips.

Chapter 3 has presented an analysis of the PWM processes for one phase leg of a voltage source inverter. This analysis has identified the fundamental and harmonic voltages produced by the modulation process itself, taking into consideration a variety of carrier and sampling alternatives that all slightly effect the phase leg switched output voltage waveform. These are all issues to do with the determination of the switched pulse *width* for the phase leg.

Once the performance of the phase leg modulation process has been determined, the next step is to consider interactions between the phase legs that are grouped together to make a complete inverter system. The overall harmonic performance of an inverter is determined both by the harmonics produced by each phase leg and the potential harmonic cancellation that may occur between the phase legs. With grouped phase legs, issues such as switched pulse *placement* within a carrier interval and switched pulse *sequence* across carrier intervals now also become important and can influence the inverter performance.

This chapter investigates the operation and performance of the simplest type of inverter — a single-phase inverter made up of two phase legs that are switched cooperatively. Later chapters then explore the operation and performance of three-phase legs combined into a three-phase inverter, which offers much more potential for variation in modulation strategies.

4.1 Topology of a Single-Phase Inverter

Figure 4.1 shows once more the general structure of a single-phase voltage source inverter. Essentially, it is made up of two single-phase legs as described in Chapter 3, connected to a common DC bus. Each phase leg is modulated in a complementary pattern by a carrier/reference waveform comparison circuit, which switches the phase leg to the upper DC rail when the reference waveform is greater than the carrier, and to the lower DC rail when the carrier waveform is greater than the reference waveform. The particular form of the carrier and reference waveforms depends on the PWM strategy that is implemented.

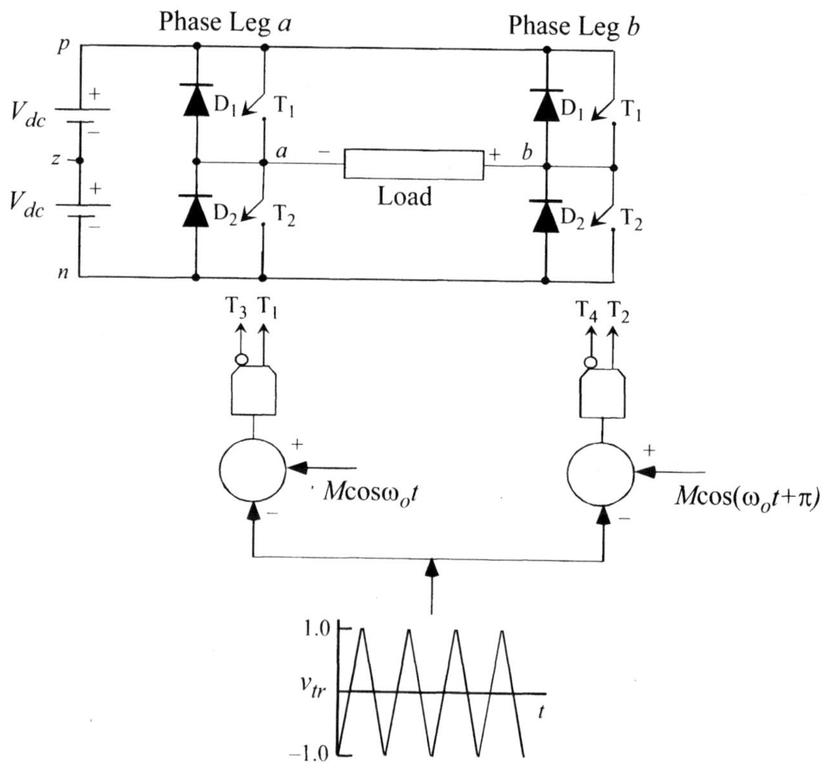

Figure 4.1 Single-phase full-bridge (H–bridge) voltage source inverter.

4.2 Three-Level Modulation of a Single-Phase Inverter

In the example shown in Figure 4.1, both phase legs use a common carrier which happens to be triangular. The two phase legs are modulated with 180° opposed reference waveforms, defined as

$$v_{az}^* = V_{dc} M \cos(\omega_o t) \tag{4.1}$$

$$v_{bz}^* = V_{dc} M \cos(\omega_o t - \pi) \tag{4.2}$$

where

M = modulation index (i.e., the normalized output voltage magnitude) with the range $0 < M < 1$

ω_o = target output angular frequency

and each phase leg reference waveform is referenced to the DC bus voltage zero midpoint.

The fundamental line-to-line (*l–l*) output voltage reference for the inverter is the difference between the two-phase leg reference voltages, and is given by

$$v_{ab}^* = v_{az}^* - v_{bz}^* = 2V_{dc} M \cos(\omega_o t) \tag{4.3}$$

This arrangement achieves three-level naturally sampled sine-triangle PWM, which has significant harmonic advantages over most other single-phase inverter modulation strategies. But it is by no means mandatory to use a common carrier for both phase legs, or even two sinusoidal reference waveforms. Many other variations have been proposed over the years with very different carriers and reference waveforms, and some have advantages in specific applications. Some of these alternatives are considered later in this chapter, but for now the analysis will be confined to continuous PWM strategies with common carrier waveforms.

Figure 4.2 shows the three-level naturally sampled PWM process for a single-phase inverter, where it can be seen how each phase leg of the inverter switches between the upper and the lower DC rails continuously over the fundamental cycle as the carrier waveform ramps above and below the reference waveform. Note that the switched *l–l* output voltage takes on values of $+2V_{dc}$ and zero during the positive period of the reference and $-2V_{dc}$ and zero during the negative period of the reference. This is why this PWM arrangement is

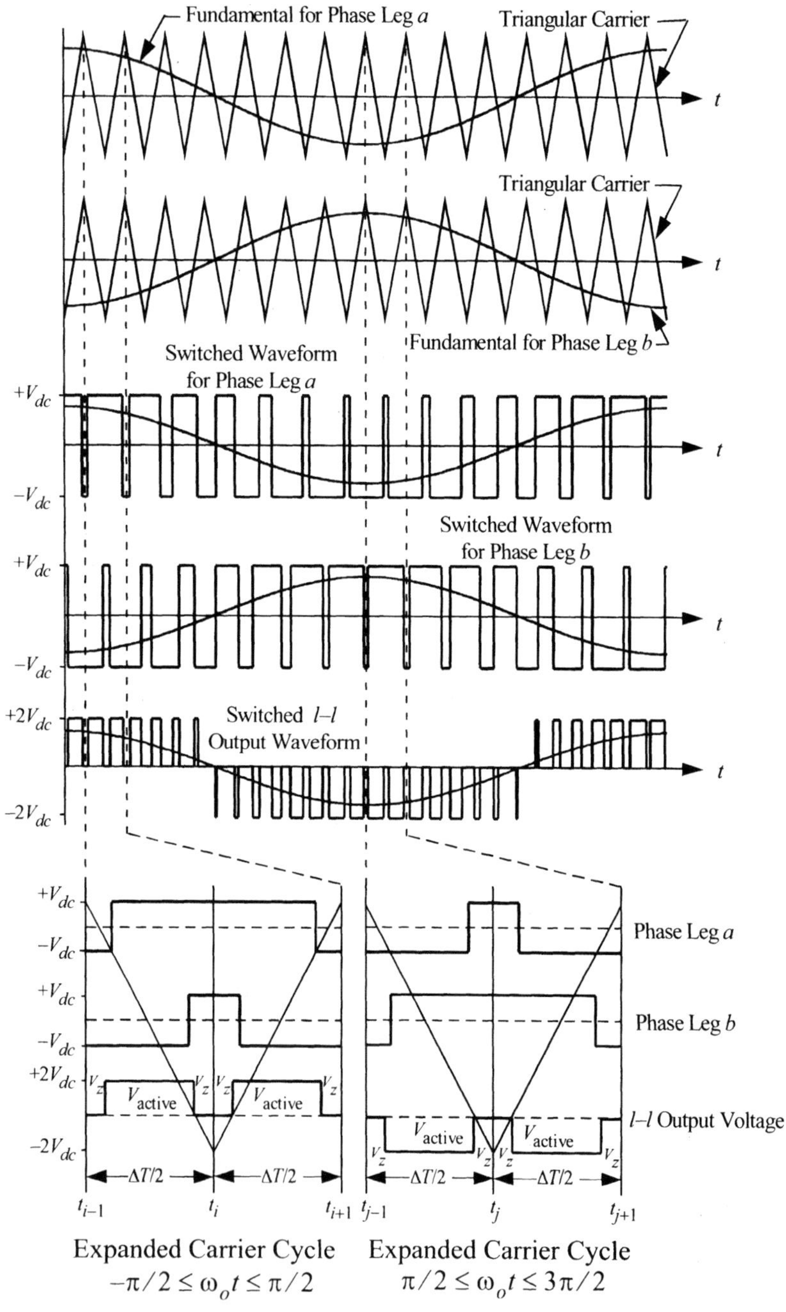

Figure 4.2 Three-level naturally sampled sine-triangle PWM process for single-phase VSI.

called three-level modulation, since the inverter output voltage switches between three levels over a complete fundamental cycle.

The switching process is illustrated in more detail in the expanded carrier period shown at the bottom of Figure 4.2. Looking at this figure, it can be seen that the switched *l–l* output voltage consists of zero voltage magnitude regions (V_Z) at the start and the end of each half-carrier period, and active voltage pulses of magnitude $\pm 2V_{dc}$ (V_{active}) which are approximately centered within the half-carrier period. It is significant to observe that the repetition frequency of the active *l–l* output pulses is twice the triangular carrier frequency (f_c) where ΔT is the period of one cycle of the triangular carrier.

The harmonic solution for *double-edge naturally sampled* PWM of a phase leg has already been established in Chapter 3 as Eq. (3.39). This solution is immediately applicable to the phase legs of the single-phase inverter by setting $\theta_o = 0$ and $\theta_o = -\pi$ for phase legs *a* and *b*, respectively, i.e.,

$$v_{az}(t) = V_{dc} + V_{dc}M\cos(\omega_o t) \qquad (4.4)$$

$$+ \frac{4V_{dc}}{\pi} \sum_{m=1}^{\infty} \sum_{n=-\infty}^{\infty} \frac{1}{m} J_n\left(m\frac{\pi}{2}M\right) \sin\left([m+n]\frac{\pi}{2}\right) \cos(m\omega_c t + n\omega_o t)$$

and

$$v_{bz}(t) = V_{dc} + V_{dc}M\cos(\omega_o t - \pi) \qquad (4.5)$$

$$+ \frac{4V_{dc}}{\pi} \sum_{m=1}^{\infty} \sum_{n=-\infty}^{\infty} \frac{1}{m} J_n\left(m\frac{\pi}{2}M\right) \sin\left([m+n]\frac{\pi}{2}\right) \cos(m\omega_c t + n[\omega_o t - \pi])$$

Note that the arbitrary carrier phase angle θ_c in both cases has been set to zero for convenience since it is the same for both phase legs. In addition, the main carrier harmonic terms, generated when $m > 0$, $n = 0$, have been incorporated into the carrier sideband terms for simplicity (this substitution must be considered individually for the analytical solution of each form of modulation, since it is not generally applicable).

The *l–l* output voltage harmonic components for the inverter are given by $v_{ab} = v_{an} - v_{bn}$ and can be readily developed from Eqs. (4.4) and (4.5) as

$$v_{ab}(t) = 2V_{dc}M\cos(\omega_o t) + \frac{8V_{dc}}{\pi}\sum_{m=1}^{\infty}\sum_{n=-\infty}^{\infty}\frac{1}{2m}J_{2n-1}(m\pi M)\cos([m+n-1]\pi) \times \cos(2m\omega_c t + [2n-1]\omega_o t) \tag{4.6}$$

Equation (4.6) shows that the odd carrier and associated sideband harmonics are completely cancelled from the *l–l* output voltage pulse train, leaving only odd sideband harmonic $(2n-1)$ terms of the even $(2m)$ carrier groups. Mathematically, this cancellation can be explained by recalling from Section 3.4.2 that when the m summation index in Eqs. (4.4) and (4.5) is *odd* (i.e., odd carrier frequency multiples), the n index must be *even* to create a non-zero $\{\sin([m+n]\pi/2)\}$ term in the last summation expression of both equations. But if n is even, the $\{\cos(m\omega_c t + n[\omega_o t - \pi])\}$ term in Eq. (4.5) creates the same harmonic components as in Eq. (4.4). These harmonic components cancel when the two phase leg voltages are subtracted, so that the *l–l* voltage can have no odd carrier harmonic or associated sideband harmonic components. Note that the indices m and n have consequently been modified in Eq. (4.6) so as to produce only *even* carrier multiples $(2m)$ with *odd* sideband harmonics $(2n-1)$ (recalling also from Section 3.4.2 that *even* harmonic sidebands are eliminated around the *even* carrier multiples by virtue of the PWM process within the phase legs).

Physically, the cancellation of the odd carrier harmonics can be explained by referring again to the expanded switching detail in Figure 4.2, where it was previously identified that there are two active output switched pulses for each triangular carrier cycle. Clearly the switching frequency of the *l–l* output is twice the carrier frequency, and thus suppression of the sideband harmonics around the odd carrier multiples is to be expected.

The theoretical phase leg a voltage harmonics are shown in Figure 4.3(a), together with the *l–l* output voltage harmonics in Figure 4.3(b), for the particular operating conditions of a carrier ratio of 21 and a modulation index M of 0.9. The cancellation of the odd carrier multiples and their associated sideband harmonics can be clearly seen. Note that all *l–l* harmonics are normalized with respect to $2V_{dc}$ to make a direct comparison with the phase leg harmonics.

Exactly the same approach can now be used to determine the *l–l* switched output voltage harmonic components for all of the combinations of carrier and

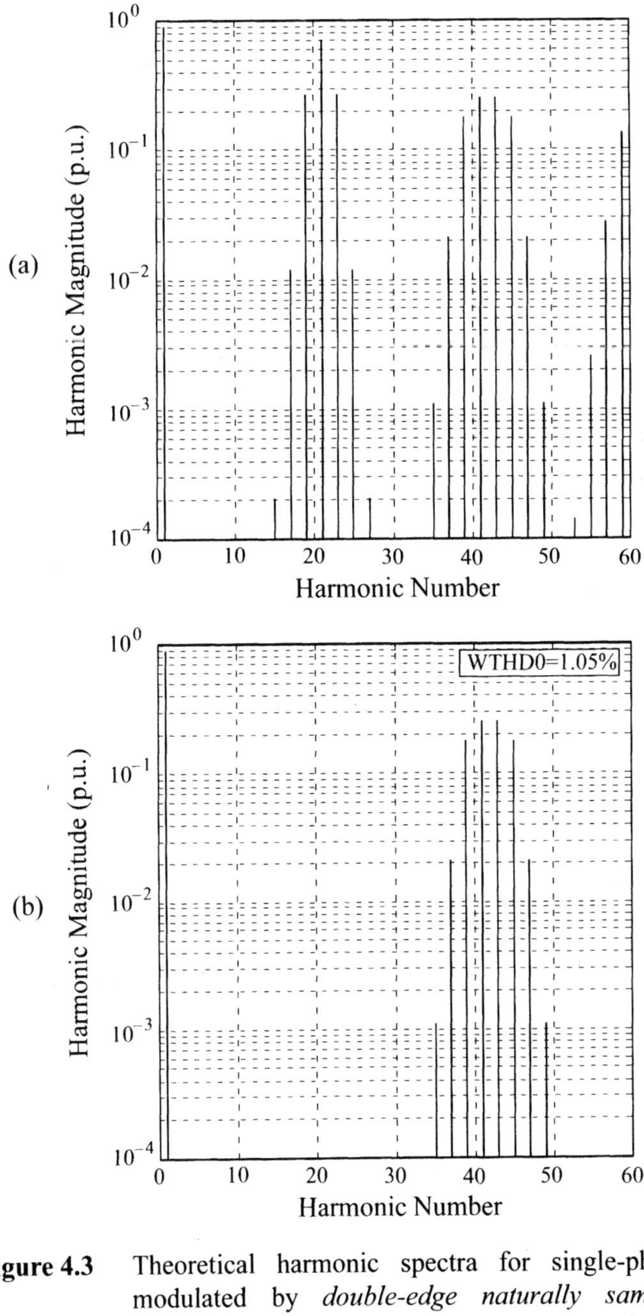

Figure 4.3 Theoretical harmonic spectra for single-phase inverter modulated by *double-edge naturally sampled* PWM: (a) phase leg a and (b) l–l output switched voltage waveforms, $M = 0.9$, $f_c/f_o = 21$.

sampling alternatives that were explored in Chapter 3. In fact, the following combinations are possible, all within the definition of three-level continuous PWM:

- Naturally sampled reference, single-edge carrier — Eq. (3.26).
- Naturally sampled reference, double-edge carrier (see above) — Eq. (3.39).
- Regularly sampled reference, single-edge carrier — Eq. (3.70).
- Symmetrical regular sampled reference, double-edge carrier — Eq. (3.78).
- Asymmetrical regular sampled reference, double-edge carrier — Eq. (3.98).

Analytical solutions for all these alternatives are

Naturally sampled reference, single-edge carrier:

$$v_{ab}(t) = 2V_{dc}M\cos(\omega_o t) + \frac{4V_{dc}}{\pi}\sum_{m=1}^{\infty}\sum_{\substack{n=-\infty \\ (n\neq 0)}}^{\infty}\frac{1}{m}J_n(m\pi M)\sin n\frac{\pi}{2}\cos(m\omega_c t + n\omega_o t) \tag{4.7}$$

Naturally sampled reference, double-edge carrier (repeated for completeness):

$$v_{ab}(t) = 2V_{dc}M\cos(\omega_o t) + \frac{8V_{dc}}{\pi}\sum_{m=1}^{\infty}\sum_{n=-\infty}^{\infty}\frac{1}{2m}J_{2n-1}(m\pi M)\cos([m+n-1]\pi) \times \cos(2m\omega_c t + [2n-1]\omega_o t) \tag{4.8}$$

Regular sampled reference, single-edge carrier:

$$v_{ab}(t) = \frac{4V_{dc}}{\pi}\sum_{n=1}^{\infty}\frac{1}{\left[n\frac{\omega_o}{\omega_c}\right]}J_n\left(n\frac{\omega_o}{\omega_c}\pi M\right)\sin n\frac{\pi}{2}\cos(n\omega_o t) \tag{4.9}$$

$$+\frac{4V_{dc}}{\pi}\sum_{m=1}^{\infty}\sum_{\substack{n=-\infty \\ (n\neq 0)}}^{\infty}\frac{1}{\left[m+n\frac{\omega_o}{\omega_c}\right]}J_n\left(\left[m+n\frac{\omega_o}{\omega_c}\right]\pi M\right)\sin n\frac{\pi}{2} \times \cos(m\omega_c t + n\omega_o t)$$

Symmetrical regular sampled reference, double-edge carrier:

$$v_{ab}(t) = \frac{8V_{dc}}{\pi}\sum_{n=1}^{\infty}\frac{J_n\left(n\frac{\omega_o}{\omega_c}\frac{\pi}{2}M\right)}{\left[n\frac{\omega_o}{\omega_c}\right]}\sin\left(n\left[1+\frac{\omega_o}{\omega_c}\right]\frac{\pi}{2}\right)\left|\sin n\frac{\pi}{2}\right|\cos(n\omega_o t) \tag{4.10}$$

$$+\frac{8V_{dc}}{\pi}\sum_{m=1}^{\infty}\sum_{\substack{n=-\infty \\ (n\neq 0)}}^{\infty}\frac{J_n\left(\left[m+n\frac{\omega_o}{\omega_c}\right]\frac{\pi}{2}M\right)}{\left[m+n\frac{\omega_o}{\omega_c}\right]}\sin\left(\left[m+n\frac{\omega_o}{\omega_c}+n\right]\frac{\pi}{2}\right)$$
$$\times\left|\sin n\frac{\pi}{2}\right|\cos(m\omega_c t+n\omega_o t)$$

Asymmetrical regular sampled reference, double-edge carrier:

$$v_{ab}(t) = \frac{8V_{dc}}{\pi}\sum_{n=1}^{\infty}\frac{1}{\left[n\frac{\omega_o}{\omega_c}\right]}J_n\left(n\frac{\omega_o}{\omega_c}\frac{\pi}{2}M\right)\sin n\frac{\pi}{2}\cos(n\omega_o t) \tag{4.11}$$

$$+\frac{8V_{dc}}{\pi}\sum_{m=1}^{\infty}\sum_{n=-\infty}^{\infty}\frac{1}{q'}J_{2n-1}\left(q'\frac{\pi}{2}M\right)\cos([m+n-1]\pi)$$
$$\times\cos(2m\omega_c t+[2n-1]\omega_o t)$$

where q' is defined as $q' = 2m+[2n-1]\omega_o/\omega_c$ and need not be an integer.

Spectral plots for phase leg *a* and the *l–l* voltages are shown for all these combinations of carrier and sampling alternatives on the following pages.

From these results, it can be seen that all harmonics where *n* is *even* are cancelled between phase legs for all carrier/sampling combinations. But what is interesting is that this apparently minor effect has a large influence on the resulting performance of the various modulation strategies. With a single-edge carrier or with symmetrical regular sampling, where the modulation process produces both *odd* and *even* sideband harmonics around each carrier multiple, significant *odd* sideband harmonics remain in the inverter output waveform around the *odd* carrier multiples despite harmonic cancellation between the phase legs. With a double-edge carrier, the effect for both natural sampling and asymmetrical regular sampling is to totally cancel the sideband harmonics around the *odd* carrier multiples from the *l–l* waveform. Clearly these latter two modulation alternatives are superior as a consequence.

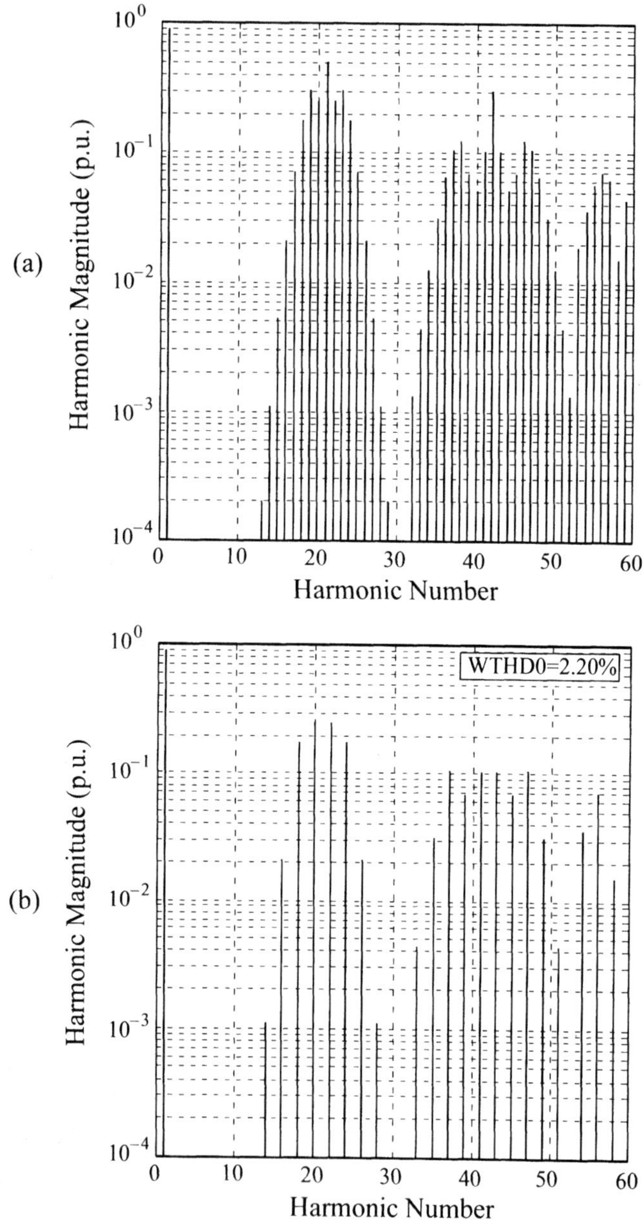

Figure 4.4 Theoretical harmonic spectra for single-phase inverter modulated by *trailing single-edge naturally sampled* PWM: (a) phase leg a and (b) l–l output switched voltage waveforms, $M = 0.9$, $f_c/f_o = 21$.

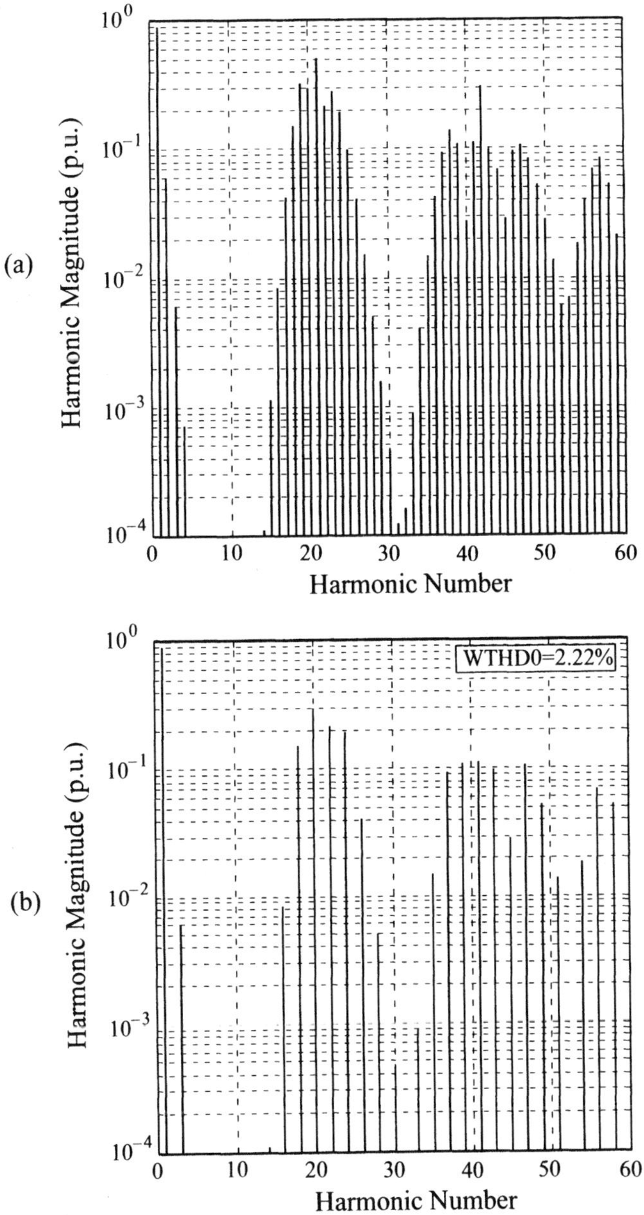

Figure 4.5 Theoretical harmonic spectra for single-phase inverter modulated by *trailing single-edge regular sampled* PWM: (a) phase leg a and (b) *l–l* output switched voltage waveforms, $M = 0.9$, $f_c/f_o = 21$.

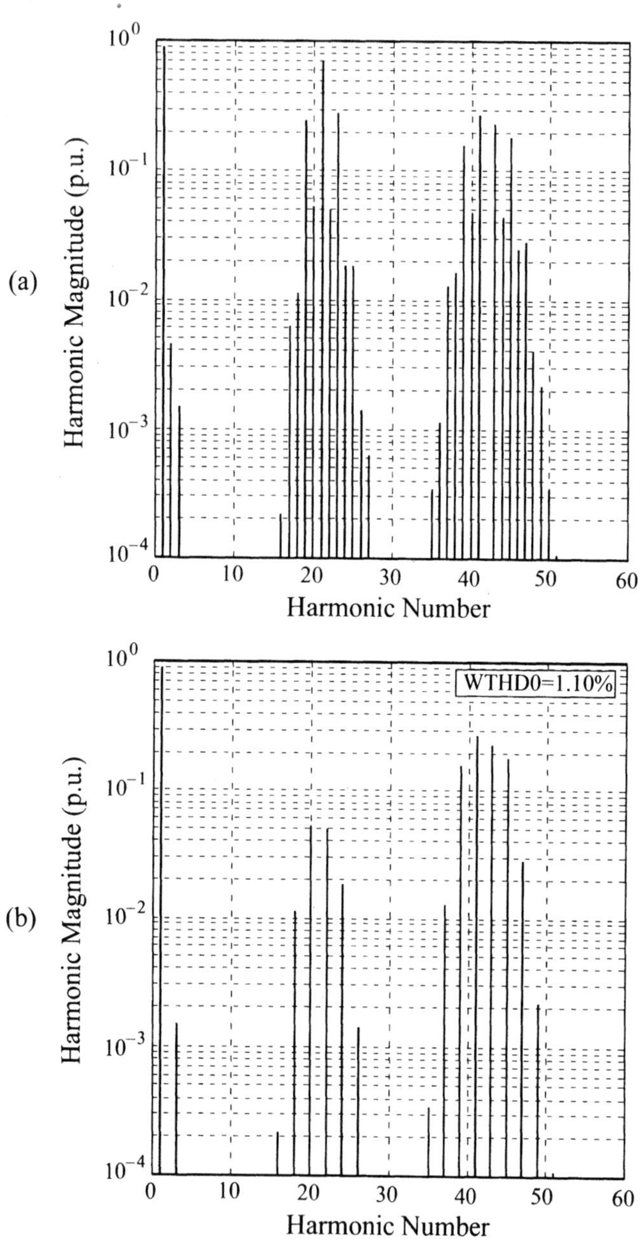

Figure 4.6 Theoretical harmonic spectra for single-phase inverter modulated by *symmetrical double-edge regular sampled* PWM: (a) phase leg a and (b) l–l output switched voltage waveforms, $M = 0.9$, $f_c/f_o = 21$.

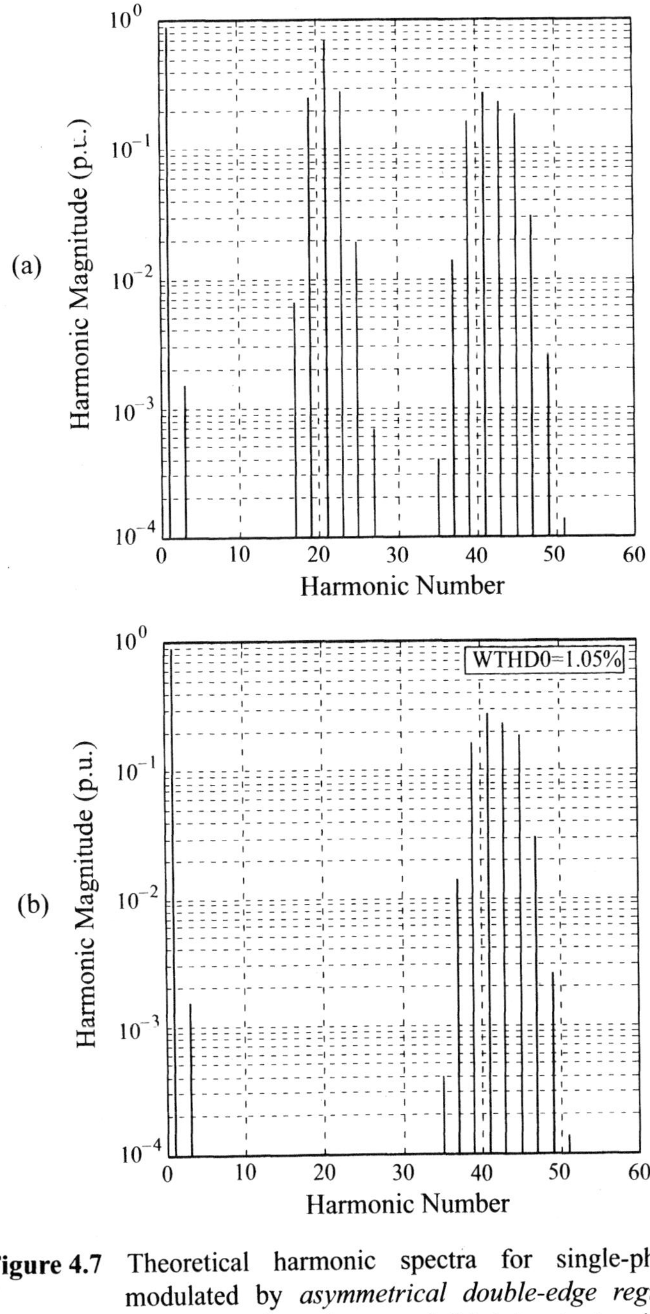

Figure 4.7 Theoretical harmonic spectra for single-phase inverter modulated by *asymmetrical double-edge regular sampled* PWM: (a) phase leg a and (b) l–l output switched voltage waveforms, $M = 0.9$, $f_c/f_o = 21$.

It should be observed also that for both symmetrical and asymmetrical regular sampling, only *odd* baseband harmonics remain in the *l–l* output voltage, since the *even* baseband harmonics produced by the symmetrical regular sampling process cancel between the phase legs. But the distinction between this cancellation and the elimination of the even harmonic with *asymmetrical* regular sampled PWM is important. Cancellation of harmonics between phase legs depends on precise modulation control implementation. The elimination of the *even* harmonics by asymmetrical regular sampling is intrinsic to the nature of the asymmetrical modulation process itself and is not so dependent on the accuracy of the modulation implementation.

It is interesting to reflect also on the physical mechanism which leads to symmetrical modulation retaining sideband harmonics around the *odd* carrier multiples despite harmonic cancellation between the phase legs. Essentially, while symmetrical sampling produces two *l–l* output pulses per carrier cycle in the same way as natural sampling, the width of these pulses is only varied once per carrier cycle. Hence some harmonic influence must be retained at all carrier multiples to reflect this sampling process. Asymmetrical sampling reassesses the switched pulse widths at twice the carrier frequency, and so does not require any sideband harmonics around the *odd* carrier multiples. This is an intrinsic limitation of *symmetrical* sampled modulation compared to *asymmetrical* modulation as identified by Boys and Handley [1].

The cancellation of the main carrier harmonics themselves between the phase legs also substantially reduces the overall WTHD of the *l–l* switched output voltage, as shown in Figure 4.8. Note that the results shown in Figure 4.8 have been deliberately presented at the very low carrier pulse ratio of 11, to highlight any differences between the various carrier/sampling alternatives. But even in this extreme case, there is essentially no difference between double-edge naturally sampled and double-edge asymmetrically regular sampled PWM. Double-edge symmetrical sampling has a little more WTHD at higher values of M, and single-edge naturally and regularly sampled PWM produce significantly more switching harmonics in general. The same trend applies for any carrier pulse ratio; but, of course, the differences between the PWM strategies become less significant at higher carrier frequencies. However, the influence of the baseband harmonics caused by sampling processes is still negligible for triangular carrier modulation strategies, even with the very low carrier ratio selected for this example.

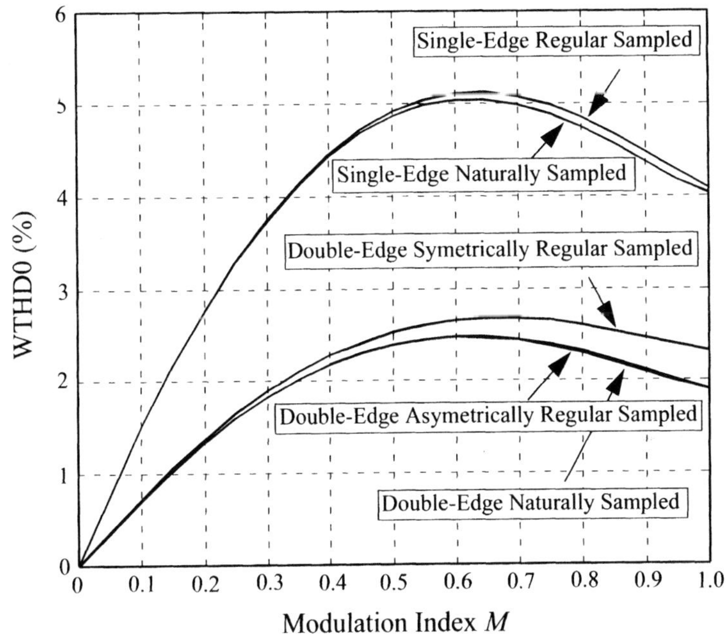

Figure 4.8 WTHD0 of *l–l* switched output voltage of a single-phase inverter for various carrier and sampling alternatives, $f_c/f_o = 11$.

4.3 Analytic Calculation of Harmonic Losses

Since the half-bridge circuit forms the basic building block for many inverter configurations, it is useful to derive closed-form solutions for the harmonic currents and consequential losses of this circuit, in order to form a basis for comparison against and between more advanced modulation methods [2].

To obtain such closed-form solutions, it is reasonable to make the assumption that the average internal EMF e_{az} of the load, measured with respect to the DC bus midpoint over an arbitrary switching interval ΔT is essentially constant. The load current can then be expected to have a high-frequency AC ripple component at the switching frequency $f_c = 1/\Delta T$. It can be further assumed that the losses in the load, while nontrivial, can be neglected for the purposes of calculating the ripple current. This approach is a good approximation for high-efficiency loads such as transformers and electrical machines.

With such an effectively inductive load, the load current will rise and fall linearly about its target value, driven by the difference between the instantaneous applied voltage from the phase leg as it switches and the internal EMF of load. This is shown in Figure 4.9, where the quantity Δi_a is the current ripple, i.e., the difference between the actual current and the reference current.

Over the interval T_1 the rise in current ripple is given by

$$\Delta i_a(t) = \frac{(V_{dc} - e_{az})}{L_\sigma} t = \frac{V_{dc}}{L_\sigma}(1-u)t \qquad \text{for } 0 \le t \le T_1 \tag{4.12}$$

where for convenience $u = e_{az}/V_{dc}$ and where L_σ corresponds to the effective inductance between the phase leg switched output and the load internal EMF. In a transformer or an induction motor this inductance corresponds to

$$L_\sigma = L_1 + \frac{L_2 L_m}{L_2 + L_m} \tag{4.13}$$

where L_1, L_2, and L_m are the primary leakage (or stator leakage), secondary leakage (or rotor leakage), and magnetizing inductances of the transformer (or induction motor) load, respectively.

At the end of T_1, the current ripple will have reached a peak value of

$$\Delta i_a(T_1) = \frac{(V_{dc} - e_{az})}{L_\sigma} T_1 = \frac{V_{dc}}{L_\sigma}(1-u)T_1 \tag{4.14}$$

During the second interval T_2, the fall in current ripple is given by

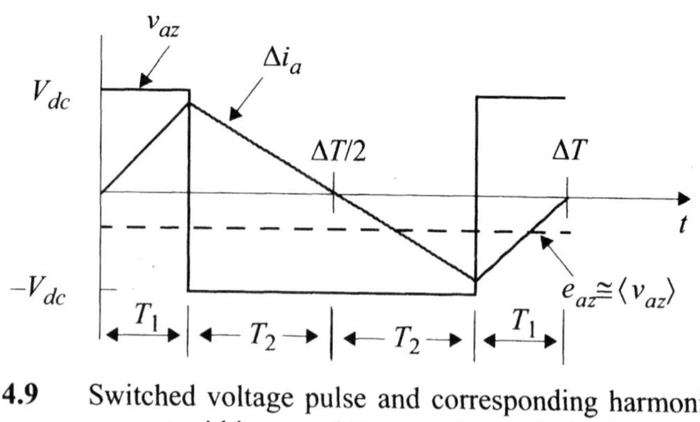

Figure 4.9 Switched voltage pulse and corresponding harmonic ripple current within an arbitrary pulse period of a single-phase half-bridge with an inductive load.

$$\Delta i_a(t) = \frac{(V_{dc} - e_{az})}{L_\sigma} T_1 - \frac{(V_{dc} + e_{az})}{L_\sigma}(t - T_1)$$

$$= \frac{V_{dc}}{L_\sigma}[(1 - u)T_1 - (1 + u)(t - T_1)] \tag{4.15}$$

for $T_1 \le t \le T_1 + T_2$.

At time $T_1 + T_2$, the current ripple will have returned to zero, so that

$$\frac{V_{dc}}{L_\sigma}[(1 - u)T_1 - (1 + u)T_2] = 0 \tag{4.16}$$

Equation (4.16) together with the knowledge that $T_1 + T_2 = \Delta T/2$, can be used to solve for T_1 and T_2 in terms of ΔT, with the result of

$$T_1 = (1 + u)\frac{\Delta T}{4} \qquad T_2 = (1 - u)\frac{\Delta T}{4} \tag{4.17}$$

The RMS copper losses due to switching over the interval $\Delta T/2$ are proportional to the average of the square of the current during this time. Thus, an expression proportional to the losses over this interval can be written as

$$\langle \Delta i_a^2 \rangle = \left(\frac{V_{dc}}{L_\sigma}\right)^2 \frac{2}{\Delta T}\left\{ \int_0^{T_1} (1 - u)^2 t^2 dt + \int_{T_1}^{T_1 + T_2} [(1 - u)T_1 - (1 + u)(t - T_1)]^2 dt \right\} \tag{4.18}$$

Equation (4.18) can be simplified by changing the integrating variable to $x = t - T_1$, so that the loss expression becomes

$$\langle \Delta i_a^2 \rangle = \left(\frac{V_{dc}}{L_\sigma}\right)^2 \frac{2}{\Delta T}\left\{ \int_0^{T_1} (1 - u)^2 t^2 dt + \int_0^{T_2} [(1 - u)T_1 - (1 + u)x]^2 dx \right\} \tag{4.19}$$

which reduces to

$$\langle \Delta i_a^2 \rangle = \left(\frac{V_{dc}}{L_\sigma}\right)^2 (1 - u^2)^2 \frac{\Delta T^2}{48} \tag{4.20}$$

The RMS value of the harmonic current ripple over any PWM interval ΔT is therefore

$$I_h = \sqrt{\langle \Delta i_a^2 \rangle} = \frac{V_{dc}}{L_\sigma}(1-u^2)\frac{\Delta T}{4\sqrt{3}} \tag{4.21}$$

since the contribution over each of the two half periods $\Delta T/2$ is the same.

The instantaneous level of harmonic current ripple varies over the complete fundamental cycle and is worst when the voltage ratio u crosses through zero (i.e., when the EMF of the load passes through zero). Figure 4.10 shows the variation in the harmonic current over a complete fundamental cycle for a modulation index of $M = 0.9$ with a highly inductive load.

It is of interest to also determine the accumulative effect of the harmonic current over one complete cycle. The harmonic power dissipated in the load is proportional to the equivalent load resistance R_e times the square of the harmonic current. Thus integrating Eq. (4.20) over a fundamental cycle gives

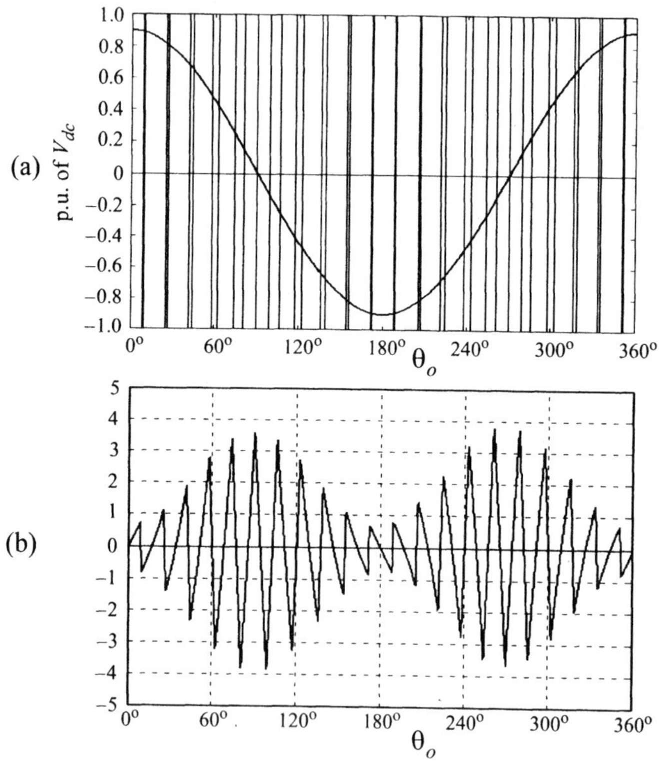

Figure 4.10 (a) Reference and switched voltages for single-phase leg (voltages referred to DC bus centerpoint) and (b) harmonic current ripple with inductive load, $M = 0.9$, $f_c/f_o = 21$.

$$P_{h,\,cu} = R_e I_{h,\,ave}^2 = R_e\left(\frac{V_{dc}}{L_\sigma}\right)^2 \frac{\Delta T^2}{48}\frac{1}{2\pi}\int_0^{2\pi}[1-u^2]^2 d\theta_o \quad (4.22)$$

In general, the average value of the applied voltage over a switching instant equals the internal EMF uV_{dc} plus the fundamental component of the reactive voltage drop across the internal inductance L_σ. For an ideal inductance, this voltage drop will be in quadrature to the load EMF, so that the applied voltage will have the same magnitude as the internal EMF (but of course phase shifted to some extent). So in terms of the overall harmonic losses, the EMF term u in Eq. (4.22) can simply be replaced by the instantaneous applied p.u. voltage at any point in the fundamental cycle, $M\cos\theta_o$, and the loss expression becomes

$$P_{h,\,cu} = R_e\left(\frac{V_{dc}}{L_\sigma}\right)^2 \frac{\Delta T^2}{48}\frac{1}{2\pi}\int_0^{2\pi}[1-M^2\cos^2\theta_o]^2 d\theta_o \quad (4.23)$$

By repeated use of the trigonometric identity, $\cos^2\varphi = (1+\cos 2\varphi)/2$, Eq. (4.23) can be expanded to an expression containing only a constant and a part containing cosine terms. Since the integral is evaluated over a complete cycle, the part involving cosine terms is zero, and the constant becomes

$$P_{h,\,cu} = R_e I_{h,\,ave}^2 = R_e\left(\frac{V_{dc}}{L_\sigma}\right)^2 \frac{\Delta T^2}{48}\left[1 - M^2 + \frac{3}{8}M^4\right] \quad (4.24)$$

The function $f(M) = 1 - M^2 + \frac{3}{8}M^4$ is plotted in Figure 4.11 and illustrates how the overall harmonic losses vary with modulation index.

Figure 4.12 shows the ripple current over one carrier interval for the more realistic case of the complete single-phase inverter of Figure 4.1, controlled by the modulation process illustrated in Figure 4.2. Note that in this case four switching events occur over the modulation interval, rather than the two that occur for the single-phase leg. As a consequence, the ripple frequency is doubled for the same ΔT. Also, the ripple current is now driven by a voltage that switches between zero and $\pm 2V_{dc}$ rather than $+V_{dc}$ and $-V_{dc}$ as before.

Figure 4.13 shows the variation in the harmonic current for a single-phase inverter over a fundamental cycle for a modulation index of $M = 0.9$ with a highly inductive load. The difference between this current ripple and that of a single-phase leg shown in Figure 4.10 can be clearly seen.

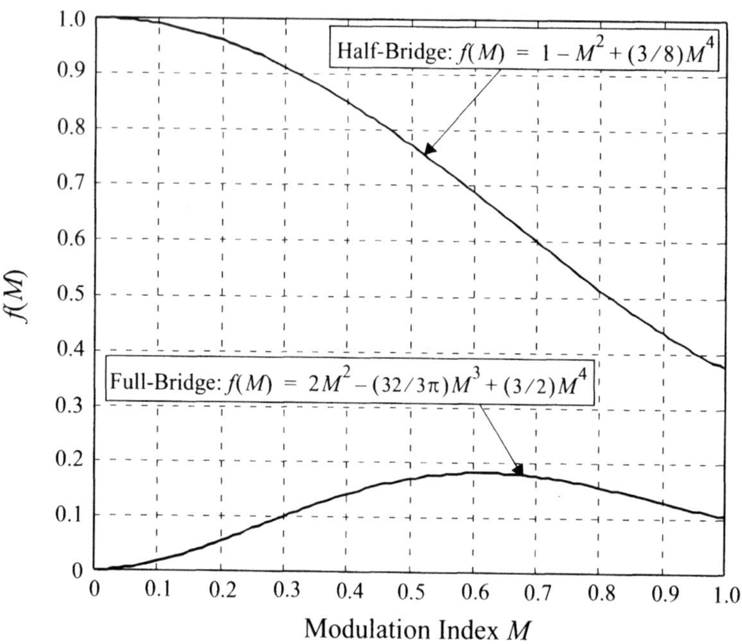

Figure 4.11 Normalized harmonic copper losses as a function of modulation index M for pure inductive load.

However, the harmonic losses as a function of modulation index for the complete single-phase inverter can still be solved in the same manner, as follows:

For $0 \le t \le T_1$

$$\Delta i_{ab}(t) = -\frac{e_{ab}}{L_\sigma}t \tag{4.25}$$

$$\Delta i_{ab}(T_1) = -\frac{e_{ab}}{L_\sigma}T_1 \tag{4.26}$$

and when $T_1 \le t \le T_1 + T_2$ (for the positive half fundamental cycle)

$$\Delta i_{ab}(t) = \left(\frac{2V_{dc} - e_{ab}}{L_\sigma}\right)(t - T_1) - \frac{e_{ab}}{L_\sigma}T_1 \tag{4.27}$$

$$\Delta i_{ab}(T_1 + T_2) = \left(\frac{2V_{dc} - e_{ab}}{L_\sigma}\right)T_2 - \frac{e_{ab}}{L_\sigma}T_1 = 0 \tag{4.28}$$

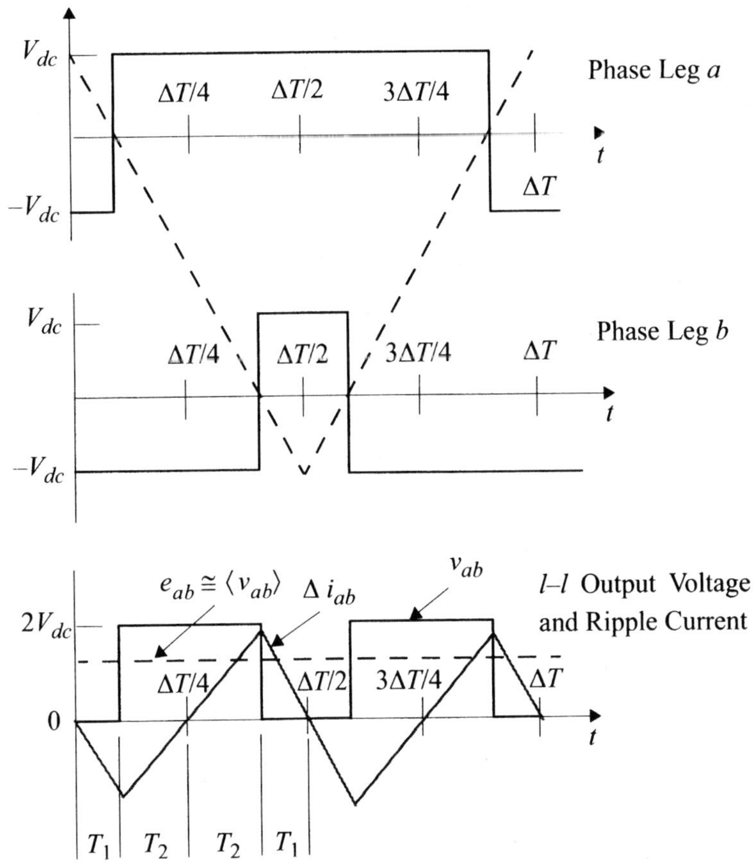

Figure 4.12 Modulation process and resulting ripple current for single-phase full-bridge VSI.

From the relationship $T_1 + T_2 = \Delta T/4$ it is straightforward using Eqs. (4.27) and (4.28) to show that $T_1 = (1-u)\Delta T/4$ and $T_2 = u\Delta T/4$ for the positive half fundamental cycle, where u is now defined as $u = e_{ab}/2V_{dc}$.

Using these relationships, the average squared value of the current ripple over the interval $\Delta T/4$ can be written in the form

$$\langle \Delta i_{ab}^2 \rangle = \left(\frac{2V_{dc}}{L_\sigma}\right)^2 \frac{4}{\Delta T}\left\{ \int_0^{T_1} u^2 t^2 dt + \int_0^{T_2} [(1-u)x - uT_1]^2 dx \right\} \quad (4.29)$$

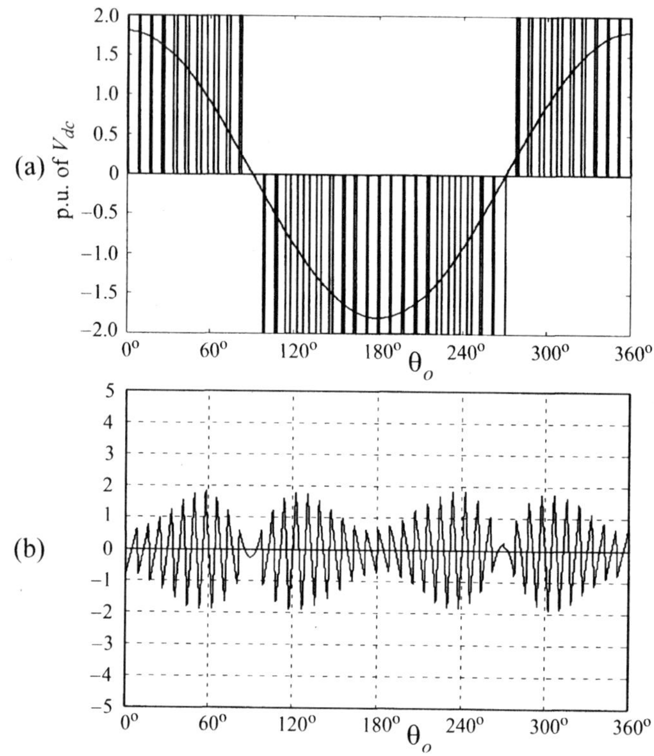

Figure 4.13 (a) Reference and switched voltages for complete single-phase inverter and (b) harmonic current ripple with inductive load, $M = 0.9$, $f_c/f_o = 21$.

where $x = t - T_1$. Equation (4.29) can be evaluated to form

$$\langle \Delta i_{ab}^2 \rangle = \left(\frac{V_{dc}}{L_\sigma}\right)^2 u^2 (1-u)^2 \frac{\Delta T^2}{12} \tag{4.30}$$

The losses due to the harmonic current ripple over a full cycle are therefore

$$P_{h,cu} = R_e \frac{1}{\pi} \int_{-\pi/2}^{\pi/2} \langle \Delta i_{ab}^2 \rangle \, d\theta_o$$

$$= R_e \left(\frac{V_{dc}}{L_\sigma}\right)^2 \frac{\Delta T^2}{12} \frac{1}{\pi} \int_{-\pi/2}^{\pi/2} (M\cos\theta_o)^2 (1 - M\cos\theta_o)^2 \, d\theta_o \tag{4.31}$$

using the same substitution of replacing u with $M\cos\theta_o$, and recognizing from Figure 4.13 that inverter operation in the negative half of the fundamental cycle produces the mirror image current ripple response compared to the positive half cycle, with identical harmonic losses.

By repeated use of standard trigonometric identities, it is again not difficult to show that Eq. (4.31) reduces to

$$P_{h,\,cu} = R_e\left(\frac{V_{dc}}{L_\sigma}\right)^2\frac{\Delta T^2}{48}\left[2M^2 - \frac{32M^3}{3\pi} + \frac{3M^4}{2}\right] \tag{4.32}$$

The square bracketed term in Eq. (4.32) has the same per unit normalized value as in Eq. (4.24) and is also plotted in Figure 4.11. However, a major difference between the two equations is the absence of the constant term in Eq. (4.32) which occurs because of the optimum use of the zero states in the modulation of the full-bridge. Also, while many other modulation algorithms can also be devised, placement of the active pulses in the center of each half period as shown in Figure 4.12 clearly achieves the minimum losses since it minimizes the peak of the current ripple.

Finally, it is important to mention that while Eqs. (4.24) and (4.32) are useful figures of merit, they become less useful at reduced carrier ratios f_c/f_o or at close to unity modulation ($M \approx 1$), since the assumption that the EMF remains constant during a switching interval becomes less accurate. It has been shown, however, that these results remain accurate to within a few percent for $f_c/f_o \geq 9$ [2] and for modulation indices below about $M = 0.95$.

Where precise accuracy is required, the exact harmonic solutions of Eqs. (4.7) to (4.11) can be used under any modulation conditions or carrier pulse ratio to determine the harmonic current ripple on a per harmonic basis.

4.4 Sideband Modulation

Examination of the terms in the spectrum of Figure 4.3(b) shows that the dominant harmonic terms in the spectrum are located at $2m\omega_c + (2n \pm 1)\omega_o$. For the case shown, the most significant of these harmonics is at $(2 \times 21)\omega_o - (2 \times 1 - 1)\omega_o = 41\omega_o$, and it would be beneficial if this term could be reduced or eliminated. Takahashi et. al. [3] have shown that by using a derivative of frequency modulation (FM) theory, the triangular carrier wave can be "dithered" or modulated in such a manner so as to cancel this term.

While difficult to represent in the linear time domain, a solution becomes possible in a time domain that is adjusted to account for the dither effect. In particular let the phase angle of the triangular carrier wave be modulated such that $\mathrm{tri}(\omega_c t')$ denotes a triangular carrier in a modified time domain t' where

$$\omega_c t' = \omega_c t + \alpha_1 \sin(\omega_o t) \tag{4.33}$$

While the two time domains are related by a nonlinear equation, a solution in the linear time domain t can be obtained if there is a one-to-one correspondence between the two times, that is, if there are no "double value" or negative solutions to Eq. (4.33). Consequently a monotonic positively increasing solution is guaranteed under the constraint

$$\frac{dt'}{dt} > 0$$

From Eq. (4.33), this constraint requires that

$$\frac{dt'}{dt} = 1 + \alpha_1 \frac{\omega_o}{\omega_c} \cos(\omega_o t) > 0 \tag{4.34}$$

Therefore the two times continue to increase as long as $\alpha_1 \frac{\omega_o}{\omega_c} < 1$.

In addition, intersections of the triangular carrier and sinusoidal reference must always occur in each carrier interval. This requirement can be satisfied if

$$\left| \frac{d[\mathrm{tri}(\omega_c t')]}{dt'} \right| > \left| \frac{dv_{an}^*}{dt'} \right| \tag{4.35}$$

where v_{an}^* is the target reference voltage in the t' domain, defined by

$$v_{an}^* = 2V_{dc} M \cos[\omega_o t + \alpha_1 \sin(\omega_o t)] \tag{4.36}$$

The slope of the triangular carrier wave is given by

$$\left| \frac{d[\mathrm{tri}(\omega_c t')]}{dt'} \right| = \frac{2V_{dc}\omega_c}{\pi} \tag{4.37}$$

while the slope of the fundamental reference is

$$\left| \frac{dv_{an}^*}{dt'} \right| = \left| \frac{dv_{an}^*}{dt} \frac{dt}{dt'} \right|$$

$$= \frac{-2V_{dc} M \sin[\omega_o t + \alpha_1 \sin\omega_o t][\omega_o + \alpha_1 \omega_o t \ \sin\omega_o t]}{1 + \alpha_1 \frac{\omega_o}{\omega_c} \cos\omega_o t} \tag{4.38}$$

As a worst case condition, let the numerator be as large and the denominator as small as possible for Eq. (4.38), in which case the limiting condition of Eq. (4.35) becomes

$$\frac{2V_{dc}\omega_c}{\pi} > \frac{2V_{dc}M(\omega_o + \alpha_1\omega_o)}{1 - \alpha_1\dfrac{\omega_o}{\omega_c}} \tag{4.39}$$

or solving for α_1,

$$\alpha_1 < \frac{\omega_c - M\pi\omega_o}{\omega_o + M\pi\omega_o} \tag{4.40}$$

The restriction is greatest for large values of M. Setting $M = 1$ gives

$$\alpha_1 < \frac{\omega_c - \pi\omega_o}{\omega_o + \pi\omega_o} \quad \text{or} \quad \alpha_1 < \frac{\omega_c/\omega_o - \pi}{1 + \pi} \tag{4.41}$$

At a typical carrier ratio of say $\omega_c/\omega_o = 21$, $\alpha_1 < 4.31$ rad or ~245°.

Consider now the behavior of the harmonic sidebands corresponding to $m = 1$ in the general solution of Eq. (4.6). These sideband terms have the form

$$v_{ab(m=1)} = \frac{8V_{dc}}{\pi} \sum_{n=-\infty}^{\infty} \frac{1}{2} J_{2n-1}(\pi M)\cos n\pi \, \cos(2\omega_c t + [2n-1]\omega_o t) \tag{4.42}$$

Replacing $\omega_o t$ with $\omega_o t + \alpha_1 \sin\omega_o t$ in this expression gives

$$v_{ab(m=1)} = \frac{4V_{dc}}{\pi} \sum_{n=-\infty}^{\infty} J_{2n-1}(\pi M)\cos n\pi \times \cos(2\omega_c t + [2n-1](\omega_o t + \alpha_1 \sin\omega_o t)) \tag{4.43}$$

The time-varying terms of Eq. (4.43) can be rearranged into the form $\cos(2\omega_c t + [2n-1]\omega_o t + [2n-1]\alpha_1 \sin\omega_o t)$, and then the complete expression can be expanded using Eq. (A2.8) to become

$$v_{ab(m=1)} = \frac{4V_{dc}}{\pi} \sum_{n=-\infty}^{\infty} J_{2n-1}(\pi M)\cos n\pi \times \sum_{k=-\infty}^{\infty} J_k([2n-1]\alpha_1)\cos(2\omega_c t + [2n-1]\omega_o t + k\omega_o t) \tag{4.44}$$

The terms from Eq. (4.44) contributing to the sideband harmonics just above and below the double carrier frequency at $2\omega_c \pm \omega_o$ can be collected by setting $[2n-1]+k = \pm 1$ whereupon $k = -2n; -2n+2$. Similarly the values of k producing harmonic terms at $2\omega_c \pm 3\omega_o$ are $k = -2n-2; -2n+4$ for the lower and upper spectral lines, respectively. For the two sideband harmonics immediately below $2\omega_c$, the precise magnitudes are therefore

$$v_{ab(2\omega_c - \omega_o)} = \frac{4V_{dc}}{\pi} \sum_{n=-\infty}^{\infty} \cos n\pi \, J_{2n-1}(\pi M) \, J_{-2n}([2n-1]\alpha_1) \tag{4.45}$$

$$v_{ab(2\omega_c - 3\omega_o)} = \frac{4V_{dc}}{\pi} \sum_{n=-\infty}^{\infty} \cos n\pi \, J_{2n-1}(\pi M) \, J_{-2n-2}([2n-1]\alpha_1) \tag{4.46}$$

The magnitudes of the two major sideband harmonic components just above $2\omega_c$ at $2\omega_c + \omega_o$ and $2\omega_c + 3\omega_o$ will be the same by virtue of the symmetry in the Bessel functions.

Figure 4.14 shows the variation of the sideband harmonic component defined by Eq. (4.45) (normalized against $2V_{dc}$ since this is the maximum value of the *l–l* output voltage) as a function of α_1 for five values of the modulation index *M*. Since the summation terms roll off rapidly, only values of *n* from –5 to 5 need be considered in the summation term. The corresponding behavior of the next sideband harmonic defined by Eq. (4.46) and located at $2\omega_c - 3\omega_o$ is illustrated in Figure 4.15.

From Figure 4.14 it can be seen that the first sideband harmonic is minimized at all modulation indices for a value of α_1 of about 3.5 rad. On the other hand, Figure 4.15 shows that the second sideband harmonic is minimized for a value of α_1 of about 0.7 rad. Clearly there will an optimum value for α_1 which minimizes the overall harmonic distortion, or WTHD, for the PWM switched waveform, somewhere between these two limits.

For the single-phase inverter, there are only four major harmonics which contribute significantly to the WTHD of the switched output voltage, at frequencies of $2\omega_c \pm \omega_o$ and $2\omega_c \pm 3\omega_o$ as can be identified from Figure 4.3(b). The WTHD contribution from these harmonics can be readily calculated from

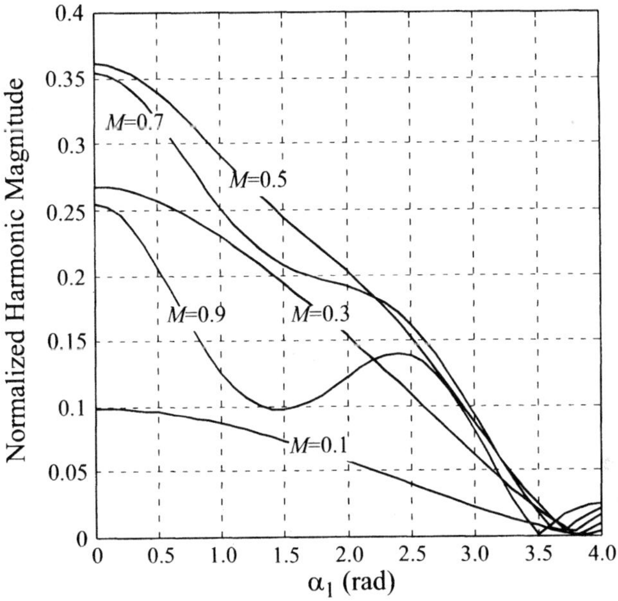

Figure 4.14 Variation of the spectral line $2\omega_c - 3\omega_o$ as a function of phase modulation amplitude α_1 for various modulation levels *M*.

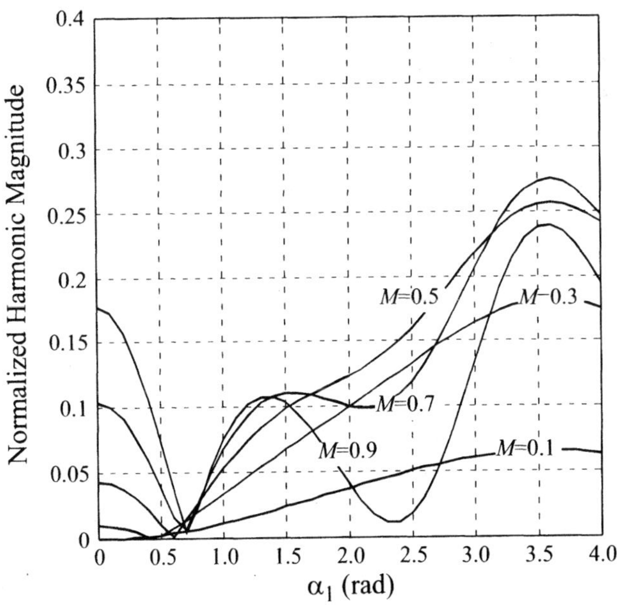

Figure 4.15 Variation of the spectral line $2\omega_c - 3\omega_o$ as a function of phase modulation amplitude α_1 for various modulation levels *M*.

$$\text{WTHD0} = \frac{\omega_o}{2V_{dc}}\sqrt{\left|\frac{v_{(2\omega_c-3\omega_o)}}{2\omega_c-3\omega_o}\right|^2+\left|\frac{v_{(2\omega_c-\omega_o)}}{2\omega_c-\omega_o}\right|^2+\left|\frac{v_{(2\omega_c+\omega_o)}}{2\omega_c+\omega_o}\right|^2+\left|\frac{v_{(2\omega_c+3\omega_o)}}{2\omega_c+3\omega_o}\right|^2} \tag{4.47}$$

where again all harmonic voltages have been normalized with respect to $2V_{dc}$.

The variation of WTHD0 with α_1 is shown in Figure 4.16, with an optimum value for α_1 of about 2.5 rad. From Eq. (4.41), this would be achievable with a carrier ratio of greater than 14, which is quite realistic. Essentially then, phase modulation of the carrier offers the potential to spread harmonic energy between the significant harmonics to an overall improvement in the harmonic distortion.

It should be realized that there is also potential to increase the magnitude of the outer sideband harmonics using this technique, so that they may become significant under some conditions. However, in this case of the single-phase inverter under three-level modulation, the next sideband harmonics at $2\omega_c \pm 5\omega_o$ are so much smaller that there is little risk of this effect occurring.

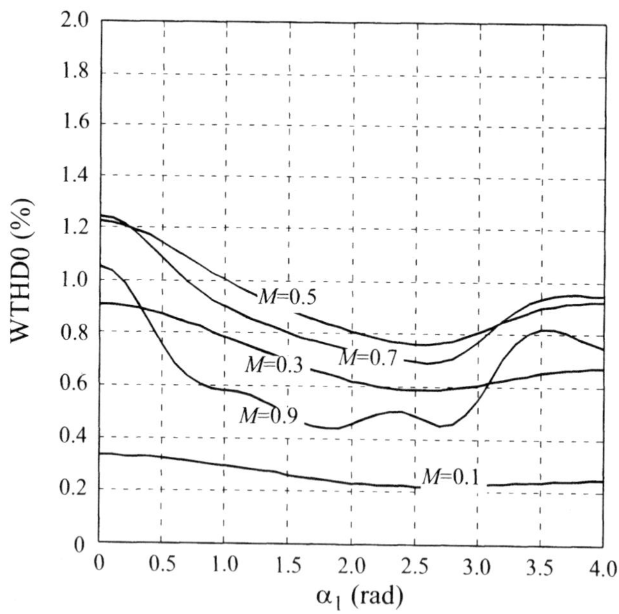

Figure 4.16 WTHD0 for *l–l* output voltage of single-phase inverter with phase modulation amplitude α_1 for various modulation levels M, $f_c/f_o = 21$.

4.5 Switched Pulse Position

Pulse *position* is a crucial factor in determining the harmonic performance of a PWM implementation, particularly for more complex converter topologies. For the single-phase inverter only a few placement variations are possible, but it is still illustrative to see the effects that these alternative placements can have on the harmonics generated. This will provide a useful introduction to the many more alternatives that are possible for pulse position for the three-phase inverters considered in the next chapter. But in all cases, the important issue is that once a pulse width has been determined by the modulation strategy, the placement of the pulse within the carrier or half carrier interval then substantially influences the harmonic performance of the inverter.

The *primary* objective of any PWM strategy is to generate a *l–l* output active pulse within each (half) carrier period which achieves the same fundamental volt-second average as the original target reference waveform over that period. This objective is satisfactorily fulfilled if the active *l–l* output pulse is placed anywhere in each half carrier period, since the position of the pulse within a half carrier period does not affect its cumulative volt-second average over that period. However, the position of the active pulse does significantly influence the harmonic performance of the switched output voltage.

This issue is illustrated in Figure 4.17, which presents the two extreme pulse placements (i.e., pulses centered and pulses placed at the edge of the half carrier period) for a 50% duty cycle switched voltage applied to an inductive load which has an internal EMF equal to half the switched voltage. While the average current is zero in both cases since the switched output volt-second average equals the load EMF, the peak current ripple when the pulses are centered in the half carrier period is clearly less than the peak current ripple when the pulses are positioned back-to-back to span across two half carrier periods. Furthermore, when the switched pulses are centered in each half carrier period, the current ripple frequency is twice that of the back-to-back pulse placement. Since the harmonic losses in the load are proportional to the RMS value of the ripple current, they are obviously also sensitive to the location of the pulses in the half carrier period, and it is not difficult to show that when the duty cycle is 50% the losses in the two extreme positions can differ by up to a factor of 4. A detailed proof of the necessity for centering the pulses in each half carrier period is given in [2].

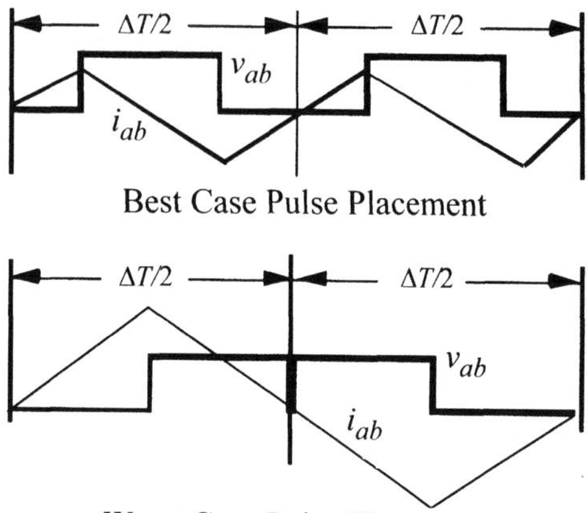

Figure 4.17 Best and worst case placement of *l–l* active voltage pulses within a half carrier period.

The conclusion to be drawn from this simple example is that modulation strategies which place their switched pulses closer to the center of each carrier period have a superior harmonic performance compared to those which do not center the pulses, irrespective of how the switched pulse width is produced. However, while centering the switched pulse in each half carrier period achieves the best harmonic response, the cost is usually that both phase legs must be independently modulated, and this is generally a more expensive implementation than some of the following alternatives. This raises the possibility of changing the position of the active *l–l* pulse within the half carrier period, to see if some implementation benefit is gained.

4.5.1 Continuous Modulation

The natural and regularly sampled PWM schemes discussed in Section 4.2 require both phase legs of the inverter to switch between the upper and lower DC rails at the carrier frequency. This approach is termed *continuous modulation*, since all devices switch continuously throughout the fundamental cycle.

The consequence of the switching patterns created by three-level *double-edge* continuous modulation for a single-phase inverter is that the *l–l* output pulse is implicitly placed approximately in the center of the half carrier period as shown in Figure 4.2. Note that in actuality, for natural sampled PWM, the switched pulse is only approximately centered since the reference wave is not a constant during the triangular carrier period, while for regular sampled PWM, the switched pulse is positioned exactly in the center of each half carrier interval for either symmetrical or asymmetrical sampling because of the sampling effect. It is this slight difference in pulse position within the carrier interval over the fundamental cycle that helps suppress the baseband harmonics of naturally sampled modulation compared to regular sampled modulation.

For regular sampled *single-edge* PWM, geometry dictates that the switched pulse width calculations are the same as for double-edge modulation. For naturally sampled PWM the switched pulse widths will vary slightly between single- and double-edge modulation, and this slightly affects the harmonic performance. But the main difference between single- and double-edge modulation is where the switched pulses are placed within each carrier period.

The analysis presented in Sections 3.4.1 and 3.6.1 show the considerable disadvantage of single-edge modulation compared to double-edge modulation, in that the phase leg switched voltages contain twice as many harmonic components as before. Essentially, the elimination of every second harmonic that occurs with double-edge modulation, achieved by the $\sin(n\pi/2)$, $\sin(m\pi/2)$, and $\sin[(m+n)\pi/2]$ terms in (3.39) and (3.98), no longer occurs. Also, the contribution of the Bessel function coefficient in reducing the sideband harmonics is less as the summation index increases, which means that the magnitude of the harmonics as well as their number is increased compared to double-edge modulation.

Figure 4.4(a) and Figure 4.5(a) particularly show the additional harmonics intrinsically created by the single-edge modulation process, with sideband harmonics occurring at every fundamental frequency interval away from each carrier multiple, rather than at every second fundamental frequency interval as is the case for double-edge modulation. And yet the switched pulse width is virtually the same for each modulation process. Pulse position is clearly a critical factor in determining the performance of a modulation system.

Furthermore the regular sampled single-edge modulation process produces larger baseband harmonic components [Figure 4.5(a)] than does the double-edge process [Figure 4.7(a)], with a particularly significant second harmonic component. While this component is eliminated from the *l–l* output voltage by cancellation between the phase legs for a single-phase inverter, this is not always the case for other topologies such as three-phase inverters.

Of course many of the carrier sideband harmonics are eliminated by cancellation between the phase legs of a single-phase inverter even with single-edge modulation, since all *even* sideband harmonics cancel in the *l–l* switched output waveform. This creates an output voltage harmonic spectrum similar to that of double-edge switching shown in Figure 4.3(b) but centered around the carrier frequency rather than around the double carrier frequency, as shown in Figure 4.4(b). However, it is always better to avoid creating harmonics within the phase leg switching pattern than to rely on accurate waveform generation to cancel harmonics between the phase legs, even disregarding the benefit of the higher effective switching frequency presented to the load by double-edge modulation.

It can also be seen that the uncanceled third harmonic low-order component is still present in Figure 4.5(b). This component is caused by the regular sampled process and does not cancel between phase legs since it is odd.

Single-edge modulation variations which incorporate *discontinuous* switching usually result in a further degradation in spectral response and are not considered further in this text since they are a straightforward extension of the concepts presented below, essentially impractical, and do not add to fundamental understanding.

4.5.2 Discontinuous Modulation

Looking again at Figure 4.2, it is straightforward to express the width of the *l–l* output voltage pulse in each half carrier period under *asymmetrical regular sampled* modulation conditions as

$$T_{l\text{-}l}(i) = \frac{\Delta T}{2} M\cos(\omega_0 t_i) \qquad 0 < |T_{l\text{-}l}(i)| < \frac{\Delta T}{2} \tag{4.48}$$

where the sampling time t_i is set at the start of each half carrier period for convenience and the sign of $T_{l\text{-}l}(i)$ determines the polarity of the output pulse. Once this pulse width is known, its position can be moved within the half carrier period to achieve the same volt-second average over this period, but with alternative switching sequences.

For example, if the two *l–l* output pulses are moved together into the middle of the full carrier period, the center V_z states shown in Figure 4.2 disappear, and it becomes unnecessary to switch both phase legs in each carrier period. This PWM strategy is known as *discontinuous* modulation, where only phase leg a is modulated in the first half of the fundamental cycle, while phase leg b is modulated in the second half of the fundamental cycle (the modulated phase leg swaps as positive and negative polarity l–l output pulses are generated in each half fundamental cycle). It should be recognized that unlike single-edge modulation, both switching edges are still modulated throughout the fundamental cycle, first by one phase leg and then by the other. But the phase legs never switch together within the same carrier period.

Discontinuous modulation can be achieved with either natural sampling using the half cycle reference waveforms for each phase leg shown in Figure 4.18, or by symmetrical or asymmetrical regular sampling with the active pulse intervals calculated using variations of Eq. (4.48) with appropriate sampling points. For all cases, the harmonic performance can be evaluated using the techniques developed in Chapter 3 for continuous modulation, with appropriate adjustments for the changed reference waveforms.

For phase leg a under *discontinuous naturally sampled* modulation, Eq. (3.9) becomes

$$A_{mn} + jB_{mn} = \frac{1}{2\pi^2} \int_{-\pi/2}^{\pi/2} \int_{-\pi M\cos y}^{\pi M\cos y} 2V_{dc} e^{j(mx+ny)} \, dx\, dy \tag{4.49}$$

where the outer integral is only evaluated for half the fundamental period since the switched output is zero with respect to the negative DC rail for the remainder of this period, and the inner integral limits reflect the revised intersection of the fundamental reference with the triangular carrier.

Equation (4.49) can now be evaluated for the various values of m and n.

For $m = n = 0$, Eq. (4.49) simplifies to

$$A_{00} + jB_{00} = \frac{V_{dc}}{\pi^2} \int_{-\pi/2}^{\pi/2} \int_{-\pi M\cos y}^{\pi M\cos y} dx\, dy$$

$$= \frac{V_{dc}}{\pi^2} \int_{-\pi/2}^{\pi/2} 2\pi M\cos y \, dy = \frac{4V_{dc}}{\pi} M \tag{4.50}$$

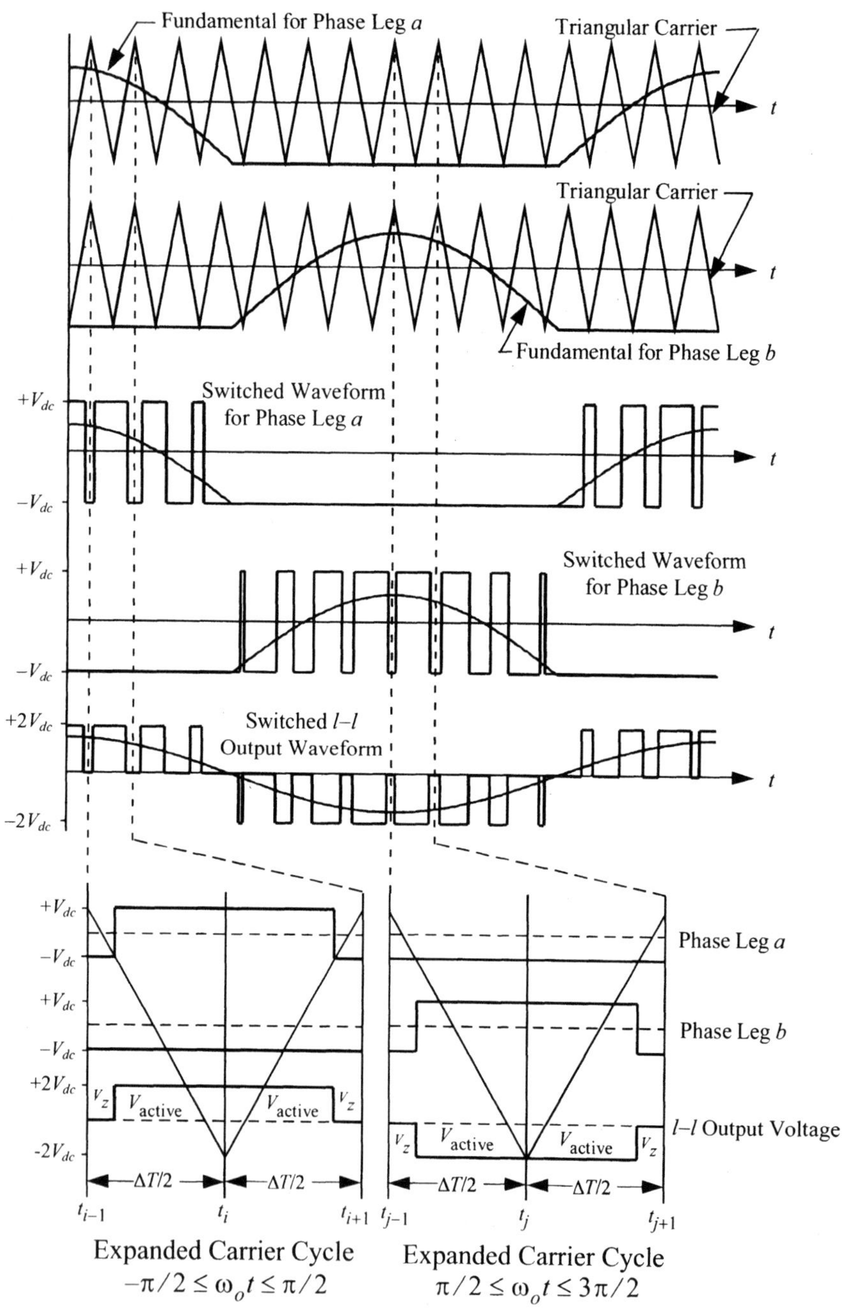

Figure 4.18 Single-phase inverter switching pattern for *discontinuous naturally sampled* PWM – both phase legs modulated.

Equation (4.50) gives the expected DC offset in the final solution caused because the negative DC bus is taken as the switched reference voltage. Unlike continuous modulation, this offset now varies with modulation index.

For $m = 0, n > 0$, Eq. (4.49) simplifies to

$$A_{0n} + jB_{0n} = \frac{V_{dc}}{\pi^2} \int_{-\pi/2}^{\pi/2} \int_{-\pi M\cos y}^{\pi M\cos y} e^{jny}\, dx\, dy$$

$$= \frac{V_{dc}}{\pi^2} \int_{-\pi/2}^{\pi/2} 2\pi M \cos y e^{jny}\, dy$$

$$= \frac{V_{dc}}{\pi} \int_{-\pi/2}^{\pi/2} M(e^{j[n+1]y} + e^{j[n-1]y})\, dy \qquad (4.51)$$

Unlike continuous modulation, Eq. (4.51) does not evaluate to zero for $n \neq 1$. Instead it integrates to

$$A_{0n} + jB_{0n} = \frac{V_{dc}M}{\pi}\left[\left.\frac{e^{j[n+1]y}}{j[n+1]}\right|_{n>0} + \left.\frac{e^{j[n-1]y}}{j[n-1]}\right|_{n>1} + \pi\Big|_{n=1}\right]_{-\pi/2}^{\pi/2}$$

$$= \frac{V_{dc}M}{\pi}\left[\left.\frac{2\sin[n+1]\frac{\pi}{2}}{[n+1]}\right|_{n>0} + \left.\frac{2\sin[n-1]\frac{\pi}{2}}{[n-1]}\right|_{n>1} + \pi\Big|_{n=1}\right]$$

$$= \frac{V_{dc}M}{\pi}\left[\pi\Big|_{n=1} - \left.\frac{4\sin[n+1]\frac{\pi}{2}}{[n+1][n-1]}\right|_{n>1}\right]$$

$$= V_{dc}M\Big|_{n=1} - \frac{4V_{dc}M}{\pi}\left.\frac{\cos n\frac{\pi}{2}}{[n+1][n-1]}\right|_{n>1} \qquad (4.52)$$

The first term of Eq. (4.52) is the same fundamental component as Eq. (3.32) in Chapter 3. The remaining terms represent even order baseband harmonics that are created because the reference waveform is only a half sinusoid, with considerable low-frequency distortion. (These additional harmonics would be expected to cancel between the phase legs to create an undistorted *l–l* output.)

For $m > 0, n = 0$, Eq. (4.49) simplifies to

$$A_{m0} + jB_{m0} = \frac{V_{dc}}{\pi^2} \int_{-\pi/2}^{\pi/2} \int_{-\pi M\cos y}^{\pi M\cos y} e^{jmx} dx\, dy$$

$$= \frac{V_{dc}}{jm\pi^2} \int_{-\pi/2}^{\pi/2} (e^{jm\pi M\cos y} - e^{-jm\pi M\cos y})\, dy$$

$$= \frac{V_{dc}}{m\pi^2} \int_{-\pi/2}^{\pi/2} 2\sin(m\pi M\cos y)\, dy \tag{4.53}$$

Using Eq. (A2.3), this becomes

$$A_{m0} + jB_{m0} = \frac{4V_{dc}}{m\pi^2} \int_{-\pi/2}^{\pi/2} \left[\sum_{k=1}^{\infty} J_k(m\pi M) \sin k\frac{\pi}{2} \cos ky \right] dy$$

$$= \frac{8V_{dc}}{m\pi^2} \sum_{k=1}^{\infty} \frac{J_k(m\pi M)\left(\sin k\frac{\pi}{2}\right)^2}{k}$$

$$= \frac{8V_{dc}}{m\pi^2} \sum_{k=1}^{\infty} \frac{1}{2k-1} J_{2k-1}(m\pi M) \tag{4.54}$$

Equation (4.54) defines the harmonics of the carrier wave itself. Unlike the carrier harmonics for continuous modulation, Eq. (4.54) defines a summation which must be evaluated for each specific value of m. However, in practice, summing over the range $0 \le k \le 10$ is usually sufficient because of the rapid roll-off in magnitude of the Bessel function $J_{2k-1}(m\pi M)$ as k increases.

For $m > 0, n \ne 0$, the inner integral of Eq. (4.49) can be evaluated, again using Eq. (A2.3), to

$$A_{mn} + jB_{mn} = \frac{V_{dc}}{jm\pi^2} \int_{-\pi/2}^{\pi/2} e^{jny} (e^{jm\pi M\cos y} - e^{-jm\pi M\cos y})\, dy$$

$$= \frac{4V_{dc}}{m\pi^2} \int_{-\pi/2}^{\pi/2} e^{jny} \left[\sum_{k=1}^{\infty} J_k(m\pi M) \sin k\frac{\pi}{2} \cos ky \right] dy$$

$$= \frac{2V_{dc}}{m\pi^2} \int_{-\pi/2}^{\pi/2} \left[\sum_{k=1}^{\infty} J_k(m\pi M) \sin k\frac{\pi}{2} (e^{j[n+k]y} + e^{j[n-k]y}) \right] dy \tag{4.55}$$

The outer integral expression then becomes, with some manipulation

$$A_{mn} + jB_{mn} = \frac{4V_{dc}}{m\pi^2} \sum_{k=1}^{\infty} J_k(m\pi M) \sin k\frac{\pi}{2} \left[\left. \frac{\sin[n+k]\frac{\pi}{2}}{[n+k]} \right|_{-n \neq k} + \left. \frac{\sin[n-k]\frac{\pi}{2}}{[n-k]} \right|_{n \neq k} + \left. \frac{\pi}{2} \right|_{\pm n = k} \right]$$

$$= \frac{2V_{dc}}{m\pi} \left[J_n(m\pi M) \sin n\frac{\pi}{2} - \frac{4}{\pi} \sum_{k=1}^{\infty} J_{2k-1}(m\pi M) \left. \frac{[2k-1] \cos n\frac{\pi}{2}}{[n+2k-1][n-2k+1]} \right|_{|n| \neq 2k-1} \right] \tag{4.56}$$

This equation defines the magnitude of the sideband harmonics of the switched waveform, and has two terms which must be added for each particular m and n. The first term has the same form as the sideband harmonic terms for continuous modulation, given by Eq. (3.38). The second term is a summation expression which once more must be evaluated for the values of m and n of interest. As before, summation over the range $0 < k < 10$ is usually sufficient because of the roll-off in the Bessel function magnitude as k increases.

The complete harmonic solution for *discontinuous naturally sampled* modulation of phase leg a can now be formed by substituting the results of Eqs. (4.50), (4.52), (4.55), and (4.56) back into Eq. (3.10), so that the time-varying switched phase leg voltage $v_{an}(t)$ can be expressed in terms of its harmonic components as

$$v_{an}(t) = \frac{2V_{dc}}{\pi}M + V_{dc}M\cos(\omega_o t) - \frac{4V_{dc}}{\pi}M\sum_{n=2}^{\infty}\frac{\cos n\frac{\pi}{2}}{[n+1][n-1]}\cos(n\omega_o t) \tag{4.57}$$

$$+\frac{8V_{dc}}{\pi^2}\sum_{m=1}^{\infty}\frac{1}{m}\sum_{k=1}^{\infty}\frac{1}{2k-1}J_{2k-1}(m\pi M)\cos(m\omega_c t)$$

$$+\frac{2V_{dc}}{\pi}\sum_{m=1}^{\infty}\frac{1}{m}\sum_{n=-\infty}^{\infty}J_n(m\pi M)\sin n\frac{\pi}{2}\cos(m\omega_c t + n\omega_o t)$$

$$-\frac{8V_{dc}}{\pi^2}\sum_{m=1}^{\infty}\frac{1}{m}\sum_{\substack{n=-\infty \\ (n\neq 0)}}^{\infty}\sum_{k=1}^{\infty}J_{2k-1}(m\pi M)\left.\frac{[2k-1]\cos n\frac{\pi}{2}}{[n+2k-1][n-2k+1]}\right|_{|n|\neq 2k-1}$$

$$\times\cos(m\omega_c t + n\omega_o t)$$

The harmonic solution for phase leg b can be found by phase shifting the fundamental reference in Eq. (4.57) by π (i.e., replacing $\omega_o t$ with $\omega_o t - \pi$), to give

$$v_{an}(t) = \frac{2V_{dc}}{\pi}M - V_{dc}M\cos(\omega_o t) - \frac{4V_{dc}}{\pi}M\sum_{n=2}^{\infty}\frac{\cos n\frac{\pi}{2}}{[n+1][n-1]}\cos(n\omega_o t) \tag{4.58}$$

$$+\frac{8V_{dc}}{\pi^2}\sum_{m=1}^{\infty}\frac{1}{m}\sum_{k=1}^{\infty}\frac{1}{2k-1}J_{2k-1}(m\pi M)\cos(m\omega_c t)$$

$$-\frac{2V_{dc}}{\pi}\sum_{m=1}^{\infty}\frac{1}{m}\sum_{n=-\infty}^{\infty}J_n(m\pi M)\sin n\frac{\pi}{2}\cos(m\omega_c t + n\omega_o t)$$

$$-\frac{8V_{dc}}{\pi^2}\sum_{m=1}^{\infty}\frac{1}{m}\sum_{\substack{n=-\infty \\ (n\neq 0)}}^{\infty}\sum_{k=1}^{\infty}J_{2k-1}(m\pi M)\left.\frac{[2k-1]\cos n\frac{\pi}{2}}{[n+2k-1][n-2k+1]}\right|_{|n|\neq 2k-1}$$

$$\times\cos(m\omega_c t + n\omega_o t)$$

Note that all the more complex expressions in Eq. (4.58) do not change sign, since they are only nonzero for even values of n. This, of course, could be anticipated, since it is expected that they will cancel between the phase legs for the *l–l* solution.

The *l–l* output voltage harmonic solution is then found by subtracting Eq. (4.58) from Eq. (4.57) to give

$$v_{ab}(t) = 2V_{dc}M\cos(\omega_o t) + \frac{4V_{dc}}{\pi}\sum_{m=1}^{\infty}\sum_{n=-\infty}^{\infty}\frac{1}{m}J_n(m\pi M)\sin n\frac{\pi}{2}\cos(m\omega_c t + n\omega_o t) \tag{4.59}$$

Equation (4.59) is very similar to the *l–l* harmonic solution for continuous modulation given by Eq. (4.6), except that the carrier sideband harmonics are clustered around multiples of $\omega_c t$ rather than $2\omega_c t$, have a different magnitude, and there are some differences in the sign of the harmonic sidebands. However, discontinuous switching halves the phase leg switching frequency for the same carrier frequency, since each phase leg is only modulated for 50% of the fundamental cycle. Thus the carrier frequency can be doubled and still achieve the same losses, and also similar THD and WTHD performances as for continuous double-edge PWM. This makes the sideband harmonics of Eq. (4.59) identical in magnitude and absolute frequency to those of Eq. (4.6), so that a harmonic performance identical to continuous PWM has been obtained, and no performance benefit has been achieved with discontinuous switching.

Figure 4.19 shows the harmonics of the switched output for phase leg a and the *l–l* output, where it can be seen that the harmonic performance of the *l–l* output is identical to that shown in Figure 4.3 for double-edge continuous naturally sampled PWM. (The carrier ratio for discontinuous modulation has been doubled for this example to achieve the same number of switch transitions for each phase leg over a complete fundamental cycle.)

While no particular benefit has been achieved in this case with discontinuous modulation applied to a single-phase inverter, the concept of combining active switched pulses together to eliminate zero intervals can be used with more complex inverter topologies and modulation strategies, where it can achieve reduced switching frequencies and sometimes harmonic benefits under particular operating conditions. This issue is considered further in Chapter 5.

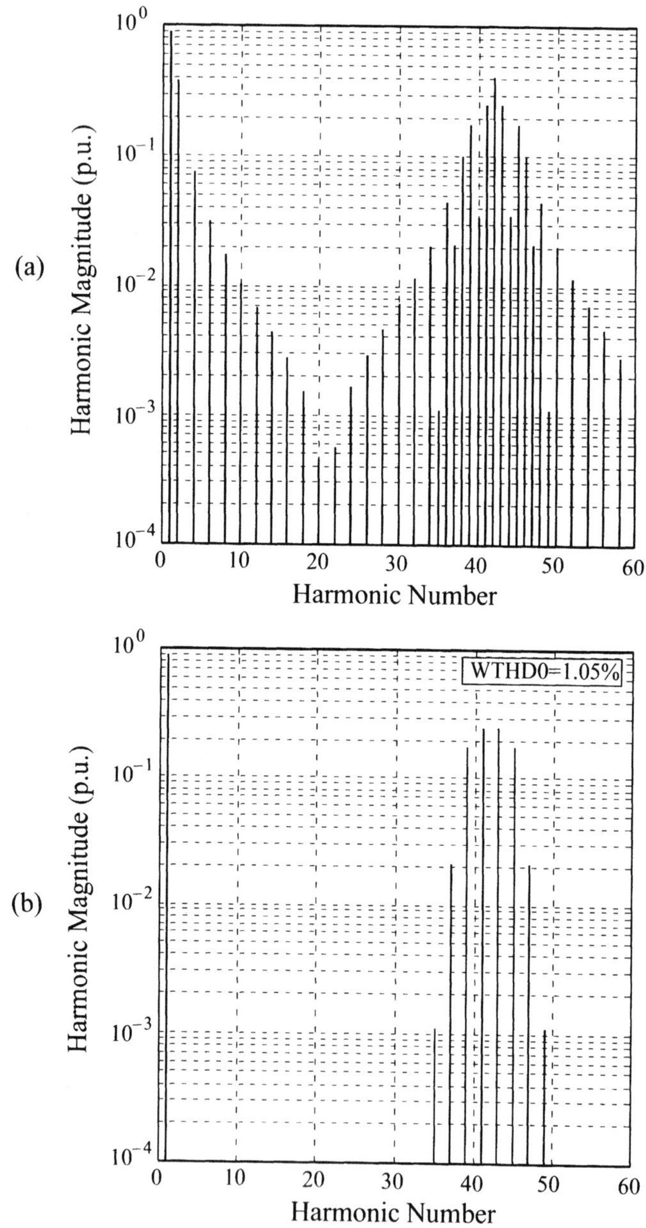

Figure 4.19 Theoretical harmonic spectra for single-phase inverter modulated by *discontinuous naturally sampled* PWM: (a) phase leg a and (b) l–l output switched voltage waveforms, $M = 0.9$, $f_c/f_o = 42$.

Similar analytical solutions for *discontinuous symmetrical* and *asymmetrical regular sampled* modulation can be found using the techniques developed in Chapter 3, by replacing y with $y = y' + (\omega_o/\omega_c)x$ in Eq. (4.49), to give

$$A_{mn} + jB_{mn} = \frac{V_{dc}}{\pi^2} \int_{-\pi/2}^{\pi/2} \int_{-\pi M \cos y'}^{\pi M \cos y'} e^{j\left(\left[m + n\frac{\omega_o}{\omega_c}\right]x + ny'\right)} dx\, dy' \tag{4.60}$$

for symmetrical sampling, and

$$A_{mn} + jB_{mn} = \frac{V_{dc}}{\pi^2} \left[\int_{-\pi/2}^{\pi/2} \int_{-\pi M \cos y_r'}^{0} e^{j\left(mx + n\left[y_r' + \frac{\omega_o}{\omega_c}x + \frac{\omega_o}{\omega_c}\frac{\pi}{2}\right]\right)} dx\, dy_r' + \int_{-\pi/2}^{\pi/2} \int_{0}^{\pi M \cos y_f'} e^{j\left(mx + n\left[y_f' + \frac{\omega_o}{\omega_c}x - \frac{\omega_o}{\omega_c}\frac{\pi}{2}\right]\right)} dx\, dy_f' \right] \tag{4.61}$$

for asymmetrical sampling.

For *discontinuous symmetrical regular sampled* modulation, after some algebraic manipulation and using the substitution of $q = m + n(\omega_o/\omega_c)$, Eq. (4.60) can be solved to yield the solution of Eq. (4.62) for phase leg a.

Examination of this solution yields some familiar terms.

The first line of Eq. (4.62) contains a DC offset term, and a series of baseband harmonics that are very similar to those of the analytical solution for *continuous asymmetrical* regular sampled modulation developed as Eq. (3.98). It should be noted that these harmonics are only odd, unlike those of *continuous symmetrical* regular sampled modulation developed as Eq. (3.78).

Line two of Eq. (4.62) contains a more complex double summation series of even baseband harmonics reflecting the half sinusoid nature of the reference, which will cancel between the two phase legs of the inverter.

Lines three, four, and five of Eq. (4.62) are the same as those of Eq. (4.57), and identify as carrier and complex even sideband harmonics that will cancel between phases, and more conventional odd sideband harmonics that again closely match those of Eq. (3.98) for *continuous asymmetrical* regular sampled modulation.

$$v_{an}(t) = \frac{2V_{dc}}{\pi}M + \frac{2V_{dc}}{\pi}\sum_{n=1}^{\infty}\frac{1}{\left[n\frac{\omega_o}{\omega_c}\right]}J_n\left(n\frac{\omega_o}{\omega_c}\pi M\right)\sin n\frac{\pi}{2}\cos(n\omega_o t) \tag{4.62}$$

$$-\frac{8V_{dc}}{\pi^2}\sum_{n=1}^{\infty}\frac{1}{\left[n\frac{\omega_o}{\omega_c}\right]}\left\{\sum_{k=1}^{\infty}J_{2k-1}\left(n\frac{\omega_o}{\omega_c}\pi M\right)\frac{[2k-1]\cos n\frac{\pi}{2}}{[n+2k-1][n-2k+1]}\Bigg|_{|n|\neq 2k-1}\right\}$$
$$\times\cos(n\omega_o t)$$

$$+\frac{8V_{dc}}{\pi^2}\sum_{m=1}^{\infty}\frac{1}{m}\sum_{k=1}^{\infty}\frac{1}{2k-1}J_{2k-1}(m\pi M)\cos(m\omega_c t)$$

$$+\frac{2V_{dc}}{\pi}\sum_{m=1}^{\infty}\sum_{n=-\infty}^{\infty}\frac{1}{q}J_n(q\pi M)\sin n\frac{\pi}{2}\cos(m\omega_c t + n\omega_o t)$$

$$-\frac{8V_{dc}}{\pi^2}\sum_{m=1}^{\infty}\sum_{\substack{n=-\infty\\(n\neq 0)}}^{\infty}\frac{1}{q}\left\{\sum_{k=1}^{\infty}J_{2k-1}(q\pi M)\frac{[2k-1]\cos n\frac{\pi}{2}}{[n+2k-1][n-2k+1]}\Bigg|_{|n|\neq 2k-1}\right\}$$
$$\times\cos(m\omega_c t + n\omega_o t)$$

As before, the solution for phase leg b can be found by replacing $\omega_o t$ with $(\omega_o t - \pi)$ in Eq. (4.62), and the l–l output voltage harmonic solution is then found by subtracting the phase leg b solution from the phase leg a solution to give

$$v_{ab}(t) = \frac{4V_{dc}}{\pi}\sum_{n=1}^{\infty}\frac{1}{\left[n\frac{\omega_o}{\omega_c}\right]}J_n\left(n\frac{\omega_o}{\omega_c}\pi M\right)\sin n\frac{\pi}{2}\cos(n\omega_o t) \tag{4.63}$$

$$+\frac{4V_{dc}}{\pi}\sum_{m=1}^{\infty}\sum_{n=-\infty}^{\infty}\frac{1}{q}J_n(q\pi M)\sin n\frac{\pi}{2}\cos(m\omega_c t + n\omega_o t)$$

Figure 4.20 shows the phase leg and l–l harmonics for this modulation strategy, with the upper and lower sideband harmonic magnitude skew that is a characteristic of sampled modulation systems, clearly visible in Figure 4.20(b).

(a)

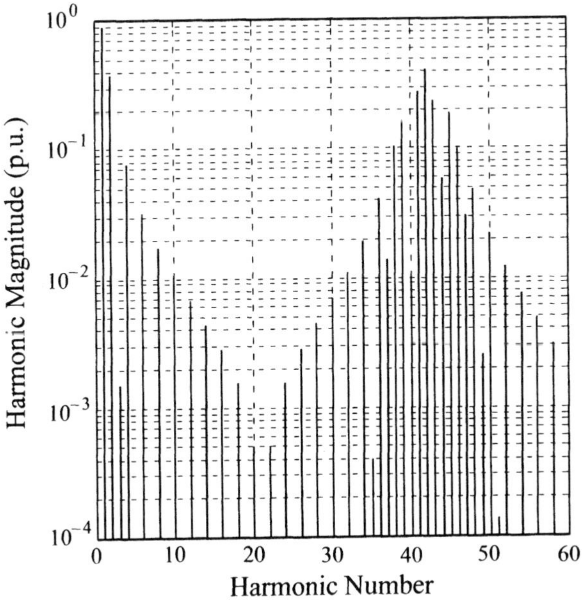

(b)

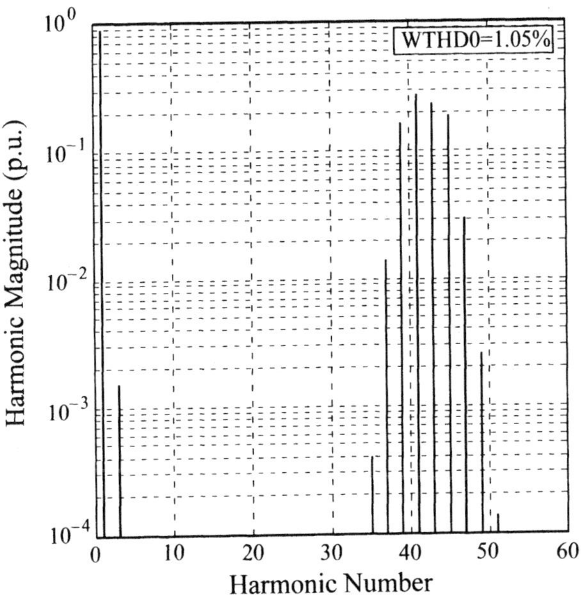

Figure 4.20 Theoretical harmonic spectra for single-phase inverter modulated by *discontinuous symmetrical regularly sampled* PWM: (a) phase leg a and (b) l–l output switched voltage waveforms, $M = 0.9$, $f_c/f_o = 42$.

In fact the *l–l* harmonic performance of this *discontinuous symmetrical* regular sampled modulation exactly matches that of *continuous asymmetrical* regular sampled modulation (Figure 4.7) once the switching frequencies are adjusted to achieve the same average number of phase leg switch transitions over a complete fundamental cycle. Hence it might be argued that for symmetrical modulation, there is a benefit in using discontinuous modulation, since it eliminates the carrier sideband harmonics that are visible in Figure 4.6. But, of course, the effective sampling frequency has increased by two for discontinuous modulation to achieve the same number of switching transitions, and so it is no surprise that the first carrier sidebands now occur at twice the continuous modulation of the first carrier frequency band. Overall, no harmonic performance benefit has been achieved in moving to discontinuous modulation for a single-phase inverter, and in practice some performance degradation might be expected since there are now many more harmonics that need to precisely match between the phase legs so they can exactly cancel on a *l–l* basis.

Similar techniques can be used to evaluate Eq. (4.61) for *discontinuous asymmetrical* regular sampled modulation. However, the solution is not included here because the result presents little additional insight or advantage. In fact, asymmetrical sampling has no particular benefit for discontinuous modulation, as the following explanation illustrates.

For continuous modulation, the *l–l* output pulses of a single-phase inverter occur at twice the carrier frequency, because of the interaction between the phase leg transitions as shown in Figure 4.2. Since asymmetrical sampling separately calculates the rising and falling switching edges of the phase leg transitions, for continuous modulation this sampling strategy effectively recalculates each *l–l* output pulse width individually. In contrast, for discontinuous modulation, since only one phase leg is switching at a time, the *l–l* output pulses occur only at the carrier frequency. (Recall that the carrier frequency is doubled for this case, so that the *l–l* output pulse frequency is the same as for continuous modulation.) Thus symmetrical sampling is sufficient to recalculate each *l–l* output pulse width individually. Asymmetrical sampling in this situation separately calculates each rising and falling edge of the *l–l* output pulses, which achieves negligible harmonic advantage while requiring the modulation system to operate twice as fast again. Figure 4.21 shows the harmonic performance for *discontinuous asymmetrical regular sampled* modulation, with only a slight change in the sideband harmonic magnitudes as anticipated.

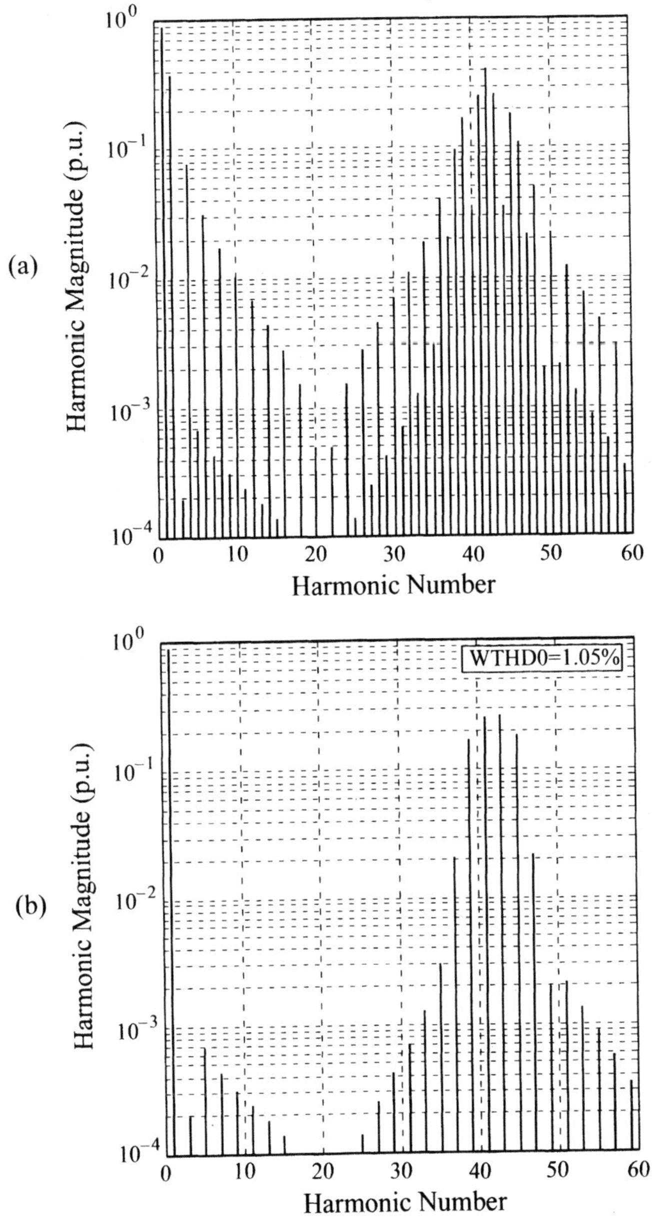

Figure 4.21 Theoretical harmonic spectra for single-phase inverter modulated by *discontinuous* asymmetrical *regularly sampled* PWM: (a) phase leg a and (b) l–l output switched voltage waveforms, $M = 0.9$, $f_c/f_o = 42$.

4.6 Switched Pulse Sequence

Once the switched pulse *width* and *position* within the carrier have been determined, a final parameter to be determined is the pulse *position* sequence across multiple carrier periods (the pulse *width* to be used across multiple carrier periods is determined directly by the modulation process). Again, this parameter has more significance for the modulation of more complex converter topologies, but it does have some relevance for single-phase inverter modulation.

For example, another perspective of double-edge modulation is to consider that it makes each phase leg switching sequence "invert" from one half carrier period to the next. For example, as shown in Figure 4.2, each phase leg starts low and switches high in the first half carrier period, and then starts high and switches low in the next half carrier period. This alternating switch sequence is the prime benefit of continuously switched double-edge modulation, since it leads to higher frequency harmonic components for the same device switching frequency.

For the single-phase inverter, the *l–l* switched output pulse produced by double-edge modulation is identical in each half carrier period, and the reversed switching sequence cannot be detected in the output switched waveform. But conceptually, the *l–l* switched output pulse can be considered to be reversed by double-edged modulation for every half carrier period, with the start becoming the finish in the next period, and the finish becoming the start. While this reversal has no significance for continuous modulation of a single-phase inverter, it is important for the harmonic performance of modulation strategies of more complex inverter topologies, as discussed in Chapter 5.

For discontinuous PWM as shown in Figure 4.18, the effect of pulse sequence is more obvious, since the active *l–l* output pulse is placed alternatively at the start and the end of each successive half carrier period. The result is a single (wider) pulse placed across the half carrier period boundary, but this wider pulse can alternatively be viewed as two half carrier pulses which happen to adjoin at the half carrier boundary. The implications of this issue for discontinuous *single-phase leg switched* modulation will now be considered.

4.6.1 Discontinuous PWM — Single-Phase Leg Switched

The discontinuous PWM variations presented in Section 4.5.2 require both phase legs to be modulated and demand a certain level of complexity for the controller implementation. Sometimes, a more cost-effective approach may be to modulate phase leg a only, and to simply switch phase leg b between the negative DC rail and the positive DC rail for each half fundamental cycle, as shown in Figure 4.22. It can be seen that for this switching arrangement, the interval of time that phase leg a is connected to the negative rail is the inactive zero interval V_z during the first half of the fundamental cycle, and the l–l output pulse interval V_{active} during the second half of the fundamental cycle. Figure 4.22 also shows the inverted sinusoidal reference waveform required for phase leg a during the negative half cycle of the reference that is required to achieve this modulation strategy.

It should be further noted that the switching arrangement shown in Figure 4.22 means that the modulation of phase leg a over each carrier interval starts with the phase leg connected to the negative DC rail. This is a common constraint for hardware timer-based PWM implementations and is highlighted here to illustrate how alternative pulse sequence arrangements can occur perhaps unintentionally with a particular PWM hardware design. Note also the two additional switch transitions required for both phase legs as the fundamental reference changes sign, which contribute nothing to the l–l switched output.

A particular consequence of this arrangement is that for the fundamental region of $\pi/2 \le \omega_o t \le 3\pi/2$, the active l–l pulse is split into two segments at each side of the carrier period with each active segment joining that of the adjoining carrier period to make a complete active pulse. This effectively causes a half carrier shift in the active pulse placement for each half of the fundamental cycle as shown in Figure 4.22(b).

The harmonic analysis of the complete switching pattern of phase leg a can be evaluated by considering it as two trains of pulses, during the intervals $-\pi/2 \le \omega_o t \le \pi/2$ and $\pi/2 \le \omega_o t \le 3\pi/2$, respectively. The analytical solution for the first train of pulses is immediately identifiable as Eq. (4.57), since the phase leg a reference waveform is the same during this interval as for two phase leg switched discontinuous modulation. For the second interval, when the fundamental target is negative, Eq. (4.49) becomes

$$A_{mn} + jB_{mn} = \frac{V_{dc}/2}{\pi^2} \int_{\pi/2}^{3\pi/2} \int_{-\pi - \pi M\cos y}^{\pi + \pi M\cos y} e^{j(mx+ny)}\, dx\, dy \qquad (4.64)$$

and can be evaluated, resulting in

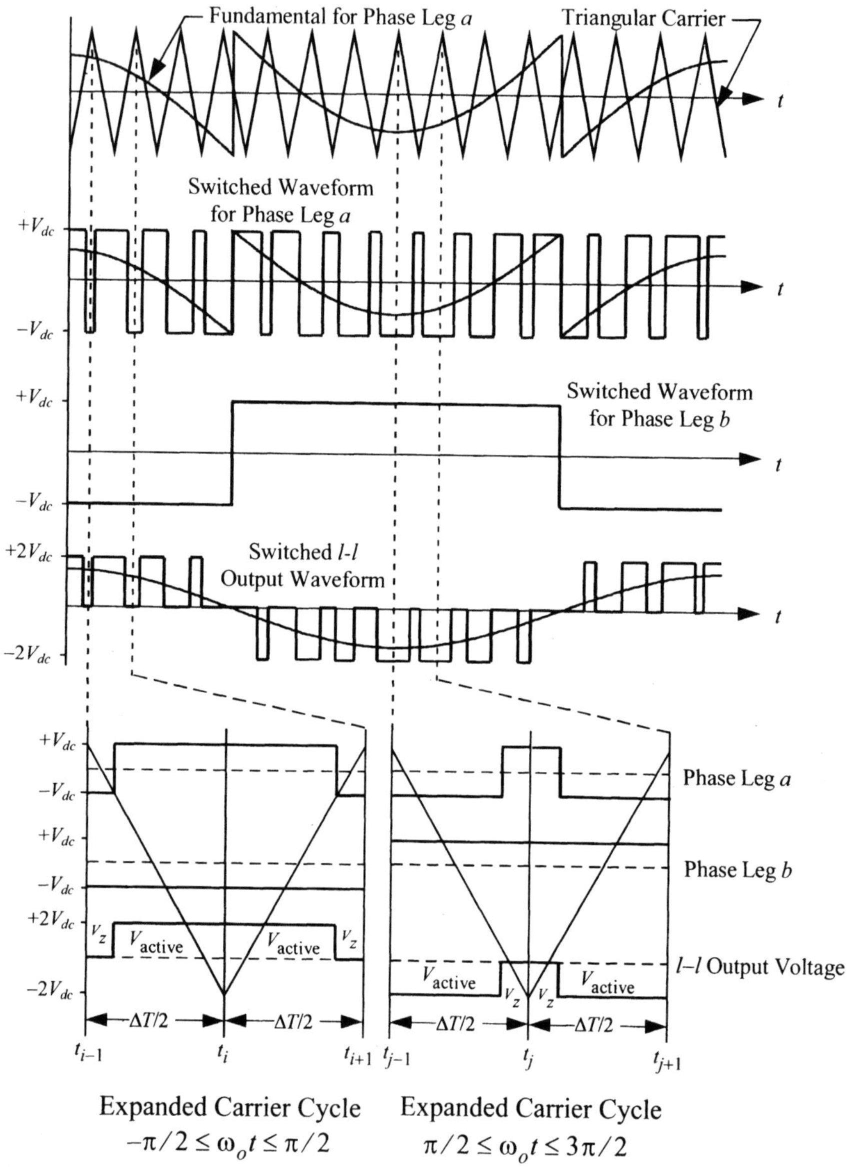

Figure 4.22 Single-phase inverter switching pattern for *discontinuous naturally sampled* PWM — phase leg a only modulated.

$$v_{an(part)}(t) = V_{dc} - \frac{2V_{dc}}{\pi}M + V_{dc}M\cos(\omega_o t) - \frac{4V_{dc}}{\pi}\sum_{n=1}^{\infty}\frac{1}{n}\sin n\frac{\pi}{2}\cos(n\omega_o t) \tag{4.65}$$

$$+ \frac{4V_{dc}}{\pi}M\sum_{n=2}^{\infty}\frac{\cos n\frac{\pi}{2}}{[n+1][n-1]}\cos(n\omega_o t)$$

$$- \frac{8V_{dc}}{\pi^2}\sum_{m-1}^{\infty}\frac{e^{jm\pi}}{m}\sum_{k=1}^{\infty}\frac{1}{2k-1}J_{2k-1}(m\pi M)\cos(m\omega_c t)$$

$$+ \frac{2V_{dc}}{\pi}\sum_{m=1}^{\infty}\sum_{n=-\infty}^{\infty}\frac{e^{jm\pi}}{m}J_n(m\pi M)\sin n\frac{\pi}{2}\cos(m\omega_c t + n\omega_o t)$$

$$+ \frac{8V_{dc}}{\pi^2}\sum_{m=1}^{\infty}\sum_{\substack{n=-\infty \\ (n\neq 0)}}^{\infty}\frac{e^{jm\pi}}{m}\left\{\sum_{k=1}^{\infty}J_{2k-1}(m\pi M)\left.\frac{[2k-1]\cos n\frac{\pi}{2}}{[n+2k-1][n-2k+1]}\right|_{|n|\neq 2k-1}\right\} \times \cos(m\omega_c t + n\omega_o t)$$

Summing Eqs. (4.57) and (4.65) gives the complete harmonic solution for phase leg a of

$$v_{an}(t) = V_{dc} + 2V_{dc}M\cos(\omega_o t) - \frac{4V_{dc}}{\pi}\sum_{n=1}^{\infty}\frac{1}{n}\sin n\frac{\pi}{2}\cos(n\omega_o t) \tag{4.66}$$

$$+ \frac{8V_{dc}}{\pi^2}\sum_{m=1}^{\infty}\frac{1-e^{jm\pi}}{m}\sum_{k=1}^{\infty}\frac{1}{2k-1}J_{2k-1}(m\pi M)\cos(m\omega_c t)$$

$$+ \frac{2V_{dc}}{\pi}\sum_{m=1}^{\infty}\sum_{n=-\infty}^{\infty}\frac{1+e^{jm\pi}}{m}J_n(m\pi M)\sin n\frac{\pi}{2}\cos(m\omega_c t + n\omega_o t)$$

$$- \frac{8V_{dc}}{\pi^2}\sum_{m=1}^{\infty}\sum_{\substack{n=-\infty \\ (n\neq 0)}}^{\infty}\frac{1-e^{jm\pi}}{m}\left\{\sum_{k=1}^{\infty}J_{2k-1}(m\pi M)\left.\frac{[2k-1]\cos n\frac{\pi}{2}}{[n+2k-1][n-2k+1]}\right|_{|n|\neq 2k-1}\right\} \times \cos(m\omega_c t + n\omega_o t)$$

The switched waveform of phase leg b is a simple square wave at the fundamental reference frequency, which is easily shown to have a harmonic solution of

$$v_{bn}(t) = V_{dc} - \frac{4V_{dc}}{\pi}\sum_{n=1}^{\infty}\frac{1}{n}\sin n\frac{\pi}{2}\cos(n\omega_o t) \tag{4.67}$$

taking into account the phase shift of the square wave.

As before, the l–l output voltage is found by subtracting phase leg b from phase leg a, i.e., Eq. (4.67) from Eq. (4.66). This clearly cancels the square-wave harmonics of Eq. (4.67), but the result equally clearly does not match Eq. (4.59), the l–l output voltage when both phase legs are discontinuously modulated. Figure 4.23 shows the harmonic performance of single-phase leg switched modulation, where it can be seen that there is considerable difference in the harmonic pattern of the switched l–l output waveform, Figure 4.23(c), compared to Figure 4.19(b). It should be noted also that in this arrangement the carrier frequency cannot usually be doubled, since phase leg a is always modulated and switches at the carrier frequency. In effect, the switching losses of both phase legs have been pushed onto phase leg a only, and the increased carrier frequency benefit of discontinuous modulation often cannot be achieved. However, in the interests of a direct harmonic comparison between single- and two-phase leg discontinuous modulation, the carrier frequency has still been doubled for the results presented here.

The reason that discontinuous modulation with only one phase leg switched does not create the same harmonics as when two phase legs are switched is because of the change in position of the active l–l output pulse for each half of the fundamental cycle. Figure 4.24 shows how this can be resolved, by inverting the carrier waveform during the second half of the fundamental cycle when the alternative phase leg a reference waveform is active.

The effect of this apparently trivial change is to change the formulation of the harmonic solution during the negative half of the fundamental cycle from Eq. (4.64) to

$$A_{mn} + jB_{mn} = \frac{V_{dc}}{\pi^2}\int_{\pi/2}^{3\pi/2}\left[\int_{-\pi}^{\pi M\cos y} e^{j(mx+ny)}\,dx + \int_{-\pi M\cos y}^{\pi} e^{j(mx+ny)}\,dx\right]dy \tag{4.68}$$

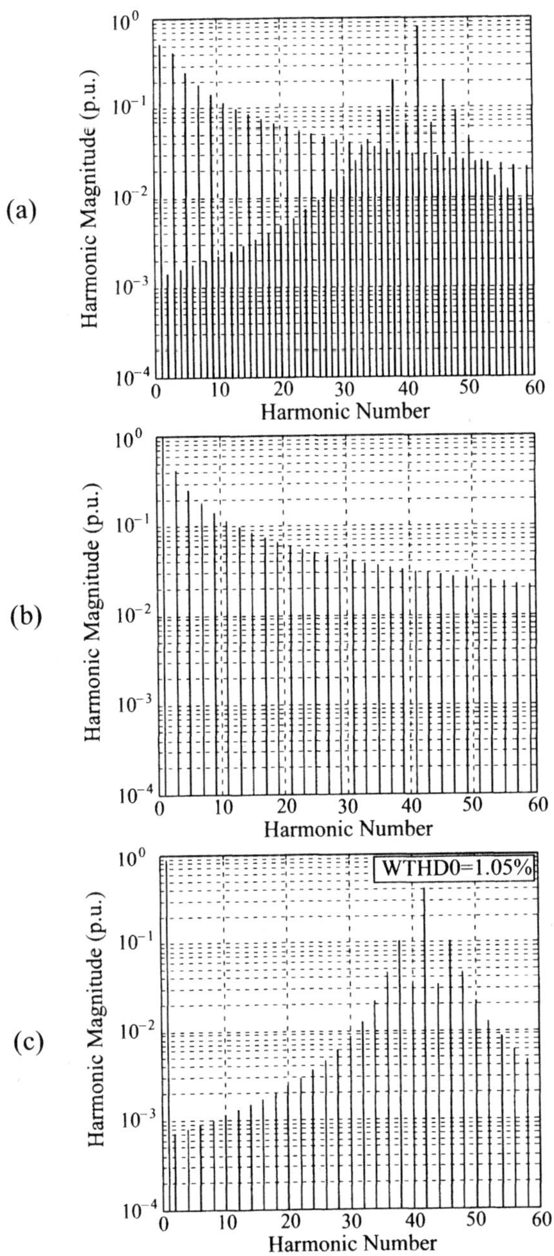

Figure 4.23 Theoretical harmonic spectra for single-phase inverter modulated by *single-phase leg switched discontinuous naturally sampled* PWM without inverted carrier: (a) phase leg *a*, (b) phase leg *b* and (c) *l–l* switched outputs, $M = 0.9$, $f_c/f_o = 42$.

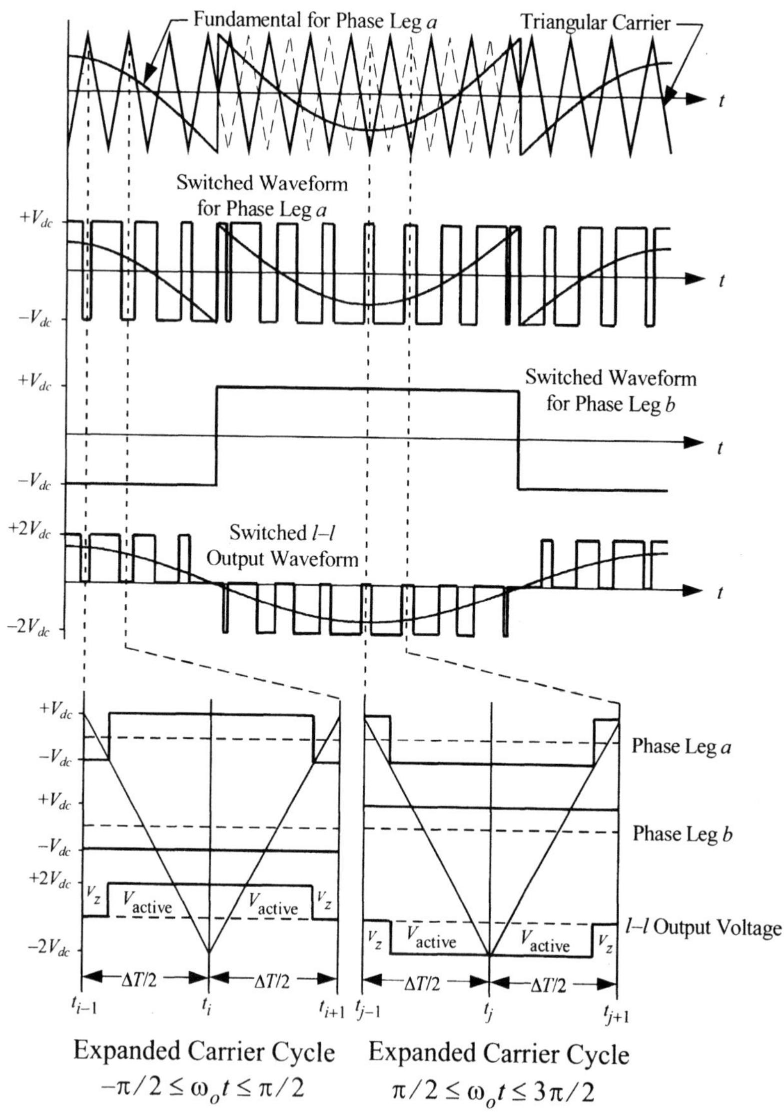

Figure 4.24 Single-phase inverter switching pattern for *discontinuous naturally sampled* PWM — phase leg a only modulated with carrier inverted while reference target is negative.

Solving Eq. (4.68) in the same manner as previously leads to a full fundamental cycle harmonic solution for phase leg *a* of

$$v_{an}(t) = V_{dc} + 2V_{dc}M\cos(\omega_o t) - \frac{4V_{dc}}{\pi}\sum_{n=1}^{\infty}\frac{1}{n}\sin n\frac{\pi}{2}\cos(n\omega_o t) \qquad (4.69)$$

$$+\frac{4V_{dc}}{\pi}\sum_{m=1}^{\infty}\sum_{n=-\infty}^{\infty}\frac{1}{m}J_n(m\pi M)\sin n\frac{\pi}{2}\cos(m\omega_c t + n\omega_o t)$$

which is clearly the same as Eq. (4.59) except for the extra baseband harmonics that will cancel out against the square harmonics of phase leg *b*. Therefore modulating a single-phase leg in this way will again achieve the *l–l* harmonic performance of Figure 4.19(b), as shown in Figure 4.25 However, it can be seen from Figure 4.24 that phase leg *a* must now start "high" at the beginning of the carrier cycle for the negative half of the reference cycle, and this may not be possible with some digital PWM hardware implementations.

This result illustrates the effect that alternative pulse sequences can have, where the harmonics in the output can vary substantially depending on the way in which the active *l–l* output pulses are moved together and placed throughout the fundamental cycle. For the single-phase inverter with matched phase leg switching frequencies, the difference in WTHD when modulating either one or both phase legs with discontinuous conduction is negligible, but for more complex inverters, poorer modulation performance can occur if care is not taken. Furthermore, it should be remembered that when the switching frequencies are matched for single-phase leg discontinuous modulation, all the switching losses reflect into phase leg *a* devices only, and this affects losses and hence PWM implementation considerations.

4.6.2 Two-Level Single-Phase PWM

Two-level single-phase modulation is included in this chapter for completeness, although it is in fact merely an alternative combination of all three factors of pulse width determination, pulse placement, and pulse sequence. Essentially, the two-level modulation strategy makes the switching of one phase leg exactly mirror the other, as shown in Figure 4.26. Both single- and double-edge modulation are quite feasible, although only double-edge modulation is illustrated here.

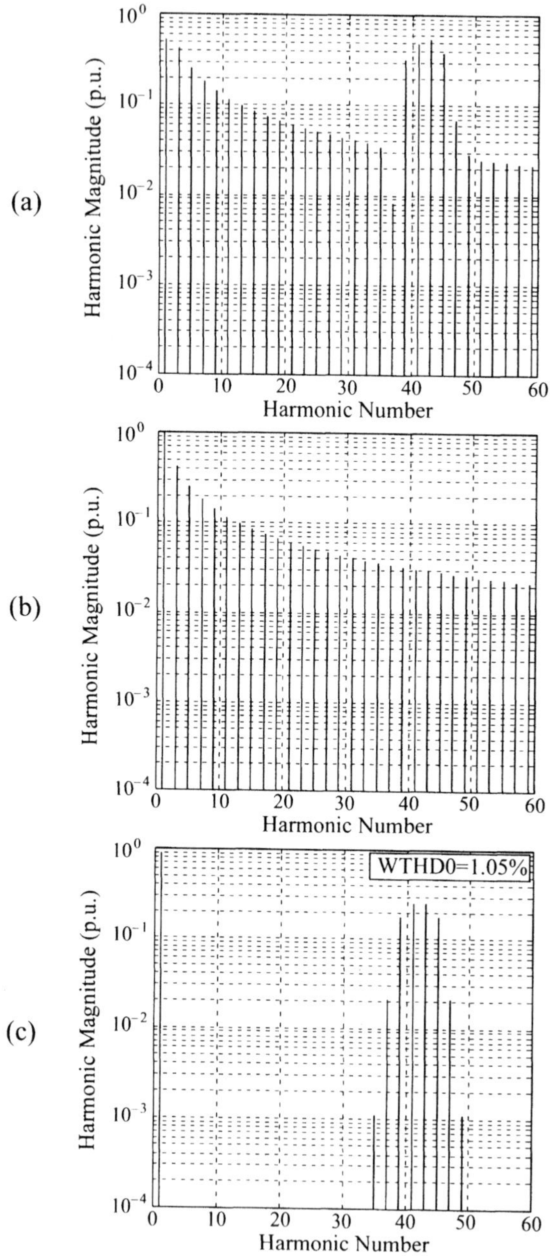

Figure 4.25 Theoretical harmonic spectra for single-phase inverter modulated by *single-phase leg switched discontinuous naturally sampled* PWM with inverted carrier: (a) phase leg *a*, (b) phase leg *b*, and (c) *l–l* switched outputs, $M = 0.9$, $f_c/f_o = 42$.

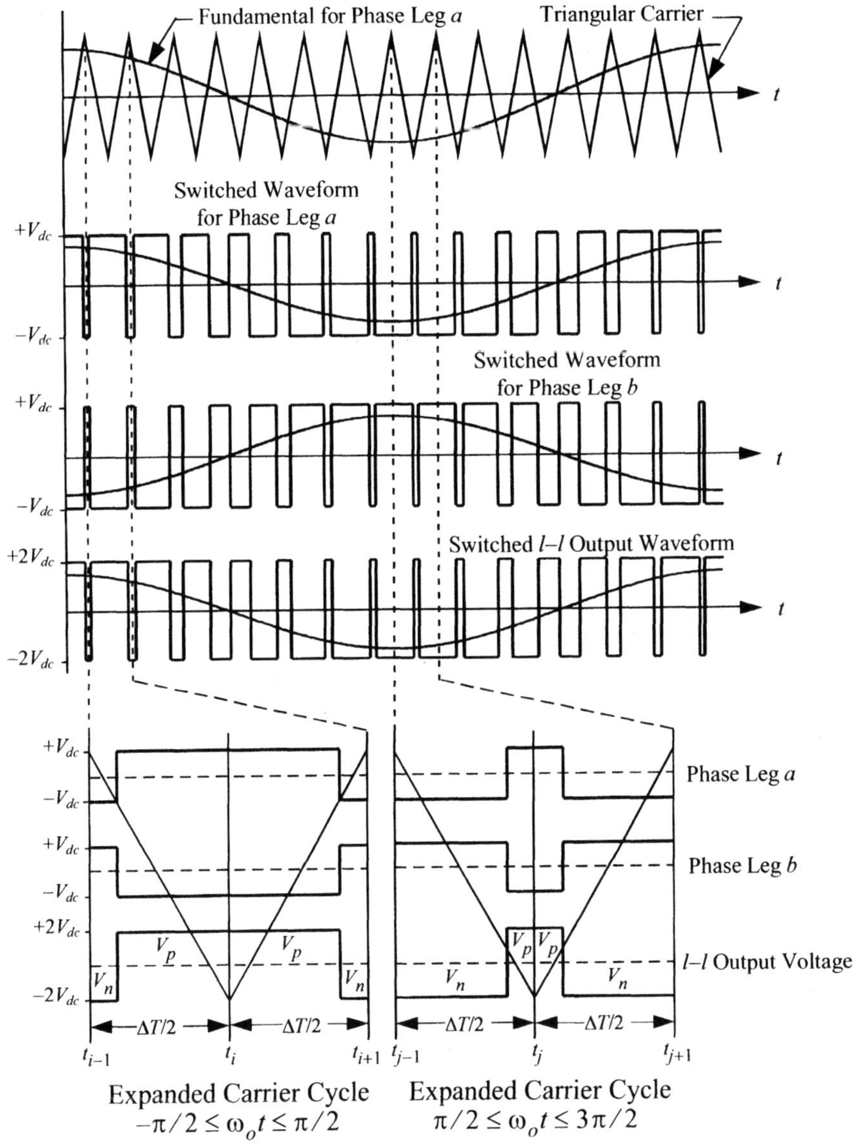

Figure 4.26 Two-level naturally sampled sine-triangle PWM process for single-phase VSI.

The *l–l* output voltage v_{ab} produced by two-level modulation is quite different from three-level modulation, since it continuously switches between positive (V_p) and negative (V_n) active pulses in each half carrier period without any zero states. One way to view this strategy is to consider that one part of the switched waveform provides the same volt-second average as the target reference waveform, while the remainder of the half carrier period contributes an effective zero state because of cancellation between the remaining positive and negative waveform segments. But, of course, this would not be expected to be a harmonically advantageous arrangement.

Since the switched output of phase leg b is the exact opposite of phase leg a, it will have identical magnitude and inverted sign harmonic components. Consequently when phase leg b is subtracted from phase leg a to create the l–l output voltage for two-level modulation, all the harmonic components of phase leg a will remain in the l–l solution without any cancellation. As a particular consequence, the *odd* carrier sideband harmonic cancellation between the phase legs that occurs for three-level modulation, will not happen for two-level modulation. And of course, this lack of cancellation will occur for any of the sampling or carrier variations considered in this chapter.

Analytical solutions for two-level modulation can be developed by reformulating the inner integration limits of Eq. (3.9) for phase leg b to suit the alternative switching arrangements, i.e.,

$$A_{mn} + jB_{mn} = \frac{V_{dc}}{\pi^2} \int_{-\pi}^{\pi} \int_{\pi M \cos y}^{\pi} e^{j(mx + ny)}\, dx\, dy \tag{4.70}$$

from Eq. (3.14) for single-edge modulation, and

$$A_{mn} + jB_{mn} = \frac{V_{dc}}{\pi^2} \int_{-\pi}^{\pi} \left[\int_{-\pi}^{-\frac{\pi}{2}(1 + M\cos y)} e^{j(mx + ny)}\, dx + \int_{\frac{\pi}{2}(1 + M\cos y)}^{\pi} e^{j(mx + ny)}\, dx \right] dy \tag{4.71}$$

from Eq. (3.29) for double-edge modulation.

These formulations reflect the change in switching pattern for phase leg b, which now starts at the positive DC bus, and switches to the negative DC bus and back over the period of the carrier unit cell.

Note that although at first glance it seems attractive, it is in fact invalid to attempt to generate analytical solutions for two-level modulation by setting $\theta_o, \theta_c = -\pi$ for phase leg b in the various single-phase leg modulation solutions of Chapter 3. This is because these solutions were obtained by assuming that the phase leg switching starts at the negative DC bus and switches to the positive DC bus and back over the unit cell period. Shifting the phase offset of the carrier does not change the way in which these unit cell switching limits are defined and thus does not correctly define the two-level modulation phase leg b switching pattern. However, it is interesting to observe that the two-level modulation phase leg b solution is correctly formulated for two-level *triangular carrier, naturally sampled* and *asymmetrical regular sampled* modulation by setting $\theta_o, \theta_c = -\pi$ in the single-phase leg solutions, while the solution for two level *sawtooth carrier* and *triangular carrier symmetrical regular sampled* modulation derived in this manner is incorrect. This result is because the changed $\{\sin([x]\pi/2)\}$ delimiters that are developed in the exact solutions can only be reflected into the $\{\cos(m[\omega_c t + \theta_c] + n[\omega_o t + \theta_o])\}$ time-varying terms for those modulation strategies that intrinsically cancel sideband harmonics within the phase leg solution. Further consideration of this issue is left for the interested reader to pursue.

Two-level modulation has the particular attraction that it can be implemented with very simple circuitry since one phase leg maintains the inverse switch state of the other. However, Figure 4.27 shows the inferior harmonic performance of *two-level single-* or *double-edge* modulation compared to *three-level double-edge* modulation. In particular, observe how the *two-level double-edge* implementation generates substantial carrier frequency and sideband harmonics in contrast to the (only) double carrier frequency and sideband harmonics generated by three-level modulation. Note also how the *two-level single-edge* implementation generates even more significant second and third fundamental harmonics because of a lack of harmonic cancellation between the two phase legs and the reduced roll-off in magnitude of the baseband harmonics that occurs with this modulation strategy.

4.7 Summary

Techniques for modulating a single-phase VSI have been explained in this chapter using new perspectives of the fundamental principles of PWM which are independent of converter topology and modulation strategy.

From this work, it has been demonstrated that there are three basic factors which affect the harmonic performance of a PWM system, viz:

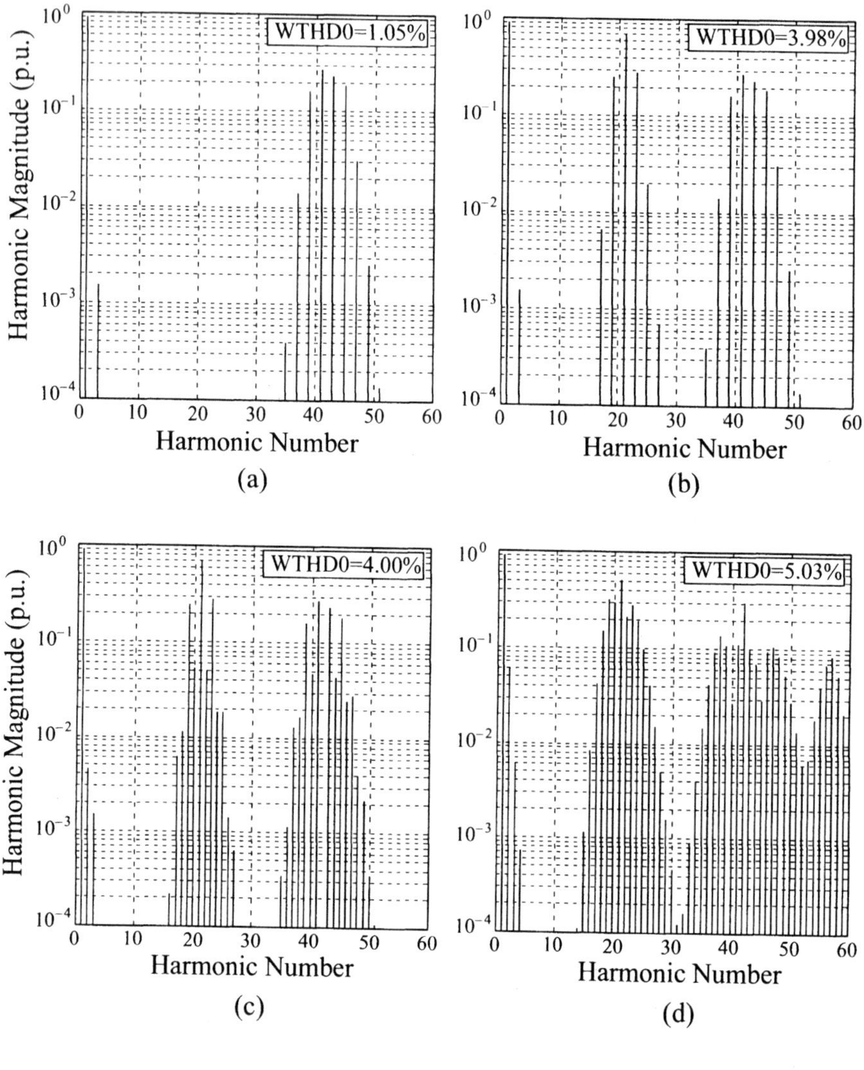

Figure 4.27 Theoretical *l–l* harmonic spectra for single-phase inverter modulated by

(a) *three level asymmetrical regular sampled* PWM,

(b) *two level asymmetrical regular sampled* PWM,

(c) *two level symmetrical regular sampled* PWM,

(d) *two level single-edge regular sampled* PWM,

$$M = 0.9,\ f_c/f_o = 21.$$

1. *Determination of the width of the active l–l output switched pulse.* This can be achieved using a naturally sampled, regular sampled, or direct modulation strategy, but the only significant difference for pulse ratios above about 11 is the presence of low-order harmonics of the fundamental generated by regular sampled PWM or direct modulation, compared to naturally sampled PWM.

2. *The position of the active pulse within the half carrier period.* Naturally sampled PWM moves the active pulse around the center of the half carrier period during the fundamental cycle, and this has harmonic advantages. Regular sampled PWM exactly places the active pulse in the center of the half carrier period. Single-edge modulation places the active pulse at the start of the carrier period, and this arrangement is usually harmonically disadvantageous.

3. *The sequence of the active pulses within and across successive half carrier periods.* The effect of this is best seen by comparing the pulse pattern generated by double-edge modulation with that generated by single-edge modulation.

The interplay between these three factors determines the harmonic performance of a particular modulation implementation. In particular, different PWM strategies can avoid creating harmonics within a phase leg's switched voltage, eliminate harmonics by cancellation between phase legs, or achieve a combination of both. In the next chapter, these three factors will be applied to a three-phase system to show how they relate to this more complex topology. From this extension, the concept of zero pulse creation and placement will be included as a fundamental parameter which determines the harmonic performance of a PWM algorithm.

References

[1] J.T. Boys and P.G. Handley, "Harmonic analysis of space vector modulated PWM waveforms," *IEE Proceedings* (*London*), vol. 137, Pt. B, no. 4, July, 1990, pp. 197–204.

[2] H.W. Van der Broeck and H.C. Skudelny, "Analytical analysis of the harmonic effects of a PWM ac drive," *IEEE Trans. on Power Electronics,* vol. 3, no. 2, March/April, 1988, pp. 216–223.

[3] I. Takahashi, S. Sekiguchi, and S. Miyairi, "Control of PWM inverter output wave with reduced low-order harmonics," *Electrical Engineering in Japan*, vol. 97, no. 3, 1977, pp. 57–63.

5

Modulation of Three-Phase Voltage Source Inverters

Chapter 3 has presented a comprehensive development of the fundamental principles of PWM that analyzes fixed-frequency open-loop modulation strategies in terms of:

- Switched pulse width determination.
- Switched pulse placement within a carrier period.
- Switched pulse sequence within and across carrier periods.

This approach offers a more integrated perspective than that presented in existing papers and textbooks, which can be used with any converter topology and modulation implementation. In Chapter 4 these principles were applied to the simple topology of a single-phase inverter to explain the concepts.

In this chapter these concepts of pulse width, position, and sequence are extended to a three-phase voltage source inverter (VSI) and are used to present a common understanding of the established fixed carrier frequency modulation strategies that have been proposed for this topology. From this development, the position of the *zero space vector* is identified as the significant parameter which differentiates the performance of various well-known PWM strategies for a VSI [1]. Later chapters will then explore the influence of the zero space vector and show how the VSI and current source inverter (CSI) topologies are linked through their common topological base so that the same integrated modulation perspective can be applied also to a CSI.

5.1 Topology of a Three-Phase Inverter (VSI)

The topology of a three-phase inverter is shown in Figure 5.1, where it can be seen that the essential difference compared to a single-phase inverter is that a third phase leg has been added, and the reference sinusoids for each phase leg are now displaced by $120°$, not $180°$, as was the case in Chapter 4.

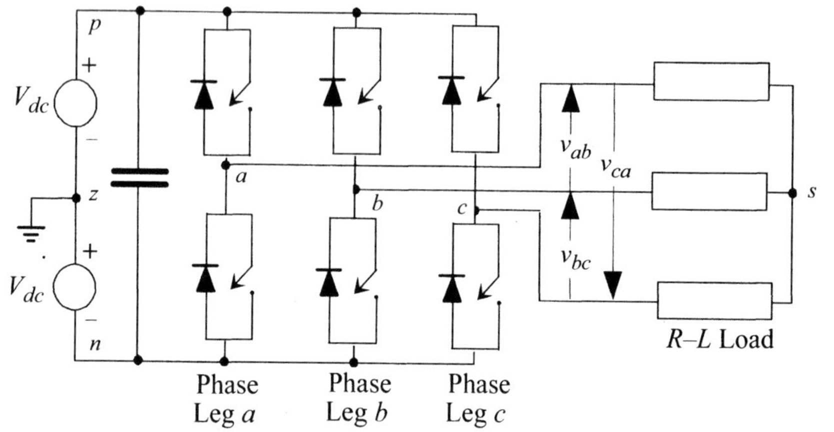

Figure 5.1 Topology of a three-phase voltage source inverter.

The fundamental modulation concepts for a three-phase inverter remain the same as for a single-phase inverter and, of course, include the variations of naturally sampled and regularly sampled PWM, continuous and discontinuous switching, and a triangular or a sawtooth carrier waveform, as discussed in Chapters 3 and 4. However, the implementation of these variations for a three-phase inverter is somewhat more complex, and the identification of pulse width, position, and sequence as the fundamental building blocks of modulation requires some development, as will now be discussed.

5.2 Three-Phase Modulation with Sinusoidal References

Figures 5.2 and 5.3 show the naturally sampled PWM process applied to a three-phase VSI, with the three sinusoidal references displaced in time by 120°, viz:

$$v_{az}^* = V_o \cos\omega_o t = MV_{dc}\cos\omega_o t \tag{5.1}$$

$$v_{bz}^* = V_o \cos(\omega_o t - 2\pi/3) = MV_{dc}\cos(\omega_o t - 2\pi/3) \tag{5.2}$$

$$v_{cz}^* = V_o \cos(\omega_o t + 2\pi/3) = MV_{dc}\cos(\omega_o t + 2\pi/3) \tag{5.3}$$

where V_o = output voltage peak magnitude, M = modulation index = V_o/V_{dc}, and the reference waveforms are defined w.r.t. the DC bus center point z.

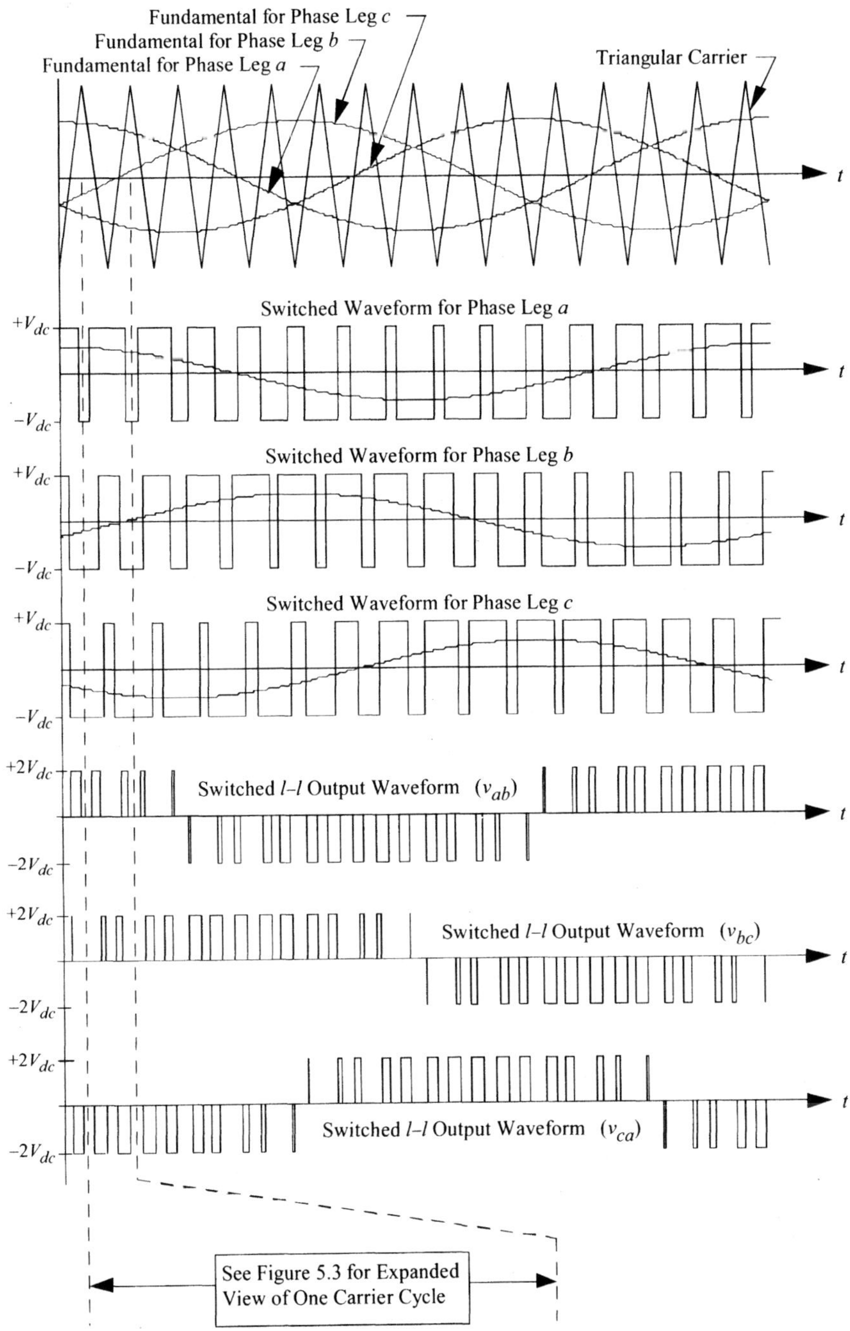

Figure 5.2 Naturally sampled sine–triangle modulation for three-phase voltage source inverter.

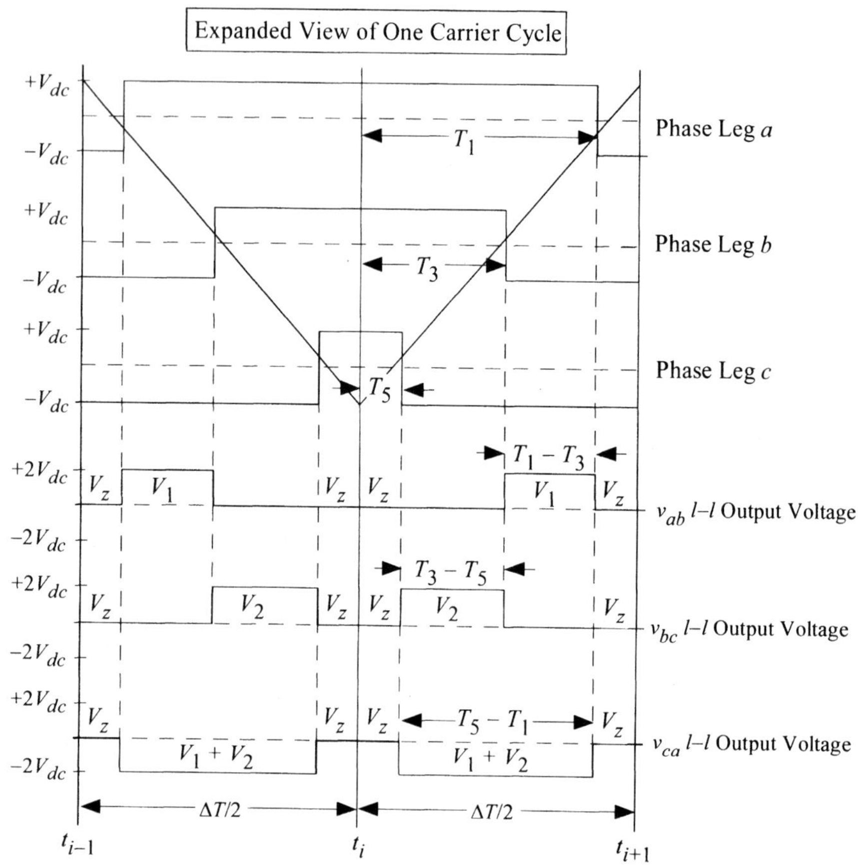

Figure 5.3 Sine–triangle modulation for three-phase voltage source inverter: expanded view of one carrier interval.

As with the single-phase inverter, the fundamental target three-phase l–l output voltages are given by the differences between the phase leg voltages, i.e.,

$$v_{ab}^* = v_{az}^* - v_{bz}^* = M\sqrt{3}V_{dc}\cos\left(\omega_o t + \frac{\pi}{6}\right) \tag{5.4}$$

$$v_{bc}^* = v_{bz}^* - v_{cz}^* = M\sqrt{3}V_{dc}\cos\left(\omega_o t - \frac{\pi}{2}\right) \tag{5.5}$$

$$v_{ca}^* = v_{cz}^* - v_{az}^* = M\sqrt{3}V_{dc}\cos\left(\omega_o t + \frac{5\pi}{6}\right) \tag{5.6}$$

Note that the maximum reference magnitude for a modulation index of $M = 1$ is only $\sqrt{3}V_{dc}$, compared to $2V_{dc}$ for a single-phase inverter, as developed in

Eq. (4.3). Note further, however, that this limit can be increased to $M = 1.15$ by changing the common mode reference definition for the target fundamental waveforms, as will be discussed later in this chapter.

From Figure 5.3 it can also be seen that, similar to the single-phase inverter, each switched *l–l* output voltage is once again made up of two pulses per carrier period with magnitude $+V_{dc}$ or $-V_{dc}$ (the polarity depends on the position in the fundamental cycle). But now these pulses are not always centered in each half carrier period, and this is the important difference between single-phase inverter modulation and three-phase inverter modulation.

The analytical harmonic solution for *double-edge naturally sampled* PWM of a three-phase inverter can be readily developed using the same strategies as has been applied to a single-phase inverter in Chapter 4, i.e., set $\theta_o = 0, -2\pi/3, 2\pi/3$ for phase legs *a, b,* and *c* respectively, i.e.,

$$v_{az}(t) = V_{dc} + V_{dc}M\cos(\omega_o t) + \frac{4V_{dc}}{\pi}\sum_{m=1}^{\infty}\sum_{n=-\infty}^{\infty}\frac{1}{m}J_n\left(m\frac{\pi}{2}M\right)\sin\left([m+n]\frac{\pi}{2}\right)\cos(m\omega_c t + n\omega_o t) \tag{5.7}$$

$$v_{bz}(t) = V_{dc} + V_{dc}M\cos\left(\omega_o t - \frac{2\pi}{3}\right) + \frac{4V_{dc}}{\pi}\sum_{m=1}^{\infty}\sum_{n=-\infty}^{\infty}\frac{1}{m}J_n\left(m\frac{\pi}{2}M\right)\sin\left([m+n]\frac{\pi}{2}\right)\cos\left(m\omega_c t + n\left[\omega_o t - \frac{2\pi}{3}\right]\right) \tag{5.8}$$

$$v_{cz}(t) = V_{dc} + V_{dc}M\cos\left(\omega_o t + \frac{2\pi}{3}\right) + \frac{4V_{dc}}{\pi}\sum_{m=1}^{\infty}\sum_{n=-\infty}^{\infty}\frac{1}{m}J_n\left(m\frac{\pi}{2}M\right)\sin\left([m+n]\frac{\pi}{2}\right)\cos\left(m\omega_c t + n\left[\omega_o t + \frac{2\pi}{3}\right]\right) \tag{5.9}$$

Subtracting these phase leg solutions from each other yields, after some manipulation, the *l–l* output voltage solution of (voltage v_{ab} only shown as representative)

$$v_{ab}(t) = \sqrt{3}V_{dc}M\cos\left(\omega_o t + \frac{\pi}{6}\right) + \frac{8V_{dc}}{\pi}\sum_{m=1}^{\infty}\sum_{n=-\infty}^{\infty}\frac{1}{m}J_n\left(m\frac{\pi}{2}M\right)\sin\left([m+n]\frac{\pi}{2}\right)\sin n\frac{\pi}{3} \times \cos\left(m\omega_c t + n\left[\omega_o t - \frac{\pi}{3}\right] + \frac{\pi}{2}\right) \tag{5.10}$$

Equation (5.10) is significantly different from the single-phase inverter *l–l* output voltage solution, Eq. (4.6), since a quite different harmonic cancellation occurs between the phase legs. In particular, there is no longer complete harmonic sideband cancellation around the *odd* carrier multiples, and instead it is the *triplen* (i.e., n is a multiple of 3) sideband harmonics that cancel on an *l–l* basis. Note also the magnitude adjustment of the sideband harmonics to allow for the magnitude contribution of the $\sin(n\pi/3)$ triplen elimination term.

Overall, the following phase leg harmonic components will not appear in the *l–l* output voltage:

- Carrier harmonics, since they are the same for all phase legs.
- Sideband harmonics with even combinations of $m \pm n$. More precisely, these harmonics are eliminated within each phase leg by the $\sin([m+n]\pi/2)$ terms in Eqs. (5.7), (5.8), and (5.9).
- Triplen sideband harmonics, where n is a multiple of 3. The phase angles of these harmonics rotate by multiples of 2π for all phase legs and hence are the same for all phase legs.

Since all *odd* sideband harmonics are eliminated in the first carrier harmonic group ($m = 1$), this means that the significant sideband harmonics in this carrier group will occur at frequencies of $\omega_c t \pm 2\omega_o t$, $\omega_c t \pm 4\omega_o t$. (Phase leg sideband harmonics for $n > 7$ are already known to be insignificant, as shown in Figure 3.12). For the second carrier harmonic group, the significant sideband harmonics will occur at $2\omega_c t \pm \omega_o t$, $2\omega_c t \pm 5\omega_o t$, and $2\omega_c t \pm 7\omega_o t$.

Figure 5.4 shows the phase leg and *l–l* harmonic components for a three-phase inverter operating under double-edge naturally sampled PWM, where the predicted significant sideband harmonics in the first and second carrier groups can be clearly identified in Figure 5.4(b). Note the increase in WTHD0 for the *l–l* voltage, from 1.05% for a single-phase inverter, as identified in Figure 4.3, to 2.04% for this three-phase system. This is a direct consequence of first carrier group sideband harmonics remaining in the *l–l* output voltage.

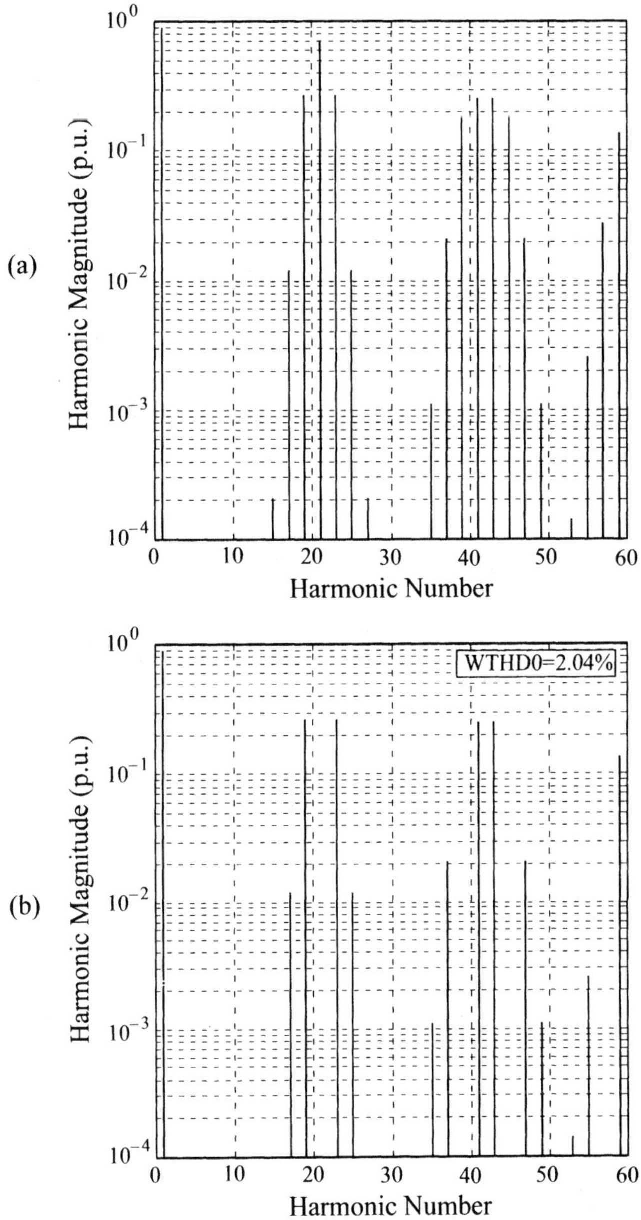

Figure 5.4 Theoretical harmonic spectra for three-phase inverter modulated by *double-edge naturally sampled* PWM: (a) phase leg a and (b) l–l output switched voltage waveforms, $M = 0.9$, $f_c/f_o = 21$.

As would be expected, a three-phase VSI can also be modulated using a *regular sampled* PWM process, with the reference value sampled either asymmetrically or symmetrically. However, the analysis of these two alternatives presented in Chapter 3 clearly shows the advantage of the asymmetrical sampling process, with half the phase leg harmonics (i.e., for $m \pm n$ even) being intrinsically suppressed. This lessens the number of harmonics remaining to potentially be cancelled between phase legs, which is obviously preferable. Hence only asymmetrically sampled PWM will be analyzed in detail in this chapter, since from the perspective of pulse width determination, symmetrical sampling is a subset variation which has poorer harmonic performance.

Figure 5.3 shows the switching times T_1, T_3, T_5, for the three-phase VSI when controlled by regular sampled PWM, which are given by

$$T_1 = \frac{\Delta T}{4} m_1(i) = \frac{\Delta T}{4}\left\{1 + M\cos\left[\frac{\omega_o(t_i + t_{i+1})}{2}\right]\right\} \tag{5.11}$$

$$T_3 = \frac{\Delta T}{4} m_3(i) = \frac{\Delta T}{4}\left\{1 + M\cos\left[\frac{\omega_o(t_i + t_{i+1})}{2} - \frac{2\pi}{3}\right]\right\} \tag{5.12}$$

$$T_5 = \frac{\Delta T}{4} m_5(i) = \frac{\Delta T}{4}\left\{1 + M\cos\left[\frac{\omega_o(t_i + t_{i+1})}{2} + \frac{2\pi}{3}\right]\right\} \tag{5.13}$$

where $m_1(i)$, $m_3(i)$, $m_5(i)$, represent the normalized on-switch time (i.e., switched to the positive rail) for each phase leg over the half carrier interval t_i to t_{i+1}.

The active *l–l* pulse widths $T_1 - T_3$, $T_3 - T_5$, $T_5 - T_1$, shown in the lower part of Figure 5.3 are now readily determined by subtracting each of the two appropriate phase leg switch on-times, whereupon

$$T_1 - T_3 = \frac{\Delta T}{4}[m_1(i) - m_3(i)] = \frac{\Delta T}{4}\left\{\sqrt{3}M\cos\left[\frac{\omega_o(t_i + t_{i+1})}{2} + \frac{\pi}{6}\right]\right\} \tag{5.14}$$

$$T_3 - T_5 = \frac{\Delta T}{4}[m_3(i) - m_5(i)] = \frac{\Delta T}{4}\left\{\sqrt{3}M\cos\left[\frac{\omega_o(t_i + t_{i+1})}{2} - \frac{\pi}{2}\right]\right\} \tag{5.15}$$

$$T_5 - T_1 = \frac{\Delta T}{4}[m_5(i) - m_1(i)] = \frac{\Delta T}{4}\left\{\sqrt{3}M\cos\left[\frac{\omega_o(t_i + t_{i+1})}{2} + \frac{5\pi}{6}\right]\right\} \tag{5.16}$$

where a negative pulse width time in these equations is interpreted to mean that a negative polarity *l–l* pulse with the same magnitude is generated by the action of the regular sampled *PWM* process.

Equations (5.14) to (5.16) confirm that the peak *l–l* output voltage obtainable from a three-phase inverter under regular sampled *PWM* is $\sqrt{3}/2$ of the available DC bus voltage $2V_{dc}$, since the $\sqrt{3}M\cos\theta$ terms can never exceed $\sqrt{3}$ for $0 \le M \le 1$. Hence each *l–l* pulse width is always less than $\sqrt{3}/2$ of the half carrier period $\Delta T/2$, and so it could be considered that the PWM process does not fully utilize the available modulation space as a consequence. This issue will be explored further in the next section of this chapter.

The analytical harmonic solution for *asymmetrical regular sampled* PWM of a three-phase inverter is similarly developed as for naturally sampled modulation, by setting $\theta_o = 0, -2\pi/3, 2\pi/3$ for phase legs *a*, *b*, and *c* respectively, into the single-phase leg solution of Eq. (3.99). The *l–l* output voltage is (voltage v_{ab} only shown as representative)

$$v_{ab}(t) = \frac{8V_{dc}}{\pi} \sum_{\substack{m=0 \\ m>0}}^{\infty} \sum_{\substack{\leftrightarrow\; n=1 \\ \leftrightarrow\; n=-\infty}}^{\infty} \frac{1}{q} J_n\left(q\frac{\pi}{2}M\right) \sin\left([m+n]\frac{\pi}{2}\right) \sin n\frac{\pi}{3} \times \cos\left(m\omega_c t + n\left[\omega_o t - \frac{\pi}{3}\right] + \frac{\pi}{2}\right) \tag{5.17}$$

where $q = m + n(\omega_o/\omega_c)$ as in Chapter 3.

Note that the maximum value of the fundamental component in Eq. (5.17) is

$$\frac{4\sqrt{3}V_{dc}}{\pi} \frac{1}{\omega_o/\omega_c} J_1\left(\frac{\omega_o}{\omega_c}\frac{\pi}{2}M\right) \cong \sqrt{3}V_{dc} \tag{5.18}$$

since

$$\frac{4\sqrt{3}}{\pi} \frac{1}{\omega_o/\omega_c} J_1\left(\frac{\omega_o}{\omega_c}\frac{\pi}{2}\right) \to 1 \qquad \text{for } \frac{\omega_c}{\omega_o} \gg 1$$

This result agrees with naturally sampled PWM, as expected.

Figure 5.5 shows the theoretical harmonic spectra for naturally and the various regular sampled PWM strategies applied to a three-phase inverter [the analytical solution for *trailing single-edge* modulation is simply derived from Eq. (3.70) and is left as an exercise for the reader to develop]. As before, the harmonic performance of these algorithms is determined by the intrinsic har-

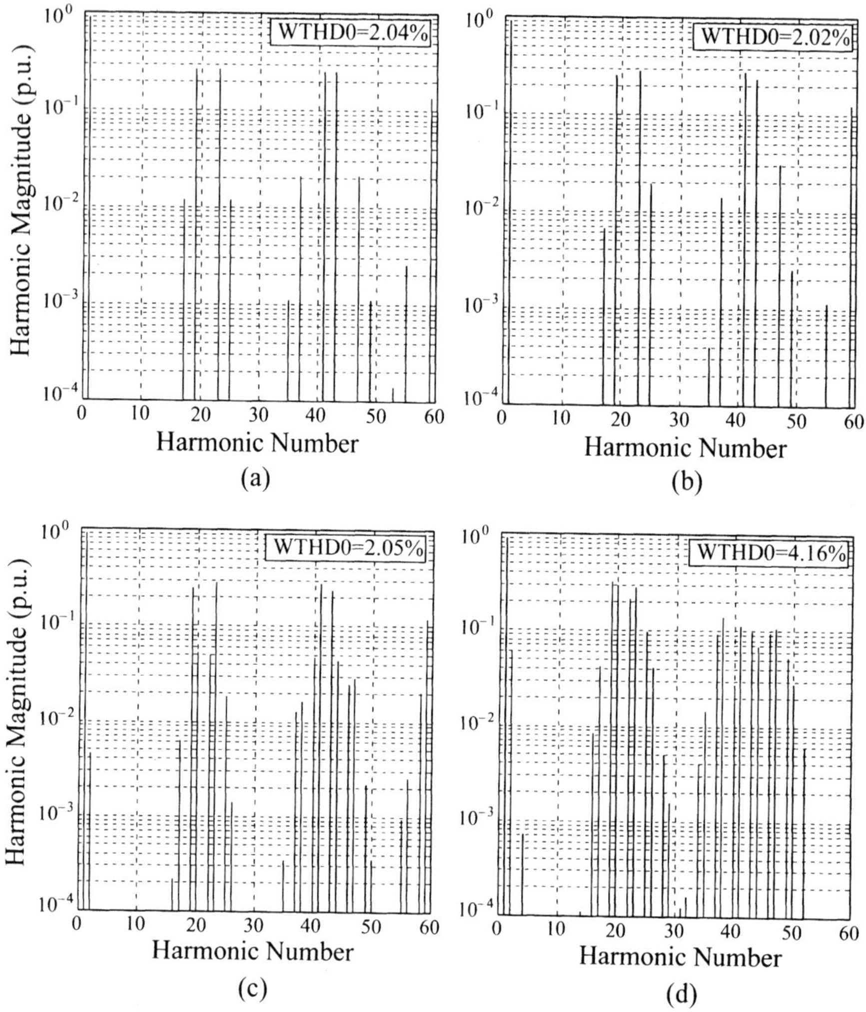

Figure 5.5 Theoretical *l–l* harmonic spectra for three-phase inverter modulated by

(a) *double-edge naturally sampled* PWM,

(b) *asymmetrical regular sampled* PWM,

(c) *symmetrical regular sampled* PWM,

(d) *trailing single-edge regular sampled PWM,*

$$M = 0.9,\ f_c/f_o = 21.$$

monic cancellation within each phase leg achieved by each modulation strategy and the harmonic cancellation that occurs between the phase legs.

From Figure 5.5, it can be seen that there are some interesting similarities and differences compared to the harmonic spectra given in Chapter 4 for the single-phase inverter, viz:

- The sideband spectra for a three-phase inverter are quite different from that of a single-phase inverter for all modulation strategies. This is because for the three-phase inverter, $m \pm 3n$ harmonics are cancelled in each carrier sideband harmonic group, instead of the $m \pm 2n$ cancellation which occurs for the single-phase system. One major consequence of this is that it is impossible to eliminate all sideband harmonics in the first carrier group for a three-phase inverter. Hence for the same switching frequency, the *l–l* output voltage WTHD for a three-phase inverter will always be greater than that of a single-phase inverter.
- As before, the only significant difference between naturally sampled modulation shown in Figure 5.5(a) and asymmetrical regular sampled modulation shown in Figure 5.5(b) is the skew in the sideband harmonic magnitudes. (Recall that this effect was identified as a theoretical consequence of the regular sampling process in Chapter 3.) However, it is interesting to note that unlike the single-phase inverter, in this case the effect achieves a slight reduction in WTHD because of the magnitude skew toward the higher sideband harmonics in the first carrier group. Although trivial for the modulation conditions used here, this effect can become more significant for very low carrier/fundamental ratios.
- The third-harmonic baseband component created by the regular sampling process shown in Figure 4.7(b) is not present in Figure 5.5(b) because all triplen harmonic components are cancelled between phase legs for a three-phase inverter. But in contrast the second-harmonic baseband component, which is created by symmetrical regular sampling as shown in Figure 4.6(a), is not cancelled in a three-phase system, as shown in Figure 5.5(c), and may significantly increase WTHD in particular implementations. This again illustrates the limitations of symmetrical sampling.
- The same effect also occurs for regular sampled single-edge modulation, However, in this case the baseband second harmonic has a magnitude of nearly 6%, and this is quite significant. Hence single-edge modulation is unattractive for use with a three-phase inverter unless the carrier/fundamental ratio is high enough to rapidly roll off the magnitude of the baseband nontriplen harmonics.

5.3 Third-Harmonic Reference Injection

A major limitation with the three-phase inverter modulation concepts presented thus far is the reduced maximum peak fundamental output line voltage of $\sqrt{3}V_{dc}$ that can be obtained compared to the available DC link voltage. This limitation has important implications for motor drive applications where it is very desirable to use a standard, rated voltage motor in variable speed drive application, for the following reasons.

It is readily shown that for a peak input *l–l* voltage of V_p, the average rectified DC link voltage will be $3V_p/\pi$ assuming a perfect zero impedance source and the rectifier operating in continuous conduction. (In practice an approximate 5% drop in the rectified DC link voltage would be typically expected at rated load due to source impedance.) The maximum peak inverter output voltage will therefore be $(\sqrt{3}/2)(3V_p/\pi)$ or 82.7% of V_p. Since the motor must run at rated volts per hertz to sustain rated load torque without overheating, this implies that the motor, when connected to the output side PWM inverter, will be capable of only about 83% of rated power, a very significant derating.

If the rectified input voltage is smoothed with a very large DC link capacitor, the DC link voltage will increase by 4.7% to V_p, and this will increase the peak inverter output voltage to $(\sqrt{3}/2)V_p$, or 86.67% of V_p. However, while this is an improvement, it is still well short of achieving the desired target of inverter output voltage equals AC source input voltage at maximum modulation index.

5.3.1 Optimum Injection Level

Beginning with Buja and Indri in 1975 [2], it has been gradually recognized that maximum modulation index of a three-phase inverter PWM system can be increased by including a common mode third-harmonic term into the target reference waveform of each phase leg [3–5]. This third-harmonic component does not affect the *l–l* fundamental output voltage, since the common mode voltages cancel between the phase legs, but it does reduce the peak size of the envelope of each phase leg voltage. Hence the modulation index M can be increased beyond $M = 1.0$ without moving into overmodulation (i.e., the region where the reference waveform magnitudes exceed the carrier peak at various times during the fundamental cycle). Overmodulation is known to produce low-frequency baseband distortion and is to be avoided if possible.

Equations (5.19), (5.20), and (5.21) define the appropriate sinusoidal target reference waveforms, which have been extended from Eqs. (5.1), (5.2), and (5.3) to include this third-harmonic component, viz:

$$v^*_{az(ref+3)} = V_{dc}[M\cos\omega_o t + M_3\cos 3\omega_o t] \tag{5.19}$$

$$v^*_{bz(ref+3)} = V_{dc}[M\cos(\omega_o t - 2\pi/3) + M_3\cos 3\omega_o t] \tag{5.20}$$

$$v^*_{cz(ref+3)} = V_{dc}[M\cos(\omega_o t + 2\pi/3) + M_3\cos 3\omega_o t] \tag{5.21}$$

By dividing through by MV_{dc}, these equations can be written in per unit form [6] as

$$v = \cos\theta_o + \gamma\cos 3\theta_o \tag{5.22}$$

where $\gamma = M_3/M$ represents a parameter to be optimized to reduce the maximum value of the function as much below unity as possible. The maximum value of Eq. (5.22) will occur where its derivative is zero, i.e.,

$$\frac{dv}{dt} = 0 = \sin\theta_o + 3\gamma\sin 3\theta_o \tag{5.23}$$

While this function obviously has either a maximum/minimum at $\theta_o = 90°$ (depending upon the polarity of γ), if it is a maximum, the value would be greater than unity, while if it is a minimum, it would not be of interest anyway. However, another maximum/minimum can be found as follows.

Recall that

$$\begin{aligned}\sin 3\theta_o &= \sin 2\theta_o\cos\theta_o + \cos 2\theta_o\sin\theta_o \\ &= 2\cos^2\theta_o\sin\theta_o + (2\cos^2\theta_o - 1)\sin\theta_o \\ &= (4\cos^2\theta_o - 1)\sin\theta_o \end{aligned} \tag{5.24}$$

Substituting this expression into Eq. (5.23) gives

$$0 = 1 + 3\gamma(4\cos^2\theta_o - 1)$$

which rearranges to

$$1 - \frac{1}{3\gamma} = 4\cos^2\theta_o$$

and finally

$$\cos\theta_o = \sqrt{\frac{3\gamma - 1}{12\gamma}} \tag{5.25}$$

Similarly

$$\begin{aligned}\cos 3\theta_o &= \cos 2\theta_o \cos\theta_o - \sin 2\theta_o \sin\theta_o \\ &= (2\cos^2\theta_o - 1)\cos\theta_o - 2\sin^2\theta_o \cos\theta_o \\ &= (4\cos^2\theta_o - 3)\cos\theta_o \end{aligned} \tag{5.26}$$

so that, from Eq. (5.25),

$$\cos 3\theta_o = -\frac{(6\gamma+1)}{6\gamma}\sqrt{\frac{3\gamma-1}{3\gamma}} \tag{5.27}$$

Substituting Eqs. (5.25) and (5.27) into (5.22) yields

$$v_{max} = -\frac{1}{3}(3\gamma-1)\sqrt{\frac{3\gamma-1}{3\gamma}} = -\frac{1}{3}(3\gamma-1)\sqrt{1-\frac{1}{3}\gamma} \tag{5.28}$$

Equation (5.28) identifies v_{max}, the maximum value of v, as a function of the parameter γ. The maximum possible value of v_{max} is obtained by setting its derivative with respect to γ equal to zero. Thus

$$\frac{dv_{max}}{d\gamma} = -\sqrt{1-1-\frac{1}{3}\gamma} - \frac{1}{6}\left(\frac{3\gamma-1}{\sqrt{1-1/3\gamma}}\right)\frac{1}{3\gamma^2} = 0 \tag{5.29}$$

With some rearrangement, this becomes

$$0 = -\sqrt{1-\frac{1}{3}\gamma}\left(1+\frac{1}{6\gamma}\right) \tag{5.30}$$

For this expression to equal zero, its numerator must be zero, which gives possible values for γ of

$$\gamma = \frac{1}{3} \text{ and } -\frac{1}{6} \tag{5.31}$$

From Eq. (5.28), the value $\gamma = \frac{1}{3}$ produces $v_{max} = 0$, which must be a minimum because the square root term cannot become smaller than zero without becoming imaginary. The value $\gamma = -\frac{1}{6}$ produces a maximum of $v_{max} = \sqrt{3}/2 = 0.866$. Recalling that $v_{az} = MV_{dc}$, this means that

$$v_{az,\,max} = \frac{\sqrt{3}}{2}MV_{dc} \tag{5.32}$$

Hence the modulation index M can now increase to $2/\sqrt{3} = 1.15$ before $v_{az,\,max}$ reaches V_{dc}. Note also that since $\gamma = M_3/M$, M_3 must consequently be increased in proportion to M and reaches a value of $M_3 = \sqrt{3}/9 = 0.192$

when $M = 2/\sqrt{3}$. Note that at this modulation index, with a large filter capacitor in the DC bus, the AC output voltage magnitude now equals the AC source input voltage magnitude.

An alternative and simpler method of establishing this result is to recognize that the third-harmonic has no effect on the value of the reference waveform expressions when $\omega_o t = [2k+1]\pi/6$, since $\cos(3[2k+1]\pi/6) = 0$ for all k. Hence M_3 can be chosen (for $k = 0$) to make the peak magnitude of the reference waveforms defined in Eqs. (5.19), (5.20), and (5.21) occur where the third-harmonic is zero, i.e., at $\omega_o t = \pi/6$. This would, in turn, assure the maximum possible value for the fundamental component. For example, when $k = 0$, the voltage $v_{az(ref+3)}$ reaches a maximum when

$$\frac{dv_{az(ref+3)}}{d\omega_o t} = -MV_{dc}\sin\omega_o t - 3M_3 V_{dc}\sin 3\omega_o t = 0 \quad (5.33)$$

which for $\omega_o t = \pi/6$ results in

$$M_3 = -\frac{M}{6}$$

Under these conditions, the maximum possible modulation index M is

$$|v_{az(ref+3)}| = \left|MV_{dc}\cos\omega_o t - \frac{M}{6}V_{dc}\cos 3\omega_o t\right| = V_{dc} \quad \text{for } \omega_o t = \frac{\pi}{6} \quad (5.34)$$

which occurs when $M = 2/\sqrt{3} = 1.155$ as before.

This analysis shows that a 15% increase in modulation index can be achieved by simply including a one-sixth third-harmonic injection into the fundamental reference waveforms. The technique is equally appropriate for naturally and regularly sampled PWM, with either a triangular or sawtooth carrier wave. Furthermore, this increase in possible modulation index is exactly that obtainable by the more elaborate method of space vector modulation which suggests that the increase in modulation index is achieved by a fundamental process separate from the *pulse width* determination. Chapter 6 identifies this process as the zero space vector placement, which controls the switched *pulse position* within each half carrier period.

While a one-sixth value of third-harmonic has been confirmed here as achieving the maximum possible increase in fundamental component, other researchers [7, 8] have proposed that the optimum third-harmonic injection component should have a magnitude of 25% of the target fundamental, arguing that this leads to reduced harmonic distortion. The cost of this option is a slight reduction in the maximum linear modulation index from 1.15 down to 1.12, since the peak of the target reference waveform no longer occurs at $\omega_o t = \pi/6$.

Figure 5.6 shows the fundamental target reference waveforms for phase leg *a* without third-harmonic injection, one-sixth magnitude third-harmonic injection, and one-quarter magnitude third-harmonic injection. The reduction in the peak magnitude of the resultant reference waveform, and hence the ability to increase the maximum linear modulation index, can be clearly seen.

5.3.2 Analytical Solution for Third-Harmonic Injection

The theoretical harmonic solution for PWM with third-harmonic injection is developed using a similar analysis technique as before, but with rather more complicated mathematics because of the additional reference component [9].

For *naturally sampled* modulation, the analysis proceeds by recognizing that the integration limits, defined by Eqs. (3.27) and (3.28) for a simple sinusoidal reference, change to include the third-harmonic reference component. Hence Eq. (3.29) becomes

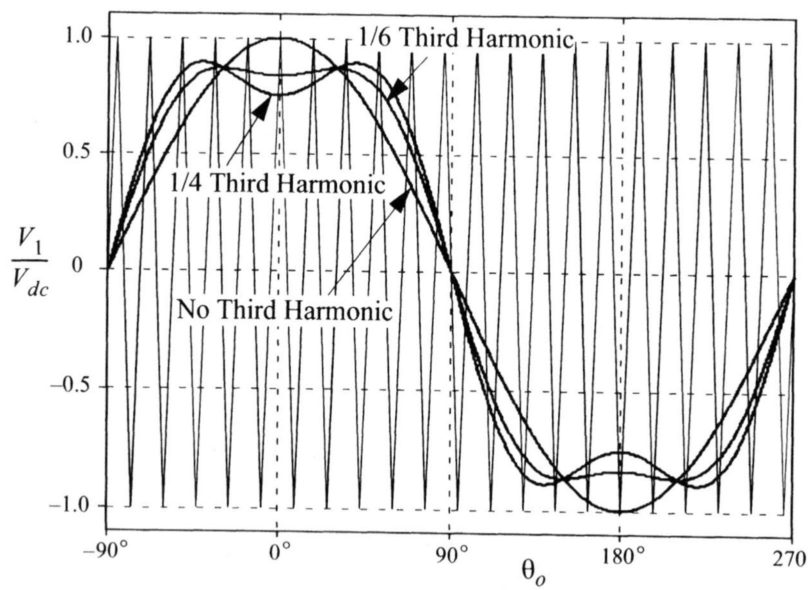

Figure 5.6 Phase leg *a* PWM reference waveforms for third-harmonic injection of zero, one-sixth, and one-quarter of the fundamental magnitude, $M = 1.0$, $f_c/f_o = 21$.

$$A_{mn} + jB_{mn} = \frac{1}{2\pi^2} \int_{-\pi}^{\pi} \int_{-\frac{\pi}{2}(1 + M\cos y + M_3\cos 3y)}^{\frac{\pi}{2}(1 + M\cos y + M_3\cos 3y)} 2V_{dc} e^{j(mx + ny)} dx\, dy \tag{5.35}$$

which can again be evaluated for the various possible values of m and n.

For $\underline{m = n = 0}$, Eq. (5.35) once more is found to result in

$$A_{00} = 2V_{dc} \qquad B_{00} = 0 \tag{5.36}$$

which gives the expected V_{dc} offset in the final solution when the negative DC bus is taken as the switched reference voltage.

For $\underline{m = 0, n > 0}$, Eq. (5.35) simplifies to

$$A_{0n} + jB_{0n} = \frac{V_{dc}}{\pi^2} \int_{-\pi}^{\pi} \int_{-\frac{\pi}{2}(1 + M\cos y + M_3\cos 3y)}^{\frac{\pi}{2}(1 + M\cos y + M_3\cos 3y)} e^{jny} dx\, dy$$

$$= \frac{V_{dc}}{\pi^2} \int_{-\pi}^{\pi} [\pi(1 + M\cos y + M_3\cos 3y)e^{jny}]\, dy$$

$$= \frac{V_{dc}}{\pi} \int_{-\pi}^{\pi} \left[e^{jny} + \frac{M}{2}(e^{j[n+1]y} + e^{j[n-1]y}) + \frac{M_3}{2}(e^{j[n+3]y} + e^{j[n-3]y}) \right] dy \tag{5.37}$$

Again using $\int_{-\pi}^{\pi} e^{jny} dy = 0$ for any non zero value of n, Eq. (5.37) reduces to

$$A_{01} + jB_{01} = \frac{V_{dc}}{\pi} \int_{-\pi}^{\pi} \frac{M}{2} dy = V_{dc} M \tag{5.38}$$

$$A_{03} + jB_{03} = \frac{V_{dc}}{\pi} \int_{-\pi}^{\pi} \frac{M_3}{2} dy = V_{dc} M_3 \tag{5.39}$$

for the case of $n = 1, 3$ only. For all other n, $A_{0n} + jB_{0n} = 0$.

Not unexpectedly, these coefficients define the fundamental and third-harmonic reference components specified in Eq. (5.19).

For $m > 0, n \neq 0$, the inner integral of Eq. (5.35) is evaluated to give

$$A_{mn} + jB_{mn} = \frac{V_{dc}}{jm\pi^2}\int_{-\pi}^{\pi} e^{jny}\left[\begin{array}{c} e^{jm\frac{\pi}{2}} e^{jm\frac{\pi}{2}M\cos y} e^{jm\frac{\pi}{2}M_3\cos 3y} \\ - e^{-jm\frac{\pi}{2}} e^{-jm\frac{\pi}{2}M\cos y} e^{-jm\frac{\pi}{2}M_3\cos 3y} \end{array}\right] dy \tag{5.40}$$

This integral expression then becomes, using Eq. (A2.1),

$$A_{mn} + jB_{mn} = \tag{5.41}$$

$$\frac{V_{dc}}{jm\pi^2}\int_{-\pi}^{\pi} e^{jny}\left[\begin{array}{l} e^{jm\frac{\pi}{2}}\left[\begin{array}{r} \left\{J_0\left(m\frac{\pi}{2}M\right) + 2\sum\limits_{k=1}^{\infty} j^k J_k\left(m\frac{\pi}{2}M\right)\cos ky\right\} \\ \times\left\{J_0\left(m\frac{\pi}{2}M_3\right) + 2\sum\limits_{h=1}^{\infty} j^h J_k\left(m\frac{\pi}{2}M_3\right)\cos 3hy\right\} \end{array}\right] \\ -e^{-jm\frac{\pi}{2}}\left[\begin{array}{r} \left\{J_0\left(m\frac{\pi}{2}M\right) + 2\sum\limits_{k=1}^{\infty} j^{-k} J_k\left(m\frac{\pi}{2}M\right)\cos ky\right\} \\ \times\left\{J_0\left(m\frac{\pi}{2}M_3\right) + 2\sum\limits_{h=1}^{\infty} j^{-h} J_k\left(m\frac{\pi}{2}M_3\right)\cos 3hy\right\} \end{array}\right] \end{array}\right] dy$$

which rearranges with some manipulation to

$$A_{mn}+jB_{mn}= \tag{5.42}$$

$$\frac{2V_{dc}}{m\pi^2}\int_{-\pi}^{\pi}\left[\begin{array}{l} J_0\left(m\frac{\pi}{2}M\right)J_0\left(m\frac{\pi}{2}M_3\right)\sin m\frac{\pi}{2}e^{jny} \\ +J_0\left(m\frac{\pi}{2}M_3\right)\sum_{k=1}^{\infty}J_k\left(m\frac{\pi}{2}M\right)\sin\left([m+k]\frac{\pi}{2}\right)\left[\begin{array}{l}e^{j[n+k]y}\\ \quad +e^{j[n-k]y}\end{array}\right] \\ +J_0\left(m\frac{\pi}{2}M\right)\sum_{h=1}^{\infty}J_h\left(m\frac{\pi}{2}M_3\right)\sin\left([m+h]\frac{\pi}{2}\right)\left[\begin{array}{l}e^{j[n+3h]y}\\ \quad +e^{j[n-3h]y}\end{array}\right] \\ +\sum_{k=1}^{\infty}\sum_{h=1}^{\infty}\left[\begin{array}{l}J_k\left(m\frac{\pi}{2}M\right)J_h\left(m\frac{\pi}{2}M_3\right)\sin\left([m+k+h]\frac{\pi}{2}\right)\\ \qquad \times\left(e^{j[n+k+3h]y}+e^{j[n-k-3h]y}\right)\end{array}\right] \\ +\sum_{k=1}^{\infty}\sum_{h=1}^{\infty}\left[\begin{array}{l}J_k\left(m\frac{\pi}{2}M\right)J_h\left(m\frac{\pi}{2}M_3\right)\sin\left([m+k+h]\frac{\pi}{2}\right)\\ \qquad \times\left(e^{j[n+k-3h]y}+e^{j[n-k+3h]y}\right)\end{array}\right] \end{array}\right]dy$$

Integrating Eq. (5.42) gives the closed-form solution for third-harmonic injected PWM of

$$A_{mn}+jB_{mn}=\frac{4V_{dc}}{m\pi}\left[\begin{array}{l} J_0\left(m\frac{\pi}{2}M_3\right)J_k\left(m\frac{\pi}{2}M\right)\sin\left([m+k]\frac{\pi}{2}\right)\Big|_{k=|n|} \\ +J_0\left(m\frac{\pi}{2}M\right)J_h\left(m\frac{\pi}{2}M_3\right)\sin\left([m+h]\frac{\pi}{2}\right)\Big|_{3h=|n|} \\ +\sum J_k\left(m\frac{\pi}{2}M\right)J_h\left(m\frac{\pi}{2}M_3\right)\sin\left([m+k+h]\frac{\pi}{2}\right)\Big|_{k+3h=|n|} \\ +\sum J_k\left(m\frac{\pi}{2}M\right)J_h\left(m\frac{\pi}{2}M_3\right)\sin\left([m+k+h]\frac{\pi}{2}\right)\Big|_{k-3h=|n|} \\ +\sum J_k\left(m\frac{\pi}{2}M\right)J_h\left(m\frac{\pi}{2}M_3\right)\sin\left([m+k+h]\frac{\pi}{2}\right)\Big|_{3h-k=|n|} \end{array}\right] \tag{5.43}$$

where the terms in the summation expressions are summed for any values of k and h over the ranges $1 \le k \le \infty$, $1 \le h \le \infty$ for which the conditional limitations are met. For example, for $n = +2$, values for (k, h) in the second summation expression would be (2, 0), (5, 1), (8, 2), (1, 1), (4, 2), (7, 3), etc. Note that in principle an infinite summation of Bessel function terms is therefore required for the harmonic coefficients to be calculated for any particular m and n. This is common with the closed-form solution of the more complex modulation strategies. However, experience suggests that only the first 10 or so terms need to be calculated, because of the rapid roll off of the Bessel function magnitudes.

It is appropriate to comment here that for integer carrier/fundamental ratios, harmonics generated in the baseband, the first carrier sidebands, and higher order carrier sidebands may have the same resultant frequency. The extent of this overlap depends on the carrier/fundamental ratio, and the rate at which the sideband magnitudes roll off (which is determined by the modulation strategy). Hence it is important to add as phasor quantities, harmonics from all harmonic group sources that have the same resultant frequency, to correctly identify any particular harmonic cancellation that may occur.

Equation (5.43) is valid for both the sideband harmonic components, and also for the carrier harmonics by making $m > 0, n = 0$.

Taken together, Eqs. (5.36), (5.38), (5.39), and (5.43) define the coefficients of the general form of the phase leg harmonic solution, Eq. (3.10), for *double-edge naturally sampled modulation with third-harmonic injection.*

Figure 5.7 shows the harmonic performance of this modulation scheme, for both a single-phase leg and the resultant *l–l* output voltage. The result is a substantial change in harmonic performance compared to the simple sinusoidal reference of Figure 5.4. In particular, the single-phase leg solution shows the expected significant third-harmonic baseband harmonic component (which of course cancels between the phase legs since it is common mode), and also a flattening and broadening of the first carrier group sideband magnitudes. Effectively, harmonic energy has been channeled from the first pair of sideband harmonics to the outer sideband pairs.

For the *l–l* output voltage, Figure 5.7(b), the expected cancellation of triplen harmonics between the phase legs has again occurred, while the remaining sideband harmonics remain flattened and broadened in magnitude

(a)

(b)

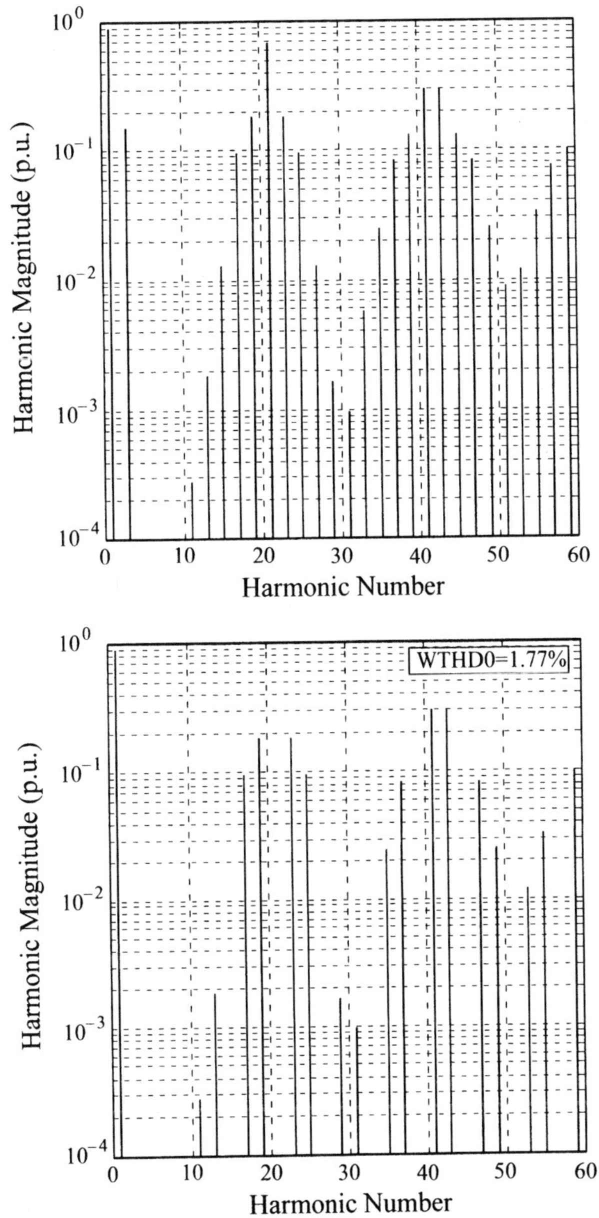

Figure 5.7 Theoretical harmonic spectra for three-phase inverter modulated by *double-edge naturally sampled* PWM *with one-sixth third-harmonic injection*: (a) phase leg a and (b) l–l output switched voltage waveforms, $M = 0.9$, $f_c/f_o = 21$.

compared to a simple sinusoidal reference, Figure 5.4(b). The result is a significant reduction in WTHD0, from 2.04 to 1.77%, that occurs entirely because of the redistribution of the sideband harmonic energy. This effect is a fundamental benefit of using references containing a third-harmonic component.

Once again, theoretical analysis of *single-edge,* and *double-edge symmetrical* and *asymmetrical regularly sampled* modulation with *third-harmonic injection* proceeds using the techniques developed in Chapter 3, by replacing y with $y = y' + (\omega_o/\omega_c)x$ in Eq. (5.35), to give phase leg solutions of

$$A_{mn} + jB_{mn} = \frac{V_{dc}}{\pi^2} \int_{-\pi}^{\pi} \int_{-\pi}^{\pi(M\cos y' + M_3 \cos 3y')} e^{j\left(mx + n\left[y' + \frac{\omega_o}{\omega_c}x\right]\right)} dx\, dy' \tag{5.44}$$

for *single-edge regularly sampled* modulation,

$$A_{mn} + jB_{mn} = \frac{V_{dc}}{\pi^2} \int_{-\pi/2}^{\pi/2} \int_{-\frac{\pi}{2}(1 + M\cos y' + M_3 \cos 3y')}^{\frac{\pi}{2}(1 + M\cos y' + M_3 \cos 3y')} e^{j\left(mx + n\left[y' + \frac{\omega_o}{\omega_c}x\right]\right)} dx\, dy' \tag{5.45}$$

for *double-edge symmetrical regularly sampled* modulation, and

$$A_{mn} + jB_{mn} = \frac{V_{dc}}{\pi^2} \left[\int_{-\pi/2}^{\pi/2} \int_{-\frac{\pi}{2}\left[1 + M\cos y_r' + M_3 \cos 3y_r'\right]}^{0} e^{j\left(mx + n\left[y_r' + \frac{\omega_o}{\omega_c}x + \frac{\omega_o}{\omega_c}\frac{\pi}{2}\right]\right)} dx\, dy_r' + \int_{-\pi/2}^{\pi/2} \int_{0}^{\frac{\pi}{2}\left[1 + M\cos y_f' + M_3 \cos 3y_f'\right]} e^{j\left(mx + n\left[y_f' + \frac{\omega_o}{\omega_c}x - \frac{\omega_o}{\omega_c}\frac{\pi}{2}\right]\right)} dx\, dy_f' \right] \tag{5.46}$$

for *double-edge asymmetrical regularly sampled* modulation.

Solutions for these three modulation variations can be obtained using similar analysis principles as presented above for naturally sampled PWM, and are as follows:

$$A_{m0} + jB_{m0} = \frac{-2jV_{dc}}{q\pi}[J_0(m\pi M)J_0(m\pi M_3) - e^{jm\pi}] \tag{5.47}$$

$$A_{mn} + jB_{mn} = \frac{-2jV_{dc}}{q\pi}\left[\left[\begin{array}{l} J_0(q\pi M_3)J_k(q\pi M)j^k\Big|_{k=|n|} \\ + J_0(q\pi M)J_h(q\pi M_3)j^h\Big|_{3h=|n|} \\ + \sum J_k(q\pi M)J_h(q\pi M_3)j^{[k+h]}\Big|_{k+3h=|n|} \\ + \sum J_k(q\pi M)J_h(q\pi M_3)j^{[k+h]}\Big|_{k-3h=|n|} \\ + \sum J_k(q\pi M)J_h(q\pi M_3)j^{[k+h]}\Big|_{3h-k=|n|} \end{array}\right]\right]_{n \neq 0}$$

for *single-edge regularly sampled* modulation, and

$$A_{mn} + jB_{mn} = \frac{4V_{dc}}{q\pi}\left[\begin{array}{l} J_0\left(q\frac{\pi}{2}M_3\right)J_k\left(q\frac{\pi}{2}M\right)\sin\left([q+k]\frac{\pi}{2}\right)\Big|_{k=|n|} \\ + J_0\left(q\frac{\pi}{2}M\right)J_h\left(q\frac{\pi}{2}M_3\right)\sin\left([q+h]\frac{\pi}{2}\right)\Big|_{3h=|n|} \\ + \sum J_k\left(q\frac{\pi}{2}M\right)J_h\left(q\frac{\pi}{2}M_3\right)\sin\left([q+k+h]\frac{\pi}{2}\right)\Big|_{k+3h=|n|} \\ + \sum J_k\left(q\frac{\pi}{2}M\right)J_h\left(q\frac{\pi}{2}M_3\right)\sin\left([q+k+h]\frac{\pi}{2}\right)\Big|_{k-3h=|n|} \\ + \sum J_k\left(q\frac{\pi}{2}M\right)J_h\left(q\frac{\pi}{2}M_3\right)\sin\left([q+k+h]\frac{\pi}{2}\right)\Big|_{3h-k=|n|} \end{array}\right] \tag{5.48}$$

for *double-edge symmetrical regularly sampled* modulation, and

$$A_{mn} + jB_{mn} = \frac{4V_{dc}}{q\pi}\left[\begin{array}{l} J_0\left(q\frac{\pi}{2}M_3\right)J_k\left(q\frac{\pi}{2}M\right)\sin\left([m+k]\frac{\pi}{2}\right)\Big|_{k=|n|} \\ +J_0\left(q\frac{\pi}{2}M\right)J_h\left(q\frac{\pi}{2}M_3\right)\sin\left([m+h]\frac{\pi}{2}\right)\Big|_{3h=|n|} \\ +\sum J_k\left(q\frac{\pi}{2}M\right)J_h\left(q\frac{\pi}{2}M_3\right)\sin\left([m+k+h]\frac{\pi}{2}\right)\Big|_{k+3h=|n|} \\ +\sum J_k\left(q\frac{\pi}{2}M\right)J_h\left(q\frac{\pi}{2}M_3\right)\sin\left([m+k+h]\frac{\pi}{2}\right)\Big|_{k-3h=|n|} \\ +\sum J_k\left(q\frac{\pi}{2}M\right)J_h\left(q\frac{\pi}{2}M_3\right)\sin\left([m+k+h]\frac{\pi}{2}\right)\Big|_{3h-k=|n|} \end{array}\right] \tag{5.49}$$

for *double-edge asymmetrical regular sampled* modulation, over the limits $m = 0, 1 \le n \le \infty$, $1 \le m \le \infty, -\infty \le n \le \infty$ for all cases and with $q = m + n(\omega_o/\omega_c)$.

Note that once again, the process of sampled modulation produces a decaying series of baseband harmonic components. Also, asymmetrical sampling continues to naturally eliminate the *even* baseband harmonic components, the *odd* sideband harmonics around the *odd* carrier multiples, and the *even* sideband harmonics around the *even* carrier multiples, within the phase leg.

Figure 5.8 shows the change in harmonic spectra of the *l–l* output voltage for the various third-harmonic injection alternatives that have been presented. It can be seen in this figure that there is a slight reduction in WTHD0 for a one-quarter magnitude compared to a one-sixth magnitude third-harmonic component. This is primarily because the sideband harmonics in the first carrier group are flattened more with the one-quarter magnitude component, to the point where all four major sideband harmonics have almost the same magnitude. This is clearly the point of optimum contribution to WTHD. However, the benefit is very slight for the modulation conditions shown, as is often the case.

More interesting is the effect of symmetrical sampling and single-edge modulation on the baseband harmonic components. It has, of course, already been identified that these modulation strategies produce additional harmonic components because of the poorer harmonic cancellation that occurs within the phase leg, including, in particular, additional baseband harmonics. When a

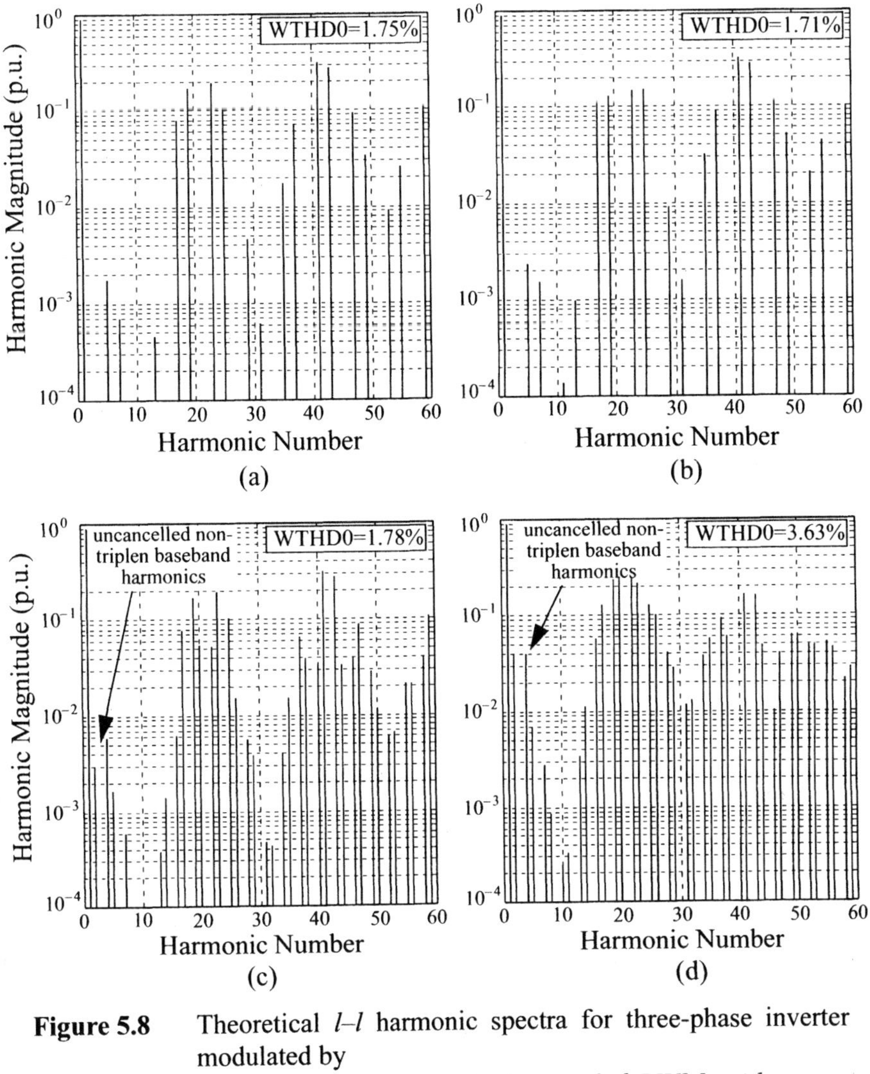

Figure 5.8 Theoretical *l–l* harmonic spectra for three-phase inverter modulated by

(a) *asymmetrical regularly sampled* PWM *with one-sixth magnitude third-harmonic,*

(b) *asymmetrical regularly sampled* PWM *with one-quarter magnitude third-harmonic,*

(c) *symmetrical regularly sampled* PWM *with one-sixth magnitude third-harmonic,*

(d) *single-edge regularly sampled PWM with one-sixth magnitude third-harmonic,*

$$M = 0.9, \; f_c/f_o = 21.$$

third-harmonic component is added to the reference, this effect is enhanced, because the fourth, fifth, ..., baseband components increase in magnitude (because of the deliberately injected third-harmonic component). Obviously the nontriplen baseband harmonics do not cancel and hence become enhanced in the *l–l* output voltage, compared to a simple sinusoidal reference. Figures 5.8(c) and 5.8(d) show this effect clearly and should be directly compared against Figures 5.5(c) and 5.5(d), respectively. However, it is interesting to note that for the modulation conditions used in these figures, the WTHD is still less for third-harmonic injection compared to a simple sinusoidal reference in all cases, primarily because of the contribution of the first major sideband harmonics around the first carrier group. Of course, this influence would be less if a more realistic carrier/fundamental ratio was used instead of the deliberately low ratio of 21 that has been used for illustration purposes in this book.

Figure 5.9 shows WTHD0 versus modulation index for the various three-phase inverter modulation strategies presented so far. Clearly third-harmonic injection into the reference has the advantages of achieving both a greater fundamental output voltage ($M = 1.15$) and a reduced harmonic distortion, compared to a simple sinusoidal reference. Asymmetrical regular sampling also provides an additional WTHD advantage, because of the shift in harmonic energy to the upper sideband harmonics around the first carrier group. However, this benefit depends on the actual carrier/fundamental ratio and is substantially enhanced for the results shown in Figure 5.9 because of the very low ratio of 11 that has been used for illustration. With a more realistic carrier ratio, the difference between natural and regular sampling is usually slight.

It should be noted also that while there is a slight WTHD benefit for injecting a one-quarter magnitude compared to a one-sixth magnitude third-harmonic, there is more benefit to be gained simply by implementing asymmetrical regular sampling in preference to symmetrical regular sampling. This is because the skew in the sideband harmonic magnitudes toward the high side, and the elimination of the even low-order baseband harmonics achieved by asymmetrical sampling, outweighs the slight evening of the sideband harmonic magnitudes achieved by the injection of a one-quarter magnitude third-harmonic. Such a result is not uncommon. Quite often it is the quality of the PWM implementation that determines the performance of a particular system, not a minimal theoretical advantage of one modulation scheme compared to another.

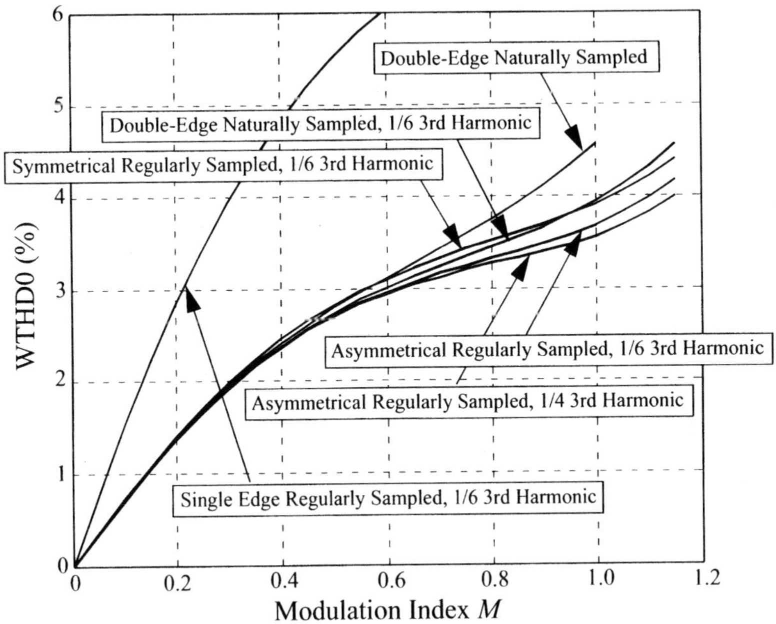

Figure 5.9 WTHD0 of *l–l* switched output voltage of a three-phase inverter for various modulation alternatives, $f_c/f_o = 11$.

5.4 Analytic Calculation of Harmonic Losses

In the same way as was done for a single-phase inverter in Section 4.3, it is useful to derive closed-form solutions for the harmonic currents and consequential losses for three-phase modulation, in order to once more allow comparison against and between the various modulation methods [10]. As before, these solutions are developed under the assumptions that the average internal EMF e_{am} of the load over an arbitrary switching interval ΔT is essentially constant, and the losses in the load can be neglected for the purposes of calculating the ripple current.

For the case of a three-phase inverter, it is convenient to assume that the inductive load is delta connected to avoid issues associated with the zero sequence component. If the load is in fact star connected, the currents which flow can always be readily calculated by means of the usual delta/wye transformation equations.

Figure 5.10 shows a sketch of a single carrier switching period with the load between output phase legs a and b. For each of the periods T_1, T_2, T_3, the current ripple through a delta load element is defined by

$$\underline{0 \le t \le T_1:} \qquad \Delta i_{ab}(t) = -\frac{e_{ab}}{L_\sigma}t \tag{5.50}$$

$$\underline{T_1 \le t \le T_1 + T_2:} \qquad \Delta i_{ab}(t) = \left(\frac{2V_{dc} - e_{ab}}{L_\sigma}\right)(t - T_1) - \frac{e_{ab}}{L_\sigma}T_1 \tag{5.51}$$

$\underline{T_1 + T_2 \le t \le T_1 + T_2 + T_3:}$

$$\Delta i_{ab}(t) = -\frac{e_{ab}}{L_\sigma}(t - T_1 - T_2) + \left(\frac{2V_{dc} - e_{ab}}{L_\sigma}\right)T_2 - \frac{e_{ab}}{L_\sigma}T_1 \tag{5.52}$$

where L_σ is now the l–l leakage inductance of one element of a delta load.

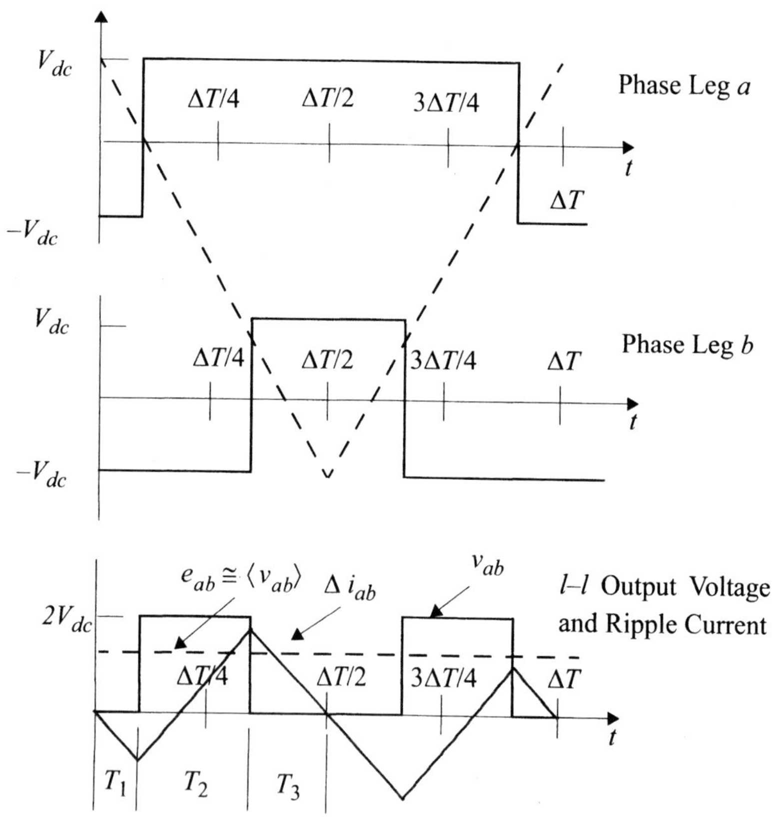

Figure 5.10 Modulation process and resulting delta element ripple current for two phase legs of three-phase VSI.

Substituting $t = T_1 + T_2 + T_3 = \Delta T/2$ into Eq. (5.52) gives

$$\Delta i_{ab}\left(\frac{\Delta T}{2}\right) = -\frac{e_{ab}}{L_\sigma}T_3 + \left(\frac{2V_{dc} - e_{ab}}{L_\sigma}\right)T_2 - \frac{e_{ab}}{L_\sigma}T_1 = 0 \qquad (5.53)$$

from which it can be shown that

$$T_2 = (u_1 - u_2)\frac{\Delta T}{4} \qquad (5.54)$$

where $u_1 = \frac{e_a}{V_{dc}}$, $u_2 = \frac{e_b}{V_{dc}}$, and $u_1 - u_2 = \frac{e_{ab}}{V_{dc}}$. Similarly, it follows that

$$T_1 = (1 - u_1)\frac{\Delta T}{4} \quad \text{and} \quad T_3 = (1 + u_2)\frac{\Delta T}{4} \qquad (5.55)$$

The average squared value of the current ripple over the interval $\Delta T/2$ can now be written in the form

$$\langle \Delta i_{ab}^2 \rangle = \left(\frac{V_{dc}}{L_\sigma}\right)^2 \frac{2}{\Delta T}\left[\begin{array}{l} \int_0^{T_1} (u_2 - u_1)^2 t^2 dt \\ + \int_0^{T_2} [(2 + u_2 - u_1)x + (u_2 - u_1)T_1]^2 dx \\ + \int_0^{T_3} [(u_2 - u_1)x' + (2 + u_2 - u_1)T_2 + (u_2 - u_1)T_1]^2 dx' \end{array}\right] \qquad (5.56)$$

where $x = t - T_1$ and $x' = t - T_1 - T_2$. Evaluating Eq. (5.56) gives

$$\langle \Delta i_{ab}^2 \rangle = \left(\frac{V_{dc}}{L_\sigma}\right)^2 \frac{1}{3}\frac{2}{\Delta T}\left[\begin{array}{l} (u_2 - u_1)^2 T_1^{\,3} \\ + \dfrac{[(2 + u_2 - u_1)T_2 + (u_2 - u_1)T_1]^3 - [(u_2 - u_1)T_1]^3}{2 + u_2 - u_1} \\ - \dfrac{[(2 + u_2 - u_1)T_2 + (u_2 - u_1)T_1]^3}{u_2 - u_1} \end{array}\right] \qquad (5.57)$$

Substituting for T_1, T_2, and T_3 using Eqs. (5.54) and (5.55) gives

$$\langle \Delta i_{ab}^2 \rangle = \left(\frac{V_{dc}}{L_\sigma}\right)^2 \frac{\Delta T^2}{96}\left[(u_2-u_1)^2(1-u_1)^3 - \frac{[(u_1-u_2)(1+u_2)]^3}{u_2-u_1} + \frac{[(u_1-u_2)(1+u_2)]^3 - [(u_2-u_1)(1-u_1)]^3}{2+u_2-u_1}\right]$$

$$= \left(\frac{V_{dc}}{L_\sigma}\right)^2 \frac{\Delta T^2}{48}(u_2-u_1)^2\left\{\frac{(1+u_2)^3+(1-u_1)^3}{(1+u_2)+(1-u_1)}\right\} \tag{5.58}$$

After some manipulation, Eq. (5.58) simplifies to

$$\langle \Delta i_{ab}^2 \rangle = \left(\frac{V_{dc}}{L_\sigma}\right)^2 \frac{\Delta T^2}{48}\{(u_2-u_1)^2+(u_2-u_1)^3+(u_2-u_1)(u_2^3-u_1^3)\} \tag{5.59}$$

The harmonic losses can then be determined by integrating Eq. (5.59) over a positive half fundamental cycle (the above development is only valid for $e_{ab} > 0$), with appropriate substitutions for u_1 and u_2 for each modulation strategy to be evaluated, to determine the (squared) harmonic current ripple, and then multiplying this result by the equivalent load resistance R_e, i.e.,

$$P_{h,\,cu} = R_e \frac{1}{\tau}\int_0^\tau \langle \Delta i_{ab}^2 \rangle \, d\theta_o \tag{5.60}$$

For a single-phase inverter, $u_1 = -u_2 = M\cos\theta_o$. Substituting this simplification into Eq. (5.59) gives

$$\langle \Delta i_{ab}^2 \rangle = \left(\frac{V_{dc}}{L_\sigma}\right)^2 \frac{\Delta T^2}{12}(M\cos\theta_o)^2(1-M\cos\theta_o)^2 \tag{5.61}$$

which is the same result as was achieved in Eq. (4.31) in Chapter 4.

For a three-phase inverter modulated with simple sinusoidal references, the phase leg reference voltages are given by $u_1 = M\cos\theta_o$ and $u_2 = M\cos(\theta_o - 2\pi/3)$. Substituting these definitions into Eq. (5.59), with θ_o replaced by θ for brevity, gives

$$\langle \Delta i_{ab}^2 \rangle = \left(\frac{V_{dc}}{L_\sigma}\right)^2 \frac{\Delta T^2}{48}\left[3M^2\cos^2\left(\theta+\frac{\pi}{6}\right) - 3\sqrt{3}M^3\cos^3\left(\theta+\frac{\pi}{6}\right) - \sqrt{3}M^4\cos\left(\theta+\frac{\pi}{6}\right)\left(\cos^3\left(\theta-\frac{2\pi}{3}\right)-\cos^3\theta\right)\right] \tag{5.62}$$

Figure 5.11 shows the variation in the delta load harmonic current for a three-phase inverter under this modulation strategy over a fundamental cycle for a modulation index of $M = 0.9$ with a highly inductive load. The asymmetry and harmonic distortion of this current compared to the harmonic load current for a single-phase inverter as shown in Figure 4.13 is obvious and reflects the changing position of the active *l–l* pulse within each carrier interval over the fundamental cycle caused by the three-phase modulation process.

The average RMS harmonic current can now be determined by integrating Eq. (5.62) over the positive half fundamental cycle of the *l–l* voltage, i.e., over the interval $-2\pi/3 < \theta_o < \pi/3$. Only the positive half cycle need be considered, since it is clear from Figure 5.11 that the current ripple during the negative half cycle is a mirror image of the positive half cycle, and hence will have the same RMS harmonic contribution. The integral expression is

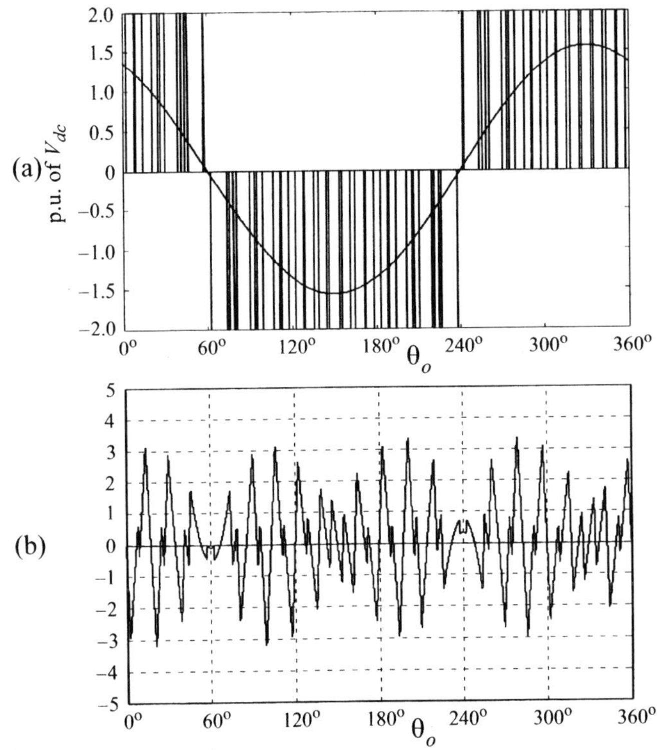

Figure 5.11 (a) Reference and switched *l–l* voltages for three-phase inverter with simple sinusoidal PWM and (b) harmonic current ripple with inductive load, $M = 0.9$, $f_c/f_o = 21$.

$$I_{ab,\,h,\,rms}^2 = \left(\frac{V_{dc}}{L_\sigma}\right)^2 \frac{\Delta T^2}{48} \frac{1}{\pi} \int_{-2\pi/3}^{\pi/3} \begin{bmatrix} 3M^2\cos^2\left(\theta + \frac{\pi}{6}\right) - 3\sqrt{3}M^3\cos^3\left(\theta + \frac{\pi}{6}\right) \\ -\sqrt{3}M^4\cos\left(\theta + \frac{\pi}{6}\right)\left(\cos^3\left(\theta - \frac{2\pi}{3}\right) - \cos^3\theta\right) \end{bmatrix} d\theta \tag{5.63}$$

which after considerable manipulation and simplification, reduces to

$$I_{ab,\,h,\,rms}^2 = \left(\frac{V_{dc}}{L_\sigma}\right)^2 \frac{\Delta T^2}{48}\left(\frac{3}{2}M^2 - \frac{4\sqrt{3}}{\pi}M^3 + \frac{9}{8}M^4\right) \tag{5.64}$$

A similar development can be done for a three-phase inverter modulated with sinusoidal references with a third-harmonic injection component. For this case, the phase leg references are defined by $u_1 = M\cos\theta_o + M_3\cos 3\theta_o$ and $u_2 = M\cos(\theta_o - 2\pi/3) + M_3\cos 3\theta_o$, with the third-harmonic component magnitude being defined by M_3. As discussed in Section 5.3.1, the two most useful levels of third-harmonic injection are $M_3 = -M/6$ and $M_3 = -M/4$.

Note that the addition of a third-harmonic component only effects the M^4 term in Eq. (5.62), since it cancels out in all the $(u_2 - u_1)$ terms in Eq. (5.59). Hence the integration and evaluation of Eq. (5.59) with third-harmonic injection is only a little more complicated and results in an average harmonic current ripple term of

$$I_{ab,\,h,\,rms}^2 = \left(\frac{V_{dc}}{L_\sigma}\right)^2 \frac{\Delta T^2}{48}\left(\frac{3}{2}M^2 - \frac{4\sqrt{3}}{\pi}M^3 + \frac{9}{8}M^4\,[1 + b + b^2]\right) \tag{5.65}$$

where $b = \dfrac{M_3}{M}$.

Figure 5.12 shows the harmonic current ripple waveform over a complete fundamental cycle for a three-phase inverter under PWM modulation with, respectively, one-sixth and one-fourth third-harmonic injection. It can be seen that the effect of the injected third-harmonic component is to improve the symmetry of the ripple current waveform and to reduce the peak excursions, by better centering the active *l–l* pulse within each carrier interval over the fundamental cycle. This in turn reduces the RMS average value of the ripple current and explains the better harmonic performance of PWM modulation with third-harmonic injection.

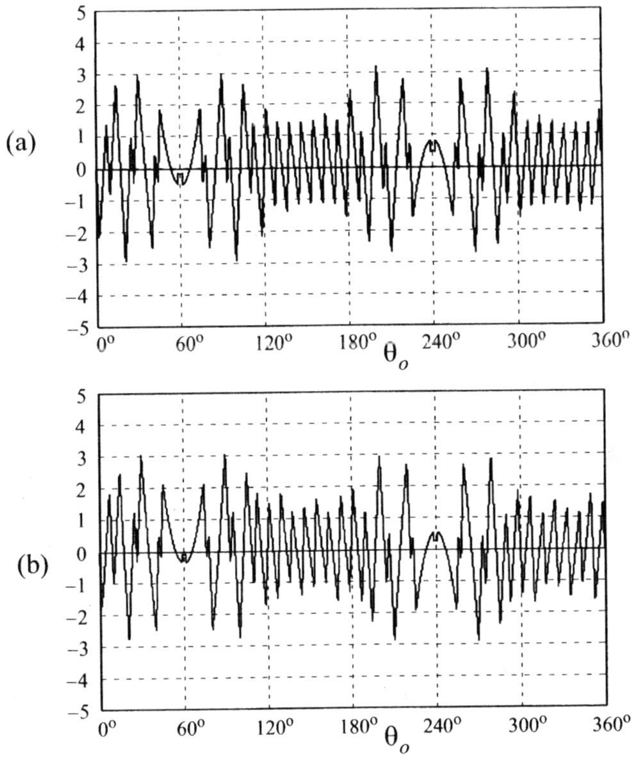

Figure 5.12 Harmonic current ripple with inductive load: (a) one-sixth third-harmonic injection and (b) one-fourth third-harmonic injection, $M = 0.9$, $f_c/f_o = 21$.

Evaluating Eq. (5.65) for the various cases gives

Simple Sinusoidal PWM

$$I_{ab,\,h,\,rms}^2 = \left(\frac{V_{dc}}{L_\sigma}\right)^2 \frac{\Delta T^2}{48}\left(\frac{3}{2}M^2 - \frac{4\sqrt{3}}{\pi}M^3 + \frac{9}{8}M^4\right) \tag{5.66}$$

One-Sixth Third-Harmonic Injection PWM

$$I_{ab,\,h,\,rms}^2 = \left(\frac{V_{dc}}{L_\sigma}\right)^2 \frac{\Delta T^2}{48}\left(\frac{3}{2}M^2 - \frac{4\sqrt{3}}{\pi}M^3 + M^4\right) \tag{5.67}$$

One-Fourth Third-Harmonic Injection PWM

$$I_{ab,\,h,\,rms}^2 = \left(\frac{V_{dc}}{L_\sigma}\right)^2 \frac{\Delta T^2}{48}\left(\frac{3}{2}M^2 - \frac{4\sqrt{3}}{\pi}M^3 + \frac{63}{64}M^4\right) \tag{5.68}$$

or in more general terms as

$$I_{ab,h,rms}^2 = \left(\frac{V_{dc}}{L_\sigma}\right)^2 \frac{\Delta T^2}{48} f(M) \tag{5.69}$$

where the function $f(M)$ is the *harmonic distortion factor* (HDF). HDF is commonly used as a figure of merit for PWM strategies that are independent of switching frequency, DC bus voltage, and load inductance.

The HDFs for the three continuous PWM strategies discussed in this chapter for a three-phase inverter are shown in Figure 5.13, where the performance superiority of the third-harmonic injected modulation strategy for a three-phase system can be clearly seen. For comparison, the HDF for a single-phase inverter taken from Chapter 4 is also shown. It is interesting to note the harmonic superiority of single-phase modulation over all three-phase alternatives, particularly at high modulation levels. But a moment's thought confirms that this is only to be expected, since single-phase three-level modulation eliminates all harmonics up to the second carrier group, in contrast to three-phase modulation which must retain some harmonics in the first carrier group.

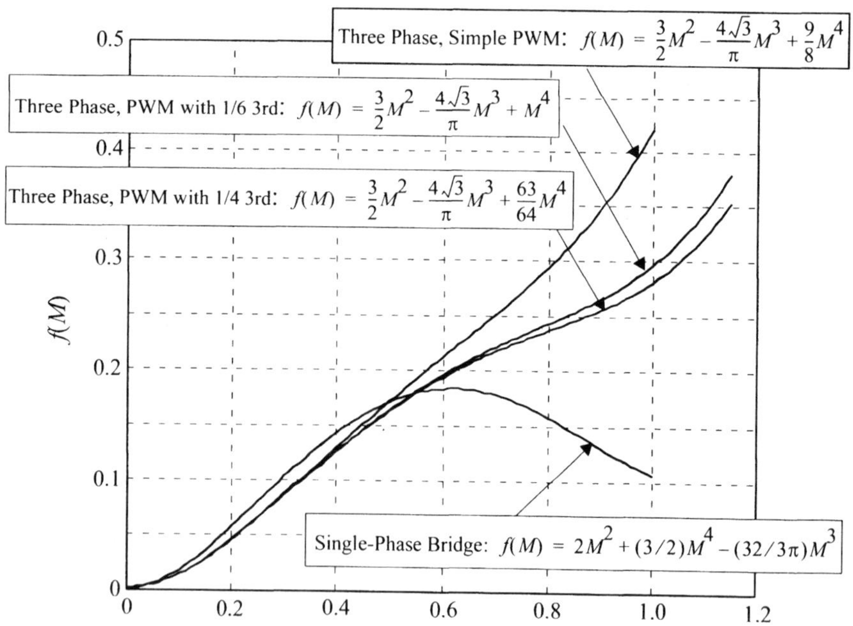

Figure 5.13 Harmonic distortion factors for different PWM modulation strategies as a function of modulation index M for pure inductive load.

The relationship between HDF and WTHD0 is readily derived as follows. By definition from Eq. (2.116)

$$\mathrm{WTHD0} = \frac{\sqrt{\sum_{n=2}^{\infty} \frac{V_n^2}{n^2}}}{V_1|_{M=1}} = \frac{\sqrt{\sum_{n=2}^{\infty} \frac{V_n^2}{n^2}}}{\sqrt{3}V_{dc}} \tag{5.70}$$

where V_1 at $M = 1$ has a magnitude of $\sqrt{3}V_{dc}$ since it is a *l–l* quantity.

Each harmonic voltage component in the *l–l* switched output voltage creates a harmonic current component of

$$I_n = \frac{V_n}{\omega_n L_\sigma} = \frac{V_n}{n}\frac{1}{\omega_o L_\sigma} \tag{5.71}$$

where all voltages and currents in Eq. (5.71) are *peak* quantities.

The RMS magnitude of the harmonic current components taken together is given by

$$I_{ab,h,rms}^2 = \sum_{n=2}^{\infty} \frac{1}{2} I_n^2 \tag{5.72}$$

where the scaling factor of $\frac{1}{2}$ inside the summation is required to convert from *peak* to RMS magnitude. Substituting Eq. (5.71) into Eq. (5.72) gives

$$I_{ab,h,rms}^2 = \sum_{n=2}^{\infty} \frac{1}{2}\left(\frac{V_n}{n}\frac{1}{\omega_o L_\sigma}\right)^2 \tag{5.73}$$

or alternatively

$$\sum_{n=2}^{\infty} \frac{V_n^2}{n^2} = 2\omega_o^2 L_\sigma^2 I_{ab,h,rms}^2 \tag{5.74}$$

Substituting this result into Eq. (5.70) gives an alternative expression for WTHD0 of

$$\mathrm{WTHD0} = \frac{\sqrt{2}\omega_o L_\sigma I_{ab,h,rms}}{\sqrt{3}V_{dc}} \tag{5.75}$$

Further substituting from Eq. (5.69) gives

$$\text{WTHD0} = \frac{\sqrt{2}\omega_o L_\sigma \sqrt{\left(\frac{V_{dc}}{L_\sigma}\right)^2 \frac{\Delta T^2}{48} f(M)}}{\sqrt{3}V_{dc}}$$

$$= \frac{\omega_o \Delta T \sqrt{f(M)}}{\sqrt{72}} \tag{5.76}$$

Finally, since

$$\Delta T = \frac{1}{f_c} \tag{5.77}$$

Eq. (5.76) simplifies to

$$\text{WTHD0} = \frac{2\pi f_o}{f_c}\sqrt{\frac{f(M)}{72}} = \frac{\pi}{f_c/f_o}\sqrt{\frac{\text{HDF}}{18}} = \frac{\pi}{p}\sqrt{\frac{\text{HDF}}{18}} \tag{5.78}$$

where f_c/f_o is often referred to as the *pulse ratio, p*.

However, it is important to appreciate that Eq. (5.78) becomes progressively less exact as the carrier/fundamental frequency ratio is reduced since $f(M)$ was derived assuming a sufficiently large pulse ratio so that the load EMF over each half carrier interval could be assumed to be constant.

Finally, Figure 5.14 shows the WTHD0 for sinusoidal PWM with one-sixth third-harmonic injection for the cases of $p = 5, 7, 10, 15, 20$, and 25. It can be recalled from Eq. (2.31) that the WTHD for simple six-step square-wave modulation (no PWM) is 4.64%. Once this result is scaled by $4/\pi$ to convert to WTHD0, it can be shown as a straight line in Figure 5.14 with a value of 5.91%. Hence, for a three-phase inverter, the carrier/fundamental frequency ratio must be at least 7 for the WTHD of a PWM system to be less than that of six-step modulation at $M = 1$. This provides a convenient reference point for the decision to use a PWM strategy versus another more complex modulation strategy in particular applications (such as very high power inverters).

5.5 Discontinuous Modulation Strategies

Looking again at Figure 5.3, it can be seen that within each half carrier interval, there is an initial *zero l–l* output voltage pulse labeled V_Z, two sequential active voltage pulses labeled V_1 and V_2, and a trailing *zero l–l* output voltage

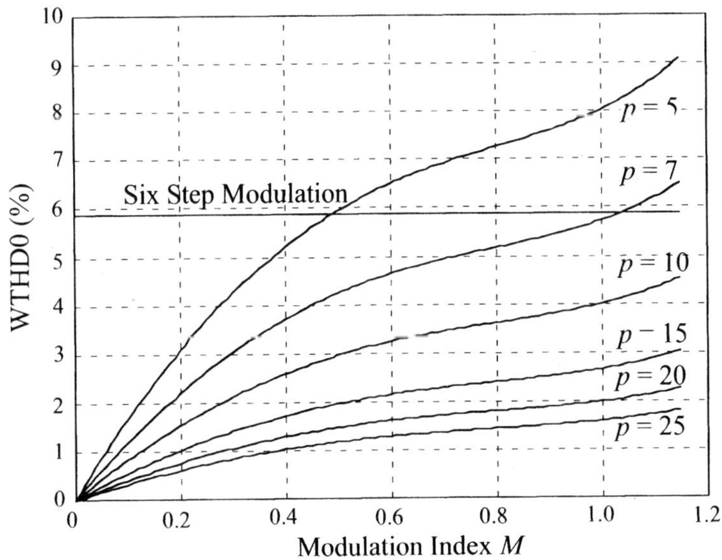

Figure 5.14 WTHD0 versus modulation index M for sinusoidal PWM with a one-sixth third-harmonic injection for increasing carrier/fundamental frequency ratios.

pulse again labeled V_Z. Clearly, as was presented in Section 4.5.2 for single-phase inverters, it is possible to move the position of the active voltage pulses around within the half carrier interval, to eliminate one zero output voltage pulse while still retaining the same average volt–seconds over the carrier interval. Modulation strategies using this concept are termed *discontinuous modulation*, and a number of possible alternatives for three-phase inverter systems have been reported over the years. However, since all these schemes essentially just rearrange the placement of the zero output voltage pulse(s) within each half carrier or carrier interval, their treatment will be deferred until Chapter 6 where they can be discussed using *space vector* concepts.

5.6 Triplen Carrier Ratios and Subharmonics

5.6.1 Triplen Carrier Ratios

It is firmly entrenched in the literature that three-phase inverter systems require a carrier that is an odd triplen multiple of the fundamental reference frequency in order to achieve optimum harmonic cancellation between the phase legs, particularly for low carrier/fundamental pulse ratios. However, the material presented in this chapter shows no justification at all for this belief, and it is clear, from the analytical solutions presented in Sections 5.2 and 5.3, that the cancellation of harmonics between phase legs is independent of the frequency ratio between the carrier and the fundamental. Carrier harmonics cancel between phase legs because they are the same, while triplen sideband harmonics cancel between the phase legs because of the 360° rotation that occurs when n = *triplen*. Other sideband harmonics do not cancel.

It is interesting to speculate on how this very widespread belief of the requirement for an *odd triplen* carrier pulse ratio may have occurred. One explanation is that early researchers [11] anticipated that the harmonics caused by the PWM process would be integer multiples of the fundamental (assuming an integer carrier ratio). This assumption would easily extend from an appreciation of the harmonics produced by SCR type systems, which always occur at simple integer multiples of the fundamental. Hence the general harmonic forms for the switched waveforms of phase legs a, b, and c may have been expected to be

$$v_{az}(t) = \sum_{n=1}^{\infty} [A_{0n}\cos(n\omega_o t) + B_{0n}\sin(n\omega_o t)] \tag{5.79}$$

$$v_{bz}(t) = \sum_{n=1}^{\infty} \left[A_{0n}\cos\left(n\left[\omega_o t - \frac{2\pi}{3}\right]\right) + B_{0n}\sin\left(n\left[\omega_o t - \frac{2\pi}{3}\right]\right)\right] \tag{5.80}$$

$$v_{cz}(t) = \sum_{n=1}^{\infty} \left[A_{0n}\cos\left(n\left[\omega_o t - \frac{2\pi}{3}\right]\right) + B_{0n}\sin\left(n\left[\omega_o t - \frac{2\pi}{3}\right]\right)\right] \tag{5.81}$$

With this formulation, it is obvious that harmonics would only cancel between the phase legs when n is a triplen integer multiple, which may well have led to the assumption that the carrier frequency had to be a triplen multiple of the fundamental to cancel between phases. In fact, of course, as has been shown, this is simply not the case.

The requirement for an *odd* carrier ratio may well derive from the fairly classical consideration that *even* harmonics represent a DC offset, or at least asymmetrical AC half cycles. Such a waveform is considered to be intrinsically unbalanced for an AC system, and should therefore be avoided if at all possible. However, once again, since the major harmonics of PWM are the sidebands that occur around the carrier multiples, there seems to be no particular reason to require an odd carrier/fundamental ratio.

The analytical solutions presented in this chapter show an important harmonic result. First, the frequency of the carrier harmonic is independent of the fundamental reference frequency, and hence always cancels between phase legs. Second, it is the "triplen multiple of n" sideband harmonics that cancel between the phase legs, with their frequencies matched irrespective of the absolute carrier frequency, and their phase shifts matched to the 120° shift of the three-phase references (times the n multiple). With this understanding, it is clear that there is no particular benefit to be gained by a triplen carrier pulse ratio.

5.6.2 Subharmonics

A further constraint that is often presented in the literature is that the carrier should be an integral ratio of the fundamental frequency to avoid *subharmonics*, particularly for low pulse ratios. Subharmonics are harmonic components with a frequency that is less than the fundamental and can cause considerable heating losses in AC motors if they are close to 0 Hz.

Once again, the analytical solutions for all forms of PWM applied to single- and three-phase inverters show that no such harmonic components are created as a result of the modulation process. The only possible way that modulation-induced harmonics would have a frequency below the fundamental is if the carrier frequency is so low that the lower sideband harmonics of the first carrier group intrude below the fundamental. This would imply that $\omega_c - 4\omega_o < \omega_o$, or $\omega_c < 5\omega_o$. While virtually no PWM system would operate with a pulse ratio this low, this result does suggest some justification for setting the minimum carrier pulse ratio to 7. At this ratio, the lowest significant sideband occurs at $\omega_c - 4\omega_o = 3\omega_o$, which may have some advantages in particular applications. But it should be noted that this apparently triplen harmonic is *not* co-phasal between the phase legs and will not cancel in the *l–l* output voltages.

It is appreciated by the authors that these conclusions are inconsistent with many reported observations of very low frequency subharmonics in drive systems using PWM-controlled inverters, particularly when the carrier /fundamental ratio is low. There are two explanations for this discrepancy between theory and observed practice.

First, it is possible that poorer quality PWM implementations were used in some of the reported work. This is perhaps more likely for earlier systems that typically used analog reference and carrier waveforms. Slight errors in these waveforms, particularly in the fundamental frequency range, would cause a variety of intermodulation effects that could well result in unwanted very low frequency harmonics. This explanation would be consistent with the significantly reduced reports of subharmonics in more recent PWM literature.

The second explanation stems from a basic assumption in the solutions presented in this book — that the DC bus voltage is exactly constant. In practice, the DC bus voltage will vary with load, and will often also have a six times AC input frequency ripple component because of the diode input rectifier that is conventionally used. This ripple frequency interacts with both the output frequency and the carrier frequency, to produce a rich spectrum of additional harmonic components. Some of these may well be at subharmonic frequencies. Further analysis of this interaction is beyond the scope of this book and is left for future research.

Figure 5.15 presents some experimental results for a three-phase inverter modulated by asymmetrical regular sampled PWM under the conditions of $M = 0.9$ at a carrier frequency of 1000 Hz. The inverter DC bus supply is provided from a 200 V switched mode power supply to provide a clean constant DC supply voltage with minimal differential or common mode AC harmonics and was operated into a very small resistive load of 200 W per phase, to minimize deadtime effects.

Comparison of these experimental results with the previous theoretical results of Figures 5.5(b) and 5.8(a) shows an extremely close match for harmonics of magnitude greater than 0.1%. Below this level, it is considered that deadtime and other practical switching effects have influenced the experimentally measured harmonics.

Figure 5.16 shows an expanded low-order spectrum of one phase leg of the inverter with a fundamental frequency of 44 Hz, where the DC bus has been alternatively supplied from a high-frequency switched mode DC power supply, and from a simple rectified single-phase AC supply.

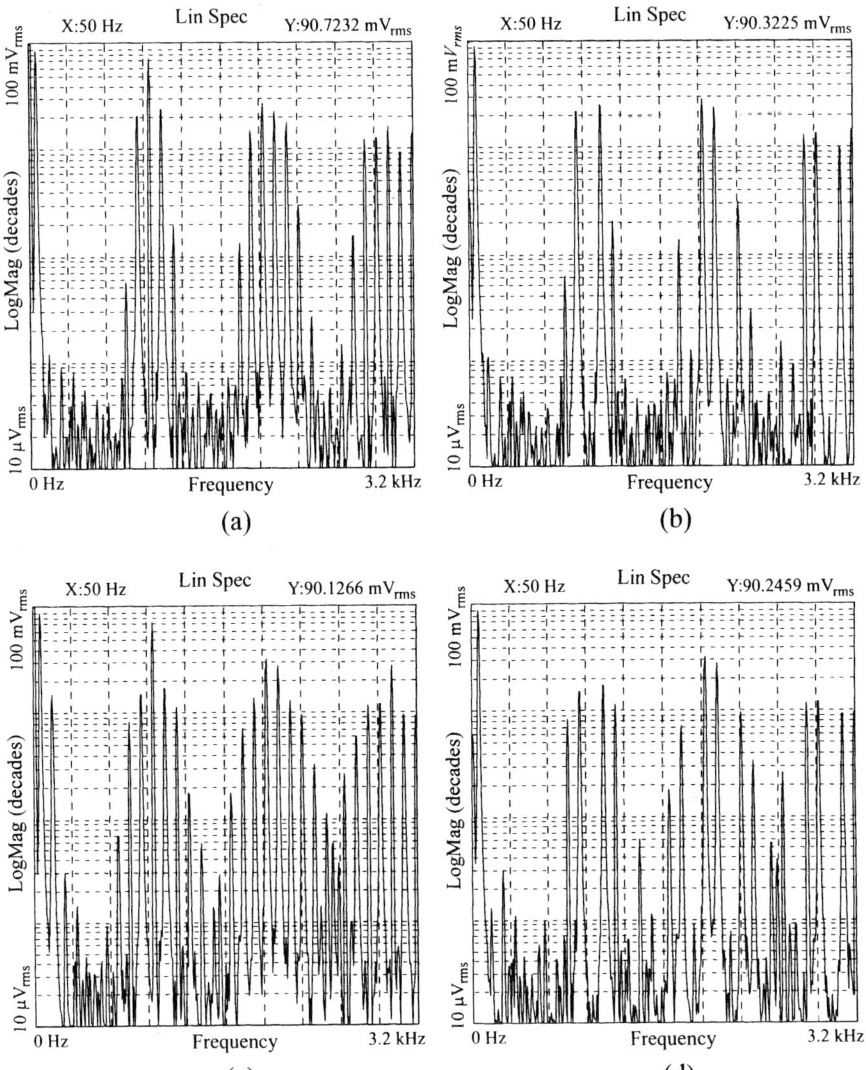

Figure 5.15 Experimental harmonic spectra for three-phase inverter modulated by *asymmetrical regularly sampled* PWM:

(a) phase voltage, simple sinusoidal reference,

(b) *l–l* voltage, simple sinusoidal reference,

(c) phase voltage, one-sixth third-harmonic injected,

(d) *l–l* voltage, onr-sixth third-harmonic injected,

$M = 0.9,\ f_o = 50$ Hz, $f_c = 1000$ Hz.

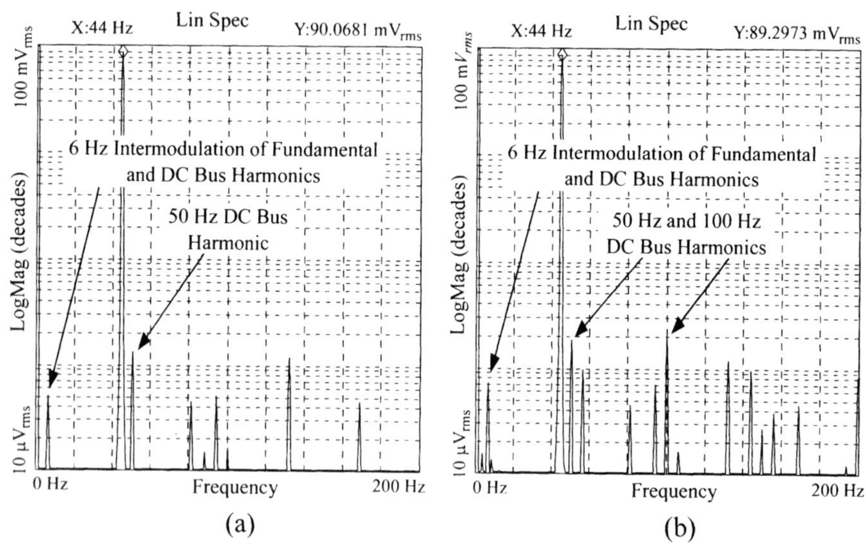

Figure 5.16 Low-frequency experimental harmonic spectra for three-phase inverter modulated by *asymmetrical regularly sampled* PWM:
(a) DC bus supplied by 2500 W DC power supply and
(b) DC bus supplied by rectified single-phase AC supply,
$M = 0.9$, $f_o = 44$ Hz, $f_c = 1000$ Hz.

In these results, it is particularly interesting to see 50 Hz and multiples of this frequency in the phase leg output spectra (note that these experimental results were measured in Australia, where the mains supply frequency is 50 Hz). Investigations showed that these harmonics were present on the DC bus itself, despite the substantial filtering that was implemented. In fact, it was also suspected that some of these harmonics came from common mode injection through the mains connection of the power supplies of the instrumentation systems and the inverter controller itself, but this was extremely difficult to precisely quantify.

However, irrespective of the source of these mains input harmonics, it is clear that their effect is to create a very low frequency harmonic of 6 Hz, most likely because of intermodulation between the target fundamental output frequency of 44 Hz and the mains supply of 50 Hz. Clearly, this harmonic has nothing to do with the carrier frequency and the modulation process. These results support the explanation that "subharmonics" as reported in the literature can be caused by incoming mains ripple on the DC bus interacting with the target output frequency, independently of any particular modulation process.

5.7 Summary

This chapter has extended the concepts of naturally and regularly sampled pulse width determination from the single-phase development of Chapter 4 to a three-phase inverter. It has shown the cancellation of the *l–l* harmonic components that occurs compared to the phase leg components, including in particular the triplen sideband harmonics. Also, while the modulation index for both regularly and naturally sampled PWM is limited to unity for a simple sinusoidal reference, it has been further shown that with the simple addition of a one-sixth third-harmonic to the modulation reference waveforms, the modulation index can be increased to 1.15.

References

[1] D.G. Holmes, "The significance of zero space vector placement for carrier-based PWM schemes," *IEEE Trans. on Industry Applications,* vol. 32, no. 5, Oct. 1996, pp. 1122–1129.

[2] G. Buja and G. Indri, "Improvement of pulse width modulation techniques," *Archiv für Elektrotechnik*, vol. 57, 1975, pp. 281–289.

[3] R. Bonert and R.S. Wu, "Improved three phase pulsewidth modulation for overmodulation," *IEEE Trans. on Industry Applications,* vol. IA–20, no. 5, Sept./Oct. 1985, pp. 1224–1228.

[4] D.A. Grant, J.A. Houldsworth, and K. Lower, "A new high-quality PWM ac drive," in *Conf. Rec. IEEE Industry Applications Society Annual Mtg.*, 1982, pp. 530–535.

[5] D.A. Grant, M. Stevens, and J.A. Houldsworth, "The effect of word length on the harmonic content of microprocessor-based PWM waveform generators," *IEEE Trans. on Industry Applications,* vol. IA–21, no. 1, Jan./Feb. 1985, pp. 218–225.

[6] S. Fukuda, H. Hasegawa, and Y. Iwaji, "PWM technique for inverter with sinusoidal output current," *IEEE Trans. on Power Electronics*, vol. 5, no. 1, Jan. 1990, pp. 54–61.

[7] S.R. Bowes and A. Midoun, "Suboptimal switching strategies for microprocessor-controlled PWM inverter drives," *IEE Proceedings (London),* vol. 132, Pt. B, no. 3., May 1985, pp. 133–148.

[8] J.T. Boys and B.E. Walton, "A loss minimised sinusoidal PWM inverter," *IEE Proceedings (London),* vol. 132, Pt. B, no. 5, Sept. 1985, pp. 260–268.

[9] D.G. Holmes, "A general analytical method for determining the theoretical harmonic components of carrier based PWM strategies," in *Conf. Rec. IEEE Industry Applications Society Annual Mtg*, St. Louis, 1998, pp. 1207–1214.

[10] H.W. Van der Broeck and H.C. Skudelny, "Analytical analysis of the harmonic effects of a PWM ac drive," *IEEE Trans. on Power Electronics,* vol. 3, no. 2, March/April, 1988, pp. 216–223.

[11] A. Schonung and H. Stemmler, "Static frequency changers with subharmonic control in conjunction with reversible variable speed AC drives," *Brown Boveri Review*, 1964, pp. 555–577.

6

Zero Space Vector Placement Modulation Strategies

In the development of three-phase naturally and regularly sampled PWM presented in Chapter 5, the placement of the switched pulses for each phase leg was explicitly defined by the modulation strategy. Consequently, there was no opportunity for variation of this placement within the basic definition of the algorithm. However, pulse placement within each half carrier period was shown to have a significant effect on both the VSI maximum output voltage (modulation index) and the harmonic performance of the modulation implementation, as demonstrated by the effect of adding a third-harmonic component to the sinusoidal reference component.

In this chapter, modulation techniques which explicitly vary pulse placement are presented and reviewed in terms of their increased modulation gain and spectral advantages. However, it is commented that all these strategies simply manipulate the placement of the inactive V_z intervals within each half carrier period. Hence they are really just pulse position variations of the common approach to PWM that has been presented already.

6.1 Space Vector Modulation

6.1.1 Principles of Space Vector Modulation

In the mid-1980s a form of PWM called *space vector modulation* (SVM) was proposed, which was claimed to offer significant advantages over natural and regular sampled PWM in terms of performance, ease of implementation, and maximum transfer ratio [1, 2, 3]. In this section, the fundamentals of SVM are presented, and SVM is identified as simply an alternative method for determining switched pulse widths. In fact, the main benefit of SVM is the explicit identification of pulse placement as an additional degree of freedom that can be exploited to achieve harmonic performance gains.

The principle of SVM is based on the fact that there are only eight possible switch combinations for a three-phase inverter. The basic inverter switch states were discussed in Chapter 1 and are shown again in Figure 6.1. Two of these states ($\overline{SV_0}$ and $\overline{SV_7}$) correspond to a short circuit on the output, while the other six can be considered to form stationary vectors in the *d–q* complex plane as shown in Figure 6.2. Note that each stationary vector corresponds to a particular fundamental angular position as shown in Figure 1.23. The magnitude of each of the six active vectors is, from Eq. (1.50),

$$V_m = \frac{4}{3}V_{dc} \tag{6.1}$$

Having identified the stationary vectors, at any point in time, an arbitrary target output voltage vector $\overline{V_o^*}$ can be formed by the summation ("averaging") of a number of these *space vectors* within one switching period $\Delta T/2$. This is shown in Figure 6.3 for a target phasor in the first 60° segment of the plane. From geometric considerations, the minimum number of active space vector components required to create any arbitrary vector on an average basis is at least two [4] but could be three or more.

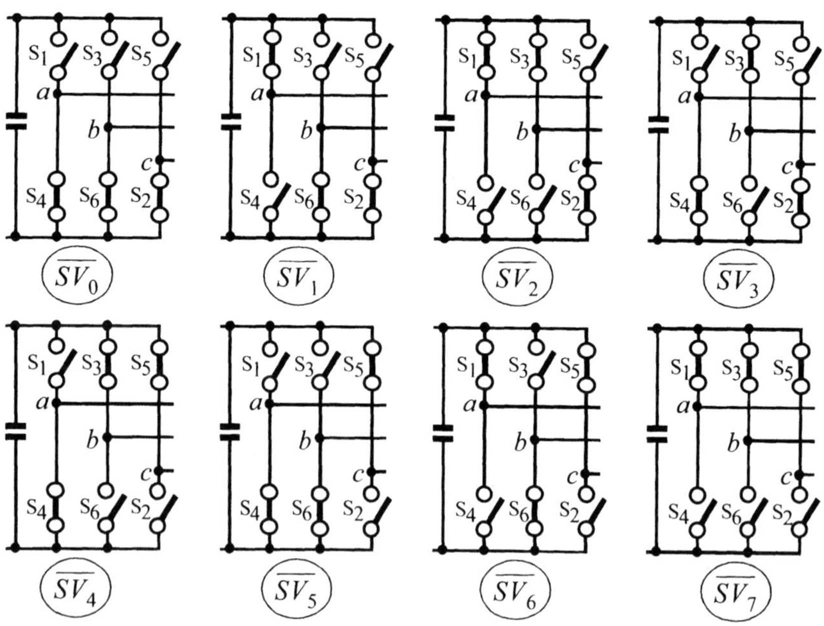

Figure 6.1 Eight possible phase leg switch combinations for a VSI.

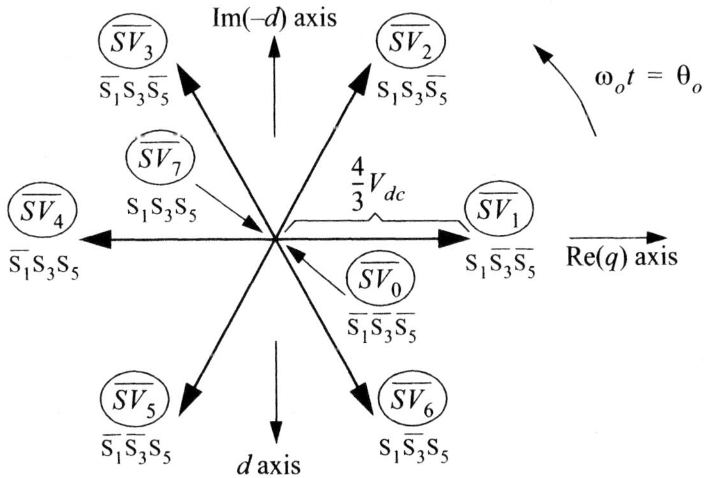

Figure 6.2 Location of eight possible stationary voltage vectors for a VSI in the d–q (Re–Im) plane, each vector has a length $(4/3)V_{dc}$.

As an example, the geometric summation shown in Figure 6.3 can then be expressed mathematically as

$$\overline{V_o^*} = V_o\angle\theta_o = \frac{T_{SV_1}}{\Delta T/2}\overline{SV_1} + \frac{T_{SV_2}}{\Delta T/2}\overline{SV_2} \tag{6.2}$$

for each switching period of $\Delta T/2$, where T_{SV_1} is the time for which space vector $\overline{SV_1}$ is selected, and T_{SV_2} is the time for which space vector $\overline{SV_2}$ is selected. In polar form (using peak voltages), Eq. (6.2) can be expressed as

$$\frac{\Delta T}{2}V_o\angle\theta_o = T_{SV_1}V_m\angle 0 + T_{SV_2}V_m\angle\pi/3 \tag{6.3}$$

or in Cartesian form

$$V_o(\cos\theta_o + j\sin\theta_o)\frac{\Delta T}{2} = T_{SV_1}V_m + T_{SV_2}V_m\left(\cos\frac{\pi}{3} + j\sin\frac{\pi}{3}\right) \tag{6.4}$$

Equating real and imaginary components gives the solution of

$$T_{SV_1} = \frac{V_o\sin\left(\frac{\pi}{3} - \theta_o\right)}{V_m\sin\frac{\pi}{3}}\frac{\Delta T}{2} \quad \text{(active time for } \overline{SV_1}\text{)} \tag{6.5}$$

$$T_{SV_2} = \frac{V_o\sin\theta_o}{V_m\sin\frac{\pi}{3}}\frac{\Delta T}{2} \quad \text{(active time for } \overline{SV_2}\text{)} \tag{6.6}$$

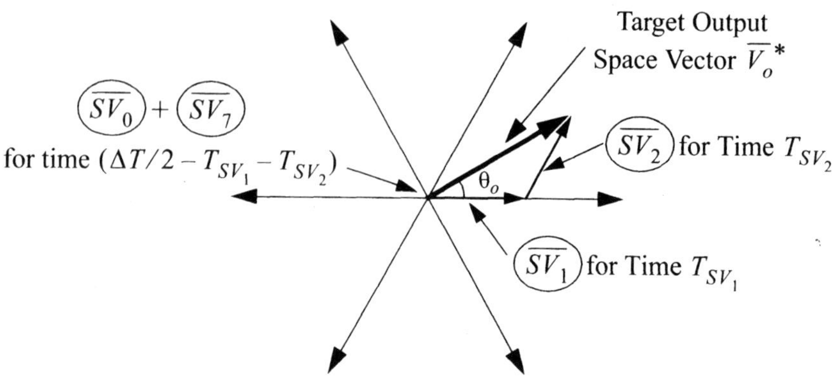

Figure 6.3 Creation of an arbitrary output target phasor by the geometrical summation of the two nearest space vectors.

It should be noted in passing that SVM is an intrinsically regular sampled process, since in essence it matches the sum of two active space vector volt–second averages over an equivalent half carrier period, to a sampled target volt–second average over the same period.

Since $0 \le T_{SV_1}, T_{SV_2} \le \Delta T/2$, the maximum possible magnitude for V_o is V_m, which can occur at $\theta_o = 0$ or $\pi/3$ radians. However, a further constraint is that the sum of the active times for the two space vectors obviously cannot exceed the half carrier period, i.e., $T_{SV_1} + T_{SV_2} \le \Delta T/2$. From simple geometry, the limiting case for this occurs at $\theta_o = \pi/6$, which means that

$$\frac{T_{SV_1} + T_{SV_2}}{\Delta T/2} = \frac{2V_o \sin\frac{\pi}{6}}{V_m \sin\frac{\pi}{3}} \le 1 \tag{6.7}$$

and this relationship constrains the maximum possible magnitude of V_o to

$$V_o = V_m \sin\frac{\pi}{3} = \frac{2}{\sqrt{3}} V_{dc} \tag{6.8}$$

Since V_o is the magnitude of the output *phase* voltage, the maximum possible *l–l* output voltage using SVM must equal

$$V_{l\text{-}l} = \sqrt{3} V_o = 2V_{dc} \tag{6.9}$$

This result is the same as for a three-phase VSI with a common mode third-harmonic injected into the reference, and again represents an increase of $2/\sqrt{3}$ or ~1.15 compared to regular sampled PWM, as developed in Section 5.3.

Note that the balance of the half carrier period is made up of any combination of the zero space vectors $\overline{SV_0}$ and $\overline{SV_7}$. However, so far no rationale has been identified for selecting a particular combination of the two zero space vectors (although equal intervals for $\overline{SV_0}$ and $\overline{SV_7}$ are common). This freedom of choice allows the placement of the space vectors to be varied anywhere within the half carrier period, which is the basis of most of the various space vector modulation alternatives that have been reported in the literature. This issue is addressed further in Section 6.6.

Table 6.1 shows the two nearest space vector components used to create an arbitrary target output phasor of any phase angle, together with appropriate expressions defining the active time intervals for each space vector. It should be noted that all the sets of space vector active times in Table 6.1 are identical in structure and simply reflect $60°$ movements of the target reference phasor from space vector segment to segment. In practice most space vector implementations take the alternative approach of redefining the target phasor reference angle every $60°$ to keep it within the bounds of $0 \le \theta_o \le \pi/3$, and then use only one set of expressions.

It is not difficult to show that the space vectors and sampling times summarized in Table 6.1 produce the desired output voltage. For example, during the period $0 \le \theta_o < \pi/3$ the target voltage is made up using $\overline{SV_1}$ and $\overline{SV_2}$. The target voltage in vector form is, from Eq. (6.2),

$$\overline{V_o}^* = \frac{T_{SV_1}}{\Delta T/2}\overline{SV_1} + \frac{T_{SV_2}}{\Delta T/2}\overline{SV_2}$$

$$= \frac{V_o}{V_{dc}}\frac{\sqrt{3}}{2}\cos\left(\theta_o + \frac{\pi}{6}\right)\overline{SV_1} + \frac{V_o}{V_{dc}}\frac{\sqrt{3}}{2}\cos\left(\theta_o - \frac{\pi}{2}\right)\overline{SV_2} \tag{6.10}$$

The component of this vector along load phase a is the average value of the phase voltage with respect to the load-neutral point, taken over the sampling interval centered at $\theta_o = \omega_o t$, i.e.,

$$\langle v_{as}\rangle = \frac{\sqrt{3}}{2}\frac{V_o}{V_{dc}}\mathrm{Re}\left[\cos\left(\theta_o + \frac{\pi}{6}\right)\overline{SV_1} + \cos\left(\theta_o - \frac{\pi}{2}\right)\overline{SV_2}\right] \tag{6.11}$$

However

$$\overline{SV_1} = \frac{4}{3}V_{dc} \quad \text{and} \quad \overline{SV_2} = \frac{4}{3}(-\bar{a}^{2})V_{dc} \tag{6.12}$$

Table 6.1 Active Space Vector Components for a VSI

$\omega_o t = \theta_o$	Space Vectors	Space Vector Active Times
$0 \le \theta_o < \frac{\pi}{3}$	$\overline{SV_1}$	$T_{SV_1} = \frac{V_o}{V_{dc}}\frac{\sqrt{3}}{2}\cos\left(\theta_o + \frac{\pi}{6}\right)\frac{\Delta T}{2}$
	$\overline{SV_2}$	$T_{SV_2} = \frac{V_o}{V_{dc}}\frac{\sqrt{3}}{2}\cos\left(\theta_o - \frac{\pi}{2}\right)\frac{\Delta T}{2}$
$\frac{\pi}{3} \le \theta_o < \frac{2\pi}{3}$	$\overline{SV_2}$	$T_{SV_2} = \frac{V_o}{V_{dc}}\frac{\sqrt{3}}{2}\cos\left(\theta_o - \frac{\pi}{6}\right)\frac{\Delta T}{2}$
	$\overline{SV_3}$	$T_{SV_3} = \frac{V_o}{V_{dc}}\frac{\sqrt{3}}{2}\cos\left(\theta_o - \frac{5\pi}{6}\right)\frac{\Delta T}{2}$
$\frac{2\pi}{3} \le \theta_o < \pi$	$\overline{SV_3}$	$T_{SV_3} = \frac{V_o}{V_{dc}}\frac{\sqrt{3}}{2}\cos\left(\theta_o - \frac{\pi}{2}\right)\frac{\Delta T}{2}$
	$\overline{SV_4}$	$T_{SV_4} = \frac{V_o}{V_{dc}}\frac{\sqrt{3}}{2}\cos\left(\theta_o - \frac{7\pi}{6}\right)\frac{\Delta T}{2}$
$\pi \le \theta_o < \frac{4\pi}{3}$	$\overline{SV_4}$	$T_{SV_4} = \frac{V_o}{V_{dc}}\frac{\sqrt{3}}{2}\cos\left(\theta_o - \frac{5\pi}{6}\right)\frac{\Delta T}{2}$
	$\overline{SV_5}$	$T_{SV_5} = \frac{V_o}{V_{dc}}\frac{\sqrt{3}}{2}\cos\left(\theta_o - \frac{3\pi}{2}\right)\frac{\Delta T}{2}$
$\frac{4\pi}{3} \le \theta_o < \frac{5\pi}{3}$	$\overline{SV_5}$	$T_{SV_5} = \frac{V_o}{V_{dc}}\frac{\sqrt{3}}{2}\cos\left(\theta_o - \frac{7\pi}{6}\right)\frac{\Delta T}{2}$
	$\overline{SV_6}$	$T_{SV_6} = \frac{V_o}{V_{dc}}\frac{\sqrt{3}}{2}\cos\left(\theta_o - \frac{11\pi}{6}\right)\frac{\Delta T}{2}$
$\frac{5\pi}{3} \le \theta_o < 2\pi$	$\overline{SV_6}$	$T_{SV_6} = \frac{V_o}{V_{dc}}\frac{\sqrt{3}}{2}\cos\left(\theta_o - \frac{3\pi}{2}\right)\frac{\Delta T}{2}$
	$\overline{SV_1}$	$T_{SV_1} = \frac{V_o}{V_{dc}}\frac{\sqrt{3}}{2}\cos\left(\theta_o - \frac{\pi}{6}\right)\frac{\Delta T}{2}$

so that Eq. (6.11) becomes

$$\langle v_{as} \rangle = \frac{2}{\sqrt{3}} V_o \left[\cos\left(\theta_o + \frac{\pi}{6}\right) + \frac{1}{2} \sin\theta_o \right] \tag{6.13}$$

which reduces to

$$\langle v_{as} \rangle = V_o \cos\theta_o \tag{6.14}$$

as expected.

6.1.2 SVM Compared to Regular Sampled PWM

While the formulation of the SVM algorithm is quite different from that of regular sampled PWM, the two algorithms are both regularly sampled systems and have the same primary objective of creating a switched output waveform which represents a given target fundamental. Clearly, there must be a relationship between the two strategies. This relationship is established by again examining Figure 5.3, where it can be seen that the output voltage pulse V_1 has the same switch conditions as space vector $\overline{SV}_1$, and the output voltage pulse V_2 has the same switch conditions as the space vector $\overline{SV}_2$. In other words, the process of regular sampled PWM *automatically* generates the two nearest space vectors to create the target output waveform.

The active *l–l* pulse widths for regular sampled PWM have been developed previously as Eqs. (5.14), (5.15), and (5.16). These clearly have the same form as the space vector intervals defined in Table 6.1, when the reference phase angle θ_o is made the same as the sampled fundamental angle $\omega_o(t_i + t_{i+1})/2$. In particular, for the region $0 \le \theta_o \le \pi/3$, which is shown in Figure 5.3,

$$T_1 - T_3 = T_{SV_1} = M\sqrt{3}\cos\left(\theta_o + \frac{\pi}{6}\right)\frac{\Delta T}{4} \tag{6.15}$$

$$T_3 - T_5 = T_{SV_2} = M\sqrt{3}\cos\left(\theta_o - \frac{\pi}{2}\right)\frac{\Delta T}{4} \tag{6.16}$$

if $V_o / V_{dc} = M$.

Equations (6.15) and (6.16) show that the active *l–l* output pulses produced by SVM and by regular sampled three-phase modulation are in fact identical. Taking account of the limits of V_0 defined by Eq. (6.8), this definition of M also confirms that SVM allows the modulation index to increase to $2/\sqrt{3}$, i.e., an increase of $\approx 15\%$ over simple regular sampled PWM. Indeed the only difference between regular sampled PWM and SVM is the position of the zero (inactive) space vectors within each half carrier period. For regular sampled PWM, the position of these zeros is prescribed by the algorithm definition given in Eqs. (5.11), (5.12), and (5.13), and there is no opportunity for variation. For SVM, the zero space vector positions are left undefined, and there is an opportunity to explore possible harmonic benefits by manipulating the zero pulse placements as noted before. However, the conventional SVM implementation [1] centers the active space vectors in each half carrier period, and splits the remaining zero space vector time equally between $\overline{SV}_0$ and $\overline{SV}_7$. This creates a space vector sequence (for $0 \leq \theta_o \leq \pi/3$) of:

$$\underbrace{\overline{SV}_0 \text{->} \overline{SV}_1 \text{->} \overline{SV}_2 \text{->} \overline{SV}_7}_{\Delta T/2} \text{->} \underbrace{\overline{SV}_7 \text{->} \overline{SV}_2 \text{->} \overline{SV}_1 \text{->} \overline{SV}_0}_{\Delta T/2} \quad (6.17)$$

which is precisely the same as the sequence produced by regular sampled PWM except for the equal spacing of the zero space vector time. A similar sequence is readily established for the other five sextants of the space vector phasor space using the two nearest space vector components for each sextant. Figure 6.4 shows the pulse pattern in the first sextant for conventional space vector PWM with equally spaced zero space vectors $\overline{SV}_0$ and $\overline{SV}_7$.

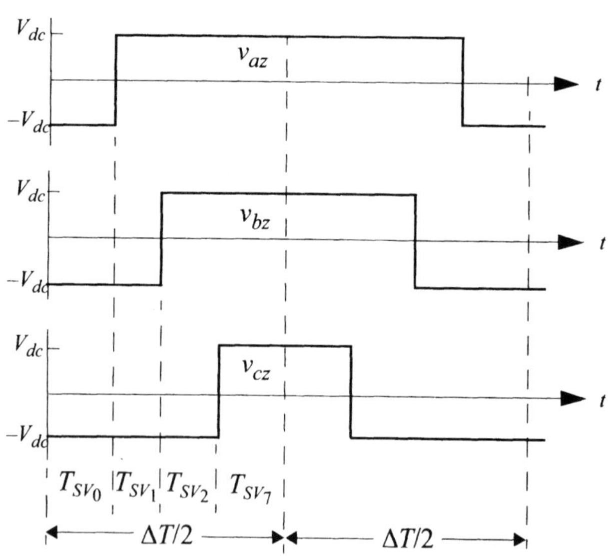

Figure 6.4 Pulse pattern of space vector modulation in the first sextant, $0 \leq \theta_o \leq \pi/3$ with centered active space vectors.

It should be noted from Eq. (6.17) that the sequence of space vectors reverses over a complete carrier interval, and this is a normal consequence of triangular carrier modulation. It is, of course, entirely possible to implement a SVM scheme without this active space vector sequence reversal, but a moment's thought identifies that such a sequence is equivalent to sawtooth carrier modulation, which has been previously identified as being suboptimal in a harmonic sense. In the same way, the sequence of Eq. (6.17) can be implemented with the active space vector periods T_{SV_1} and T_{SV_2} recalculated every half carrier period (equivalent to asymmetric regular sampled PWM) or once per entire ΔT (equivalent to symmetric regular sampled PWM). Each of these variations would be expected to achieve a different harmonic performance similar to carrier-based PWM, with reverse sequence SVM calculated every half carrier period anticipated to be the harmonically superior alternative. Hence, this is the implementation that should be compared against carrier-based PWM to identify any benefits of explicit zero space vector placement.

6.2 Phase Leg References for Space Vector Modulation

It can be recalled from Chapter 5 the magnitude and harmonic benefits that were achieved by simply adding a third-harmonic component to the sinusoidal phase leg reference voltages. Since both regular sampled PWM and SVM produce identical mean output voltages, one could expect that a similar component could be added to sinusoidal phase leg reference voltages to achieve SVM. This component can be found by examining Figure 6.4 in more detail.

From Figure 6.4, the reference (average) values of the three-phase leg voltages over the interval $\Delta T/2$, expressed in terms of the midpoint of the DC link, are

$$\langle v_{az}\rangle = V_{dc}\left(\frac{T_{SV_1}}{\Delta T/2}+\frac{T_{SV_2}}{\Delta T/2}\right) \tag{6.18}$$

$$\langle v_{bz}\rangle = V_{dc}\left(-\frac{T_{SV_1}}{\Delta T/2}+\frac{T_{SV_2}}{\Delta T/2}\right) \tag{6.19}$$

$$\langle v_{cz}\rangle = V_{dc}\left(-\frac{T_{SV_1}}{\Delta T/2}-\frac{T_{SV_2}}{\Delta T/2}\right) \tag{6.20}$$

Substituting from Table 6.1, these average voltages can be expressed as

$$\langle v_{az}\rangle = \frac{\sqrt{3}}{2}V_o\left[\cos\left(\theta_o+\frac{\pi}{6}\right)+\cos\left(\theta_o-\frac{\pi}{2}\right)\right] \tag{6.21}$$

$$\langle v_{bz}\rangle = \frac{\sqrt{3}}{2}V_o\left[-\cos\left(\theta_o+\frac{\pi}{6}\right)+\cos\left(\theta_o-\frac{\pi}{2}\right)\right] \tag{6.22}$$

$$\langle v_{cz}\rangle = \frac{\sqrt{3}}{2}V_o\left[-\cos\left(\theta_o+\frac{\pi}{6}\right)-\cos\left(\theta_o-\frac{\pi}{2}\right)\right] \tag{6.23}$$

Since the maximum value of V_o is set by Eq. (6.8), V_o in general can be expressed as $V_o = MV_{dc}$ where $0 \le M \le 2/\sqrt{3}$. The line-to-DC midpoint voltages can therefore be written as

$$\langle v_{az}\rangle = \frac{\sqrt{3}}{2}MV_{dc}\left[\cos\left(\theta_o+\frac{\pi}{6}\right)+\cos\left(\theta_o-\frac{\pi}{2}\right)\right] \tag{6.24}$$

$$\langle v_{bz}\rangle = \frac{\sqrt{3}}{2}MV_{dc}\left[-\cos\left(\theta_o+\frac{\pi}{6}\right)+\cos\left(\theta_o-\frac{\pi}{2}\right)\right] \tag{6.25}$$

$$\langle v_{cz}\rangle = \frac{\sqrt{3}}{2}MV_{dc}\left[-\cos\left(\theta_o+\frac{\pi}{6}\right)-\cos\left(\theta_o-\frac{\pi}{2}\right)\right] \tag{6.26}$$

These equations reduce to

$$\langle v_{az}\rangle = \frac{\sqrt{3}}{2}MV_{dc}\cos\left(\theta_o-\frac{\pi}{6}\right) \tag{6.27}$$

$$\langle v_{bz}\rangle = \frac{3}{2}MV_{dc}\cos\left(\theta_o-\frac{2\pi}{3}\right) \tag{6.28}$$

$$\langle v_{cz}\rangle = \frac{\sqrt{3}}{2}MV_{dc}\cos\left(\theta_o+\frac{5\pi}{6}\right) \tag{6.29}$$

for the period $0 \le \theta_o \le \pi/3$.

Once the solution for all three-phases is known over one 60° interval, a complete solution can be readily determined for an entire cycle using half cycle and three-phase symmetry concepts as developed in Appendix 3. In particular, for the interval $\pi/3 \le \theta_o + \pi/3 \le 2\pi/3$,

$$\left\langle v_{az}\left(\theta_o+\frac{\pi}{3}\right)\right\rangle = -\langle v_{bz}(\theta_o)\rangle \tag{6.30}$$

$$\left\langle v_{bz}\left(\theta_o+\frac{\pi}{3}\right)\right\rangle = -\langle v_{cz}(\theta_o)\rangle \tag{6.31}$$

$$\langle v_{cz}\left(\theta_o + \frac{\pi}{3}\right)\rangle = -\langle v_{az}(\theta_o)\rangle \tag{6.32}$$

The solutions for the remaining four 60° sextants are easily obtained by progressively increasing the arguments of Eqs. (6.30) to (6.32) and are summarized in Table 6.2.

The resulting solution for phase leg a is plotted in Figure 6.5, where $\langle v_{az}\rangle$ is a space vector equivalent phase leg reference voltage that can be used in either a naturally or regular sampled PWM system to produce output voltages with the same average low-frequency content. This voltage can be thought of as consisting of two quantities: $\langle v_{as}\rangle$ the sinusoidal phase-to-load neutral voltage which can be as large as $(2/\sqrt{3})V_{dc}$ when $M = 2/\sqrt{3}$ and $\langle v_{sz}\rangle$ the three-times-fundamental-frequency triangular voltage between load-neutral and DC midpoint that is made up of odd triplen harmonics and is injected into $\langle v_{az}\rangle$. Hence it is evident that the space vector concept adds third-harmonic quantities to the basic regular sampled PWM method in much the same manner as does simple third-harmonic injection.

Table 6.2 Phase Leg Reference Voltage Waveforms for SVM

60° sextant	Phase Leg a $\langle v_{az}\rangle/V_{dc}$	Phase Leg b $\langle v_{bz}\rangle/V_{dc}$	Phase Leg c $\langle v_{cz}\rangle/V_{dc}$
$\frac{2\pi}{3} \le \theta_o \le \pi$	$\frac{\sqrt{3}}{2}M\cos\left(\theta_o + \frac{\pi}{6}\right)$	$\frac{\sqrt{3}}{2}M\cos\left(\theta_o - \frac{5\pi}{6}\right)$	$\frac{3}{2}M\cos\left(\theta_o + \frac{2\pi}{3}\right)$
$\frac{\pi}{3} \le \theta_o \le \frac{2\pi}{3}$	$\frac{3}{2}M\cos\theta_o$	$\frac{\sqrt{3}}{2}M\sin\theta_o$	$-\frac{\sqrt{3}}{2}M\sin\theta_o$
$0 \le \theta_o \le \frac{\pi}{3}$	$\frac{\sqrt{3}}{2}M\cos\left(\theta_o - \frac{\pi}{6}\right)$	$\frac{3}{2}M\cos\left(\theta_o - \frac{2\pi}{3}\right)$	$\frac{\sqrt{3}}{2}M\cos\left(\theta_o + \frac{5\pi}{6}\right)$
$-\frac{\pi}{3} \le \theta_o \le 0$	$\frac{\sqrt{3}}{2}M\cos\left(\theta_o + \frac{\pi}{6}\right)$	$\frac{\sqrt{3}}{2}M\cos\left(\theta_o - \frac{5\pi}{6}\right)$	$\frac{3}{2}M\cos\left(\theta_o + \frac{2\pi}{3}\right)$
$-\frac{2\pi}{3} \le \theta_o \le -\frac{\pi}{3}$	$\frac{3}{2}M\cos\theta_o$	$\frac{\sqrt{3}}{2}M\sin\theta_o$	$-\frac{\sqrt{3}}{2}M\sin\theta_o$
$-\pi \le \theta_o \le -\frac{2\pi}{3}$	$\frac{\sqrt{3}}{2}M\cos\left(\theta_o - \frac{\pi}{6}\right)$	$\frac{3}{2}M\cos\left(\theta_o - \frac{2\pi}{3}\right)$	$\frac{\sqrt{3}}{2}M\cos\left(\theta_o + \frac{5\pi}{6}\right)$

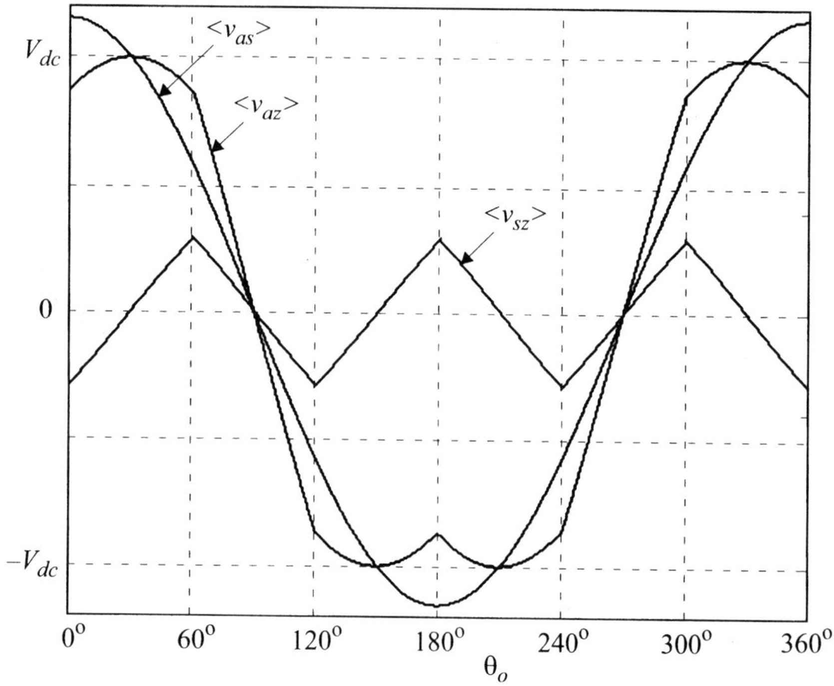

Figure 6.5 Mean phase leg output voltage and its two components under SVM, referred to z (midpoint of the DC bus), $M = 2/\sqrt{3}$.

6.3 Naturally Sampled SVM

As developed in Section 6.1, conventional SVM intrinsically assumes that it is a regularly sampled process. Usually, the active space vector magnitudes and placement are determined by a microprocessor, but, of course, this does somewhat limit the switching frequency that can be achieved in a practical system and high-speed digital signal processors (DSPs) must usually be used to obtain an acceptable performance.

In Section 6.2, an equivalent phase leg reference waveform was developed that will achieve SVM by centering the active space vectors in each half carrier period. This reference waveform can be generated in an analog sine–triangle modulator by adding a continuously varying offset voltage to the set of three-phase voltage references which centers their envelope around zero at all times.

A simple circuit to implement this arrangement is shown in Figure 6.6, where the three-phase reference waveforms are rectified to produce their envelope magnitude, this magnitude is scaled by 0.5, and is then added as an offset to each voltage reference as it is passed on to a conventional sine–triangle modulator [5]. The mathematical expression for this implementation is

$$v_{az(\text{comp})} = v_{az}^* - \frac{\max(v_{az}^*, v_{bz}^*, v_{cz}^*) + \min(v_{az}^*, v_{bz}^*, v_{cz}^*)}{2} \tag{6.33}$$

where the "max/min" expression centers the sinusoidal references defined in Eqs. (5.1), (5.2), and (5.3) around zero at all times. [It should also be noted that Eq. (6.33) can be implemented equally easily using regular sampled modulation techniques. In the authors' opinion, this is often the most effective way to implement a digital SV modulation system.]

This equivalent *naturally sampled* SVM concept relates to *regularly sampled* SVM in exactly the same way that naturally sampled PWM relates to regularly sampled PWM, as described in Chapter 3. Thus the theoretical differences between the two techniques would be expected to be that

- Naturally sampled modulation balances the low-side/high-side carrier sideband harmonic components.
- Naturally sampled modulation does not create low-order fundamental components in the phase leg harmonic spectrum. This means that a single-edge implementation of the analog SVM, where the triangular carrier is replaced by a sawtooth carrier, would not produce low-order fundamental harmonic components.

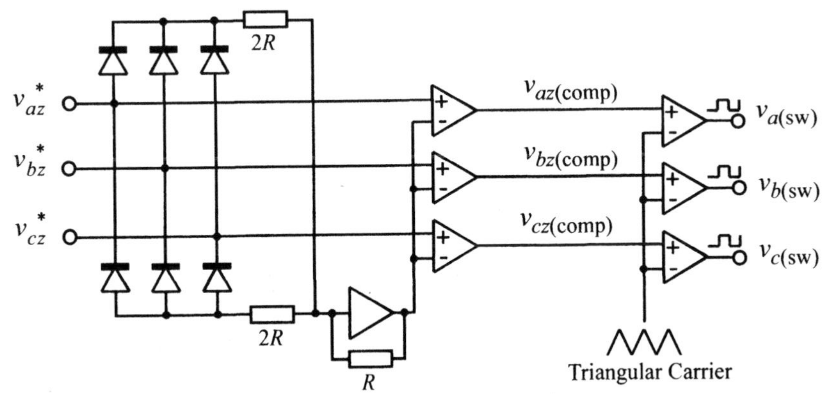

Figure 6.6 Circuit for implementation of analog space vector modulation.

These differences are shown in Figure 6.7, where the harmonic spectra for regular sampled SVM are compared against naturally sampled SVM for both double- and single-edge modulation conditions. In particular, the uncanceled second and fourth low-order harmonics caused by single-edge regular sampled PWM, shown in Figure 6.7(c), are clearly not produced by the naturally sampled SVM equivalent process shown in Figure 6.7(d). Note that these results were generated using the analytic solutions presented in Section 6.4 for double-edge modulation, and by time-based switched waveform simulations for single-edge modulation.

The major benefit of the analog SVM concept is that the switching frequency can be set to any arbitrary maximum frequency limited only by the inverter phase leg switching capability. Also, since the strategy does not require the injection of a specific third-harmonic offset, it is particularly effective for situations where the demanded inverter phasor voltage may abruptly change from computational cycle to cycle and determination of the third-harmonic is difficult. Typical applications would be a motor drive system or an active filter system [6].

Finally, it should be noted that as with any analog sine–triangle comparison modulation implementation, it may be necessary to include slew rate limiting for the final reference signal, which is compared against the triangular carrier, to limit the slope to less than that of the carrier. Otherwise, multiple switch transitions can occur within one half carrier period. However, this is a normal consideration for any analog sine–triangle modulation implementation

6.4 Analytical Solution for SVM

The analytical solution for SVM can now be developed by extending the techniques that have been applied to single-phase and three-phase inverters with continuous reference waveforms in the previous chapters, first for naturally sampled SVM and then in turn for symmetrical and asymmetrical sampled SVM [7]. Essentially, this extension involves re-defining the integral limits of the basic double Fourier integration expression Eq. (3.9) to take account of the fact that the phase leg reference waveform is no longer continuous but is now made up of six segments across a complete fundamental cycle. Hence the outer integral term becomes a summation of six integral terms, each spanning 60° of the fundamental, i.e.,

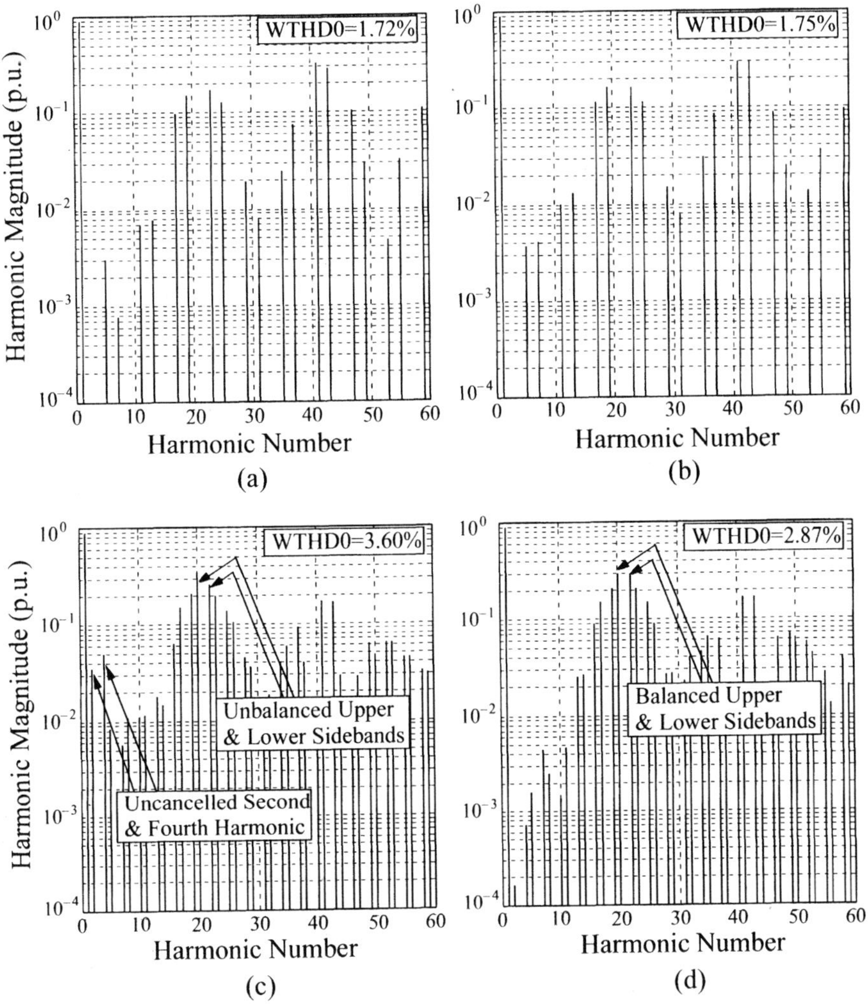

Figure 6.7 Theoretical *l–l* harmonic spectra for three-phase inverter modulated by
(a) *asymmetrical regular sampled double-edge* SVM,
(b *naturally sampled double-edge* SVM,
(c) *regular sampled single-edge* SVM,
(d) *naturally sampled single-edge* SVM,
$M = 0.9,\ f_c/f_o = 21.$

$$A_{mn} + jB_{mn} = \frac{1}{2\pi^2}\sum_{i=1}^{6}\int_{y_s(i)}^{y_e(i)}\int_{x_r(i)}^{x_f(i)} 2V_{dc}e^{j(mx+ny)}\,dx\,dy \quad (6.34)$$

where the outer and inner integral limits of Eq. (6.34) are defined in Table 6.3.

With these limits identified, Eq. (6.34) can now be evaluated for various possible values of m and n.

For $m = n = 0$, Eq. (6.34) simplifies to

$$A_{00} + jB_{00} = \frac{V_{dc}}{\pi^2}\sum_{i=1}^{6}\int_{y_s(i)}^{y_e(i)}\int_{x_r(i)}^{x_f(i)} dx\,dy \quad (6.35)$$

which not unexpectedly evaluates to

$$A_{00} + jB_{00} = 2V_{dc} \quad (6.36)$$

Table 6.3 Outer and Inner Double Fourier Integral Limits for SVM

i	$y_s(i)$	$y_e(i)$	$x_r(i)$ (rising edge of switched waveform)	$x_f(i)$ (falling edge of switched waveform)
1	$\frac{2\pi}{3}$	π	$-\frac{\pi}{2}\left[1 + \frac{\sqrt{3}}{2}M\cos\left(y + \frac{\pi}{6}\right)\right]$	$\frac{\pi}{2}\left[1 + \frac{\sqrt{3}}{2}M\cos\left(y + \frac{\pi}{6}\right)\right]$
2	$\frac{\pi}{3}$	$\frac{2\pi}{3}$	$-\frac{\pi}{2}\left[1 + \frac{3}{2}M\cos y\right]$	$\frac{\pi}{2}\left[1 + \frac{3}{2}M\cos y\right]$
3	0	$\frac{\pi}{3}$	$-\frac{\pi}{2}\left[1 + \frac{\sqrt{3}}{2}M\cos\left(y - \frac{\pi}{6}\right)\right]$	$\frac{\pi}{2}\left[1 + \frac{\sqrt{3}}{2}M\cos\left(y - \frac{\pi}{6}\right)\right]$
4	$-\frac{\pi}{3}$	0	$-\frac{\pi}{2}\left[1 + \frac{\sqrt{3}}{2}M\cos\left(y + \frac{\pi}{6}\right)\right]$	$\frac{\pi}{2}\left[1 + \frac{\sqrt{3}}{2}M\cos\left(y + \frac{\pi}{6}\right)\right]$
5	$-\frac{2\pi}{3}$	$-\frac{\pi}{3}$	$-\frac{\pi}{2}\left[1 + \frac{3}{2}M\cos y\right]$	$\frac{\pi}{2}\left[1 + \frac{3}{2}M\cos y\right]$
6	$-\pi$	$-\frac{2\pi}{3}$	$-\frac{\pi}{2}\left[1 + \frac{\sqrt{3}}{2}M\cos\left(y - \frac{\pi}{6}\right)\right]$	$\frac{\pi}{2}\left[1 + \frac{\sqrt{3}}{2}M\cos\left(y - \frac{\pi}{6}\right)\right]$

since the DC offset would be expected once more to be V_{dc} for a symmetrical reference waveform such as is shown in Figure 6.5.

For $m = 0, n > 0$, Eq. (6.34) simplifies to

$$A_{0n} + jB_{0n} = \frac{V_{dc}}{\pi^2}\sum_{i=1}^{6}\int_{y_s(i)}^{y_e(i)}\int_{x_r(i)}^{x_f(i)} e^{jny}\,dx\,dy$$

$$= \frac{V_{dc}}{\pi^2}\sum_{i=1}^{6}\int_{y_s(i)}^{y_e(i)} e^{jny}\,[x_f(i) - x_r(i)]\,dy \qquad (6.37)$$

Substituting for $x_r(i)$ and $x_f(i)$ from Table 6.3 gives, after some manipulation and recalling that $\int_{-\pi}^{\pi} e^{jny}\,dy = 0$ for any nonzero value of n,

$$A_{0n} + jB_{0n} = \frac{\sqrt{3}MV_{dc}}{4\pi}\left[\begin{array}{l} \displaystyle\sum_{i=1,4}\int_{y_s(i)}^{y_e(i)}\left[e^{j\left([n+1]y+\frac{\pi}{6}\right)} + e^{j\left([n-1]y-\frac{\pi}{6}\right)}\right]dy \\ \displaystyle +\sum_{i=2,5}\sqrt{3}\int_{y_s(i)}^{y_e(i)}[e^{j[n+1]y} + e^{j[n-1]y}]\,dy \\ \displaystyle +\sum_{i=3,6}\int_{y_s(i)}^{y_e(i)}\left[e^{j\left([n+1]y-\frac{\pi}{6}\right)} + e^{j\left([n-1]y+\frac{\pi}{6}\right)}\right]dy \end{array}\right]$$

(6.38)

For $n = 1$, Eq. (6.38) further simplifies to

$$A_{01} + jB_{01} = \frac{\sqrt{3}MV_{dc}}{4\pi}\left[\begin{array}{l} \displaystyle\sum_{i=1,4}\int_{y_s(i)}^{y_e(i)}\left[e^{j\left(2y+\frac{\pi}{6}\right)} + e^{-j\frac{\pi}{6}}\right]dy\ dy \\ \displaystyle + \sum_{i=2,5}\sqrt{3}\int_{y_s(i)}^{y_e(i)}[e^{j2y} + 1]\,dy \\ \displaystyle + \sum_{i=3,6}\int_{y_s(i)}^{y_e(i)}\left[e^{j\left(2y-\frac{\pi}{6}\right)} + e^{j\frac{\pi}{6}}\right]dy \end{array}\right] \tag{6.39}$$

which integrates, with substitution for $y_s(i)$ and $y_e(i)$ from Table 6.3, to

$$A_{01} + jB_{01} = \frac{\sqrt{3}MV_{dc}}{4\pi}\left[\begin{array}{l} \displaystyle\left[\frac{e^{j\left(2y+\frac{\pi}{6}\right)}}{j2} + ye^{-j\frac{\pi}{6}}\right]\Bigg|_{\frac{2\pi}{3}}^{\pi} + \sqrt{3}\left[\frac{e^{j2y}}{j2} + y\right]\Bigg|_{\frac{\pi}{3}}^{\frac{2\pi}{3}} \\ \displaystyle + \left[\frac{e^{j\left(2y-\frac{\pi}{6}\right)}}{j2} + ye^{j\frac{\pi}{6}}\right]\Bigg|_{0}^{\frac{\pi}{3}} + \left[\frac{e^{j\left(2y+\frac{\pi}{6}\right)}}{j2} + ye^{-j\frac{\pi}{6}}\right]\Bigg|_{-\frac{\pi}{3}}^{0} \\ \displaystyle + \sqrt{3}\left[\frac{e^{j2y}}{j2} + y\right]\Bigg|_{-\frac{2\pi}{3}}^{-\frac{\pi}{3}} + \left[\frac{e^{j\left(2y-\frac{\pi}{6}\right)}}{j2} + ye^{j\frac{\pi}{6}}\right]\Bigg|_{-\pi}^{-\frac{2\pi}{3}} \end{array}\right] \tag{6.40}$$

Equation (6.40) reduces, after some simplification, to

$$A_{01} + jB_{01} = MV_{dc} \tag{6.41}$$

Equation (6.41) is immediately recognizable as the target reference waveform magnitude, which is the object of the modulation process. It should further be noted that nothing in this development has restricted the maximum value of M, so that the modulation limit remains at $2/\sqrt{3}$ as identified already for SVM.

For $n > 1$, again with substitution for $y_s(i)$ and $y_e(i)$ from Table 6.3, Eq. (6.38) integrates to form

$$A_{0n} + jB_{0n} = \frac{\sqrt{3}MV_{dc}}{4\pi}\left[\begin{array}{l} \left[\dfrac{e^{j\left([n+1]y+\frac{\pi}{6}\right)}}{j[n+1]} + \dfrac{e^{j\left([n-1]y-\frac{\pi}{6}\right)}}{j[n-1]}\right]\Bigg|_{\frac{2\pi}{3}}^{\pi} \\ + \sqrt{3}\left[\dfrac{e^{j[n+1]y}}{j[n+1]} + \dfrac{e^{j[n-1]y}}{j[n-1]}\right]\Bigg|_{\frac{\pi}{3}}^{\frac{2\pi}{3}} \\ + \left[\dfrac{e^{j\left([n+1]y-\frac{\pi}{6}\right)}}{j[n+1]} + \dfrac{e^{j\left([n-1]y+\frac{\pi}{6}\right)}}{j[n-1]}\right]\Bigg|_{0}^{\frac{\pi}{3}} \\ + \left[\dfrac{e^{j\left([n+1]y+\frac{\pi}{6}\right)}}{j[n+1]} + \dfrac{e^{j\left([n-1]y-\frac{\pi}{6}\right)}}{j[n-1]}\right]\Bigg|_{-\frac{\pi}{3}}^{0} \\ + \sqrt{3}\left[\dfrac{e^{j[n+1]y}}{j[n+1]} + \dfrac{e^{j[n-1]y}}{j[n-1]}\right]\Bigg|_{-\frac{2\pi}{3}}^{-\frac{\pi}{3}} \\ + \left[\dfrac{e^{j\left([n+1]y-\frac{\pi}{6}\right)}}{j[n+1]} + \dfrac{e^{j\left([n-1]y+\frac{\pi}{6}\right)}}{j[n-1]}\right]\Bigg|_{-\pi}^{-\frac{2\pi}{3}} \end{array}\right] \tag{6.42}$$

which can be expressed more compactly as

$$A_{0n} + jB_{0n}\Big|_{n>1} = \frac{\sqrt{3}MV_{dc}}{\pi}\left[\begin{array}{l}\dfrac{1}{n+1}\left[\begin{array}{l}\sin\left([n+1]\dfrac{\pi}{6}\right)\cos\left([n+1]\dfrac{\pi}{2}\right)\\ \qquad\times\left\{\sqrt{3}+2\cos\left([n+1]\dfrac{\pi}{3}+\dfrac{\pi}{6}\right)\right\}\end{array}\right]\\ +\dfrac{1}{n-1}\left[\begin{array}{l}\sin\left([n-1]\dfrac{\pi}{6}\right)\cos\left([n-1]\dfrac{\pi}{2}\right)\\ \qquad\times\left\{\sqrt{3}+2\cos\left([n-1]\dfrac{\pi}{3}-\dfrac{\pi}{6}\right)\right\}\end{array}\right]\end{array}\right] \tag{6.43}$$

Expressing this result under a common denominator produces

$$A_{0n} + jB_{0n}\Big|_{n>1} = \frac{\sqrt{3}MV_{dc}}{\pi(n^2-1)}\left[\begin{array}{l}\sin\left([n+1]\dfrac{\pi}{6}\right)\cos\left([n+1]\dfrac{\pi}{2}\right)\left(2\sin\left(n\dfrac{\pi}{3}\right)-\sqrt{3}\right)\\ +\sin\left([n-1]\dfrac{\pi}{6}\right)\cos\left([n-1]\dfrac{\pi}{2}\right)\left(2\sin\left(n\dfrac{\pi}{3}\right)+\sqrt{3}\right)\end{array}\right] \tag{6.44}$$

It can be readily verified from this expression that only the triplen baseband harmonics are nonzero. Equation (6.44) reduces even further, if it is now rephrased to take account of the fact that the nontriplen harmonics are zero, to

$$A_{0n} + jB_{0n} = \frac{3\sqrt{3}MV_{dc}}{\pi(n^2-1)}\sin\left(n\frac{\pi}{6}\right)\sin\left(n\frac{\pi}{2}\right) \qquad n = 3, 9, 15\ldots \tag{6.45}$$

Thus it is clear that Eq. (6.43) defines the triplen baseband harmonic components that make up the triangular common mode offset voltage shown in Figure 6.5, similar to the third-harmonic injected component defined by Eq. (5.39) in Chapter 5.

For $m > 0$, the inner integral of Eq. (6.34) evaluates to

$$A_{mn} + jB_{mn} = \frac{V_{dc}}{jm\pi^2}\sum_{i=1}^{6}\int_{y_s(i)}^{y_e(i)} e^{jny}[e^{jmx_f(i)} - e^{jmx_r(i)}]\,dy \tag{6.46}$$

Substituting for $x_r(i)$ and $x_f(i)$ from Table 6.3 gives

$$A_{mn}+jB_{mn} = \frac{V_{dc}}{jm\pi^2}\left[\begin{array}{l} \displaystyle\sum_{i=1,4}\int_{y_s(i)}^{y_e(i)} e^{jny}\left[\begin{array}{l} e^{jm\frac{\pi}{2}}e^{jm\frac{\pi}{2}\frac{\sqrt{3}}{2}M\cos\left(y+\frac{\pi}{6}\right)} \\ -e^{-jm\frac{\pi}{2}}e^{-jm\frac{\pi}{2}\frac{\sqrt{3}}{2}M\cos\left(y+\frac{\pi}{6}\right)}\end{array}\right]dy \\ \displaystyle+\sum_{i=2,5}\int_{y_s(i)}^{y_e(i)} e^{jny}\left[\begin{array}{l} e^{jm\frac{\pi}{2}}e^{jm\frac{\pi}{2}\frac{3}{2}M\cos y} \\ -e^{-jm\frac{\pi}{2}}e^{-jm\frac{\pi}{2}\frac{3}{2}M\cos y}\end{array}\right]dy \\ \displaystyle+\sum_{i=3,6}\int_{y_s(i)}^{y_e(i)} e^{jny}\left[\begin{array}{l} e^{jm\frac{\pi}{2}}e^{jm\frac{\pi}{2}\frac{\sqrt{3}}{2}M\cos\left(y-\frac{\pi}{6}\right)} \\ -e^{-jm\frac{\pi}{2}}e^{-jm\frac{\pi}{2}\frac{\sqrt{3}}{2}M\cos\left(y-\frac{\pi}{6}\right)}\end{array}\right]dy \end{array}\right] \tag{6.47}$$

Equation (6.47) can be rearranged, using Eq. (A2.1), to become

$$A_{mn}+jB_{mn} = \frac{2V_{dc}}{m\pi^2}\left\{\sum_{i=1,4}\int_{y_s(i)}^{y_e(i)}\Pi_1\,dy+\sum_{i=2,5}\int_{y_s(i)}^{y_e(i)}\Pi_2\,dy+\sum_{i=3,6}\int_{y_s(i)}^{y_e(i)}\Pi_3\,dy\right\} \tag{6.48}$$

where the terms within the integrals are given by

$$\Pi_1 = e^{jny}\left[\begin{array}{l} J_0\left(m\frac{\sqrt{3}\pi}{4}M\right)\sin m\frac{\pi}{2} \\ \displaystyle+2\sum_{k=1}^{\infty}J_k\left(m\frac{\sqrt{3}\pi}{4}M\right)\sin\left([m+k]\frac{\pi}{2}\right)\cos\left(k\left[y+\frac{\pi}{6}\right]\right)\end{array}\right] \tag{6.49}$$

$$\Pi_2 = e^{jny}\left[\begin{array}{l} J_0\left(m\frac{3\pi}{4}M\right)\sin m\frac{\pi}{2} \\ \displaystyle+2\sum_{k=1}^{\infty}J_k\left(m\frac{3\pi}{4}M\right)\sin\left([m+k]\frac{\pi}{2}\right)\cos ky\end{array}\right] \tag{6.50}$$

$$\Pi_3 = e^{jny}\left[\begin{array}{l} J_0\left(m\frac{\sqrt{3}\pi}{4}M\right)\sin m\frac{\pi}{2} \\ \quad + 2\sum_{k=1}^{\infty} J_k\left(m\frac{\sqrt{3}\pi}{4}M\right)\sin\left([m+k]\frac{\pi}{2}\right)\cos\left(k\left[y-\frac{\pi}{6}\right]\right)\end{array}\right] \tag{6.51}$$

Terms Π_1, Π_2, and Π_3 can be readily rearranged into the more easily integrable forms of

$$\Pi_1 = J_0\left(m\frac{\sqrt{3}\pi}{4}M\right)\sin m\frac{\pi}{2}e^{jny} + \sum_{k=1}^{\infty} J_k\left(m\frac{\sqrt{3}\pi}{4}M\right)\sin\left([m+k]\frac{\pi}{2}\right)\left[\begin{array}{l} e^{j\left([n+k]y+k\frac{\pi}{6}\right)} \\ \quad + e^{j\left([n-k]y-k\frac{\pi}{6}\right)}\end{array}\right] \tag{6.52}$$

$$\Pi_2 = J_0\left(m\frac{3\pi}{4}M\right)\sin m\frac{\pi}{2}e^{jny} + \sum_{k=1}^{\infty} J_k\left(m\frac{3\pi}{4}M\right)\sin\left([m+k]\frac{\pi}{2}\right)\left[\begin{array}{l} e^{j[n+k]y} \\ \quad + e^{j[n-k]y}\end{array}\right] \tag{6.53}$$

$$\Pi_3 = J_0\left(m\frac{\sqrt{3}\pi}{4}M\right)\sin m\frac{\pi}{2}e^{jny} + \sum_{k=1}^{\infty} J_k\left(m\frac{\sqrt{3}\pi}{4}M\right)\sin\left([m+k]\frac{\pi}{2}\right)\left[\begin{array}{l} e^{j\left([n+k]y-k\frac{\pi}{6}\right)} \\ \quad + e^{j\left([n-k]y+k\frac{\pi}{6}\right)}\end{array}\right] \tag{6.54}$$

so that Eq. (6.48) can now be integrated to

$$A_{mn}+jB_{mn} = \frac{2V_{dc}}{m\pi^2}\left[\Lambda_1\Big|_{\frac{2\pi}{3}}^{\pi} + \Lambda_2\Big|_{\frac{\pi}{3}}^{\frac{2\pi}{3}} + \Lambda_3\Big|_{0}^{\frac{\pi}{3}} + \Lambda_1\Big|_{-\frac{\pi}{3}}^{0} + \Lambda_2\Big|_{-\frac{2\pi}{3}}^{-\frac{\pi}{3}} + \Lambda_3\Big|_{-\pi}^{-\frac{2\pi}{3}}\right] \tag{6.55}$$

where

$$\Lambda_1 = \left[\begin{array}{l} J_0\left(m\frac{\sqrt{3}\pi}{4}M\right)\sin m\frac{\pi}{2}\frac{e^{jny}}{jn} + J_n\left(m\frac{\sqrt{3}\pi}{4}M\right)\sin\left([m+n]\frac{\pi}{2}\right)e^{-jn\frac{\pi}{6}}y\Bigg|_{k=|n|} \\ + \sum_{k=1}^{\infty} J_k\left(m\frac{\sqrt{3}\pi}{4}M\right)\sin\left([m+k]\frac{\pi}{2}\right)\left[\begin{array}{l} \dfrac{e^{j\left([n+k]y+k\frac{\pi}{6}\right)}}{j[n+k]}\Bigg|_{k\neq -n} \\ + \dfrac{e^{j\left([n-k]y-k\frac{\pi}{6}\right)}}{j[n-k]}\Bigg|_{k\neq n} \end{array}\right] \end{array}\right] \tag{6.56}$$

$$\Lambda_2 = \left[\begin{array}{l} J_0\left(m\frac{3\pi}{4}M\right)\sin m\frac{\pi}{2}\frac{e^{jny}}{jn} + J_n\left(m\frac{3\pi}{4}M\right)\sin\left([m+n]\frac{\pi}{2}\right)y\Bigg|_{k=|n|} \\ + \sum_{k=1}^{\infty} J_k\left(m\frac{3\pi}{4}M\right)\sin\left([m+k]\frac{\pi}{2}\right)\left[\begin{array}{l} \dfrac{e^{j[n+k]y}}{j[n+k]}\Bigg|_{k\neq -n} \\ + \dfrac{e^{j[n-k]y}}{j[n-k]}\Bigg|_{k\neq n} \end{array}\right] \end{array}\right] \tag{6.57}$$

$$\Lambda_3 = \left[\begin{array}{l} J_0\left(m\frac{\sqrt{3}\pi}{4}M\right)\sin m\frac{\pi}{2}\frac{e^{jny}}{jn} + J_n\left(m\frac{\sqrt{3}\pi}{4}M\right)\sin\left([m+n]\frac{\pi}{2}\right)e^{jn\frac{\pi}{6}}y\Bigg|_{k=|n|} \\ + \sum_{k=1}^{\infty} J_k\left(m\frac{\sqrt{3}\pi}{4}M\right)\sin\left([m+k]\frac{\pi}{2}\right)\left[\begin{array}{l} \dfrac{e^{j\left([n+k]y-k\frac{\pi}{6}\right)}}{j[n+k]}\Bigg|_{k\neq -n} \\ + \dfrac{e^{j\left([n-k]y+k\frac{\pi}{6}\right)}}{j[n-k]}\Bigg|_{k\neq n} \end{array}\right] \end{array}\right] \tag{6.58}$$

With considerable manipulation, a final solution for Eq. (6.46) can now be developed as

$$A_{mn}+jB_{mn}= \tag{6.59}$$

$$\frac{8V_{dc}}{m\pi^2}\left[\begin{array}{l}\left.\frac{\pi}{6}\sin\left([m+n]\frac{\pi}{2}\right)\left\{J_n\left(m\frac{3\pi}{4}M\right)+2\cos n\frac{\pi}{6}J_n\left(m\frac{\sqrt{3}\pi}{4}M\right)\right\}\right. \\ \left.+\frac{1}{n}\sin m\frac{\pi}{2}\cos n\frac{\pi}{2}\sin n\frac{\pi}{6}\left\{J_0\left(m\frac{3\pi}{4}M\right)-J_0\left(m\frac{\sqrt{3}\pi}{4}M\right)\right\}\right|_{n\neq 0} \\ +\sum_{\substack{k=1\\(k\neq -n)}}^{\infty}\left[\begin{array}{l}\frac{1}{[n+k]}\sin\left([m+k]\frac{\pi}{2}\right)\cos\left([n+k]\frac{\pi}{2}\right)\sin\left([n+k]\frac{\pi}{6}\right) \\ \times\left\{J_k\left(m\frac{3\pi}{4}M\right)+2\cos\left([2n+3k]\frac{\pi}{6}\right)J_k\left(m\frac{\sqrt{3}\pi}{4}M\right)\right\}\end{array}\right] \\ +\sum_{\substack{k=1\\(k\neq n)}}^{\infty}\left[\begin{array}{l}\frac{1}{[n-k]}\sin\left([m+k]\frac{\pi}{2}\right)\cos\left([n-k]\frac{\pi}{2}\right)\sin\left([n-k]\frac{\pi}{6}\right) \\ \times\left\{J_k\left(m\frac{3\pi}{4}M\right)+2\cos\left([2n-3k]\frac{\pi}{6}\right)J_k\left(m\frac{\sqrt{3}\pi}{4}M\right)\right\}\end{array}\right]\end{array}\right]$$

It should be noted that this solution is also valid for the condition of $m > 0, n = 0$. In addition, while Eq. (6.59) contains infinite summations for each value of m and n, in practice, summation over the range $1 \le k \le 10$ is usually sufficient to calculate harmonic magnitudes to an acceptable level of accuracy.

Equations (6.36), (6.41), (6.43), and (6.59) together define the magnitudes of the harmonics for any value of m and n for *naturally sampled space vector modulation* with the active space vectors centered in each half carrier interval.

Once again, theoretical analysis of *double-edge symmetrical* and *asymmetrical regular sampled* space vector modulation proceeds by replacing y with $y = y' + (\omega_o/\omega_c)x$ in Eq. (6.34), to give

$$A_{mn}+jB_{mn} = \frac{V_{dc}}{\pi^2}\sum_{i=1}^{6}\int_{y_s(i)}^{y_e(i)}\int_{x_r(i)}^{x_f(i)} e^{j\left(\left[m+n\frac{\omega_o}{\omega_c}\right]x+ny'\right)}dx\,dy' \tag{6.60}$$

for *double-edge symmetrical sampled* modulation, and

$$A_{mn} + jB_{mn} = \frac{V_{dc}}{\pi^2} \sum_{i=1}^{6} \left[\int_{y_s(i)}^{y_e(i)} \int_{x_r(i)}^{0} e^{j\left(mx + n\left[y_r' + \frac{\omega_o}{\omega_c}x + \frac{\omega_o}{\omega_c}\frac{\pi}{2}\right]\right)} dx\, dy_r' + \int_{y_s(i)}^{y_e(i)} \int_{0}^{x_f(i)} e^{j\left(mx + n\left[y_f' + \frac{\omega_o}{\omega_c}x - \frac{\omega_o}{\omega_c}\frac{\pi}{2}\right]\right)} dx\, dy_f' \right] \quad (6.61)$$

for *double-edge asymmetrical sampled* modulation.

Solutions for these two modulation variations can be obtained using similar analysis principles as have been developed above for naturally sampled SVM. However, the detailed working is not presented here since it follows almost exactly the processes used for the solution of naturally sampled modulation, and little point is served by the repetition. Detailed development of the solutions presented below is left as an exercise for the reader.

For $m = n = 0$, the solution of Eq. (6.36) continues to be valid, since all sampled reference waveforms have a DC offset of V_{dc}.

For all other harmonics, the solution process leading to Eq. (6.59) must be used, since the exponential terms remain in Eqs. (6.60) and (6.61) irrespective of whether m is zero or nonzero. Not unexpectedly, the results have a very similar form to naturally SVM, with substitutions of q for m as appropriate for the two regular sampled variations. [The parameter q was developed in Chapter 3 as $q = m + n(\omega_o/\omega_c)$.]

For *symmetrical regularly sampled SVM*, the harmonic magnitudes are given by

$$
A_{mn}+jB_{mn}= \tag{6.62}
$$

$$
\frac{8V_{dc}}{m\pi^2}\left[\begin{aligned}
&\left.\begin{aligned}&\frac{\pi}{6}\sin\left([q+n]\frac{\pi}{2}\right)\left\{J_n\left(q\frac{3\pi}{4}M\right)+2\cos n\frac{\pi}{6}J_n\left(q\frac{\sqrt{3}\pi}{4}M\right)\right\}\\
&+\frac{1}{n}\sin q\frac{\pi}{2}\cos n\frac{\pi}{2}\sin n\frac{\pi}{6}\left\{J_0\left(q\frac{3\pi}{4}M\right)-J_0\left(q\frac{\sqrt{3}\pi}{4}M\right)\right\}\end{aligned}\right|_{n\neq 0}\\
&+\sum_{\substack{k=1\\(k\neq -n)}}^{\infty}\left[\begin{aligned}&\frac{1}{[n+k]}\sin\left([q+k]\frac{\pi}{2}\right)\cos\left([n+k]\frac{\pi}{2}\right)\sin\left([n+k]\frac{\pi}{6}\right)\\
&\times\left\{J_k\left(q\frac{3\pi}{4}M\right)+2\cos\left([2n+3k]\frac{\pi}{6}\right)J_k\left(q\frac{\sqrt{3}\pi}{4}M\right)\right\}\end{aligned}\right]\\
&+\sum_{\substack{k=1\\(k\neq n)}}^{\infty}\left[\begin{aligned}&\frac{1}{[n-k]}\sin\left([q+k]\frac{\pi}{2}\right)\cos\left([n-k]\frac{\pi}{2}\right)\sin\left([n-k]\frac{\pi}{6}\right)\\
&\times\left\{J_k\left(q\frac{3\pi}{4}M\right)+2\cos\left([2n-3k]\frac{\pi}{6}\right)J_k\left(q\frac{\sqrt{3}\pi}{4}M\right)\right\}\end{aligned}\right]
\end{aligned}\right]
$$

For *asymmetrical regularly sampled SVM*, the harmonic magnitudes are given by

$$A_{mn} + jB_{mn} = \tag{6.63}$$

$$\frac{8V_{dc}}{m\pi^2}\left[\begin{aligned}&\left[\frac{\pi}{6}\sin\left([m+n]\frac{\pi}{2}\right)\left\{J_n\left(q\frac{3\pi}{4}M\right)+2\cos n\frac{\pi}{6}J_n\left(q\frac{\sqrt{3}\pi}{4}M\right)\right\}\right.\\&\left.+\frac{1}{n}\sin m\frac{\pi}{2}\cos n\frac{\pi}{2}\sin n\frac{\pi}{6}\left\{J_0\left(q\frac{3\pi}{4}M\right)-J_0\left(q\frac{\sqrt{3}\pi}{4}M\right)\right\}\right]\Bigg|_{n\neq 0}\\&+\sum_{\substack{k=1\\(k\neq -n)}}^{\infty}\left[\begin{aligned}&\frac{1}{[n+k]}\sin\left([m+k]\frac{\pi}{2}\right)\cos\left([n+k]\frac{\pi}{2}\right)\sin\left([n+k]\frac{\pi}{6}\right)\\&\times\left\{J_k\left(q\frac{3\pi}{4}M\right)+2\cos\left([2n+3k]\frac{\pi}{6}\right)J_k\left(q\frac{\sqrt{3}\pi}{4}M\right)\right\}\end{aligned}\right]\\&+\sum_{\substack{k=1\\(k\neq n)}}^{\infty}\left[\begin{aligned}&\frac{1}{[n-k]}\sin\left([m+k]\frac{\pi}{2}\right)\cos\left([n-k]\frac{\pi}{2}\right)\sin\left([n-k]\frac{\pi}{6}\right)\\&\times\left\{J_k\left(q\frac{3\pi}{4}M\right)+2\cos\left([2n-3k]\frac{\pi}{6}\right)J_k\left(q\frac{\sqrt{3}\pi}{4}M\right)\right\}\end{aligned}\right]\end{aligned}\right]$$

It is commented again that the summation expressions defining the harmonic component magnitudes for SVM are quite complex and must be carefully evaluated to calculate accurate values for particular harmonics.

Figure 6.8 shows the phase leg and *l–l* harmonic components for *naturally sampled* SVM, where the triplen harmonic common mode injection created by centering the active space vectors can be clearly seen in the phase leg harmonics. Figures 6.9 and 6.10 show the comparative harmonic performance for *symmetrical* and *asymmetrical regularly sampled* SVM, respectively.

Once more, it can be seen that for symmetrical regular sampled SVM, the $\sin([q+n]\pi/2)$ and $\sin([q+k]\pi/2)$ terms in Eq. (6.62) lead to additional harmonic terms compared to naturally sampled and asymmetrical regular sampled modulation. Also, the baseband components created by the symmetrical sampling process can be clearly seen as low-magnitude second, fourth and fifth fundamental harmonics in the *l–l* voltage shown in Figure 6.9(b). However, as expected, only the fifth harmonic component is present in the *l–l* voltage of asymmetrical regularly sampled SVM (Figure 6.10) because of the even baseband harmonic cancellation achieved intrinsically by this sampling strategy.

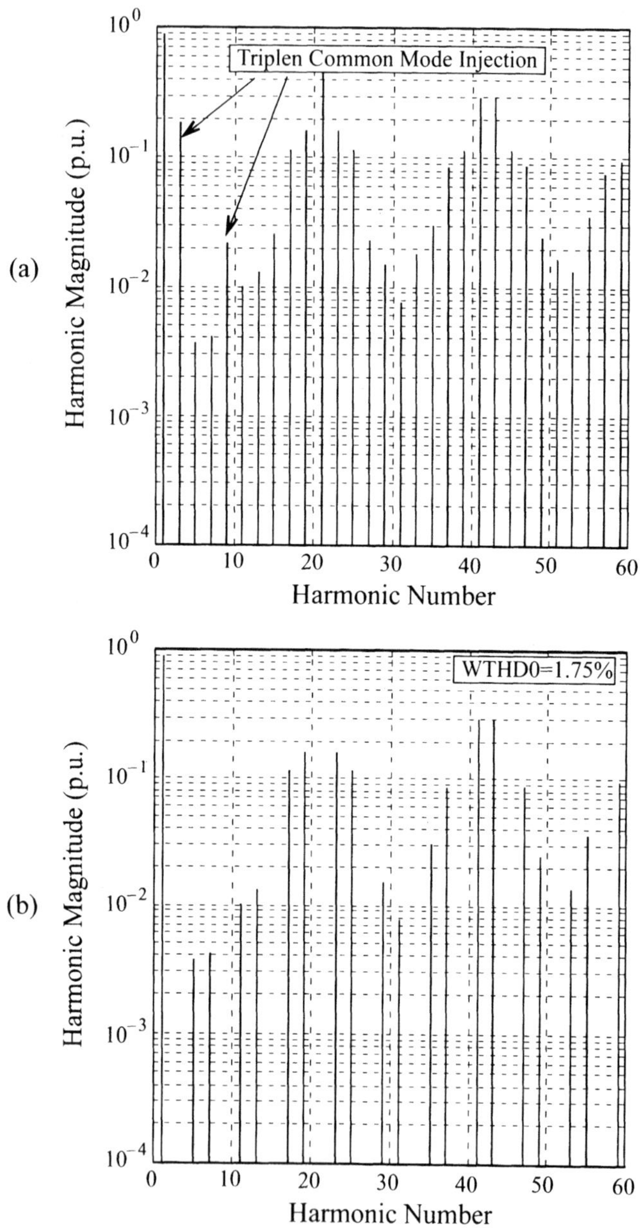

Figure 6.8 Theoretical harmonic spectra for three-phase inverter modulated by *naturally sampled* SVM: (a) phase leg a and (b) l–l output switched voltage waveforms, $M = 0.9$, $f_c/f_o = 21$.

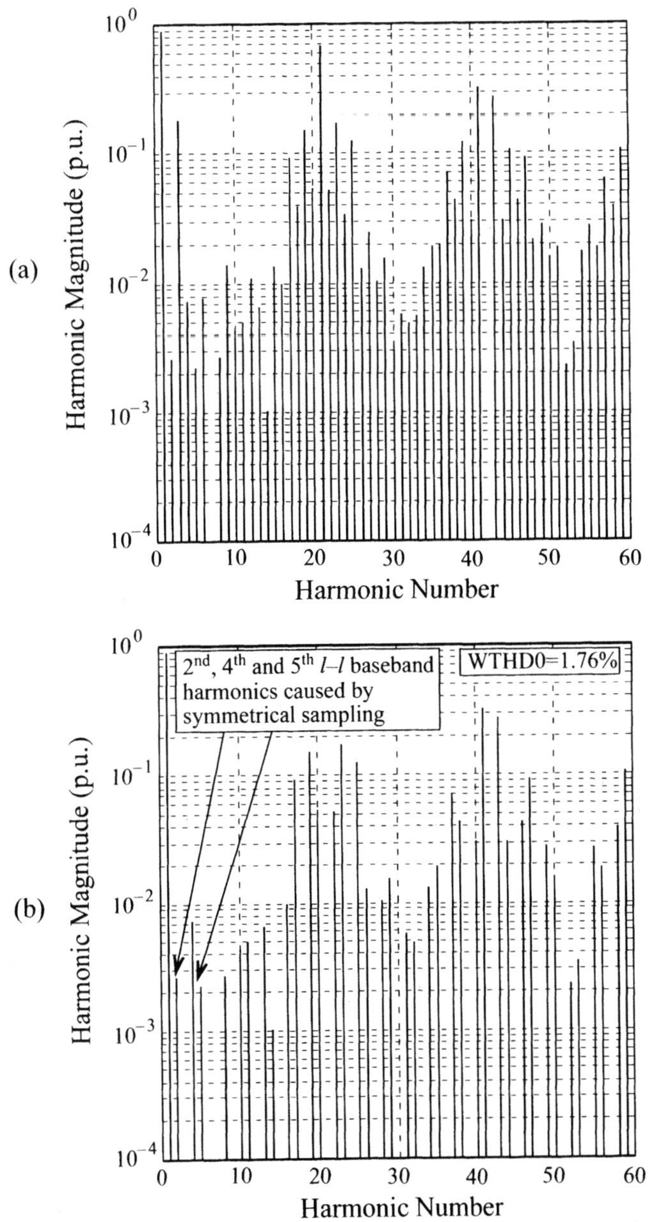

Figure 6.9 Theoretical harmonic spectra for three-phase inverter modulated by *double-edge symmetrical regularly sampled* SVM: (a) phase leg a and (b) l–l output switched voltage waveforms, $M = 0.9$, $f_c/f_o = 21$.

(a)

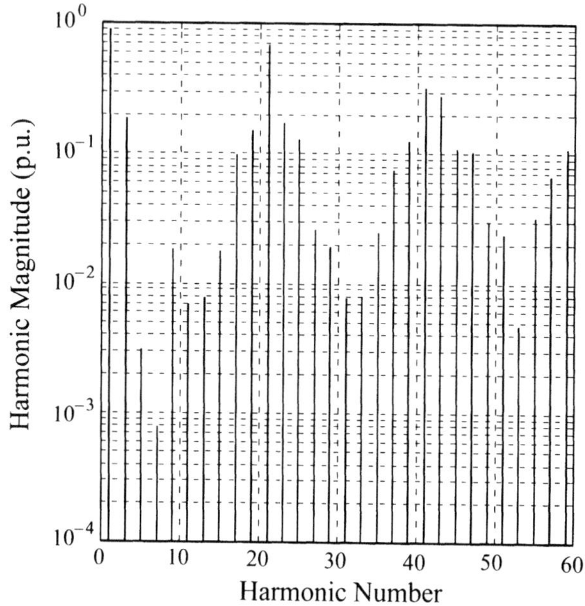

(b)

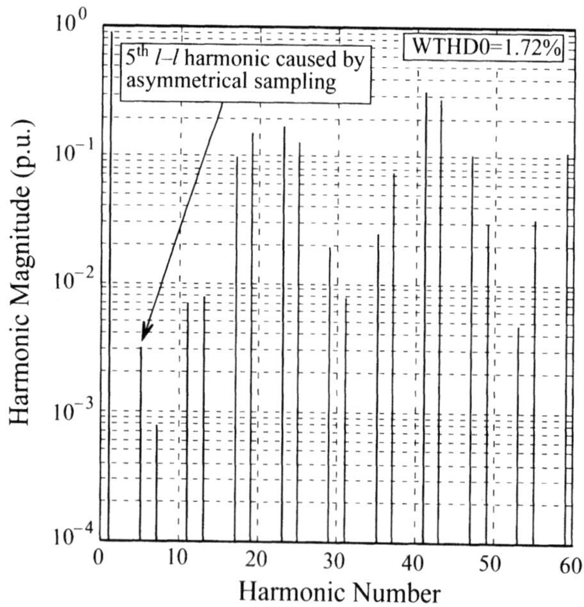

Figure 6.10 Theoretical harmonic spectra for three-phase inverter modulated by *double-edge asymmetrical regularly sampled* PWM: (a) phase leg *a* and (b) *l–l* output switched voltage waveforms, $M = 0.9$, $f_c/f_o = 21$.

The main benefit of SVM compared to simple sinusoidal or third-harmonic injected PWM can be seen by contrasting Figure 5.5(b) for a simple sinusoidal reference, and Figure 5.8(a) for one-sixth third-harmonic injected modulation, against Figure 6.10(b). Clearly, the effect of SVM is to transfer harmonic energy into the outer sideband harmonics and to some extent to the sidebands of the second carrier group, and reduce the inner sideband harmonics as a consequence. This reduces the overall WTHD of the switched output waveform.

To further illustrate this effect, Figure 5.8(b) shows that with a one-quarter third-harmonic injection, the magnitudes of the four major sidebands of the first carrier group are almost identical. This explains the improved WTHD performance of this modulation strategy even over SVM. But, of course, these distinctions become less significant at higher carrier and sideband harmonic frequencies, where the influence on the load ripple current is greatly reduced.

It should be noted also that the rate of decay of the sideband harmonics is much slower than for PWM with either a sinusoidal or a third-harmonic injected reference. This occurs because the common mode injection component for SVM contains multiple triplen harmonics, and the higher order triplen components increase the magnitude of the outer sideband harmonics. The effect of this is twofold.

First, when calculating the magnitude of a particular harmonic frequency, it is important to add all harmonics that have the same resultant frequency as phasor quantities. Harmonic contributions can come from the baseband, and sidebands from various carrier groups, and the wider sideband spread of SVM increases the potential for harmonic groups to overlap and unexpectedly cancel. For example, there is no seventh harmonic in the symmetrically sampled SVM result of Figure 6.9(a), and this is because of phasor cancellation between the baseband component and the first sideband group component.

Second, the increased lower tail of the first carrier group sidebands increases the possibility of noticeable harmonic intrusion into (and perhaps below) the fundamental frequency region, particularly for low fundamental/carrier ratios. In general, this is not a concern for practical systems, but it may need to be kept in mind for applications with special modulation requirements.

Table 6.4 confirms these conclusions by summarizing the normalized magnitudes of the significant harmonics for each of the four major PWM strategies. Table 6.4 also includes an estimate of the harmonic distortion produced by the major sideband harmonics of the first carrier group (single carrier harmonic distortion, or SCHD) and the second carrier group (double carrier harmonic distortion, or DCHD), calculated as

$$\text{SCHD or DCHD} = \sqrt{\sum_{n=i}^{n=j} (h_n)^2} \qquad \begin{array}{l} i = \text{start of harmonic region} \\ j = \text{end of harmonic region} \end{array}$$

and where h_n is the harmonic magnitude normalized to V_{dc} as before.

The overall shift in harmonic energy to the outer sidebands (and also a little up to the second carrier group sidebands) because of the improved space vector component placement is clearly shown by these partial harmonic distortion parameters. Note also that the total harmonic distortion calculated using the significant harmonics over the carrier and double carrier regions is almost identical for all four modulation strategies. This result confirms that the harmonics listed in Table 6.4 are the only significant components for a three-phase VSI PWM implementation, since these are the components which are summed in the THD given in Table 6.4.

Table 6.4 ***l–l*** **Harmonic Components for Asymmetrical Regularly Sampled PWM and SVM,** $f_c/f_o = 21$, $M = 0.9$

Harmonic	Sinusoidal PWM (%)	PWM + 1/6 3rd Harmonic (%)	PWM + 1/4 3rd Harmonic (%)	SVM (%)
$\omega_c - 4\omega_o$	0.66	7.98	11.65	9.80
$\omega_c - 2\omega_o$	25.05	17.04	12.91	15.09
$\omega_c + 2\omega_o$	28.38	19.50	14.85	17.26
$\omega_c + 4\omega_o$	1.94	10.75	15.19	12.92
SCHD (1st carrier group)	37.91	29.15	27.45	28.08
$2\omega_c - 5\omega_o$	1.39	7.10	9.11	7.39
$2\omega_c - \omega_o$	27.57	31.53	32.18	31.88
$2\omega_c + \omega_o$	23.46	27.36	28.03	27.72
$2\omega_c + 5\omega_o$	3.05	9.35	11.59	10.22
DCHD (2nd carrier group)	36.35	43.37	45.15	44.09
THD ($\sqrt{\text{SCHD}^2 + \text{DCHD}^2}$)	52.52	52.25	52.84	52.28
WTHD0 ($n = 15$)	0.00	0.04	0.05	0.11
WTHD0 ($n = 60$)	2.02	1.75	1.72	1.72

6.5 Harmonic Losses for SVM

A closed-form solution for the harmonic current produced by a SVM scheme can now be developed using a simple extension of the solution process presented for three-phase modulation in Chapter 5, as follows.

Equation (5.59) in Chapter 5 is a general expression which defines the square of the harmonic ripple current over one carrier interval for any arbitrary instantaneous values of the phase leg reference voltages. The overall harmonic losses for any particular modulation scheme can then be calculated by integrating this equation over a positive half fundamental cycle. For the modulation schemes presented in Chapter 5, the phase leg reference voltages are continuous functions, and the integration can be done over one set of limits, as shown in Eq. (5.63). For space vector modulation, the phase leg reference waveforms are separate sinusoidal segments in each 60° sextant, as listed in Table 6.2. Substituting $u_1 = v_{az}$ and $u_2 = v_{bz}$ into Eq. (5.59) from this table (with θ_o again replaced by θ for brevity) then creates separate expressions for the squared harmonic ripple current in each sextant over the positive half fundamental cycle, as follows:

$$\langle \Delta i_{ab}^2 \rangle = \left(\frac{V_{dc}}{L_\sigma}\right)^2 \frac{\Delta T^2}{48} \left[\begin{array}{l} 3M^2\cos^2\left(\theta+\frac{\pi}{6}\right) - 3\sqrt{3}M^3\cos^3\left(\theta+\frac{\pi}{6}\right) \\ -\frac{9}{8}M^4\cos\left(\theta+\frac{\pi}{6}\right)\left[\begin{array}{l} 3\sqrt{3}\cos^3\left(\theta-\frac{2\pi}{3}\right) \\ -\cos^3\left(\theta-\frac{\pi}{6}\right) \end{array}\right] \end{array} \right]_{0<\theta<\frac{\pi}{3}} \tag{6.64}$$

$$\langle \Delta i_{ab}^2 \rangle = \left(\frac{V_{dc}}{L_\sigma}\right)^2 \frac{\Delta T^2}{48} \left[\begin{array}{l} 3M^2\cos^2\left(\theta+\frac{\pi}{6}\right) - 3\sqrt{3}M^3\cos^3\left(\theta+\frac{\pi}{6}\right) \\ -\frac{9}{8}M^4\cos\left(\theta+\frac{\pi}{6}\right)\left[\cos^3\left(\theta-\frac{5\pi}{6}\right) - \cos^3\left(\theta+\frac{\pi}{6}\right)\right] \end{array} \right]_{-\frac{\pi}{3}<\theta<0} \tag{6.65}$$

$$\langle \Delta i_{ab}^2 \rangle = \left(\frac{V_{dc}}{L_\sigma}\right)^2 \frac{\Delta T^2}{48} \left[\begin{array}{l} 3M^2\cos^2\left(\theta+\frac{\pi}{6}\right) - 3\sqrt{3}M^3\cos^3\left(\theta+\frac{\pi}{6}\right) \\ -\frac{9}{8}M^4\cos\left(\theta+\frac{\pi}{6}\right)[\sin^3\theta - 3\sqrt{3}\cos^3\theta] \end{array} \right]_{-\frac{2\pi}{3}<\theta<-\frac{\pi}{3}} \tag{6.66}$$

Not unexpectedly, the M^2 and M^3 terms in these equations are identical to each other and to those in Eq. (5.62). This is because they involve only *l–l* reference quantities, and these must be the same for any modulation strategy for a given fundamental *l–l* output. Hence the integrated solution coefficients for these terms developed in Eqs. (5.66), (5.67), and (5.68) are still valid. However, the M^4 term must be integrated separately in three parts, as follows:

$$I^2_{ab,\,h,\,rms}\Big|_{M^4} = -\left(\frac{V_{dc}}{L_\sigma}\right)^2 \frac{\Delta T^2}{48}\frac{9}{8}\frac{M^4}{\pi} \tag{6.67}$$

$$\times\left[\begin{aligned}&\int_0^{\frac{\pi}{3}} \cos\left(\theta+\frac{\pi}{6}\right)\left[3\sqrt{3}\cos^3\left(\theta-\frac{2\pi}{3}\right)-\cos^3\left(\theta-\frac{\pi}{6}\right)\right]d\theta\\ &+\int_{-\frac{\pi}{3}}^{0} \cos\left(\theta+\frac{\pi}{6}\right)\left[\cos^3\left(\theta-\frac{5\pi}{6}\right)-\cos^3\left(\theta+\frac{\pi}{6}\right)\right]d\theta\\ &+\int_{-\frac{2\pi}{3}}^{-\frac{\pi}{3}} \cos\left(\theta+\frac{\pi}{6}\right)[\sin^3\theta-3\sqrt{3}\cos^3\theta]\,d\theta\end{aligned}\right]$$

Using standard trigonometric substitutions of

$$\begin{aligned}\sin^3\theta &= \frac{3}{4}\sin\theta-\frac{1}{4}\sin 3\theta\\ \cos^3\theta &= \frac{3}{4}\cos\theta+\frac{1}{4}\cos 3\theta\\ \cos a\cos b &= \frac{1}{2}(\cos[a+b]+\cos[a-b])\\ \cos a\sin b &= \frac{1}{2}(\sin[a+b]-\sin[a-b])\end{aligned} \tag{6.68}$$

Eq. (6.67) rearranges to

$$I_{ab,h,rms}^2\Big|_{M^4} = \left(\frac{V_{dc}}{L_\sigma}\right)^2 \frac{\Delta T^2}{48}\frac{9}{8} M^4$$

$$\times\frac{1}{\pi}\left[\begin{aligned}&\int_0^{\pi/3}\frac{1}{4}\cos\left(\theta+\frac{\pi}{6}\right)\left[\begin{aligned}&9\sqrt{3}\cos\left(\theta+\frac{\pi}{3}\right)-3\sqrt{3}\cos 3\theta\\&\qquad+3\cos\left(\theta-\frac{\pi}{6}\right)+\sin 3\theta\end{aligned}\right]d\theta\\&+\int_{-\pi/3}^{0} 2\cos^4\left(\theta+\frac{\pi}{6}\right)d\theta\\&+\int_{-2\pi/3}^{-\pi/3}\frac{1}{4}\cos\left(\theta+\frac{\pi}{6}\right)\left[\begin{aligned}&-3\sin\theta+\sin 3\theta\\&\qquad+9\sqrt{3}\cos\theta+3\sqrt{3}\cos 3\theta\end{aligned}\right]d\theta\end{aligned}\right]$$

$$= \left(\frac{V_{dc}}{L_\sigma}\right)^2 \frac{\Delta T^2}{48}\frac{9}{8} M^4 \tag{6.69}$$

$$\times\frac{1}{8\pi}\left[\begin{aligned}&\int_0^{\pi/3}\left[\begin{aligned}&-3\sqrt{3}\left\{3\sin 2\theta-3\cos\frac{\pi}{6}+\cos\left(4\theta+\frac{\pi}{6}\right)+\cos\left(2\theta-\frac{\pi}{6}\right)\right\}\\&\qquad+3\cos 2\theta+3\cos\frac{\pi}{3}+\sin\left(4\theta+\frac{\pi}{6}\right)+\sin\left(2\theta-\frac{\pi}{6}\right)\end{aligned}\right]d\theta\\&+\int_{-\pi/3}^{0}\left[4+8\cos\left(2\theta+\frac{\pi}{3}\right)+2+2\cos\left(4\theta+\frac{2\pi}{3}\right)\right]d\theta\\&+\int_{-2\pi/3}^{-\pi/3}\left[\begin{aligned}&-3\sin\left(2\theta+\frac{\pi}{6}\right)+3\sin\frac{\pi}{6}+\sin\left(4\theta+\frac{\pi}{6}\right)+\sin\left(2\theta-\frac{\pi}{6}\right)\\&+3\sqrt{3}\left[\begin{aligned}&3\cos\left(2\theta+\frac{\pi}{6}\right)+3\cos\frac{\pi}{6}\\&\qquad+\cos\left(4\theta+\frac{\pi}{6}\right)+\cos\left(2\theta-\frac{\pi}{6}\right)\end{aligned}\right]\end{aligned}\right]d\theta\end{aligned}\right]$$

Equation (6.69) integrates with some manipulation to

$$I^2_{ab,\,h,\,rms}\Big|_{M^4} = \left(\frac{V_{dc}}{L_\sigma}\right)^2 \frac{\Delta T^2}{48}\,\frac{9}{8}\,M^4\left(\frac{3}{2}-\frac{9}{8}\frac{\sqrt{3}}{\pi}\right) \tag{6.70}$$

so that the complete closed-form solution for the harmonic current produced by a SVM scheme is

$$I^2_{ab,\,h,\,rms} = \left(\frac{V_{dc}}{L_\sigma}\right)^2 \frac{\Delta T^2}{48}\left[\frac{3}{2}M^2 - \frac{4\sqrt{3}}{\pi}M^3 + \frac{9}{8}\left(\frac{3}{2}-\frac{9}{8}\frac{\sqrt{3}}{\pi}\right)M^4\right] \tag{6.71}$$

Evaluating the coefficient of the M^4 term gives a result of 0.9897, which is very close to the result for sinusoidal PWM with third-harmonic injection, Eqs. (5.67) and (5.68) in Chapter 5.

6.6 Placement of the Zero Space Vector

In Section 4.5 it was concluded that the best harmonic performance for a PWM algorithm is achieved when the active *l–l* output pulses are placed in the middle of each half carrier period. This was verified for a single-phase VSI, where it was determined that an implicit consequence of three-level PWM was to place the *l–l* pulses in this manner. For a three-phase VSI, it is not physically possible to place all three sets of *l–l* output pulses centrally in each half carrier period, since the switching process implicitly generates sequential *l–l* output pulses as shown in Figure 5.3. Consequently the placement of the *l–l* output voltage pulses for a three-phase system is constrained by the need to balance the pulse position of all three sets of *l–l* output pulses, so that the harmonic components of all three *l–l* output voltages match.

Further examination of Figure 5.3 shows that the two active pulses V_1 and V_2 always directly follow each other, since the transition between them is a single-phase leg state change. Hence the process of pulse position for a three-phase VSI becomes simply the process of placing the total active pulse interval $V_1 + V_2$ within each half carrier period. Of course, the phase leg switching could be arranged so that an extra zero state is inserted between the two active space vector components of Figure 5.3, but this requires additional switch transitions within the half carrier period. Such a strategy is clearly suboptimal, since it increases the switching frequency of the phase legs and moves the active *l–l* output pulses (which are the space vector components) further from the center of the half carrier period, rather than closer as is desired.

Figures 6.11 and 6.12 show how the active *l–l* output pulses are distributed within each half carrier period over a complete fundamental cycle, for:

- Simple sinusoidal regular sampled PWM.
- PWM + one-sixth magnitude third-harmonic injection.
- PWM + one-quarter magnitude third-harmonic injection.
- SVM with the active space vectors centered in each half carrier period.

It should be noted that the carrier/fundamental ratio for Figures 6.11 and 6.12 has been changed to 18 (i.e., 10° fundamental steps between asymmetrical sampled points) for clarity of illustration. This change does not affect the validity of the conclusions drawn below, since the relative influence of pulse position on harmonic performance is the same at any carrier frequency.

The effect that the different modulation strategies have on the position of the active space vectors can be seen by looking at the region $\pi/6 \le \theta_o \le 2\pi/3$. For all four modulation strategies, the total active space vector period at $\theta_o = \pi/6$ and $\theta_o = \pi/2$ is identically placed in the center of the half carrier period. While all three-phase VSI pulse position variations add common mode triplen components to the target reference waveform, the triplen components are zero at $(2n+1)\pi/6$ multiples of the fundamental phasor angle.

Note also that the total active space vector interval has its maximum value at the $(2n+1)\pi/6$ multiples of the fundamental phasor angle, since this is the point where equal amounts of the two nearest space vectors are required to make the target space vector. This interval is shown as $SV_{\min}$ on Figure 6.11 and is the limiting factor for the maximum possible modulation index of $M = 1.15$ as discussed earlier. However, while the total active space vector interval is centered in the half carrier period at $\theta_o = \pi/6$ and $\theta_o = \pi/2$, the two individual space vector components within this interval reverse their position in each successive half carrier period. This creates a carrier frequency oscillation for each space vector component, and, since the *l–l* output voltage pulses are made up from these components, one would expect to see incomplete odd carrier multiple sideband cancellation and consequential odd carrier multiple sideband harmonics in the *l–l* output switched waveform, as indeed occurs.

Furthermore, at $\theta_o = \pi/3$ and $\theta_o = 2\pi/3$ there is a clear difference between the modulation strategies, as follows:

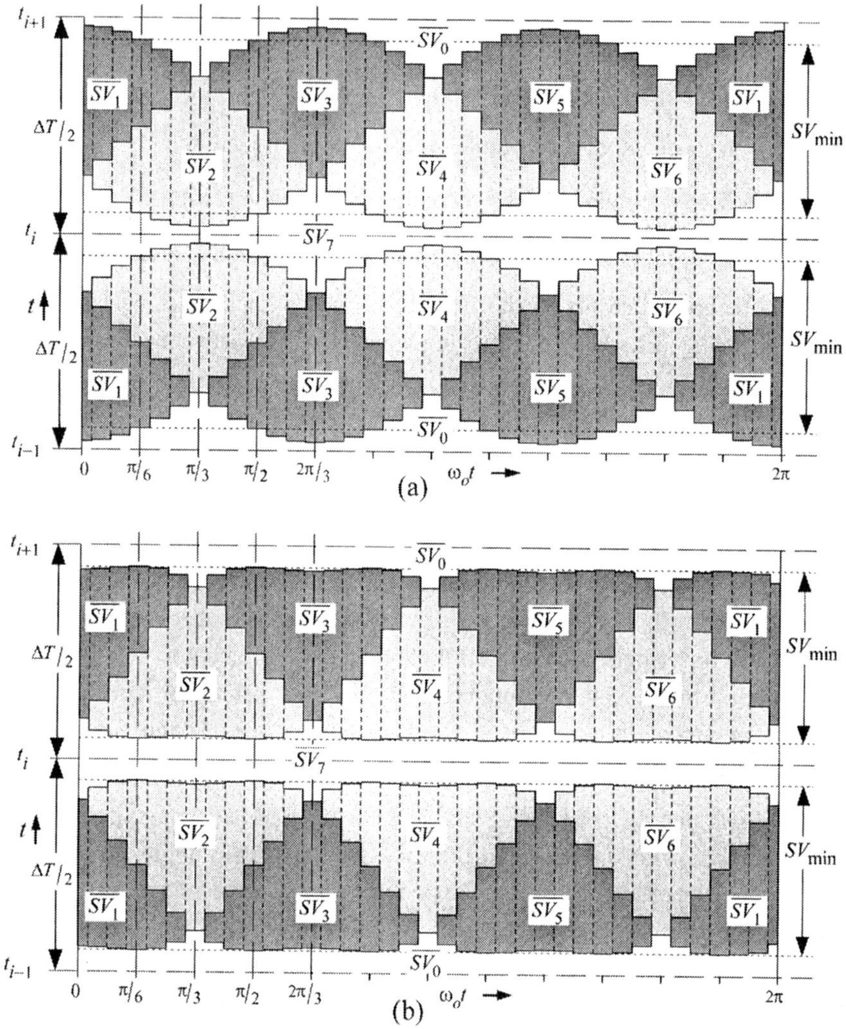

Figure 6.11 Space vector placement for (a) regular sampled PWM and (b) regular sampled PWM + one-sixth magnitude third-harmonic (modulation index $M = 0.9$).

- Regular sampled PWM without third-harmonic injection [Figure 6.11(a)] places one edge of the total active space vector interval significantly outside the SV_{min} region, first on one side and then on the other side in each successive half carrier period. This placement not only limits the maximum modulation index to $M = 1.0$, but also increases the

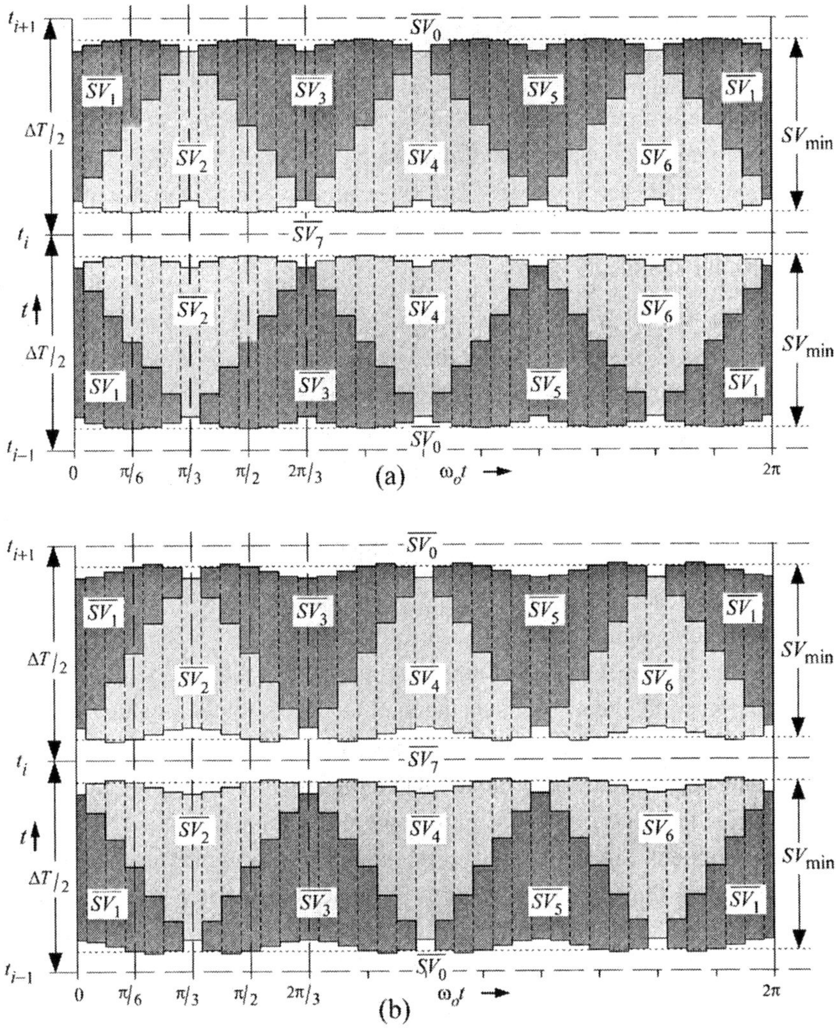

Figure 6.12 Space vector placement for (a) SVM with active pulses centered and (b) regular sampled PWM + one-quarter magnitude third-harmonic (modulation index $M = 0.9$).

oscillatory movement of the space vector component(s) within each successive half carrier period. This additional movement increases the magnitude of the odd carrier multiple sideband harmonics since pulses in successive half carrier periods are not separated as far apart as they would be if they were centrally placed in the half carrier period.

- Regular sampled PWM with one-sixth magnitude third-harmonic injection [Figure 6.11(b)] restricts the placement of the total active space vector interval to within the $SV_{\min}$ region, which allows the maximum modulation index to increase to $M = 1.15$. However, there is still some oscillation of the position of the space vector component(s) in each successive half carrier period. However, at least the magnitude of the odd carrier multiple sideband harmonics should be reduced compared to regular sampled PWM.
- SVM with centered active pulses [Figure 6.12(a)] continues this concept, by explicitly centering the vector component(s) in every half carrier period. This cancels the odd carrier sidebands at this point in the fundamental cycle, but of course the space vector oscillation at other parts of the fundamental cycle still remains so that the total odd carrier sideband cancellation will not be complete. However, the harmonic performance clearly will improve under this strategy.
- Regular sampled PWM with one-quarter magnitude third-harmonic injection [Figure 6.12(b)] has a further effect. This strategy centers the total active space vector interval at every $\pi/6$ step in the fundamental phase angle, but also offsets the total interval slightly outside the $SV_{\min}$ region in between these points. The effect of this is to bias the intermediate space vector placements so that the larger space vector component (i.e., the longer pulse) is more centered at the expense of the smaller component. Since the larger component is more significant, the result is closer to the (unreachable) ideal of exactly centering every *l–l* output pulse in each half carrier period. This should improve the harmonic performance of this strategy even compared to SVM, although there is a slight consequential loss in maximum modulation index down to $M = 1.12$ because the space vector interval is now placed slightly outside the $SV_{\min}$ region for some of the fundamental cycle.

Indeed, taking this concept further, it should be possible to position the space vector components in each half carrier period to optimize the reduction in odd carrier sideband harmonic components by biasing the position of the space vectors in favor of the larger space vector component. This work has been done by [8–10], and in all three cases the conclusion reached is that the optimal space vector pulse position is almost exactly that achieved by a one-quarter third-harmonic injection strategy.

6.7 Discontinuous Modulation

The previous sections of this chapter have established that the zero space vector contributes as the fundamental difference between PWM strategies, and the various alternative implementations simply vary the position of the space vector active pulses within each half carrier period. From this understanding, further alternative modulation strategies can now be identified where the active space vectors in successive half carrier intervals are moved to join together, and the intermediate zero space vector consequently disappears. These strategies have the advantage of eliminating one switching transition in each half carrier interval, which allows the switching frequency to increase by a nominal value of 3/2 accordingly for the same inverter losses. This may improve the harmonic performance of the inverter by virtue of the reduced influence of higher frequency switching harmonics. However, any particular benefit will depend on the precise harmonic performance of these alternative *discontinuous* modulation strategies.

6.7.1 120° Discontinuous Modulation

The most straightforward approach to achieve discontinuous switching is to move the active space vectors in Figure 6.11 to eliminate one of the two space vectors, similar to the removal of the center V_Z states described in Section 4.5.2 for the single-phase VSI. The only difference is that for the three-phase topology, each phase leg in turn is now continuously locked to the upper or lower DC rail (whichever is chosen) for one-third of the fundamental cycle (120°). This variation is termed 120° *discontinuous modulation*, and the consequential space vector placement is shown in Figure 6.13, with either the $\overline{SV_0}$ or the $\overline{SV_7}$ zero vector eliminated depending on which DC rail is used as the clamping reference.

Examination of Figure 6.13(a) (DPWMMIN strategy) shows that space vectors $\overline{SV_2}$, $\overline{SV_4}$, $\overline{SV_6}$, have been concatenated into one pulse per carrier period, while space vectors $\overline{SV_1}$, $\overline{SV_3}$, $\overline{SV_5}$, remain as separate pulses in each half carrier period. This procedure is done in a similar manner as the V_{active} pulses which were combined for the single-phase inverter as shown in Figure 4.18. However, the arrangement has a complex effect on the *l–l* output switched waveform. Considering the *l–l* voltage v_{ab}, Figure 6.1 shows that only space vectors $\overline{SV_1}$, $\overline{SV_3}$, $\overline{SV_4}$, $\overline{SV_6}$, lead to active pulses between phase legs *a* and *b* (the other space vectors create an *l–l* short circuit across *ab*). Also,

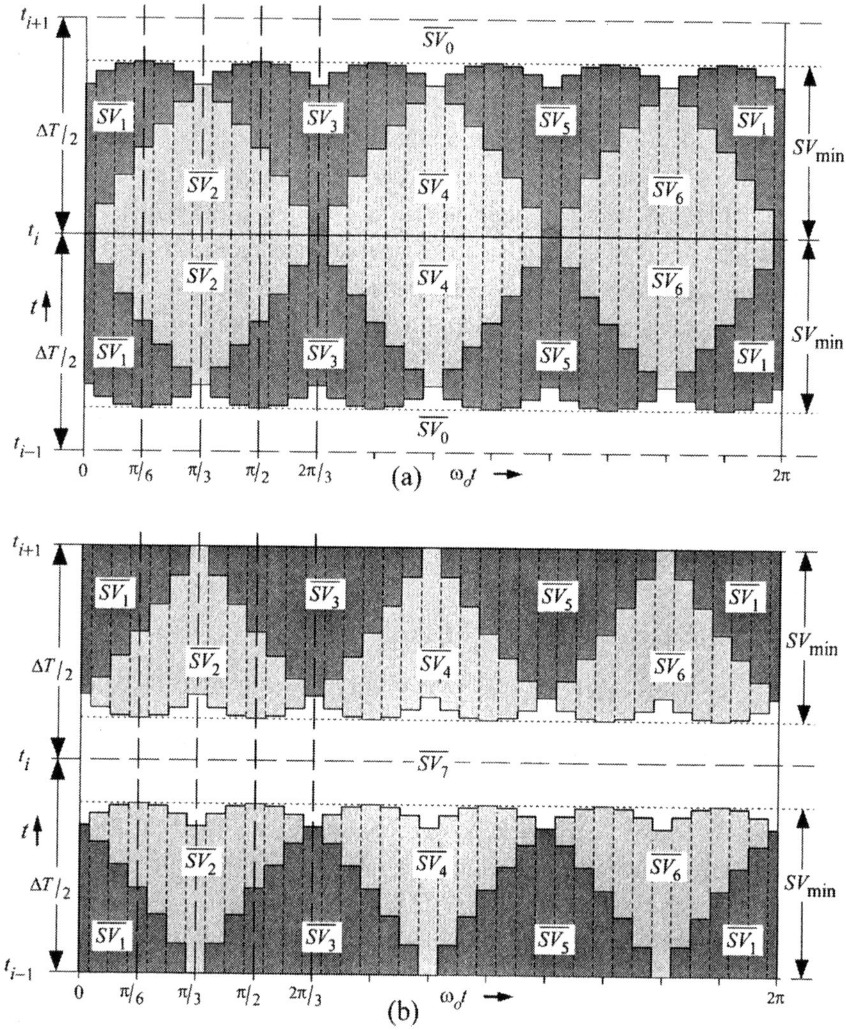

Figure 6.13 Space vector placement for (a) DPWMMIN with active pulses moved to lower DC rail and (b) DPWMMAX with active pulses moved to upper DC rail.

it has just been determined that 120° discontinuous modulation concatenates space vectors $\overline{SV_4}$, $\overline{SV_6}$ into one pulse per carrier period, while space vectors $\overline{SV_1}$, $\overline{SV_3}$ remain as two pulses per carrier period. Hence all *l–l* voltages will be asymmetrical across a fundamental cycle. Figure 6.14 shows the switched phase leg voltages and v_{ab} voltage for this strategy, where the asymmetry in the *l–l* voltage can be clearly seen.

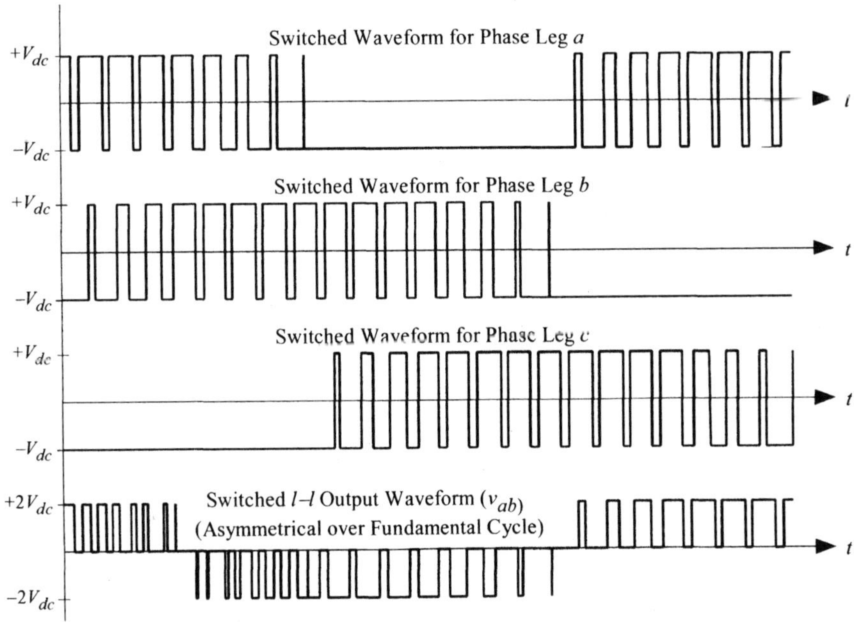

Figure 6.14 Phase leg and l–l switched output voltages for DPWMMIN $120°$ *discontinuously switched* PWM, $M = 0.9$, $f_c/f_o = 24$.

Simply from this asymmetry, it would be expected that the harmonic performance of $120°$ discontinuous PWM will be suboptimal compared to continuous modulation strategies. However, it can also be identified that to maintain the same effective phase leg switching frequency as for continuous modulation, the space vector calculation frequency (i.e., twice the carrier frequency) should be increased by approximately 50%, since each phase leg only switches during two-thirds of each fundamental cycle. It is the trade-off between these two effects that offers potential advantages for $120°$ discontinuous PWM under some modulation conditions.

A further limitation with this strategy is that one device of each phase leg is always turned off during its $120°$ unmodulated region, while the other device is always conducting. Hence conduction losses are not shared equally across the two devices in each phase leg. However, this limitation can be avoided by clamping to alternate DC buses every half cycle (i.e., cycle between the DPWMMIN and the DPWMMAX strategies).

6.7.2 60° and 30° Discontinuous Modulation

An improved discontinuous modulation strategy that inherently balances the switching losses is to place the space vectors to alternately eliminate zero space vectors $\overline{SV_0}$ and $\overline{SV_7}$ for successive 60° segments of the fundamental cycle. This arrangement is called 60° *discontinuous modulation*, since each phase leg is now unmodulated for only 60° at a time (alternately switched to the upper or to the lower DC rail). It has the particular benefit that the *l–l* switched voltages are now symmetrical, as shown in Figure 6.15. The space vector arrangement for this modulation strategy is shown in Figure 6.16(a).

This type of discontinuous switching pattern is termed DPWM1 and centers the nonswitching periods for each phase leg symmetrically around the positive and negative peaks of its fundamental reference voltage. This is the best position for a resistive load, since the peaks of the line currents follow the peaks of the fundamental voltages. Therefore each phase leg does not switch just when the current is at its maximum value, and this obviously minimizes switching losses.

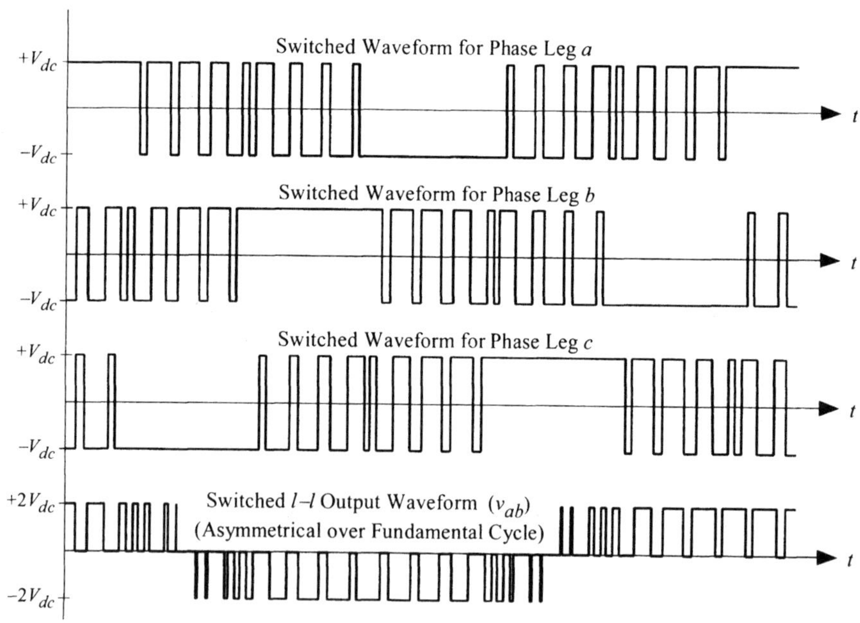

Figure 6.15 Phase leg and *l–l* switched output voltages for DPWM1 60° discontinuously switched PWM, $M = 0.9$, $f_c/f_o = 24$.

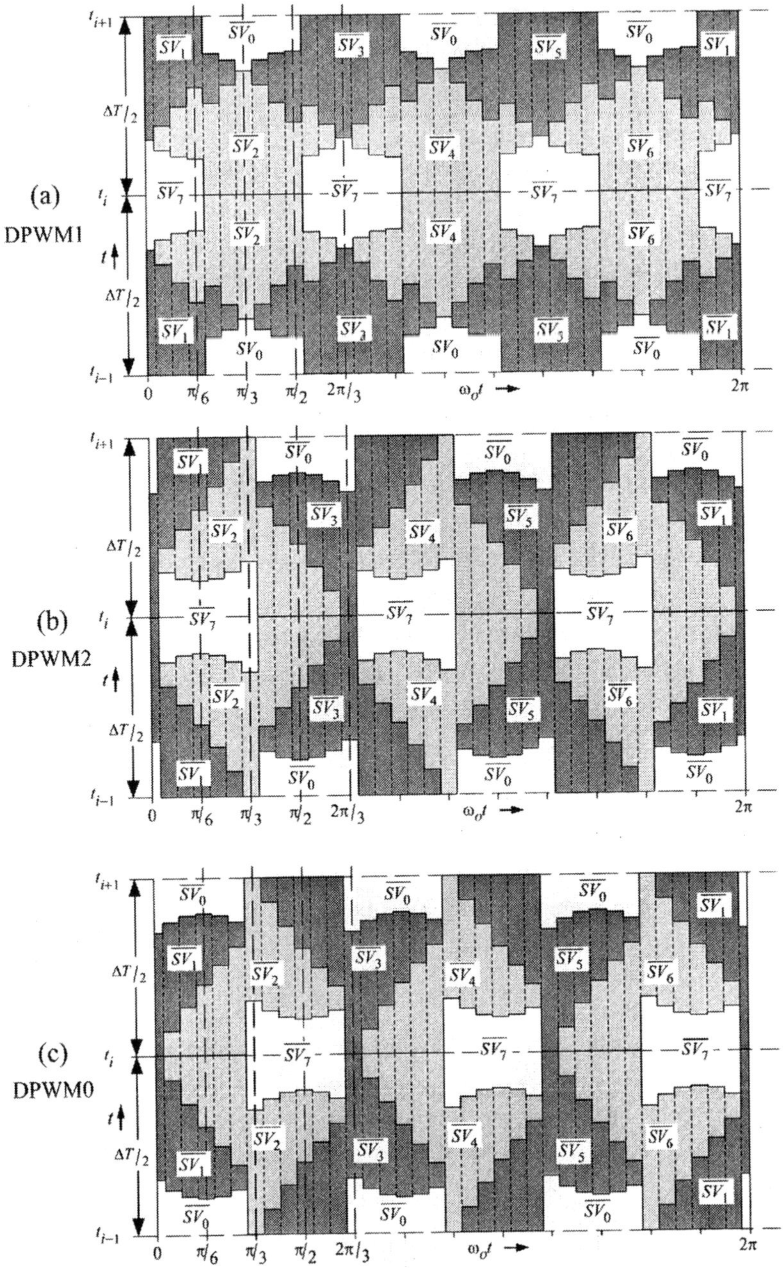

Figure 6.16 Space vector placement for 60° *discontinuous* PWM: (a) DPWM1 (zero clamp at voltage peak), (b) DPWM2 (30° lagging clamp), and (c) DPWM0 (30° leading clamp).

A little thought suggests further that it is feasible to place each 60° nonswitching period anywhere within the 120° region where the appropriate phase leg reference voltage is the maximum/minimum of the three-phase set. For example, the 60° "clamp to $+V_{dc}$" nonswitching period for phase leg a can be placed anywhere in the region $-\pi/3 \leftrightarrow \pi/3$, since the a phase reference voltage is more positive than the other two phase reference voltages during this period. Similarly, the "clamp to $-V_{dc}$" nonswitching period for phase leg a can be placed anywhere in the region $2\pi/3 \leftrightarrow 4\pi/3$, since this is the region where the a phase fundamental voltage is more negative than the other two-phase reference voltages. This freedom allows alternative 60° discontinuous PWM strategies to be considered which minimize the switching losses for loads which are not unity power factor.

For a *lagging* power factor (pf) load, it is clearly preferable to *retard* the non-switching period by up to a maximum of 30° (pf of 0.866 *lag*), depending on the load current power factor. This is the discontinuous modulation strategy DPWM2 shown in Figure 6.16(b). For a leading power factor load, the non-switching period can be *advanced* by up to 30° (pf of 0.866 lead). This is discontinuous modulation strategy DPWM0 shown in Figure 6.16(c). It should be noted however, that incorporating this advance/retard of the nonswitching period reintroduces asymmetry into the *l–l* voltage, as shown in Figure 6.17.

A final variation [11] is to eliminate zero space vectors so as to clamp the phase legs to the opposite DC rails in each 60° segment, as shown in Figure 6.18. This strategy is sometimes termed 30° *discontinuous modulation,* since it requires the clamping phase leg to change every 30°. Figure 6.19 shows the switching pattern for this modulation strategy, which is again symmetrical.

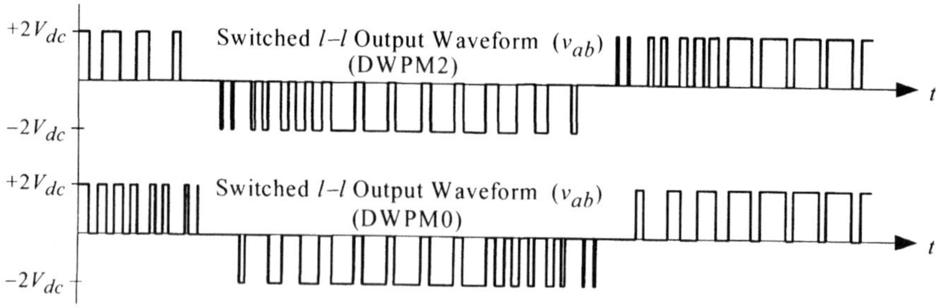

Figure 6.17 Phase leg and *l–l* switched output voltages for DPWM2 and DPWM0 60° discontinuously switched PWM, $M = 0.9$, $f_c/f_o = 24$.

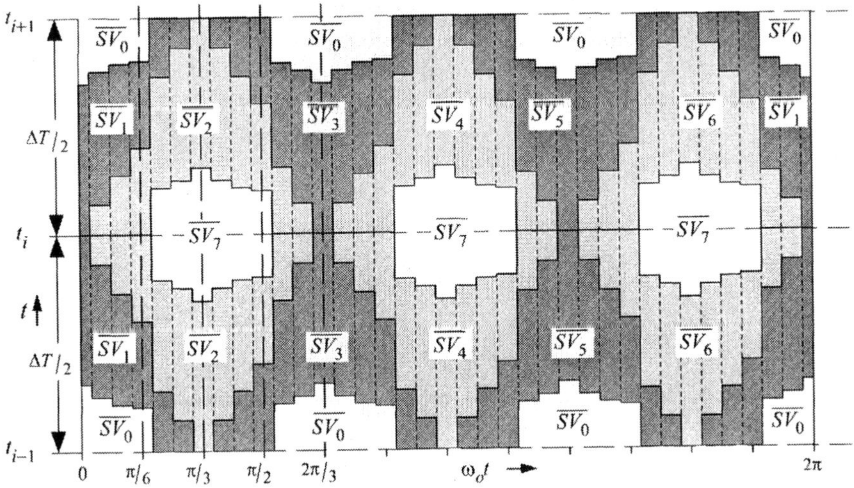

Figure 6.18 Space vector placement for DPWM3 30° *discontinuous* PWM.

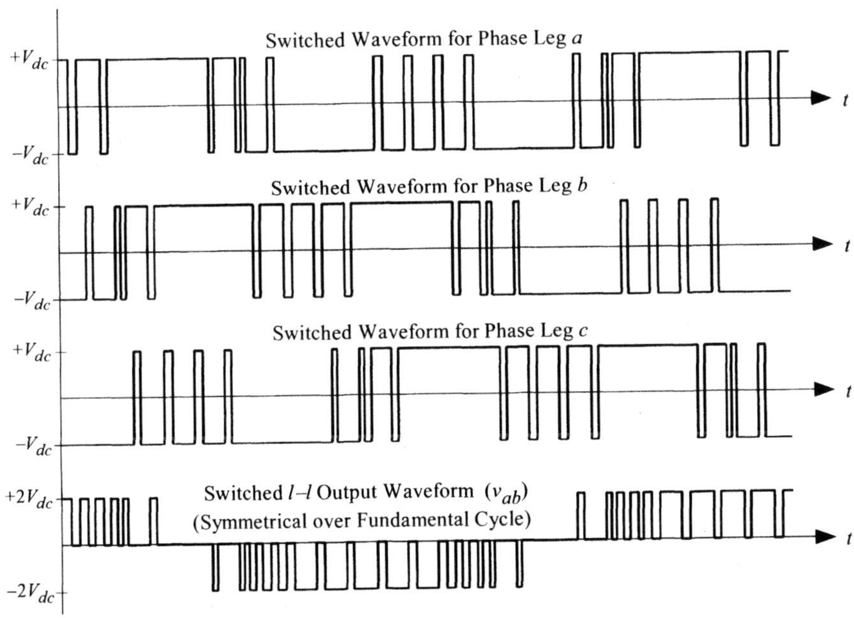

Figure 6.19 Phase leg and *l–l* switched output voltages for DPWM3 30° discontinuously switched PWM, $M = 0.9$, $f_c/f_o = 24$.

Figure 6.20 shows the sections of the three-phase voltage envelopes that are clamped to the DC rails for the various discontinuous modulation strategies.

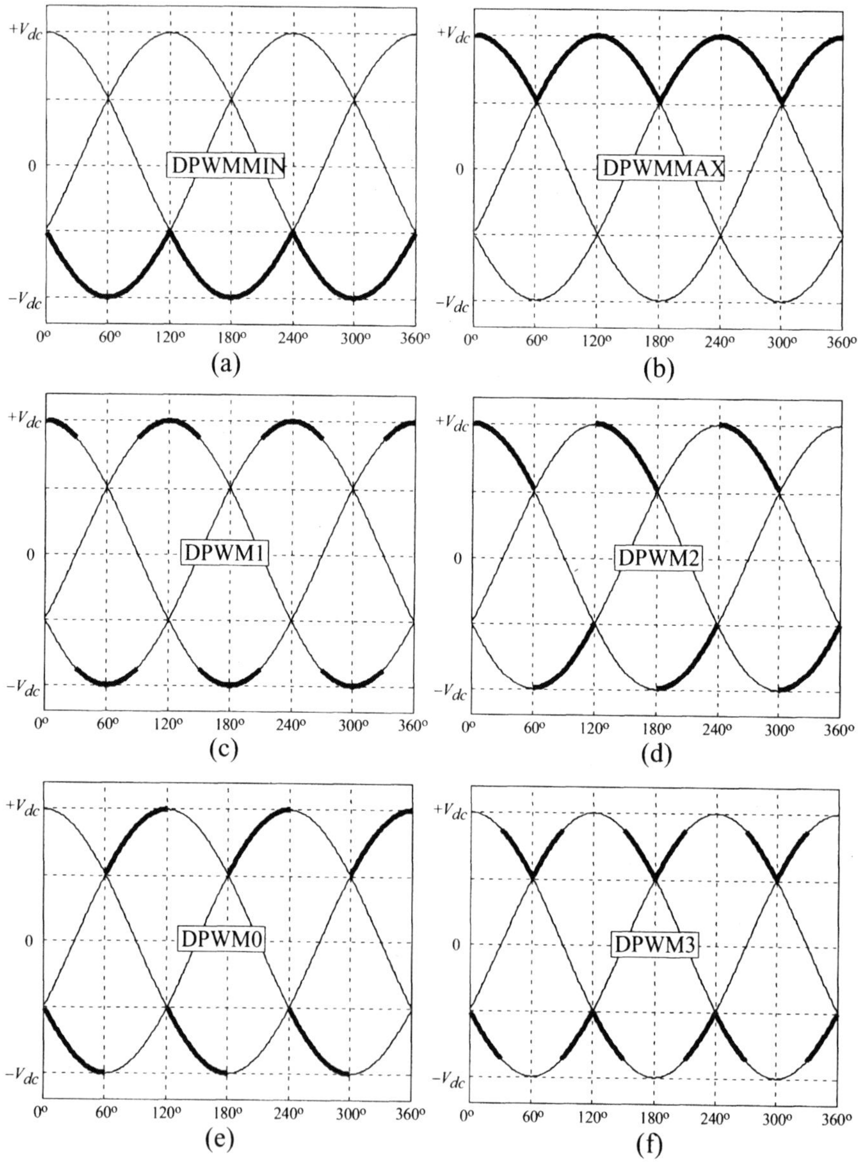

Figure 6.20 Clamped waveform sections for discontinuous PWM: (a) DPWMMIN (clamp to lower DC rail), (b) DPWMMAX (clamp to upper DC rail), (c) DPWM1 (clamp at voltage peaks), (d) DPWM2 (30° lagging clamp), (e) DPWM0 (30° leading clamp), and (f) DPWM3 (clamp to opposite DC rails).

6.8 Phase Leg References for Discontinuous PWM

It is useful to consider how the 120°, 60°, and 30° discontinuous modulation strategies achieve the proper mean sinusoidal commanded voltages across the load. Consider, for example, the 120° DPWMMIN case illustrated in Figure 6.13. During operation in the first sextant of Figure 6.2, $0 \le \theta \le \pi/3$, the phase leg switched voltages have the pulse patterns shown in Figure 6.21.

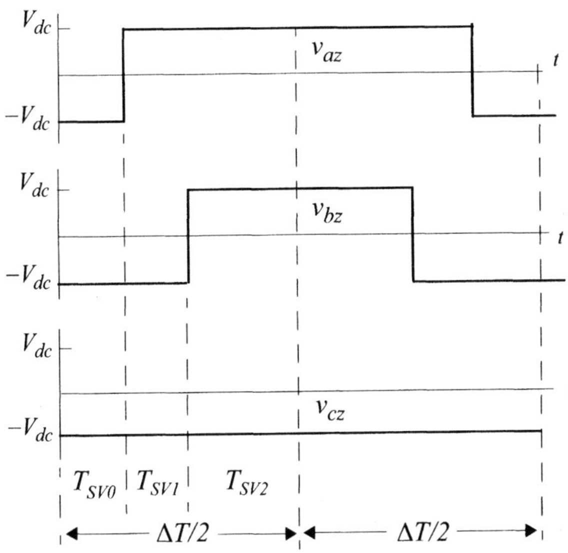

Figure 6.21 Pulse pattern for 120° DPWMMIN discontinuous PWM in the first sextant, $0 \le \theta_o \le \pi/3$.

The three-phase leg voltages can now be expressed as

$$\langle v_{az}\rangle = V_{dc}\frac{2}{\Delta T}(T_{SV1} + T_{SV2} - T_{SV0}) \tag{6.72}$$

$$\langle v_{bz}\rangle = V_{dc}\frac{2}{\Delta T}(-T_{SV1} + T_{SV2} - T_{SV0}) \tag{6.73}$$

$$\langle v_{cz}\rangle = V_{dc}\frac{2}{\Delta T}(-T_{SV1} - T_{SV2} - T_{SV0}) \tag{6.74}$$

However, since

$$T_{SV1} + T_{SV2} + T_{SV0} = \frac{\Delta T}{2}$$

Eqs. (6.72) to (6.74) become

$$\langle v_{az}\rangle = 2V_{dc}\left(\frac{2}{\Delta T}T_{SV1}+\frac{2}{\Delta T}T_{SV2}-\frac{1}{2}\right) \tag{6.75}$$

$$\langle v_{bz}\rangle = 2V_{dc}\left(\frac{2}{\Delta T}T_{SV2}-\frac{1}{2}\right) \tag{6.76}$$

$$\langle v_{cz}\rangle = -V_{dc} \tag{6.77}$$

From Table 6.1

$$T_{SV1} = \frac{V_o}{V_{dc}}\frac{\sqrt{3}}{2}\cos\left(\theta_o+\frac{\pi}{6}\right)\frac{\Delta T}{2} \tag{6.78}$$

$$T_{SV2} = \frac{V_o}{V_{dc}}\frac{\sqrt{3}}{2}\cos\left(\theta_o-\frac{\pi}{2}\right)\frac{\Delta T}{2} \tag{6.79}$$

Substituting Eqs. (6.78) and (6.79) into Eqs. (6.75) and (6.76), and knowing that $V_o = MV_{dc}$, the three-phase leg reference voltages become

$$\langle v_{az}\rangle = V_{dc}\left[\sqrt{3}M\cos\left(\theta_o-\frac{\pi}{6}\right)-1\right]\Bigg|_{0\le\theta_o\le 2\pi/3} \tag{6.80}$$

$$\langle v_{bz}\rangle = V_{dc}\left[\sqrt{3}M\cos\left(\theta_o-\frac{\pi}{2}\right)-1\right]\Bigg|_{0\le\theta_o\le 2\pi/3} \tag{6.81}$$

$$\langle v_{cz}\rangle = -V_{dc}\big|_{0\le\theta_o\le\pi/3} \tag{6.82}$$

Once the solution for the first 120° interval has been obtained, the solution for the remaining intervals can be obtained in the same manner. A similar strategy can also be used to determine the phase leg reference waveforms for the other discontinuous modulation strategies.

Tables 6.5 and 6.6 summarize the reference waveform segments for DPWMMIN and DWPMMAX 120° discontinuous modulation strategies, respectively. Table 6.7 summarizes the reference waveform segments for the DPWM1 60° discontinuous modulation strategy, while Tables 6.8 and 6.9 summarize the reference waveform segments for the 30° phase-shifted variations of DPWM0 and DPWM2. Table 6.10 lists the reference waveform segments for the DPWM3 30° discontinuous modulation strategy. In all cases, it can be seen how the phase leg reference segments are in fact sections of the required *l–l* voltage, referenced to the phase leg which is clamped to either the upper or the lower DC rail voltage.

Table 6.5 Phase Leg Reference Voltage Waveforms for 120° DPWMMIN Discontinuous PWM

60° Sextant	Phase Leg a $\langle v_{az}\rangle/V_{dc}$	Phase Leg b $\langle v_{bz}\rangle/V_{dc}$	Phase Leg c $\langle v_{cz}\rangle/V_{dc}$
$\frac{2\pi}{3} \le \theta_o \le \pi$	-1	$-1-\sqrt{3}M\cos\left(\theta_o+\frac{\pi}{6}\right)$	$-1+\sqrt{3}M\cos\left(\theta_o+\frac{5\pi}{6}\right)$
$0 \le \theta_o \le \frac{2\pi}{3}$	$-1-\sqrt{3}M\cos\left(\theta_o+\frac{5\pi}{6}\right)$	$-1+\sqrt{3}M\sin\theta_o$	-1
$-\frac{2\pi}{3} \le \theta_o \le 0$	$-1+\sqrt{3}M\cos\left(\theta_o+\frac{\pi}{6}\right)$	-1	$-1-\sqrt{3}M\sin\theta_o$
$-\pi \le \theta_o \le -\frac{2\pi}{3}$	-1	$-1-\sqrt{3}M\cos\left(\theta_o+\frac{\pi}{6}\right)$	$-1+\sqrt{3}M\cos\left(\theta_o+\frac{5\pi}{6}\right)$

Table 6.6 Phase Leg Reference Voltage Waveforms for 120° DPWMMAX Discontinuous PWM

60° Sextant	Phase Leg a $\langle v_{az}\rangle/V_{dc}$	Phase Leg b $\langle v_{bz}\rangle/V_{dc}$	Phase Leg c $\langle v_{cz}\rangle/V_{dc}$
$\frac{\pi}{3} \le \theta_o \le \pi$	$+1+\sqrt{3}M\cos\left(\theta_o+\frac{\pi}{6}\right)$	$+1$	$+1-\sqrt{3}M\sin\theta_o$
$-\frac{\pi}{3} \le \theta_o \le \frac{\pi}{3}$	$+1$	$+1-\sqrt{3}M\cos\left(\theta_o+\frac{\pi}{6}\right)$	$+1+\sqrt{3}M\cos\left(\theta_o+\frac{5\pi}{6}\right)$
$-\pi \le \theta_o \le -\frac{\pi}{3}$	$+1-\sqrt{3}M\cos\left(\theta_o+\frac{5\pi}{6}\right)$	$+1+\sqrt{3}M\sin\theta_o$	$+1$

Figure 6.22 shows the reference voltages for phase leg a and its two components, the target sinusoid and the offset voltage required to achieve discontinuous modulation, for the cases of DPWMMIN, DPWM1, and DPWM3 at the maximum modulation condition of $M = 2/\sqrt{3}$. Figure 6.23 shows the phase leg reference waveforms for a reduced modulation index of $M = 0.9$, where the discontinuous nature of the switching process can be seen more clearly.

Figure 6.24 shows variation of the 60° DPWM1 discontinuous switching scheme as the modulation index M changes. While complicated in shape, since the equations describing the functions are relatively simple sections of sinusoids, they can be readily implemented in real-time digital form.

Table 6.7 Phase Leg Reference Voltage Waveforms for 60° DPWM1 Discontinuous PWM

60° Sextant	Phase Leg a $\langle v_{az}\rangle/V_{dc}$	Phase Leg b $\langle v_{bz}\rangle/V_{dc}$	Phase Leg c $\langle v_{cz}\rangle/V_{dc}$
$\frac{5\pi}{6} \le \theta_o \le \pi$	-1	$-1-\sqrt{3}M\cos\left(\theta_o+\frac{\pi}{6}\right)$	$-1+\sqrt{3}M\cos\left(\theta_o+\frac{5\pi}{6}\right)$
$\frac{\pi}{2} \le \theta_o \le \frac{5\pi}{6}$	$+1+\sqrt{3}M\cos\left(\theta_o+\frac{\pi}{6}\right)$	$+1$	$+1-\sqrt{3}M\sin\theta_o$
$\frac{\pi}{6} \le \theta_o \le \frac{\pi}{2}$	$-1-\sqrt{3}M\cos\left(\theta_o+\frac{5\pi}{6}\right)$	$-1+\sqrt{3}M\sin\theta_o$	-1
$-\frac{\pi}{6} \le \theta_o \le \frac{\pi}{6}$	$+1$	$+1-\sqrt{3}M\cos\left(\theta_o+\frac{\pi}{6}\right)$	$+1+\sqrt{3}M\cos\left(\theta_o+\frac{5\pi}{6}\right)$
$-\frac{\pi}{2} \le \theta_o \le -\frac{\pi}{6}$	$-1+\sqrt{3}M\cos\left(\theta_o+\frac{\pi}{6}\right)$	-1	$-1-\sqrt{3}M\sin\theta_o$
$-\frac{5\pi}{6} \le \theta_o \le -\frac{\pi}{2}$	$+1-\sqrt{3}M\cos\left(\theta_o+\frac{5\pi}{6}\right)$	$+1+\sqrt{3}M\sin\theta_o$	$+1$
$-\pi \le \theta_o \le -\frac{5\pi}{6}$	-1	$-1-\sqrt{3}M\cos\left(\theta_o+\frac{\pi}{6}\right)$	$-1+\sqrt{3}M\cos\left(\theta_o+\frac{5\pi}{6}\right)$

Table 6.8 Phase Leg Reference Voltage Waveforms for 60° DPWM0 Discontinuous PWM (30° Leading Clamp)

60° Sextant	Phase Leg a $\langle v_{az}\rangle/V_{dc}$	Phase Leg b $\langle v_{bz}\rangle/V_{dc}$	Phase Leg c $\langle v_{cz}\rangle/V_{dc}$
$\frac{2\pi}{3} \le \theta_o \le \pi$	-1	$-1-\sqrt{3}M\cos\left(\theta_o+\frac{\pi}{6}\right)$	$-1+\sqrt{3}M\cos\left(\theta_o+\frac{5\pi}{6}\right)$
$\frac{\pi}{3} \le \theta_o \le \frac{2\pi}{3}$	$+1+\sqrt{3}M\cos\left(\theta_o+\frac{\pi}{6}\right)$	$+1$	$+1-\sqrt{3}M\sin\theta_o$
$0 \le \theta_o \le \frac{\pi}{3}$	$-1-\sqrt{3}M\cos\left(\theta_o+\frac{5\pi}{6}\right)$	$-1+\sqrt{3}M\sin\theta_o$	-1
$-\frac{\pi}{3} \le \theta_o \le 0$	$+1$	$+1-\sqrt{3}M\cos\left(\theta_o+\frac{\pi}{6}\right)$	$+1+\sqrt{3}M\cos\left(\theta_o+\frac{5\pi}{6}\right)$
$-\frac{2\pi}{3} \le \theta_o \le -\frac{\pi}{3}$	$-1+\sqrt{3}M\cos\left(\theta_o+\frac{\pi}{6}\right)$	-1	$-1-\sqrt{3}M\sin\theta_o$
$-\pi \le \theta_o \le -\frac{2\pi}{3}$	$+1-\sqrt{3}M\cos\left(\theta_o+\frac{5\pi}{6}\right)$	$+1+\sqrt{3}M\sin\theta_o$	$+1$

Table 6.9 Phase Leg Reference Voltage Waveforms for 60° DPWM2 Discontinuous PWM (30° Lagging Clamp)

60° Sextant	Phase Leg a $\langle v_{az}\rangle/V_{dc}$	Phase Leg b $\langle v_{bz}\rangle/V_{dc}$	Phase Leg c $\langle v_{cz}\rangle/V_{dc}$
$\frac{2\pi}{3} \le \theta_o \le \pi$	$+1+\sqrt{3}M\cos\left(\theta_o+\frac{\pi}{6}\right)$	$+1$	$+1-\sqrt{3}M\sin\theta_o$
$\frac{\pi}{3} \le \theta_o \le \frac{2\pi}{3}$	$-1-\sqrt{3}M\cos\left(\theta_o+\frac{5\pi}{6}\right)$	$-1+\sqrt{3}M\sin\theta_o$	-1
$0 \le \theta_o \le \frac{\pi}{3}$	$+1$	$+1-\sqrt{3}M\cos\left(\theta_o+\frac{\pi}{6}\right)$	$+1+\sqrt{3}M\cos\left(\theta_o+\frac{5\pi}{6}\right)$
$-\frac{\pi}{3} \le \theta_o \le 0$	$-1+\sqrt{3}M\cos\left(\theta_o+\frac{\pi}{6}\right)$	-1	$-1-\sqrt{3}M\sin\theta_o$
$-\frac{2\pi}{3} \le \theta_o \le -\frac{\pi}{3}$	$+1-\sqrt{3}M\cos\left(\theta_o+\frac{5\pi}{6}\right)$	$+1+\sqrt{3}M\sin\theta_o$	$+1$
$-\pi \le \theta_o \le -\frac{2\pi}{3}$	-1	$-1-\sqrt{3}M\cos\left(\theta_o+\frac{\pi}{6}\right)$	$-1+\sqrt{3}M\cos\left(\theta_o+\frac{5\pi}{6}\right)$

6.9 Analytical Solutions for Discontinuous PWM

The analytical solution for 120° discontinuous modulation can now be developed using the same principles as those used for SVM in Section 6.4. In particular, Eq. (6.34) still holds, but with a different set of limits for y_s and y_e, as listed in Table 6.11, to reflect the discontinuous reference waveform segments appropriate for this modulation strategy. In fact, since there are only two active waveform segments, Eq. (6.34) reduces to

$$A_{mn}+jB_{mn} = \frac{1}{2\pi^2}\sum_{i=1}^{2}\int_{y_s(i)}^{y_e(i)}\int_{x_r(i)}^{x_f(i)} 2V_{dc}e^{j(mx+ny)}\,dx\,dy \qquad (6.83)$$

where the outer and inner integral limits of Eq. (6.83) are defined in Table 6.11. With these limits identified, Eq. (6.83) can now be written explicitly as

Table 6.10 Phase Leg Reference Voltage Waveforms for 30° DPWM3 Discontinuous PWM

60° Sextant	Phase Leg a $\langle v_{az}\rangle / V_{dc}$	Phase Leg b $\langle v_{bz}\rangle / V_{dc}$	Phase Leg c $\langle v_{cz}\rangle / V_{dc}$
$\frac{5\pi}{6} \le \theta_o \le \pi$	$+1+\sqrt{3}M\cos\left(\theta_o+\frac{\pi}{6}\right)$	$+1$	$+1-\sqrt{3}M\sin\theta_o$
$\frac{2\pi}{3} \le \theta_o \le \frac{5\pi}{6}$	-1	$-1-\sqrt{3}M\cos\left(\theta_o+\frac{\pi}{6}\right)$	$-1+\sqrt{3}M\cos\left(\theta_o+\frac{5\pi}{6}\right)$
$\frac{\pi}{2} \le \theta_o \le \frac{2\pi}{3}$	$-1-\sqrt{3}M\cos\left(\theta_o+\frac{5\pi}{6}\right)$	$-1+\sqrt{3}M\sin\theta_o$	-1
$\frac{\pi}{3} \le \theta_o \le \frac{\pi}{2}$	$+1+\sqrt{3}M\cos\left(\theta_o+\frac{\pi}{6}\right)$	$+1$	$+1-\sqrt{3}M\sin\theta_o$
$\frac{\pi}{6} \le \theta_o \le \frac{\pi}{3}$	$+1$	$+1-\sqrt{3}M\cos\left(\theta_o+\frac{\pi}{6}\right)$	$+1+\sqrt{3}M\cos\left(\theta_o+\frac{5\pi}{6}\right)$
$0 \le \theta_o \le \frac{\pi}{6}$	$-1-\sqrt{3}M\cos\left(\theta_o+\frac{5\pi}{6}\right)$	$-1+\sqrt{3}M\sin\theta_o$	-1
$-\frac{\pi}{6} \le \theta_o \le 0$	$-1+\sqrt{3}M\cos\left(\theta_o+\frac{\pi}{6}\right)$	-1	$-1-\sqrt{3}M\sin\theta_o$
$-\frac{\pi}{3} \le \theta_o \le -\frac{\pi}{6}$	$+1$	$+1-\sqrt{3}M\cos\left(\theta_o+\frac{\pi}{6}\right)$	$+1+\sqrt{3}M\cos\left(\theta_o+\frac{5\pi}{6}\right)$
$-\frac{\pi}{2} \le \theta_o \le -\frac{\pi}{3}$	$+1-\sqrt{3}M\cos\left(\theta_o+\frac{5\pi}{6}\right)$	$+1+\sqrt{3}M\sin\theta_o$	$+1$
$-\frac{2\pi}{3} \le \theta_o \le -\frac{\pi}{2}$	$-1+\sqrt{3}M\cos\left(\theta_o+\frac{\pi}{6}\right)$	-1	$-1-\sqrt{3}M\sin\theta_o$
$-\frac{5\pi}{6} \le \theta_o \le -\frac{2\pi}{3}$	-1	$-1-\sqrt{3}M\cos\left(\theta_o+\frac{\pi}{6}\right)$	$-1+\sqrt{3}M\cos\left(\theta_o+\frac{5\pi}{6}\right)$
$-\pi \le \theta_o \le -\frac{5\pi}{6}$	$+1-\sqrt{3}M\cos\left(\theta_o+\frac{5\pi}{6}\right)$	$+1+\sqrt{3}M\sin\theta_o$	$+1$

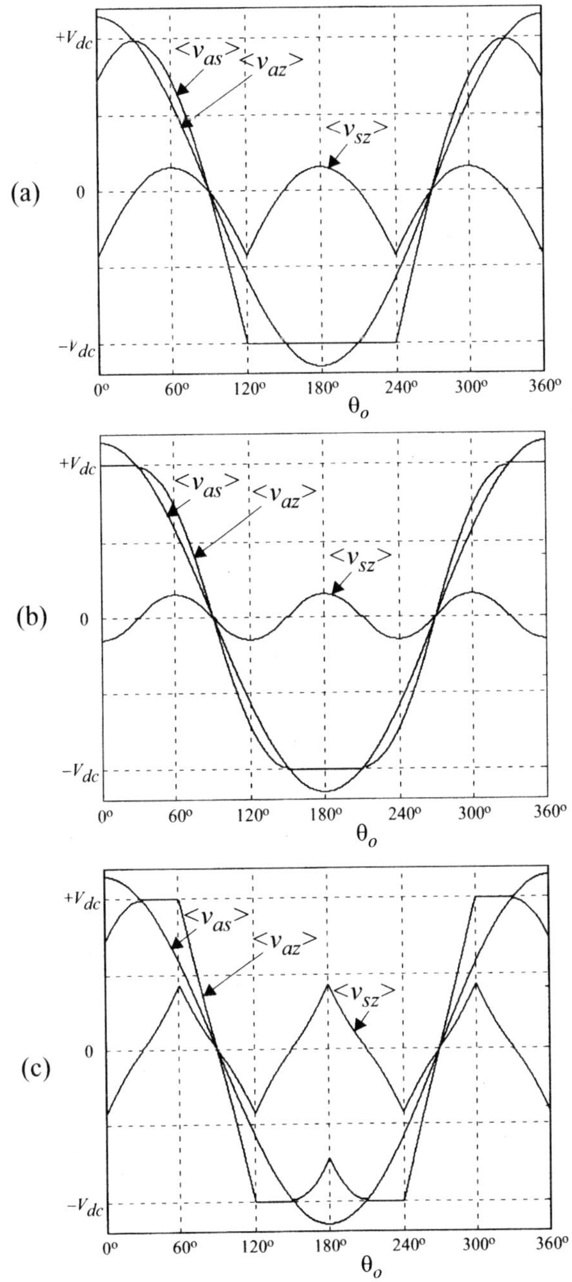

Figure 6.22 Phase leg *a* mean reference voltage and its two components for (a) 120° DPWMMIN, (b) 60° DPWM1, and (c) 30° DPWM3, referred to *z* (midpoint of the DC bus), $M = 2/\sqrt{3}$.

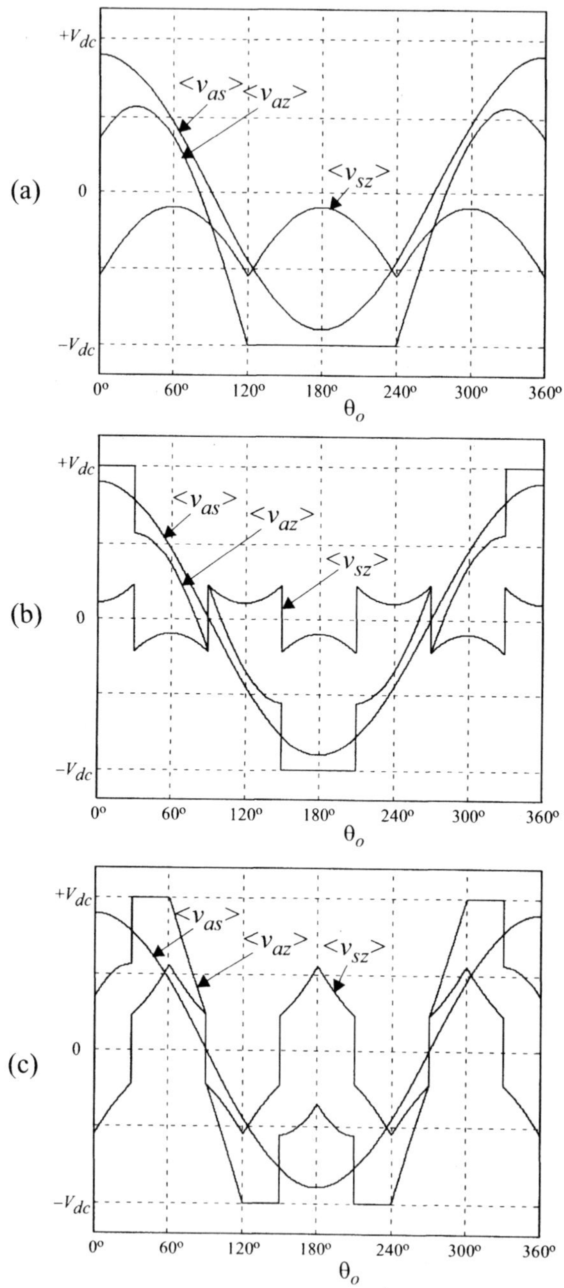

Figure 6.23 Phase leg *a* mean reference voltage and its two components for (a) 120° DPWMMIN, (b) 60° DPWM1, and (c) 30° DPWM3, referred to *z* (midpoint of the DC bus), $M = 0.9$.

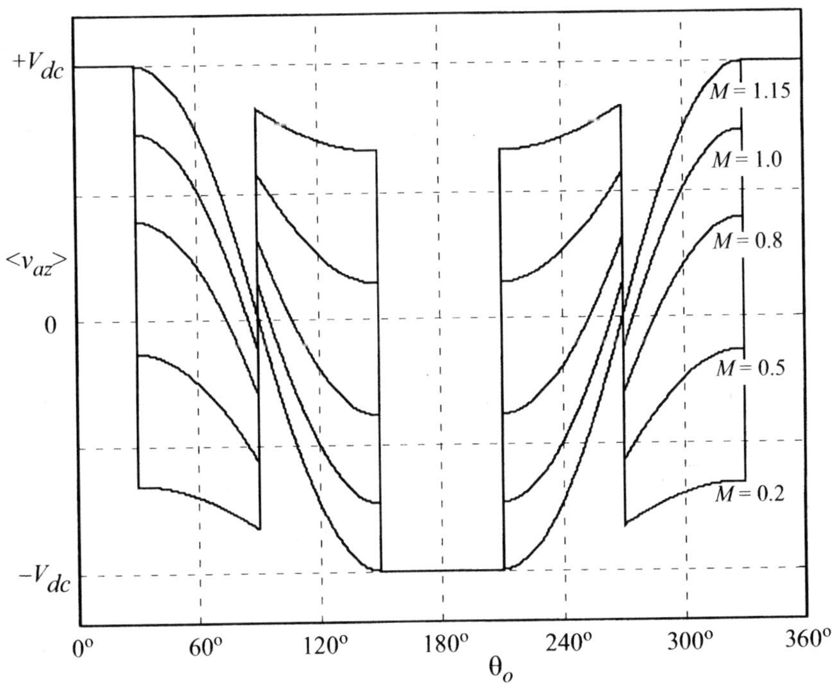

Figure 6.24 Variation of the mean reference voltage $<v_{az}>$ with modulation index, 60° DWPM1 discontinuous modulation.

Table 6.11 Outer and Inner Double Fourier Integral Limits for 120° DPWMMIN Discontinuous Modulation

i	$y_s(i)$	$y_e(i)$	$x_r(i)$ (rising edge of switched waveform)	$x_f(i)$ (falling edge of switched waveform)
1	$\frac{2\pi}{3}$	π	0 (always low)	0 (always low)
2	0	$2\pi/3$	$-\frac{\pi}{2}\sqrt{3}M\cos\left(y-\frac{\pi}{6}\right)$	$\frac{\pi}{2}\sqrt{3}M\cos\left(y-\frac{\pi}{6}\right)$
3	$-\frac{2\pi}{3}$	0	$-\frac{\pi}{2}\sqrt{3}M\cos\left(y+\frac{\pi}{6}\right)$	$\frac{\pi}{2}\sqrt{3}M\cos\left(y+\frac{\pi}{6}\right)$
4	$-\pi$	$-\frac{2\pi}{3}$	0 (always low)	0 (always low)

$$A_{mn} + jB_{mn} = \frac{1}{2\pi^2}\left[\int_{0}^{\frac{2\pi}{3}} \int_{-\frac{\pi}{2}\sqrt{3}M\cos\left(y-\frac{\pi}{6}\right)}^{\frac{\pi}{2}\sqrt{3}M\cos\left(y-\frac{\pi}{6}\right)} 2V_{dc}e^{j(mx+ny)}\,dx\,dy + \int_{-\frac{2\pi}{3}}^{0} \int_{-\frac{\pi}{2}\sqrt{3}M\cos\left(y+\frac{\pi}{6}\right)}^{\frac{\pi}{2}\sqrt{3}M\cos\left(y+\frac{\pi}{6}\right)} 2V_{dc}e^{j(mx+ny)}\,dx\,dy\right] \tag{6.84}$$

For $m = n = 0$, Eq. (6.84) simplifies to

$$\begin{aligned}
A_{00} + jB_{00} &= \frac{V_{dc}}{\pi^2}\left[\int_{0}^{\frac{2\pi}{3}} \int_{-\frac{\pi}{2}\sqrt{3}M\cos\left(y-\frac{\pi}{6}\right)}^{\frac{\pi}{2}\sqrt{3}M\cos\left(y-\frac{\pi}{6}\right)} dx\,dy + \int_{-\frac{2\pi}{3}}^{0} \int_{-\frac{\pi}{2}\sqrt{3}M\cos\left(y+\frac{\pi}{6}\right)}^{\frac{\pi}{2}\sqrt{3}M\cos\left(y+\frac{\pi}{6}\right)} dx\,dy\right] \\
&= \frac{V_{dc}}{\pi^2}\left[\int_{0}^{\frac{2\pi}{3}} \pi\sqrt{3}M\cos\left(y-\frac{\pi}{6}\right)dy + \int_{-\frac{2\pi}{3}}^{0} \pi\sqrt{3}M\cos\left(y+\frac{\pi}{6}\right)dy\right] \\
&= \frac{\sqrt{3}MV_{dc}}{\pi}\left(\sin\frac{\pi}{2} + \sin\frac{\pi}{6} + \sin\frac{\pi}{6} + \sin\frac{\pi}{2}\right) = \frac{3\sqrt{3}MV_{dc}}{\pi}
\end{aligned} \tag{6.85}$$

so that the DC offset component is

$$A_{00} + jB_{00} = \frac{3\sqrt{3}MV_{dc}}{\pi} \tag{6.86}$$

Note that the DC offset is no longer constant, but varies with modulation index, as would be expected from the shape of the reference waveform.

For $m = 0, n > 0$, Eq. (6.84) simplifies to

$$A_{0n} + jB_{0n} = \frac{V_{dc}}{\pi^2}\left[\int_0^{\frac{2\pi}{3}} \int_{-\frac{\pi}{2}\sqrt{3}M\cos\left(y-\frac{\pi}{6}\right)}^{\frac{\pi}{2}\sqrt{3}M\cos\left(y-\frac{\pi}{6}\right)} e^{jny}\,dx\,dy + \int_{-\frac{2\pi}{3}}^{0} \int_{-\frac{\pi}{2}\sqrt{3}M\cos\left(y+\frac{\pi}{6}\right)}^{\frac{\pi}{2}\sqrt{3}M\cos\left(y+\frac{\pi}{6}\right)} e^{jny}\,dx\,dy\right] \tag{6.87}$$

Evaluating the inner integral of Eq. (6.87) gives

$$A_{0n} + jB_{0n} = \frac{V_{dc}}{\pi^2}\left[\int_0^{\frac{2\pi}{3}} e^{jny}\pi\sqrt{3}M\cos\left(y-\frac{\pi}{6}\right)dy + \int_{-\frac{2\pi}{3}}^{0} e^{jny}\pi\sqrt{3}M\cos\left(y+\frac{\pi}{6}\right)dy\right] \tag{6.88}$$

which then rearranges to

$$A_{0n} + jB_{0n} = \frac{\sqrt{3}MV_{dc}}{2\pi}\left[\begin{array}{l}\displaystyle\int_0^{\frac{2\pi}{3}} [e^{j([n+1]y-\pi/6)} + e^{j([n-1]y+\pi/6)}]\,dy \\ \displaystyle + \int_{-\frac{2\pi}{3}}^{0} [e^{j([n+1]y+\pi/6)} + e^{j([n-1]y-\pi/6)}]dy\end{array}\right] \tag{6.89}$$

For $n = 1$, Eq. (6.89) further simplifies to

$$A_{01} + jB_{01} = \frac{\sqrt{3}MV_{dc}}{2\pi}\left[\int_0^{\frac{2\pi}{3}} [e^{j(2y-\pi/6)} + e^{j\pi/6}]dy + \int_{-\frac{2\pi}{3}}^{0} [e^{j(2y+\pi/6)} + e^{-j\pi/6}]dy\right] \tag{6.90}$$

which integrates to

$$A_{01} + jB_{01} = \frac{\sqrt{3}MV_{dc}}{2\pi}\left[\left[\frac{e^{j(2y-\pi/6)}}{j2} + e^{j\pi/6}\right]\Bigg|_0^{\frac{2\pi}{3}} + \; + \left[\frac{e^{j(2y+\pi/6)}}{j2} + e^{-j\pi/6}\right]\Bigg|_{-\frac{2\pi}{3}}^{0}\right] \tag{6.91}$$

Equation (6.91) again reduces, not unexpectedly, to

$$A_{01} + jB_{01} = MV_{dc} \tag{6.92}$$

which is once more immediately recognizable as the target reference waveform magnitude. Furthermore, as with SVM, nothing in this development has restricted the maximum value of M, so that the modulation limit remains at $2/\sqrt{3}$.

For $n > 1$, Eq. (6.89) integrates to

$$A_{0n} + jB_{0n}\Big|_{n>1} = \frac{\sqrt{3}MV_{dc}}{2\pi}\left[\left[\frac{e^{j([n+1]y-\pi/6)}}{j[n+1]} + \frac{e^{j([n-1]y+\pi/6)}}{j[n-1]}\right]\Bigg|_0^{\frac{2\pi}{3}} + \left[\frac{e^{j([n+1]y+\pi/6)}}{j[n+1]} + \frac{e^{j([n-1]y-\pi/6)}}{j[n-1]}\right]\Bigg|_{-\frac{2\pi}{3}}^{0}\right] \tag{6.93}$$

which can be evaluated to give

$$A_{0n} + jB_{0n}\Big|_{n>1} = \frac{2\sqrt{3}MV_{dc}}{\pi}\left[\frac{1}{n+1}\sin\left([n+1]\frac{\pi}{3}\right)\cos\left([2n+1]\frac{\pi}{6}\right) + \frac{1}{n-1}\sin\left([n-1]\frac{\pi}{3}\right)\cos\left([2n-1]\frac{\pi}{6}\right)\right] \tag{6.94}$$

As with SVM the two sinusoid product terms of Eq. (6.94) are nonzero only for odd triplen values of n, i.e., $n = 3, 9, 15, \ldots$, and can be simplified to

$$A_{0n} + jB_{0n} = -\frac{2\sqrt{3}MV_{dc}}{\pi(n^2-1)}\left\{\cos n\frac{2\pi}{3} + \frac{1}{2}\right\} \qquad n = 3, 9, 15, \ldots \tag{6.95}$$

so that it only defines triplen baseband harmonic components that will cancel in the l–l voltage.

For $m > 0$, the inner integral of Eq. (6.84) can be evaluated to give

$$A_{mn} + jB_{mn} = \frac{V_{dc}}{jm\pi^2}\left[\begin{array}{l}\displaystyle\int_0^{\frac{2\pi}{3}} e^{jny}\left(e^{jm\frac{\pi}{2}\sqrt{3}M\cos\left(y-\frac{\pi}{6}\right)} - e^{-jm\frac{\pi}{2}\sqrt{3}M\cos\left(y-\frac{\pi}{6}\right)}\right)dy \\ \displaystyle + \int_{-\frac{2\pi}{3}}^{0} e^{jny}\left(e^{jm\frac{\pi}{2}\sqrt{3}M\cos\left(y+\frac{\pi}{6}\right)} - e^{-jm\frac{\pi}{2}\sqrt{3}M\cos\left(y+\frac{\pi}{6}\right)}\right)dy\end{array}\right] \tag{6.96}$$

Equation (6.96) can be rearranged, using Eq. (A2.1), to become

$$A_{mn} + jB_{mn} = \frac{2V_{dc}}{jm\pi^2}\left[\begin{array}{l}\displaystyle\int_0^{\frac{2\pi}{3}} e^{jny}\sum_{k=1}^{\infty}(j^k - j^{-k})J_k\left(m\frac{\sqrt{3}\pi}{2}M\right)\cos\left(k\left[y-\frac{\pi}{6}\right]\right)dy \\ \displaystyle + \int_{-\frac{2\pi}{3}}^{0} e^{jny}\sum_{k=1}^{\infty}(j^k - j^{-k})J_k\left(m\frac{\sqrt{3}\pi}{2}M\right)\cos\left(k\left[y+\frac{\pi}{6}\right]\right)dy\end{array}\right]$$

$$= \frac{4V_{dc}}{m\pi^2}\left[\begin{array}{l}\displaystyle\int_0^{\frac{2\pi}{3}} e^{jny}\sum_{k=1}^{\infty}\sin k\frac{\pi}{2}J_k\left(m\frac{\sqrt{3}\pi}{2}M\right)\cos\left(k\left[y-\frac{\pi}{6}\right]\right)dy \\ \displaystyle + \int_{-\frac{2\pi}{3}}^{0} e^{jny}\sum_{k=1}^{\infty}\sin k\frac{\pi}{2}J_k\left(m\frac{\sqrt{3}\pi}{2}M\right)\cos\left(k\left[y+\frac{\pi}{6}\right]\right)dy\end{array}\right] \tag{6.97}$$

which can be further rearranged to

$$A_{mn}+jB_{mn} = \frac{2V_{dc}}{m\pi^2}\left[\begin{array}{l}\displaystyle\int_0^{\frac{2\pi}{3}}\sum_{k=1}^{\infty}\sin k\frac{\pi}{2}J_k\left(m\frac{\sqrt{3}\pi}{2}M\right)\left[\begin{array}{l}e^{j\left([n+k]y-k\frac{\pi}{6}\right)}\\ +e^{j\left([n-k]y+k\frac{\pi}{6}\right)}\end{array}\right]dy\\ \displaystyle+\int_{-\frac{2\pi}{3}}^{0}\sum_{k=1}^{\infty}\sin k\frac{\pi}{2}J_k\left(m\frac{\sqrt{3}\pi}{2}M\right)\left[\begin{array}{l}e^{j\left([n+k]y+k\frac{\pi}{6}\right)}\\ +e^{j\left([n-k]y-k\frac{\pi}{6}\right)}\end{array}\right]dy\end{array}\right] \tag{6.98}$$

and then integrated to give

$$A_{mn}+jB_{mn} = \frac{2V_{dc}}{m\pi^2}\left[\Lambda_1\Big|_0^{\frac{2\pi}{3}}+\Lambda_2\Big|_{-\frac{2\pi}{3}}^{0}\right] \tag{6.99}$$

where

$$\Lambda_1 = \left[\begin{array}{l}J_n\left(m\frac{\sqrt{3}\pi}{2}M\right)\sin n\frac{\pi}{2}\,e^{jn\frac{\pi}{6}}\,y\Big|_{k=|n|}\\ \displaystyle+\sum_{k=1}^{\infty}J_k\left(m\frac{\sqrt{3}\pi}{2}M\right)\sin k\frac{\pi}{2}\left[\frac{e^{j\left([n+k]y-k\frac{\pi}{6}\right)}}{j[n+k]}\Bigg|_{k\neq -n}+\frac{e^{j\left([n-k]y+k\frac{\pi}{6}\right)}}{j[n-k]}\Bigg|_{k\neq n}\right]\end{array}\right] \tag{6.100}$$

$$\Lambda_2 = \left[\begin{array}{l}J_n\left(m\frac{\sqrt{3}\pi}{2}M\right)\sin n\frac{\pi}{2}\,e^{-jn\frac{\pi}{6}}\,y\Big|_{k=|n|}\\ \displaystyle+\sum_{k=1}^{\infty}J_k\left(m\frac{\sqrt{3}\pi}{2}M\right)\sin k\frac{\pi}{2}\left[\frac{e^{j\left([n+k]y+k\frac{\pi}{6}\right)}}{j[n+k]}\Bigg|_{k\neq -n}+\frac{e^{j\left([n-k]y-k\frac{\pi}{6}\right)}}{j[n-k]}\Bigg|_{k\neq n}\right]\end{array}\right] \tag{6.101}$$

A final solution for Eq. (6.99) can now be developed as

$$A_{mn}+jB_{mn}= \tag{6.102}$$

$$\frac{8V_{dc}}{m\pi^2}\left[\begin{array}{l}\frac{\pi}{3}\sin n\frac{\pi}{2}\cos n\frac{\pi}{6}J_n\left(m\frac{\sqrt{3}\pi}{2}M\right)\\[2ex] +\displaystyle\sum_{\substack{k=1\\(k\neq -n)}}^{\infty}\frac{1}{[n+k]}\sin\left([n+k]\frac{\pi}{3}\right)\sin k\frac{\pi}{2}\cos\left([2n+k]\frac{\pi}{6}\right)J_k\left(m\frac{\sqrt{3}\pi}{2}M\right)\\[2ex] +\displaystyle\sum_{\substack{k=1\\(k\neq n)}}^{\infty}\frac{1}{[n-k]}\sin\left([n-k]\frac{\pi}{3}\right)\sin k\frac{\pi}{2}\cos\left([2n-k]\frac{\pi}{6}\right)J_k\left(m\frac{\sqrt{3}\pi}{2}M\right)\end{array}\right]$$

As with SVM, this solution is valid also for the condition of $m > 0, n = 0$.

Equations (6.86), (6.92), (6.94), and (6.102) together define the magnitudes of the harmonics for any value of m and n for *naturally sampled* 120° *DPWM-MIN discontinuous* modulation.

An analytical solution for *symmetrically* or *asymmetrically sampled* 120° *discontinuous* modulation can be developed once more using the substitution of $y = y' + (\omega_o/\omega_c)x$ into Eq. (6.84) and following through the evaluation using the principles presented already for continuous modulation strategies.

Without presenting the working detail, the harmonic magnitudes for *asymmetrically sampled* 120° *DPWMMIN discontinuous* modulation are given by

$$A_{mn}+jB_{mn} = \frac{8V_{dc}}{q\pi^2}\left[\begin{array}{l} \frac{\pi}{3}\sin\left(\left[n-n\frac{\omega_o}{\omega_c}\right]\frac{\pi}{2}\right)\cos n\frac{\pi}{6}J_n\left(q\frac{\sqrt{3}\pi}{2}M\right) \\ +\frac{1}{2n}\sin\left(n\frac{2\pi}{3}\right)\sin\left(n\frac{\omega_o}{\omega_c}\frac{\pi}{2}\right)\left(1-J_0\left(q\frac{\sqrt{3}\pi}{2}M\right)\right)\Bigg|_{n\neq 0} \\ +\sum_{\substack{k=1\\(k\neq -n)}}^{\infty}\left[\begin{array}{r}\frac{1}{[n+k]}\sin\left([n+k]\frac{\pi}{3}\right)\sin\left(\left[k-n\frac{\omega_o}{\omega_c}\right]\frac{\pi}{2}\right) \\ \times\cos\left([2n+k]\frac{\pi}{6}\right)J_k\left(m\frac{\sqrt{3}\pi}{2}M\right)\end{array}\right] \\ +\sum_{\substack{k=1\\(k\neq n)}}^{\infty}\left[\begin{array}{r}\frac{1}{[n-k]}\sin\left([n-k]\frac{\pi}{3}\right)\sin\left(\left[k-n\frac{\omega_o}{\omega_c}\right]\frac{\pi}{2}\right) \\ \times\cos\left([2n-k]\frac{\pi}{6}\right)J_k\left(m\frac{\sqrt{3}\pi}{2}M\right)\end{array}\right] \end{array}\right] \tag{6.103}$$

where q again equals $m+n(\omega_o/\omega_c)$.

This solution is valid for all values of m and n, except $m = n = 0$, where Eq. (6.86) still applies.

Solutions for 60° DPWM0, DPWM1, DPWM2, and 30° DPWM3 discontinuous modulation can be developed using similar principles as presented above but are left for the reader to develop as an extension to this text.

6.10 Comparison of Harmonic Performance

Figure 6.25 shows the theoretical harmonic spectra for DPWMMIN 120° discontinuous modulation at a modulation index of $M = 0.9$. Figure 6.26 shows the spectra for DPWM1 60° discontinuous modulation at the same modulation index. Figure 6.27 shows the *l–l* spectra for the discontinuous modulation strategies of DPWM0, DPWM1, DPWM2, and DPWM3. Note that in all cases the carrier-to-fundamental ratio has been increased to 30, to at least partially reflect the increase in switching frequency that is possible for discontinuous modulation strategies because of the reduced number of switch transitions per fundamental cycle. This issue is discussed further in Section 6.11.

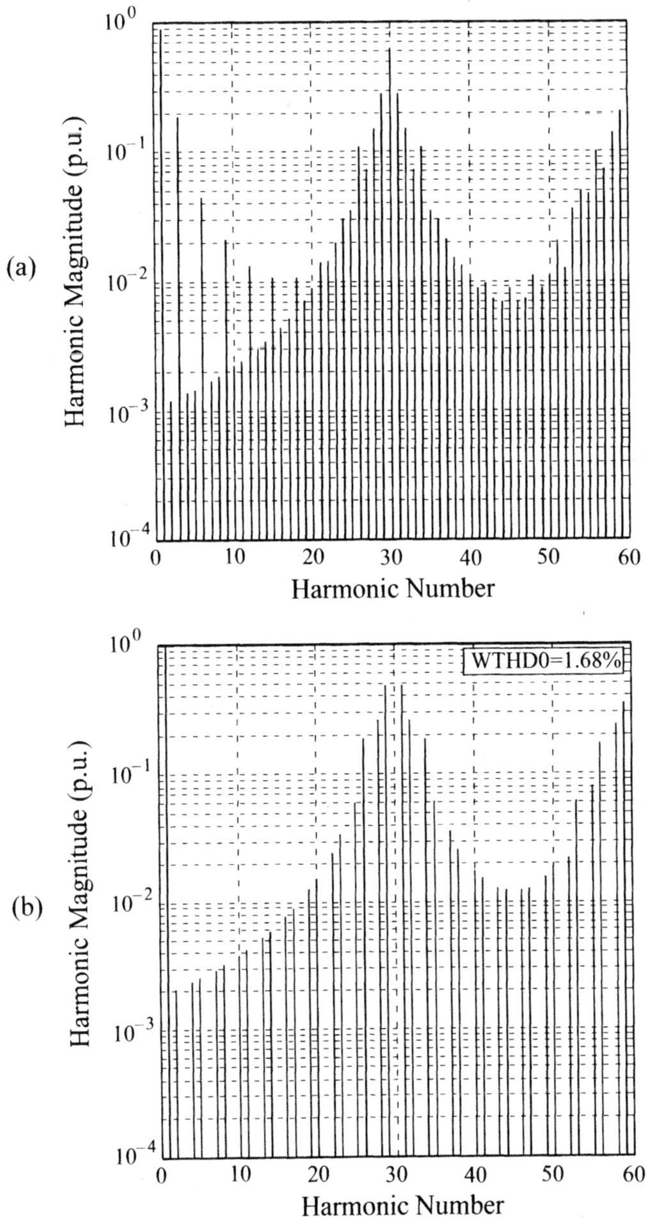

Figure 6.25 Theoretical harmonic spectra for three-phase inverter modulated by *naturally sampled* 120° *DPWMMIN discontinuous* PWM: (a) phase leg a and (b) l–l output switched voltage waveforms, $M = 0.9$, $f_c/f_o = 30$.

(a)

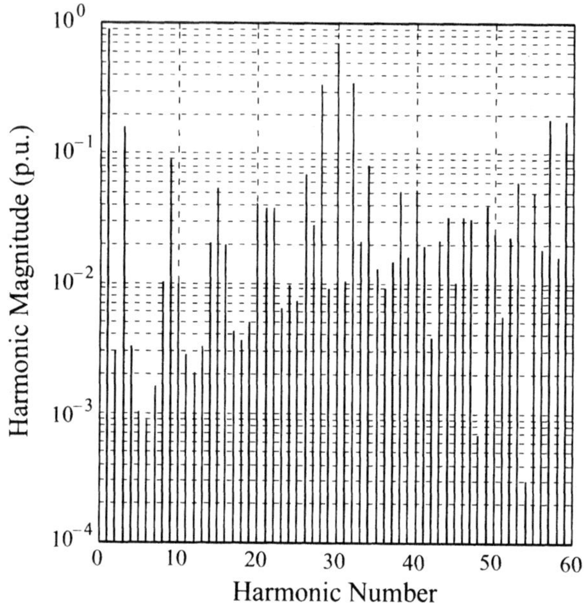

(b)

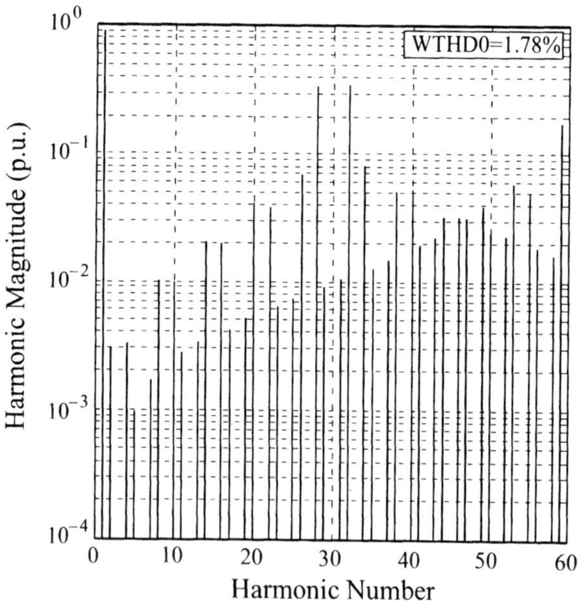

Figure 6.26 Theoretical harmonic spectra for three-phase inverter modulated by *naturally sampled* 60° *DPWM1* PWM: (a) phase leg *a* and (b) *l–l* output switched voltage waveforms, $M = 0.9$, $f_c/f_o = 30$, spectra derived from time-based switching simulation.

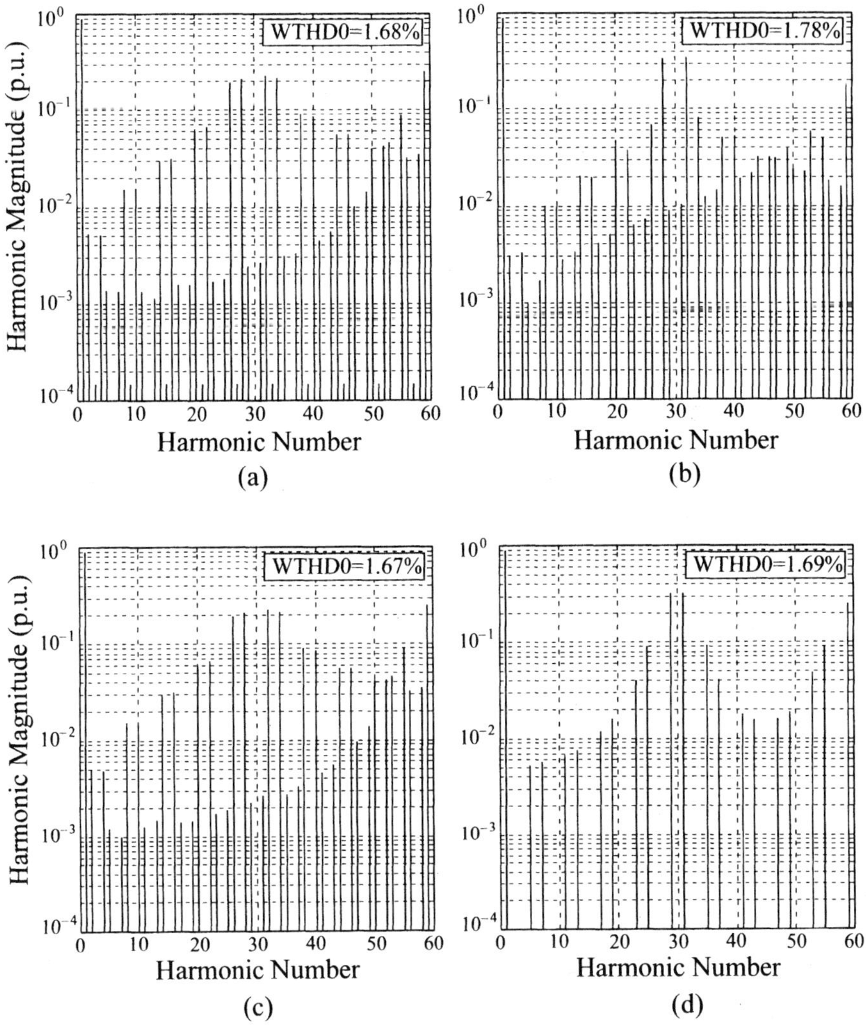

Figure 6.27 Theoretical *l–l* harmonic spectra for three-phase inverter modulated by

(a) *naturally sampled* 60° *DPWM0* PWM,
(b) *naturally sampled* 60° *DPWM1* PWM,
(c) *naturally sampled* 60° *DPWM2* PWM,
(d) *naturally sampled* 30° *DPWM3* PWM,

$M = 0.9$, $f_c/f_o = 30$, spectra derived from time-based switching simulation.

It can be particularly noted from these plots that the harmonic sideband cancellations (e.g., the cancellation of odd harmonics around the first carrier multiple) that is a feature of continuous modulation strategies no longer occurs. In addition, the roll-off in magnitude of the sideband harmonic components is much slower than for continuous modulation. In the first instance, it would be expected that these additional harmonic components would cause an increased WTHD compared to continuous modulation strategies, and indeed this is the case (at least for the same switching frequency, as discussed in the next section). Finally, it is illuminating to identify baseband triplen harmonics in the phase leg voltages for discontinuous modulation, in Figures 6.25(a) and 6.26(a). These triplen harmonics are to be expected, since they are required for PWM strategies to modulate beyond $M = 1.0$, up to the theoretical maximum of $M = 1.15$.

6.11 Harmonic Losses for Discontinuous PWM

From the previous section, it should be clear that discontinuous switching patterns lead to a suboptimal harmonic performance compared to continuous modulation arrangements in terms of the number and magnitude of carrier sidebands that are generated. However, the advantage of these modulation strategies arises from the reduction in the number of switch transitions per phase leg that can be achieved over each fundamental cycle. This reduction makes it possible to increase the carrier frequency for each variation by approximately 3/2 compared to continuously switched regular or naturally sampled PWM, while still maintaining the same number of device switch transitions for each phase leg over a fundamental cycle. (It should be noted that the exact multiplier varies slightly for each discontinuous modulation alternative, because of extra switch transitions that are required to change the clamped DC bus and the position of the DC bus clamping interval.) For higher modulation indices, the increased switching frequency leads then to an improvement in the WTHD performance of discontinuous modulation compared to continuously switched modulation despite the additional sideband harmonics that are generated.

This advantage can be illustrated by calculating the RMS ripple current in the same manner as was done in Section 6.5 for space vector modulation, using the reference waveform segments listed in Tables 6.5 through to 6.10 for discontinuous modulation strategies DPWMMIN, DPWMMAX, DPWM1, DPWM0, DPWM2, and DPWM3, respectively.

The resultant closed-form solutions [12] for the harmonic current produced by the various modulation strategies are (note that previous results already presented for continuous modulation strategies are reproduced here also for completeness):

<u>*Simple sinusoidal PWM*</u>

$$I_{ab,h,rms}^2 = \left(\frac{V_{dc}}{L_\sigma}\right)^2 \frac{\Delta T^2}{48}\left[\frac{3}{2}M^2 - \frac{4\sqrt{3}}{\pi}M^3 + \frac{9}{8}M^4\right] \tag{6.104}$$

<u>*One-sixth third-harmonic injection PWM*</u>

$$I_{ab,h,rms}^2 = \left(\frac{V_{dc}}{L_\sigma}\right)^2 \frac{\Delta T^2}{48}\left[\frac{3}{2}M^2 - \frac{4\sqrt{3}}{\pi}M^3 + M^4\right] \tag{6.105}$$

<u>*One-quarter third-harmonic injection PWM*</u>

$$I_{ab,h,rms}^2 = \left(\frac{V_{dc}}{L_\sigma}\right)^2 \frac{\Delta T^2}{48}\left[\frac{3}{2}M^2 - \frac{4\sqrt{3}}{\pi}M^3 + \frac{63}{64}M^4\right] \tag{6.106}$$

<u>*Space vector PWM*</u>

$$I_{ab,h,rms}^2 = \left(\frac{V_{dc}}{L_\sigma}\right)^2 \frac{\Delta T^2}{48}\left[\frac{3}{2}M^2 - \frac{4\sqrt{3}}{\pi}M^3 + \frac{9}{8}\left(\frac{3}{2} - \frac{9}{8}\frac{\sqrt{3}}{\pi}\right)M^4\right] \tag{6.107}$$

<u>*DPWMMIN, DPWMMAX, DPWM0, DPWM2 discontinuous PWM*</u>

$$I_{ab,h,rms}^2 = \left(\frac{V_{dc}}{L_\sigma}\right)^2 \frac{\Delta T^2}{48}\left[6M^2 - \frac{35\sqrt{3}}{2\pi}M^3 + \left(\frac{27}{8} + \frac{81}{64}\frac{\sqrt{3}}{\pi}\right)M^4\right] \tag{6.108}$$

<u>*DPWM1 discontinuous PWM*</u>

$$I_{ab,h,rms}^2 = \left(\frac{V_{dc}}{L_\sigma}\right)^2 \frac{\Delta T^2}{48}\left[6M^2 - \left(\frac{45}{2\pi} + \frac{4\sqrt{3}}{\pi}\right)M^3 + \left(\frac{27}{8} + \frac{27}{32}\frac{\sqrt{3}}{\pi}\right)M^4\right] \tag{6.109}$$

<u>*DPWM3 discontinuous PWM*</u>

$$I_{ab,h,rms}^2 = \left(\frac{V_{dc}}{L_\sigma}\right)^2 \frac{\Delta T^2}{48}\left[6M^2 + \left(\frac{45}{2\pi} - \frac{31\sqrt{3}}{\pi}\right)M^3 + \left(\frac{27}{8} + \frac{27}{16}\frac{\sqrt{3}}{\pi}\right)M^4\right] \tag{6.110}$$

The harmonic distortion factors for the various modulation strategies, defined as $f(M)$ in Eq. (5.69), are compared in Figure 6.28(a). It is clear from this figure that, as expected, the discontinuous modulation strategies produce greater harmonic distortion and harmonic losses than continuous modulation strategies for the same switching frequency.

However, since with discontinuous switching the inverter phase legs only switch for two-thirds of the fundamental carrier period, the switching fre-

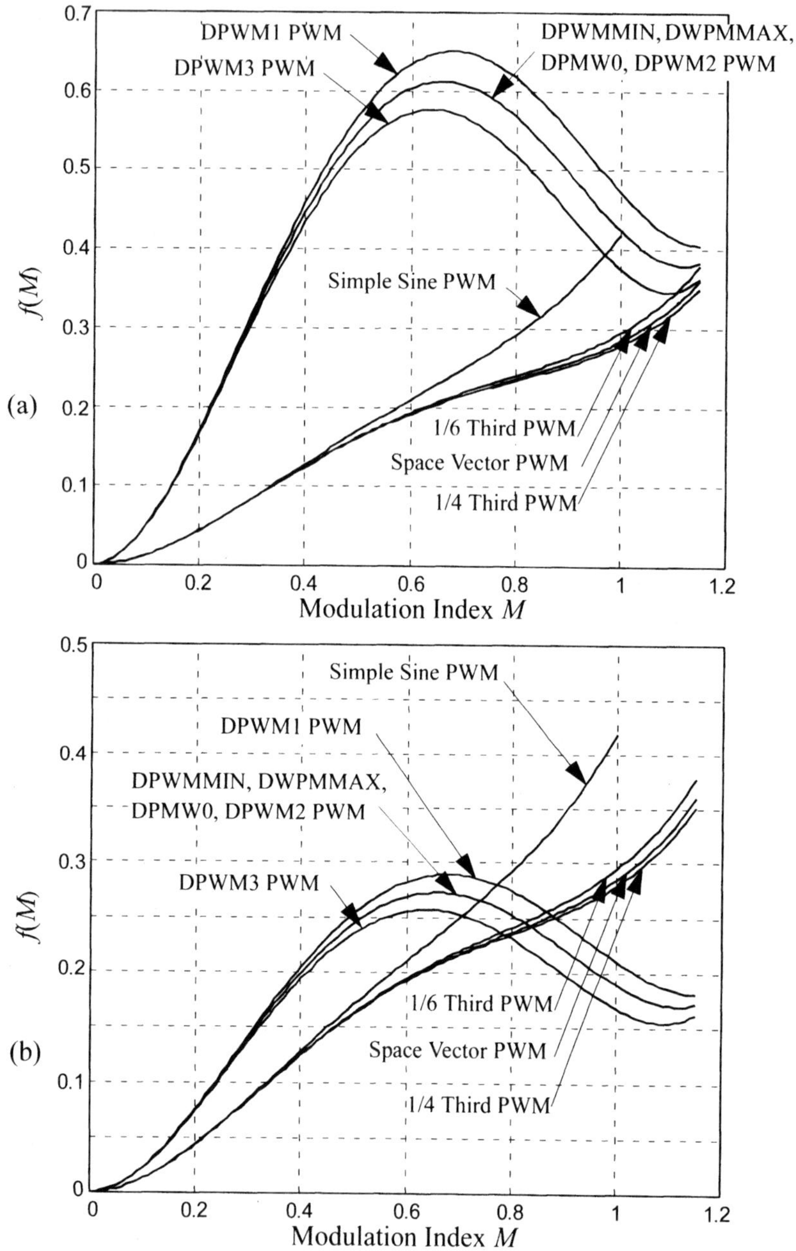

Figure 6.28 Harmonic distortion factors (HDF) for different PWM modulation strategies as a function of modulation index M: (a) same switching frequency and (b) discontinuous strategy switching frequencies increased by 3/2.

quency of these strategies can be increased by this amount for the same switching losses. Hence the HDFs for discontinuous modulation should be scaled by $(2/3)^2$ for a direct comparison with continuous modulation. These results are shown in Figure 6.28(b), where it can be seen that discontinuous modulation strategies now become harmonically advantageous at high modulation indices.

From a practical point of view, device losses are the key parameter for comparing switching strategies. Switching losses essentially vary proportionally to the magnitude of the instantaneous load current, for an inverter operating with a constant DC bus voltage. Hence for a unity power factor load, the switching losses for a continuously switching phase leg over one half of a fundamental cycle are

$$P_{loss(\text{continuous})} = V_e I_m \frac{1}{\pi} \int_{-\pi/2}^{\pi/2} \cos\theta \, d\theta \tag{6.111}$$

where V_e is an equivalent DC voltage for determining the switching losses. Equation (6.111) readily works out to

$$P_{loss(\text{continuous})} = \frac{2}{\pi} V_e I_m \tag{6.112}$$

i.e., the average value of a sine wave, as one would expect.

For the case of DPWM1 60° discontinuous modulation, switching losses do not occur over the region $-\pi/6 \le \theta \le \pi/6$, so that

$$P_{loss(\text{DPWM1})} = V_e I_m \frac{1}{\pi} \left(\int_{-\pi/2}^{-\pi/6} \cos\theta \, d\theta + \int_{\pi/6}^{\pi/2} \cos\theta \, d\theta \right) \tag{6.113}$$

which can be evaluated to give

$$P_{loss(DPWM1)} = \frac{1}{\pi} V_e I_m = 0.5 P_{loss(\text{continuous})} \tag{6.114}$$

or one half of the switching losses of continuous modulation. In a similar manner it can be determined that

$$P_{loss(\text{PWMMIN,PWMMAX})} = \frac{2}{\pi}\left(1 - \frac{\sqrt{3}}{4}\right) V_e I_m = 0.567 P_{loss(\text{continuous})} \tag{6.115}$$

$$P_{loss(\text{DPWM3})} = \frac{2}{\pi}\left(\frac{3}{2} - \frac{\sqrt{3}}{2}\right) V_e I_m = 0.634 P_{loss(\text{continuous})} \tag{6.116}$$

When the power factor of the load is not unity, as is invariably the case with practical loads, the above calculations will change since the peak of the load current no longer coincides with the peak of the target fundamental waveform. As identified previously, discontinuous strategies DPWM0 and DPWM2 offer some benefit in this case, since they slew the DC bus clamping intervals by 30° leading and 30° lagging, respectively. Hence, they will achieve the minimum switching loss condition for loads with power factors of 0.866 (leading or lagging as appropriate). Obviously, it is possible to go further and design a 60° discontinuous modulation strategy that would clamp for 60° anywhere within the region $+60^\circ$ to -60° around the peak of the phase leg reference voltage. This would allow minimum loss discontinuous switching to be achieved for loads with power factors of between 0.866 leading and 0.866 lagging, which covers the range of many typical loads. Finally, it is also possible to modulate with 60° discontinuous switching outside of this power factor range, with the cost of some loss of maximum modulation index [12].

This analysis also suggests that it is possible to increase the switching frequency of discontinuous modulation strategies by more than 3/2, and still have the same overall losses. However, the calculations are only approximate since they are based on the harmonic current loss expressions which in turn assume an infinite switching frequency. Also, the harmonic losses for discontinuous modulation increase away from the minimum loss condition as the power factor of the load varies from the optimum value, and an allowance must be maintained for this mismatch. Finally, discontinuous switching strategies require an additional switching transition as the DC bus clamp varies from positive to negative, and this has not been considered in the calculations so far. For a practical modulation system operating at realistically low switching frequencies, all of these issues must be taken into account when comparing continuous versus discontinuous modulation strategies. Nevertheless, discontinuous modulation does offer considerable advantages at high modulation indexes.

6.12 Single-Edge SVM

The effect of single-edge modulation for a SVM harmonic injection strategy is similar to that using the regular sampled technique but with an increased fourth harmonic and decreased second-harmonic low-order component compared to Figure 5.8(d). Detailed results for this variation are not presented here since they contribute little additional fundamental understanding regarding the effects of single-edge modulation. Similarly, discontinuous single-edge modulation for a three-phase VSI is not explored further, except to comment that it simply produces a degraded spectral response compared to the previous schemes described.

6.13 Switched Pulse Sequence

It has been shown that the major parameter which defines all three-phase VSI modulation alternatives is the *zero space vector* pulse position. However, the sequence of the active space vectors within a half carrier period and across multiple half carrier periods can also effect the harmonic performance of the modulation algorithms, similar to the effect described for a single-phase VSI in Section 4.6.1.

In general, the pulse sequence for a three-phase VSI is constrained by switching considerations. Once an active space vector has been selected, the minimum switching transition is to switch to the next nearest space vector when appropriate, and then to a zero space vector for continuous modulation, or back to the previous space vector for discontinuous modulation. While it is theoretically feasible to insert an additional zero between the two active space vectors, as mentioned in Section 6.6, this increases the phase leg switching frequency for no harmonic benefit, and is therefore of little practical value.

The effect of this switching constraint is that the normal switched pulse sequence for a three-phase VSI operating under continuous conduction is

$$\overline{SV_7} \to \overline{SV_i} \to \overline{SV_j} \to \overline{SV_0} \to \overline{SV_0} \to \overline{SV_j} \to \overline{SV_i} \to \overline{SV_7} \qquad (6.117)$$

$$|\longleftarrow \Delta T/2 \longrightarrow|\;|\longleftarrow \Delta T/2 \longrightarrow|$$

where $\overline{SV_i}$ and $\overline{SV_j}$ represent the two nearest stationary space vector components.

For discontinuous conduction, when the $\overline{SV_0}$ zero space vector is eliminated, for example, the switched pulse sequence is

$$\overline{SV_7} \to \overline{SV_i} \to \overline{SV_j} \to \overline{SV_j} \to \overline{SV_i} \to \overline{SV_7} \qquad (6.118)$$

$$|\longleftarrow \Delta T/2 \longrightarrow|\;|\longleftarrow \Delta T/2 \longrightarrow|$$

Single-edge modulation creates the following pulse sequences:

$$\overline{SV}_7 \to \overline{SV}_i \to \overline{SV}_j \to \overline{SV}_0 \to \overline{SV}_7 \to \overline{SV}_i \to \overline{SV}_j \to \overline{SV}_0$$

$$|\longleftarrow \Delta T/2 \longrightarrow|\;|\longleftarrow \Delta T/2 \longrightarrow|$$

for continuous modulation, and

$$\overline{SV}_7 \to \overline{SV}_i \to \overline{SV}_j \to \overline{SV}_7 \to \overline{SV}_i \to \overline{SV}_j$$

$$|\longleftarrow \Delta T/2 \longrightarrow|\;|\longleftarrow \Delta T/2 \longrightarrow|$$

for single-edge discontinuous modulation ($\overline{SV}_0$ eliminated).

Some variations on these sequences have been proposed [12] which achieve limited harmonic benefits at best when their effect on phase leg switching frequencies are considered. However, the intrinsic switching constraints of the VSI topology leave little scope for significant variations in switch sequence.

The other major pulse sequence issue for a three-phase VSI occurs with 60° discontinuous modulation at the space vector boundaries, where the zero space vector usage changes between $\overline{SV}_0$ and $\overline{SV}_7$. Figure 6.16(a) shows the common variation, where the position of the zero space vector shifts by one half carrier period as the zero vector usage changes from $\overline{SV}_0$ to $\overline{SV}_7$. The shift occurs because it is usually easier to build a modulator which maintains a constant switching sequence across successive half carrier periods. Hence, for example, at time t_i, either two- or three-phase legs are always switched to the positive DC rail, while at time t_{i+1} either two- or three-phase legs are always switched to the negative DC rail (the third phase leg will be switched to the opposite DC rail if the inverter is currently producing an active space vector pulse).

The alternative approach would be to arrange the inverter switching so that the zero space vectors are always placed across the same half carrier period transitions. For example, this would mean changing the switching sequence to move say $\overline{SV}_7$ in Figure 6.16(a) to always occur at time t_{i+1}. This could be achieved by inverting all phase legs simultaneously at each 60° space vector transition boundary. One approach to achieve this would be to invert the carrier waveform for every second sextant. But, of course, this would result in a consequential increase in modulation controller complexity.

Figure 6.29 shows the harmonic spectra for these two alternative placements of the zero space vector in successive sextants, and they are clearly quite different. For the case where the switching sequence shifts the zero space vector one half carrier period in each successive sextant, the spectrum is more spread out, with flatter carrier sideband peaks. For the case where the switching sequence does not shift the zero space vector placement, the harmonic spectra more resembles 120° discontinuous modulation with larger low-order sideband peaks but a faster roll-off. There is also better cancellation of harmonics between phase legs for this sequence.

The selection of the "best" of these two alternatives is not straightforward, since based on the WTHD, there is virtually no difference between the two approaches for the modulation conditions analyzed. But equally, one sequence may achieve a lower WTHD for other modulation indices and/or carrier frequencies. This would need to be checked by more extensive simulation studies for a practical implementation, operating with the particular modulation conditions of interest. Overall, pulse sequence is an important factor in a modulation implementation and needs to be considered carefully, particularly for discontinuous switching arrangements.

6.14 Summary

This chapter has further extended the concepts of pulse width determination, pulse position, and pulse sequence for a three-phase VSI. The pulse placement concepts which have been presented offer a clear physical explanation as to why the differences between the alternative modulation strategies occur. It has been shown that the third-harmonic injection and space vector modulation methods achieve essentially the same pulse placement objective, indirectly via third-harmonic injection and by direct calculation in the case of space vector modulation. The extension has identified the placement of the zero space vector as the major factor in the implementation of a particular PWM algorithm and has demonstrated how the same basic concepts can be found in all major PWM variations that have been proposed in the literature.

In this context, discontinuous modulation can be regarded as a pulse placement variation of continuous modulation concepts, achieved by the elimination of selected zero space vectors for sections of the fundamental cycle. This elimination allows a higher effective carrier frequency to be implemented for the

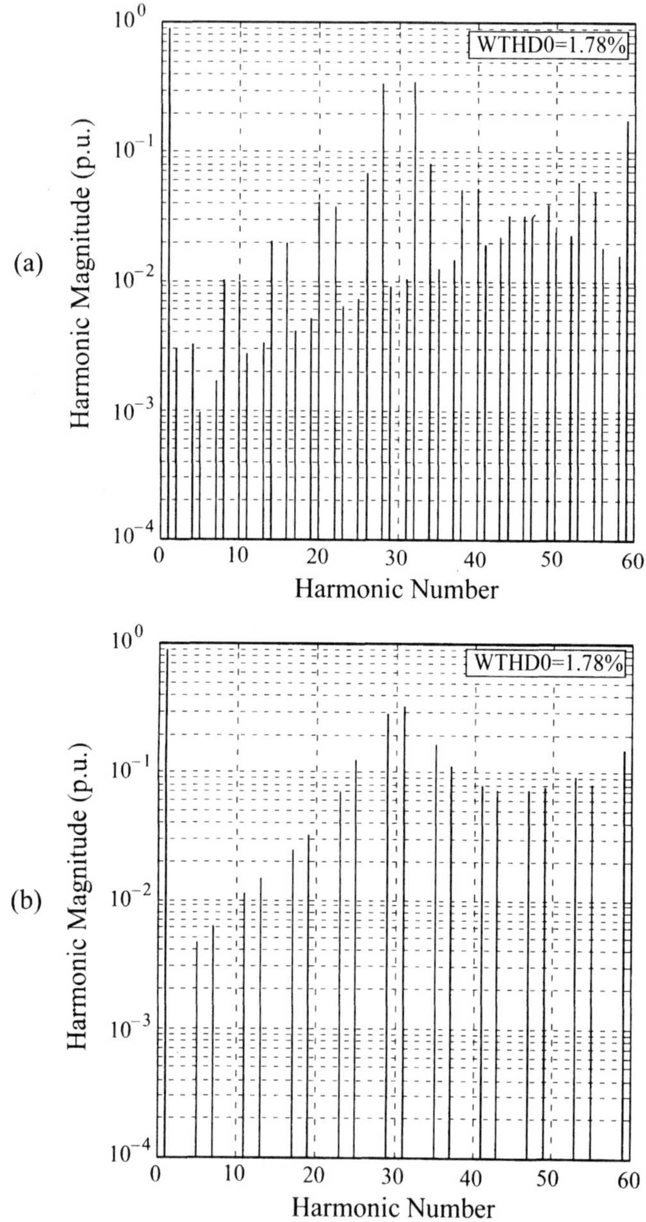

Figure 6.29 Simulated *l–l* harmonic spectra for three-phase inverter modulated by *regularly sampled DPWM1 discontinuous* PWM: (a) zero space vector position alternating every 60° and (b) zero space vector position constant in successive carrier periods, $M = 0.9$, $f_c/f_o = 30$.

same number of phase leg switching transitions over the complete fundamental cycle, which achieves an improved WTHD performance for higher modulation indices. Conceptually, discontinuous conduction becomes advantageous when the oscillatory placement of continuously modulated active space vector components across two half carrier periods causes more carrier sideband harmonic distortion than is caused by the more numerous, but higher frequency, sideband harmonics of discontinuous modulation.

References

[1] J. Holtz and S. Stadtfeld, "A predictive controller for the stator current vector of ac machines fed from a switched voltage source," in *Conf. Rec. IPEC Conf.*, Tokyo, 1983, pp. 1665–1675.

[2] H.W. Van der Broeck, H. Skudelny, and G. Stanke, "Analysis and realization of a pulse width modulator based on voltage space vectors," *IEEE Trans. on Industry Applications.*, vol. 24, no. 1, Jan./Feb., 1988, pp. 142–150.

[3] O. Ogasawara, H. Akagi, and A. Nabae, "A novel PWM scheme of voltage source inverters based on space vector theory," in *Conf. Rec. European Power Electronics Conf. (EPE)*, Aachen, 1989, pp. 1197–1202.

[4] H.W. Van der Broeck and H.C. Skudelny, "Analytical analysis of the harmonic effects of a PWM ac drive," *IEEE Trans. on Power Electronics,* vol. 3, no. 2, March/April 1988, pp. 216–223.

[5] M. Depenbrock, "Pulsewidth control of a 3-phase inverter with non-sinusoidal phase voltages," in *Conf Rec. IEEE Industry Applications Society Int. Semiconductor Power Converter Conf.*, 1997, pp. 399–403.

[6] S. Bhattacharya, D.G. Holmes, and D.M. Divan, "Optimizing three phase current regulators for low inductance loads," in *Conf. Rec. IEEE Industry Applications Society Annual Mtg.*, Orlando, 1995, pp. 2357–2366.

[7] J.F. Moynihan, M.G. Egan, and J.M.D. Murphy, "Theoretical spectra of space-vector-modulated waveforms," *IEE Proceedings (London) Electr. Power Applications*, vol. 145, no. 1, Jan. 1998, pp. 17–24.

[8] S.R. Bowes and A. Midoun, "Suboptimal switching strategies for microprocessor–controlled PWM inverter drives," *IEE Proceedings (London),* vol. 132, Pt. B, no. 3., May 1985, pp. 133–148.

[9] J.T. Boys and B.E. Walton, "A loss minimised sinusoidal PWM inverter," *IEE Proceedings (London),* vol. 132, Pt. B, no. 5, Sept. 1985, pp. 260–268.

[10] S. Fukuda and Y. Iwaji, "A single-chip microprocessor-based PWM technique for sinusoidal inverters," in *Conf. Rec. IEEE Industry Applications Society Annual Mtg.*, Pittsburgh, 1988, pp. 921–926.

[11] J.W. Kolar, H. Ertl, and F.C. Zach, "Calculation of the passive and active component stress of three phase PWM converter system with high pulse rate," in *Conf. Rec. European Power Electronics* (*EPE*), Aachen, 1989, pp. 1303–1311.

[12] H.W. Van der Broeck, "Analysis of the harmonics in voltage fed converter drives caused by PWM schemes with discontinuous switching operation," in *Conf. Rec. European Power Electronics Conf.* (*EPE*), Florence, 1991, pp. 3:261–3:266.

7

Modulation of Current Source Inverters

The majority of this work on pulse width modulation relates to the control of a voltage source/stiff inverter (VSI), since this topology has a number of characteristics which make it more attractive for many applications. In particular, a VSI requires a fixed DC voltage supply which is easily created from an AC source using a simple rectifier, and is also suitable for open-loop modulation for low-performance variable speed drive systems without requiring expensive feedback transducers. The major modulation strategies for a VSI are analog natural sampled sine-triangle PWM, regular sampled PWM, and more recently space vector modulation. Each approach has particular advantages and disadvantages, which have already been documented.

In contrast, there has been considerably less research work into the control and modulation of the alternative current source/stiff inverter (CSI) topology. This topology has the benefit of implicit short circuit protection, and direct capability of reverse power flow back to an AC supply when the current source is based around a thyristor controlled rectifier (as is often the case). However, variable speed drive operation of an induction motor using a CSI can only be achieved using a closed-loop control system. Nevertheless, the CSI topology has considerable potential in drive systems particularly in high-power applications, provided a modulation performance similar to a VSI can be achieved [1]. However, this advantage has not been widely exploited to date.

In this chapter, a general method of adapting any VSI modulation strategy to suit a CSI is presented. The method works by treating the modulation process simply as a way of generating active and zero switching states, and then mapping these states to their equivalent CSI switch pattern. With this approach, all of the existing knowledge and experience relating to modulation and control of a VSI is readily applicable to a CSI, without requiring any additional research or development of specific modulation algorithms. Much of the material presented comes from the research work of Dr. D.N. Zmood [2].

7.1 Three-Phase Modulators as State Machines

It has been shown that conventional three-phase two-level inverters can only change between a limited number of switch combinations, which can be identified as six active states and two null states for a VSI. In this same manner, examination of the switch states for the CSI again reveals six active states plus three null states in which short circuits occur in the three legs of the CSI. Each of the active states directly map to six stationary output phasors, spaced equally at 60° intervals around the complex plane, as shown in Figure 7.1.

In terms of a state machine, the modulation objectives for a three-phase inverter are:

- The identification of the active states required to achieve a target output phasor.
- The placement and sequencing of these active states in conjunction with additional nulls as required.

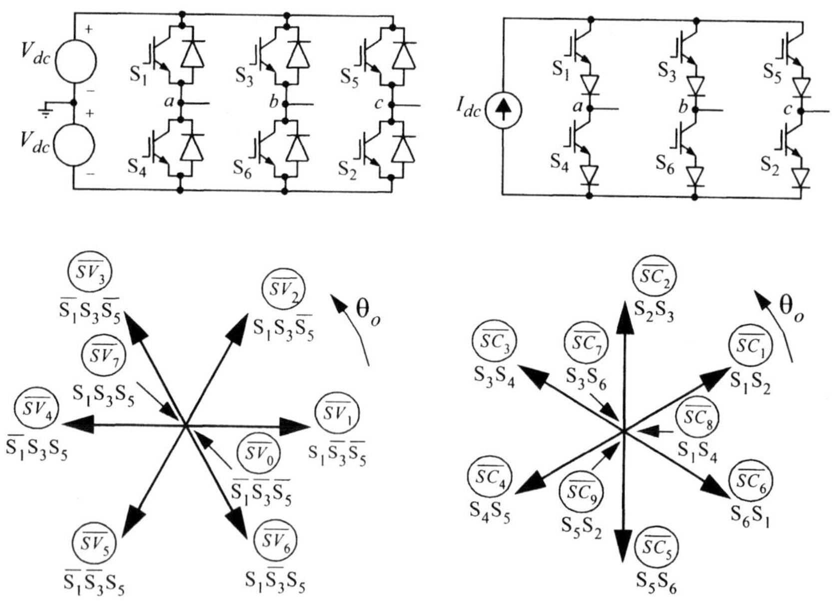

(a) Three-Phase VSI (b) Three-Phase CSI

Figure 7.1 Topologies and space vector states for three-phase VSI and CSI configurations.

The well-known "standard" PWM strategies achieve these objectives either implicitly or explicitly when applied to VSI control. For example, naturally or regular sampled three-phase VSI modulators both generate six active states and two null states, while 120° discontinuous three-phase VSI modulators generate six active states and one null state.

Using the state concepts developed above, a CSI can be directly controlled by any VSI modulation strategy if the active states created by the modulator are mapped to the six stationary space vectors, and hence to the switch combinations associated with these phasors, as follows.

$$\text{State 1} \Leftrightarrow \overline{SV_1} \Leftrightarrow \overline{SC_1} \qquad \text{State 2} \Leftrightarrow \overline{SV_2} \Leftrightarrow \overline{SC_2}$$

$$\text{State 3} \Leftrightarrow \overline{SV_3} \Leftrightarrow \overline{SC_3} \qquad \text{State 4} \Leftrightarrow \overline{SV_4} \Leftrightarrow \overline{SC_4}$$

$$\text{State 5} \Leftrightarrow \overline{SV_5} \Leftrightarrow \overline{SC_5} \qquad \text{State 6} \Leftrightarrow \overline{SV_6} \Leftrightarrow \overline{SC_6}$$

However, it should be noted that the resultant output current for a CSI will be displaced 30° compared to the VSI modulator reference waveforms because of the phase shift in the state phasors used by the CSI. Of course, this can be easily compensated by phase shifting the reference inputs to the modulator.

Next, it is necessary to determine how the CSI null states should relate to the modulator state outputs, which is done as follows:

For a VSI, the identification of the minimum switch transition null states by the modulation process is often implicit (especially for sine-triangle modulation) and may not be identifiable as a separate part of the modulation process. This is particularly so because the output of each phase leg sine-triangle comparison is a binary quantity (i.e., on or off), and this maps to the VSI phase leg "either upper or lower" switching requirements on a one-to-one relationship.

For a CSI, there is no direct mapping between the binary output of a sine-triangle phase leg modulator and the inverter phase legs because each phase leg can adopt one of three states (either upper on, lower on, or off). Hence an explicit minimum switch criterion must be developed to determine how the three null states of a CSI should be used.

From Figure 7.1, the CSI switch logic control signals which relate to each CSI state are defined by the simple logical combinations of

$$S_{1(csi)} = SC_6 + SC_1 + SC_7 \tag{7.1}$$

$$S_{2(csi)} = SC_1 + SC_2 + SC_9 \tag{7.2}$$

$$S_{3(csi)} = SC_2 + SC_3 + SC_8 \tag{7.3}$$

$$S_{4(csi)} = SC_3 + SC_4 + SC_7 \tag{7.4}$$

$$S_{5(csi)} = SC_4 + SC_5 + SC_9 \tag{7.5}$$

$$S_{6(csi)} = SC_5 + SC_6 + SC_8 \tag{7.6}$$

where SC_7, SC_8, SC_9 are the three null states corresponding to the shorting of switched pairs $S_{1(csi)}$–$S_{4(csi)}$, $S_{3(csi)}$–$S_{6(csi)}$, and $S_{5(csi)}$–$S_{2(csi)}$, respectively. Note that in these equations, $SC_x, x = 1, \ldots, 8$, denote logic variables that simply identify when the CSI is in a particular space vector state.

The CSI switch control logic signals can also be defined in terms of switch outputs from a VSI modulator, i.e.,

$$S_{1(csi)} = S_{6(csi)} * S_{1(vsi)} + SC_7 \tag{7.7}$$

$$S_{2(csi)} = S_{1(csi)} * S_{2(vsi)} + SC_9 \tag{7.8}$$

$$S_{3(csi)} = S_{2(csi)} * S_{3(vsi)} + SC_8 \tag{7.9}$$

$$S_{4(csi)} = S_{3(csi)} * S_{4(vsi)} + SC_7 \tag{7.10}$$

$$S_{5(csi)} = S_{4(csi)} * S_{5(vsi)} + SC_9 \tag{7.11}$$

$$S_{6(csi)} = S_{5(csi)} * S_{6(vsi)} + SC_8 \tag{7.12}$$

which is similar to the approach developed by [1].

The identification of when a null state is required can now be determined either directly from the state modulator (i.e., by identifying when an active state is not required) or indirectly derived from the outputs of a VSI modulator using the logic identity

$$\text{null} = S_{1(vsi)} \bullet S_{3(vsi)} \bullet S_{5(vsi)} + \overline{S_{1(vsi)}} \bullet \overline{S_{3(vsi)}} \bullet \overline{S_{5(vsi)}} \tag{7.13}$$

where the overbar on these symbols designate the logic complement of the phase leg switch state. This result can be mapped to one of the three real CSI nulls based on the last known active state, according to Table 7.1.

This approach ensures the same (minimum) number of switching transitions as for an equivalent VSI modulator and suits modulation strategies which

Table 7.1 CSI Zero States for Minimum Switching Frequency, Equivalent Triangular Carrier

Last Active State	Last VSI Switch State	Null State
$\overline{SC_1}$	$S_{1(vsi)} \bullet S_{6(vsi)} \bullet S_{2(vsi)}$	$\overline{SC_9}$
$\overline{SC_2}$	$S_{1(vsi)} \bullet S_{3(vsi)} \bullet S_{2(vsi)}$	$\overline{SC_8}$
$\overline{SC_3}$	$S_{4(vsi)} \bullet S_{3(vsi)} \bullet S_{2(vsi)}$	$\overline{SC_7}$
$\overline{SC_4}$	$S_{4(vsi)} \bullet S_{3(vsi)} \bullet S_{5(vsi)}$	$\overline{SC_9}$
$\overline{SC_5}$	$S_{4(vsi)} \bullet S_{6(vsi)} \bullet S_{5(vsi)}$	$\overline{SC_8}$
$\overline{SC_6}$	$S_{1(vsi)} \bullet S_{6(vsi)} \bullet S_{5(vsi)}$	$\overline{SC_7}$

exit from the null state back to the same "last active state". Such strategies essentially use an explicit or implicit triangular carrier waveform, which reverses the sequence of the active states in each half carrier cycle.

For example, the active states required in the first sextant of CSI modulation are $\overline{SC_6}$ and $\overline{SC_1}$, and the state sequence for continuous switching modulation would be

$$\overline{\text{null}_1} \to \overline{SC_6} \to \overline{SC_1} \to \overline{\text{null}_2} \to \overline{SC_1} \to \overline{SC_6} \to \overline{\text{null}_1} \quad (7.14)$$

From Table 7.1, the correct null states for minimum switching with this state sequence are

$$\overline{SC_7} \to \overline{SC_6} \to \overline{SC_1} \to \overline{SC_9} \to \overline{SC_1} \to \overline{SC_6} \to \overline{SC_7} \quad (7.15)$$

Discontinuous switching removes one null state, so that the state sequence would become

$$\overline{SC_7} \to \overline{SC_6} \to \overline{SC_1} \to \overline{SC_6} \to \overline{SC_7} \quad (7.16)$$

Similar sequences can be established for all sextants of modulation using Table 7.1. However, the approach used in Table 7.1 does not take advantage of the additional reduction in switching transitions that can be achieved with a CSI by virtue of the extra null state that is available. An alternative approach is to identify the null state to be used from the sextant of the reference waveforms, according to Table 7.2. This approach allows an equivalent sawtooth carrier to be used without the penalty of extra switching from one null state to another that occurs for a VSI.

Table 7.2 CSI Zero States for Minimum Switching Frequency, Equivalent Saw Tooth Carrier

Sextant	Last Active State #1	Last Active State #2	null
$-\pi/6 \rightarrow \pi/6$	$\overline{SC_6}$	$\overline{SC_1}$	$\overline{SC_7}$
$-\pi/6 \rightarrow \pi/2$	$\overline{SC_1}$	$\overline{SC_2}$	$\overline{SC_9}$
$\pi/2 \rightarrow 5\pi/6$	$\overline{SC_2}$	$\overline{SC_3}$	$\overline{SC_8}$
$5\pi/6 \rightarrow 7\pi/6$	$\overline{SC_3}$	$\overline{SC_4}$	$\overline{SC_7}$
$7\pi/6 \rightarrow 3\pi/2$	$\overline{SC_4}$	$\overline{SC_5}$	$\overline{SC_9}$
$3\pi/2 \rightarrow -\pi/6$	$\overline{SC_5}$	$\overline{SC_6}$	$\overline{SC_8}$

For example, continuous sawtooth modulation for a VSI produces the state sequence over one carrier interval of

$$\ldots \overline{SV_0} \rightarrow \overline{SV_6} \rightarrow \overline{SV_1} \rightarrow \overline{SV_7} \rightarrow \overline{SV_0} \rightarrow \ldots \tag{7.17}$$

where the transition from null $\overline{SV_7}$ to null $\overline{SV_0}$ in the next carrier interval requires extra switch transitions on two phase legs and hence five switch transitions overall per carrier period. Discontinuous modulation reduces this to four switch transitions because one null state is no longer required.

The equivalent sawtooth modulation for a CSI using the null definitions of Table 7.2 produces the state sequence over one carrier interval of

$$\ldots \overline{SC_7} \rightarrow \overline{SC_6} \rightarrow \overline{SC_1} \rightarrow \overline{SC_7} \rightarrow \ldots \tag{7.18}$$

where the transition from null to null in the next carrier interval, of course, requires no switch transitions. Hence the total number of switch transitions is only three over the carrier interval irrespective of whether continuous or discontinuous modulation is used (the only difference is the placement of the active states within the carrier period). Consequently the sawtooth carrier frequency can be set to twice that of a triangular carrier (which requires six switch transitions), and this can give significant harmonic benefits at higher modulation indices as discussed previously.

A further advantage of this general approach to CSI modulation is that saturation is implicitly managed in the same way as for VSI modulation. Irrespective of the modulation algorithm that is implemented, saturation of the modulator still simply creates a sequence of active states over the carrier period, and these states map immediately to their CSI counterparts.

Finally, the concept of third-harmonic injection is still relevant with the modulation mapping concept presented here. The prime function of third-harmonic injection strategies is to move the position of the active states within each carrier interval to achieve maximum modulation index and decreased low-order harmonics, and these requirements apply equally to CSI modulation.

7.2 Naturally Sampled CSI Space Vector Modulator

From the ideas of zero vector placement in the carrier interval the analog equivalent of regular sampled SVM was developed previously for use with a VSI. Since it has already been shown how any VSI modulator can be mapped to control a CSI, the concept of an analog Space Vector modulator for a CSI is obvious, and needs no further development. Essentially, the modulator uses the envelope of the three-phase reference waveforms to develop an offset which centers the active states in the middle of each half carrier interval, and the harmonic benefits of this placement are equally applicable to a VSI or a CSI.

7.3 Experimental Confirmation

The generalized modulation approach for a CSI presented here has been confirmed experimentally by taking conventional VSI modulation systems and mapping the switch outputs from the modulators through a programmable logic device (PLD) using Eqs. (7.7) to (7.12) and Table 7.1 to control the CSI switches. The crossover delay in the VSI modulators was set to zero for these experiments, and it was assumed that IGBT switching delays in the CSI would provide sufficient overlap for current commutation. The experimental converter had ratings of 10 A, 5 kW, and the DC current source was supplied from a 10 A current regulated power supply through a 80 mH air cored inductor.

Figure 7.2 shows the switched line currents and associated harmonics for a single-phase CSI controlled by a naturally sampled VSI modulator. The carrier frequency was 1050 Hz, and the elimination of the carrier frequency sideband components can be clearly seen in Figure 7.2(b), as would be expected for a three-level single-phase modulation system.

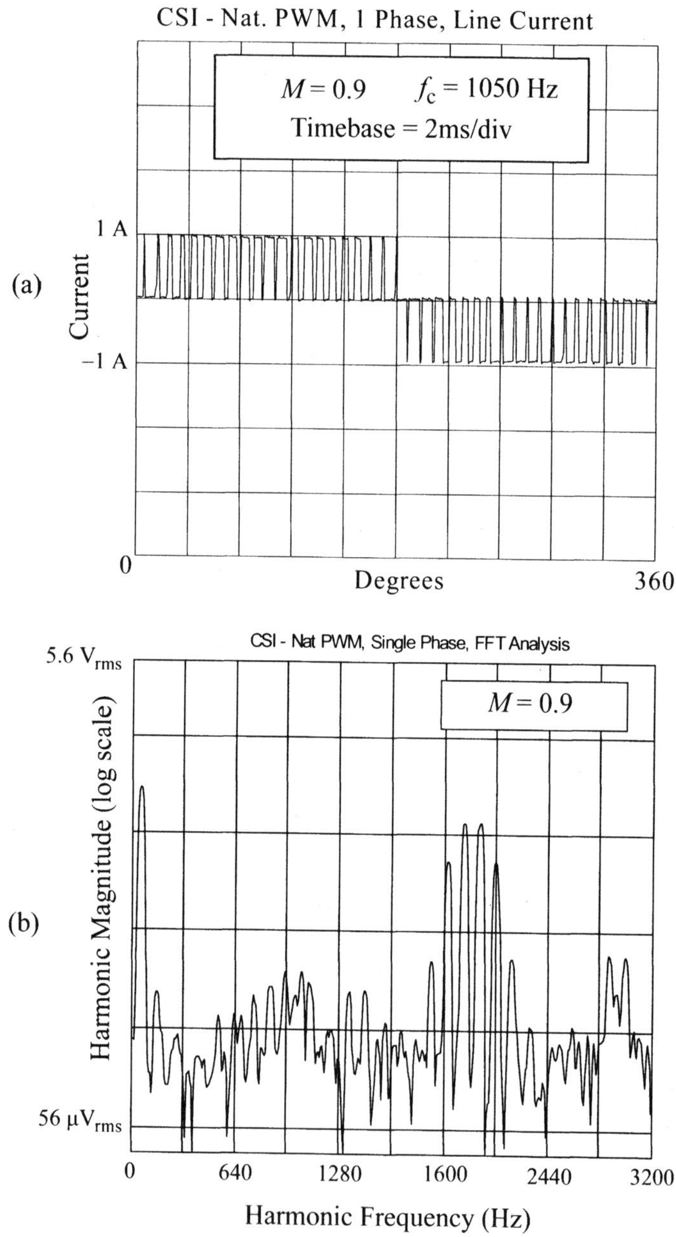

Figure 7.2 (a) Experimental switched current for single-phase CSI controlled by naturally sampled VSI modulation and (b) measured resulting spectrum, $M = 0.9$, $f_c = 50$ Hz [3].

Figure 7.3 shows the switched line currents and associated harmonics for a three-phase CSI controlled by a regular sampled VSI modulator without third-harmonic injection. The carrier frequency was 1050 Hz, and the spectrum shows the well-known two major sideband harmonics either side of the (suppressed) carrier frequency.

Figures 7.4 and 7.5 show the harmonic spectrum of the line current for a three-phase CSI controlled by regular sampled PWM with third-harmonic injection, and by analog space vector modulation, respectively.

Figures 7.3(b), 7.4, and 7.5, show in particular the expected reduction of the two major carrier sideband components in favor of the wider and higher frequency sidebands, as the modulation strategy moves from regular sampled PWM, through third-harmonic injection, to centered space vector modulation. These results match exactly the results obtained previously using the same modulators to directly control a VSI, and confirm that the mapping concept does achieve exactly the same modulation performance for a CSI as for a VSI controlled by the same PWM strategy.

Figure 7.6 shows the harmonic spectrum of the line current for 60° discontinuous modulation, with a modulation calculation frequency adjusted to maintain the same number of switch transitions per phase leg. The result is the expected wider and more complex sideband progression, but the weighted THD performance will still be improved because of the higher frequencies of the harmonic components.

7.4 Summary

This chapter has described a generalized approach to the modulation control of a CSI, which enables all types of fixed-frequency VSI modulation strategies to be readily applied to a CSI. The approach used is to map the active states of the VSI modulator to equivalent active states for the CSI. Then CSI null states are chosen for the periods when no active state is required, selected so as to minimize the number of switch transitions so the carrier frequency can be maximized for a given amount of device switching losses. The result is a series of PWM modulators for a CSI which achieve exactly the same harmonic performance as their equivalent VSI controller.

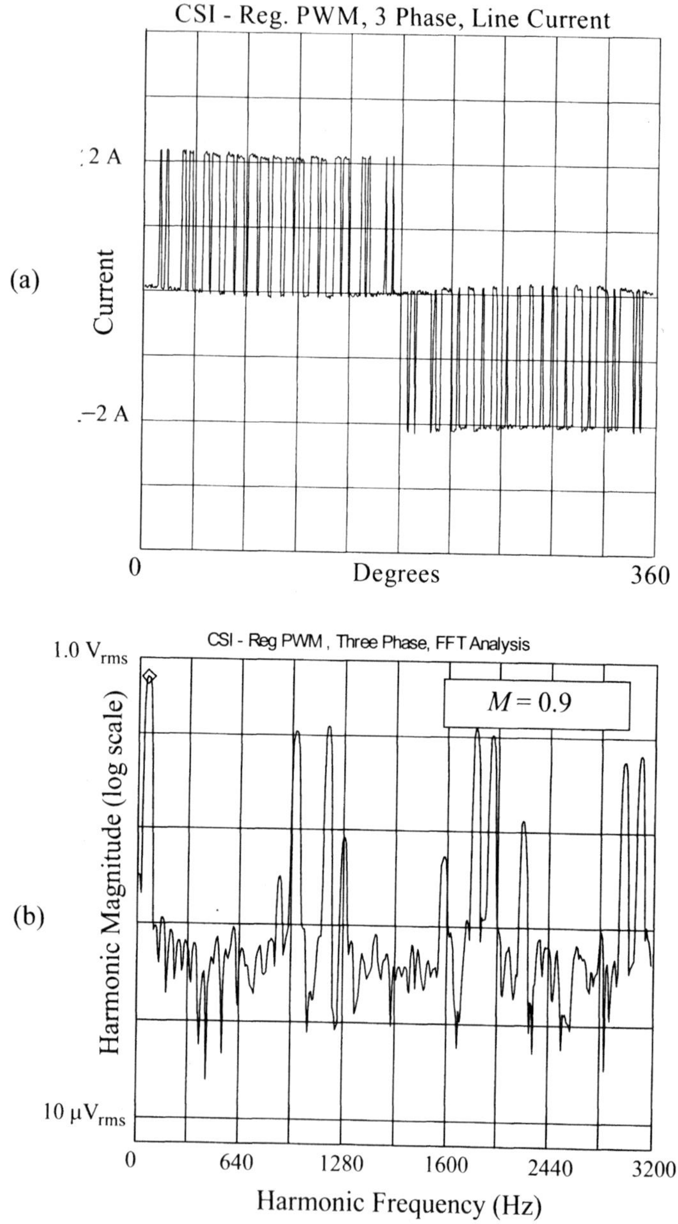

Figure 7.3 (a) Experimental switched current for three-phase CSI controlled by regular sampled sine-triangle modulation without third-harmonic injection and (b) corresponding experimental harmonic spectrum [3].

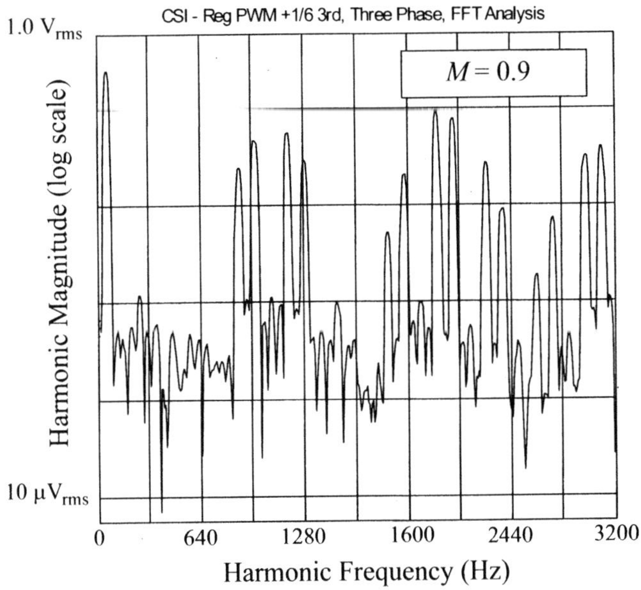

Figure 7.4 Experimental harmonic spectrum for three-phase CSI controlled by regular sampled sine-triangle modulation with third-harmonic injection, $M = 0.9$ [3].

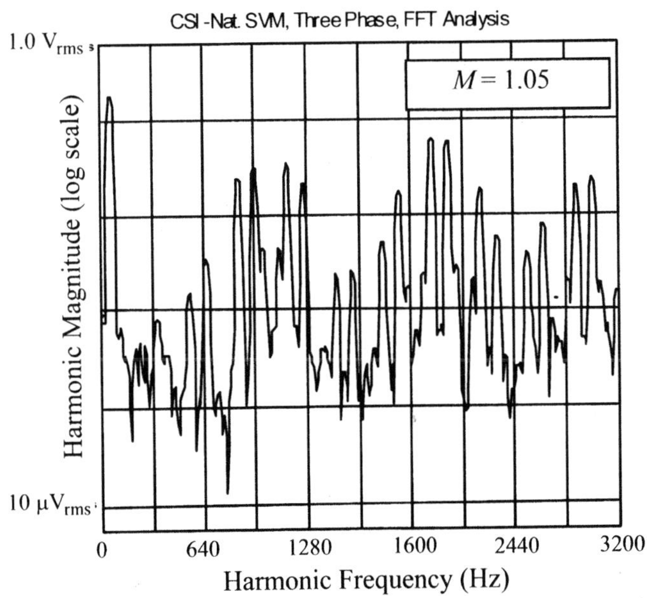

Figure 7.5 Experimental harmonic spectrum for three-phase CSI controlled by natural space vector modulation, $M = 1.05$ [3].

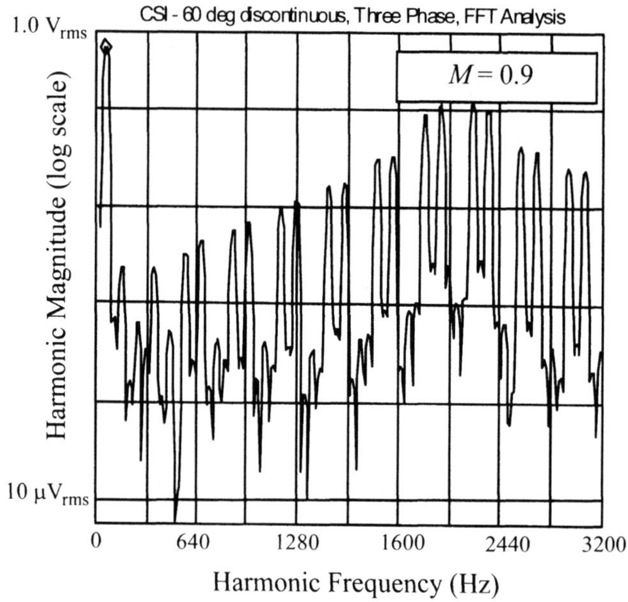

Figure 7.6 Experimental harmonic spectrum for three-phase CSI controlled by 60° discontinuous modulation, $M = 0.9$ [3].

The concepts presented here have been verified experimentally by mapping the output of a number of VSI modulation systems to obtain switch control signals for a test CSI. The harmonic spectra of the resultant switched output currents exactly match their VSI counterparts and hence confirm the viability of the approach.

References

[1] G. Joos, G. Moschopoulos, and P.D. Ziogas, "A high performance current source inverter," in *Conf. Rec. IEEE Power Electronics Specialists Conf. (PESC)*, Cambridge MA, 1991, pp. 123–130.

[2] D.N. Zmood, "A systematic development of improved linear regulators for sinusoidal power converters," Ph.D. Thesis, Monash University, Australia, 2002.

[3] D.N. Zmood and D.G. Holmes, "Generalised approach to the modulation of current source inverters," in *Conf. Rec. IEEE Power Electronics Specialist Conf. (PESC)*, Fukuoka, 1998, pp. 739–745.

8

Overmodulation of an Inverter

It has been shown in Chapters 5 and 6, that by either adding a third-harmonic component to natural or regular sampled PWM or by adopting the space vector modulation strategy, the range of linear control of the output fundamental component can be extended by a factor of 1.15 from $V_1 = V_{dc}$ to $V_1 = (2/\sqrt{3})V_{dc} = 1.15V_{dc}$. However, it has also been shown in Eq. (1.7) that for a six-step square-wave controlled inverter, the magnitude of the output fundamental voltage is $V_1 = (4/\pi)V_{dc} = 1.273V_{dc}$. Increasing the output voltage of a PWM-controlled inverter from $1.15V_{dc}$ to the limit of $1.273V_{dc}$ is achieved by entering the nonlinear region of *overmodulation*, within which the modulation controller gain (the ratio between the target reference and the actual inverter output voltage) decreases from 1 to 0 as switched pulses progressively disappear.

The reader might question why it is necessary to enter this region at all since the maximum modulation index could simply be restricted to the end of the linear control region, that is, $V_1 = 1.15V_{dc}$. However, it can be recalled that if a diode rectifier is operating in continuous conduction, its average DC voltage is given by $2V_{dc} = 1.35V_p/\sqrt{2}$. Hence the maximum inverter peak *l–n* output voltage will be limited to

$$V_1 = (2/\sqrt{3})V_{dc} = (2/\sqrt{3})1.35V_p/(2\sqrt{2}) = 0.955(V_p/\sqrt{3}) \qquad (8.1)$$

and a motor rated at the input line voltage would still not reach rated power even with third-harmonic injection or space vector modulation. On the other hand if the inverter can reach square-wave operation, then the peak output voltage will be $V_1 = (4/\pi)1.35V_p/(2\sqrt{2}) = 1.053(V_p/\sqrt{3})$, which is, ideally, 5% greater than the peak *l–n* input voltage. While this additional 5% capability is not typically offered to a customer, it is an important reserve for making up the losses in the drive electronics. On the other hand, if the inverter limits at 95% of rated voltage, a nontrivial 5% (or more) of the power is inaccessible from motor that is rated at the voltage of the AC input, a significant disadvantage from an applications point of view.

8.1 The Overmodulation Region

The region of operation between the loss of linear control ($M = 1.15$) and complete loss of control ($M = 1.27$) is called the *overmodulation region*. When overmodulation occurs, the modulation index M exceeds the triangle wave as shown in Figure 8.1. Note that when $M > 1$, or $M > 1.15$ as appropriate, the actual resultant fundamental component does not linearly follow M, and the controller is said to have saturated. Consequently, the shape of the output voltage waveform is only partially under control. Since the modulator effectively loses control of the output waveform during the saturation intervals, the output waveform becomes progressively distorted and includes low-frequency harmonics of the fundamental component. The final result is the "six-step" waveform corresponding to simple square-wave operation of each phase leg. The process of dropping pulses from the output waveform is termed *overmodulation* and will now be examined in detail.

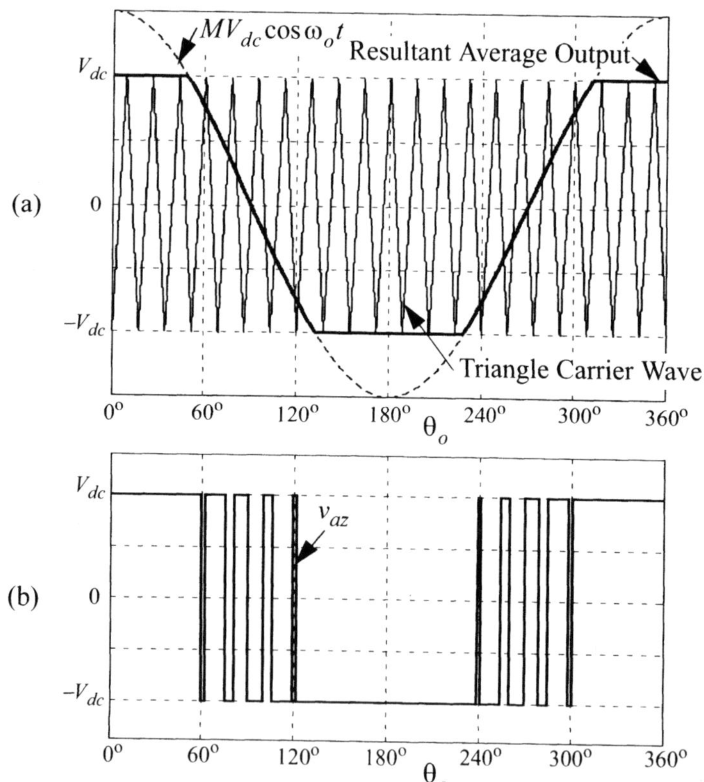

Figure 8.1 Overmodulated naturally sampled PWM: (a) reference and carrier waves, and (b) switched phase leg output, $M = 1.5$.

8.2 Naturally Sampled Overmodulation of One Phase Leg of an Inverter

It is useful to begin again with the modulation of one phase leg of an inverter, discussed previously in Chapter 3. The solution process in general involves evaluation of the double Fourier integral expression

$$C_{mn} = A_{mn} + jB_{mn} = \frac{1}{2\pi^2}\int_{-\pi}^{\pi}\int_{-\pi}^{\pi} f(x, y)\, e^{j(mx + ny)}\, dx\, dy \tag{8.2}$$

where $f(x, y)$ defines the value of the switched waveform as the modulation process proceeds and C_{mn} defines the magnitude of the $(m\omega_c + n\omega_o)^{th}$ harmonic component of this waveform.

To define $f(x, y)$ in the overmodulation region it is first necessary to define the carrier and reference functions in terms of the time-dependent variables x and y shown in Figure 8.2. Referring to this figure, the triangular carrier waveform is defined by

$$f(x) = -1 - \frac{2x}{\pi} \qquad \text{for} \quad -\pi \le x < 0 \tag{8.3}$$

$$f(x) = -1 + \frac{2x}{\pi} \qquad \text{for} \quad 0 \le x < \pi \tag{8.4}$$

while the reference waveform is defined by

$$f(y) = M\cos y \tag{8.5}$$

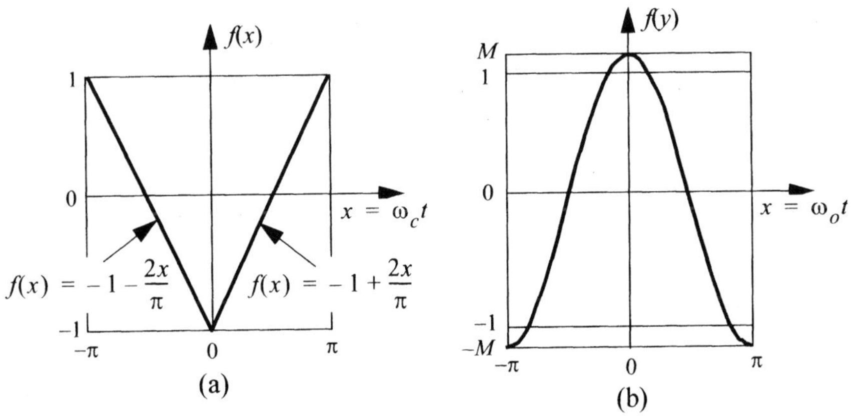

Figure 8.2 Triangle carrier waveform and fundamental reference waveform in the x and y subvariable space for $M \ge 1.0$.

Combining Eqs. (8.3) to (8.5), and using the knowledge that the phase leg switches from zero to $+2V_{dc}$ only when the reference sine wave is greater than the carrier, leads to the following expressions for $f(x, y)$.

During the interval $-\pi \le x < 0$ when the triangle wave is decreasing,

$$f(x, y) = 2V_{dc} \text{ when } M\cos y > -1 - \frac{2x}{\pi} \quad \left[\text{i.e., } x < -\frac{\pi}{2}(1 + M\cos y)\right] \tag{8.6}$$

and otherwise equals zero.

During the interval $0 \le x \le \pi$ when the triangle wave is increasing,

$$f(x, y) = 2V_{dc} \text{ when } M\cos y > -1 + \frac{2x}{\pi} \quad \left[\text{i.e., } x < \frac{\pi}{2}(1 + M\cos y)\right] \tag{8.7}$$

and otherwise equals zero.

Equations (8.6) and (8.7) lead to the unit cell contour plot for overmodulation shown in Figure 8.3. Note that the point of intersection of the reference waveform with the y axis is where

$$\frac{\pi}{2}(1 + M\cos\psi) = \pi \tag{8.8}$$

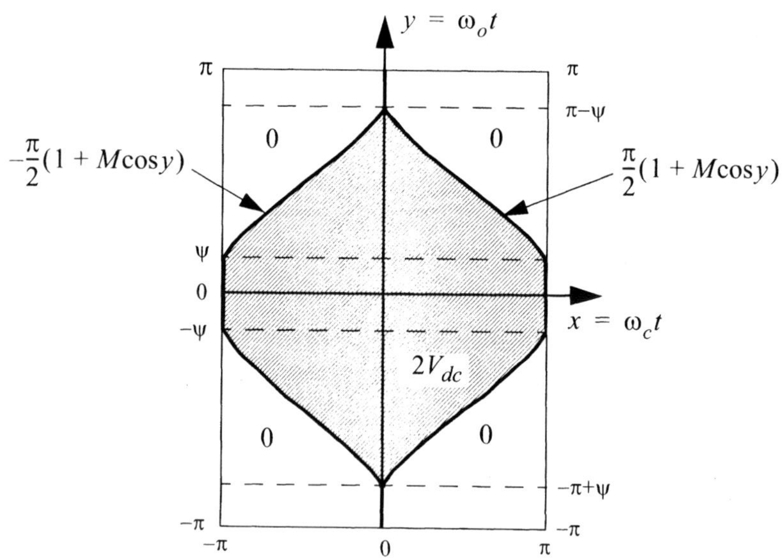

Figure 8.3 Unit cell for double-edge naturally sampled two-level PWM under overmodulation conditions.

in which case

$$\psi = \cos^{-1}\left(\frac{1}{M}\right) \tag{8.9}$$

The general form of the Fourier integral, Eq. (8.2), can now be written as

$$C_{mn} = \frac{V_{dc}}{\pi^2}\left[\int_{-\pi+\psi}^{-\psi}\int_{-\frac{\pi}{2}(1+M\cos y)}^{\frac{\pi}{2}(1+M\cos y)} e^{j(mx+ny)}\,dx\,dy + \int_{-\psi}^{\psi}\int_{-\pi}^{\pi} e^{j(mx+ny)}\,dx\,dy + \int_{\psi}^{\pi-\psi}\int_{-\frac{\pi}{2}(1+M\cos y)}^{\frac{\pi}{2}(1+M\cos y)} e^{j(mx+ny)}\,dx\,dy\right] \tag{8.10}$$

which can be solved using the solution process described in Appendix 4 to determine the magnitude of the harmonic components for various index values of m and n. The complete solution for *naturally sampled* overmodulation of one inverter phase leg can then be written as

$$\begin{aligned} v_{az}(t) = {} & V_{dc} + \frac{V_{dc}}{\pi}[4\sin\psi + M(\pi - 2\psi - \sin 2\psi)]\cos\omega_o t \\ & + \frac{V_{dc}}{\pi}\sum_{n=2}^{\infty}\left[\frac{4}{[2n-1]}\sin([2n-1]\psi) - M\left[\frac{\sin 2n\psi}{n} + \frac{\sin(2[n-1]\psi)}{[n-1]}\right]\right]\cos([2n-1]\omega_o t) \\ & + \frac{4V_{dc}}{\pi^2}\sum_{m=1}^{\infty}\frac{1}{m}\sin m\frac{\pi}{2}\left[J_0\left(m\frac{\pi}{2}M\right)(\pi - 2\psi) - 2\sum_{k=1}^{\infty}\frac{1}{k}J_{2k}\left(m\frac{\pi}{2}M\right)\cos k\pi\sin 2k\psi\right]\cos m\omega_c t \\ & + \sum_{m=1}^{\infty}\sum_{\substack{n=-\infty \\ (n\neq 0)}}^{\infty} C_{mn}\cos(m\omega_c t + n\omega_o t) \end{aligned} \tag{8.11}$$

where C_{mn} is given by

$$C_{mn} = \frac{4V_{dc}}{m\pi^2}\left[\begin{array}{l} -\frac{1}{n}J_0\left(m\frac{\pi}{2}M\right)\sin m\frac{\pi}{2}(1+\cos n\pi)\sin n\psi \\ +J_n\left(m\frac{\pi}{2}M\right)\sin\left([m+n]\frac{\pi}{2}\right)\left[\pi - 2\psi - \frac{1}{n}\sin 2n\psi\right] \\ -\sum_{\substack{k=1 \\ k\neq|n|}}^{\infty}\left[\begin{array}{l} J_k\left(m\frac{\pi}{2}M\right)\sin\left([m+k]\frac{\pi}{2}\right)\{1+\cos([n+k]\pi)\} \\ \qquad\times\left[\frac{\sin([n+k]\psi)}{[n+k]}+\frac{\sin([n-k]\psi)}{[n-k]}\right]\end{array}\right]\end{array}\right] \tag{8.12}$$

and ψ is defined by

$$\psi = \left\{\begin{array}{ll} \cos^{-1}\left(\frac{1}{M}\right) & \text{for } M > 1 \\ 0 & \text{for } M \leq 1 \end{array}\right\} \tag{8.13}$$

Not unexpectedly, the solution for the voltage with respect to the DC bus center point z is simply Eq. (8.11) without the V_{dc} offset term. Also, Eq. (8.11) readily devolves back to the solution for linear modulation of Eq. (3.39) for $\psi = 0$, since all $(\sin n\psi)$ terms become zero for any n.

Figure 8.4 shows the spectral plot corresponding to Eq. (8.11) for the conditions of $M = 1.5$ and a carrier/fundamental ratio of $f_c/f_o = 21$. Note that the calculation of these harmonics must be done carefully, allowing for phasor summation between the harmonic components of the various baseband and sideband groups. The result can be compared against Figure 3.12, and it can be seen how significant baseband harmonics have now been introduced into the spectra as a consequence of the distortion of the overmodulation process. It would be expected that these baseband harmonic components would develop toward the harmonic pattern of a square wave as the modulation index continues to increase.

Also, it is interesting to observe how the magnitude of the first carrier harmonic is smaller for the increased modulation index. But this is explained by a moment's examination of the analytical solutions for linear modulation, Eq. (3.39), and for overmodulation, Eqs. (8.11) and (8.12). In both cases, the magnitude of the carrier harmonic includes a $J_0\left(m\frac{\pi}{2}M\right)$ term, and, from Figure

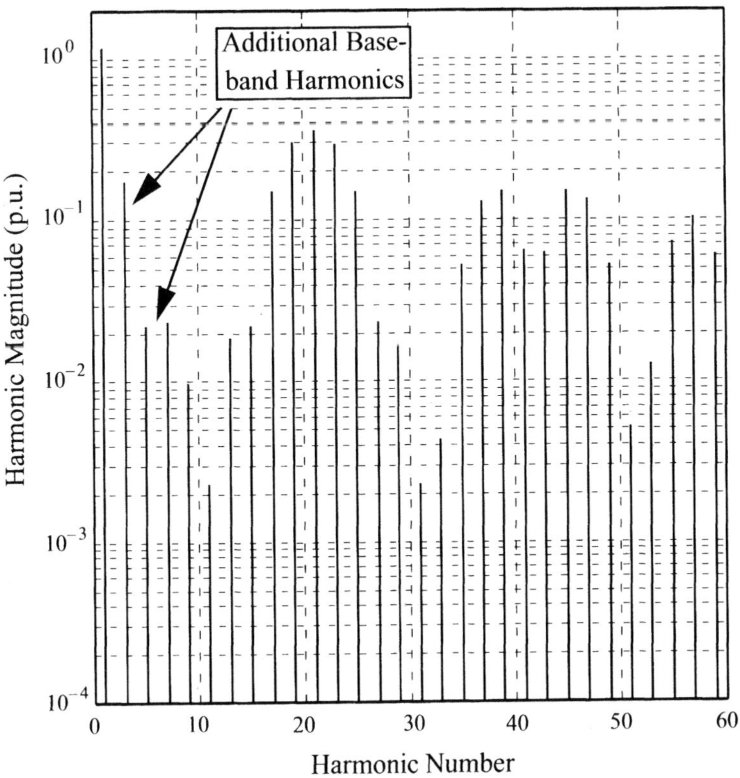

Figure 8.4 Harmonic components for one inverter phase leg with *double-edge naturally sampled* PWM under *overmodulation* conditions, $M = 1.5$, $f_c/f_o = 21$.

A2.1, the magnitude of this term increases for smaller values of M. Hence decreased modulation indices always lead to increased carrier harmonic magnitudes.

Finally, it is important to note that overmodulation does not create any sideband harmonics at new sideband frequencies, and that in particular all even sidebands for odd carrier sideband groups and all odd sidebands for even carrier sideband groups continue to be eliminated. Thus the influence of the overmodulation process on carrier and sideband harmonics is only to vary the magnitude of harmonics that are already present to some degree in the switched output waveform. Hence sideband harmonic filter systems designed to work in the linear modulation region will continue to be effective under overmodulation conditions.

8.3 Regular Sampled Overmodulation of One Phase Leg of an Inverter

Analytical solutions can be developed for symmetrical and asymmetrical regular sampled PWM under overmodulation conditions using the same techniques as were applied to naturally sampled PWM in the previous section. However, the detailed solution process is quite lengthy and is developed in Appendix 4 with only the results presented here for reference.

The complete solution for *symmetrical regular sampled* overmodulation of one inverter phase leg is

$$v_{az}(t) = V_{dc} + \sum_{n=1}^{\infty} C_{0n} \cos n\omega_o t \tag{8.14}$$

$$+ \frac{4V_{dc}}{\pi^2} \sum_{m=1}^{\infty} \frac{1}{m} \sin m\frac{\pi}{2} \left[\begin{array}{l} J_0\left(m\frac{\pi}{2}M\right)(\pi - 2\psi) \\ -2\displaystyle\sum_{k=1}^{\infty} \frac{1}{k} J_{2k}\left(m\frac{\pi}{2}M\right) \cos k\pi \sin 2k\psi \end{array} \right] \cos m\omega_c t$$

$$+ \sum_{m=1}^{\infty} \sum_{\substack{n=-\infty \\ (n \neq 0)}}^{\infty} C_{mn} \cos(m\omega_c t + n\omega_o t)$$

where C_{0n} is given by

$$C_{0n} = \frac{4V_{dc}}{\left[n\frac{\omega_o}{\omega_c}\right]^2 \pi} \left[\begin{array}{l} \frac{1}{n}\left[\sin\left(n\frac{\omega_o}{\omega_c}\pi\right) - J_0\left(n\frac{\omega_o}{\omega_c}\frac{\pi}{2}M\right)\sin\left(n\frac{\omega_o}{\omega_c}\frac{\pi}{2}\right)(1 + \cos n\pi)\right]\sin n\psi \\ \quad + J_n\left(n\frac{\omega_o}{\omega_c}\frac{\pi}{2}M\right)\sin\left(\left[n\frac{\omega_o}{\omega_c} + n\right]\frac{\pi}{2}\right)\left[\pi - 2\psi - \frac{1}{n}\sin 2n\psi\right] \\ \quad - \displaystyle\sum_{\substack{k=1 \\ k \neq |n|}}^{\infty} \left[\begin{array}{l} J_k\left(n\frac{\omega_o}{\omega_c}\frac{\pi}{2}M\right)\sin\left(\left[n\frac{\omega_o}{\omega_c} + k\right]\frac{\pi}{2}\right) \\ \quad \times \{1 + \cos([n+k]\pi)\} \\ \quad \quad \times \left[\frac{\sin([n+k]\psi)}{[n+k]} + \frac{\sin([n-k]\psi)}{[n-k]}\right] \end{array} \right] \end{array} \right] \tag{8.15}$$

and C_{mn} is given by

$$C_{mn} = \frac{4V_{dc}}{q\pi^2}\left[\begin{array}{l} \frac{1}{n}\left[\sin(q\pi) - J_0\left(q\frac{\pi}{2}M\right)\sin q\frac{\pi}{2}(1+\cos n\pi)\right]\sin n\psi \\ \quad + J_n\left(q\frac{\pi}{2}M\right)\sin\left([q+n]\frac{\pi}{2}\right)\left[\pi - 2\psi - \frac{1}{n}\sin 2n\psi\right] \\ \quad - \sum_{\substack{k=1 \\ k\neq|n|}}^{\infty}\left[\begin{array}{l} J_k\left(q\frac{\pi}{2}M\right)\sin\left([q+k]\frac{\pi}{2}\right)\{1+\cos([n+k]\pi)\} \\ \qquad \times\left[\frac{\sin([n+k]\psi)}{[n+k]} + \frac{\sin([n-k]\psi)}{[n-k]}\right] \end{array}\right] \end{array}\right] \tag{8.16}$$

with $q = m + n(\omega_o/\omega_c)$ and ψ defined by Eq. (8.13).

The complete solution for *asymmetrical regular sampled* overmodulation of one inverter phase leg is

$$v_{az}(t) = V_{dc} + \sum_{n=1}^{\infty} C_{0n}\cos n\omega_o t \tag{8.17}$$

$$+ \frac{4V_{dc}}{\pi^2}\sum_{m=1}^{\infty}\frac{1}{m}\sin m\frac{\pi}{2}\left[\begin{array}{l} J_0\left(m\frac{\pi}{2}M\right)(\pi - 2\psi) \\ -2\sum_{k=1}^{\infty}\frac{1}{k}J_{2k}\left(m\frac{\pi}{2}M\right)\cos k\pi \sin 2k\psi \end{array}\right]\cos m\omega_c t$$

$$+ \sum_{m=1}^{\infty}\sum_{\substack{n=-\infty \\ (n\neq 0)}}^{\infty} C_{mn}\cos(m\omega_c t + n\omega_o t)$$

where C_{0n} is given by

$$C_{0n} = \frac{4V_{dc}}{\left[n\frac{\omega_o}{\omega_c}\right]\pi^2}\left[\begin{array}{l}\frac{1}{n}\sin\left(n\frac{\omega_o}{\omega_c}\frac{\pi}{2}\right)(1-\cos n\pi)\sin n\psi \\ \quad + J_n\left(n\frac{\omega_o}{\omega_c}\frac{\pi}{2}M\right)\sin n\frac{\pi}{2}\left[\pi - 2\psi - \frac{1}{n}\sin 2n\psi\right] \\ \quad - \sum_{\substack{k=1\\k\neq|n|}}^{\infty}\left[\begin{array}{l}J_k\left(n\frac{\omega_o}{\omega_c}\frac{\pi}{2}M\right)\sin k\frac{\pi}{2}\{1+\cos([n+k]\pi)\} \\ \quad \times\left[\frac{\sin([n+k]\psi)}{[n+k]} + \frac{\sin([n-k]\psi)}{[n-k]}\right]\end{array}\right]\end{array}\right] \tag{8.18}$$

and C_{mn} is given by

$$C_{mn} = \frac{4V_{dc}}{q\pi^2}\left[\begin{array}{l}\frac{1}{n}\left[\begin{array}{l}\sin\left(n\frac{\omega_o}{\omega_c}\frac{\pi}{2}\right)(\cos m\pi - \cos n\pi) \\ -J_0\left(q\frac{\pi}{2}M\right)\sin m\frac{\pi}{2}(1+\cos n\pi)\end{array}\right]\sin n\psi \\ \quad + J_n\left(q\frac{\pi}{2}M\right)\sin\left([m+n]\frac{\pi}{2}\right)\left[\pi - 2\psi - \frac{1}{n}\sin 2n\psi\right] \\ \quad - \sum_{\substack{k=1\\k\neq|n|}}^{\infty}\left[\begin{array}{l}J_k\left(q\frac{\pi}{2}M\right)\sin\left([m+k]\frac{\pi}{2}\right)\{1+\cos([n+k]\pi)\} \\ \quad \times\left[\frac{\sin([n+k]\psi)}{[n+k]} + \frac{\sin([n-k]\psi)}{[n-k]}\right]\end{array}\right]\end{array}\right] \tag{8.19}$$

with $q = m + n(\omega_o/\omega_c)$ and ψ once more defined by Eq. (8.13). Spectral plots for these two solutions for the conditions of $M = 1.5$ and a carrier/fundamental ratio of $f_c/f_o = 21$ are shown in Figure 8.5. From this figure, similar differences between naturally sampled and regular sampled modulation as for linear modulation, can be observed.

First, as before, symmetrical sampling creates even harmonics both in the baseband region and in the sideband regions. These harmonics occur because the noninteger $n(\omega_o/\omega_c)$ terms in the arguments of the various sinusoidal terms of the symmetrically sampled solution preclude $\sin(n\pi/2)$ cancellation for even n (as before, this does not occur for asymmetrical sampling).

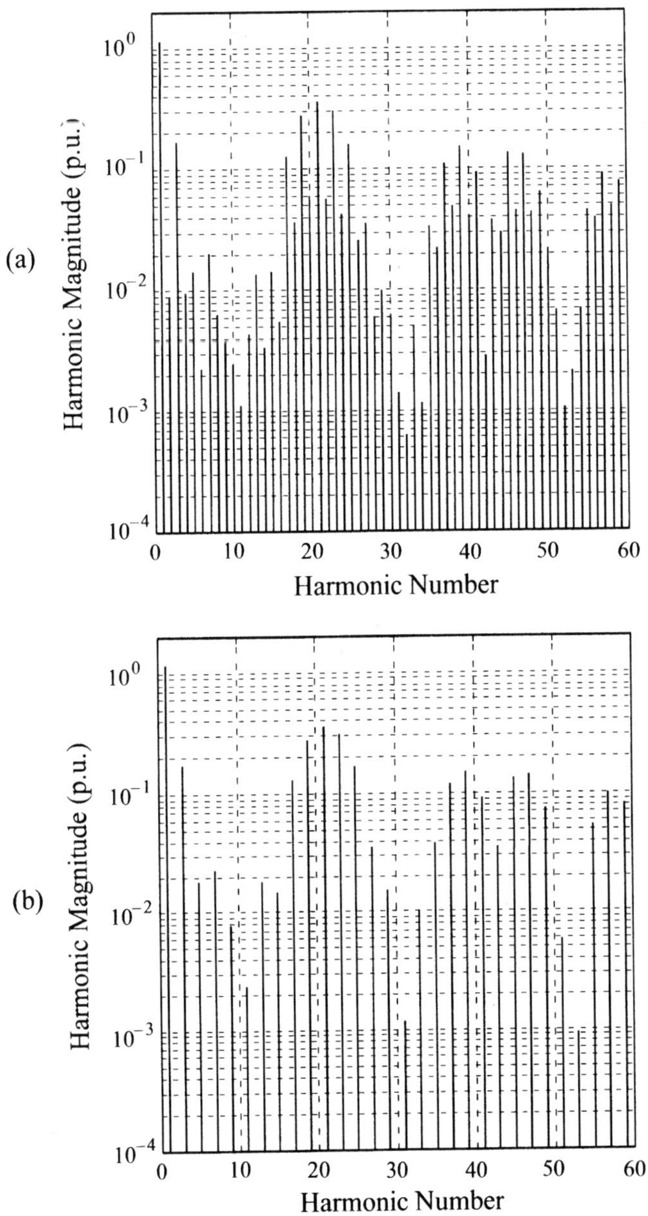

Figure 8.5 Theoretical harmonic spectra for one inverter phase leg under *overmodulation* conditions: (a) *symmetrically sampled* PWM and (b) *asymmetrically sampled* PWM, $M = 1.5$, $f_c/f_o = 21$.

Second, asymmetrically sampled modulation again produces almost the same harmonic spectra as for naturally sampled modulation, but with a slight transfer of energy from the lower harmonic terms to the higher harmonic terms. This shift occurs for both the sideband harmonics and the additional baseband harmonics created by the overmodulation process.

8.4 Naturally Sampled Overmodulation of Single- and Three-Phase Inverters

Using the same approach as was developed for linear modulation in Chapter 4, the *l–l* voltage of a single-phase inverter under overmodulation conditions can be obtained by simply again recognizing, for the second phase leg, that

$$v_{bz}^{*} = V_{dc} M \cos(\omega_o t - \pi) \tag{8.20}$$

whereupon

$$v_{ab}^{*}(t) = v_{az}^{*}(\omega_o t) + v_{az}^{*}(\omega_o t - \pi) \tag{8.21}$$

Substituting from Eq. (8.11) and simplifying to remove all even harmonic terms that cancel between the two phase legs, gives

$$\begin{aligned} v_{ab}(t) &= \frac{2V_{dc}}{\pi}[4\sin\psi + M(\pi - 2\psi - \sin 2\psi)]\cos\omega_o t \\ &+ \frac{2V_{dc}}{\pi}\sum_{n=2}^{\infty}\left[\begin{array}{l} \dfrac{4}{[2n-1]}\sin([2n-1]\psi) \\ \quad - M\left[\dfrac{\sin 2n\psi}{n} + \dfrac{\sin(2[n-1]\psi)}{[n-1]}\right] \end{array}\right]\cos([2n-1]\omega_o t) \\ &+ \sum_{m=1}^{\infty}\sum_{n=-\infty}^{\infty} C_{mn}\cos(2m\omega_c t + [2n-1]\omega_o t) \end{aligned} \tag{8.22}$$

where C_{mn}, with the simplifying substitution of $2m$ for m and $2n-1$ for n (since all other sideband harmonics cancel between the two phase legs), is given by

$$C_{mn} = \frac{8V_{dc}}{2m\pi^2}\left[\begin{array}{l} J_{2n-1}(m\pi M)\cos([m+n-1]\pi)\left[\pi - 2\psi - \dfrac{\sin(2[2n-1]\psi)}{[2n-1]}\right] \\ + \displaystyle\sum_{\substack{k=1 \\ 2k-1 \neq |2n-1|}}^{\infty} \left[\begin{array}{l} J_{2k-1}(m\pi M)\cos([m+k]\pi) \\ \times \quad \dfrac{\sin(2[n+k-1]\psi)}{[n+k-1]} + \dfrac{\sin(2[n-k]\psi)}{[n-k]} \end{array}\right] \end{array}\right] \tag{8.23}$$

and ψ is once more defined by Eq. (8.13).

A spectral plot of Eq. (8.22) is shown in Figure 8.6(a) and should be compared against Fig. 4.3(b) which shows the spectral response of a single-phase inverter for linear modulation conditions. From these plots, it can be seen that, while the sideband harmonic magnitudes have changed, the sideband harmonics in the first carrier group again cancel in the *l–l* bridge output voltage and only occur around the even (second) carrier multiples. Essentially, the only harmonic difference for overmodulation conditions is the additional baseband harmonic components caused by the nonlinearity of the overmodulation process.

A similar approach for a three-phase inverter can be taken by recognizing that

$$v_{bz}(\omega_o t) = v_{az}\left(\omega_o t - \frac{2\pi}{3}\right) \tag{8.24}$$

$$v_{cz}(\omega_o t) = v_{az}\left(\omega_o t + \frac{2\pi}{3}\right) \tag{8.25}$$

so that

$$v_{ab}(\omega_o t) = v_{az}(\omega_o t) - v_{az}\left(\omega_o t - \frac{2\pi}{3}\right) \tag{8.26}$$

$$v_{bc}(\omega_o t) = v_{az}\left(\omega_o t - \frac{2\pi}{3}\right) - v_{az}\left(\omega_o t + \frac{2\pi}{3}\right) \tag{8.27}$$

$$v_{ca}(\omega_o t) = v_{az}\left(\omega_o t + \frac{2\pi}{3}\right) - v_{az}(\omega_o t) \tag{8.28}$$

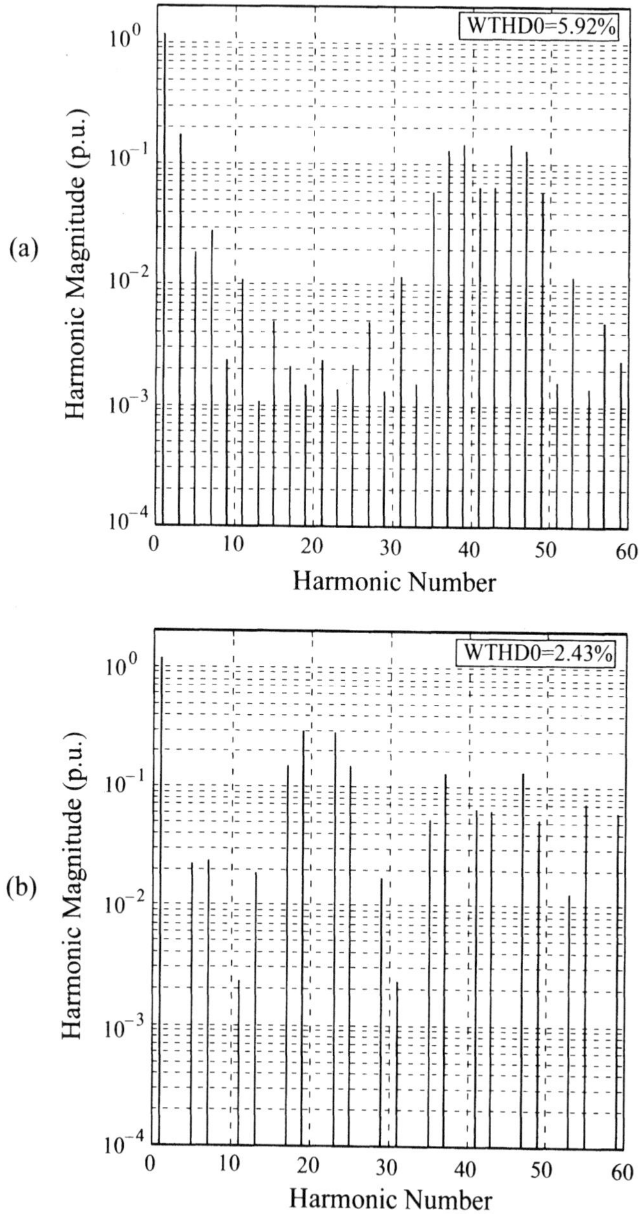

Figure 8.6 Theoretical harmonic spectra for inverter bridges with *double-edge naturally sampled* PWM under *overmodulation* conditions: (a) single-phase bridge *l–l* voltage and (b) three-phase bridge *l–l* voltage, $M = 1.5$, $f_c/f_o = 21$.

The solution process then follows the same pattern as for the single-phase case and is not presented here since the result is lengthy and requires careful use of Eqs. (8.11) and (8.12) to achieve correct cancellation. The result is shown in Figure 8.6(b) and should be compared against Figure 5.4(b). Once again, the same harmonic sideband components as for linear modulation are present, with the expected additional baseband components caused by the overmodulation process. It can also be observed how the third-harmonic baseband component cancels between the phase legs for the three-phase case.

It is interesting in passing to note that the WTHD0 for the three-phase inverter is less than that of the single-phase bridge for these overmodulation conditions. This is because the additional (odd) baseband harmonics are a major harmonic contributor to WTHD0 during overmodulation, and the third-harmonic component of this distortion cancels between the phase legs for the three-phase inverter, but does not cancel for a single-phase inverter.

Similar results can be obtained for sampled modulation, with even baseband and sideband harmonics remaining for symmetrical modulation, and a slight magnitude shift to the higher harmonics for asymmetrical modulation, as would be expected. However, little point is served here in pursing these variations further, and the interested reader is directed to Eqs. (8.14) through (8.19) as a starting point if such solutions are required. It should also be noted that it is usually easier to calculate the harmonic components of each phase leg of an inverter separately and numerically subtract them (as complex phasors) to calculate the *l–l* harmonics, rather than developing simplified analytical expressions along the lines of Eq. (8.22).

Finally, it is obvious that overmodulation of a three-phase inverter is equally possible using a phase reference waveform that includes a third-harmonic component, except of course the nonlinear distortion will not begin until M exceeds $2/\sqrt{3} = 1.15$ rather than $M > 1.0$. As could be expected, the harmonic responses for overmodulation with a third-harmonic injected reference under the various sampling alternatives are very similar to those shown above, except with the anticipated flattening of the sideband harmonic magnitudes as occurred under linear modulation conditions. However, there is a difference in the roll-off rate of modulation gain (the ratio between the target reference fundamental magnitude and the actual inverter output voltage) for the alternatives of simple sinusoidal reference waveforms and more complex reference waveforms, and this issue will now be addressed in Section 8.5.

8.5 PWM Controller Gain during Overmodulation

8.5.1 Gain with Sinusoidal Reference

Examination of Eq. (8.11) shows that the amplitude of the fundamental component of the overmodulated waveform for one inverter phase leg is

$$V_{az(1)} = \frac{V_{dc}}{\pi}[4\sin\psi + M(\pi - 2\psi - \sin 2\psi)] \tag{8.29}$$

or, explicitly in terms of M,

$$V_{az(1)} = \frac{V_{dc}}{\pi}\left\{4\sin\left[\cos^{-1}\left(\frac{1}{M}\right)\right] + M\left\{\pi - 2\cos^{-1}\left(\frac{1}{M}\right) - \sin\left[2\cos^{-1}\left(\frac{1}{M}\right)\right]\right\}\right\} \tag{8.30}$$

During overmodulation conditions, the effective modulation index M' can be defined as

$$M' = \frac{V_{az(1)}}{V_{dc}} \tag{8.31}$$

and the gain achieved by a naturally sampled PWM controller under overmodulation conditions can then be written as

$$G(M) = \frac{M'}{M} = \frac{1}{\pi}\left\{\frac{4}{M}\sin\left[\cos^{-1}\left(\frac{1}{M}\right)\right] + \pi - 2\cos^{-1}\left(\frac{1}{M}\right) - \sin\left[2\cos^{-1}\left(\frac{1}{M}\right)\right]\right\} \tag{8.32}$$

This expression for the change in the modulation gain can be obtained in alternative fashion if the frequency of the triangle wave is assumed to be much higher than the commanded frequency, so that only the average output voltage across each carrier period need be considered. Then, if ψ denotes the fundamental angle below which the commanded voltage $v_{az}^* = MV_{dc}\cos\theta_o$ exceeds the triangle carrier magnitude (so that v_{az} clamps to V_{dc}), the fundamental component of the output can be expressed as the first term of a Fourier series, using Eq. (A1.2) and quarter cycle symmetry, as

$$V_{az(1)} = \frac{4}{\pi}\left(\int_0^{\psi} V_{dc}\cos\theta_o \, d\theta_o + \int_{\psi}^{\pi/2} MV_{dc}\cos^2\theta_o \, d\theta_o\right) \tag{8.33}$$

Equation (8.33) can be integrated to give

$$V_{az(1)} = \frac{4V_{dc}}{\pi}\left(\sin\psi + \frac{M}{2}\left[\frac{\pi}{2} - \psi - \frac{\sin 2\psi}{2}\right]\right)$$

$$= \frac{V_{dc}}{\pi}(4\sin\psi + M[\pi - 2\psi - \sin 2\psi]) \tag{8.34}$$

which is the same as Eq. (8.29). Also, since ψ has been defined as the angle where

$$V_{dc} = MV_{dc}\cos\psi \tag{8.35}$$

this means that $\psi = \cos^{-1}\left(\frac{1}{M}\right)$ as before.

Substituting Eq. (8.35) into Eq. (8.34) gives

$$V_{az(1)} = \frac{V_{dc}}{\pi}(4M\sin\psi\cos\psi + M[\pi - 2\psi - \sin 2\psi])$$

$$= \frac{MV_{dc}}{\pi}(\pi - 2\psi + \sin 2\psi) \tag{8.36}$$

Hence the effective modulation index M' can be written as

$$M' = \frac{V_{az(1)}}{V_{dc}} = \frac{M}{\pi}(\pi - 2\psi + \sin 2\psi) \tag{8.37}$$

and the modulation gain G can be written as

$$G(M) = \frac{M'}{M} = \frac{1}{\pi}(\pi - 2\psi + \sin 2\psi) \tag{8.38}$$

Since the modulation gain is still described in terms of an intermediate, saturation-dependent, variable ψ, it would be useful to eliminate this quantity. This can be done by recalling that

$$\sin 2\psi = 2\sin\psi\cos\psi \tag{8.39}$$

and since $\cos\psi = \frac{1}{M}$, this means that $\sin\psi = \sqrt{1 - \left(\frac{1}{M}\right)^2}$. Hence $\sin 2\psi$ can be expressed as

$$\sin 2\psi = 2\frac{1}{M}\sqrt{1 - \left(\frac{1}{M}\right)^2} \tag{8.40}$$

Also, if $\psi = \cos^{-1}\left(\frac{1}{M}\right)$, this means that $\pi/2 - \psi = \sin^{-1}\left(\frac{1}{M}\right)$.

Substituting these results into Eqs. (8.37) and (8.38) gives the effective modulation index and gain of a saturating modulation controller as

$$M' = \frac{2}{\pi}\left\{M\sin^{-1}\left(\frac{1}{M}\right) + \sqrt{1-\left(\frac{1}{M}\right)^2}\right\} \tag{8.41}$$

$$G = \frac{2}{\pi}\left\{\sin^{-1}\left(\frac{1}{M}\right) + \frac{1}{M}\sqrt{1-\left(\frac{1}{M}\right)^2}\right\} \tag{8.42}$$

Figure 8.7 shows the variation of the effective modulation index and modulation gain against the modulation index M. Note that the value of M' corresponding to $M = 1.5$, which was used for the calculation of the spectra of Figures 8.4 to 8.6, is $M' = 1.171$. Note also that since M' approaches the limiting value of $4/\pi = 1.273$ in an asymtotic fashion as the modulation index M approaches infinity, it appears in the first instance that this point will never be reached with finite amplitude control signals. However, since the frequency of the triangle carrier wave is also finite, the intersections of the carrier and the fundamental sine wave reference will be eliminated discretely as the modulation index increases. Eventually the final PWM pulse will disappear when the maximum slope of the sine wave reference exceeds the slope of the triangular carrier wave.

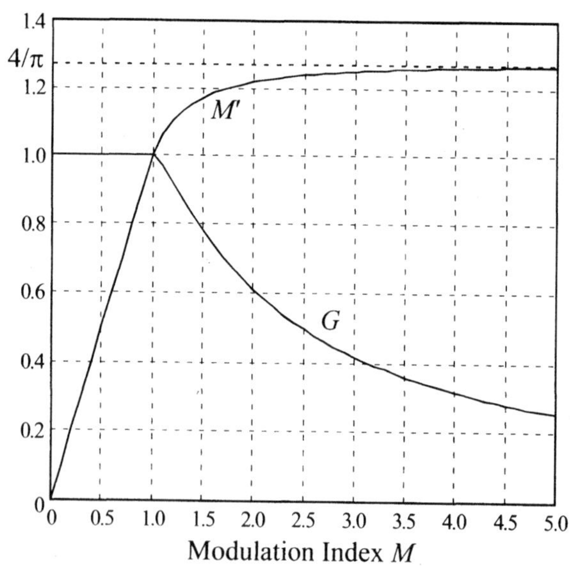

Figure 8.7 Effective modulation index M' and gain G as a function of the modulation index M, for naturally sampled sine–triangle modulation.

For example, for a "constant-volts-per-Hz" variable speed drive system, if the period of the nominal rated frequency is T_{rated}, then, for a carrier frequency of $f_c = 100 f_{rated}$, the triangle waveform changes from −1 p.u. to +1 p.u. in 1/200th of T_{rated}. The magnitude of the sinusoidal wave reference waveform at any output frequency f_o is defined by $M\cos(2\pi t/T_o) = M\cos(2\pi f_o t)$, which has a maximum slope of $2\pi M f_o$. For this type of system, the modulation index also varies proportionally to the fundamental frequency, i.e., $M = f_o/f_{rated}$. Hence the maximum slope of the reference becomes $2\pi f_o^2/f_{rated}$. Intersections between the carrier and the reference waveforms no longer occur when

$$\frac{2}{T_{rated}/200} = 400 f_{rated} < 2\pi \frac{f_o^2}{f_{rated}} \tag{8.43}$$

Hence

$$f_o = \sqrt{\frac{400}{2\pi}} f_{rated} = 7.98 f_{rated}$$

and the phase leg output switched waveform reverts to a square wave when the output frequency (and the modulation index) reach eight times the nominal rated value for the system.

The need to allow a control variable to reach, for example, eight times its nominal value is clearly a drawback in either an analog or digital implementation of the PWM controller. For example, in a fixed point digital controller, most of the computation will be done over only one-eighth of the range that must be reserved for the integer control variable, thereby increasing computational inaccuracies.

8.5.2 Gain with Space Vector Reference

A similar approach can be used to determine the effect of overmodulation on the gain of more complex PWM strategies such as those using space vector reference waveforms. Recall from Chapter 6 that space vector modulation is equivalent to naturally sampled PWM with the sinusoidal waveform replaced by the clamped space vector reference waveform shown in Figure 6.5. Overmodulation in this case begins to occur when M reaches 1.15 or $2/\sqrt{3}$, as shown in Figure 8.8(a). While $2/\sqrt{3} \le M \le 4/3$, the clamped overmodulated reference waveform exceeds the DC bus voltage limits twice in each of the positive and negative half cycles. When M becomes greater than $4/3$, overmodulation moves into a second region of operation where the clamped reference waveform exceeds the DC bus voltage limits only once in each half cycle, as shown in Figure 8.8(b).

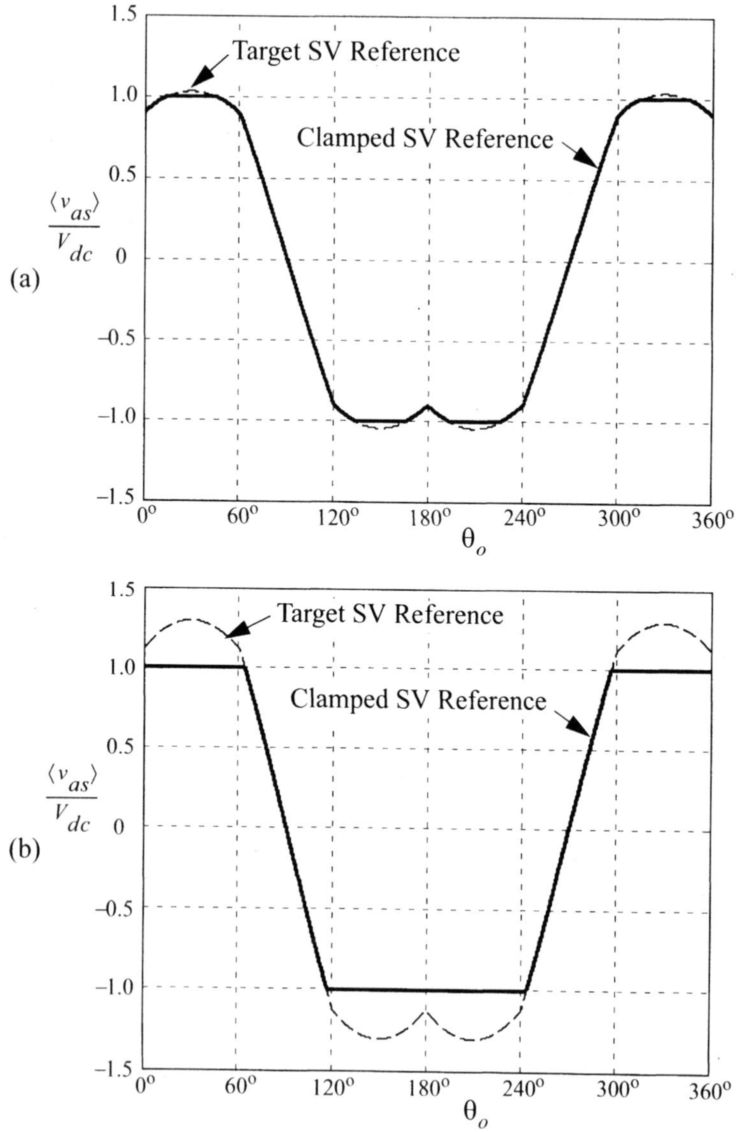

Figure 8.8 Target and clamped reference waveforms for naturally sampled space vector PWM in the overmodulation region: (a) Region 1, $M = 1.2$ and (b) Region 2, $M = 1.5$.

Figure 8.9 shows the trajectory of the output voltage vector $\boldsymbol{v}_{dqs}$ in the Re–Im or d–q plane during overmodulation, where it can be seen to be clamped to the space vector hexagon boundary as the vector rotates. When $M > 4/3$ the trajectory of $\boldsymbol{v}_{dqs}$ becomes entirely limited to motion on the hexagon.

Using the alternative approach presented in the last section, it can be shown that during Region 1 when $2/\sqrt{3} \le M \le 4/3$, the effective modulation index becomes [1]

$$M' = -\frac{1}{2}M + \frac{3}{\pi}M\sin^{-1}\left(\frac{2}{\sqrt{3}M}\right) + \frac{2\sqrt{3}}{\pi}\sqrt{1-\left(\frac{2}{\sqrt{3}M}\right)^2} \tag{8.44}$$

from which it can be determined that

$$\frac{2}{\sqrt{3}} \le M' < 1.218 \tag{8.45}$$

During Region 2 when $M \ge 4/3$, the effective modulation index is given by

$$M' = \left(\frac{3}{\pi}\right)M\sin^{-1}\left(\frac{2}{\sqrt{3}M}\right) + \frac{2}{\pi}\sqrt{1-\left(\frac{2}{3M}\right)^2} \tag{8.46}$$

A plot of M' versus M is shown in Figure 8.10.

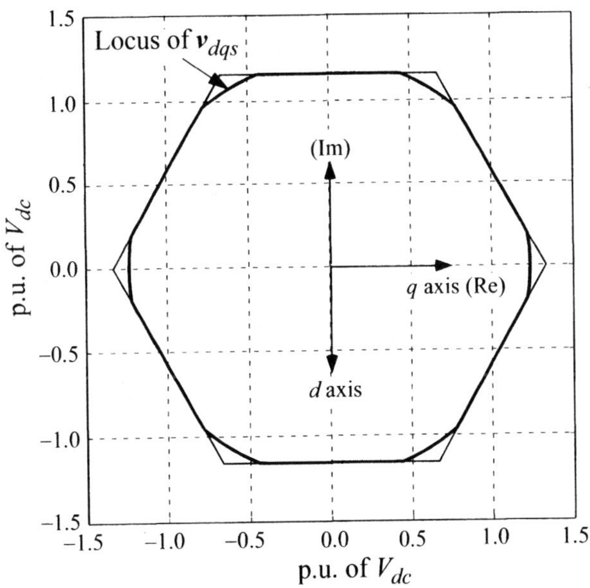

Figure 8.9 Waveforms for SVM operation in the overmodulation range, Region 1, $M = 1.22$.

(a)

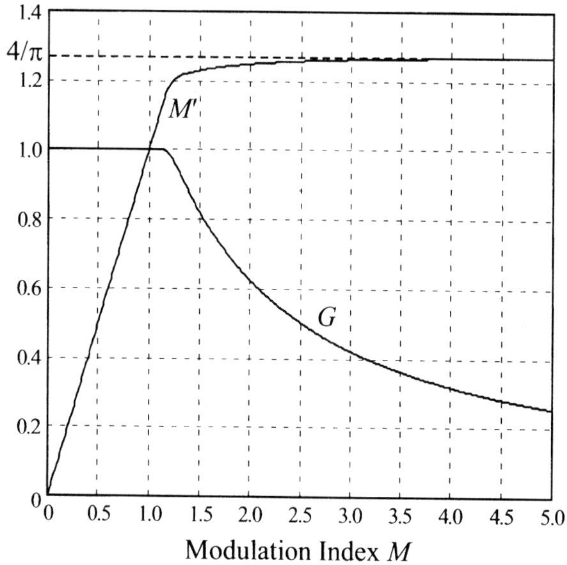

(b)

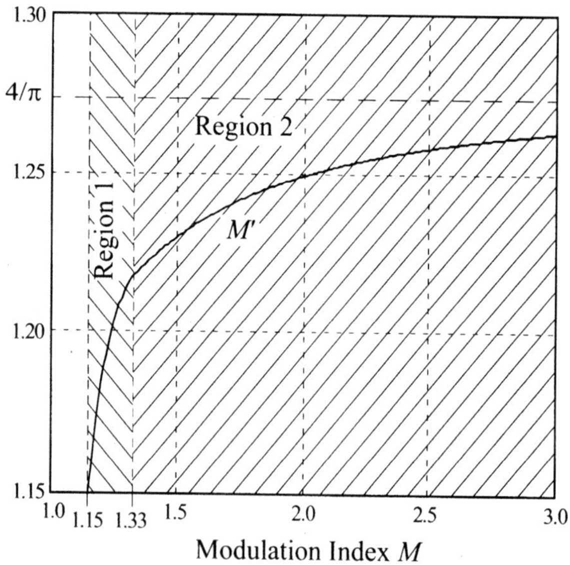

Figure 8.10 Effective modulation index M' and gain G as a function of the modulation index M, for naturally sampled SV modulation: (a) full trace and (b) expanded trace showing effective modulation index variation in Regions 1 and 2.

The corresponding gain for the two cases is readily obtained by taking

$$G = \frac{M'}{M} \tag{8.47}$$

and is plotted in Figure 8.10(a), for naturally sampled space vector PWM in the overmodulation region.

8.5.3 Gain with 60° Discontinuous Reference

Discontinuous PWM, the benefits of which were discussed in Chapter 6, can also be extended to the overmodulation region. Figure 8.11 shows the target and clamped reference waveforms for the case of 60° discontinuous switching, where it can be seen that only two intersections occur per half cycle for a modulation index beyond $M = 1$. Also, the output voltage reverts to a square wave when M exceeds $4/\sqrt{3}$ because of the straight sides of the reference waveform. Hence, for $1 \le M \le 4/\sqrt{3}$, the effective modulation index M' is [2]

$$M' = -\frac{4}{\pi} + \left(\frac{\sqrt{3}}{\pi} - \frac{1}{2}\right)M + \frac{4}{\sqrt{3}\pi M} + \frac{3}{\pi}M\sin^2\left(\frac{2}{\sqrt{3}M}\right) + \frac{2\sqrt{3}}{\pi}\sqrt{1 - \left(\frac{2}{\sqrt{3}M}\right)^2} \tag{8.48}$$

while $M' = 4/\pi$ for $M > 4/\sqrt{3}$.

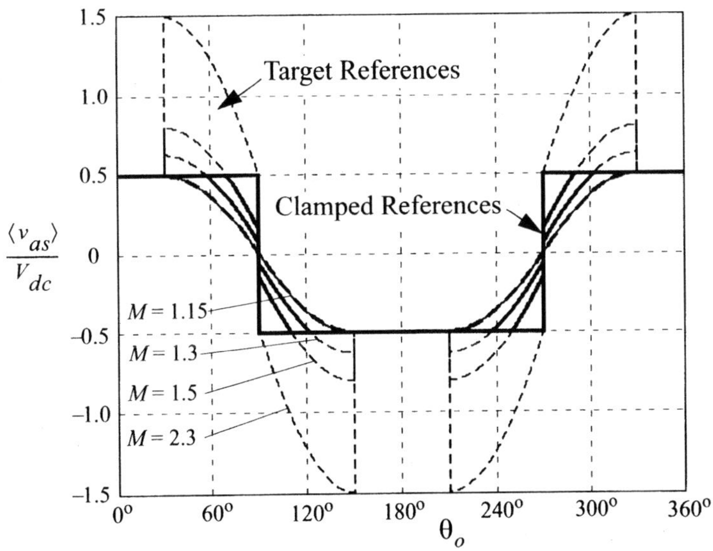

Figure 8.11 Target and clamped reference waveforms for naturally sampled 60° discontinuous PWM in the overmodulation region.

Figure 8.12 shows the effective modulation index and gain for all three modulation strategies in the overmodulation region.

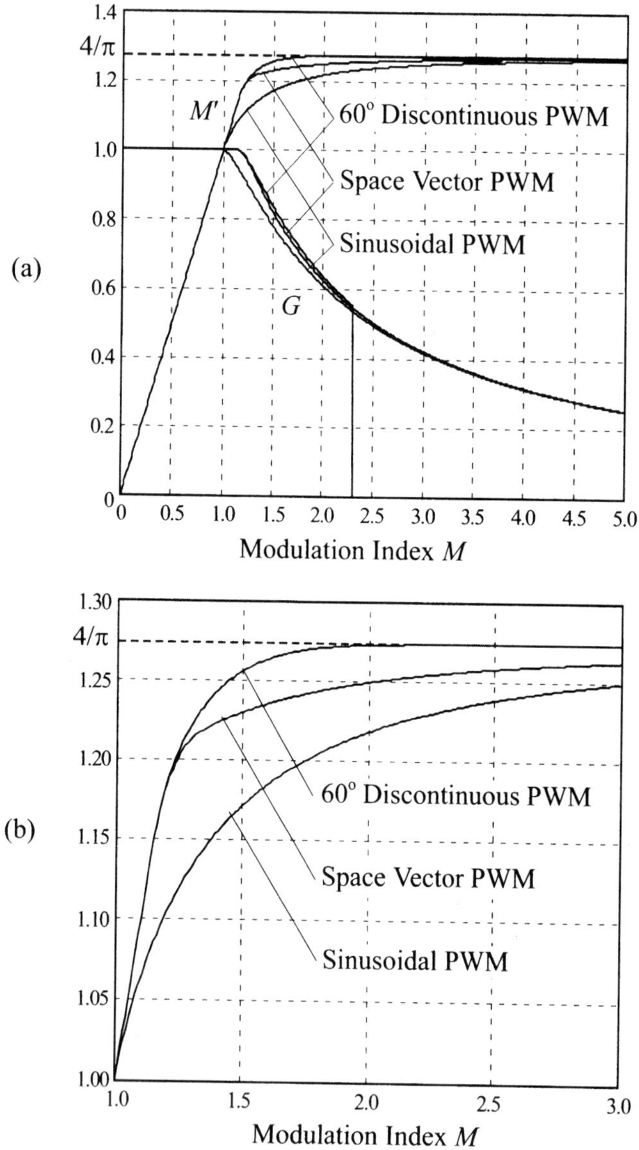

Figure 8.12 Effective modulation index M' and gain G as a function of the modulation index M, for three alternative PWM strategies: (a) full trace and (b) expanded trace.

8.5.4 Compensated Modulation

In many applications the loss of gain in the overmodulation region can cause deterioration of the overall performance of a system, since inverter characteristics often play a significant role in establishing the performance of an AC drive. For example, the constant-volts-per-hertz control strategy requires a PWM algorithm that delivers a fundamental voltage to the motor whose magnitude varies linearly with supply frequency, to ensure stable operation with a desired power factor and efficiency. An obvious approach is to simply vary the modulation index with frequency, i.e., $M = k\omega_o$. However, this strategy breaks down when the inverter controller enters the overmodulation condition, and the modulator gain is reduced.

The problem can be overcome by compensating for the loss of gain by inserting an inverse gain block into the modulation controller, to linearize its overall transfer function [3]. This inverse gain is plotted in Figure 8.13 and is essentially the gain plot shown in Figure 8.12 with its axes reversed. Note that, in principle, the inverse gains for naturally sampled PWM and space vector PWM must approach infinity as the commanded modulation index M^* approaches $4/\pi$. However, as discussed in Section 8.5.1, a gain of 10 to 50, depending on the application, is usually sufficient to ensure that all switching

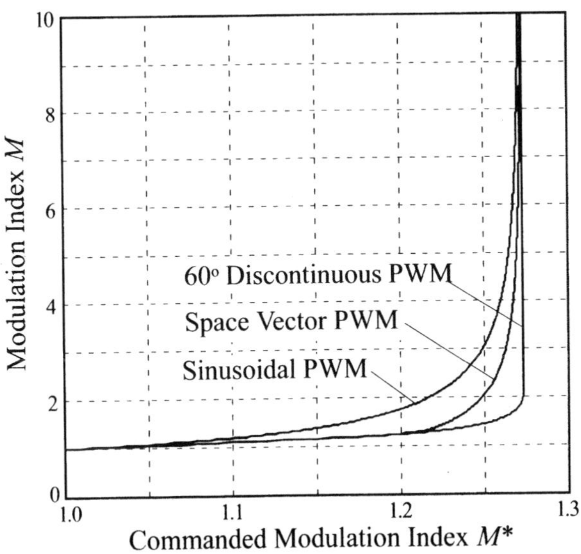

Figure 8.13 Inverse gain function required to compensate a commanded modulation index M^* for overmodulation gain loss.

pulses are dropped. Also, because of the straight sides achieved by the algorithm, 60° discontinuous modulation requires a gain compensation of only slightly more than two. This is an important advantage over the other two PWM methods since large excursions in gain limit numerical accuracy and require fast slewing capability, which can be difficult to accomplish in fixed point digital systems.

Figure 8.14 shows a simple PWM control system which has been modified to include inverse gain compensation. It can be seen how the commanded modulation index, M^*, is first adjusted to decouple the modulation process from DC bus voltage variations, by scaling the modulation index as the bus voltage varies from its nominal setpoint V_{dc0} [4]. Then, the inverse gain block adjusts the decoupled modulation index to allow for overmodulation, and the resultant corrected modulation index M is fed to the sine–triangle modulation process. Finally, the switched phase leg output is low pass filtered to produce a low-frequency average output voltage.

Figure 8.15 shows the low-frequency average phase voltages produced by a PWM controller using gain compensation, for the three modulation strategies considered. As expected, both space vector and 60° discontinuous modulation remain linear up to a commanded modulation index of $2/\sqrt{3} = 1.15$. Beyond this limit, all PWM strategies move eventually to six-step operation, but the resultant voltage waveshape for each strategy differs during this progression, as shown. Also, the actual modulation indices required to achieve the commanded modulation indices for each strategy can be read from Figure 8.13.

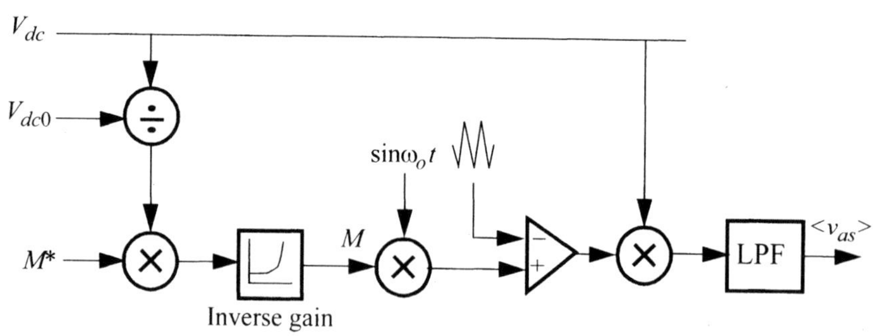

Figure 8.14 Block diagram of compensated PWM controller including DC bus voltage decoupling and overmodulation compensation, V_{dc0} = rated DC link voltage, $\langle v_{as}\rangle$ = low frequency average phase output voltage.

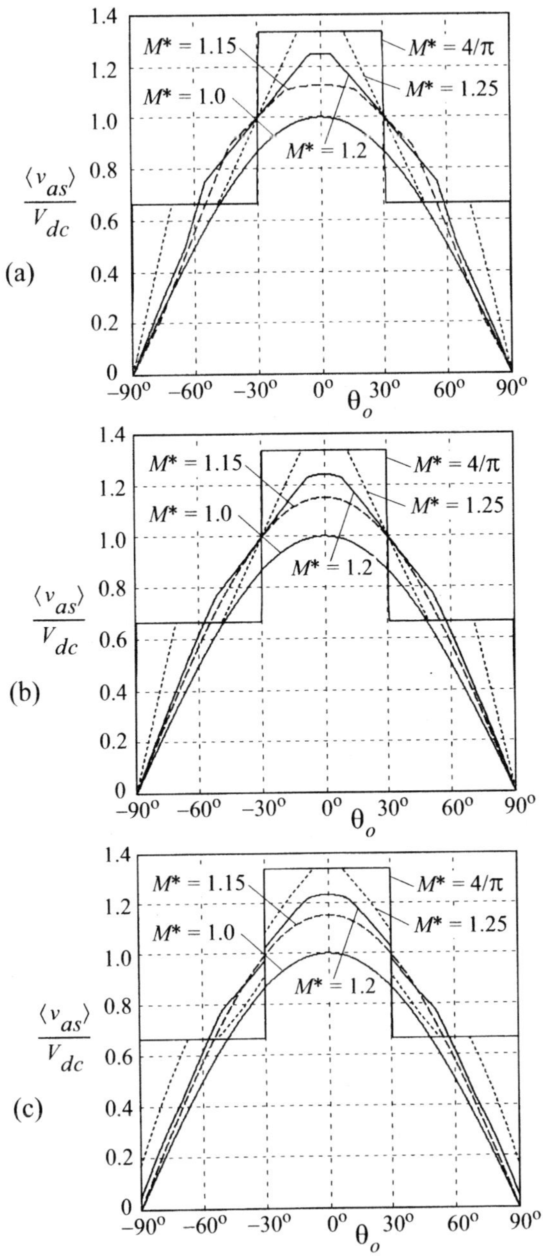

Figure 8.15 Resultant line to load-neutral reference voltages after modulation gain compensation, for naturally sampled (a) sinusoidal, (b) space vector, and (c) 60° discontinuous PWM. Commanded modulation indices M^* are 1, 1.5, 1.2, 1.25, and 1.273.

8.6 Space Vector Approach to Overmodulation

The space vector method of PWM operates by representing the three target phase output voltages as a target reference vector $\overline{V}_o^*$ located within a hexagon drawn between six stationary space vectors in a complex vector space (the stationary space vectors correspond to the six active states of a three-phase inverter). Then the sextant of the hexagon containing the reference vector is determined, and finally a PWM pattern is formed by appropriately dwelling at the two points of the hexagon nearest the reference voltage vector (plus dwelling for any remaining time at the origin) over a clock cycle $\Delta T/2$. This produces the same average voltages for all three phases as the original targets.

The dwell times for the first sextant are, from Table 6.1,

$$T_{SV_1} = \frac{V_o^*}{V_{dc}}\frac{\sqrt{3}}{2}\cos\left(\theta_o + \frac{\pi}{6}\right)\frac{\Delta T}{2} \tag{8.49}$$

$$T_{SV_2} = \frac{V_o^*}{V_{dc}}\frac{\sqrt{3}}{2}\cos\left(\theta_o - \frac{\pi}{2}\right)\frac{\Delta T}{2} \tag{8.50}$$

$$T_{SV_{0/7}} = \Delta T - T_{SV_1} - T_{SV_2} \tag{8.51}$$

where the states corresponding to 1, 2, and 0/7 are shown in Figure 6.1. The trajectory of the reference voltage vector is circular in the steady state and is located within the space vector hexagon until the modulation index reaches the value of $2/\sqrt{3}$. At this point the on-duration of the zero vector, Eq. (8.51), becomes zero and the reference voltage vector trajectory touches the hexagon at the angles $\theta_o = (2k+1)(\pi/6)$ where $k = 0,...,5$. If the modulation index is further increased, the on-duration time of the zero vector becomes negative and hence meaningless. Therefore, the modulation index cannot be increased beyond $2/\sqrt{3} = 1.15$ while maintaining a sinusoidal reference voltage.

Once the limit of linear sinusoidal space vector modulation has been reached, there is still the possibility of programming a higher modulation index by modifying the angle and magnitude of the reference vector using a preprocessor [1]. This preprocessor uses a set of nonlinear functions to modify the original circular track of the reference vector so as to follow the boundary of the space vector hexagon whenever the reference vector points outside the hexagon. Essentially, a distorted modulation index is defined, $M'(\theta_o)$, which functionally depends upon the spatial command angle θ_o.

Figure 8.16 shows this effect for one 60° sector of the hexagon. The dashed circle indicates the trajectory of the desired target reference voltage vector $\overline{V}_o^*$. Whenever this target vector is within the space vector hexagon, i.e., near the outer corners of a sector on the circle segments *a–b* and *c–d*, the inverter controller selects the switching space vectors using Eqs. (8.49) to (8.51). However, when the target reference trajectory passes outside the hexagon, the space vector switching times must be modified to make the actual average space vector trajectory follow the locus of the hexagon itself, i.e., along the straight line segment *b–c*. The solid line $\overline{V}_o'$ is the trajectory of the modified reference vector generated by the preprocessor to take account of this requirement.

The switching times during the *b–c* traverse of the modified reference vector are proportional to the space vector lengths shown in Figure 8.16. If the angles defined by the vertices *aOb* and *cOd* equal α, then the real and imaginary parts of $\overline{V}_o'$ are given by

$$V_o' \cos\theta_o = V_1 + V_2 \cos 60° \qquad (8.52)$$

$$V_o' \sin\theta_o = V_2 \sin 60° \qquad (8.53)$$

for $\alpha < \theta_o < \frac{\pi}{3} - \alpha$.

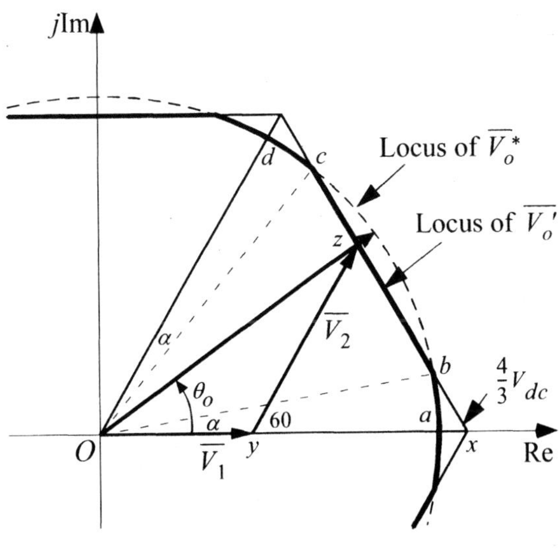

Figure 8.16 Overmodulation of space vector method when the modulation index varies over the range $2/\sqrt{3} < M < 4/3$.

Hence

$$V_1 + \frac{1}{2}V_2 = V_o' \cos\theta_o \tag{8.54}$$

and

$$V_2 = \frac{2}{\sqrt{3}} V_o' \sin\theta_o \tag{8.55}$$

Substituting Eq. (8.55) into Eq. (8.54) gives

$$V_1 = V_o' \left(\cos\theta_o - \frac{1}{\sqrt{3}}\sin\theta_o\right) \tag{8.56}$$

Now, *xyz* forms an equilateral triangle, and so all sides have the same length of V_2. Hence it follows that

$$V_1 + V_2 = \frac{4}{3}V_{dc} \tag{8.57}$$

Combining Eqs. (8.55), (8.56) and (8.57) gives

$$V_o' = \frac{4}{3}\frac{V_{dc}}{\cos\theta_o + \frac{1}{\sqrt{3}}\sin\theta_o} \tag{8.58}$$

Finally, Eqs. (8.55) and (8.56) can be expressed as

$$\frac{V_1}{V_{dc}} = \frac{4}{3}\frac{(\sqrt{3}\cos\theta_o - \sin\theta_o)}{(\sqrt{3}\cos\theta_o + \sin\theta_o)} \tag{8.59}$$

and

$$\frac{V_2}{V_{dc}} = \frac{4}{3}\frac{2\sin\theta_o}{\sqrt{3}\cos\theta_o + \sin\theta_o} \tag{8.60}$$

for $\alpha < \theta_o < \frac{\pi}{3} - \alpha$.

While these equations are valid for only the first sextant (the sextant shown in Figure 8.16), it is clear that similar equations can be easily developed for the remaining five sextants.

Since the per unit space vector switching intervals are equal to the per unit voltages, they can be written as

$$\frac{T_{SV_1}}{\Delta T/2} = \frac{V_1}{4/3V_{dc}} \quad \text{and} \quad \frac{T_{SV_2}}{\Delta T/2} = \frac{V_2}{4/3V_{dc}} \tag{8.61}$$

which can be rearranged to give

$$T_{SV_1} = \left(\frac{\sqrt{3}\cos\theta_o - \sin\theta_o}{\sqrt{3}\cos\theta_o + \sin\theta_o}\right)\frac{\Delta T}{2} \tag{8.62}$$

$$T_{SV_2} = \left(\frac{2\sin\theta_o}{\sqrt{3}\cos\theta_o + \sin\theta_o}\right)\frac{\Delta T}{2} \tag{8.63}$$

$$T_{SV_{0/7}} = 0 \tag{8.64}$$

for $\alpha < \theta_o < \frac{\pi}{3} - \alpha$. Note that T_{SV_2} can be more readily computed in a practical system using

$$T_{SV_2} = \Delta T/2 - T_{SV_1} \tag{8.65}$$

For small excursions of the target vector beyond the hexagon, a further refinement is to recover the volt–seconds lost as the modified reference vector tracks segment *b–c,* by tracking above segments *a–b* and *c–d* to compensate. This allows the correct fundamental component to still be maintained for limited overmodulation excursions of the target reference vector [1].

Figure 8.17(a) shows the low-frequency (average) phase voltage waveform for space vector modulation up to a modulation index of $M = 4/3$. When M reaches this limit, the distorted modulation index $M'(\theta_o)$ becomes completely constrained by the space vector hexagon, so that $\alpha = 0°$, and hence $T_{SV_1} = \Delta T/2$ and $T_{SV_2} = 0$ at $\theta_o = 0°$, while $T_{SV_1} = 0$ and $T_{SV_2} = \Delta T/2$ at $\theta_o = 60°$. The amplitude of the fundamental component of the phase voltage at this point is $1.212 V_{dc}$.

As the modulation index increases beyond this point, a smooth transition to square-wave operation is accomplished by changing gradually from tracing the continuous hexagon, to switching in a discrete six-step sequence. This is achieved by dwelling progressively longer at the six discrete hexagon vertices as M increases beyond 4/3, as shown in Figure 8.17(b). Commonly, the progression to *six-step* is arbitrarily constrained to take place over the range $1.33 < M < 1.45$. The dwell interval, α_p, is the period of fundamental angular time where the inverter stays switched to each of the two hexagon vertices of each $60°$ sextant interval, and increases linearly from 0 to $30°$ as M increases from 1.33 to 1.45.

The magnitude variation of the actual fundamental component of the phase voltage (the distorted or *effective* modulation index M') as M progresses from 1.15 to 1.45 is shown in Figure 8.18. It should be noted that the variation is roughly linear. However, if necessary, the nonlinearity can be corrected by an inverse function as discussed in Section 8.5.4.

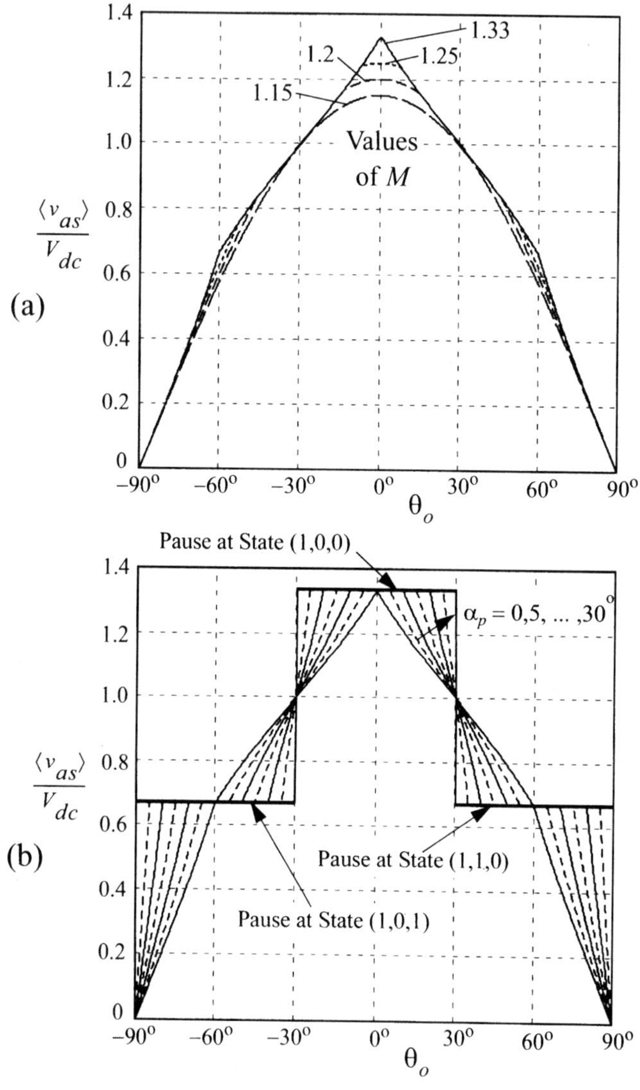

Figure 8.17 Equivalent line to load-neutral reference voltage for naturally sampled space vector modulation: (a) M increasing from 1.15 to 1.33 and (b) α_p increasing from 0 to 30° as M is increased from 1.33 to 1.45. Dwell angle α_p corresponds to fraction of each 60° sextant interval locked to one space vector state.

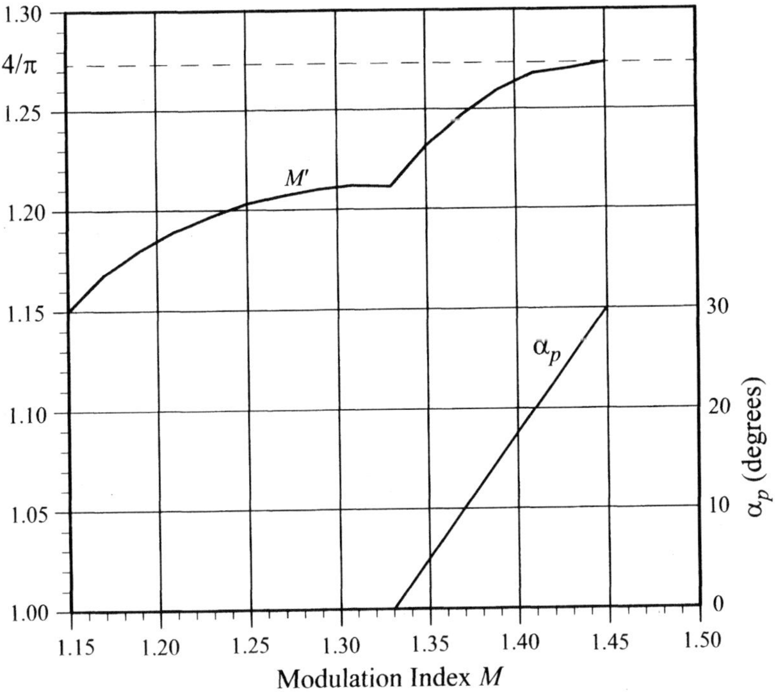

Figure 8.18 Effective modulation index M' as a function of modulation index M within the overmodulation region using modulation scheme of Figure 8.17.

Table 8.1 summarizes the strategies to be used for space vector modulation in the different regions of overmodulation.

Table 8.1 Space Vector Modulation Strategies for Various Levels of Overmodulation

$0 < M < 1.15$	$M'(\theta_o) = M = \dfrac{\lvert \overline{V}_o^* \rvert}{V_{dc}}$	Linear modulation
$1.15 < M < 1.33$	$M'(\theta_o) = \dfrac{\lvert \overline{V}_o'^* \rvert}{V_{dc}} = f(M, \theta_o)$	Trace hexagon boundary using Eqs. (8.59) and (8.60)
$1.33 < M < 1.45$	$M'(\theta_o) = f(\alpha_p)$	Transition from hexagon trace to six-step modulation
$1.45 < M$	$M'(\theta_o) = 4/\pi$	Six-step modulation

8.7 Summary

This chapter has presented a detailed evaluation of the overmodulation operation of PWM inverter systems. The overmodulation region of PWM provides a useful extension of the operating range of an inverter without requiring an increased DC link voltage. However, the modulation gain of all modulation systems becomes nonlinear in this region, and the switched output voltage includes low-order baseband harmonic components, as well as the expected sideband harmonics caused by the switching process.

The general form of gain roll-off is the same for all modulation strategies, but the rate varies for different approaches. Simple PWM with a sinusoidal reference has the most rapid gain roll-off and also becomes nonlinear above a modulation index of $M = 1$. Third-harmonic, space vector, and the discontinuous reference modulation strategies are linear until M exceeds 1.15, with the $60°$ discontinuous PWM strategy then having the least gain roll-off until the limiting output voltage of $4V_{dc}/\pi$ is reached. This makes this strategy attractive in fixed-point digital implementations where the numerical gain range is an issue. Direct space vector modulation offers an alternative approach that achieves a similar roughly linear gain response in the overmodulation region, by distorting the space vector modulation command vector.

References

[1] J. Holtz, A.M. Khambadkone, and W. Lotzkat, "On continuous control of PWM inverters in the overmodulation range including the six-step mode," *IEEE Trans. on Power Electronics*, vol. 8, no. 4, Oct. 1993, pp. 546–553.

[2] A.M. Hava, "Carrier based PWM voltage source inverter in the overmodulation range," Ph.D. Thesis, University of Wisconsin, 1998.

[3] .R.J. Kerkman, D. Leggate, B.J. Seibel, and T.M. Rowan, "Operation of PWM voltage source inverters in the overmodulation region," *IEEE Trans on Industrial Electronics,* vol. 41, no. 1, Feb. 1996, pp. 132–141.

[4] A.B. Plunkett and T.A. Lipo, "Source impedance effects in the control of inverter-induction motor drives," *World Electrotechnical Congress*, Moscow, USSR, June 21–25, 1977.

9

Programmed Modulation Strategies

It has already been learned from Chapter 3 that modulation techniques fall into several classes. These alternatives are:

- Modulation using naturally sampled sine wave–modulating wave intersections
- Regular sampling defined by sine wave–modulating wave intersections
- Direct modulation

It has also been shown that modulation involves the exact positioning of the pulses within a clock cycle so as to achieve a minimal distortion waveform. However, up to this point, all three modulation methods have been "clocked". That is, the basic carrier interval (mark-space interval) has been held fixed. In this chapter this constraint will be removed so that the individual pulses will no longer be constrained within a clock pulse window but can vary in width and position over a much wider interval (typically set by cycle symmetry requirements). The overall approach to define the switching times is then based on the minimization of a suitable objective function which typically represents system losses.

It is a common feature to all these methods that the optimization computation process is done off-line on a personal or mainframe computer. The result of the computation is a set of switching angles which are functions of the modulation index *M*. The switching angles are stored in the memory of a PWM controller or an EPROM. The stored angles are accessed in real time to determine the optimized switching angles. Since these angles always are synchronized to the fundamental component, the harmonic spectrum is free from subharmonic components. The switching of pulses always operates in synchronism, so that the number of pulses per cycle must be changed in discrete fashion as the frequency decreases in order to maintain a good quality waveform. Because the computational effort needed to compute the switching angles increases greatly with the number of switching angles to be calculated, these methods are generally combined with natural or regular sampling methods to complement these optimal methods over the lower end of the fundamental frequency range. While benefits exist, the overhead in time, effort, and computing resources frequently prevent their use in many lower cost applications. However, optimized methods can be combined with regular sampling to produce a low-cost approximation to the methods described in this chapter [1].

9.1 Optimized Space Vector Modulation

Space vector modulation, described in Section 6.1, is based on converting a reference vector $\overline{V_o^*}(t)$ into a sequence of discrete sampled values $\overline{V_o^*}(t_s)$ with sampling instants t_s regularly spaced at $\Delta T/2$ intervals. Each sampled reference value is then approximated by selecting a sequence of three switching state vectors in closest proximity to $\overline{V_o^*}(t_s)$ to achieve the same time-averaged value over the interval $\Delta T/2$. For example, when the vector $\overline{V_o^*}(t_s)$ is located between switching state vectors $\overline{SV_1}$, $\overline{SV_2}$, and $\overline{SV_0}$(or $\overline{SV_7}$) the time-averaged value is

$$\overline{V_o^*}(t_s) = \frac{T_{SV_1}}{\Delta T/2}\overline{SV_1} + \frac{T_{SV_2}}{\Delta T/2}\overline{SV_2} \tag{9.1}$$

while the remaining time in the interval

$$T_{SV_Z} = \Delta T/2 - T_{SV_1} - T_{SV_2} \tag{9.2}$$

is spent on a combination of the null voltage vectors $\overline{SV_0}$ and $\overline{SV_7}$. Within the interval ΔT the minimum number of switching events of the inverter is achieved by the sequence

$$\ldots \Rightarrow \overline{SV_0} \Rightarrow \overline{SV_1} \Rightarrow \overline{SV_2} \Rightarrow \overline{SV_7} \Rightarrow \overline{SV_2} \Rightarrow \overline{SV_1} \Rightarrow \overline{SV_0} \Rightarrow \ldots$$

When normal space vector modulation is used, the duration of all of the half-intervals $\Delta T/2$ is assumed as constant. Assuming that the switching frequency is synchronized to the fundamental cycle, N such intervals can be formed over one period. That is,

$$\frac{\Delta T}{2} = \frac{1}{Nf_1} \tag{9.3}$$

where $N = 2p$ and p is the *pulse number* (the number of on–off transitions per cycle of each inverter switch). The pulse number is, in turn, related to the switching frequency f_s by

$$f_s = pf_o \tag{9.4}$$

The current flowing over any interval k, $1<k<N$, is, approximately,

$$L_\sigma \frac{d\bar{i}_k}{dt} = \bar{V}_{o,k} - \bar{e}_k$$

where L_σ and $\bar{e}_k$ represent the motor load leakage inductance [transient inductance defined by Eq. (2.23)] and the load induced EMF, respectively. This expression can be arranged into the form

$$L_\sigma \frac{d\bar{i}_{1,k}}{dt} + L_\sigma \frac{d\bar{i}_{h,k}}{dt} = (\bar{V}^*_{o,k} - \bar{e}_k) + (\bar{V}_{o,k} - \bar{V}^*_{o,k}) \tag{9.5}$$

where $\bar{i}_k = \bar{i}_{1,k} + \bar{i}_{h,k}$ and the subscript h indicates the harmonic component.

The first term on both sides of Eq. (9.5) represents the voltage producing the desired fundamental component of output current while the second term represents the undesired harmonic component

$$L_\sigma \frac{d\bar{i}_{h,k}}{dt} = \bar{V}_{o,k} - \bar{V}^*_{o,k} \tag{9.6}$$

where

$$\bar{V}^*_{o,k} = MV_{dc}e^{j\theta_k} \quad \text{and} \quad \theta_k = \omega_o t_k \tag{9.7}$$

In the first sextant the inverter voltage over every first $\Delta T/2$ half-interval is made up of the three subintervals (see Section 6.1) of

$$\begin{aligned}
&1.\ t_k \le t \le t_k + T_{SV_1} && \bar{V}_{o,k_1} = \overline{SV}_1 = V_{dc}e^{j0} \\
&2.\ t_k + T_{SV_1} \le t \le t_k + T_{SV_1} + T_{SV_2} && \bar{V}_{o,k_2} = \overline{SV}_2 = V_{dc}e^{j\frac{\pi}{3}} \\
&3.\ t_k + T_{SV_1} + T_{SV_2} \le t \le t_k + \Delta T/2 && \bar{V}_{o,k_3} = \overline{SV}_7 = 0
\end{aligned} \tag{9.8}$$

The times T_{SV_1} and T_{SV_2} are determined from

$$T_{SV_1}\overline{SV}_1 + T_{SV_2}\overline{SV}_2 = \frac{\Delta T}{2}\,\bar{V}^*_{o,k}(t_k) \tag{9.9}$$

resulting in Eqs. (6.5) and (6.6), that is,

$$T_{SV_1} = \frac{V_o \sin\left(\frac{\pi}{3} - \theta_o\right)}{V_m \sin\frac{\pi}{3}} \frac{\Delta T}{2} \tag{9.10}$$

$$T_{SV_2} = \frac{V_o \sin(\theta_o)}{V_m \sin\frac{\pi}{3}} \frac{\Delta T}{2} \tag{9.11}$$

The alternate time intervals are made up of the sequence $\overline{SV_2} \Rightarrow \overline{SV_1} \Rightarrow \overline{SV_0}$ and can be written out explicitly in similar fashion.

The harmonic component of each interval has a period equal to $\Delta T/2$. If it is assumed that current which flows within each interval is independent of the current flowing in the previous or subsequent intervals k–1 or k+1, the RMS current for the *kth* interval is simply

$$I_{h,k} = \sqrt{\frac{2}{\Delta T} \int_{t_k}^{t_k + \Delta T/2} \left|\bar{i}_{h,k}\right|^2 dt} \tag{9.12}$$

where

$$\bar{i}_{h,k} = \frac{\overline{V}_{o,k} - \overline{V}^*_{o,k}}{L_\sigma}(t - t_k) + I_{h,k}(t_k) \tag{9.13}$$

Over a full cycle comprising N intervals, the RMS current is therefore

$$I_h = \sqrt{\frac{2}{\Delta T} \sum_{k=1}^{N} \int_{t_k}^{t_k + \Delta T/2} \left|\bar{i}_{h,k}\right|^2 dt} \tag{9.14}$$

The duration of the half-interval $\Delta T/2$, thus far considered as constant, can now be treated as a variable [2]. Replacing the constant $\Delta T/2$ by the variable T_k, N independent variables are produced which can be used to define an optimization problem by assuming that the harmonic current waveform defined by Eq. (9.14) is piecewise linear. The optimization is subject to the constraint that the number of intervals per cycle remains constant, that is

$$\sum_{k=1}^{N} T_k = p\,\Delta T = N\frac{\Delta T}{2} = \frac{1}{f_o} \tag{9.15}$$

During the subcycle T_k one can write

$$I_{h,k} = \sqrt{\frac{1}{T_k} \int_{t_k}^{t_k + T_k} \left|\bar{i}_{h,k}\right|^2 dt} \tag{9.16}$$

Assuming operation in the first sextant, the three states chosen are $\overline{SV}_1$, $\overline{SV}_2$, and $\overline{SV}_0$. Since the forcing function for each interval is assumed constant, the current changes linearly during each of the three subintervals. Also, since the harmonic current is periodic with period T_k, it can be assumed that the initial current at the beginning of each subcycle t_k is zero, i.e.,

$$I_{h,k}(t_k) = 0 \tag{9.17}$$

The square of the RMS harmonic current for the first of the three subintervals is then given by

$$I_{h,k_1}^2 = \frac{1}{T_k}\int_0^{T_{k_1}} \left|\frac{\overline{V}_{o,k_1} - \overline{e}_k}{L_\sigma}\right|^2 t^2\, dt \tag{9.18}$$

$$= \frac{T_{k_1}^3}{3T_k}\left|\frac{\overline{V}_{o,k_1} - \overline{e}_k}{L_\sigma}\right|^2 \tag{9.19}$$

If the period of the interval T_k is changed to kT_k, the corresponding subintervals change by the same value, i.e., from T_{k_1} to kT_{k_1}, etc. The square of the RMS value of the current during this modified interval is therefore

$$I_{h,k_1(mod)}^2 = k^2\frac{T_{k_1}^3}{3T_k}\left|\frac{\overline{V}_{o,k_1} - \overline{e}_k}{L_\sigma}\right|^2 \tag{9.20}$$

It can be shown that the square of the RMS value for the current in the remaining two subintervals similarly changes by k^2. Hence, the RMS value of the current of the overall interval increases in proportion to the duration of the interval itself. One can therefore say, in essence, that

$$I_{h,k} = \frac{T_k}{T_0} I_{h,k_0} \tag{9.21}$$

where I_{h,k_0} is the normalized reference value of $I_{h,k}$ given by

$$I_{h,k_0} = \sqrt{\frac{1}{T_0}\int_{t_k}^{t_k+T_0} \left|\overline{i}_{h,k}\right|^2 dt} \tag{9.22}$$

where $T_0 = \Delta T/2$.

The optimum set of variables T_k can be found by minimizing the RMS ripple current over a complete period NT_0, expressed by

$$I_h = \sqrt{\frac{1}{NT_0}\int_0^{NT_0} i_h^2\, dt} \tag{9.23}$$

$$= \sqrt{\frac{1}{NT_0}\sum_{k=1}^{N}\int_{t_k}^{t_k+T_k}\left(\frac{T_k}{T_0}I_{h,k_0}\right)^2 dt} \tag{9.24}$$

$$= \sqrt{\frac{1}{NT_0}\sum_{k=1}^{N}\left(\frac{T_k}{T_0}I_{h,k_0}\right)^2 T_k} \tag{9.25}$$

$$= \sqrt{\frac{1}{N}\sum_{k=1}^{N}\frac{T_k^3}{T_0^3}\, I_{h,k_0}^2} \tag{9.26}$$

where N is the number of subcycles within a fundamental period, T_k is the optimal time duration of the kth subcycle, and I_{h,k_0} is the harmonic current in the subcycle k before optimization when all subcycles have the time duration T_0.

Squaring Eq. (9.26), and making use of the constraint Eq. (9.15) gives

$$I_h^2 = \frac{1}{NT_0^3}\left[\sum_{k=1}^{N-1} T_k^3 I_{h,k_0}^2 + \left(NT_0 - \sum_{k=1}^{N-1} T_k\right)^3 I_{h,k_0}^2\right] \tag{9.27}$$

To minimize I_h^2, the derivatives with respect to an arbitrary member j of the set of intervals N must be set to zero [2], hence

$$\frac{\partial I_h^2}{\partial T_j} = \frac{3}{NT_0^3}\left[T_j^2 I_{h,j_0}^2 - \left(NT_0 - \sum_{k=1}^{N-1} T_k\right)^2 I_{h,k_0}^2\right] = 0 \tag{9.28}$$

The difference between derivatives of the $j\,th$ member and any other kth member (other than the last or $N\,th$ member) is therefore

$$T_j^2 I_{h,j_0}^2 - T_k^2 I_{h,k_0}^2 = 0 \tag{9.29}$$

so that for any subinterval $1 \le j \le N-1$

$$I_{h,j_0} = \frac{T_k}{T_j} I_{h,k_0} \tag{9.30}$$

However the optimal current of interval j is related to the nonoptimal reference value by Eq. (9.21), viz.

$$I_{h,j} = \frac{T_j}{T_0} I_{h,j_0} \tag{9.31}$$

$$I_{h,k} = \frac{T_k}{T_0} I_{h,k_0} \tag{9.32}$$

From Eqs. (9.31) and (9.32) it is clear that Eq. (9.30) can only be valid if

$$I_{h,j} = I_{h,k} \tag{9.33}$$

Hence a minimum RMS value for the overall switched waveform is reached only when the RMS values of the harmonic components of each subcycle are identical. Defining this optimum value as $I_{h,\min}$, the optimal value of the time duration in the *kth* interval is

$$T_k = T_0 \frac{I_{h,\min}}{I_{h,k_0}} \tag{9.34}$$

where I_{h,k_0} is given by Eq. (9.22). However, since

$$\sum_{k=1}^{N} T_k = NT_0 \tag{9.35}$$

then, from Eq. (9.34),

$$NT_0 = T_0 I_{h,\min} \sum_{k=1}^{N} \frac{1}{I_{h,k_0}} \tag{9.36}$$

or

$$I_{h,\min} = \frac{N}{\sum_{k=1}^{N} \frac{1}{I_{h,k_0}}} \tag{9.37}$$

Hence, the optimum time duration of the *kth* interval can be expressed as

$$T_k = \frac{NT_0}{I_{h,k_0} \sum_{k=1}^{N} \frac{1}{I_{h,k_0}}} \tag{9.38}$$

It is interesting to note that all of the terms on the right-hand side of Eq. (9.38) are known or can be readily calculated by means of Eq. (9.22). Therefore the optimum times have now been determined in closed analytical or numerical form so that iteration procedures are not required.

Evaluation of Eq. (9.38) over a one-sixth of a fundamental period is shown in Figure 9.1 for the case where $N = 36$ ($f_s/f_1 = 18$) and modulation index $M = 0.9$. Note that the optimum switching intervals are relatively longer when the command voltage vector is located near zero or $\pi/3$ since the voltage error is a minimum at these points (i.e., where the command vector voltage is near the stationary voltage vectors in Figure 6.2). Conversely, the intervals are shorter in the intermediate region of $\pi/6$ where the error voltage is a maximum.

Since the exact location of the sampling instants is arbitrary, the pulse durations shown in Figure 9.1 are one of an infinite number of possibilities, each differing only by a phase shift corresponding to the instant where the initial sampling instant occurs. [In Figure 9.1 the initial sampling instant is taken when the phase of the vector $\overline{V}_o^*$ with respect to the Re axis (or phase a axis) is zero.] If the chain of synchronized cycles are displaced by small increments, the discrete values T_k can be represented by a continuous function $T_k[\arg(\overline{V}_o^*)]$ as illustrated in Figures 9.2 and 9.3. Optimum values are differ-

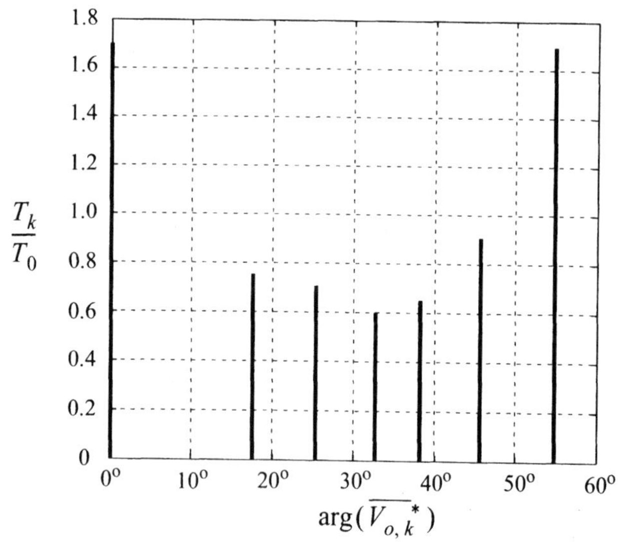

Figure 9.1 Durations of optimal switching intervals versus phase of the sampled space vector command voltage $\overline{V}_{o,k}^*$, $f_s/f_1 = 18$, $M = 0.9$.

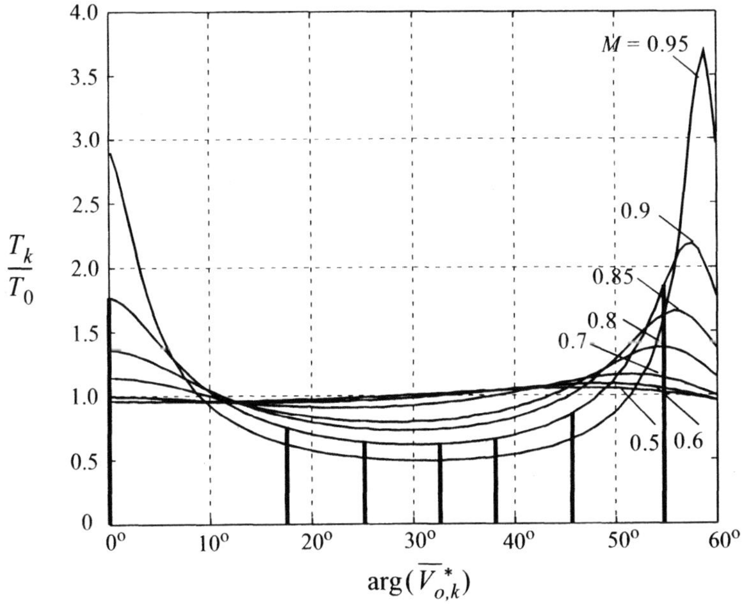

Figure 9.2 T_k/T_0 plotted as a continuous function of the phase angle of the space vector command voltage for the up-stroke interval ($SV_1 \Rightarrow SV_2 \Rightarrow SV_7$) during sextant #1; points computed in Figure 9.1 shown as vertical bars.

ent for the up-stroke and down-stroke conditions corresponding to switch state transitions $SV_1 \Rightarrow SV_2 \Rightarrow SV_7$ and $SV_2 \Rightarrow SV_1 \Rightarrow SV_0$, respectively. Note that the maximum values are reached (longest pulse width) when $\arg(\overline{V}^*_o)$ is near $\pi/3$ (and also π and $5\pi/3$) for the "up-stroke" condition where the inverter switches change from the negative to the positive bus, and when $\arg(\overline{V}^*_o)$ is near 0 (and $2\pi/3$, $4\pi/3$) for the "down-stroke" condition where the inverter switches change from the positive to the negative bus. Minimum values (shortest pulses) occur near $\pi/6$ radians and also $\pi/2$, $5\pi/6$, etc., since the function repeats every $\pi/3$ radians. It can also be seen that as the modulation index M becomes smaller, the excursions of T_k from T_0 become progressively smaller since the variations in the amplitude of the error voltage vector as a function of the vector position become smaller for decreasing M. Note that since Figure 9.3 can obviously be obtained from Figure 9.2 data by simply reading the digitally stored data in reverse order, only one set of curves needs to be stored in computer memory. A block diagram of the controller to implement this optimal form of SVM is shown in Figure 9.4.

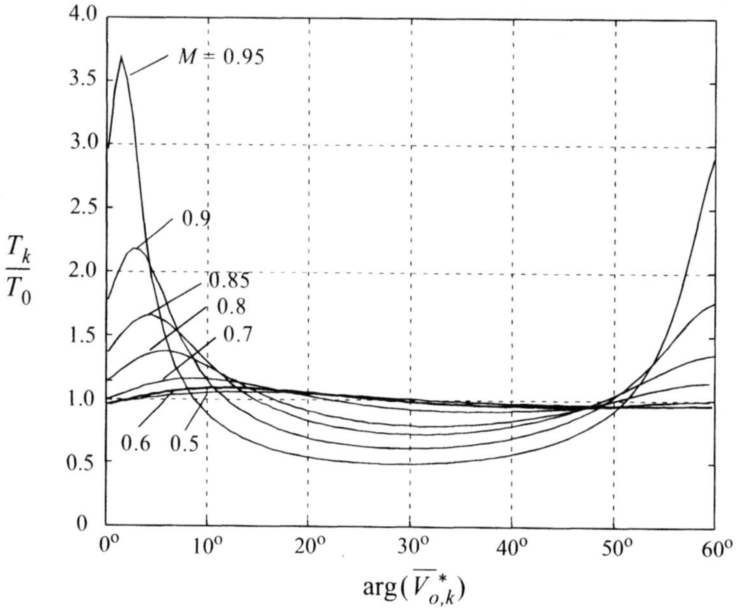

Figure 9.3 T_k/T_0 plotted as a continuous function of the phase angle of the space vector command voltage for the down-stroke interval ($SV_2 \Rightarrow SV_1 \Rightarrow SV_0$) within sextant #1.

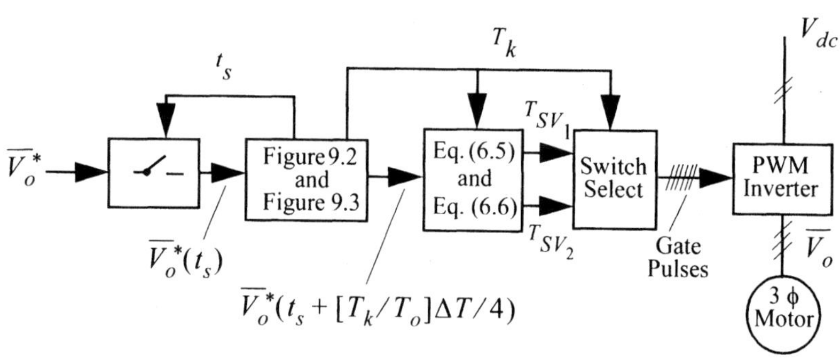

Figure 9.4 Block diagram of the optimized space vector modulator assuming operation in sextant #1 ($0 \le \arg(\overline{V}^*_{o,k}) \le \pi/3$).

It should be recalled that sampled systems such as that presented in Figure 9.4 introduce a sample delay of one-half the basic pulse width, which is often not corrected as long as the pulse period ΔT is constant as is the case for conventional fixed-frequency modulation. (This delay is typically not corrected because it is small enough to have a negligible effect on the dynamics of the system.) However, in this case the pulse widths near the $0°$ and $60°$ points have become considerably extended to more than three times the smallest subcycle duration over each $60°$ sextant. This variation in sample delay would cause a low-frequency harmonic distortion if not corrected. In Figure 9.4 the commanded voltage $\overline{V}_o^*(t_s)$ has been corrected to $\overline{V}_o^*(t_s + [T_k/T_o]\Delta T/4)$ by simple linear interpolation to avoid this error.

Figures 9.5 and 9.6 show a comparison of modulation using conventional space vector modulation against using optimized SVM. In Figure 9.5, the trajectory of the space vector corresponding to the harmonic component of current is shown over the first 60° of the switching hexagon for the case of 10 switching intervals. Because of symmetry, the behavior of the function in the remaining five 60° portions is identical. The current locus for each of the switching intervals forms a closed triangle. A circular locus of amplitude 0.175 is sketched in dotted lines so that the two modulation methods can be compared. The triangle sizes of the optimized case are clearly much more uniform.

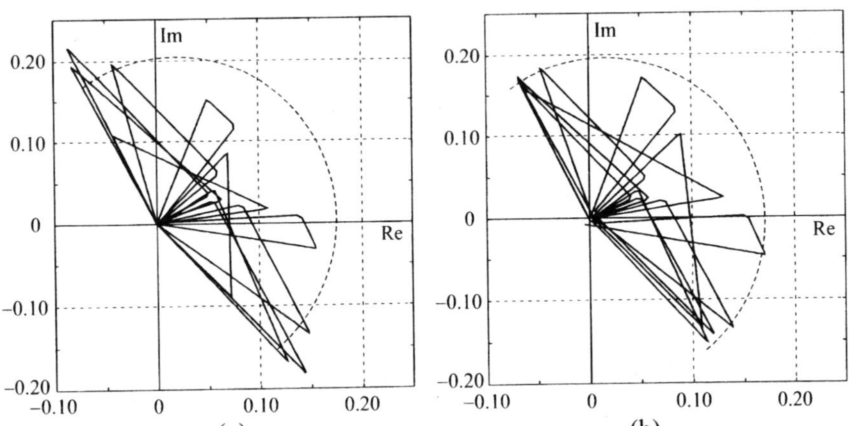

Figure 9.5 Current error space vector vs. angular position of the voltage command vector from the real axis, $\theta = \arg(\overline{V}_o^*)$ for (a) conventional SVM and (b) optimal SVM, pulse number $p = 30$, modulation index $M = 0.8$.

In Figure 9.6 the amplitude of the current error is plotted as a function of the angular location of the command voltage $\overline{V}_o^*$. It can be seen that the ripple current amplitude pulsates with a frequency of six times the fundamental component for conventional SVM, while the current ripple is nearly constant for the optimized SVM strategy. While arbitrary normalized units have been used to compute the current ripple, the ratio of the two values for $I_{h,rms}$ of $0.112/0.126 = 0.888$ will be the same regardless of the inductance. Therefore a 21% reduction in losses can be anticipated for this case ($p = 30$).

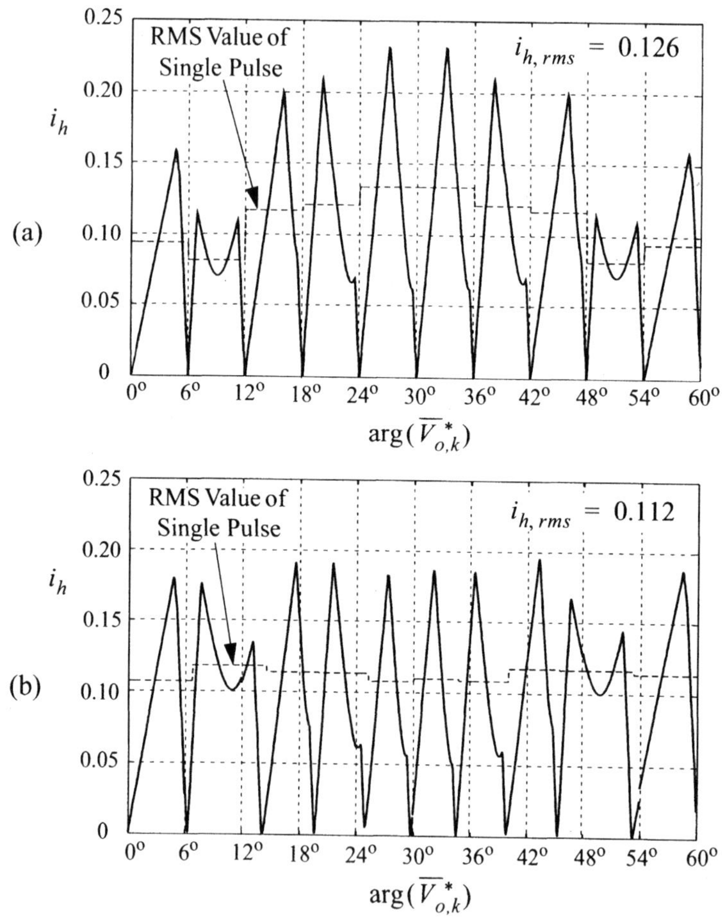

Figure 9.6 Amplitude of current error and incremental RMS value versus angular position of the voltage command vector from the real axis, $\theta = \arg(\overline{V}_o^*)$ for (a) conventional SVM and (b) optimal SVM, pulse number $p = 30$, modulation index $M = 0.8$.

By careful examination of Figure 9.6(b) it can be seen that the RMS value per pulse obtained for the optimized SVM is not quite constant. This is a result of the approximations made in the derivation (piecewise linear variation of the current amplitude). This assumption is clearly violated, particularly during switching intervals 2 and 9. Nonetheless, the approximation is within a few percent of optimal which is generally close enough. If an exact answer is required, the problem then to be solved constitutes a constrained optimization problem (the constraint being that the RMS values of the current in all of the switching cycles must be the same). The exact, nonlinear variation of the incremental current must then be taken into account. Hence the term *optimized SVM* has been used for this type of controller in the context that the switching periods have been "optimized" [2]. However, it is not, in the classical sense, a truly "optimal" PWM strategy, but is instead a "nearly optimal" strategy.

Figure 9.7 compares WTHD0 for normal SVM and optimized SVM. Very little improvement is obtained below $M = 0.5$, but the improvement reaches a factor of 2 as M approaches 1.15.

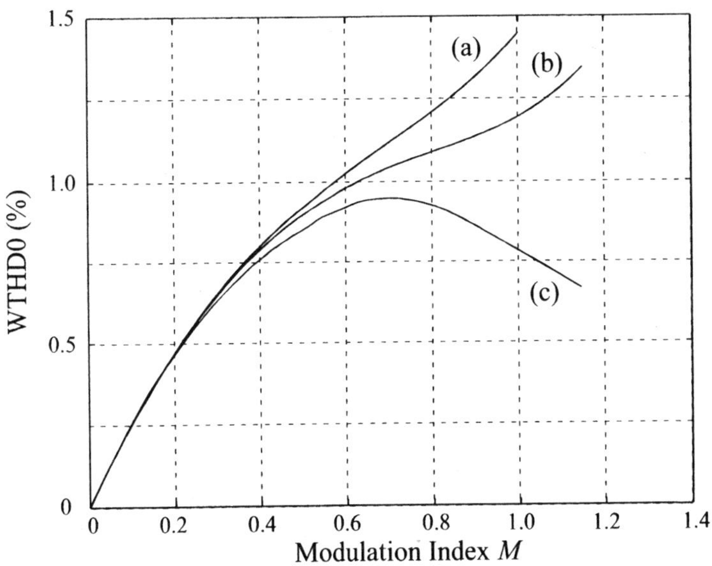

Figure 9.7 WTHD0 versus modulation index M for space vector and optimized space vector operation, $p = 30$: (a) conventional naturally sampled modulation, (b) space vector modulation (naturally sampled equivalent), and (c) optimized space vector modulation.

9.2 Harmonic Elimination PWM

Whereas the previous method of optimization relaxes the concept of regular timing pulses by allowing the half carrier interval to vary about its nominal value T_0, the concept of clocking the pulses can also be eliminated completely. In this case the switching events are allowed to roam freely over the period of the fundamental component consistent only with the need to maintain waveform symmetry. One of the earliest forms of PWM [3], termed *harmonic elimination*, seeks to progressively eliminate the low-order harmonics beginning with the largest (fifth) harmonic consistent with the number of pulses available to be manipulated [4,5].

In general, the two-level output of each phase leg can be switched an arbitrary number of times per cycle as shown in Figure 9.8, in order to vary the fundamental component of voltage applied to the load and also to reduce the harmonic content of the output voltage waveform. Figure 9.8 shows two generalized output waveforms with N notches per half cycle ($2N+1$ switching events). For both waveforms, switching events are always assumed to occur at the half cycle points, 0, 180°, 360°, resulting from the basic square-wave switching and not being part of the notching process. Hence, waveform A begins its positive half cycle with a positive-going voltage transition (a noncontrollable event since the basic square-wave voltage always goes positive at this point), while waveform B begins with a negative-going transition (also noncontrollable). Correspondingly, each negative half cycle begins with noncontrollable opposite polarity transitions.

Furthermore, N can be odd or even for either waveform. However, when N is *odd* in the case of waveform A, or *even* in the case of waveform B, a notch with the opposite polarity to the target fundamental peak will be necessarily located in the center of each half cycle, i.e., symmetrically located around the 90° and 270° points of the fundamental sinusoid. In either case this results in unacceptable solutions, so that, for practical purposes N must be *even* for waveform A and *odd* for waveform B.

Note that with no notches, it is evident that the resultant square wave contains all odd harmonics $h = 1, 3, 5, 7, 9, \ldots$. In addition, if it is assumed that three such half-bridges will be connected to form a three-phase system and, furthermore, that the load will ultimately be connected in wye, then only the non-triplen harmonics result in currents flowing in the load.

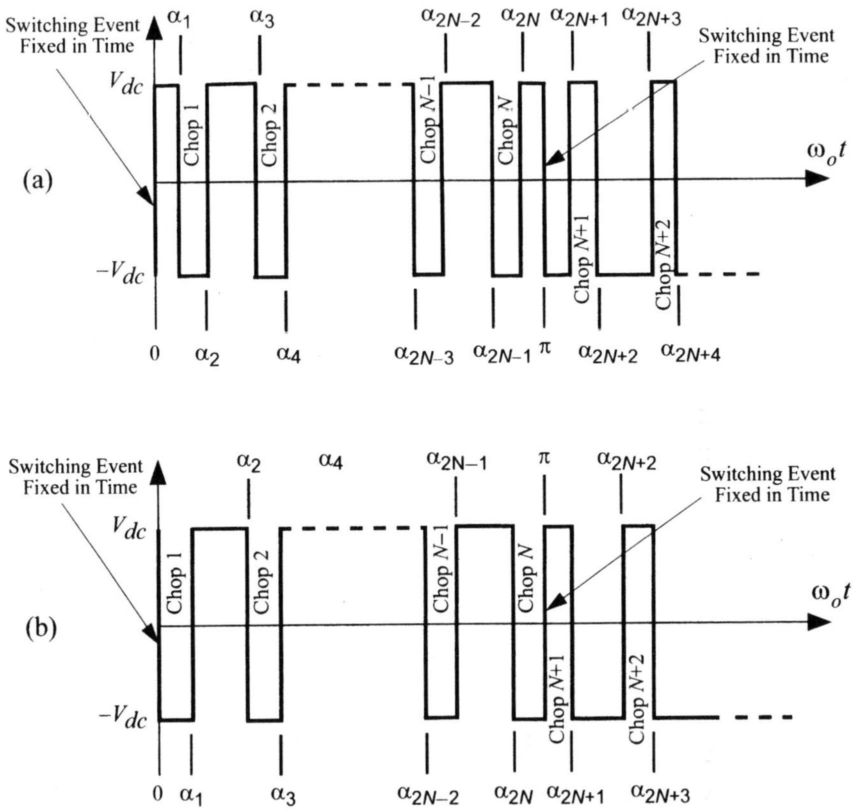

Figure 9.8 Generalized output waveform of half-bridge inverter showing $2N$ notches per cycle: (a) waveform type A, positive half cycle having positive going initial step and (b) waveform type B, positive half cycle with negative going initial step.

If $\alpha_1, \alpha_2, ..., \alpha_{2N}$ define the angular positions of the N notches shown in Figure 9.8, then the waveforms can be represented in general by a Fourier series as

$$f(\omega t) = \sum_{n=1}^{\infty} a_n \sin n\omega t + b_n \cos n\omega t \tag{9.39}$$

where, utilizing half-wave symmetry,

$$a_n = \frac{2}{\pi}\int_0^{\pi} f(\omega t) \sin n\omega t \; d\omega t \tag{9.40}$$

$$b_n = \frac{2}{\pi}\int_0^{\pi} f(\omega t)\cos n\omega t \, d\omega t \tag{9.41}$$

Substituting for $f(\omega t)$ in Eq. (9.40) and, using the half-wave symmetry property, then for waveform A of Figure 9.8(a)

$$a_n = \frac{2}{\pi}V_{dc}\left[\sum_{k=0}^{2N-1}(-1)^k \int_{\alpha_k}^{\alpha_{k+1}} \sin n\omega t \, d\omega t + \int_{\alpha_{2N}}^{\pi} \sin n\omega t \, d\omega t\right] \tag{9.42}$$

where, necessarily as identified before, $\alpha_0 = 0$.

For waveform B,

$$a_n = \frac{2}{\pi}V_{dc}\left[\sum_{k=0}^{2N-1}(-1)^{k+1} \int_{\alpha_k}^{\alpha_{k+1}} \sin n\omega t \, d\omega t - \int_{\alpha_{2N}}^{\pi} \sin n\omega t \, d\omega t\right] \tag{9.43}$$

Evaluating the integrals gives

$$a_n = (\pm 1)\frac{4}{n\pi}V_{dc}\sum_{k=0}^{2N}(-1)^k[\cos n\alpha_k - \cos(n\alpha_{k+1})] \tag{9.44}$$

$$= (\pm 1)\frac{4V_{dc}}{n\pi}\left[1 + \sum_{k=1}^{2N}(-1)^k \cos n\alpha_k\right] \tag{9.45}$$

where +1 is used for waveform of type A (for even values of N) and −1 is used for waveform type B (odd values of N).

Similarly,

$$b_n = (\pm 1)\frac{4V_{dc}}{n\pi}\left[\sum_{k=1}^{2N}(-1)^k \sin(n\alpha_k)\right] \tag{9.46}$$

Equations (9.45) and (9.46) form a set of equations in $2N$ unknowns. Hence given values for the α's for any particular case, the harmonic spectral components a_h and b_h can be readily solved. The number of unknowns can be reduced if one now assumes quarter-wave symmetry, so that

$$\alpha_k = \pi - \alpha_{2N-k+1} \qquad k = 1, 2, \ldots, N \tag{9.47}$$

Hence the sine terms in Eq. (9.46) occur in pairs, so that

$$b_n = 0 \qquad \text{for all } n \tag{9.48}$$

From Eq. (9.47), it is also apparent that

$$\cos(n\alpha_k) = \cos[n(\pi - \alpha_{2N-k+1})] \quad k = 1, 2, \ldots, N \tag{9.49}$$

or, simply

$$\cos(nh\alpha_k) = -\cos(n\alpha_{2N-k+1}) \qquad k = 1, 2, \ldots, N \tag{9.50}$$

Substituting Eq. (9.50) into Eq. (9.45), one obtains finally,

$$a_n = (\pm 1)\frac{4V_{dc}}{n\pi}\left[1 + 2\sum_{k=1}^{N}(-1)^k \cos(n\alpha_k)\right] \tag{9.51}$$

The set of equations given by Eq. (9.51) are nonlinear since they are trigonometric functions of the variables α_k. One convenient approach to solving these equations is to use an n-dimensional form of Newton's method [4,5]. For this method, the N equations of Eq. (9.51) can be written as the vector

$$\frac{4V_{dc}}{\pi}\begin{bmatrix} \frac{1}{n_1}\left[1 + 2\sum_{k=1}^{N}(-1)^k\cos(n_1\alpha_k)\right] \\ \frac{1}{n_2}\left[1 + 2\sum_{k=1}^{N}(-1)^k\cos(n_2\alpha_k)\right] \\ \vdots \\ \frac{1}{n_N}\left[1 + 2\sum_{k=1}^{N}(-1)^k\cos(n_N\alpha_k)\right] \end{bmatrix} = \begin{bmatrix} F_1(\alpha_1, \alpha_2, \ldots, \alpha_N) \\ F_2(\alpha_1, \alpha_2, \ldots, \alpha_N) \\ \vdots \\ F_N(\alpha_1, \alpha_2, \ldots, \alpha_N) \end{bmatrix} = \boldsymbol{F}(\alpha_1, \alpha_2, \ldots, \alpha_N) \tag{9.52}$$

where the n's are harmonic numbers to be eliminated. This equation can be linearized about a nominal value by taking the slope and multiplying this result by an incremental value around the nominal value. In the case of an N-dimensional variable, this is equivalent to

$$\boldsymbol{F}(\alpha_1, \alpha_2, \ldots, \alpha_N) = \boldsymbol{F}(\alpha_{1,0}, \alpha_{2,0}, \ldots, \alpha_{N,0}) + grad\boldsymbol{F}(\alpha_1, \alpha_2, \ldots, \alpha_N)\Big|_{\alpha = \alpha_0}^{\Delta\alpha} \tag{9.53}$$

where

$$gradF(\alpha_1, \alpha_2, \ldots, \alpha_N) = \begin{bmatrix} \frac{\partial F_1}{\partial \alpha_1} & \cdots & \frac{\partial F_1}{\partial \alpha_N} \\ \vdots & \vdots & \vdots \\ \frac{\partial F_N}{\partial \alpha_1} & \cdots & \frac{\partial F_N}{\partial \alpha_N} \end{bmatrix} \tag{9.54}$$

and

$$\Delta\alpha = \begin{bmatrix} \alpha_1 - \alpha_{1,0} \\ \vdots \\ \alpha_N - \alpha_{N,0} \end{bmatrix} \tag{9.55}$$

The *grad* function is called the *gradient* of the vector $\boldsymbol{F}$. The solution is obtained in the same manner as the scalar form for Newton's equation. A suitable starting point is guessed and the gradient computed from the known functions given by Eq. (9.52). Assuming that the function is linear, the function is projected toward the origin and the intercept on the axes is located. These values are then used to fix a new operating point. The slope is again computed and the process repeated until convergence is obtained (hopefully). Formally, this iteration is described by solving Eq. (9.53) as

$$\begin{bmatrix} \alpha_1 \\ \vdots \\ \alpha_N \end{bmatrix} = -\left[grad\left(\boldsymbol{F}(\alpha_1, \alpha_2, \ldots, \alpha_N)\Big|_{\alpha = \alpha_0}\right)\right]^{-1} \boldsymbol{F}(\alpha_1, \alpha_2, \ldots, \alpha_N)\Big|_{\alpha = \alpha_0} + \begin{bmatrix} \alpha_{1,0} \\ \vdots \\ \alpha_{N,0} \end{bmatrix} \tag{9.56}$$

The process of convergence is illustrated along a single axis of the N-dimensional axes shown in Figure 9.9.

Should convergence fail, other more sophisticated solution techniques are available. As a practical matter, many excellent algorithms are, for example, available in the MATLAB programming language. A low-cost solution is also available via numerical methods available in standard mathematical libraries such as IMSL [6] or NAG [7].

Equation (9.51) can be solved to make any particular a_n equal to zero, i.e., the harmonic component corresponding to *any* harmonic n can be eliminated. However, since even harmonics are readily set to zero by simply maintaining

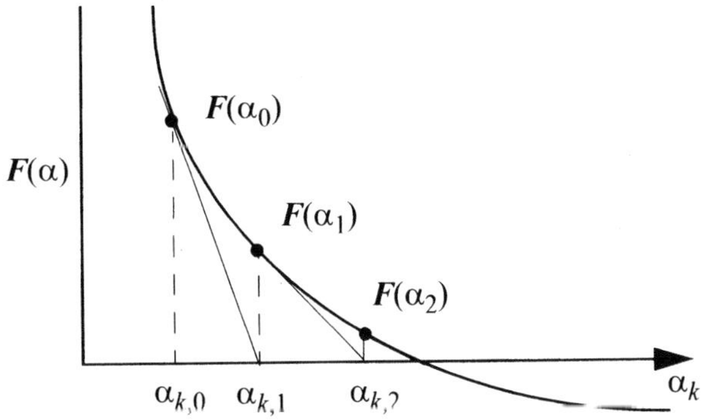

Figure 9.9 Illustration of Newton's method.

symmetry in the waveform, only odd nontriplen pairs of harmonics are typically set to zero by use of Eq. (9.56).

For example, for $N = 2$, the two lowest nontriplen harmonics of the basic square wave that should be eliminated to minimize WTHD are the fifth and seventh. In this case, $\alpha_1 = 16.247°$ and $\alpha_2 = 22.069°$ for waveform type A. A sketch of the resulting waveform together with the harmonic content of the waveform for $N = 2$ are shown in Figure 9.10 and Table 9.1. Note that while the selected harmonics are eliminated, the harmonic just above the largest eliminated harmonic increases as the number of transitions increases. In effect this is a natural consequence of the sampling principle.

Examining Figure 9.10 it can be seen that 10 switching events per phase leg result in an effective switching frequency of five times the fundamental. The fifth and seventh harmonics have now been eliminated by the optimized switching angle. (In PWM terms, these are the carrier harmonic and the sideband at the switching frequency plus twice the fundamental.) The lower sideband at the switching frequency minus twice the fundamental is eliminated by virtue of it being a triplen harmonic. The sidebands at the switching frequency plus/minus the fundamental are even harmonics which are eliminated by arranging a symmetrical waveform with half and quarter cycle symmetry. The harmonic at twice the sampling frequency is also zero since it is even, leaving only the sidebands at the switching frequency plus/minus the fundamental (10 ± 1) as significant. However, the ninth harmonic is also triplen and does not appear across the load so that, finally, the eleventh harmonic becomes the first important spectral contribution from the point of view of the load.

Table 9.1 Harmonic Magnitudes for Elimination of 5''' and 7''' Harmonic, $N = 2$

Order of Harmonic	Absolute Magnitude of Harmonic Coefficient	Magnitude of Harmonic (% of Fundamental)
1	1.1879	100.00
3	0.2070	17.43
5	0.00	0.00
7	0.00	0.00
9	0.1086	9.14
11	0.2412	21.31
13	0.3223	27.13
15	0.3084	25.96
17	0.2030	17.09
19	0.0514	4.33
21	0.0825	6.94

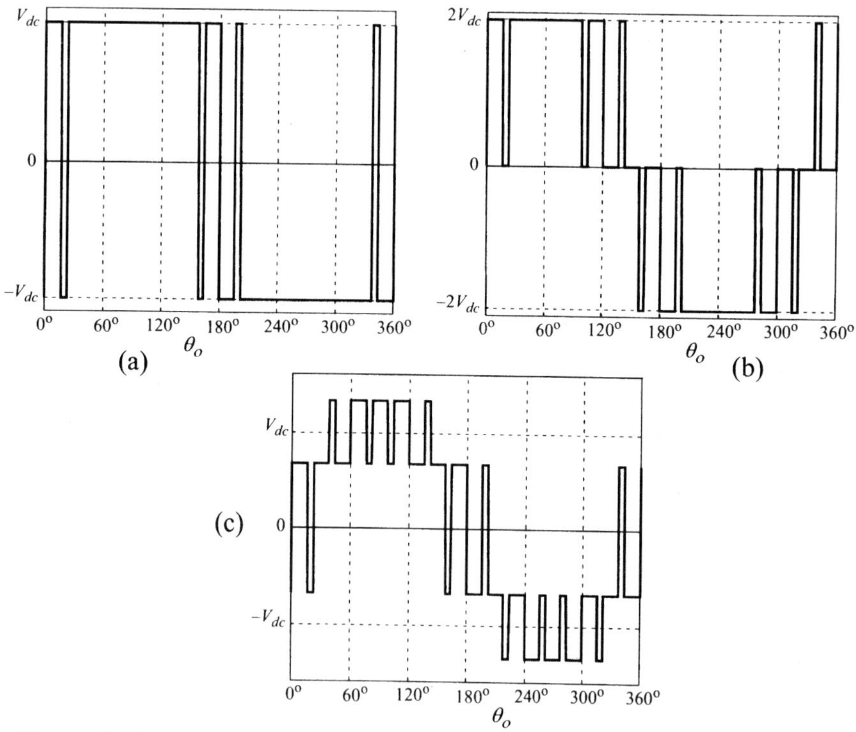

Figure 9.10 Solution and waveform for elimination of the 5th and 7th harmonics using waveform type A, $\alpha_1 = 16.247°$, $\alpha_2 = 22.069°$: (a) Line-to-DC midpoint voltage v_{az}, (b) line-to-line voltage v_{ab}, and (c) line to load-neutral voltage v_{as}.

It can be noted from Table 9.1 that if the two free values of α are used to eliminate the fifth and seventh harmonics, the fundamental component is fixed and cannot be varied. While this may be suitable for passive, constant frequency loads, use of such an inverter for an AC motor drive requires that the fundamental voltage be controlled in a manner such that the volts/hertz remain constant. Assuming that the switching pattern is not altered as the frequency changes, this implies that the fundamental voltage must also be changed. Hence, one of the constraint equations denoted by Eq. (9.51), namely the equation for a_1, must be altered to the form

$$a_1 = (\pm 1)\frac{4V_{dc}}{\pi}\left[1 + 2\sum_{k=1}^{2N}(-1)^k \cos\alpha_k\right] = V_1 \tag{9.57}$$

or, as a per unit of V_{dc},

$$a_{1,(pu)} = (\pm 1)\frac{4}{\pi}\left[1 + 2\sum_{k=1}^{2N}(-1)^k \cos\alpha_k\right] = M \tag{9.58}$$

Again the +1 and –1 are taken for even and odd N, respectively. Since Eq. (9.58) must be satisfied, one less harmonic can be eliminated than previously. Because Eq. (9.50) forms an additional constraint, the switching angles α_k now become a function of the per unit voltage and the solution becomes more complex.

Figure 9.11 shows the simple case where a single notch is introduced per half cycle using waveform A. In this case all nontriplen odd harmonics remain, and the single free switching angle is used to vary the fundamental component of voltage. Figures 9.11(c) and 9.11(d) show two harmonic distortion curves.

In Figure 9.11(c) the WTHD is expressed as a per unit of the fundamental component of voltage as defined by Eq. (2.22), namely

$$\text{WTHD} = \frac{\sqrt{\sum_{n=2}^{\infty}\left(\frac{V_n}{n}\right)^2}}{V_1} \tag{9.59}$$

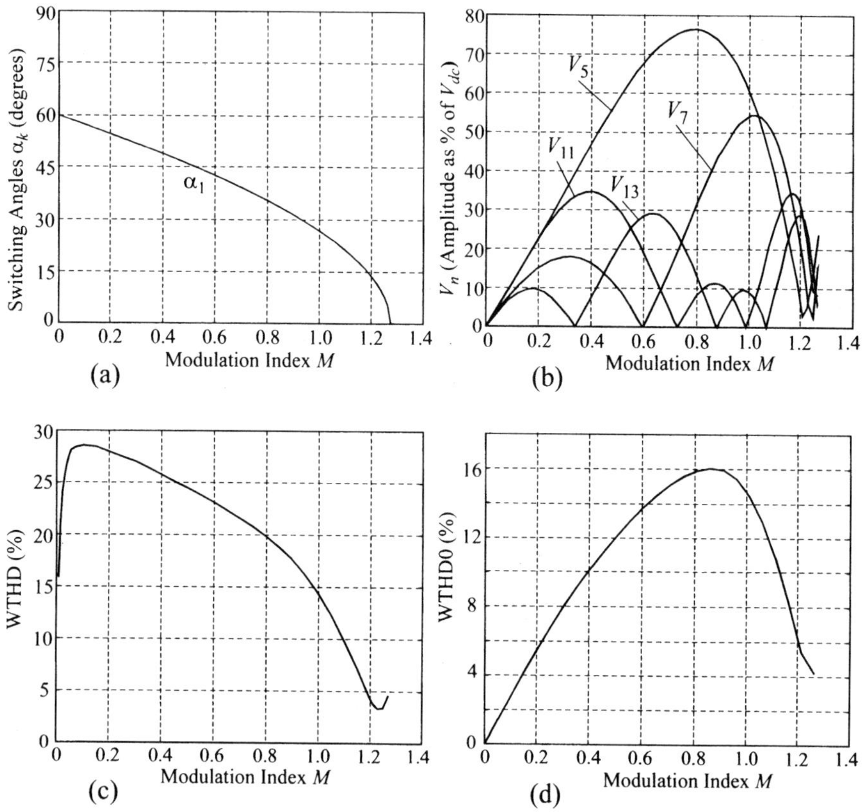

Figure 9.11 (a) Switching angle α_1 vs. modulation index M, (b) resulting 5th, 7th, 11th, and 13th harmonics, (c) WTHD, and (d) WTHD0. Type A waveform used with one chop per half cycle, harmonics shown in per unit of V_{dc}, fundamental voltage proportional to M.

In the case being examined only nontriplen odd harmonics exist since the waveform is half-wave symmetrical. If both V_n and V_1 are expressed as a per unit of V_{dc}, then Eq. (9.59) can be written as

$$\text{WTHD} = \frac{1}{M}\sqrt{\sum_{k=1}^{\infty}\left[\left(\frac{V_{6k-1,(pu)}}{6k-1}\right)^2 + \left(\frac{V_{6k+1,(pu)}}{6k+1}\right)^2\right]} \quad (9.60)$$

Equation (9.60) is important as a figure of merit for inverter drives which operate with *constant-volts-per-hertz*. In such cases the per unit fundamental component of voltage and per unit frequency are identical so that their ratio always equals unity. The modulation index M has already been defined in such a manner as to represent the per unit fundamental component of voltage and, hence, its use in Eq. (9.60). Since the fundamental component of voltage reaches zero at zero frequency, it is clear that the WTHD becomes singular when $V_1 = 0$.

In other applications, such as uninterruptible power supplies, the voltage V_1 is varied while the frequency is kept constant. In such cases the WTHD can best be expressed in terms of a fixed reference voltage. Using the peak voltage obtained with $M = 1.0$, i.e., V_{dc}, Chapter 2 has defined

$$\mathrm{WTHD0} = \sqrt{\sum_{k=1}^{\infty}\left[\left(\frac{V_{6k-1(pu)}}{6k-1}\right)^2 + \left(\frac{V_{6k+1(pu)}}{6k+1}\right)^2\right]} \tag{9.61}$$

in which case,

$$\mathrm{WTHD} = \frac{\mathrm{WTHD0}}{M} \tag{9.62}$$

Figure 9.11(d) shows the behavior of this function.

Figure 9.12 shows the two switching angles needed to eliminate the fifth harmonic on the output while again producing the desired fundamental output voltage. The angles α_1 and α_2 are plotted as a function of the desired fundamental output voltage in per unit of the maximum possible fundamental voltage for square-wave operation (i.e., in terms of the modulation index M). Switching angles and other data for two possible solutions are plotted. The solid line solution retains both switching angles within the region $0 \le \alpha \le \pi/3$ corresponding to 60° discontinuous switching as described in Chapter 6. Switching of this type suggests that each inverter phase is clamped to the positive or negative DC rail for 60° (during the region 60°–120° and 240°–300° in the case of the waveform shown in Figure 9.8). Hence, such switching can be considered as a form of 60° discontinuous switching. When the angles occupy the entire half cycle, by comparison, the solution corresponds to a form of continuous switching.

In both cases waveform A is used (positive-going pulse at the zero crossing of the fundamental component). The dashed line gives the solution for continuous switching over the 90° and therefore, by symmetry, switching over the entire 360° period. Also shown are the next three most important harmonics (i.e., the seventh, eleventh and thirteenth) as a function of M for both cases.

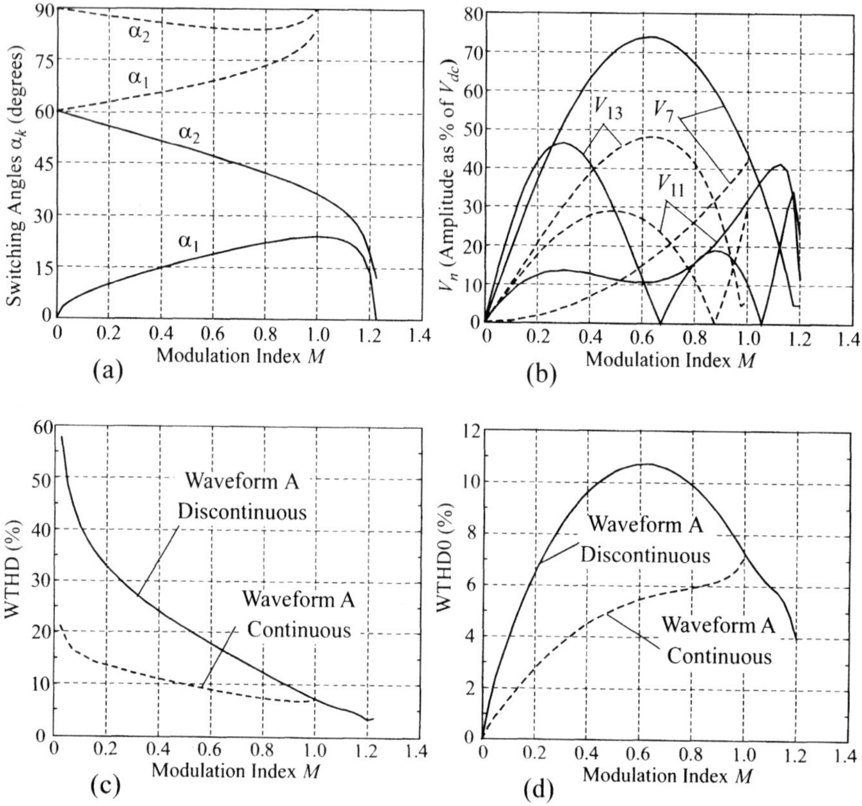

Figure 9.12 (a) Switching angles α_1, α_2 for two chops per half cycle, (b) resulting 7th, 11th, and 13th harmonics, (c) WTHD, and (d) WTHD0. Both type A waveforms, solid and dashed lines show discontinuous and continuous switching, respectively, 5th harmonic eliminated, remaining harmonics shown in per unit of V_{dc}, fundamental voltage proportional to M.

Note that the seventh harmonic component is dominant which can be described as follows. The pulse rate for this case is $P = 2N+1 = 5$. Since the fifth harmonic has been eliminated and even sideband harmonics are zero due to the symmetry of the waveform, the dominant sidebands are $5 \pm 2 = 3$ and 7. The third harmonic is eliminated because it is common mode, and thus the seventh harmonic becomes dominant. The WTHD normalized for variable frequency, constant V/Hz and for fixed-frequency operation are given in Figures 9.12(c) and 9.12(d). In both cases the continuous switching case is superior to the 60°

discontinuous case. Unfortunately, the solution terminates near $M = 1$ and cannot be readily used in the overmodulation region.

It is important to note that the solutions for the switching angles α_k terminate when any α_i is equal either to 0 or $\pi/2$. Hence, if one value of α_k is always known when the modulation index reaches its maximum value, then M_{max} can be found by letting one α_k equal zero (or $\pi/2$) and treating M as a variable instead. Solutions of the resulting set of equations yield the remaining values of α_k plus the maximum value of M for a particular solution. The number of solutions has been found to be surprisingly large [8]. Table 9.2 is taken from this reference showing the maximum values of α_k and M for $N = 2$ through 10. These values of α_k can be used as the starting point to *back calculate* the intermediate values of α_k from $M = M_{max}$ back to $M = 0$ by assuming an additional value of α_k equal to either 0 or $\pi/2$. The resulting set of α_k's may or may not generate a range of solutions over a feasible range of the parameter M. For example, when $N = 3$, the four possible points which could generate a solution are as shown in Table 9.3. It has been shown in [8] that the number of solutions obtained in this manner follow the sequence shown in Table 9.4. It can also be noted from Table 9.2 that only one of the solution set of points contains values of α_k no greater than $\pi/3$. Hence, one and only one solution corresponding to 60° switching is possible for each value of N.

Figures 9.13 through 9.16 give additional results for three chops through to five chops per half cycle. Only the significant solutions for $N = 3$, 4, and 5 are shown. It should be mentioned that the nonlinear nature of the system equations prevents a decisive assessment of all possible solutions. While the solutions found in [8] are numerous, they are not exhaustive since at least one additional solution has been found for $N = 5$, as shown in Figure 9.17. Since the α_k's for this solution do not terminate at 0 or $\pi/2$, it cannot be predicted by the approach used to develop Table 9.2. Its existence suggests the presence of possible additional solutions.

Note that when $N > 2$ a pattern emerges showing that angles depart from equally spaced points along the ordinate which can also be used as the starting points for the solution algorithm. Reference [9] illustrates a method of solving these equations by identifying these existing symmetries and defining a set of straight lines which approximate the exact solution. These existing symmetries and straight-line approximations are of considerable benefit in the implementation of this switching strategy in hardware [10].

Table 9.2 Maximum Values of α_k and Modulation Index M for Feasible Solutions, Number of Chops per Half Cycle $N = 2$ to 10 [8]

N	M_{max}	1	2	3	4	5	6	7	8	9
2	1.22	0.227								
	0.98	1.448				Maximum Delay Angle α (radians)				
3	1.1884	0.2836	0.3852							
	1.16690	0.1780	1.5448							
4	1.1779	0.1526	0.4258	0.4845						
	1.1733	0.1647	0.2578	1.5511						
5	1.1704	0.1841	0.2809	0.5394	0.5736					
	1.1690	0.1717	0.2631	1.4845	1.5057					
	1.1668	0.1186	0.3020	0.3671	0.6051	0.6283				
6	1.0330	0.1483	0.2700	0.8423	0.8854	1.5345				
	1.1663	0.1136	0.2757	0.3269	1.4546	1.4751				
	1.0217	0.1883	0.2644	1.2062	1.2526	1.5328				
	1.1638	0.1362	0.2212	0.4030	0.4474	0.6654	0.6807			
	1.1596	0.1057	0.2574	0.3208	0.7459	0.7546	1.5557			
7	1.1636	0.1339	0.2173	0.4277	0.4598	1.4108	1.4257			
	1.1596	0.1061	0.2612	0.3156	1.3392	1.3479	1.5557			
	1.1621	0.0969	0.2341	0.2963	0.4694	0.5021	0.7034	0.7148		
8	1.1613	0.1004	0.1715	0.3018	0.3490	0.7252	0.7345	1.5578		
	1.1620	0.0952	0.2263	0.2805	0.4758	0.5069	1.3777	1.3888		
	1.1612	0.1008	0.1721	0.3093	0.3469	1.3598	1.3691	1.5577		
	1.1606	0.1081	0.1825	0.3213	0.3675	0.5323	0.5561	0.7409	0.7490	
9	1.1603	0.1041	0.1763	0.3047	0.3478	0.7519	0.7591	1.5078	1.5234	
	1.1606	0.1074	0.1812	0.3321	0.3713	0.5335	0.5571	1.3449	1.3530	
	1.1602	0.1036	0.1753	0.3132	0.3497	1.3349	1.3420	1.5073	1.5229	
	1.1597	0.0818	0.1912	0.2486	0.3832	0.4192	0.5751	0.5937	0.7657	0.7721
	1.0383	0.0908	0.1764	0.2950	0.3580	0.7117	0.7421	0.9272	0.9501	1.5487
	1.1595	0.0797	0.1846	0.2398	0.3529	0.3786	0.7706	0.7766	1.4900	1.5066
10	1.0269	0.0906	0.1782	0.3199	0.3510	0.9267	0.9516	1.3506	1.3828	1.5477
	1.1597	0.0811	0.1882	0.2409	0.3862	0.4212	0.5761	0.5946	1.3219	1.3282
	1.0345	0.1125	0.1741	0.2940	0.3578	0.7116	0.7426	1.1438	1.1673	1.5484
	1.1595	0.0791	0.1818	0.2331	0.3568	0.3810	1.3176	1.3236	1.4898	1.5063
	1.0228	0.1146	0.1764	0.3193	0.3505	1.1422	1.1679	1.3500	1.3830	1.5474

Table 9.3 Possible Solution End Points for $N = 3$, α_k in Degrees

α_1	α_2	α_3	Solution
0.00	16.35	22.07	Type A
16.35	22.07	90.00	Type B*
0.00	10.20	88.51	None
10.20	88.51	90.00	Type A

* Range restricted to $1.172 \leq M \leq 1.188$.

Table 9.4 Number of Solutions vs. Number of Chops per Half Cycle N [8]

N	2	3	4	5	6	7	8	9	10	11	12	13	14	...
Solutions	3	3	3	4	6	6	6	8	12	12	12	16	24	...

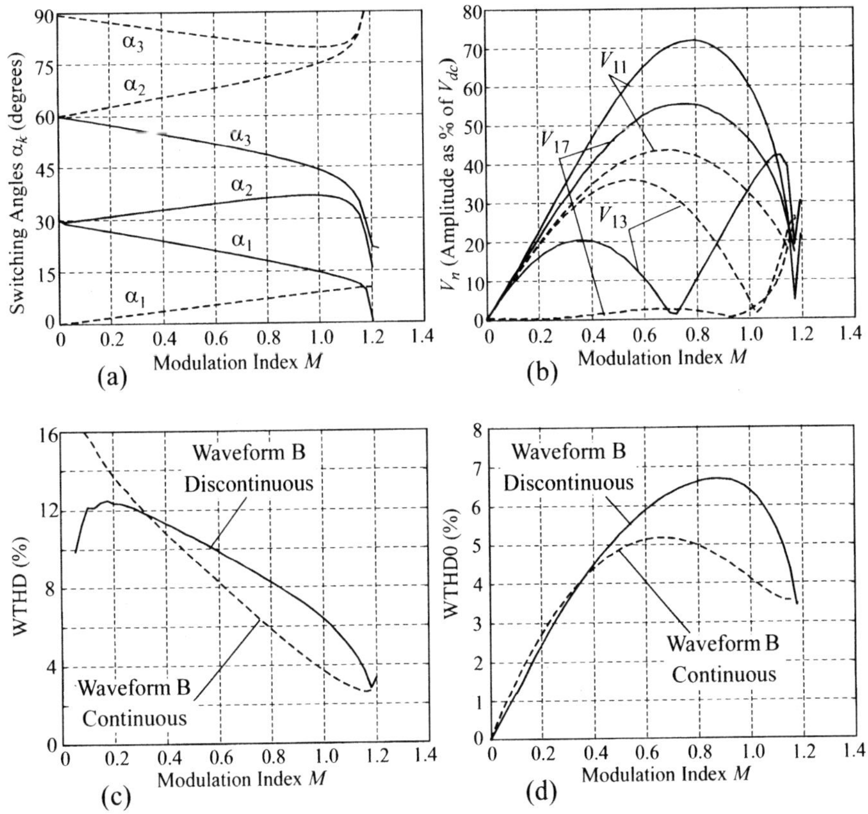

Figure 9.13 (a) Switching angles α_1, α_2, α_3 for three chops per half cycle, (b) resulting 11th, 13th, and 17th harmonics, (c) WTHD, and (d) WTHD0. Both type B waveforms, solid and dashed lines show discontinuous and continuous switching, respectively, 5th and 7th harmonics eliminated, remaining harmonics shown as per unit of V_{dc}.

Figure 9.18 shows a plot of the WTHD for both types of waveforms with two different chopping frequencies, $N = 3$ and 6. Note that type A and type B curves cross near a modulation index of 0.8. The type B curve proves superior for low voltages (low frequencies) for N odd and type A for N even. However, it is apparent that the complexity of the solutions make the use of harmonic elimination difficult to use particularly at low frequencies when the number of chops per quarter cycle becomes very large. The process of sorting out the best

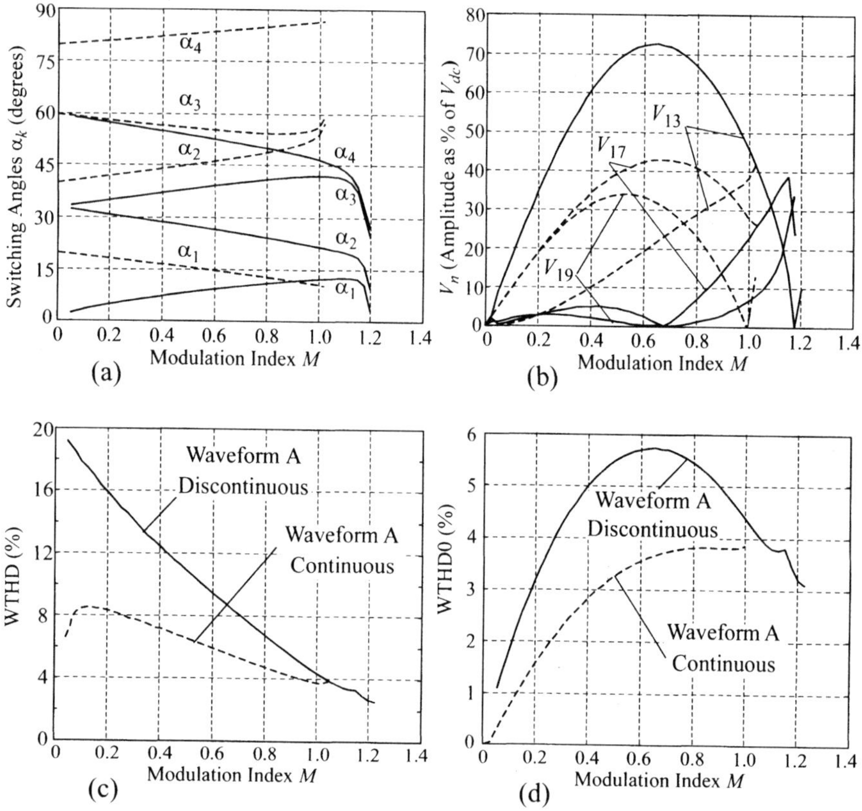

Figure 9.14 (a) Switching angles α_1, α_2, α_3, α_4, for four chops per half cycle, (b) resulting 13th, 17th, and 19th harmonics, (c) WTHD, and (d) WTHD0. Both type A waveforms, solid and dashed lines show discontinuous and continuous switching, respectively, 5th, 7th, and 11th harmonics eliminated, remaining harmonics shown as per unit of V_{dc}.

solution from the large number of possibilities when N is large becomes nearly an impossible task and has never been reported in the literature. Generally speaking, harmonic elimination becomes most valuable during the pulse dropping mode ($2/\sqrt{3} \le M \le \pi/4$). A possible modulation strategy for a converter limited to a switching frequency of 1000 Hz with a fundamental frequency at $M = 1.27$ of 89 Hz is shown in Figure 9.19.

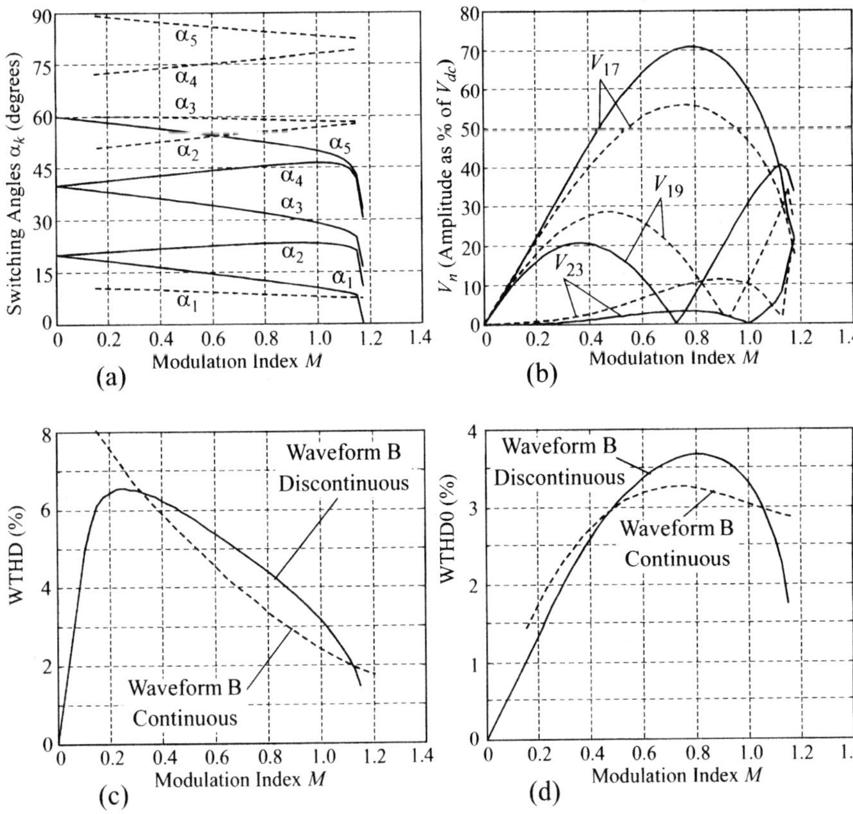

Figure 9.15 (a) Switching angles α_1, ..., α_5 for five chops per half cycle, (b) resulting 17th, 19th, and 23rd harmonics, (c) WTHD, and (d) WTHD0. Both type B waveforms, solid and dashed lines show discontinuous and continuous switching, respectively, 5th, 7th, 11th, and 13th harmonics eliminated, remaining harmonics shown as per unit of V_{dc}.

9.3 Performance Index for Optimality

Pulsewidth modulation is most often used in forced commutated inverters to supply variable speed AC motors where the ultimate purpose of the modulation is to create, to as high an accuracy as possible, a sinusoidal current flowing in the motor stator winding. Rather than deal indirectly with the problem by attempting to create a sinusoidal voltage by progressively eliminating harmon-

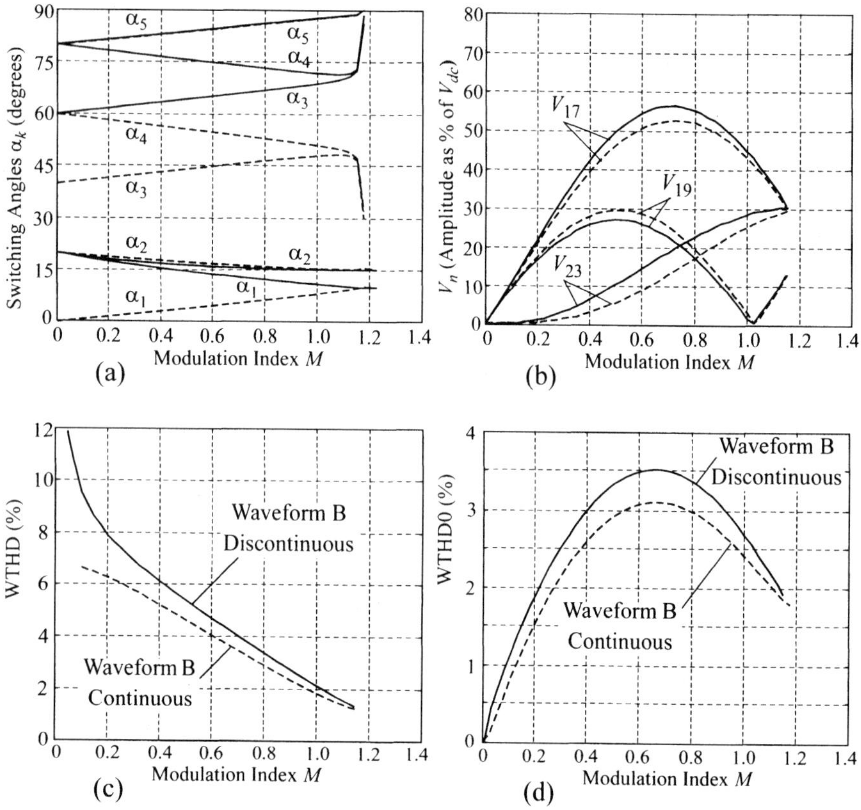

Figure 9.16 (a) Two additional groups of switching angles α_1, ..., α_5 for five chops per half cycle, (b) resulting 17th, 19th, and 23rd harmonics, (c) WTHD, and (d) WTHD0. Both type B waveforms, solid and dashed lines show discontinuous and continuous switching, respectively, 5th, 7th, 11th, and 13th harmonics eliminated, remaining harmonics shown as per unit of V_{dc}.

ics, another approach is to simply use the distortion in the current as a figure of merit and to attempt to minimize this function [11,12]. This problem is, of course, more difficult, since the current waveform is not known *a priori* without current sensors or without some knowledge of the load. Nonetheless, considerable headway can be made with a few simple assumptions. In addition, with the arrival of powerful signal processing chips, the possibility of on-line optimization appears to be on the horizon.

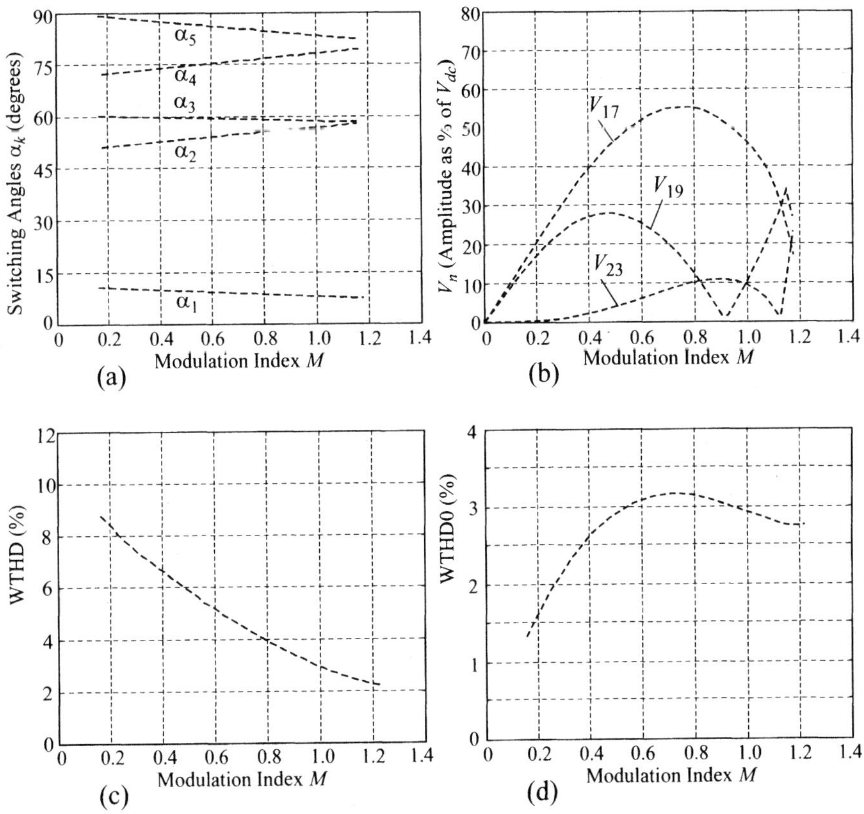

Figure 9.17 (a) Yet another undocumented group of switching angles α_1, ..., α_5 for five chops per half cycle, (b) resulting 17th, 19th, and 23rd harmonics, (c) WTHD, and (d) WTHD0. Type B waveform, 5th, 7th, 11th, and 13th harmonics eliminated, remaining harmonics shown in per unit of V_{dc}.

Assume that the load under consideration consists of an AC motor. It is well known that the impedance to harmonics above the fundamental is limited principally by the leakage inductance of the motor. Recall from Section 2.2 that if the leakage inductance is not frequency dependent and if the voltages applied to the machine are a symmetric three-phase set with no DC component, then the RMS current in one of the motor phases can be expressed as

$$I_{rms} = \sqrt{\frac{1}{2}\sum_{n=2}^{\infty}\left(\frac{V_n}{n\omega_o L_\sigma}\right)^2} \tag{9.63}$$

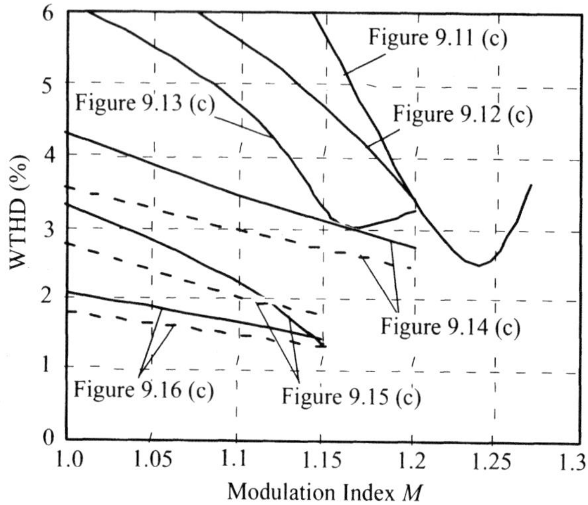

Figure 9.18 Weighted total harmonic distortion for $N = 1$ through $N = 5$ with type A and type B voltage waveforms.

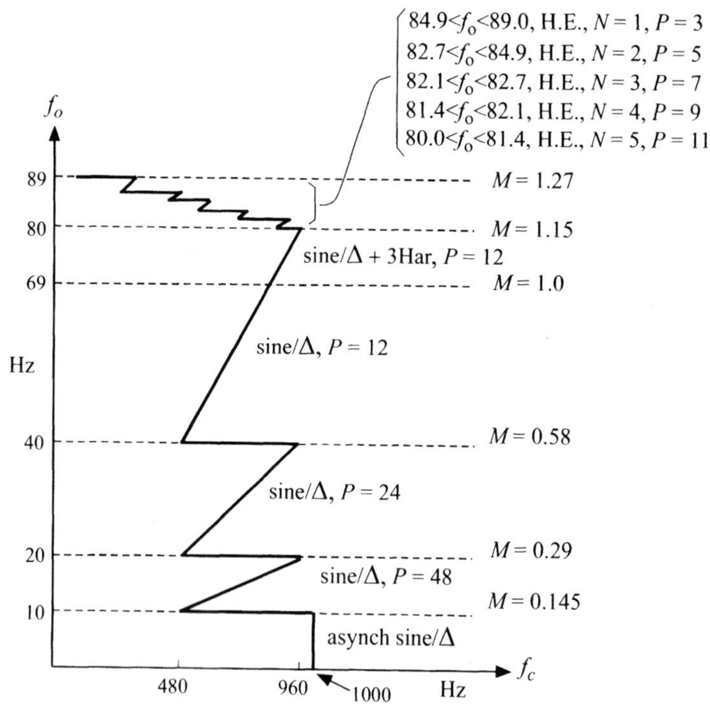

Figure 9.19 Modulation strategy to reach square wave mode using harmonic elimination (H.E.) in the overmodulation region.

where

I_{rms} = RMS value of the sum of the harmonic currents

V_n = peak value of the harmonic n of the output voltage

ω_o = fundamental angular frequency

L_σ = nominal sum of the motor stator plus rotor leakage inductance, or,

$$L_\sigma = L_1 + \frac{L_2 L_m}{L_2 + L_m} \tag{9.64}$$

with

L_1, L_2 = stator and rotor leakage inductance per phase, respectively, and

L_m = magnetizing inductance per phase

This equation can be written as

$$\frac{\omega_o L_\sigma I_{rms}}{V_{1,\,rms}} = \text{WTHD} = \frac{1}{V_1}\sqrt{\sum_{n=2}^{\infty}\left(\frac{V_n}{n}\right)^2} \tag{9.65}$$

where V_1 and $V_{1,rms}$ are the peak and RMS fundamental component of the voltage. Hence, to minimize the RMS value of the harmonic currents it is sufficient to minimize the harmonic voltages expressed in per unit weighted by their harmonic order. With motor loads, V_1 generally changes in direct proportion to frequency, so that Eq. (9.65) tends to infinity as V_1 approaches zero, making this expression unwieldy for deep modulation indices. However, if one normalizes Eq. (9.65) and chooses instead as the reference voltage, the value of V_1 obtained when $M = 1$, then

$$\frac{\omega_o L_\sigma I_{rms}}{V_1|_{M=1}} = \text{WTHD0} = \frac{1}{V_{dc}}\sqrt{\sum_{n=2}^{\infty}\left(\frac{V_n}{n}\right)^2} \tag{9.66}$$

Equation (9.66) constitutes the *performance index* against which various optimizing PWM methods can now be compared. While other factors could readily be used which incorporate skin effect, etc. as demonstrated in Chapter 2, Eq. (9.66) will be used initially here for simplicity. The use of more elaborate performance indices will be discussed toward the end of this chapter.

9.4 Optimum PWM

While the concept is straightforward, the process of minimizing the performance index given by Eq. (9.66) is very difficult since the expression is, in fact, again a function of the switching angles (i.e., same as the α's discussed in Section 9.2). Figures 9.20 and 9.21 show the optimum switching angles computed for the case of two switching angles ($N = 2$ chops per half cycle) as a function of modulation index. In this case the two switching angles allow the performance index to be minimized in addition to realizing the proper fundamental component of voltage. Two solutions are again obtained, one for type A and one for type B switching. In order to obtain the overall minimum WTHD,

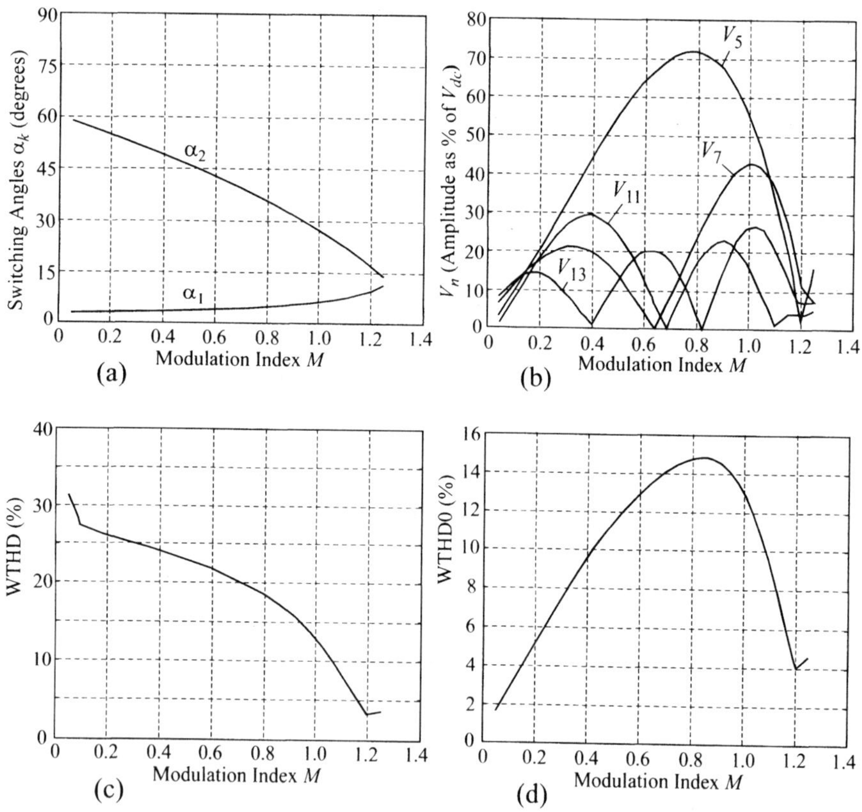

Figure 9.20 Optimum switching for chop number $N = 2$, type A switching: (a) Switching angles α_1, α_2, (b) 5th, 7th, 11th, and 13th harmonic amplitudes, (c) WTHD, and (d) WTHD0.

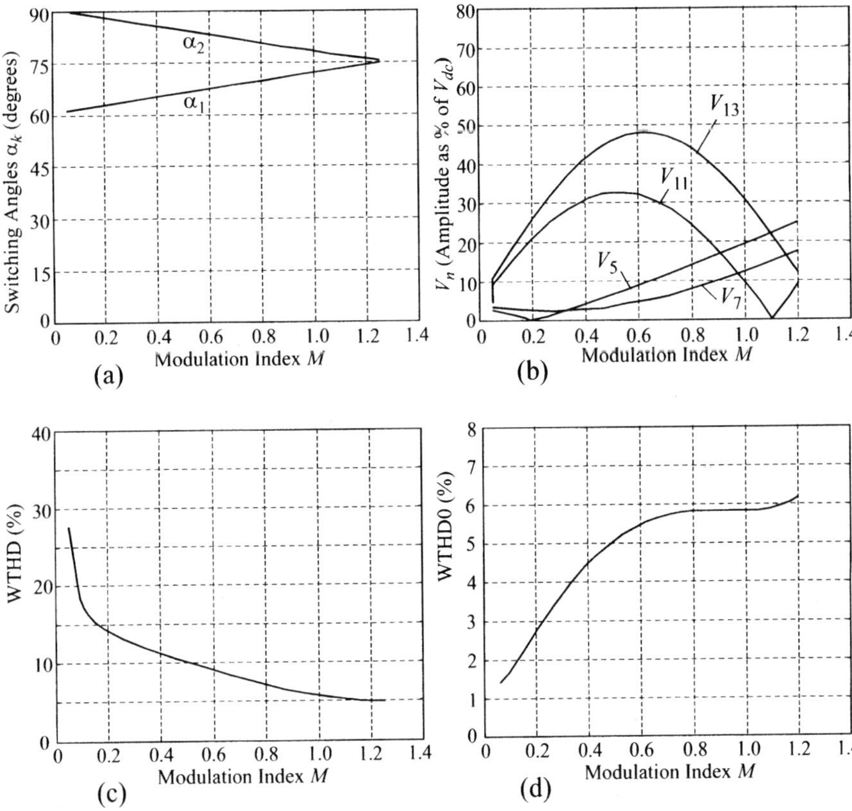

Figure 9.21 Optimum switching for chop number $N = 2$, type B switching: (a) Switching angles α_1, α_2, (b) 5th, 7th, 11th, and 13th harmonic amplitudes, (c) WTHD, and (d) WTHD0.

the two solutions must be combined as shown in Figure 9.22 and require the switching angles to be abruptly changed near $M = 1.15$.

As the number of switching angles increases, the possible number of solutions again increases rapidly. In this case the number of solutions becomes more difficult to predict than for harmonic elimination. The solution space must be carefully examined to extract all possible local minima. Figure 9.23 shows the solution space for $N = 3$ when type A switching is used. The WTHD is plotted as a contour plot versus switching angles α_1 and α_2 for the particular case of $M = 0.5$. The value of α_3 at any location on the coordinate system is implied by Eq. (9.58). Three minima can be located on the contour plot as

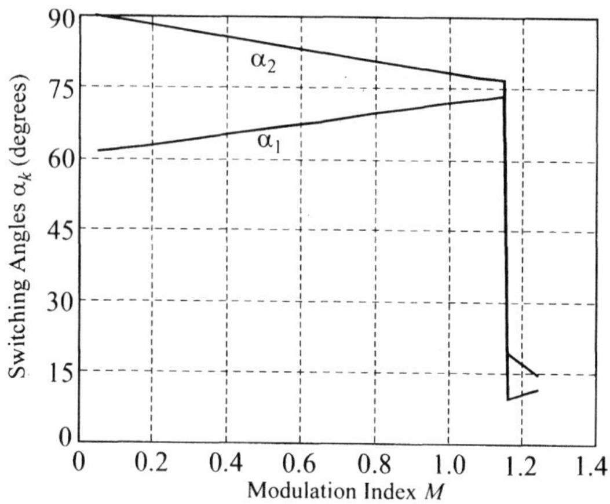

Figure 9.22 Switching angles α_1, α_2 for overall minimum WTHD with $N = 2$.

shown. Being on the boundary of the contour, minima A1 can be verified to be a special case in which $\alpha_3 = 90°$, so that the solution degenerates to the $N = 2$ case. When tracked over the full range of modulation index, Figures 9.24 and 9.25 result for points A2 and A3, respectively.

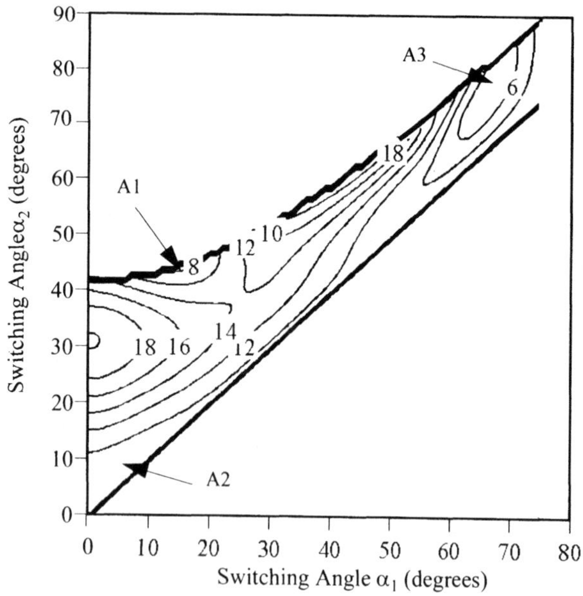

Figure 9.23 Contour plot of WTHD0 as a function of α_1 and α_2. Type A switching, chops per half cycle $N = 3$, modulation index $M = 0.5$.

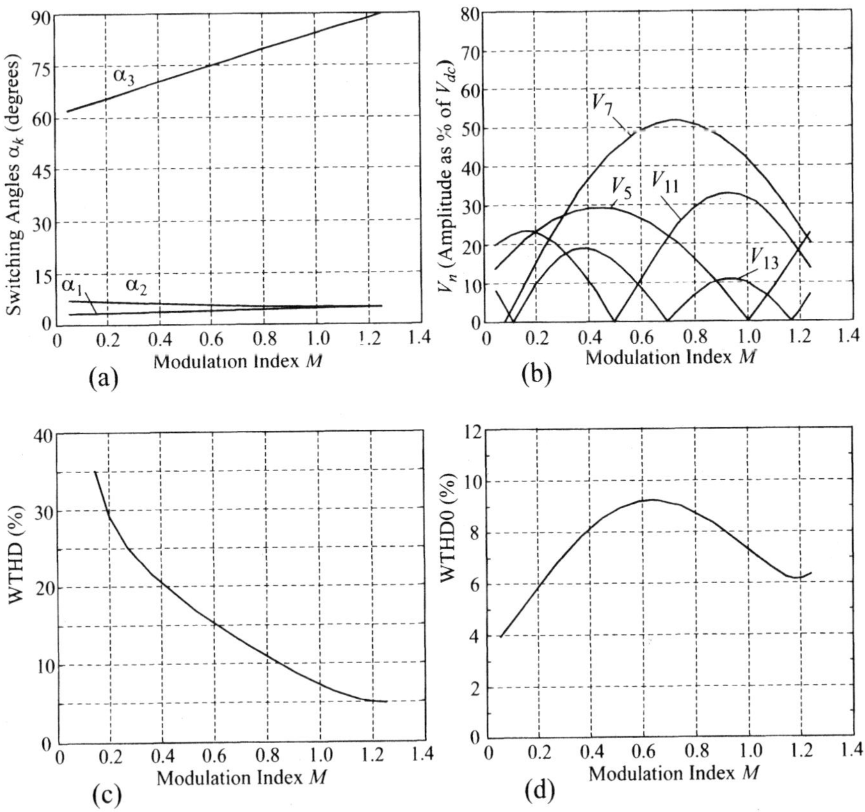

Figure 9.24 Optimum switching for chop number $N = 3$, type A switching, minima A2: (a) Switching angles α_1, α_2, α_3, (b) 5th, 7th, 11th, and 13th harmonic amplitudes, (c) WTHD, and (d) WTHD0.

A contour plot can again be computed for type B modulation. In this case the contour shown in Figure 9.26 is obtained for the case of $M = 0.5$. Three local minima can now be located designated by B1, B2 and B3. Each minimum represents a possible solution. Using the values of α_k corresponding to minima B1 ($\alpha_1 = 5°$, $\alpha_2 = 67°$, $\alpha_3 = 85°$), as the initial search point, the solution can now be traced as a function of the modulation index *M*. The result is shown in Figure 9.27. The process can be repeated for the two remaining minima B2 ($\alpha_1 = 8.7°$, $\alpha_2 = 17°$, $\alpha_3 = 47°$) and B3 ($\alpha_1 = 24°$, $\alpha_2 = 37°$, $\alpha_3 = 53°$) resulting in Figures 9.28 and 9.29. Note that minima B3 does not exist for values of *M* greater than 1.16, a situation which frequently occurs in this type of analysis.

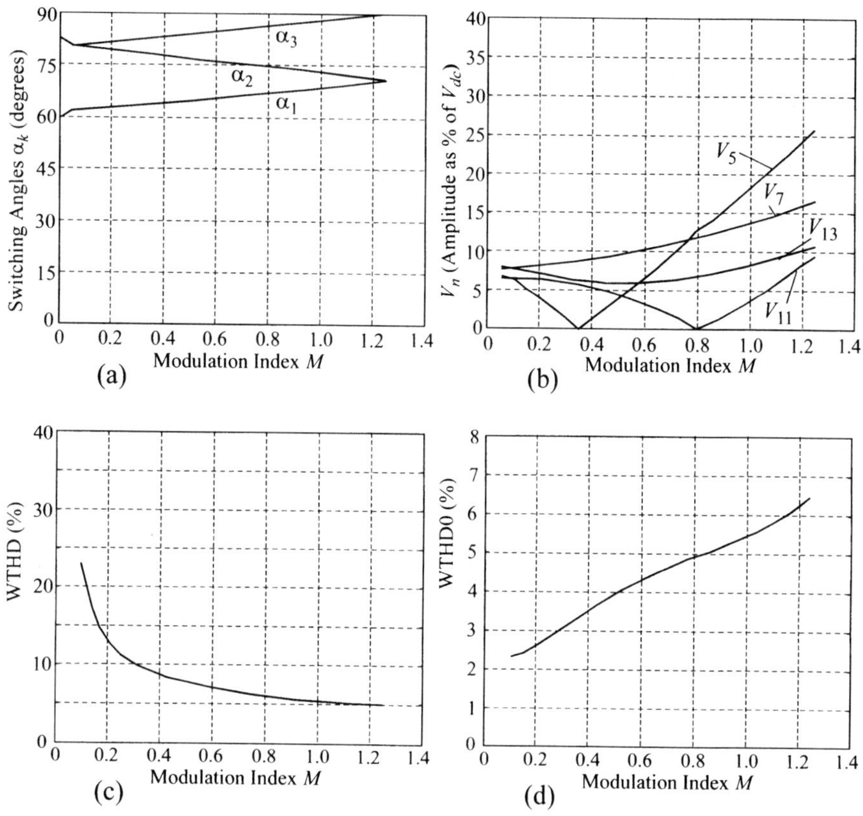

Figure 9.25 Optimum switching for chop number $N = 3$, type A switching, solution A3: (a) Switching angles α_1, α_2, α_3, (b) 5th, 7th, 11th, and 13th harmonic amplitudes, (c) WTHD, and (d) WTHD0.

The four solutions can now be combined to form the overall optimum solution shown in Figure 9.30. The solid line of Figure 9.30 shows the overall minimum obtained by using type A switching from (roughly) $M = 0.1$ to 0.5, type B1 switching from $M = 0.5$ to 1.1, and finally type B2 switching from $M = 1.1$ to 1.27. Type B3 switching is not used. If it is desired to operate with a modulation index below 0.1, a condition rarely if ever needed, switching pattern B2 would again be optimal. It is interesting to note that the optimal algorithm naturally transitions from continuous modulation (type B1 switching) to discontinuous modulation (type B2 switching) at $M = 1.1$.

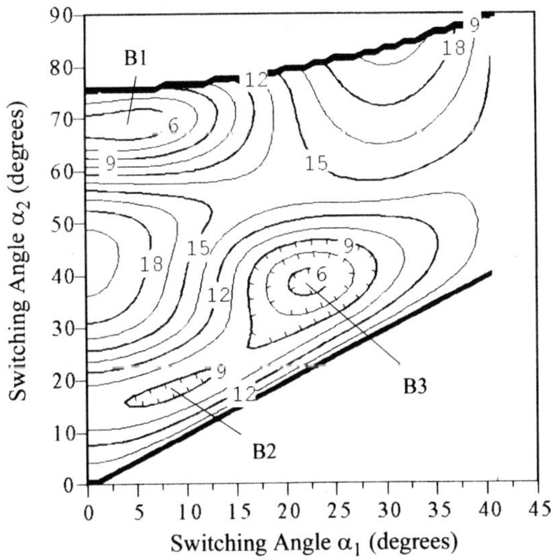

Figure 9.26 Contour plot of WTHD0 as a function of α_1 and α_2. Type B switching, chops per half cycle $N = 3$, modulation index $M = 0.5$.

The process of identifying solutions at local minima for pulse numbers of $N = 4$, 5 and larger is obtained in much the same manner as has been demonstrated. However, as the number of switching events per quarter cycle increases, the number of local minima also increases. Also, the searchable space for these minima also increases making the guarantee of finding all local minima difficult. For example, when $N = 5$, the searchable space is four dimensional (the fifth switching angle is constrained to realize the required fundamental component). Eight solutions have been shown to exist in Ref. [13] of which four are needed to assimilate the overall optimal solution.

9.5 Minimum-Loss PWM

It has already been mentioned that an optimum PWM technique should minimize the harmonic losses in an inductive load or in an induction motor with constant parameters. If the harmonic losses of interest are simple copper losses and if the WTHD is known, one can readily find this loss by solving Eq. (9.63) for I_{rms} and then taking

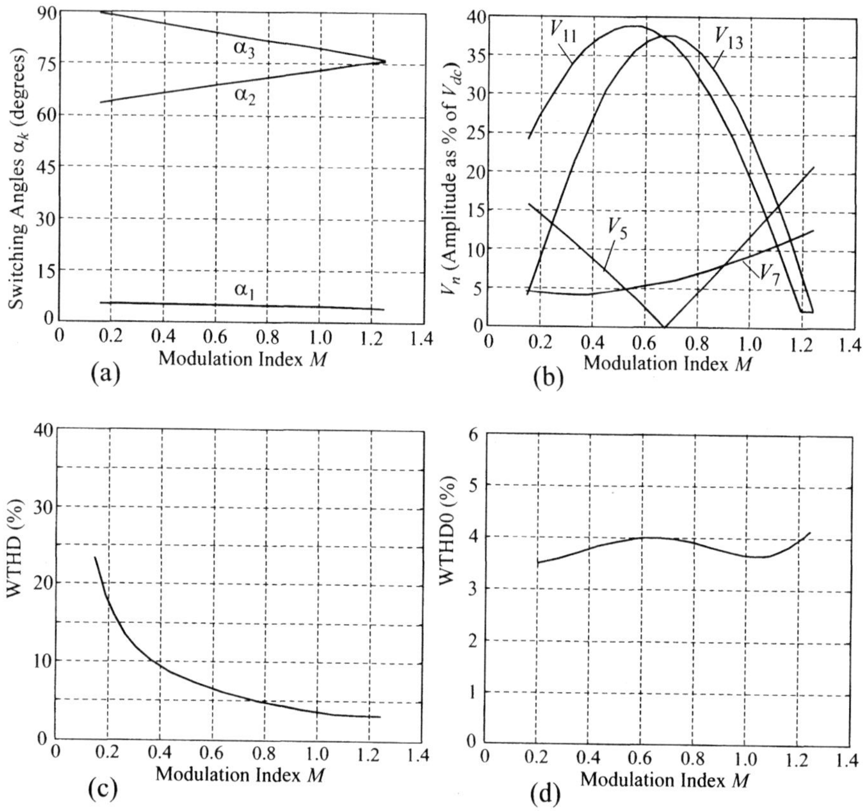

Figure 9.27 Optimum switching for chop number $N = 3$, type B switching, solution B1: (a) Switching angles α_1, α_2, α_3, (b) 5th, 7th, 11th, and 13th, (c) WTHD, and (d) WTHD0.

$$\begin{aligned} P_{1(cu)} &= 3I_{rms}^2 r_1 \\ &= 3\text{WTHD1}^2 I_{1,\,inrush}^2 r_1 \\ &= \text{WTHD1}^2 P_{1,\,inrush} \\ &= \text{WTHD1}^2 \left[3\left(\frac{V_1}{\omega_o L_\sigma}\right)^2 r_1\right] \end{aligned} \tag{9.67}$$

where r_1 is the induction motor stator winding resistance and

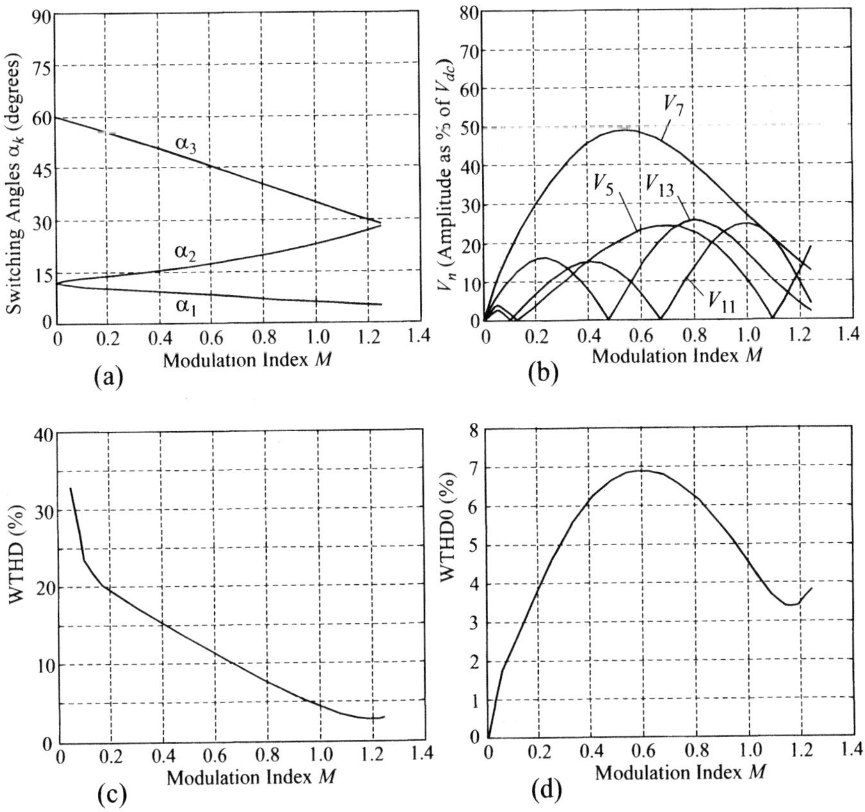

Figure 9.28 Optimum switching for chop number $N = 3$, type B switching, solution B2: (a) Switching angles α_1, α_2, α_3, (b) 5th, 7th, 11th, and 13th harmonic amplitudes, (c) WTHD, and (d) WTHD0.

$$\text{WTHD1} = \sqrt{\sum_{n=2}^{\infty} \frac{1}{n^2}\left(\frac{V_n}{V_1}\right)^2} = \frac{\text{WTHD01}}{M} \tag{9.68}$$

It is important to observe that the purely load-dependent quantities, stator resistance and leakage inductance, have been isolated from the purely source-dependent quantity outside the square brackets in Eq. (9.67). Hence the quantity outside the square brackets can be expressed as the loss factor for stator copper $\rho_{1(cu)}$ without frequency dependence, as

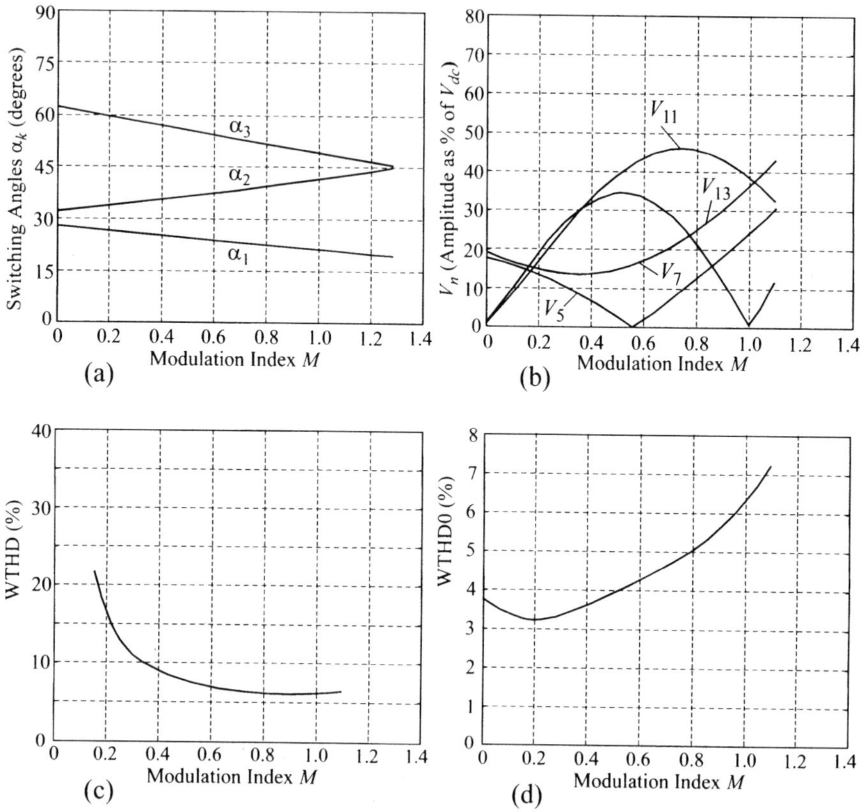

Figure 9.29 Optimum switching for chop number $N = 3$, type B switching, solution B3: (a) Switching angles α_1, α_2, α_3, (b) 5th, 7th, 11th, and 13th harmonic amplitudes, (c) WTHD, and (d) WTHD0.

$$\rho_{1(cu)} = \frac{\text{WTHD01}^2}{M^2} = \frac{1}{M^2}\left[\sum_{n=2}^{\infty} \frac{1}{n^2}\left(\frac{V_n}{V_{dc}}\right)^2\right] \tag{9.69}$$

In practice, it can be recalled that skin effect can have a significant influence on both the stator and the rotor copper losses due to the relatively large rotor bars. The stator resistive losses (in per unit of stator power dissipated during the inrush period) can be expressed approximately as a function of frequency by taking the square of Eq. (2.117), to give

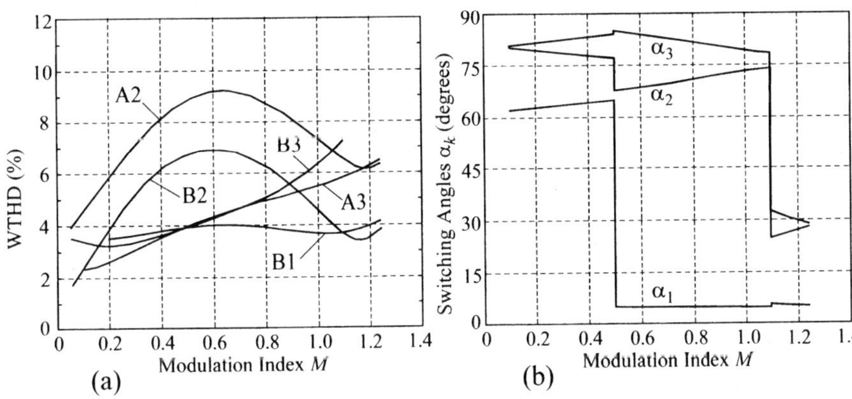

Figure 9.30 Composite optimum solution for pulse number $N = 3$ using type A3, B1, and B2 switching: (a) Overall minimum WTHD0 (solid line) and (b) corresponding switching angles.

$$\frac{P_{1(cu)}}{P_{1,\,inrush}} = \frac{\text{WTHD01}^2}{M^2} = \rho_{1(cu)}$$

$$= \frac{1}{M^2} \sum_{n\,=\,3k\pm 1}^{n_{0b}} \frac{1}{n^2}\left(\frac{V_n}{V_{dc}}\right)^2 + \left(\frac{\omega_o}{\omega_{0b}}\right) \sum_{n\,>\,n_{0b}}^{\infty} \frac{1}{n}\left(\frac{V_n}{V_{dc}}\right)^2 \frac{4}{\left[\sqrt{n\frac{\omega_o}{\omega_{0b}}} + 1.5\right]^2} \tag{9.70}$$

Likewise, the rotor copper losses as a per unit of inrush current are obtained by squaring Eq. (2.118) to give

$$\frac{P_{2(cu)}}{P_{2,\,inrush}} = \rho_{2(cu)} = \frac{\text{WTHD02}^2}{M^2} \tag{9.71}$$

$$= \frac{W1 + W2 + W3}{M^2} \tag{9.72}$$

where $W1$, $W2$, and $W3$ are defined by Eqs. (2.119) to (2.121).

Similar loss factors can be derived to account for other loss components, for example, iron loss. In this case the problem becomes exceedingly complicated since the iron losses have many components. However, theoretical and experimental investigations seem to have confirmed that the increase in core loss due to time harmonic main fluxes is negligible [14] since the stator harmonic currents in a squirrel cage induction motor are essentially opposed by equal and opposite rotor harmonic currents resulting in the main air gap flux remaining nearly sinusoidal. The core loss due to nonsinusoidal spatial variation of flux density, (space harmonics) is also small. However, the end-leakage and skew-leakage fluxes, which normally contribute only to the stray load loss, also produce an appreciable core loss at the harmonic frequencies since these flux components tend to enter the rotor perpendicular to the plane of the laminations. If an unskewed rotor is used, the dominant such component is the end-leakage loss. Alger, Angst, and Davies [15] indicate that these losses are proportional to the frequency times current squared. In effect, the equivalent resistance corresponding to end-leakage loss can then be considered as a linear function of frequency, and the power loss due to end leakage can be written as

$$P_{ew} = 3I_n^2 r_{ew0}\frac{nf_o}{f_{0b}} \tag{9.73}$$

where r_{ew0} is the equivalent end winding resistance measured at rated frequency f_{0b}. The corresponding loss factor for end winding leakage flux becomes

$$\frac{P_{ew}}{P_{i,\,inrush}} = \rho_{ew} = \mathrm{WTHD}_i^2 = \left(\frac{\mathrm{WTHD0}_i}{M}\right)^2 \tag{9.74}$$

$$= \frac{1}{M^2}\sum_{n=2}^{\infty}\frac{1}{n}\left(\frac{V'_n}{V_{dc}}\right)^2 \tag{9.75}$$

Finally, the stray load losses are given generally by

$$P_{sll} = K_{sll}\sum_{n=2}^{\infty}(I_n)^x\,(nf_1)^y \tag{9.76}$$

where the exponents x and y depend upon the machine construction. It has been determined [16] that this expression is estimated sufficiently well by setting $x = 2$ and $y = 1.5$, so that the power loss term becomes

$$P_{sll} = K_{sll}\sum_{n=2}^{\infty}I_n^2\,(nf_1)^{1.5} \tag{9.77}$$

The form of this equation suggests the definition of an equivalent stray load loss factor of the type

$$R_{sll,\,n} = K_{sll0} f_o^{1.5} \left(\frac{nf_o}{f_{0n}}\right)^{1.5} = r_{sll0} \left(\frac{nf_o}{f_{0b}}\right)^{1.5} \tag{9.78}$$

In per unit the power dissipated in the stray load loss is

$$\frac{P_{sll}}{P_{sll,\,inrush}} = \rho_{sll} = \text{WTHD}_{sll}^2 = \left(\frac{\text{WTHD0}_{sll}}{M}\right)^2 \tag{9.79}$$

$$= \frac{1}{M^2} \sum_{n=2}^{\infty} \frac{1}{\sqrt{n}} \left(\frac{V'_n}{V_{dc}}\right)^2 \tag{9.80}$$

In general, the total per unit power loss produced by harmonic current flow in the machine can be expressed by

$$P_{harmonic\ loss} = \sum_{n=2}^{\infty} (P_{1(cu)} + P_{2(cu)} + P_{ew} + P_{sll}) \tag{9.81}$$

Equation (9.81) can be written as

$$P_{harmonic\ loss} = \left[\rho_{1(cu)} + \left(\frac{r_2}{r_1}\right)\rho_{2(cu)} + \left(\frac{r_{ew0}}{r_1}\right)\rho_{ew} + \left(\frac{r_{sll0}}{r_1}\right)\rho_{sll}\right] P_{1,\,inrush} \tag{9.82}$$

Hence, it is therefore possible to define an overall harmonic loss coefficient for any machine as

$$\rho_h = \rho_{1(cu)} + \gamma_{2(cu)}\rho_{2(cu)} + \gamma_{ew}\rho_{ew} + \gamma_{sll}\rho_{sll} \tag{9.83}$$

where

$$\gamma_{2(cu)} = \frac{r_2}{r_1} \tag{9.84}$$

and so forth for γ_{ew} and γ_{sll}. Consequently, an overall effective weighted total harmonic distortion can be defined as

$$\text{WTHD}_{eff} = \frac{\text{WTHD0}_{eff}}{M} = \sqrt{\rho_{1(cu)} + \gamma_{2(cu)}\rho_{2(cu)} + \gamma_{ew}\rho_{ew} + \gamma_{s11}\rho_{s11}} \tag{9.85}$$

The per unit quantities $\gamma_{2(cu)}$, γ_{ew}, and γ_{sll} can be determined by test or by direct computation using finite elements. Hence, the harmonic loss for any particular induction machine can be minimized by choosing the switching angles to minimize WTHD_{eff} in Eq. (9.85).

The following three figures show an example of such an optimization with the parameters: $\gamma_{2(cu)} = 1.0$, $\gamma_{ew} = 0.2$, $\gamma_{sll} = 0.1$ for the case of $N = 3$. Figure 9.31 and Figure 9.32 are maps of WTHD_{eff} for the condition of $M = 0.5$ with type A and type B switching. The contours of these figures can be compared with Figures 9.23 and 9.26 to observe the effect of the frequency-dependent parameters. The distortion has again been normalized to the DC link voltage. It can be noted that the contours are not greatly affected by the frequency-dependent parameters and incorporation of iron and stray loss. It can also be observed that the amplitudes of the corresponding figures are different since the WTHD0 in Figures 9.23 and 9.26 pertains to the harmonic distortion related to the copper losses in the stator *or* the rotor while WTHD0_{eff} in Figures 9.31 and 9.32 accounts for stator *and* rotor copper losses as well as stray and iron losses. For a more representative comparison, WTHD0 in Figures 9.23 and 9.26 should be modified by $1 + r_i/r_1$ (= 2 in the case under consideration). In this case the WTHD can be estimated to have roughly doubled (four times increase in the losses). Figure 9.33 shows the overall switching angles and harmonic distortion and can be compared with Figure 9.30.

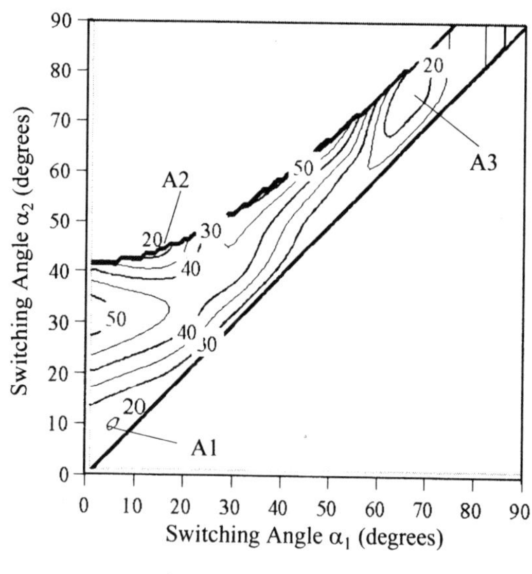

Figure 9.31 Contour plot of WTHD0_{eff} as a function of α_1 and α_2. Type A switching, chops per half cycle $N = 3$, modulation index $M = 0.5$.

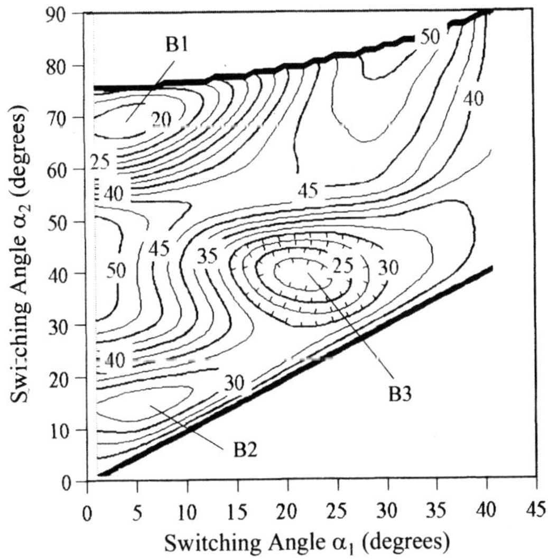

Figure 9.32 Contour plot of $WTHD0_{eff}$ as a function of α_1 and α_2. Type B switching, chops per half cycle $N = 3$, modulation index $M = 0.5$.

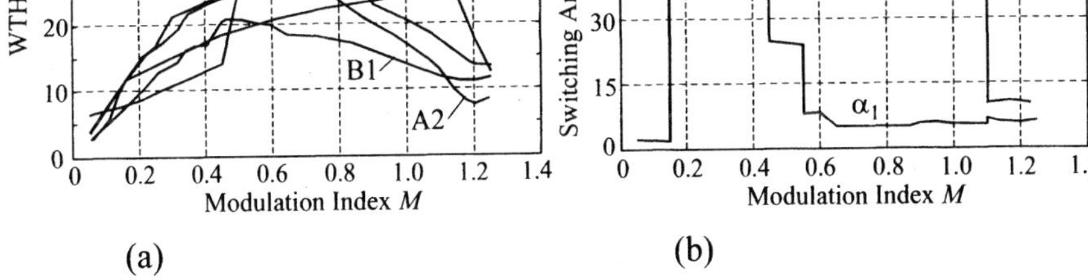

Figure 9.33 Composite optimum solution for pulse number $N = 3$, using type A2, A3, B1, and B3 switching: (a) Overall minimum $WTHD0_{eff}$ and (b) corresponding switching angles.

9.6 Summary

It has been shown in this chapter that considerable reduction in losses can be derived by varying the pulse interval ΔT during the course of a cycle. Savings in losses are particularly large during the overmodulation region when the inverter approaches square-wave operation. It should be noted that the word "optimum" must be used with extreme caution since it always implies an associated set of assumptions which sometimes are forgotten. For example, all of the optimum switching angles calculated as part of Section 9.4 assume constant motor parameters, which in actuality vary with frequency as illustrated in Section 9.5. Furthermore, both of these sections have made the implicit assumption that the optimum waveform must have no even harmonics, i.e., it is assumed that the waveform is half-wave symmetric. In reality, allowing for the presence of a small amount of even harmonics could produce switching angles which realize a still smaller value of the performance index. The true "optimum" clearly remains an elusive target which will certainly occupy the interest of researchers for years to come.

References

[1] S.R. Bowes and A. Midoun, "Suboptimal switching strategies for microprocessor-controlled PWM inverter drives," *IEE Proceedings (London)*, vol. 132, Pt. B, no. 3., May 1985, pp. 133–148.

[2] J. Holtz and B. Beyer, "Optimal pulsewidth modulation for ac servos and low-cost industrial drives," *IEEE Trans. on Industry Applications*, vol. 30, no. 4, July/Aug. 1994, pp. 1039–1047.

[3] F.G. Turnbull, "Selected harmonic reduction in static DC-AC inverters," *IEEE Trans. Communication and Electronics*, vol. 83, July 1964, pp. 374–378.

[4] H.S. Patel and R.G. Hoft, "Generalized techniques of harmonic elimination and voltage control in thyristor inverters: Part I — harmonic elimination," *IEEE Trans. on Industry Applications*, vol. IA–9, no. 3, May/June 1973, pp. 310–317.

[5] H.S. Patel and R.G. Hoft, "Generalized techniques of harmonic elimination and voltage control in thyristor inverters: Part II — voltage control techniques," *IEEE Trans. on Industry Applications, vol.* IA–10, no. 5, Sept./Oct 1974, pp. 666–673.

[6] "IMSL – International mathematical and statistical library," Visual Numerics, Inc., Suite 270, 2000 Crow Canyon Place, San Ramon CA 94583, http://www.vni.com.

[7] "NAG," Numerical Algorithms Group Ltd., Wilkinson House, Jordan Hill Rd., Oxford, OX28DR, United Kingdom, http://www.nag.co.uk.

[8] T. Kato, "Precise PWM waveform analysis of inverter for selected harmonic elimination," in *Conf. Rec. Industry Applications Society Industry Applications Society Annual Mtg*, Denver, 1986, pp. 611–616.

[9] P.N. Enjeti and J.F. Lindsay, "Solving nonlinear equations of harmonic elimination PWM in power control," *IEE Electronics Letters*, vol. 23, no.12, June 1987, pp. 656–657

[10] Q. Jiang, D.G. Holmes, and D.B. Giesner, "A method of linearising optimal PWM switching strategies to enable their computation on-line in real-time," in *Conf. Rec. IEEE Industry Applications Society Annual Mtg*., Dearborn, 1991, pp. 819–825.

[11] G.S. Buja and G.B. Indri, "Optimal pulse width modulation for feeding AC motors," *IEEE Trans. on Industry Applications,* vol. IA–13, no. 1, Jan./Feb. 1977, pp. 38–44.

[12] G.S. Buja, "Optimum output waveforms in PWM inverters," *IEEE Trans. on Industry Applications*, vol. IA–16, no. 6, Nov./Dec. 1980, pp. 830–836.

[13] F. Zach and H. Ertl, "Efficiency optimal control for ac drives with PWM inverters," *IEEE Trans. on Industry Applications,* vol. IA–21, no. 4, July/Aug. 1985, pp. 987–1000.

[14] B.J. Chalmers and B.R. Indri, "Optimal motor losses due to nonsinusoidal supply waveforms," *IEE Proc.* (*London*), vol 115, no. 12, Dec. 1968, pp. 1777–1782.

[15] P.L. Alger, G. Angst, and E.J. Davies, "Stray-load losses in polyphase induction machines," *AIEE Trans. Power Applications & Systems*, vol. 78, pt. III–A, June 1949, pp. 349–357.

[16] J.M.D. Murphy and M.G. Egan, "An analysis of induction motor performance with optimum PWM waveforms," in *Conf Rec. Int. Conf. on Electrical Machines* (*ICEM*), Athens, 1980, pp. 642–656.

10

Programmed Modulation of Multilevel Converters

Semiconductor switch ratings have limited the application of power converters rated in the tens to hundreds of megawatts. Large inverters operating at these power levels in the medium voltage range (2000 to 13.8 kV) have traditionally been the domain of gate turn off (GTO) thyristors. However, their switching speed is severely limited compared to IGBTs so that the carrier frequency of a GTO inverter is generally only a few hundred hertz. Higher switching frequencies can be achieved by replacing each of the slower switches in such a converter by a series string of faster IGBT switches so that each individual IGBT shares the impressed DC link voltage with others in the string during its off state. However, the operation of each of the series switches must be carefully matched to turn on and off in concert with the other series switches to prevent transient overvoltages. Since device characteristics vary with device current, temperature, and aging, the solution of this problem requires sophisticated adaptive gate control algorithms.

Multilevel converters, introduced in Sections 1.7 to 1.8.4, offer an attractive alternative for higher power applications. These converters also use higher speed switching devices but avoid the problems of linking them in series by connecting single devices between multiple DC voltage levels. They are more complex to modulate because of the number of switching alternatives that are available, but they do also provide the benefit of reduced harmonic components in the switched output voltage. The modulation and control of multilevel converters will be considered in this and in the next two chapters.

10.1 Multilevel Converter Alternatives

Multilevel converters were initiated by the invention of the so-called neutral point-clamped inverter of Nabae et al. [1]. At present, the two most common multilevel inverter structures are the *diode-clamped* inverter [2] (Nabae's converter topology renamed) and the *cascaded* inverter [3]. The diode-clamped inverter uses one DC bus subdivided into a number of voltage levels by a series string of capacitors. The voltages across the individual switches of the inverter are clamped by diodes at the voltage level of only one of the series string of capacitors. In contrast the cascaded inverter is made up from series-connected single-phase full-bridge inverters, each with their own isolated DC bus.

The most common type of diode-clamped inverter is shown in Figure 10.1 for one phase leg. If L is the number of levels, the diode-clamped inverter places a stress of V_{dc} across any nonconducting transistor. However, the voltage stress across the diodes can reach $V_{dc}(L-1)$ often necessitating the use of a series string of diodes to block the required voltage. A number of alternative connections have been considered to avoid the buildup of large voltages across any single device. Two possibilities are shown in Figures 10.2 and 10.3. Figure 10.4 shows a further approach using bidirectional switches. While impractical at the present time, the development of symmetrical GTO-type devices could result in this circuit becoming advantageous.

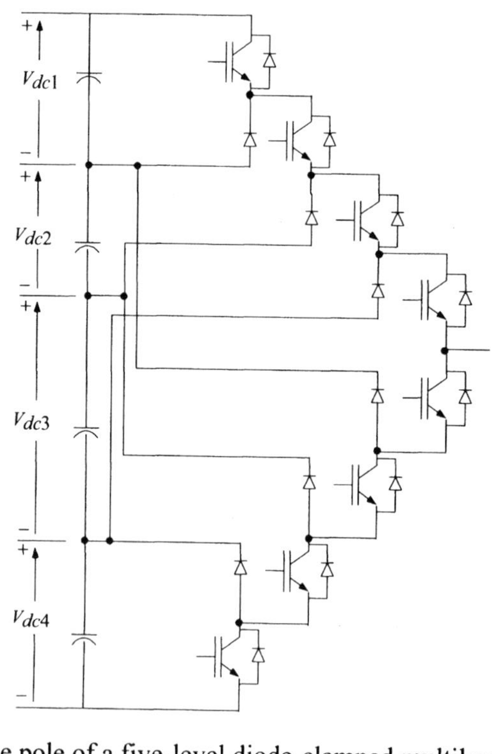

Figure 10.1 One pole of a five-level diode-clamped multilevel converter.

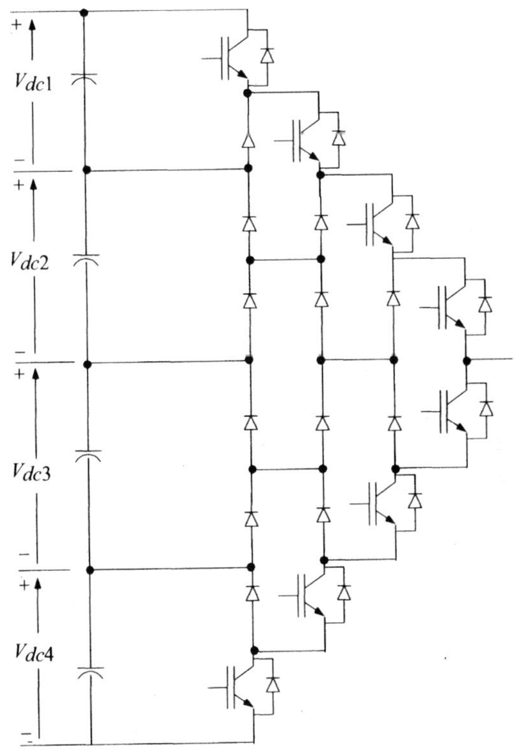

Figure 10.2 Single phase leg of alternate five-level inverter with reduced stress on clamp diodes.

The major difficulty associated with control of the diode-clamped inverter is the balancing of the capacitor voltages when the link is supplied by a single rectifier. This problem can clearly be alleviated by supplying each link capacitor with its individual, isolated DC supply. However, the cost associated with this option is considerable as the number of levels increases. Another alternative is to use an active bridge or chopper to control the inner voltages of the link (which are nominally floating). An example is shown in Figure 10.5 for a single pole of a four-level inverter. However, the voltage range can only be somewhat increased with this method of control.

The topology of the cascaded inverter concept for a seven-level case, previously introduced in Section 1.8.3, is shown again in Figure 10.6. In this case nine isolated DC power supplies are required. Although the problem of balancing the link capacitors does not then exist in principle, it is at the expense of a

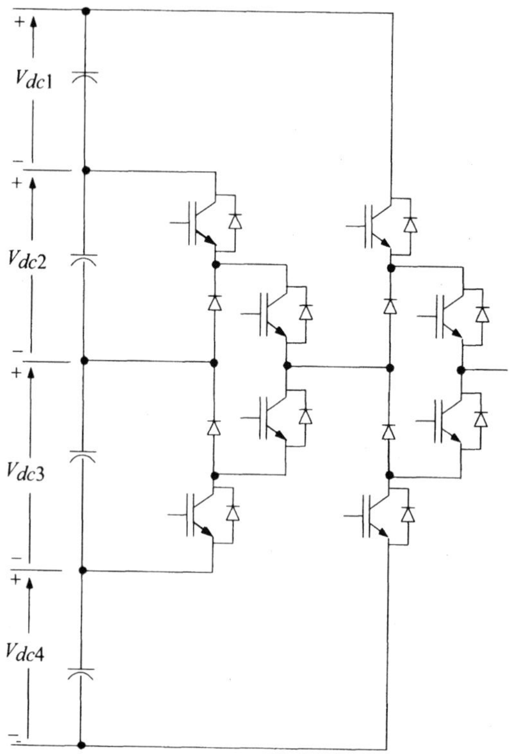

Figure 10.3 One phase leg of a nested three-level inverter producing five levels but requiring only four clamp diodes.

complicated transformer if the DC voltages are derived from a single three-phase AC supply.

Combinations of the diode-clamped and cascade converters are also possible. Figure 10.7 shows one implementation which combines a three-level NPC with a single-cascade inverter bridge. In this case six levels can be obtained. Four independent power supplies are required.

10.2 Block Switching Approaches to Voltage Control

In general, as suggested in Figures 10.1 and 10.6, the DC voltage level magnitudes need not be identical but can be adjusted as desired. Furthermore, in applications where the output voltage amplitude and frequency are relatively fixed, PWM need not be used to synthesize the output voltage since a sufficiently small WTHD can be obtained by simple fundamental frequency

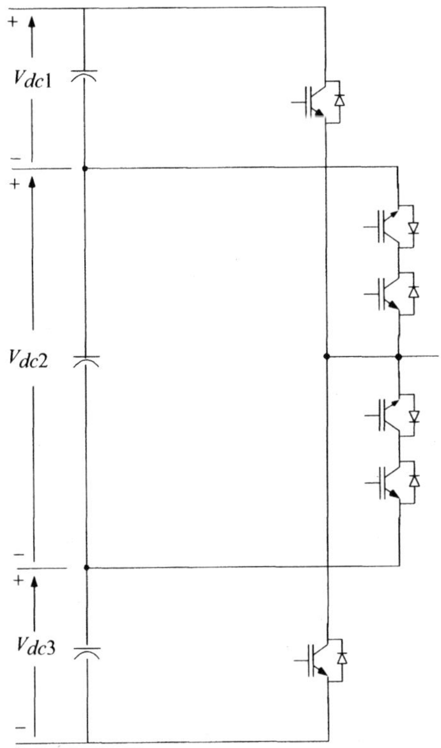

Figure 10.4 One phase leg of a four-level inverter realized with bidirectional switches.

switching, or *block switching*, of the semiconductors. Flexibility in the instant of application of the voltage blocks as well as its amplitude allows a means to synthesize an output voltage by choosing appropriate switching angles and DC voltage amplitudes as illustrated in Figure 10.8 for a seven-level inverter.

Two possibilities for voltage output synthesis can be proposed:

1. The DC voltages remain equal: $V_{dc1} = V_{dc2} = V_{dc3} = \cdots = V_{dc}$. The angles α_i are adjusted so as to adjust the fundamental voltage component and minimize the WTHD [4].

2. The voltage and angles are both adjusted to minimize the WTHD [5].

It should be mentioned that while the waveform of Figure 10.8 applies equally well to both diode-clamped and cascaded inverters, the inner voltages of the block-switched diode-clamped inverter cannot be maintained constant unless the load is purely reactive (i.e., a VAR control application). Otherwise

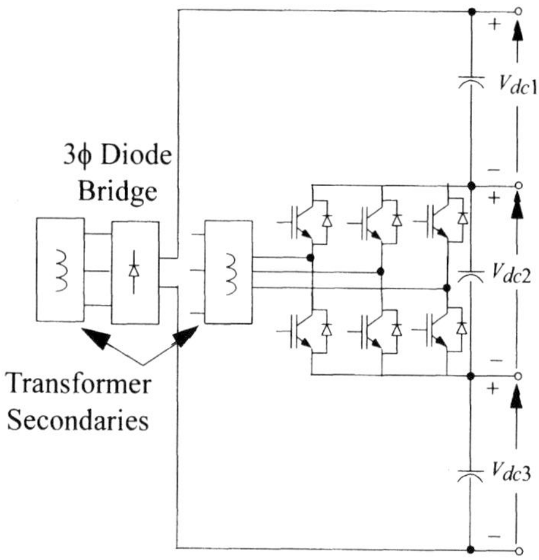

Figure 10.5 Active control of link capacitor voltages of a four-level inverter.

these inner capacitors must have auxiliary charging means not explicitly shown in Figure 10.1. It has been shown that balance can be maintained by using redundant states, i.e., pairs of states in which the current is directed into or out

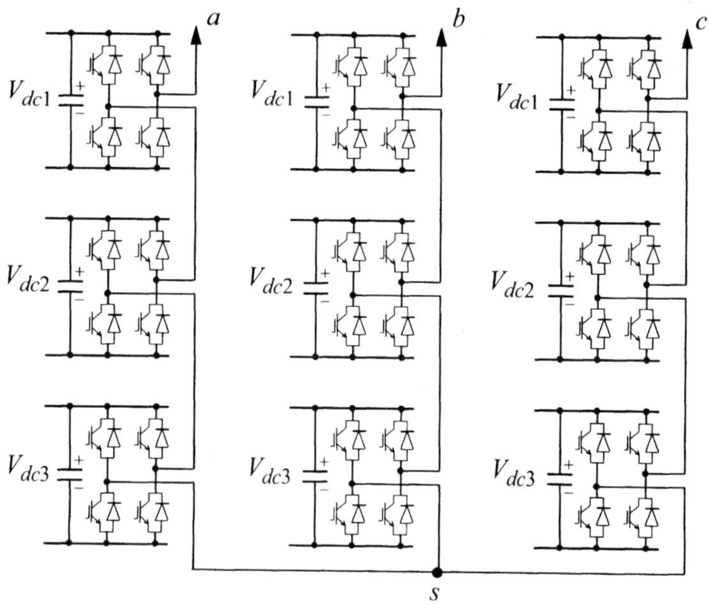

Figure 10.6 Seven-level cascaded inverter.

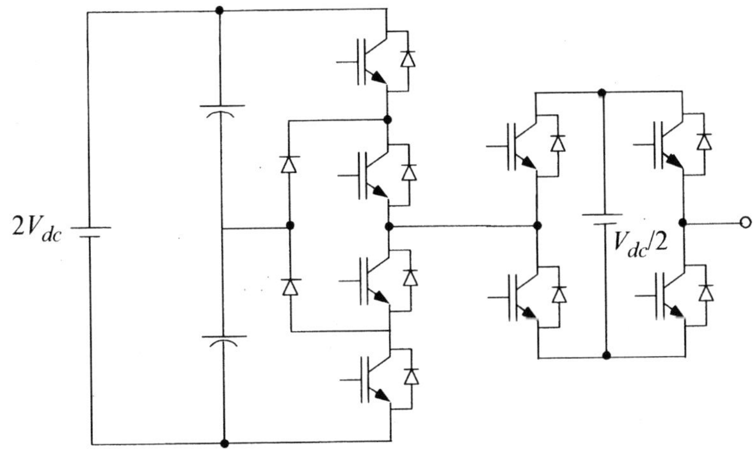

Figure 10.7 One phase leg of a hybrid three-level NPC cascade inverter.

of an inner capacitor node. However, the balance of the capacitor voltages is lost when the modulation index approaches unity since there is insufficient time to use the redundant voltage states to sustain capacitor voltage balance. Various solutions to this problem have been proposed in the literature [6-14].

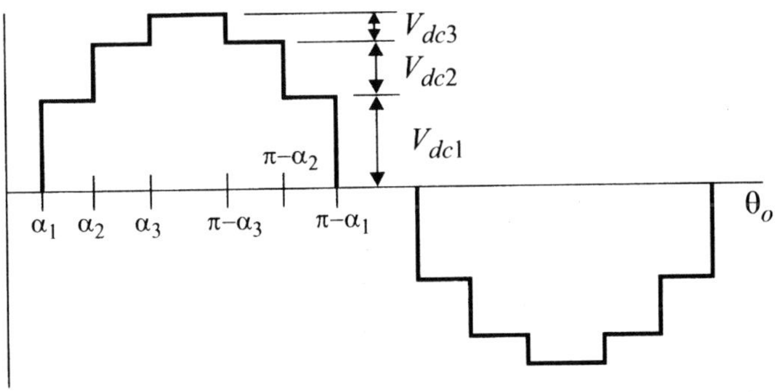

Figure 10.8 Voltage waveform of seven-level inverter using level switching.

10.3 Harmonic Elimination Applied to Multilevel Inverters

10.3.1 Switching Angles for Harmonic Elimination Assuming Equal Voltage Levels

For the case of equal voltage levels the Fourier coefficients of the output voltage are, for an L level inverter:

L odd

$$V_{az,n} = \frac{4}{n\pi}[V_{dc}\cos(n\alpha_1) + V_{dc}\cos(n\alpha_2) + \cdots + V_{dc}\cos(n\alpha_{(L-1)/2})] \quad (10.1)$$

L even

$$V_{az,n} = \frac{4}{n\pi}\left[\frac{V_{dc}}{2} + V_{dc}\cos(n\alpha_1) + \cdots + V_{dc}\cos(n\alpha_{L/2-1})\right] \quad (10.2)$$

The placement of α's for a seven-level cascade inverter are shown in Figure 10.9. The solution for even L is valid only for the diode-clamped arrangement.

To optimize these switching angles, either harmonic elimination or optimum PWM can be pursued. For example, if harmonic elimination is chosen for a seven-level inverter, then either a maximum number of three harmonics can

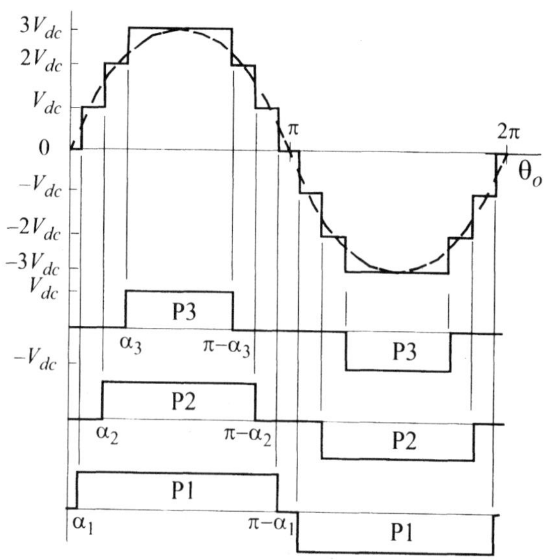

Figure 10.9 Switching diagram for seven-level cascade inverter.

be eliminated, and the resulting value of M is predetermined, or if M is specified then one fewer harmonic can be eliminated. In the first case the values of M and WTHD0 (excluding triplen harmonics since they cancel between phase legs) are shown in Table 10.1 for three to seven levels. It should be noted that when the number of levels is even, the number of capacitors on the DC link is odd. In this case, for purposes of visualization, the center capacitor can be split into two equivalent capacitors of double value to establish the link midpoint z.

For the second case where M is specified, the switching angles and WTHD as a function of M are shown in Figures 10.10 and 10.11 for five and seven levels, respectively. No solutions exist below $M = 0.35$ and 0.5 for five- and seven-level inverters, respectively. It should be noted that in a practical application the frequency of the inverter would be fixed and therefore the output voltage would be expected to vary only by a relatively small amount near the rated point. Switched voltage plots of the optimum solutions corresponding to the switching angles taken from Figures 10.10 and 10.11, for $M = 1.0$, are shown in Figures 10.12 and 10.13, respectively.

10.3.2 Equalization of Voltage and Current Stresses

It is apparent from Figure 10.9 that block modulation results in unequal losses in the semiconductor switches unless corrective action is taken. The voltage and current stresses within each single-phase bridge of a cascade inverter can be equalized as illustrated in Figure 10.14 [15]. The voltage and current stresses among the individual bridges can be accomplished by sequential swapping of the voltage pulses P1, P2, and P3 identified in Figure 10.9, as

Table 10.1 Optimum Switching Angles α_i for Harmonic Elimination When Maximum Possible Number of Harmonics Are Eliminated ($V_5 = 0$ for all cases)

	M	WTHD0 (%)	α_1 (deg)	α_2 (deg)	α_3 (deg)	V_7	V_{11}	V_{13}	V_{17}
Harmonic elimination, 3 levels	1.211	1.94	18.0	–	–	0.107	0.110	0.058	0.044
Harmonic elimination, 4 levels	1.200	1.13	24.0	–	–	0.058	0.030	0.076	0.058
Harmonic elimination, 5 levels	1.180	0.95	5.14	30.9	–	0	0.086	0.056	0.035
Harmonic elimination, 6 levels	1.176	0.45	12.3	33.6	–	0	0.036	0.009	0.037
Harmonic elimination, 7 levels	1.172	0.38	7.10	15.9	36.2	0	0	0.042	0.019

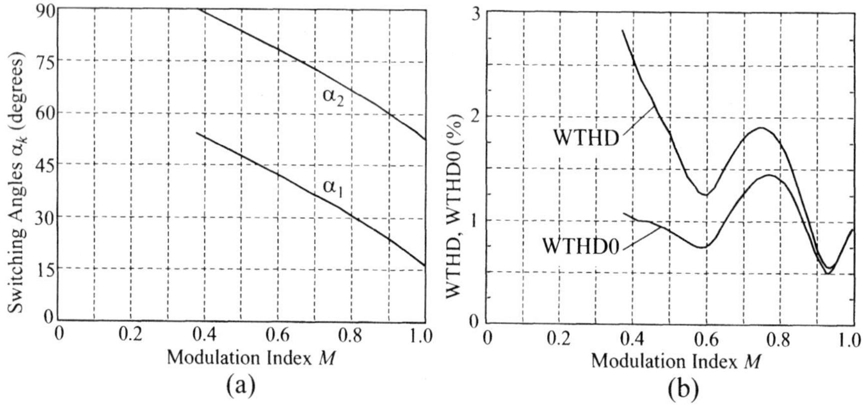

Figure 10.10 (a) Optimum switching angles α_i, $i = 1,2$ for a five-level inverter eliminating the 5th harmonic and (b) the resulting modulation indices WTHD and WTHD0.

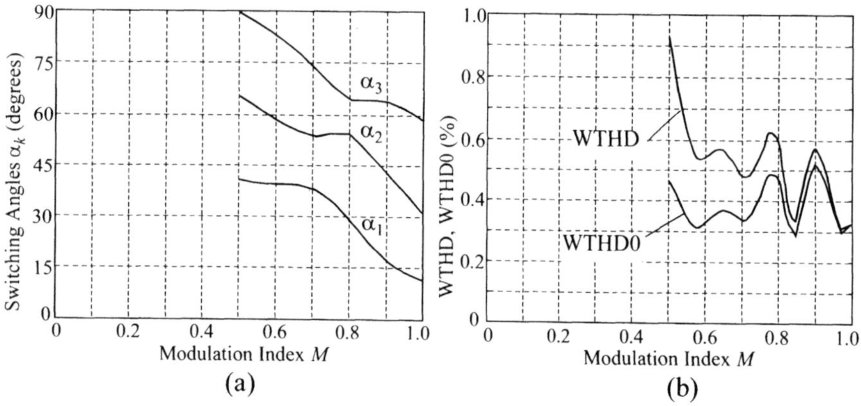

Figure 10.11 (a) Optimum switching angles α_i, $i = 1,2,3$ for a seven-level inverter eliminating the 5th and 7th harmonics and (b) the resulting modulation indices WTHD and WTHD0.

shown in Figure 10.15. Alternatively, the pulses can be programmed as shown in Figure 10.16. In this case two of the half cycle pulse widths are $\pi - \alpha_1 - \alpha_3$ while the pulse width of the third is $\pi - 2\alpha_2$. Since α_2 is roughly halfway between α_1 and α_3, satisfactory balance among the three phases can be achieved. If desired, interleaving of the two pulse widths can again be introduced to ensure balanced device losses.

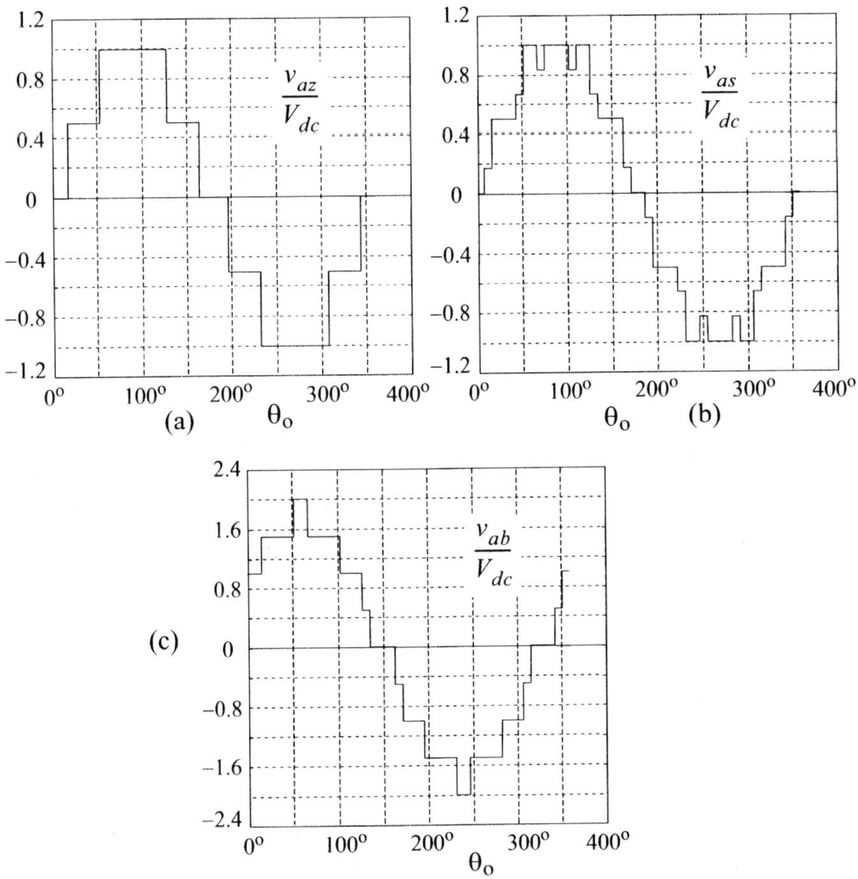

Figure 10.12 Time domain voltage waveforms for five-level inverter eliminating the 5th harmonic for modulation index $M = 1.0$: (a) line-to-DC midpoint voltage v_{az}, (b) line to load-neutral voltage v_{as}, and (c) line-to-line voltage v_{ab}.

10.3.3 Switching Angles for Harmonic Elimination Assuming Unequal Voltage Levels

If minimum harmonic distortion block switching is pursued, then the number of harmonics that can be eliminated is again equal to the number of levels $(L-1)/2$ if L is odd and $(L/2-1)$ if L is even. The bus voltages on the positive and negative side of the center point ground must be symmetrical to eliminate even harmonics. The voltage Fourier coefficients of the output voltage are then, for an L level inverter (L odd),

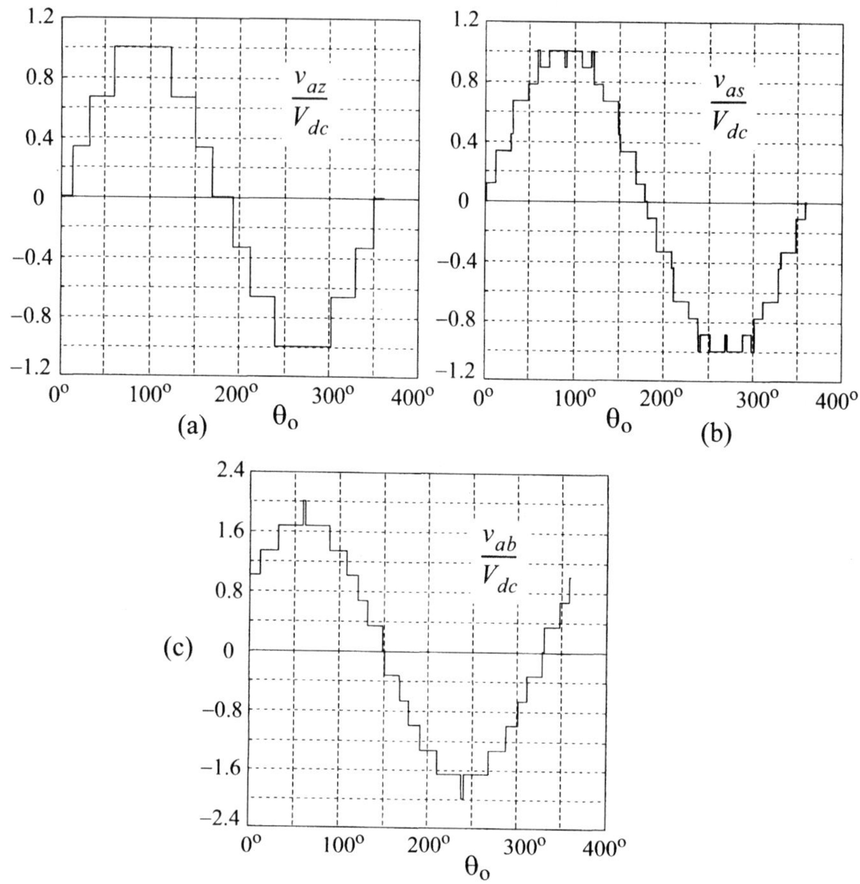

Figure 10.13 Time domain voltage waveforms for seven-level inverter eliminating the 5th and 7th harmonics for modulation index $M = 1.0$: (a) line-to-DC midpoint voltage v_{az}, (b) line to load-neutral voltage v_{as}, and (c) line-to-line voltage v_{ab}.

$$V_{az,\,n} = \frac{4}{n\pi}[V_{dc1}\cos(n\alpha_1) + V_{dc2}\cos(n\alpha_2) + \cdots + V_{dcn}\cos(n\alpha_{(L-1)/2})] \tag{10.3}$$

where the voltages are constrained by

$$V_{dc1} + V_{dc2} + \cdots + V_{dc(L-1)/2} = \frac{(L-1)}{2}V_{dc} \tag{10.4}$$

If L is even (again valid only for the diode-clamped arrangement) then the harmonics are given by

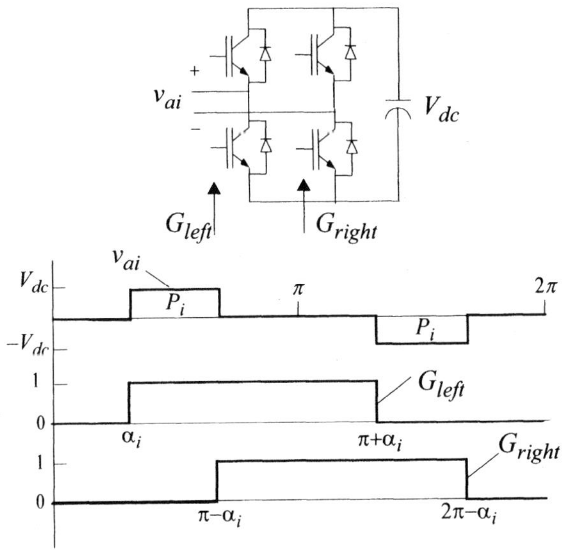

Figure 10.14 Single-bridge waveform and switching method to balance stresses, G_{left}, G_{right} represent gate control logic, i.e., "0" when lower device is on, "1" when upper device is on.

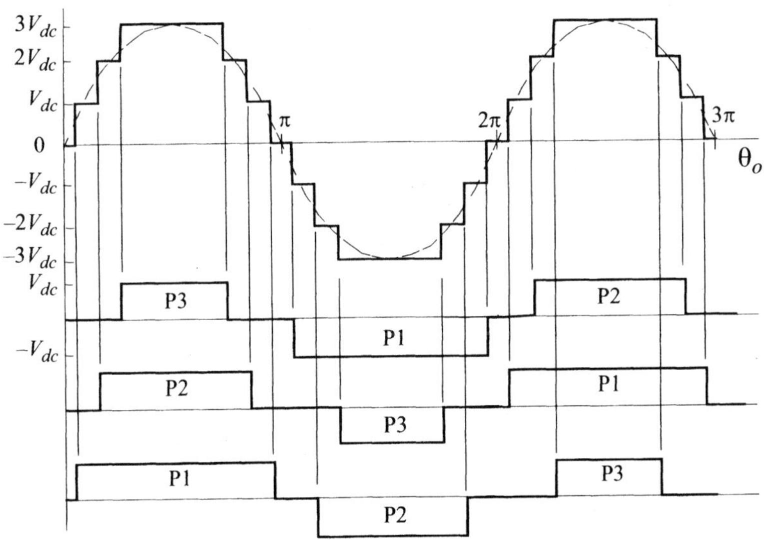

Figure 10.15 Interleaving of voltage pulses to balanced stress among cascaded inverters.

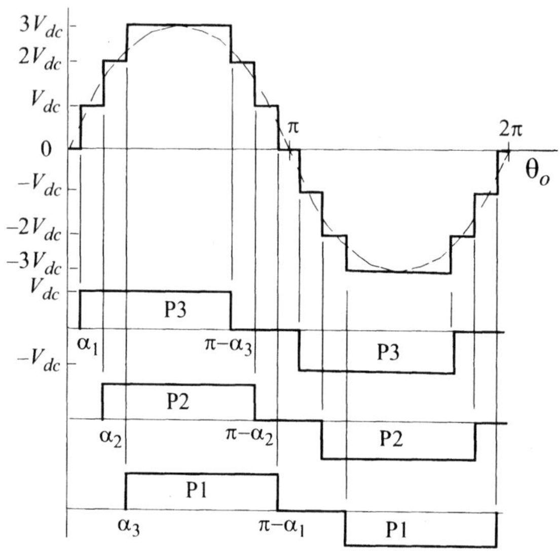

Figure 10.16 Alternate arrangement of pulses for seven-level cascaded inverter.

$$V_{az,n} = \frac{4}{n\pi}\left[\frac{V_{dc0}}{2} + V_{dc1}\cos(n\alpha_1) + \cdots + V_{dcn}\cos(n\alpha_{L/2-1})\right] \tag{10.5}$$

where the link voltages are constrained by

$$\frac{V_{dc0}}{2} + V_{dc1} + V_{dc2} + \cdots + V_{dc(L/2-1)} = \frac{(L-1)}{2}V_{dc} \tag{10.6}$$

It is apparent that since both the switching angles and values of the bus voltages are variables, more harmonics can be eliminated than if only the switching angles were varied. In general, when L is odd, the number of harmonics which can be eliminated are $[(L-1)/2+(L-3)/2]$ or L–2 harmonics. When L is even, $[(L/2-1)+(L/2-1)]$ or L–2 harmonics can be eliminated. One fewer harmonic can be eliminated if the amplitude of the voltage fundamental component (modulation index M) is specified. However, this solution can be considered as impractical since the DC link voltages which are normally set to fixed values must now vary with M to achieve this result.

Table 10.2 shows the results when the maximum number of harmonics are eliminated. Since the voltage levels of the individual cascaded inverters are now unequal, it is apparent that in this case the interleaving procedure of Figure 10.15 cannot be used to equalize the power supplied by the converters. This limitation can be considered as a significant drawback to this approach.

Table 10.2 Optimum Switching Angles α_i and Bus Voltages V_{dci} in Per Unit for Harmonic Elimination When Maximum Possible Number of Harmonics are Eliminated

	M	WTHD (%)	α_1 (deg)	α_2 (deg)	α_3 (deg)	V_{dc0} (p.u.)	V_{dc1} (p.u.)	V_{dc2} (p.u.)	V_{dc3} (p.u.)
Harmonic elimination, 4 levels	1.18	0.98	30	–	–	1.39	0.804	–	–
Harmonic elimination, 5 levels	1.173	0.47	10.97	35.24	–	–	1.268	0.732	–
Harmonic elimination, 6 levels	1.167	0.42	20.0	40.0	–	1.527	1.133	0.603	–
Harmonic elimination, 7 levels	1.032	0.92	16.44	34.59	58.99	–	1.269	049	0.682

10.4 Minimum Harmonic Distortion

A second useful criterion for selecting the switch angles of a multilevel inverter is to employ an optimization algorithm to minimize the WTHD, thus minimizing the losses in a motor load. If the DC voltage levels are assumed as equal, then only the switch angles become variables. When the number of levels equal three or four, only one switch angle can be varied, so that the "optimum" solution becomes the "only" solution. Solutions for five-, six- and seven- level inverters are given in Figures 10.17, 10.18, and 10.19.

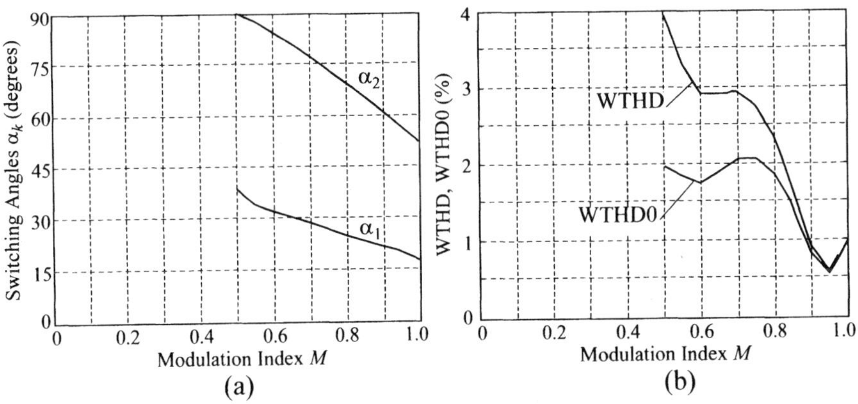

Figure 10.17 (a) Optimum switching angles α_1 and α_2 using minimum WTHD as the performance index for a five-level inverter and (b) the resulting minimum modulation indices WTHD and WTHD0.

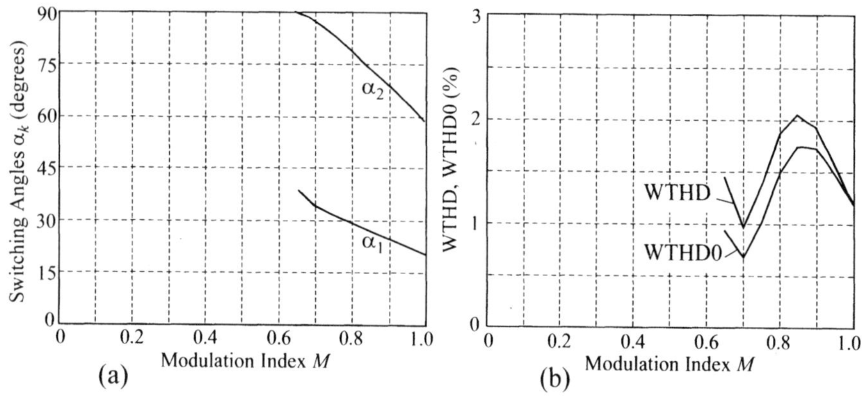

Figure 10.18 (a) Optimum switching angles α_1 and α_2 using minimum WTHD as the performance index for a six-level inverter and (b) the resulting minimum modulation indices WTHD and WTHD0.

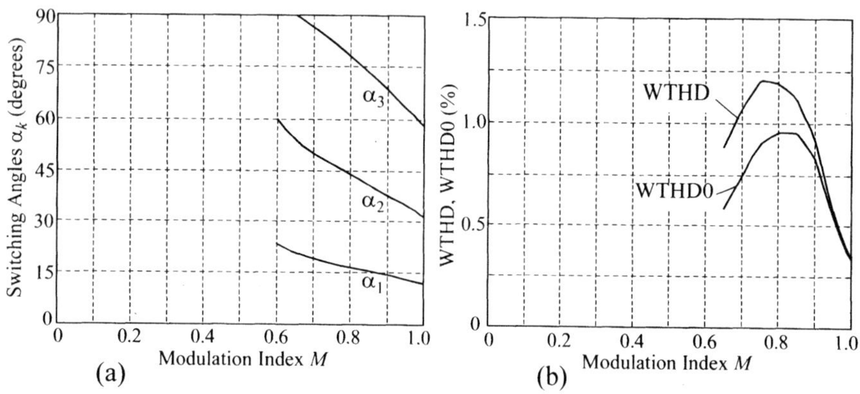

Figure 10.19 (a) Optimum switching angles α_1, α_2, and α_2 using minimum WTHD as the performance index for a seven-level inverter and (b) the resulting minimum modulation indices WTHD and WTHD0

When unequal DC bus voltage levels are permitted, nontrivial solutions can be obtained for four levels and greater. Again, the DC voltages can now be selected to realize an optimal condition for a given fundamental voltage. But once selected, this value, practically speaking, must be held constant since treating these voltages as variables would imply expensive adjustable voltage DC supplies for each level. Optimal solutions for two representative values of modulation index, $M = 1.0$ and $M = 1.1$, are summarized in Table 10.3.

Table 10.3 Minimum WTHD for Two Values of Modulation Index *M* When Both Switc Angles and Bus Voltages are Varied

$M = 1$	WTHD (%)	α_1 (deg)	α_2 (deg)	α_3 (deg)	V_{dc0} (p.u.)	V_{dc1} (p.u.)	V_{dc2} (p.u.)	V_{dc3} (p.u.)
5 levels	0.820	18.8	53.2	–	–	0.532	0.468	–
6 levels	0.630	25.39	55.67		0.350	0.429	0.397	–
7 levels	0.340	14.19	34.74	58.66	–	0.382	0.311	0.307
$M = 1.1$								
5 levels	1.32	16.7	42.5	–	–	0.575	0.425	–
6 levels	0.485	24.5	51.9	–	0.546	0.486	0.241	–
7 levels	0.430	9.54	35.7	59.9	–	0.507	0.376	0.117

10.5 Summary

This chapter has explored the modulation benefits of utilizing additional DC levels beyond the two levels provided by the conventional three-phase inverter topology that has been examined up to this point. It has been shown that the simple addition of multiple switched voltage levels has a dramatic effect on the reduction of WTHD. It can be recalled from Chapter 2 (Table 2.1) that the WTHD of a simple square-wave inverter is 4.64% and that progression to three-, four-, and five-level inverters results in a reduction in WTHD to 1.6, 1.12, and 0.987%, respectively. It has been further shown that these values can be reduced to 1.53, 0.89, and 0.75%, respectively, when the DC source voltage levels are optimized (Table 10.1) and to 0.82, 0.63, and 0.34% (Table 10.2) when both the DC levels and switching instants are optimized. Hence significantly reduced values of WTHD can be achieved with multilevel block modulation techniques without necessarily resorting to pulse width modulation. Unfortunately, however, the approach only appears to be practical when the output voltage target range is relatively small [e.g., for an uninterruptible power supply (UPS) application]. When the output voltage amplitude demand is wide and the slew rate is relatively rapid (e.g. in an AC motor drive) pulse width modulation of multilevel converters again becomes almost a necessity, and hence this approach will now be explored in the next chapter.

References

[1] A. Nabae, I. Takahashi, and H. Akagi, "A new neutral-point-clamped PWM inverter," *IEEE Trans. on Industry Applications*, vol. IA–17, no. 5, Sept./Oct. 1981, pp. 518–523.

[2] P. Bhagwat and V.R. Stefanovic, "Generalized structure of a multilevel PWM inverter," *IEEE Trans. on Industry Applications*, vol. 19, no. 6, Nov./Dec. 1983, pp. 1057–1069.

[3] P.W. Hammond, "A new approach to enhance power quality for medium voltage AC drives," *IEEE Trans. on Industry Applications*, vol. 33, no. 1, Jan/Feb. 1997, pp. 202–208.

[4] F.Z. Peng, J.S. Lai, J.W. McKeever, and J. Van Coevering, "A multilevel voltage-source inverter with separate DC sources for static VAR generation," *IEEE Trans. on Industry Applications*, vol. 32, no. 5, Sept./Oct. 1996, pp. 1130–1138.

[5] Q. Jiang and T.A. Lipo, "Switching angle and DC link voltage optimization for multilevel cascade inverters," *Electric Machines and Power Systems,* vol. 28, July 2000, pp. 605–612.

[6] M. Marchesoni, M. Mazzucchelli, and P. Tenca, "About the DC-link capacitors voltage balance in multi-point clamped converters," in *Conf. Rec. Int. Conf. on Industrial Electronics, Control and Instrumentation* (*IECON*), Aachen, 1998, pp. 548–553.

[7] M. Marchesoni, M. Mazzucchelli, F. Robinson, and P. Tenca, "Analysis of DC-link capacitor voltage balance in AC-DC-AC diode-clamped multilevel converters," in *Conf. Record European Power Electronics Conf.* (*EPE*), Lausanne, 1999, in CD ROM.

[8] M.C. Klabunde, Y. Zhao, and T.A. Lipo, "Current Control of a 3–Level Rectifier/Inverter Drive System," in *Conf. Rec. IEEE Industry Applications Society Annual Mtg.*, Denver, 1994, pp. 859–866.

[9] H.L. Liu, N.S. Choi, and G.H. Cho, "Space vector PWM for three-level inverter with DC-link voltage balancing," in *Conf. Rec. Int. Conf. on Industrial Electronics, Control and Instrumentation* (*IECON*), Kobe, 1991, pp. 197–203.

[10] S. Ogasawara and H. Akagi, "Analysis of variation of neutral point potential in neutral-point-clamped voltage source PWM inverters," in *Conf. Rec. IEEE Industry Applications Society Annual Mtg.*, Toronto, 1993, pp. 965–970.

[11] Y.-H. Lee, R.-Y. Kim, and D.-S. Hyun, "A novel SVPWM strategy considering DC-link balancing for a multi-level voltage source inverter," in *Conf. Rec. IEEE Applied Power Electronics Conf.* (*APEC*), Dallas, 1999, pp. 509–514.

[12] D.H. Lee, S.R. Lee, and F.C. Lee, "An analysis of the midpoint balance for neutral-point clamped three-level VSI," in *Conf. Rec. IEEE Power Electronics Specialists Conf.* (*PESC*), Fukuoka, 1998, pp. 193–199.

[13] N. Celanovic and D. Boroyevich, "Comprehensive study of neutral-point voltage balancing problem in three-level neutral-point-clamped voltage source PWM inverters," *IEEE Trans. on Power Electronics,* vol. 15, 2000, pp. 242–249.

[14] C. Keller, R. Jakob, and S. Salama, "Topology and balance control of medium voltage multilevel drives," in *Conf. Rec. European Power Electronics Conf. (EPE)*, Graz, 2001, in CD–ROM.

[15] L.M. Tolbert, F.Z. Peng, and T.G. Habetler, "Multilevel converters for large electric drives," *IEEE Trans. on Industry Applications*, vol. 35, no. 1, Jan./Feb. 1999, pp. 36–43.

11

Carrier-Based PWM of Multilevel Inverters

The era of multilevel inverters began in the 1980s with the development of the three-level diode-clamped inverter by Nabae, Takahashi, and Akagi in 1981 [1]. Proposals for diode-clamped and cascade multilevel inverters with higher number of DC voltage levels appeared shortly thereafter [2,3]. The first treatment of modulation principles for a multilevel inverter also dates to the 1980s. A major contribution to the categorization of the numerous modulation possibilities, the disposition process by Carrara et al. [4] in 1992, then set the stage for the detailed assessment of modulation alternatives that has appeared more recently in the research literature.

This chapter extends to multilevel inverters the principles of carrier-based PWM that have already been developed for two-level inverters. Much of the material presented comes from the research work of Dr. B.P McGrath, and a substantial part of the theoretical underpinnings of the chapter have been derived from his Ph.D. thesis [5].

11.1 PWM of Cascaded Single-Phase H-Bridges

Multilevel cascaded inverter topologies as shown in Figure 10.6 use series-connected single-phase H-bridge inverters to achieve an increased voltage range and reduced output harmonics. In the previous work of Chapter 4, the analytical solution for three-level naturally sampled modulation of one single-phase H-bridge was developed as Eq. (4.6), adapted here as

$$v_{ab}(t) = V_{dc}M\cos(\omega_o t) + \frac{4V_{dc}}{\pi}\sum_{m=1}^{\infty}\sum_{n=-\infty}^{\infty}\frac{1}{2m}J_{2n-1}(m\pi M)\cos([m+n-1]\pi) \times \cos(2m\omega_c t + [2n-1]\omega_o t) \quad (11.1)$$

by scaling $2V_{dc}$ to V_{dc} to suit the more usual bus voltage definition for single-phase bridges cascaded into a multilevel inverter.

It can be recalled that Eq. (11.1) was developed by subtracting two expressions corresponding to the switched voltages of phase legs *a* and *b* with respect to the DC neutral point. It can be further recalled that around each *odd* carrier multiple group (ω_c, $3\omega_c$, $5\omega_c$, etc.), these expressions contain only even sideband harmonics, and these sidebands cancel despite the $180°$ phase shift of the fundamental of phase leg *b* with respect to phase leg *a*. Hence, sideband harmonics of the line output voltage of a single-phase H-bridge only exist at *even* carrier multiple groups ($2\omega_c$, $4\omega_c$, $6\omega_c$, etc.).

A further cancellation can now be obtained by appropriately phase shifting the remaining harmonics of several series-connected single-phase H-bridges. This modulation process is denoted as *phase-shifted cascaded* PWM or PSCPWM. The essential principle of PSCPWM is to retain sinusoidal reference waveforms for the two phase legs of each H-bridge inverter that are phase shifted by $180°$ and to then phase shift the carriers of each bridge to achieve additional harmonic sideband cancellation around the even carrier multiple groups. To illustrate, Figure 11.1 illustrates the carrier and reference arrangements for the two single-phase H-bridges that are connected in series in the phase leg of a five-level cascaded inverter structure.

Optimum harmonic cancellation is achieved by phase shifting each carrier by $(i-1)\pi/N$, where i is the ith converter, N is the number of series-connected single-phase inverters per phase leg, and $N = (L-1)/2$ where L is the number of switched DC levels that can be achieved in each phase leg. This can be shown as follows.

Assuming natural sampling and equal DC voltages of V_{dc} across each single-phase bridge, the overall cascaded inverter phase leg to DC link midpoint voltage is given by

$$v_{az}(t) = \sum_{i=1}^{N} v_{ab}^{i}(t) \tag{11.2}$$

where v_{ab}^{i} is the line output voltage across each single-phase bridge *i*.

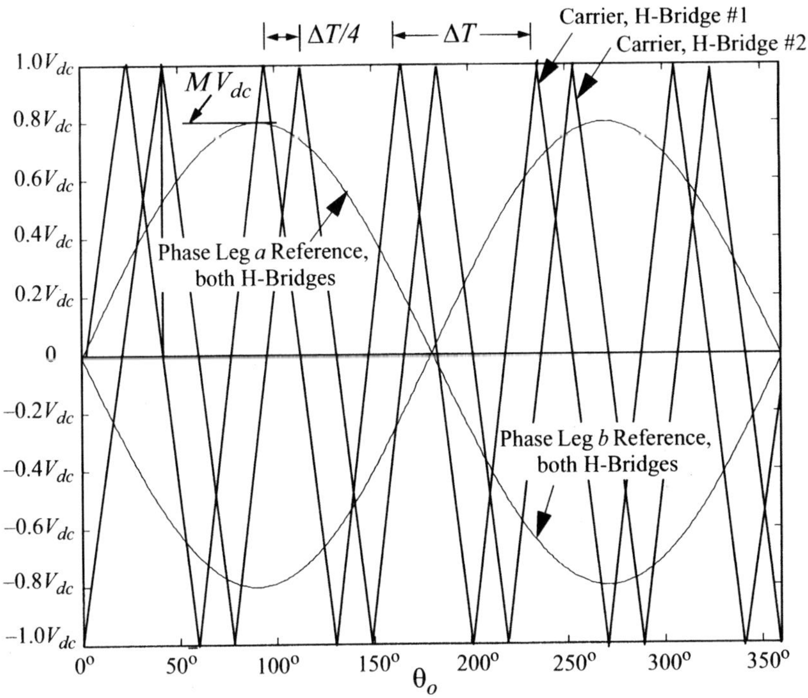

Figure 11.1 Phase-shifted carrier PWM (PSCPWM) strategy for five-level cascaded inverter showing the sinusoidal reference waveforms and carriers for the phase legs of the two cascaded single-phase bridges, $M = 0.8$, $f_c/f_o = 5$.

Substituting from Eq. (11.1) gives

$$v_{az}(t) = NV_{dc}M\cos(\omega_o t) \tag{11.3}$$

$$+\frac{4V_{dc}}{\pi}\sum_{m=1}^{\infty}\sum_{n=-\infty}^{\infty}\left[\begin{array}{l}\frac{1}{2m}J_{2n-1}(m\pi M)\cos([m+n-1]\pi)\\ \times\sum_{i=1}^{N}\cos\left\{2m\left(\omega_c t+\frac{[i-1]\pi}{N}\right)+[2n-1]\omega_o t\right\}\end{array}\right]$$

Since

$$\sum_{i=1}^{N}\cos\left(2m\omega_c t+[2n-1]\omega_o t+\frac{2m[i-1]\pi}{N}\right) = 0 \tag{11.4}$$

for all $m \neq kN,\ k = 1, 2, 3, \ldots$, the only phase leg harmonics remaining across the cascaded bridges will be sideband harmonic components centered around the $2N^{th}$ carrier multiples, or

$$v_{az}(t) = NV_{dc}M\cos(\omega_o t) + \frac{4V_{dc}}{\pi}\sum_{m=1}^{\infty}\sum_{n=-\infty}^{\infty}\frac{1}{2m}J_{2n-1}(Nm\pi M)\cos([Nm+n-1]\pi) \times \cos(2Nm\omega_c t + [2n-1]\omega_o t) \tag{11.5}$$

Therefore, for two cascaded H-bridges with carrier phase shifts of 90°, harmonic cancellation up to sidebands around multiples of $4f_c$ will be achieved. For three cascaded bridges with carrier phase shifts of 60°, harmonic cancellation up to sidebands around multiples of $6f_c$ will be achieved, and so on. It should be noted that this cancellation is not dependent on the carrier/fundamental frequency ratio and is valid even for very low pulse ratios. Of course, in addition, triplen sideband harmonic cancellation around the $2N^{th}$ carrier and its multiples will also occur between the phase legs as a separate process because of the 120° phase shift that occurs between the fundamental reference components of a three-phase system. Figure 11.2 illustrates the theoretical spectra for a single-phase cascaded inverter consisting of one, two, and three series-connected H-bridges, respectively. The successive cancellation of sideband groups is clearly evident.

For cascaded H-bridges controlled by two-level modulation, the carrier phase shift required between the cascaded inverters becomes $2(i-1)\pi/N$ because of the lesser harmonic elimination that is achieved by this modulation strategy within each single-phase bridge. Hence, this strategy will only achieve harmonic elimination up to the Nth carrier multiple and is therefore not recommended for multilevel cascaded inverters since it only achieves half the harmonic benefits of a three-level modulation strategy.

The solution for the voltage across a cascaded multilevel inverter phase leg with the single-phase bridges modulated by either symmetrical and asymmetrical regular sampled PWM with phase-shifted carriers can be developed in the same manner as has been done for the naturally sampled case. However, it can be recalled from Chapter 4 that for symmetrical regular sampled PWM, the odd carrier group sideband harmonics do not fully cancel, as shown in Figure 4.6. Hence cascaded bridges modulated under this strategy can only eliminate

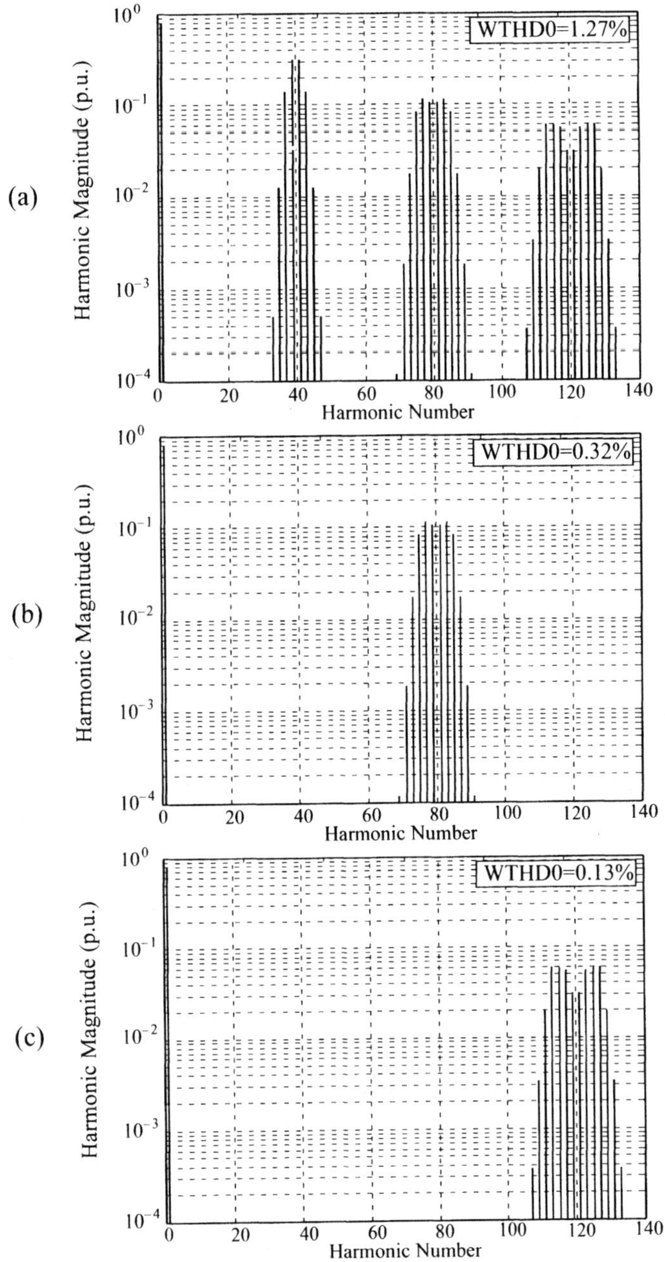

Figure 11.2 Theoretical spectra for phase leg of a cascaded inverter with *naturally sampled* PWM for (a) one H-bridge (3-level modulation), (b) two H-bridges (5-level modulation), and (c) three H-bridges (7-level modulation), $M = 0.8$, $f_c/f_o = 20$.

harmonics up to the Nth carrier multiple using a phase shift of $2(i-1)\pi/N$ (the same as for two-level PWM). In contrast, cascaded bridges modulated by asymmetrical regular sampled PWM will eliminate harmonics up to the $2N^{th}$ carrier multiple if a carrier phase shift of $(i-1)\pi/N$ is used, i.e., the same as for naturally sampled PWM. Thus only this strategy will be explored further.

The analytical solution for cascaded inverters modulated by asymmetrical regular sampled PWM is

$$v_{az}(t) = \frac{4NV_{dc}}{\pi}\sum_{n=1}^{\infty}\frac{1}{\left[n\frac{\omega_o}{\omega_c}\right]}J_n\left(n\frac{\omega_o}{\omega_c}\frac{\pi}{2}M\right)\sin n\frac{\pi}{2}\cos(n\omega_o t) \tag{11.6}$$

$$+\frac{4NV_{dc}}{\pi}\sum_{m=1}^{\infty}\sum_{n=-\infty}^{\infty}\frac{1}{q'}J_{2n-1}\left(q'\frac{\pi}{2}M\right)\cos([Nm+n-1]\pi) \times \cos(2Nm\omega_c t+[2n-1]\omega_o t)$$

where

$$q' = 2Nm+[2n-1]\omega_o/\omega_c \tag{11.7}$$

Figure 11.3 shows the progression of harmonic cancellation from a single-phase bridge (three-level modulation), two cascaded bridges (five-level modulation), and three cascaded bridges (seven-level modulation), modulated using asymmetrical regular sampled PWM. Complete cancellation of the switching harmonics up to the $2N^{th}$ carrier group sideband harmonics is again clearly evident.

For a three-phase multilevel inverter made up of cascaded bridges, the *l–l* voltage is again found by taking the voltage difference between phase legs with fundamental components that are phase displaced by 120°. The analytical *l–l* solution for naturally sampled modulation is given by

$$v_{ab}(t) = \sqrt{3}NV_{dc}M\cos\left(\omega_o t+\frac{\pi}{6}\right)$$
$$+\frac{8V_{dc}}{\pi}\sum_{m=1}^{\infty}\sum_{n=-\infty}^{\infty}\frac{1}{2m}J_{2n-1}(Nm\pi M)\cos([Nm+n]\pi)\sin\left([2n-1]\frac{\pi}{3}\right) \times \sin\left(2Nm\omega_c t+[2n-1]\left[\omega_o t-\frac{\pi}{3}\right]\right) \tag{11.8}$$

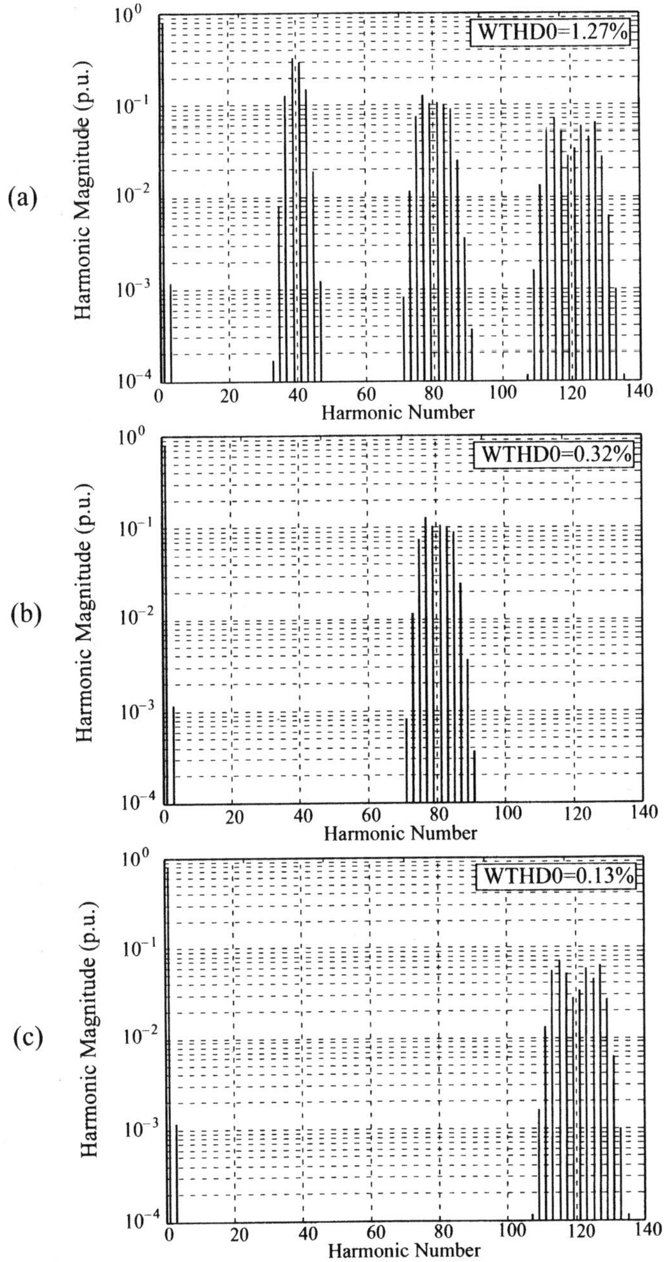

Figure 11.3 Theoretical spectra for phase leg of a cascaded inverter with *asymmetrical regular sampled* PWM for (a) one H-bridge (3-level modulation), (b) two H-bridges (5-level modulation), and (c) three H-bridges (7-level modulation), $M = 0.8$, $f_c/f_o = 20$.

while the *l–l* solution for regular sampled PWM is given by

$$v_{ab}(t) = \frac{8NV_{dc}}{\pi}\sum_{n=1}^{\infty}\frac{1}{\left[n\frac{\omega_o}{\omega_c}\right]}J_n\left(n\frac{\omega_o}{\omega_c}\frac{\pi}{2}M\right)\sin n\frac{\pi}{2}\sin n\frac{\pi}{3}\cos\left(n\left[\omega_o t-\frac{\pi}{3}\right]+\frac{\pi}{2}\right) \tag{11.9}$$

$$+\frac{8NV_{dc}}{\pi}\sum_{m=1}^{\infty}\sum_{n=-\infty}^{\infty}\frac{1}{q'}J_{2n-1}\left(q'\frac{\pi}{2}M\right)\cos([Nm+n]\pi)\sin\left([2n-1]\frac{\pi}{3}\right)\times\sin\left(2Nm\omega_c t+[2n-1]\left[\omega_o t-\frac{\pi}{3}\right]\right)$$

Figures 11.4 and 11.5 show the progression of *l–l* harmonic cancellation from a single-phase bridge (three-level modulation), two cascaded bridges (five-level modulation), and three cascaded bridges (seven-level modulation), modulated using natural and asymmetrical regular sampled PWM, respectively. Complete cancellation of the switching harmonics up to the $2N^{th}$ carrier group sideband harmonics is again clearly evident, together with the expected cancellation of the triplen harmonics from the $2N^{th}$ carrier group sidebands.

Experimental phase leg waveforms of a five- and a seven-level cascaded inverter are shown in Figures 11.6 and 11.7. Corresponding measured spectra are shown in Figures 11.8 and 11.9 These spectra compare directly with the theoretical predictions of Figures 11.3(b) and 11.3(c), respectively, and the match is very close. The only significant differences are the low-magnitude baseband harmonics in the experimental spectra, which are caused by practical implementation issues in the experimental switched output voltages [5]. Such issues are, of course, inevitable with any physical inverter and typically cause slight low-order harmonic distortion compared to the theoretical ideal.

It should be noted that to achieve effective harmonic sideband cancellation for a cascaded inverter system, it is essential that all single-phase bridges have exactly the same DC bus voltage (to within a few percent). Also, for regularly sampled implementations, it is critical to sample the reference waveform for each single-phase bridge at the peak and base of its particular (phase-shifted) triangular carrier. In other words, the reference waveform sample point must be phase shifted by the appropriate triangular carrier phase shift for each individual bridge. Otherwise, harmonics across the multiple bridges will not have the correct magnitude and phase relationships to achieve proper cancellation.

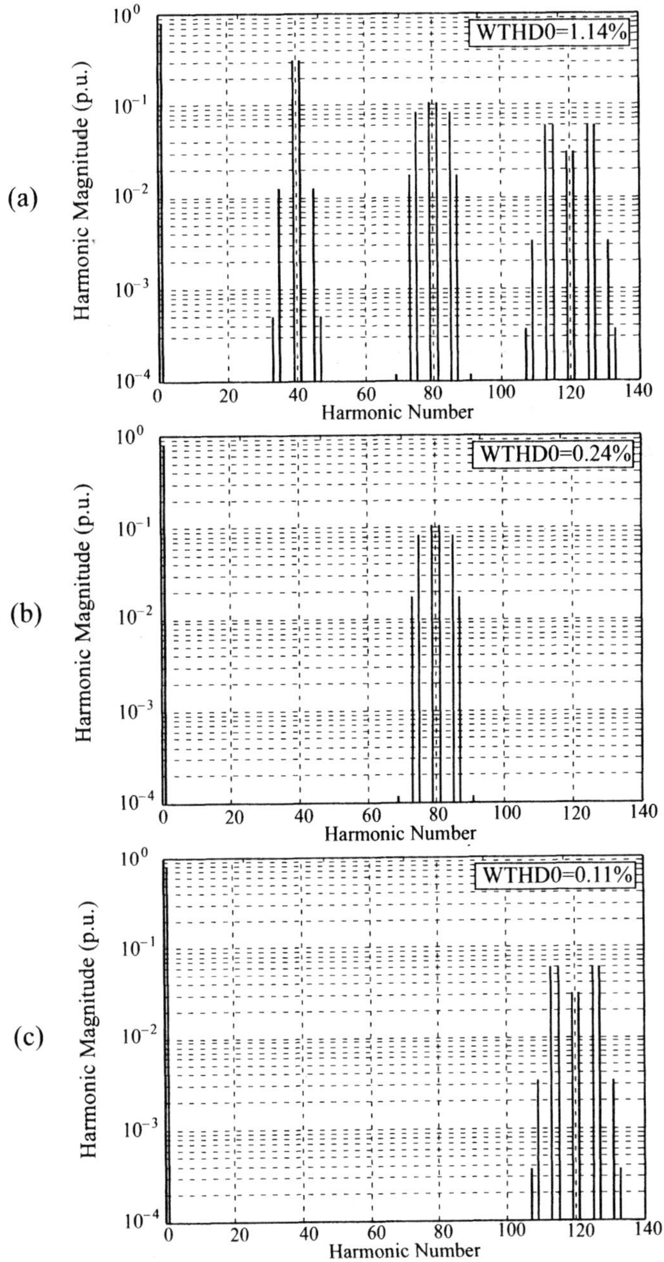

Figure 11.4 Theoretical *l–l* spectra for three-phase cascaded inverter with *naturally sampled* PWM for (a) one H-bridge (3-level modulation), (b) two H-bridges (5-level modulation), and (c) three H-bridges (7-level modulation), $M = 0.8$, $f_c/f_o = 20$.

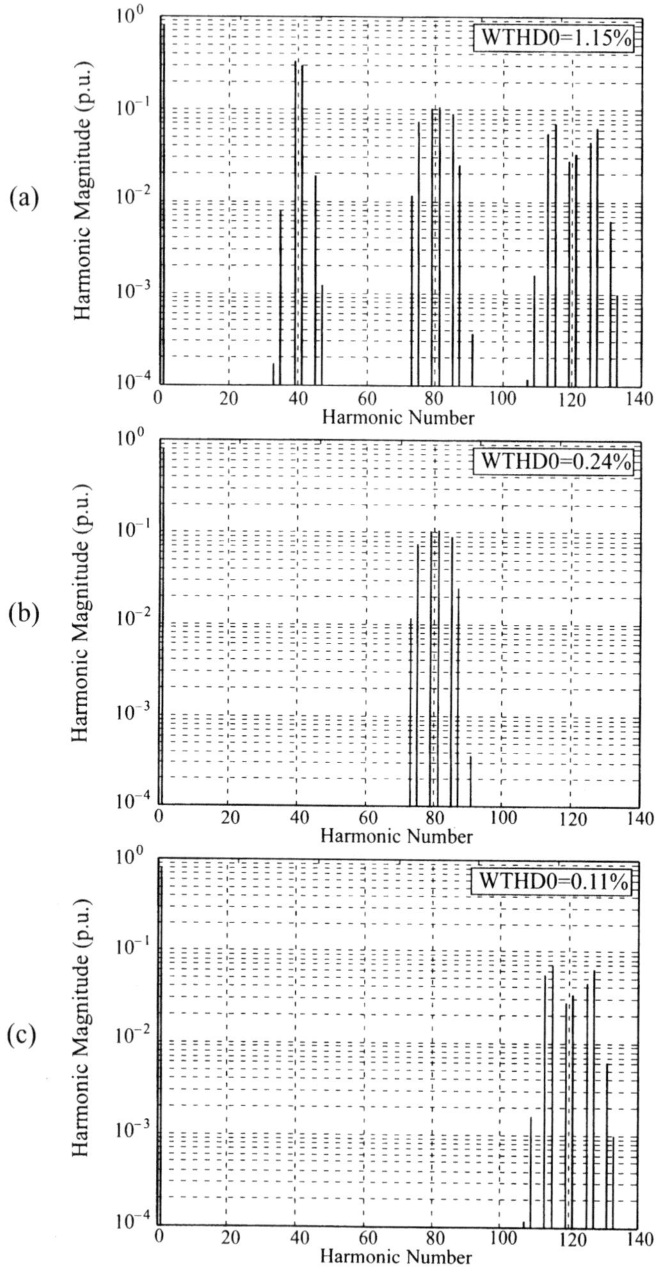

Figure 11.5 Theoretical *l–l* spectra for three-phase cascaded inverter with *asymmetrical regular sampled* PWM for (a) one H-bridge (3-level modulation), (b) two H-bridges (5-level modulation), and (c) three H-bridges (7-level modulation), $M = 0.8$, $f_c/f_o = 20$.

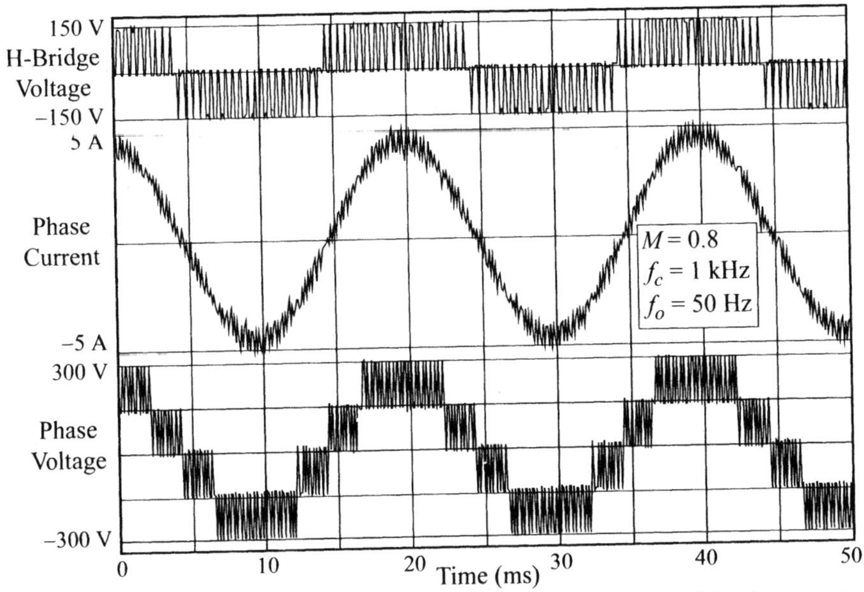

Figure 11.6 Experimental switched phase leg voltages and load current for two cascaded bridges, asymmetric regular sampled PWM, $M = 0.8$, $f_o = 50Hz$, $f_c/f_o = 20$ [6].

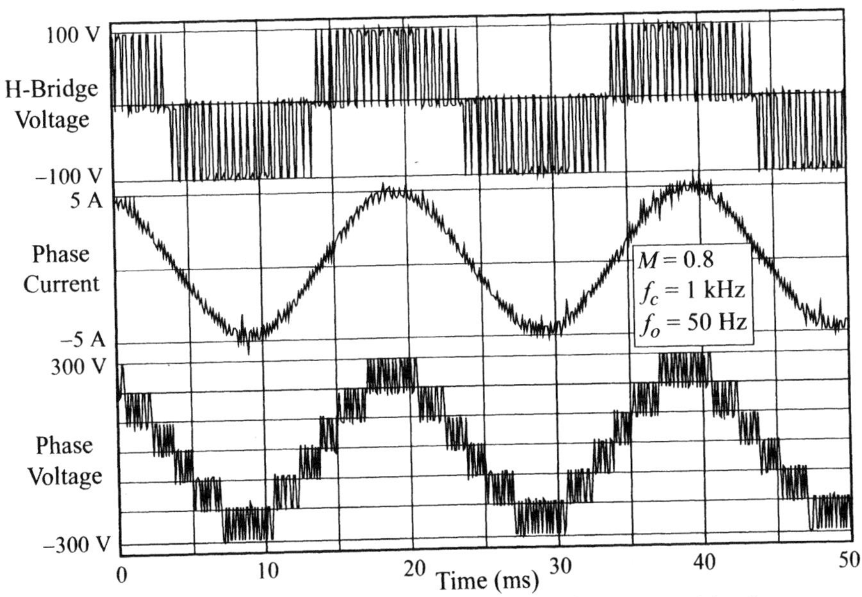

Figure 11.7 Experimental switched phase leg voltages and load current for three cascaded bridges, asymmetric regular sampled PWM, $M = 0.8$, $f_o = 50Hz$, $f_c/f_o = 20$ [6].

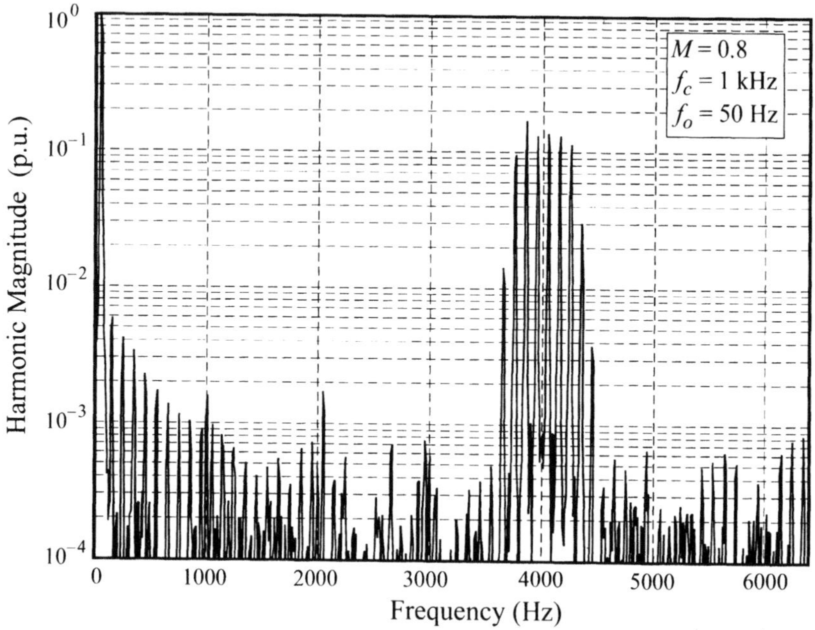

Figure 11.8 Experimental phase leg voltage spectrum for two cascaded bridges (5-level), $M = 0.8$, $f_o = 50Hz$, $f_c/f_o = 20$ [6].

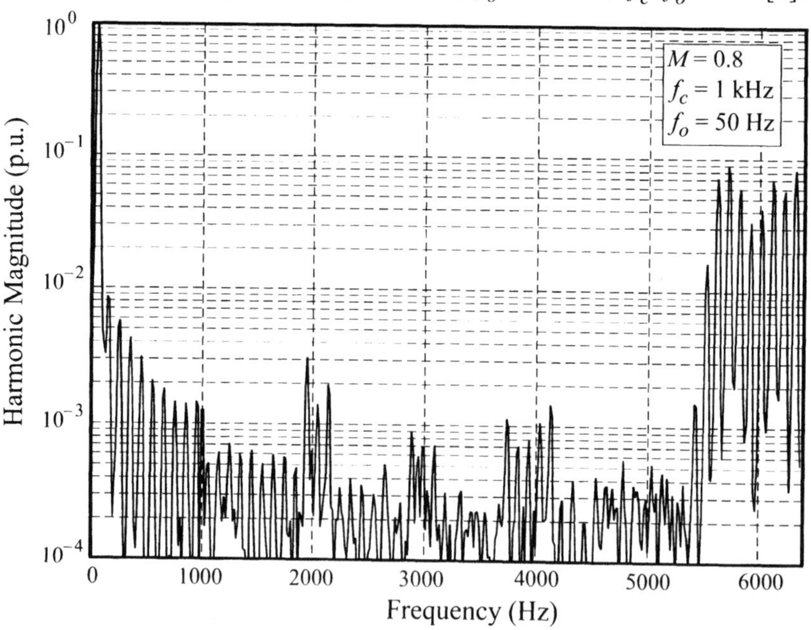

Figure 11.9 Experimental phase leg voltage spectrum for three cascaded bridges (7-level), $M = 0.8$, $f_o = 50Hz$, $f_c/f_o = 20$ [6].

11.2 Overmodulation of Cascaded H-Bridges

Utilizing the results obtained in Chapter 8, it is also possible to determine the behavior of multilevel cascaded H-bridges in the overmodulation region for both the naturally sampled and asymmetric regular sampled cases [7].

For the naturally sampled case

$$
\begin{aligned}
v_{az}(t) = {} & \frac{NV_{dc}}{\pi}[4\sin\psi + M(\pi - 2\psi - \sin 2\psi)]\cos\omega_o t \\
& + \frac{NV_{dc}}{\pi}\sum_{n=2}^{\infty}\left[\begin{array}{l} \dfrac{4}{[2n-1]}\sin([2n-1]\psi) \\ -M\left[\dfrac{\sin 2n\psi}{n} + \dfrac{\sin(2[n-1]\psi)}{[n-1]}\right]\end{array}\right]\cos([2n-1]\omega_o t) \\
& + \frac{4NV_{dc}}{\pi^2}\sum_{m=1}^{\infty}\sum_{n=-\infty}^{\infty}\frac{1}{2Nm}C_{mn}\cos(2Nm\omega_c t + [2n-1]\omega_o t)
\end{aligned}
\tag{11.10}
$$

where C_{mn} is given by

$$
C_{mn} = \left[\begin{array}{l} J_{2n-1}(Nm\pi M)\cos([Nm+n-1]\pi)\left(\pi - 2\psi - \dfrac{\sin(2[2n-1]\psi)}{[2n-1]}\right) \\ \\ + \displaystyle\sum_{\substack{k=1 \\ 2k-1 \neq |2n-1|}}^{\infty} J_{2k-1}(Nm\pi M)\cos[(Nm+k)\pi] \\ \qquad\qquad \times\left\{\dfrac{\sin(2[n+k-1]\psi)}{[n+k-1]} + \dfrac{\sin(2[n-k]\psi)}{[n-k]}\right\} \end{array}\right]
\tag{11.11}
$$

and ψ is once more defined by

$$
\psi = \left\{\begin{array}{ll} \cos^{-1}\left(\dfrac{1}{M}\right) & \text{for} \quad M > 1 \\ 0 & \text{for} \quad M \leq 1 \end{array}\right\}
\tag{11.12}
$$

For the asymmetric regular sampled case the phase leg voltage spectrum for an N-level cascaded inverter is

$$v_{az}(t) = \frac{4NV_{dc}}{\pi^2}\sum_{n=1}^{\infty} C_{0n}\cos([2n-1]\omega_o t) + \frac{4NV_{dc}}{\pi^2}\sum_{m=1}^{\infty}\sum_{n=-\infty}^{\infty} C_{mn}\cos(2mN\omega_c t + [2n-1]\omega_o t) \tag{11.13}$$

where

$$C_{0n} = \frac{1}{[2n-1]\frac{\omega_o}{\omega_c}}\left[\begin{aligned} &\frac{2}{[2n-1]}\sin([2n-1]\psi)\sin\left([2n-1]\frac{\omega_o}{\omega_c}\frac{\pi}{2}\right) \\ &+ J_{2n-1}\left([2n-1]\frac{\omega_o}{\omega_c}\frac{\pi}{2}M\right)\cos([n-1]\pi) \\ &\qquad\times\left[\pi - 2\psi - \frac{1}{[2n-1]}\sin(2[2n-1]\psi)\right] \\ &+ \sum_{\substack{k=1 \\ 2k-1 \neq |2n-1|}}^{\infty} J_{2k-1}\left([2n-1]\frac{\omega_o}{\omega_c}\frac{\pi}{2}M\right)\cos k\pi \\ &\qquad\times\left[\frac{\sin(2[n+k-1]\psi)}{[n+k-1]} + \frac{\sin(2[n-k]\psi)}{[n-k]}\right] \end{aligned}\right] \tag{11.14}$$

and

$$C_{mn} = \frac{1}{q'}\left[\begin{aligned} &\frac{2}{[2n-1]}\sin([2n-1]\psi)\sin\left([2n-1]\frac{\omega_o}{\omega_c}\frac{\pi}{2}\right) \\ &+ J_{2n-1}\left(q'\frac{\pi}{2}M\right)\cos([Nm+n-1]\pi) \\ &\qquad\times\left[\pi - 2\psi - \frac{1}{[2n-1]}\sin(2[2n-1]\psi)\right] \\ &+ \sum_{\substack{k=1 \\ 2k-1 \neq |2n-1|}}^{\infty} J_{2k+1}\left[q'\frac{\pi}{2}M\right]\cos([Nm+k]\pi) \\ &\qquad\times\left[\frac{\sin(2[n+k-1]\psi)}{[n+k-1]} + \frac{\sin(2[n-k]\psi)}{[n-k]}\right] \end{aligned}\right] \tag{11.15}$$

and where q' is defined by

$$q' = 2Nm + [2n-1]\omega_o/\omega_c \tag{11.16}$$

While these equations are quite complex, upon examination it becomes apparent that the complexity is primarily related to the magnitude of the harmonic terms, and that new harmonic components are only created in the baseband region with frequencies of $[2n-1]\omega_o t$. These harmonics can be readily identified as the low-order odd harmonics created by the overmodulation process as the effective reference waveform approaches a square wave in a similar fashion as identified for two-level modulation in Chapter 8. The important consequence of this characteristic is that the harmonic cancellation of the sideband harmonics that is achieved by cascading H-bridges into a multilevel inverter is not affected by overmodulation. Hence, one would expect the same harmonic benefits of cascaded bridges to be achieved for any modulation index.

Figure 11.10 shows the phase leg spectra for a five-level cascaded inverter with the two sampling methods for a modulation index of 1.2. Cancellation of sideband groups up to the fourth carrier multiple again occurs in the same manner as in the linear modulation region.

11.3 PWM Alternatives for Diode-Clamped Multilevel Inverters

Diode-clamped circuits make up the other major category of multilevel inverters. Carrier strategies similar to those employed for conventional two-level three-phase inverters can be readily extended to the three-phase multilevel case. Again, various types of single-edged (i.e., sawtooth) waveforms could be considered. However, it has already been shown in previous chapters that the benefits of harmonic cancellation are always superior when a double-edged triangular carrier is utilized, and hence only triangular carrier modulation of multilevel inverters will be considered further.

Double-edge modulation can be easily achieved by increasing the number of triangular carriers to $L-1$ where L is the number of voltage levels comprising the DC link [4]. These carriers are arranged so that they fully occupy contiguous bands in the range of $-(L-1)V_{dc}/2$ to $(L-1)V_{dc}/2$ (for L odd). A single sinusoidal reference is then compared with these carriers to determine the switched voltage level to be attained. Three alternative PWM strategies with differing phase relationships have been developed in the literature [4].

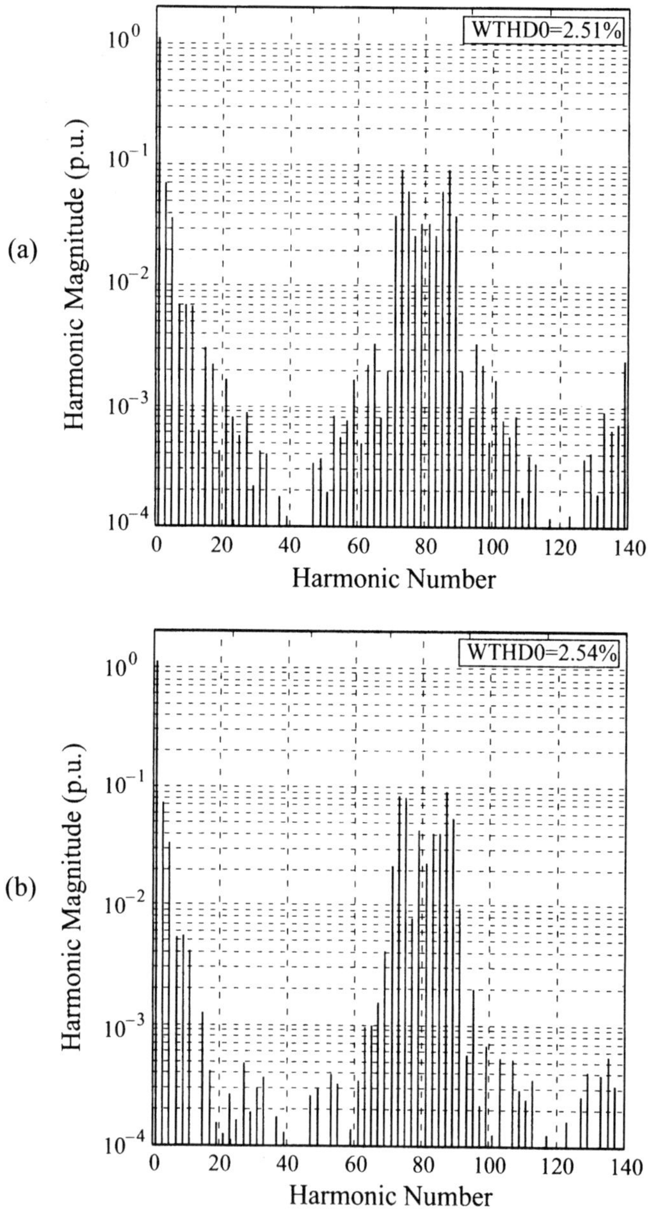

Figure 11.10 Theoretical spectra for phase leg of a five-level cascaded inverter overmodulated under (a) naturally sampled and (b) asymmetric regularly sampled modulation, $M = 1.2$, $f_c/f_o = 20$.

- Alternative phase opposition disposition (APOD), where carriers in adjacent bands are phase shifted by $180°$.
- Phase opposition disposition (POD), where the carriers above the reference zero point are out of phase with those below zero by $180°$.
- Phase disposition (PD), where all carriers are in phase across all bands.

Figure 11.11 illustrates the carrier and reference arrangements for five-level modulation for these alternatives. It is generally accepted that the PD strategy gives rise to the lowest harmonic distortion for the *l–l* voltage [4,8]. It can also be noted that the APOD and POD strategies are equivalent for three-level converters.

11.4 Three-Level Naturally Sampled PD PWM

11.4.1 Contour Plot for Three-Level PD PWM

In contrast to the cascaded topology, the multilevel diode-clamped converter analytical solution is not easily constructed from previous approaches. However, it has been shown in Eqs. (3.9) and (3.10) that any switched waveform can be described by

$$f(t) = \frac{A_{00}}{2} + \sum_{n=1}^{\infty} [A_{0n}\cos(n[\omega_o t + \theta_o]) + B_{0n}\sin(n[\omega_o t + \theta_o])] \tag{11.17}$$
$$+ \sum_{m=1}^{\infty} [A_{m0}\cos(m[\omega_c t + \theta_c]) + B_{m0}\sin(m[\omega_c t + \theta_c])]$$
$$+ \sum_{m=1}^{\infty} \sum_{\substack{n=-\infty \\ (n \neq 0)}}^{\infty} \begin{bmatrix} A_{mn}\cos(m[\omega_c t + \theta_c] + n[\omega_o t + \theta_o]) \\ + B_{mn}\sin(m[\omega_c t + \theta_c] + n[\omega_o t + \theta_o]) \end{bmatrix}$$

where

$$C_{mn} = A_{mn} + jB_{mn} = \frac{1}{2\pi^2} \int_{-\pi}^{\pi} \int_{-\pi}^{\pi} f(x, y) e^{j(mx + ny)} \, dx \, dy \tag{11.18}$$

and where $x = \omega_c t$, $y = \omega_o t$.

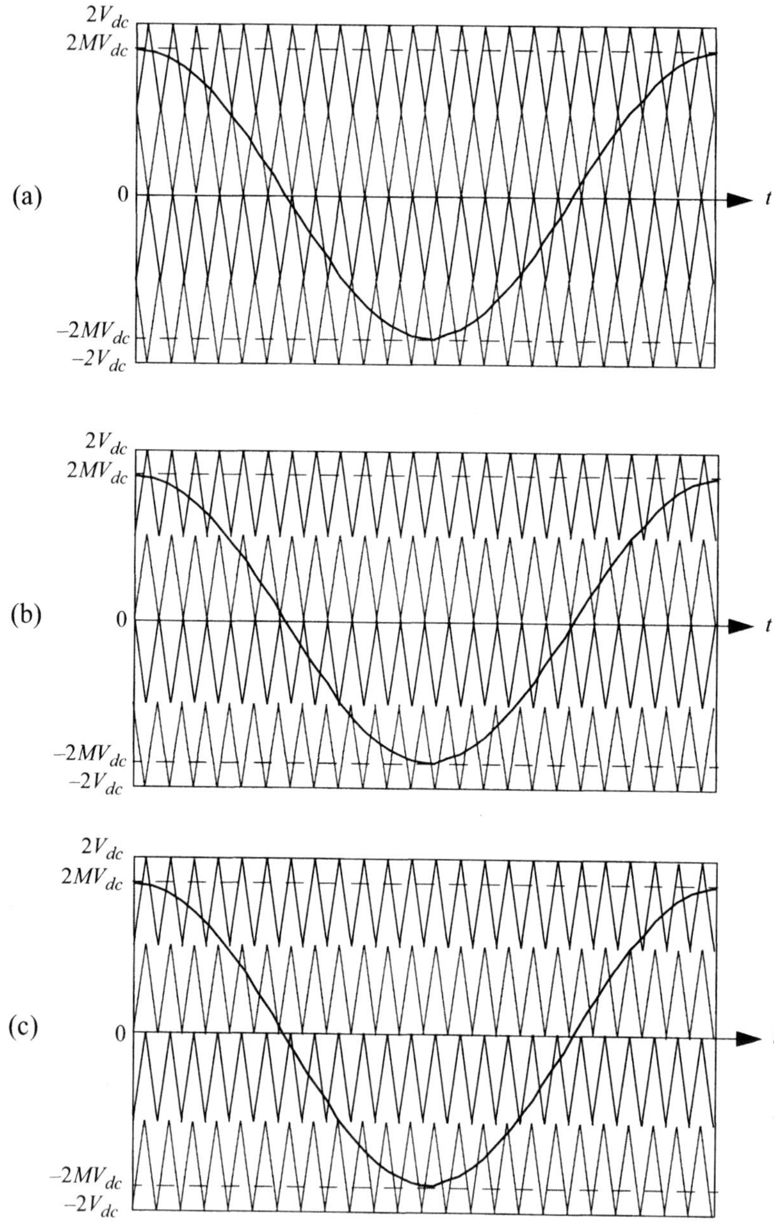

Figure 11.11 Five-level diode-clamped inverter PWM modulation options using a triangular carrier wave: (a) APOD, (b) POD, and (c) PD; $\omega_c = 24\omega_o$.

To evaluate Eq. (11.18), the function $f(x, y)$ must be properly defined in the x, y plane. In a multilevel inverter the values that the function $f(x, y)$ can take on is determined by the number of levels of the converter. In two-level modulation the available levels are $+V_{dc}$ and $-V_{dc}$. In a three-level inverter the available levels will be $+V_{dc}$, 0, and $-V_{dc}$. The boundaries between levels will be determined by the particular modulation rules. For example, the modulation rules for the PD technique, when the number of levels $N = 3$, are:

- The $N - 1 = 2$ carriers are arranged so that every carrier is in phase.
- The converter is switched to $+V_{dc}$ when the reference is greater than both carriers.
- The converter is switched to zero when the reference is greater than the lower carrier but less than the upper carrier.
- The converter is switched to the $-V_{dc}$ level when the reference is less than both carriers.

Figure 11.12 shows the defined reference input for the case of natural sampling. This waveform could obviously be revised for regular sampling if desired, but is only considered here as naturally sampled for simplicity. Figure 11.13 shows the carrier and reference waveforms in the normalized frequency space. Table 11.1 defines equations for $f(x, y)$ based on the above modulation rules for various regions of x during a carrier cycle. Figure 11.14 shows a contour plot for $f(x, y)$ over a complete cycle of x and y, based on the definitions given in Table 11.1.

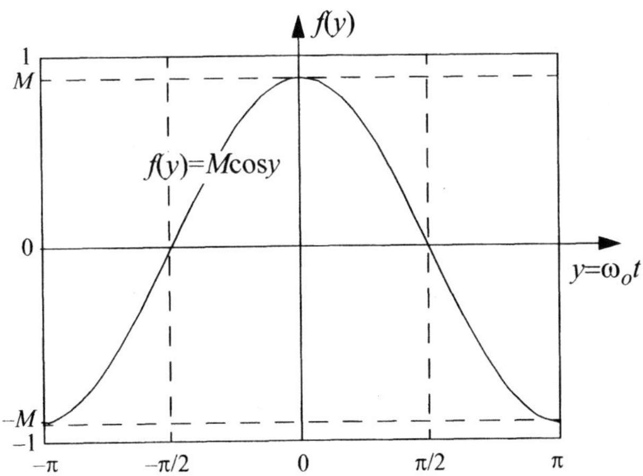

Figure 11.12 Reference waveform for natural sampled PWM.

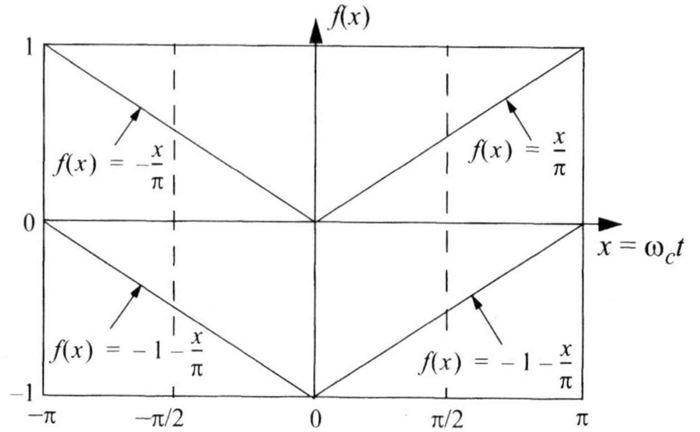

Figure 11.13 Three-level PD PWM Carrier waveforms.

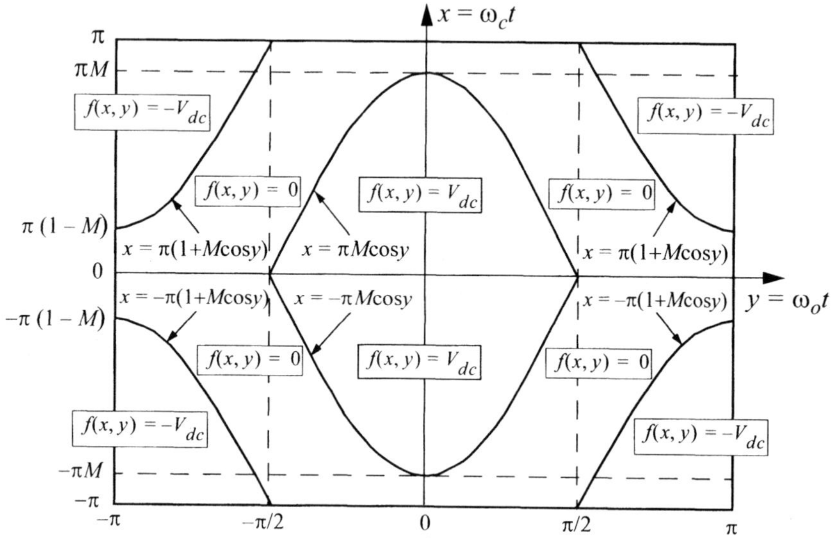

Figure 11.14 Three-level PD PWM contour plot of $f(x, y)$, natural sampling.

Table 11.1 Three-Level PD PWM Switching Function, $f(x,y)$, $x = \omega_c t$, $y = \omega_o t$

$f(x,y)$	When $-\pi \le x \le 0$	When $0 < x \le \pi$
$+V_{dc}$	$M\cos y > -\frac{x}{\pi}$	$M\cos y > \frac{x}{\pi}$
0	$-1-\frac{x}{\pi} < M\cos y < -\frac{x}{\pi}$	$-1+\frac{x}{\pi} < M\cos y < \frac{x}{\pi}$
$-V_{dc}$	$M\cos y < -1-\frac{x}{\pi}$	$M\cos y < -1+\frac{x}{\pi}$

11.4.2 Double Fourier Series Harmonic Coefficients

From this plot the double Fourier series integral of Eq. (11.18) can now be formulated as

$$C_{mn} = \frac{V_{dc}}{2\pi^2}\left[\begin{array}{l} \displaystyle\int_{-\frac{\pi}{2}}^{\frac{\pi}{2}} \int_{-\pi M\cos y}^{\pi M\cos y} e^{j(mx+ny)}\, dx\, dy \\ \displaystyle -\int_{-\pi}^{-\frac{\pi}{2}} \left\{ \int_{-\pi}^{-\pi(1+M\cos y)} e^{j(mx+ny)}\, dx + \int_{\pi(1+M\cos y)}^{\pi} e^{j(mx+ny)}\, dx \right\} dy \\ \displaystyle -\int_{\frac{\pi}{2}}^{\pi} \left\{ \int_{-\pi}^{-\pi(1+M\cos y)} e^{j(mx+ny)}\, dx + \int_{\pi(1+M\cos y)}^{\pi} e^{j(mx+ny)}\, dx \right\} dy \end{array}\right] \tag{11.19}$$

Equation (11.19) can be simplified by noting that for a periodic function $f(x)$

$$\begin{aligned} \int_{-\pi}^{-\pi-\alpha\pi} f(x)\, dx + \int_{\pi+\alpha\pi}^{\pi} f(x)\, dx &= \int_{-\pi+2\pi}^{-\pi-\alpha\pi+2\pi} f(x)\, dx + \int_{\pi+\alpha\pi}^{\pi} f(x)\, dx \\ &= \int_{\pi}^{\pi-\alpha\pi} f(x)\, dx + \int_{\pi+\alpha\pi}^{\pi} f(x)\, dx \\ &= \int_{\pi+\alpha\pi}^{\pi-\alpha\pi} f(x)\, dx \end{aligned} \tag{11.20}$$

where, in this context, $\alpha = M\cos y$. Hence Eq. (11.19) reduces to

$$C_{mn} = \frac{V_{dc}}{2\pi^2}\left[\int_{-\frac{\pi}{2}}^{\frac{\pi}{2}}\int_{-\pi M\cos y}^{\pi M\cos y} e^{j(mx+ny)}\,dx\,dy - \int_{-\pi}^{-\frac{\pi}{2}}\int_{\pi(1+M\cos y)}^{\pi(1-M\cos y)} e^{j(mx+ny)}\,dx\,dy - \int_{\frac{\pi}{2}}^{\pi}\int_{\pi(1+M\cos y)}^{\pi(1-M\cos y)} e^{j(mx+ny)}\,dx\,dy\right] \tag{11.21}$$

A similar simplification can be accomplished by operating on the y variable of the second two terms with $\alpha = -\frac{1}{2}$, to give

$$C_{mn} = \frac{V_{dc}}{2\pi^2}\left[\int_{-\frac{\pi}{2}}^{\frac{\pi}{2}}\int_{-\pi M\cos y}^{\pi M\cos y} e^{j(mx+ny)}\,dx\,dy - \int_{\frac{\pi}{2}}^{\frac{3\pi}{2}}\int_{\pi(1+M\cos y)}^{\pi(1-M\cos y)} e^{j(mx+ny)}\,dx\,dy\right] \tag{11.22}$$

Reversing the limits of integration of the x variable of the second term then yields

$$C_{mn} = \frac{V_{dc}}{2\pi^2}\left[\int_{-\frac{\pi}{2}}^{\frac{\pi}{2}}\int_{-\pi M\cos y}^{\pi M\cos y} e^{j(mx+ny)}\,dx\,dy + \int_{\frac{\pi}{2}}^{\frac{3\pi}{2}}\int_{\pi(1-M\cos y)}^{\pi(1+M\cos y)} e^{j(mx+ny)}\,dx\,dy\right] \tag{11.23}$$

Equation (11.23) can be compressed further if changes of variables are chosen for the second term such that

$$y = y' + \pi \qquad \text{and} \qquad x = x' + \pi \tag{11.24}$$

Substituting these variable changes into the second term of Eq. (11.23) gives

$$\int_{\frac{\pi}{2}}^{\frac{3\pi}{2}}\int_{\pi(1-M\cos y)}^{\pi(1+M\cos y)} e^{j(mx+ny)}\,dx\,dy = e^{j[m+n]\pi}\int_{-\frac{\pi}{2}}^{\frac{\pi}{2}}\int_{\pi M\cos y'}^{-\pi M\cos y'} e^{j(mx'+ny')}\,dx'\,dy' \tag{11.25}$$

which yields the final result for C_{mn} of

$$C_{mn} = \frac{V_{dc}}{2\pi^2}\{1 - \cos([n+m]\pi)\} \int_{-\frac{\pi}{2}}^{\frac{\pi}{2}} \int_{-\pi M\cos y}^{\pi M\cos y} e^{j(mx+ny)}\, dx\, dy \quad (11.26)$$

where the primes have been dropped on the substituted variables since they are simply dummy variables of integration. This expression can now be evaluated for various values of m and n.

11.4.3 Evaluation of the Harmonic Coefficients

For $m = n = 0$, since $1 - \cos 0 = 0$, clearly Eq. (11.26) reduces to

$$C_{00} = 0 \quad (11.27)$$

For $m = 0, n > 0$, Eq. (11.26) simplifies to

$$C_{0n} = \frac{V_{dc}}{2\pi^2}(1 - \cos n\pi) \int_{-\frac{\pi}{2}}^{\frac{\pi}{2}} \int_{-\pi M\cos y}^{\pi M\cos y} e^{jny}\, dx\, dy$$

$$= \frac{V_{dc}}{2\pi^2}(1 - \cos n\pi) \int_{-\frac{\pi}{2}}^{\frac{\pi}{2}} 2\pi M\cos y\, e^{jny}\, dy$$

$$= \frac{MV_{dc}}{\pi}(1 - \cos n\pi) \int_{-\frac{\pi}{2}}^{\frac{\pi}{2}} \frac{e^{j[n+1]y} + e^{j[n-1]y}}{2}\, dy \quad (11.28)$$

Since $(1 - \cos n\pi)$ is zero for all even values of n, Eq. (11.28) can only integrate to nonzero values for odd values of n. Note also that $1 - \cos n\pi = 2$ for all odd values of n.

For $n = 1$, Eq. (11.28) can be integrated to form

$$C_{01} = \frac{MV_{dc}}{\pi}\left\{\frac{(e^{j\pi} - e^{-j\pi})}{2j} + \frac{\pi}{2} - \left(-\frac{\pi}{2}\right)\right\}$$

$$= \frac{MV_{dc}}{\pi}(\sin(\pi) + \pi) = MV_{dc} \tag{11.29}$$

which is the fundamental component.

For $n > 1$ and n odd, Eq. (11.28) becomes

$$C_{0n} = \frac{MV_{dc}}{\pi}\left\{\frac{e^{j[n+1]\frac{\pi}{2}} - e^{-j[n+1]\frac{\pi}{2}}}{j[n+1]} + \frac{e^{j[n-1]\frac{\pi}{2}} - e^{-j[n-1]\frac{\pi}{2}}}{j[n-1]}\right\}$$

$$= \frac{MV_{dc}}{\pi}\left\{\frac{2\sin\left([n+1]\frac{\pi}{2}\right)}{[n+1]} + \frac{2\sin\left([n-1]\frac{\pi}{2}\right)}{[n-1]}\right\} = 0 \quad n > 1 \tag{11.30}$$

Thus there are no baseband harmonics at multiples of the fundamental component.

For $m > 0, n = 0$, Eq. (11.26) simplifies to

$$C_{m0} = \frac{V_{dc}}{2\pi^2}(1 - \cos m\pi)\int_{-\frac{\pi}{2}}^{\frac{\pi}{2}}\int_{-\pi M\cos y}^{\pi M\cos y} e^{jmx}\,dx\,dy$$

$$= \frac{V_{dc}}{2jm\pi^2}(1 - \cos m\pi)\int_{-\frac{\pi}{2}}^{\frac{\pi}{2}}(e^{jm\pi M\cos y} - e^{-jm\pi M\cos y})\,dy$$

$$= \frac{V_{dc}}{m\pi^2}(1 - \cos m\pi)\int_{-\frac{\pi}{2}}^{\frac{\pi}{2}}\sin(m\pi M\cos y)\,dy \tag{11.31}$$

Using Eq. (A2.3) this expression becomes (with the k summation start index changed to 1)

$$C_{m0} = \frac{2V_{dc}}{m\pi^2}(1-\cos m\pi)\sum_{k=1}^{\infty}\cos([k-1]\pi)J_{2k-1}(m\pi M)\int_{-\frac{\pi}{2}}^{\frac{\pi}{2}}\cos([2k-1]y)\,dy \tag{11.32}$$

which integrates to form

$$C_{m0} = \frac{2V_{dc}}{m\pi^2}(1-\cos m\pi)\sum_{k=1}^{\infty}J_{2k-1}(m\pi M)\cos([k-1]\pi)\left\{\frac{2\sin\left([2k-1]\frac{\pi}{2}\right)}{[2k-1]}\right\}$$

$$= \frac{4V_{dc}}{m\pi^2}(1-\cos m\pi)\sum_{k=1}^{\infty}J_{2k-1}(m\pi M)\cos([k-1]\pi)\left\{\frac{\sin\left(k\pi-\frac{\pi}{2}\right)}{[2k-1]}\right\}$$

$$= \frac{4V_{dc}}{m\pi^2}(1-\cos m\pi)\sum_{k=1}^{\infty}J_{2k-1}(m\pi M)\left\{\frac{\cos^2[k-1]\pi}{[2k-1]}\right\}$$

$$= -\frac{4V_{dc}}{m\pi^2}(1-\cos m\pi)\sum_{k=1}^{\infty}\frac{J_{2k-1}(m\pi M)}{[2k-1]} \tag{11.33}$$

This result defines the harmonics of the carrier wave itself and will only be nonzero when m is odd because of the $(1-\cos m\pi)$ term.

For $m > 0, n \neq 0$, the inner integral of Eq. (11.26) can be evaluated as

$$C_{mn} = \frac{V_{dc}}{2jm\pi^2}\{1-\cos([n+m]\pi)\}\int_{-\frac{\pi}{2}}^{\frac{\pi}{2}} e^{jny}(e^{jm\pi M\cos y}-e^{-jm\pi M\cos y})\,dy$$

$$= \frac{V_{dc}}{m\pi^2}\{1-\cos([n+m]\pi)\}\int_{-\frac{\pi}{2}}^{\frac{\pi}{2}} e^{jny}\sin(m\pi M\cos y)\,dy \tag{11.34}$$

Using Eq. (A2.3), this result becomes

$$C_{mn} = \frac{2V_{dc}}{m\pi^2}\{1-\cos([n+m]\pi)\}\int_{-\frac{\pi}{2}}^{\frac{\pi}{2}} e^{jny}\left[\sum_{k=1}^{\infty} J_k(m\pi M)\sin k\frac{\pi}{2}\cos ky\right]dy$$

$$= \frac{V_{dc}}{m\pi^2}\{1-\cos([n+m]\pi)\}\int_{-\frac{\pi}{2}}^{\frac{\pi}{2}}\left\{\sum_{k=1}^{\infty} J_k(m\pi M)\sin k\frac{\pi}{2}\left[\begin{array}{l} e^{j[n+k]y} \\ \quad + e^{j[n-k]y}\end{array}\right]\right\}dy \tag{11.35}$$

The integral expression can then be evaluated, with some manipulation, to

$$C_{mn} = \frac{2V_{dc}}{m\pi^2}\{1-\cos([n+m]\pi)\}\sum_{k=1}^{\infty} J_k(m\pi M)\sin k\frac{\pi}{2}\left\{\begin{array}{l} \left.\dfrac{\sin[n+k]\frac{\pi}{2}}{[n+k]}\right|_{-n\neq k} \\ + \left.\dfrac{\sin[n-k]\frac{\pi}{2}}{[n-k]}\right|_{n\neq k} \\ + \left.\dfrac{\pi}{2}\right|_{\pm n = k}\end{array}\right\}$$

$$= \frac{V_{dc}}{m\pi}\{1-\cos([n+m]\pi)\} \tag{11.36}$$

$$\times\left[\begin{array}{l} J_n(m\pi M)\sin n\frac{\pi}{2} \\ \quad + \dfrac{4}{\pi}\displaystyle\sum_{k=1}^{\infty} J_{2k-1}(m\pi M)\left.\frac{[2k-1]\cos n\frac{\pi}{2}}{[2k-1+n][2k-1-n]}\right|_{|n|\neq 2k-1}\end{array}\right]$$

Note that the summation expression results in a nonzero value only for even values of n because of the $\cos n\frac{\pi}{2}$ term, while the $J_n(m\pi M)\sin n\frac{\pi}{2}$ term is nonzero only for odd values of n. As a consequence, because of the $1-\cos[n+m]\pi$ term, even sideband harmonics (n even) only exist around the odd carrier multiples (m odd), and odd sideband harmonics (n odd) only exist around the even carrier multiples (m even). Also, the summation limitation $|n| \neq 2k-1$ is implicitly satisfied and need not be explicitly defined.

The complete harmonic solution for *three-level PD* modulation of phase leg a can now be formed by substituting the results of Eqs. (11.29), (11.33), and (11.36) back into Eq. (11.17), so that the time-varying switched phase leg voltage $v_{az}(t)$ can be expressed in terms of its harmonic components as

$$v_{az}(t) = V_{dc}M\cos(\omega_o t) \tag{11.37}$$

$$+\frac{8V_{dc}}{\pi^2}\sum_{m=1}^{\infty}\frac{1}{2m-1}\sum_{k=1}^{\infty}\frac{J_{2k-1}([2m-1]\pi M)}{[2k-1]}\cos([2m-1]\omega_c t)$$

$$+\frac{2V_{dc}}{\pi}\sum_{m=1}^{\infty}\frac{1}{2m}\sum_{n=-\infty}^{\infty}J_{2n+1}(2m\pi M)\cos n\pi\,\cos(2m\omega_c t+[2n+1]\omega_o t)$$

$$+\frac{8V_{dc}}{\pi^2}\sum_{m=1}^{\infty}\frac{1}{2m-1}\sum_{\substack{n=-\infty\\(n\neq 0)}}^{\infty}\sum_{k=1}^{\infty}\frac{J_{2k-1}([2m-1]\pi M)[2k-1]\cos n\pi}{[2k-1+2n][2k-1-2n]}\times\cos([2m-1]\omega_c t+2n\omega_o t)$$

Note that the phase voltage v_{az} is again taken with respect to the midpoint z of the DC bus.

As in Chapter 5, the line voltage can be found by taking

$$v_{ab}(t) = v_{az}(t) - v_{bz}(t) \tag{11.38}$$

where $v_{bz}(t)$ is obtained using Eq. (11.37) with θ_o set to $-2\pi/3$ instead of 0.

It should also be mentioned that a corresponding solution for asymmetric regular sampling can be developed by using the same approach employed in previous chapters. In this case the parameter m in Eq. (11.26) is replaced by q, which is defined as before as $q = m + n(\omega_o/\omega_c)$. The derivation then proceeds in the same manner as above.

11.4.4 Spectral Performance of Three-Level PD PWM

Phase and line voltage plots of the resulting spectrum for *three-level naturally sampled PD* modulation for the conditions of $M = 0.8$, $f_c/f_o = 20$ are shown in Figure 11.15. A number of significant features can be noted from these plots.

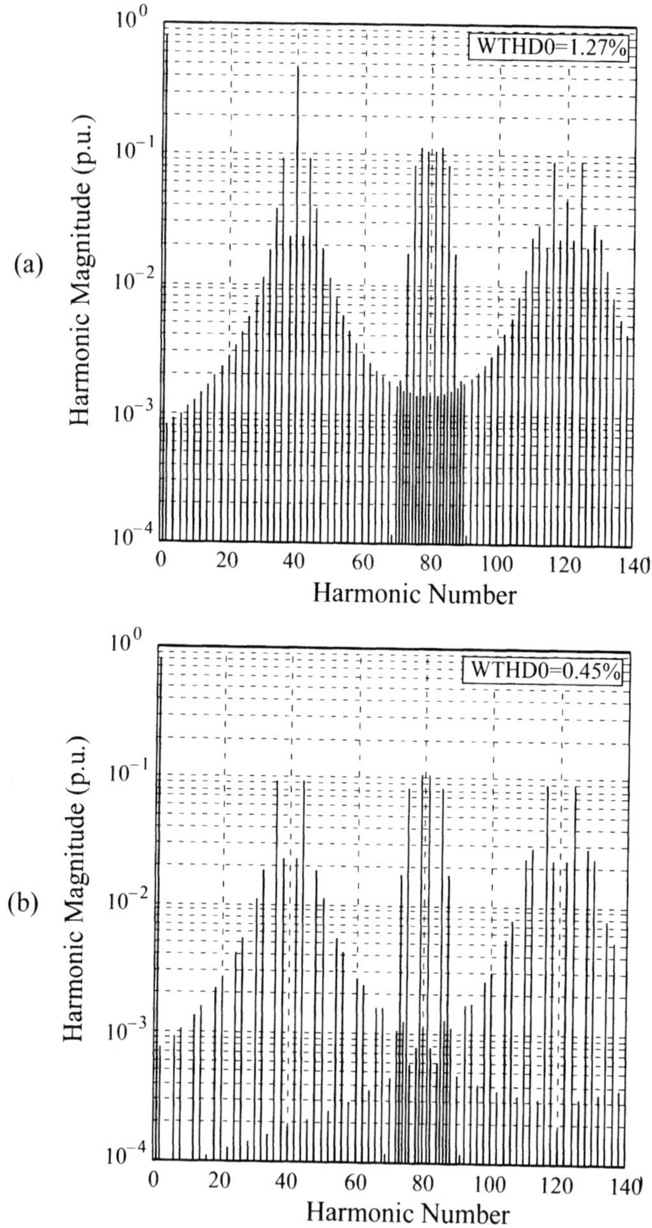

Figure 11.15 Theoretical harmonic spectra for three-level three-phase inverter modulated by *PD naturally sampled* PWM: (a) phase leg a and (b) l–l output switched voltage waveforms, $M = 0.8$, $f_c/f_o = 40$.

First, a major feature of the phase leg spectrum is the substantial first carrier harmonic. This is in contrast to the phase leg spectrum for cascaded H-bridges shown in Figure 11.2, where no carrier harmonics are present. It is this feature in particular that gives PD modulation its superior *l–l* performance, since this carrier harmonic is a common mode component across the phase legs of a three-phase inverter, and hence cancels in the *l–l* output voltage. Second, the odd sidebands around the even carrier multiples, and the even sidebands around the odd carrier multiples, can be easily seen in Figure 11.15(a), exactly as predicted by the analytical solution. Third, as with all carrier-based PWM strategies, only the triplen sideband harmonics (i.e., multiples of three away from the carrier multiples) cancel in the *l–l* voltage. As before, this cancellation is independent of the absolute carrier frequency, once again confirming that integer/triplen carrier to fundamental ratios have no influence on the harmonic performance of the modulation algorithm.

Finally, the substantial spread of the lower sideband harmonics of the first carrier group into the baseband region should be noticed. It is important to observe that these harmonics are sideband harmonics of the switching frequency and are not baseband harmonics deriving from the command frequency. Hence these harmonics are unaffected by the carrier frequency, except for their attenuation. Consequently, for low carrier frequencies they may still be significant at the fundamental frequency, and this should be allowed for in a practical system.

11.5 Three-Level Naturally Sampled APOD or POD PWM

For phase opposition disposition (POD) modulation all carrier waveforms above zero are in phase and are 180° out of phase with those below zero. Alternative phase opposition disposition (APOD) specifies that every carrier waveform is out of phase with its neighbors by 180°. It is clear that when three-level PWM is specified, the carriers of the two algorithms are identical.

The carrier functions for this case are shown in Figure 11.16. Combining the carrier waveform in *x* space and the sine wave command in *y* space from Figure 11.12 generates the switching conditions of Table 11.2 and the *x*,*y* space contour plot shown in Figure 11.17.

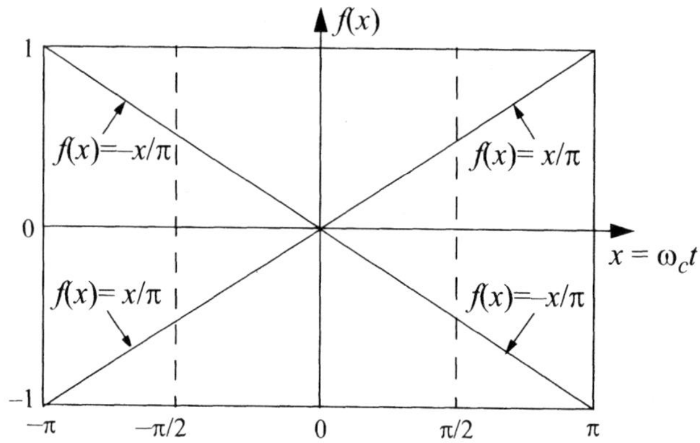

Figure 11.16 Three-level APOD/POD PWM carrier waveforms.

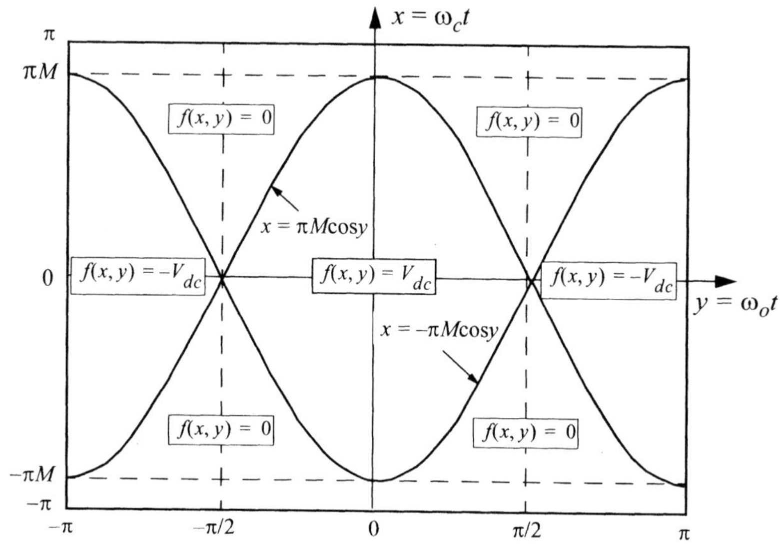

Figure 11.17 Three-level APOD/POD contour plot of $f(x,y)$, natural sampling.

Table 11.2 Three-Level APOD/POD PWM Switching Function $f(x,y)$, $x = \omega_c t$, $y = \omega_o t$

$f(x,y)$	When $-\pi \le x \le 0$	When $0 < x \le \pi$
$+V_{dc}$	$M\cos y > -\frac{x}{\pi}$	$M\cos y > \frac{x}{\pi}$
0	$\frac{x}{\pi} < M\cos y < -\frac{x}{\pi}$	$-\frac{x}{\pi} < M\cos y < \frac{x}{\pi}$
$-V_{dc}$	$M\cos y < \frac{x}{\pi}$	$M\cos y < -\frac{x}{\pi}$

With this definition of $f(x,y)$, the double Fourier series integral of Eq. (11.18) can be formulated as

$$C_{mn} = \frac{V_{dc}}{2\pi^2}\left[\int_{-\frac{\pi}{2}}^{\frac{\pi}{2}}\int_{-\pi M\cos y}^{\pi M\cos y} e^{j(mx+ny)}\,dx\,dy - \int_{-\pi}^{-\frac{\pi}{2}}\int_{\pi M\cos y}^{-\pi M\cos y} e^{j(mx+ny)}\,dx\,dy - \int_{\frac{\pi}{2}}^{\pi}\int_{\pi M\cos y}^{-\pi M\cos y} e^{j(mx+ny)}\,dx\,dy\right] \tag{11.39}$$

This expression for the coefficient C_{mn} can now be reduced using Eq. (11.20) with $\alpha = -\frac{1}{2}$ to give

$$C_{mn} = \frac{V_{dc}}{2\pi^2}\left[\int_{-\frac{\pi}{2}}^{\frac{\pi}{2}}\int_{-\pi M\cos y}^{\pi M\cos y} e^{j(mx+ny)}\,dx\;dy - \int_{\frac{\pi}{2}}^{\frac{3\pi}{2}}\int_{\pi M\cos y}^{-\pi M\cos y} e^{j(mx+ny)}\,dx\;dy\right]$$

$$= \frac{V_{dc}}{2\pi^2}\left[\int_{-\frac{\pi}{2}}^{\frac{\pi}{2}}\int_{-\pi M\cos y}^{\pi M\cos y} e^{j(mx+ny)}\,dx\;dy + \int_{\frac{\pi}{2}}^{\frac{3\pi}{2}}\int_{-\pi M\cos y}^{\pi M\cos y} e^{j(mx+ny)}\,dx\;dy\right]$$

$$= \frac{V_{dc}}{2\pi^2}\left[\int_{-\frac{\pi}{2}}^{\frac{3\pi}{2}}\int_{-\pi M\cos y}^{\pi M\cos y} e^{j(mx+ny)}\,dx\,dy\right] \tag{11.40}$$

The solution of Eq. (11.40) for various values of m and n now follows the same procedure as for the detailed PD example presented previously.

The complete harmonic solution for *three-level APOD and POD* modulation of phase leg a is

$$v_{az}(t) = V_{dc}M\cos(\omega_o t) \tag{11.41}$$

$$+\frac{2V_{dc}}{\pi}\sum_{m=1}^{\infty}\frac{1}{m}\sum_{\substack{n=-\infty\\(n\neq 0)}}^{\infty} J_{2n+1}(m\pi M)\cos n\pi\ \cos(m\omega_c t+[2n+1]\omega_o t)$$

A plot of the spectrum for this case is shown in Figure 11.18 for the phase and *l–l* voltages. This result should be compared against the spectrum for PD modulation shown in Figure 11.15. In particular, it can be seen that APOD and POD modulation do not produce a first carrier harmonic. Instead the first carrier group harmonic energy is channelled into the sidebands around the first carrier harmonic. Consequently, since only the triplen sidebands away from the carrier cancel in a three-phase system, APOD and PD modulation retain more harmonic energy in the *l–l* voltages compared to PD modulation. This is reflected in an increased WTHD result for APOD and POD modulation.

11.6 Overmodulation of Three-Level Inverters

Overmodulation occurs when the commanded amplitude of the sinusoidal voltage exceeds the amplitude of the triangle wave carrier as illustrated in Figure 11.19. The modified contour plot of $f(x, y)$ for a three-level inverter overmodulated under PD PWM is shown in Figure 11.20.

From this plot the double Fourier series integral can now be formulated as

(a)

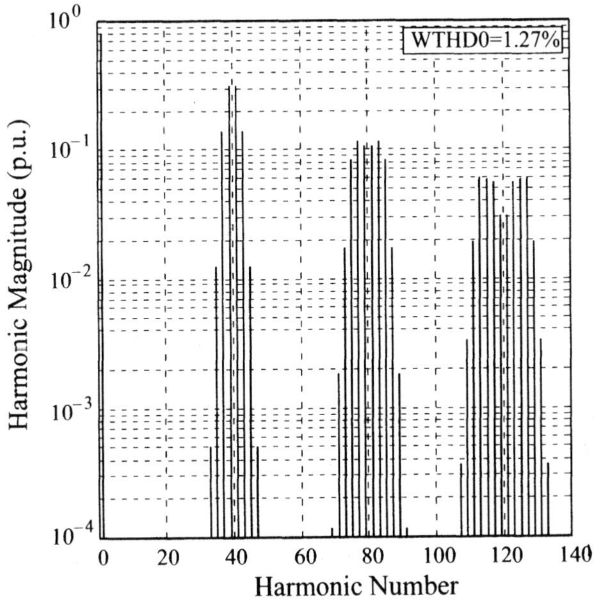

(b)

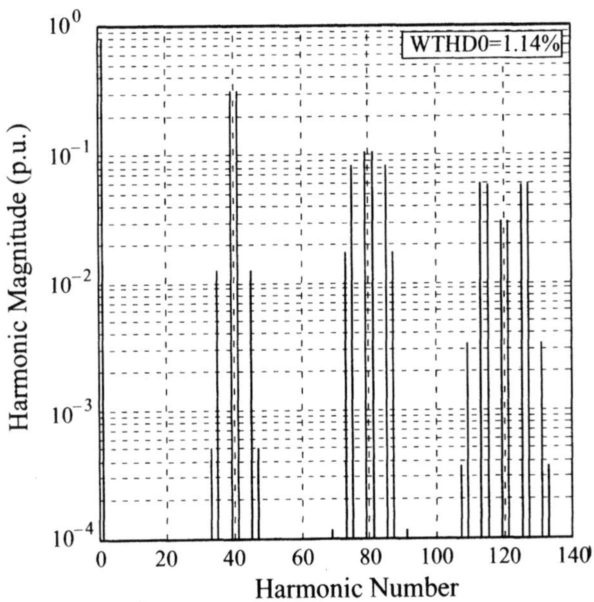

Figure 11.18 Theoretical harmonic spectra for three-level three-phase inverter modulated by *APOD/POD naturally sampled* PWM: (a) phase leg *a* and (b) *l–l* output switched voltage waveforms, $M = 0.8$, $f_c/f_o = 40$.

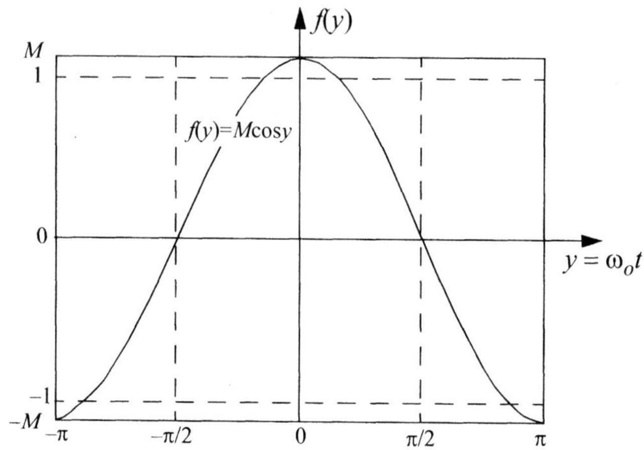

Figure 11.19 Reference waveform for natural sampled PWM under overmodulation conditions.

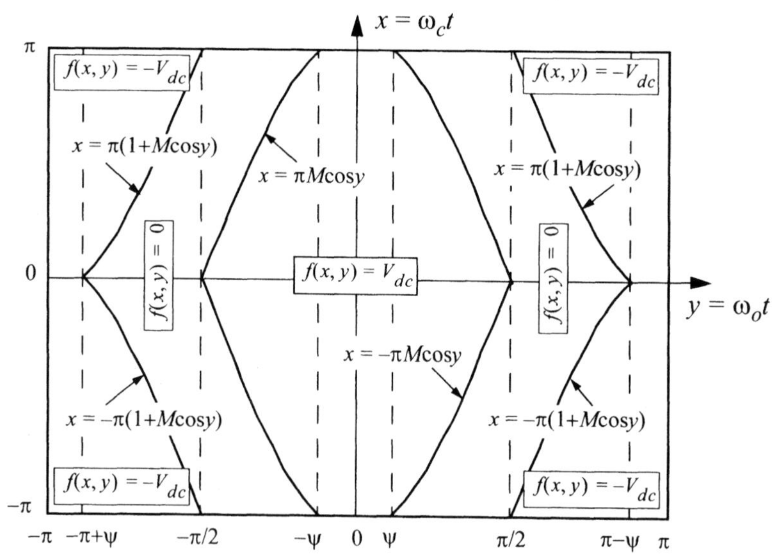

Figure 11.20 Contour plot of $f(x,y)$ for three-level PD PWM under overmodulation conditions ($M > 1$), $\psi = \cos^{-1}\left(\frac{1}{M}\right)$.

$$C_{mn} = \frac{V_{dc}}{2\pi^2}\left[\begin{aligned}&\int_{-\frac{\pi}{2}}^{-\psi}\int_{-\pi M\cos y}^{\pi M\cos y} e^{j(mx+ny)}\,dx\,dy + \int_{\psi}^{\frac{\pi}{2}}\int_{-\pi M\cos y}^{\pi M\cos y} e^{j(mx+ny)}\,dx\,dy \\ &+\int_{-\psi}^{\psi}\int_{-\pi}^{\pi} e^{j(mx+ny)}\,dx\,dy \\ &-\int_{-\pi+\psi}^{-\frac{\pi}{2}}\left\{\int_{-\pi}^{-\pi(1+M\cos y)} e^{j(mx+ny)}\,dx + \int_{\pi(1+M\cos y)}^{\pi} e^{j(mx+ny)}\,dx\right\}dy \\ &-\int_{\frac{\pi}{2}}^{\pi-\psi}\left\{\int_{-\pi}^{-\pi(1+M\cos y)} e^{j(mx+ny)}\,dx + \int_{\pi(1+M\cos y)}^{\pi} e^{j(mx+ny)}\,dx\right\}dy \\ &-\int_{-\pi}^{-\pi+\psi}\int_{-\pi}^{\pi} e^{j(mx+ny)}\,dx\,dy - \int_{\pi-\psi}^{\pi}\int_{-\pi}^{\pi} e^{j(mx+ny)}\,dx\,dy\end{aligned}\right] \tag{11.42}$$

Using periodicity conditions as before, Eq. (11.42) eventually reduces to

$$C_{mn} = \frac{V_{dc}}{2\pi^2}(1-\cos[n+m]\pi) \times \left[\begin{aligned}&\int_{-\frac{\pi}{2}}^{-\psi}\int_{-\pi M\cos y}^{\pi M\cos y} e^{j(mx+ny)}\,dx\,dy + \int_{\psi}^{\frac{\pi}{2}}\int_{-\pi M\cos y}^{\pi M\cos y} e^{j(mx+ny)}\,dx\,dy \\ &+\int_{-\psi}^{\psi}\int_{-\pi}^{\pi} e^{j(mx+ny)}\,dx\,dy\end{aligned}\right] \tag{11.43}$$

which can again be evaluated for the various values of m and n.

Using the same solution principles as before, the harmonic solution for *three-level PD* modulation of phase leg a under overmodulation conditions is

$$v_{az}(t) = \frac{V_{dc}}{\pi}\{M(\pi - 2\psi - \sin 2\psi) + 4\sin\psi\}\cos\omega_o t \tag{11.44}$$

$$+ \frac{V_{dc}}{\pi}\sum_{n=2}^{\infty}\left[\frac{4}{2n-1}\sin([2n-1]\psi) - M\left\{\frac{\sin 2n\psi}{n} + \frac{\sin(2[n-1]\psi)}{[n-1]}\right\}\right] \times \cos([2n-1]\omega_o t)$$

$$+ \frac{8V_{dc}}{\pi^2}\sum_{m=1}^{\infty}\frac{1}{2m-1}\sum_{k=1}^{\infty}\frac{1}{2k-1}\, J_{2k-1}([2m-1]\pi M)\cos k\pi \times \{\cos k\pi + \sin([2k-1]\psi)\} \times \cos([2m-1]\omega_c t)$$

$$+ \frac{2V_{dc}}{\pi^2}\sum_{m=1}^{\infty}\frac{1}{2m}\sum_{n=-\infty}^{\infty}\left[\begin{array}{l} J_{2n+1}(2m\pi M)\cos n\pi\left(\pi - 2\psi - \dfrac{\sin(2[2n+1]\psi)}{[2n+1]}\right) \\ + \displaystyle\sum_{\substack{k=1 \\ (k\neq -n) \\ (k \neq n+1)}}^{\infty}\left[\begin{array}{l} J_{2k-1}(2m\pi M)\cos k\pi \\ \times\left[\dfrac{\sin(2[k+n]\psi)}{[k+n]} + \dfrac{\sin(2[k-n-1]\psi)}{[k-n-1]}\right]\end{array}\right]\end{array}\right] \times \cos(2m\omega_c t + [2n+1]\omega_o t)$$

$$+ \frac{4V_{dc}}{\pi^2}\sum_{m=1}^{\infty}\frac{1}{2m-1}\sum_{\substack{n=-\infty \\ (n\neq 0)}}^{\infty}\sum_{k=1}^{\infty}\left[\begin{array}{l} J_{2k-1}([2m-1]\pi M)\cos k\pi \\ \times\left[\dfrac{\cos([n-k]\pi) + \sin([2k-1-2n]\psi)}{[2k-1-2n]} + \dfrac{\cos([n+k]\pi) + \sin([2k-1+2n]\psi)}{[2k-1+2n]}\right]\end{array}\right] \times \cos([2m-1]\omega_c t + 2n\omega_o t)$$

where $\psi = \cos^{-1}\left(\frac{1}{M}\right)$ and, of course, $M > 1$.

As with cascaded H-bridges, the solution to an overmodulated diode-clamped multilevel inverter seems very complex. However, on examination it becomes apparent that the complexity is again primarily related only to the magnitude of the harmonic terms, and that new harmonic components are only created in the baseband region with frequencies of $[2n-1]\omega_o t$ [line 2 of Eq. (11.44)]. Once again, these harmonics can be readily identified as the low-order odd harmonics created by the overmodulation process as the effective reference waveform approaches a square wave. Apart from this, the only effect of overmodulation is to change the magnitude of the switching harmonics, not their frequency.

Figure 11.21 shows the spectra for a three-level diode-clamped inverter for a modulation index of 1.2. Cancellation of the carrier harmonic and the triplen sideband harmonics for the *l–l* voltage again occurs in the same manner as in the linear modulation region. Note in particular the cancellation of the third-harmonic in the *l–l* voltage compared to the phase voltage, so that the first significant baseband harmonic under overmodulation conditions is the fifth.

11.7 Five-Level PWM for Diode-Clamped Inverters

11.7.1 Five-level Naturally Sampled PD PWM

The technique for the evaluation of the double Fourier coefficients C_{mn} can be readily extended to higher odd levels. The carrier and fundamental waveforms for five-level phase disposition PWM are shown in Figure 11.22. The resulting switching function conditions are shown in Table 11.3 and the corresponding contour plot for $f(x, y)$ is shown in Figure 11.23.

Table 11.3 Five-Level PD PWM Switching Function $f(x,y)$, $x = \omega_c t$, $y = \omega_o t$

$f(x,y)$	When $-\pi \le x \le 0$	When $0 < x \le \pi$
$+2V_{dc}$	$M\cos y > \frac{1}{2} - \frac{x}{2\pi}$	$M\cos y > \frac{1}{2} + \frac{x}{2\pi}$
$+V_{dc}$	$-\frac{x}{2\pi} < M\cos y < \frac{1}{2} - \frac{x}{2\pi}$	$\frac{x}{2\pi} < M\cos y < \frac{1}{2} + \frac{x}{2\pi}$
0	$-\frac{1}{2} - \frac{x}{2\pi} < M\cos y < -\frac{x}{2\pi}$	$-\frac{1}{2} + \frac{x}{2\pi} < M\cos y < \frac{x}{2\pi}$
$-V_{dc}$	$-1 - \frac{x}{2\pi} < M\cos y < -\frac{1}{2} - \frac{x}{2\pi}$	$-1 + \frac{x}{2\pi} < M\cos y < -\frac{1}{2} + \frac{x}{2\pi}$
$-2V_{dc}$	$M\cos y < -1 - \frac{x}{2\pi}$	$M\cos y < -1 + \frac{x}{2\pi}$

(a)

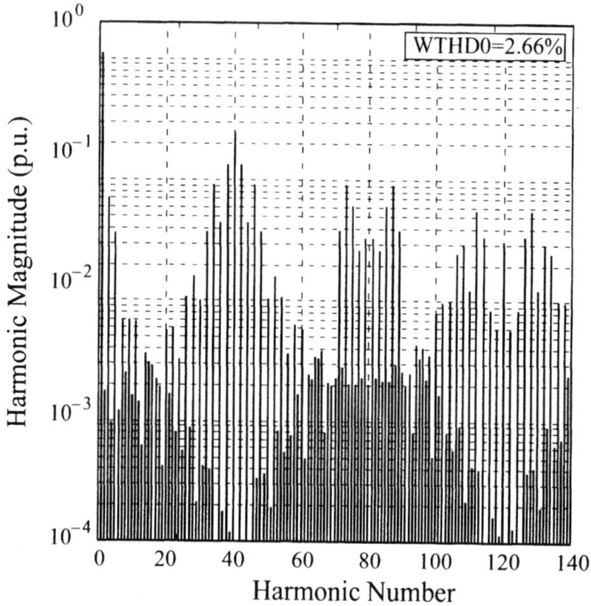

(b)

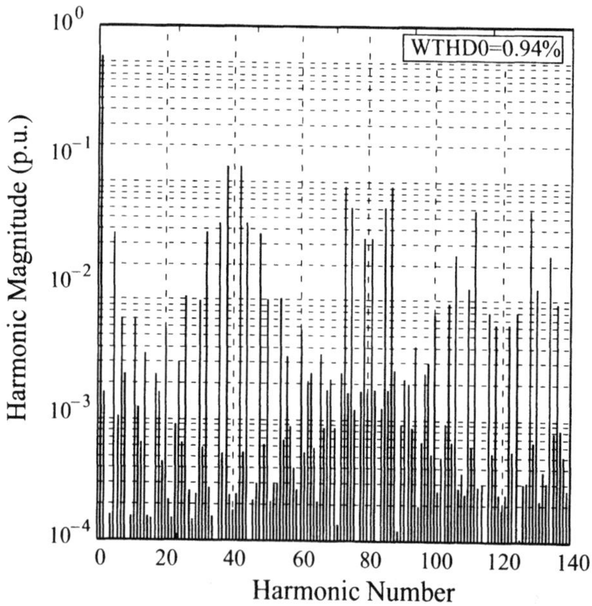

Figure 11.21 Theoretical harmonic spectra for three-level three-phase inverter overmodulated by *PD naturally sampled* PWM: (a) phase leg a and (b) l–l output switched voltage waveforms, $M = 1.2$, $f_c/f_o = 40$.

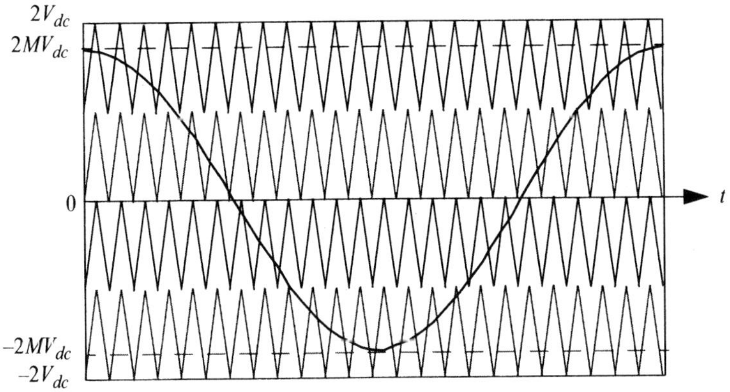

Figure 11.22 Five-level PD PWM carrier and fundamental waveform arrangements.

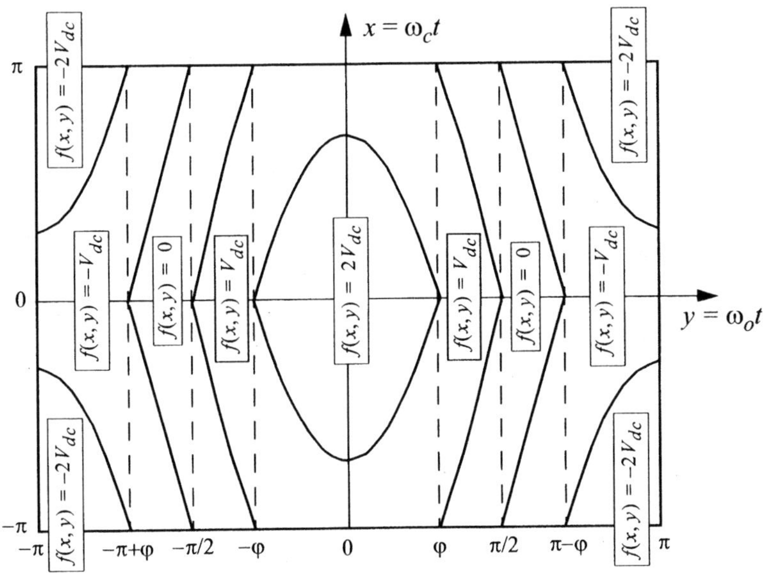

Figure 11.23 Five-level PD PWM contour plot of $f(x,y)$, natural sampling, $\varphi = \cos^{-1}\left(\frac{1}{2M}\right)$, $M \geq 0.5$. (Note that modulation reduces to three-level for $M < 0.5$.)

Using a similar process as before, the resulting solution for five-level PD PWM is

$$
\begin{aligned}
v_{az}(t) &= 2V_{dc}M\cos(\omega_o t) \qquad (11.45)\\
&+ \frac{8V_{dc}}{\pi^2}\sum_{m=1}^{\infty}\frac{1}{2m-1}\sum_{k=1}^{\infty}\frac{1}{2k-1}\,J_{2k-1}([2m-1]2\pi M)\\
&\qquad \times\{1+2\sin([2k-1]\varphi)\cos k\pi\}\\
&\qquad \times\cos([2m-1]\omega_c t)\\
&+ \frac{2V_{dc}}{\pi}\sum_{m=1}^{\infty}\frac{1}{2m}\sum_{n=-\infty}^{\infty}J_{2n+1}(4m\pi M)\cos n\pi\;\cos(2m\omega_c t+[2n+1]\omega_o t)\\
&+ \frac{4V_{dc}}{\pi^2}\sum_{m=1}^{\infty}\frac{1}{2m-1}\sum_{\substack{n=-\infty\\(n\neq 0)}}^{\infty}\sum_{k=1}^{\infty}\left[\begin{array}{l} J_{2k-1}([2m-1]2\pi M)\cos k\pi\\ \times\left[\dfrac{\cos([n-k]\pi)+2\sin([2k-1-2n]\varphi)}{[2k-1-2n]} + \dfrac{\cos([n+k]\pi)+2\sin([2k-1+2n]\varphi)}{[2k-1+2n]}\right]\end{array}\right]\\
&\qquad \times\cos([2m-1]\omega_c t+2n\omega_o t)
\end{aligned}
$$

where $\varphi = \cos^{-1}\left(\frac{1}{2M}\right)$, $M \geq 0.5$ (to maintain five-level modulation).

The phase leg voltage and *l–l* voltage spectra for five-level PD PWM with a modulation index $M = 0.8$ are shown in Figure 11.24. Note that the apparent second fundamental harmonic in Figure 11.24(b) is in fact a sideband harmonic from the first carrier group which intrudes a long way below the first carrier frequency. This low-order intrusion of the first carrier group sideband harmonics for PD modulation is an issue which is intrinsic to the modulation strategy and must be appreciated and allowed for when low carrier/fundamental ratios are implemented. In particular, it suggests that an odd carrier ratio may be appropriate for this modulation strategy when using low carrier ratios.

11.7.2 Five-Level Naturally Sampled APOD PWM

The carrier and fundamental waveforms for five-level APOD PWM are shown in Figure 11.25. The resulting switching conditions are shown in Table 11.4 and the contour plot for $f(x, y)$ is shown in Figure 11.26.

Using a similar approach as before, the harmonic solution for *five-level APOD* modulation of phase leg a is

(a)

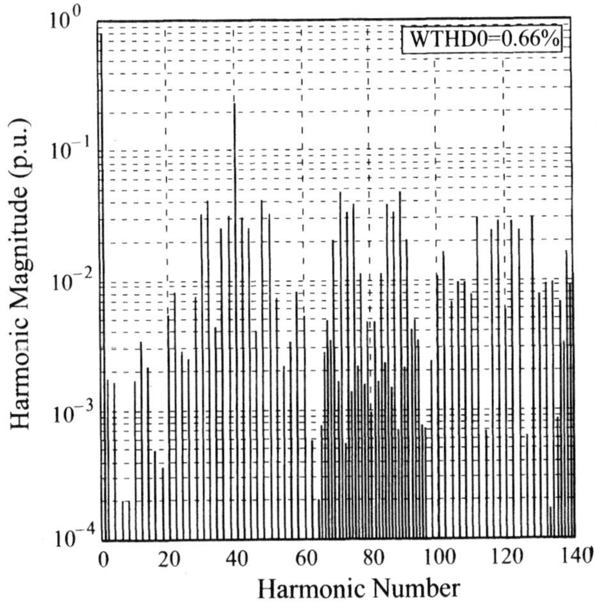

(b)

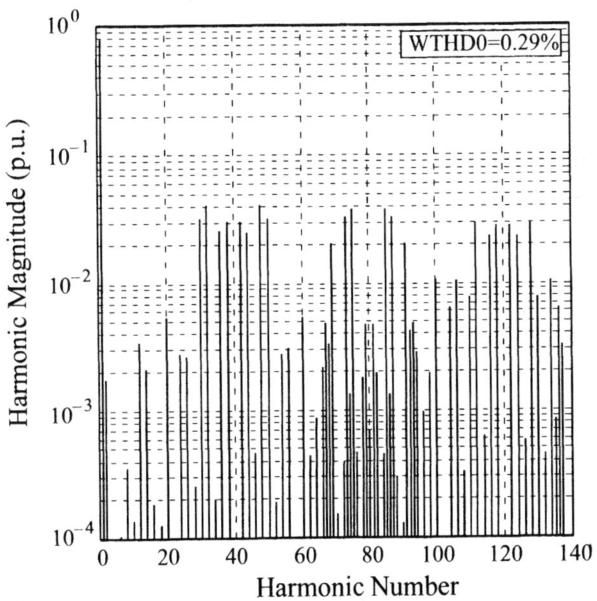

Figure 11.24 Theoretical harmonic spectra for five-level three-phase inverter modulated by *PD naturally sampled* PWM: (a) phase leg *a* and (b) *l–l* output switched voltage waveforms, $M = 0.8$, $f_c/f_o = 40$.

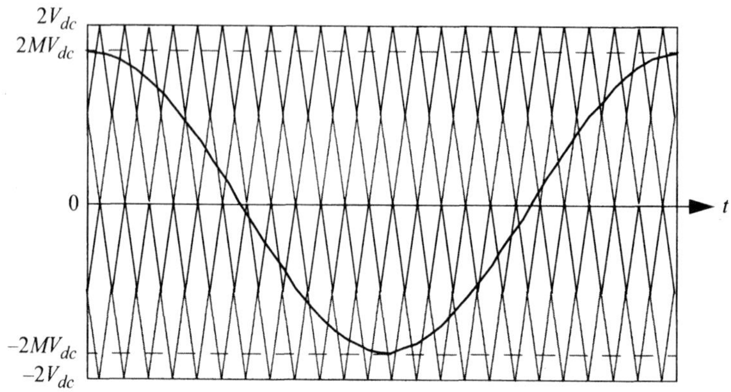

Figure 11.25 Five-level APOD PWM carrier and fundamental waveform arrangements.

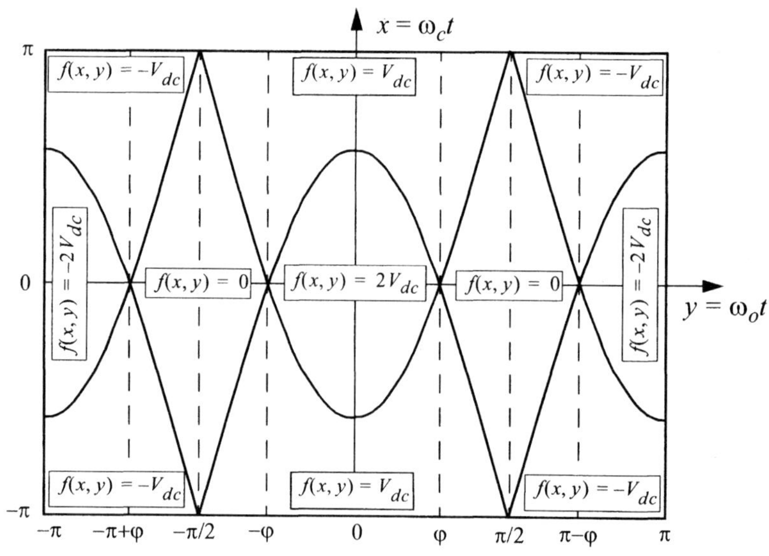

Figure 11.26 Five-level APOD PWM contour plot of $f(x,y)$, natural sampling, $\varphi = \cos^{-1}\left(\frac{1}{2M}\right)$, $M \geq 0.5$. (Note that modulation reduces to three-level for $M < 0.5$.)

$$v_{az}(t) = 2MV_{dc}\cos\omega_o t \tag{11.46}$$

$$+ \frac{2V_{dc}}{\pi}\sum_{m=1}^{\infty}\sum_{\substack{n=-\infty \\ (n \neq 0)}}^{\infty}\frac{\cos([m+n]\pi)}{m}J_{2n-1}(2m\pi M)\cos(m\omega_c t + [2n+1]\omega_o t)$$

Table 11.4 Five-Level APOD PWM Switching Function $f(x,y)$, $x = \omega_c t$, $y = \omega_o t$

$f(x,y)$	When $-\pi \le x \le 0$	When $0 < x \le \pi$
$2V_{dc}$	$M\cos y > \frac{1}{2} - \frac{x}{2\pi}$	$M\cos y > \frac{1}{2} + \frac{x}{2\pi}$
V_{dc}	$-\frac{x}{2\pi} < M\cos y < \frac{1}{2} - \frac{x}{2\pi}$	$\frac{1}{2} - \frac{x}{2\pi} < M\cos y < \frac{1}{2} + \frac{x}{2\pi}$
0	$\frac{x}{2\pi} < M\cos y < -\frac{x}{2\pi}$	$-\frac{x}{2\pi} < M\cos y < \frac{x}{2\pi}$
$-V_{dc}$	$-1 - \frac{x}{2\pi} < M\cos y < \frac{x}{2\pi}$	$1 + \frac{x}{2\pi} < M\cos y < -\frac{x}{2\pi}$
$-2V_{dc}$	$M\cos y < -1 + \frac{x}{2\pi}$	$M\cos y < -1 - \frac{x}{2\pi}$

Figure 11.27 shows the phase leg voltage and *l–l* voltage spectra for five-level APOD PWM with a modulation index $M = 0.8$.

Comparing the PD and APOD results for both three- and five-level modulation, it can be seen that the *l–l* voltage WTHD0 for PD (Figures 11.15 and 11.24) is nearly one-half that for APOD (Figures 11.18 and 11.27) even though the phase voltage WTHDs are almost the same. This clearly shows the superior *l–l* harmonic performance of the PD strategy, primarily because the phase leg spectrum for PD modulation has its most significant harmonic in the first carrier component, in contrast to APOD modulation which only generates carrier sideband components. Thus, PD modulation places significant harmonic energy into the carrier components for each phase leg and then takes advantage of common mode cancellation between the inverter phase legs to eliminate this carrier from the final *l–l* output voltage. Consequently, the harmonic sidebands, which of course do not fully cancel between the three-phase legs, have less energy.

It is also instructive to note that the five-level solution for cascaded inverters under PSCPWM [$N = 2$, Eq. (11.5) and Figure 11.2(b)] has exactly the same harmonic magnitudes as for five-level APOD PWM [Eq. (11.46) and Figure 11.27(a)]. In fact, the only difference is that the first group of sideband harmonics are centered around the carrier frequency for APOD PWM, while they are centered around the fourth multiple of the carrier frequency for cascaded inverters. However, since the total number of switch transitions for PSCPWM is exactly four times the number of switch transitions for APOD

(a)

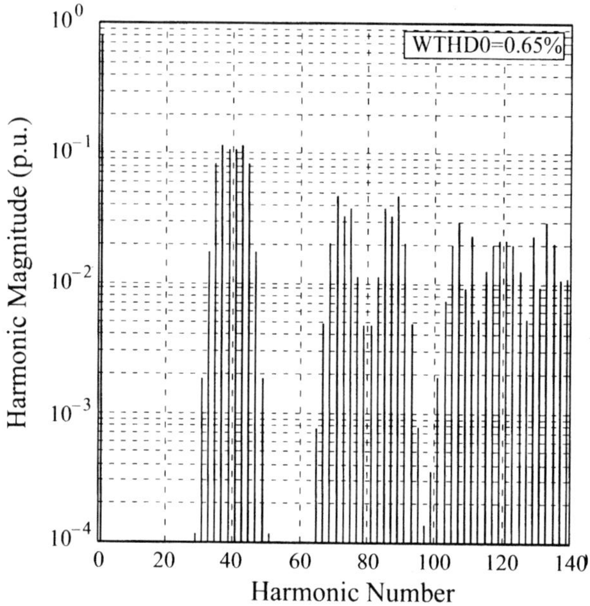

(b)

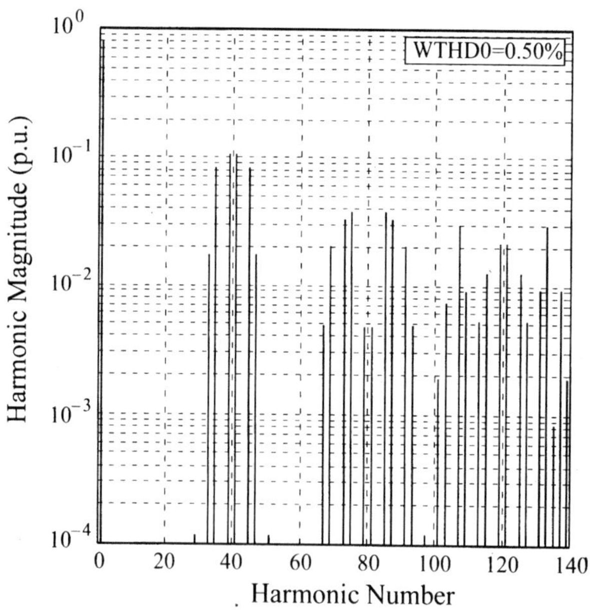

Figure 11.27 Theoretical harmonic spectra for five-level three-phase inverter modulated by *APOD naturally sampled* PWM: (a) phase leg a and (b) l–l output switched voltage waveforms, $M = 0.8$, $f_c/f_o = 40$.

PWM, the frequencies of the harmonic components become the same when the number of switch transitions are normalized across both topologies. This important observation that APOD and PSCPWM are identical, which does in fact generalize to any (odd) L-level inverter, will be used again later in this chapter.

11.7.3 Five-Level POD PWM

The switching waveforms and conditions needed to achieve phase opposition disposition PWM for a five-level inverter are shown in Figure 11.28 and Table 11.5, while the resulting contour plot is given in Figure 11.29.

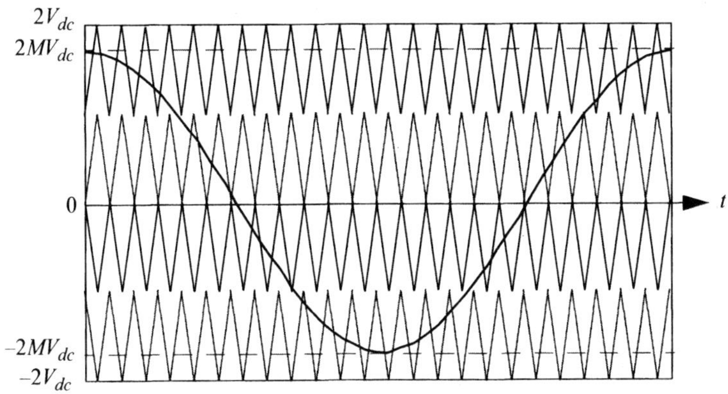

Figure 11.28 Five-level POD PWM carrier and fundamental arrangements.

Table 11.5 Five-Level POD PWM Switching Function $f(x,y)$, $x = \omega_c t$, $y = \omega_o t$

$f(x,y)$	When $-\pi \le x \le 0$	When $0 < x \le \pi$
$+2V_{dc}$	$M\cos y > -\frac{1}{2} - \frac{x}{2\pi}$	$M\cos y > \frac{1}{2} + \frac{x}{2\pi}$
V_{dc}	$-\frac{x}{2\pi} < M\cos y < -\frac{1}{2} - \frac{x}{2\pi}$	$\frac{x}{2\pi} < M\cos y < -\frac{1}{2} + \frac{x}{2\pi}$
0	$\frac{x}{2\pi} < M\cos y < -\frac{x}{2\pi}$	$-\frac{x}{2\pi} < M\cos y < \frac{x}{2\pi}$
$-V_{dc}$	$-\frac{1}{2} + \frac{x}{2\pi} < M\cos y < \frac{x}{2\pi}$	$-\frac{1}{2} - \frac{x}{2\pi} < M\cos y < -\frac{x}{2\pi}$
$-2V_{dc}$	$M\cos y < -\frac{1}{2} + \frac{x}{2\pi}$	$M\cos y < -\frac{1}{2} - \frac{x}{2\pi}$

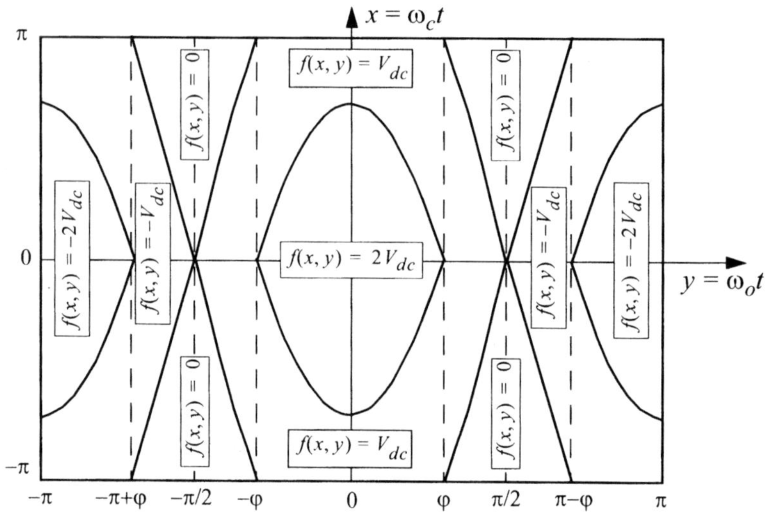

Figure 11.29 Five-level POD PWM contour plot of $f(x,y)$, natural sampling, $\varphi = \cos^{-1}\left(\frac{1}{2M}\right)$, $M \geq 0.5$. (Note that modulation reduces to three-level for $M < 0.5$.)

The harmonic solution for *five-level POD* modulation of phase leg *a* is

$$
\begin{aligned}
v_{az}(t) &= 2V_{dc}M\cos(\omega_o t) \qquad (11.47)\\
&+ \frac{2V_{dc}}{\pi}\sum_{m=1}^{\infty}\frac{1}{2m}\sum_{n=-\infty}^{\infty}J_{2n+1}(4m\pi M)\cos n\pi\,\cos(2m\omega_c t + [2n+1]\omega_o t)\\
&+ \frac{4V_{dc}}{\pi^2}\sum_{m=1}^{\infty}\frac{1}{2m-1}\sum_{\substack{n=-\infty\\(n\neq 0)}}^{\infty}\left[\begin{aligned}&J_{2n+1}([2m-1]2\pi M)\cos n\pi\\&\qquad\times\left\{\frac{\pi}{2} - 2\varphi - \frac{\sin(2[2n+1]\varphi)}{[2n+1]}\right\}\\&+\sum_{\substack{k=1\\(k\neq -n)\\(k\neq n+1)}}^{\infty}J_{2k-1}([2m-1]2\pi M)\cos k\pi\\&\qquad\times 2\left[\frac{\sin([2k+2n]\varphi)}{[2k+2n]} + \frac{\sin([2k-2n-2]\varphi)}{[2k-2n-2]}\right]\end{aligned}\right]\\
&\qquad\times\cos([2m-1]\omega_c t + [2n+1]\omega_o t)
\end{aligned}
$$

where $\varphi = \cos^{-1}\left(\frac{1}{2M}\right)$, $M \geq 0.5$ (to maintain five-level modulation).

Figure 11.30 shows the line-to-neutral voltage and l–l voltage for five-level POD PWM with a modulation index $M = 0.8$. It can been seen from these results that POD PWM has the worst l–l voltage WTHD of the three multilevel diode-clamped modulation strategies that have been considered. This is because, while both APOD and POD do not channel harmonic energy into the first carrier harmonic, the APOD strategy still places more harmonic energy into the triplen sideband harmonics than POD. These triplen sideband harmonics cancel on a l–l basis, hence improving the performance of the APOD strategy compared to the POD strategy.

11.8 PWM of Higher Level Inverters

Analytic solutions for PWM of higher level inverters can be derived using the same principles as described above, but the algebra is quite tedious [5,6,9]. As an example, the general phase leg solution for multilevel PD PWM under both linear and overmodulation conditions, adapted from [5], is

$$v_{az}(t) = \frac{N'V_{dc}}{\pi}\{M(\pi - 2\psi - \sin 2\psi) + 4\sin\psi\}\cos(\omega_o t) \tag{11.48}$$

$$+\frac{N'V_{dc}}{\pi}\sum_{n=2}^{\infty}\left\{\frac{4}{2n-1}\sin([2n-1]\psi) - M\left\{\frac{\sin 2n\psi}{n} + \frac{\sin(2[n-1]\psi)}{[n-1]}\right\}\right\} \times \cos([2n-1]\omega_o t)$$

$$+\frac{8V_{dc}}{\pi^2}\sum_{m=1}^{\infty}\frac{1}{2m-1}\sum_{k=1}^{\infty}C_{m0}\cos([2m-1]\omega_c t)$$

$$+\frac{2V_{dc}}{\pi^2}\sum_{m=1}^{\infty}\frac{1}{2m}\sum_{n=-\infty}^{\infty}C_{m_{even}n_{odd}}\cos(2m\omega_c t + [2n+1]\omega_o t)$$

$$+\frac{4V_{dc}}{\pi^2}\sum_{m=1}^{\infty}\frac{1}{2m-1}\sum_{\substack{n=-\infty \\ (n\neq 0)}}^{\infty}\sum_{k=1}^{\infty}C_{m_{odd}n_{even}}\cos([2m-1]\omega_c t + 2n\omega_o t)$$

where

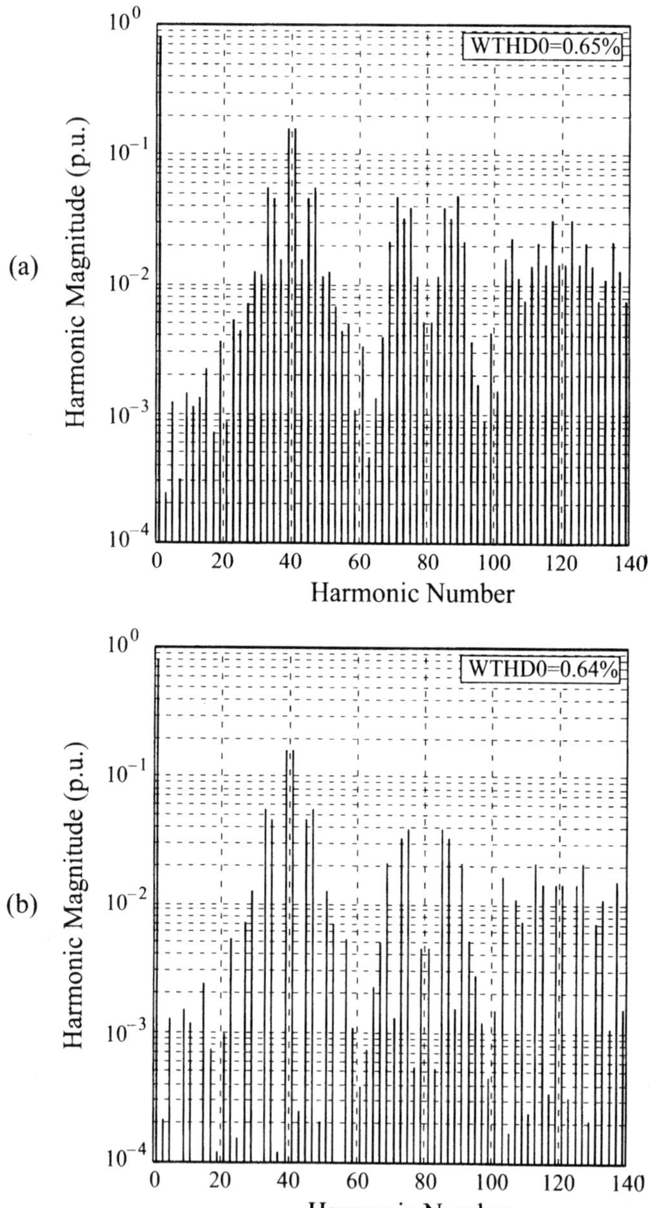

Figure 11.30 Theoretical harmonic spectra for five-level three-phase inverter modulated by *POD naturally sampled* PWM: (a) phase leg *a* and (b) *l–l* output switched voltage waveforms, $M = 0.8$, $f_c/f_o = 40$.

$$C_{m0} = \frac{1}{2k-1} J_{2k-1}([2m-1]N'\pi M)\cos k\pi \times \left[\begin{array}{l} \cos k\pi - \cos N'\pi \sin([2k-1]\psi) \\ \quad -2\displaystyle\sum_{h=1}^{N'-1} \sin\left([2k-1]\cos\left(\frac{h}{N'M}\right)^{-1}\right)\cos h\pi \end{array}\right] \tag{11.49}$$

$$C_{m_{even}n_{odd}} = J_{2n+1}(2mN'\pi M)\cos n\pi \left\{\pi - 2\psi - \frac{\sin(2[2n+1]\psi)}{[2n+1]}\right\} + \sum_{\substack{k=1 \\ (k \neq -n) \\ (k \neq n+1)}}^{\infty} J_{2k-1}(2mN'\pi M)\cos k\pi \times \left[\frac{\sin(2[k+n]\psi)}{[k+n]} + \frac{\sin(2[k-n-1]\psi)}{[k-n-1]}\right] \tag{11.50}$$

$$C_{m_{odd}n_{even}} = J_{2k-1}([2m-1]N'\pi M)\cos k\pi \times \left\{\begin{array}{l} \dfrac{\left[\begin{array}{l} \cos([n-k]\pi) - \cos N'\pi \sin([2k-1-2n]\psi) \\ -2\displaystyle\sum_{h=1}^{N'-1} \sin\left\{[2k-2n-1]\cos^{-1}\left(\frac{h}{N'M}\right)\right\}\cos h\pi \end{array}\right]}{[2k-1-2n]} \\ + \dfrac{\left[\begin{array}{l} \cos([n+k]\pi) + \cos N'\pi \sin([2k-1+2n]\psi) \\ -2\displaystyle\sum_{h=1}^{N'-1} \sin\left\{[2k-1+2n]\cos^{-1}\left(\frac{h}{N'M}\right)\right\}\cos h\pi \end{array}\right]}{[2k-1+2n]} \end{array}\right\} \tag{11.51}$$

and $$\psi = \left\{\begin{array}{lll} \cos^{-1}\left(\frac{1}{M}\right) & \text{for} \quad M > 1 & \text{(overmodulation)} \\ 0 & \text{for} \quad M \leq 1 & \text{(linear region)} \end{array}\right\}$$

$N' = \dfrac{N-1}{2}$ where N is the number of voltage levels of the diode-clamped inverter.

The reduction of this general solution to the three- and five-level solutions for multilevel PD PWM presented earlier as Eqs. (11.37), (11.44), and (11.45) is left as an exercise for the reader.

Figure 11.31 shows the phase leg and *l–l* output voltage spectra for a seven-level diode-clamped inverter modulated by PD PWM. Once again, the strong first carrier harmonic can be seen in the phase leg voltage spectrum and cancels in the *l–l* voltage since it is a common mode component.

In contrast, the general phase leg solution for multilevel APOD PWM under both linear and overmodulation conditions is, again adapted from [5],

$$v_{az}(t) = \frac{N'V_{dc}}{\pi}\{M(\pi - 2\psi - \sin 2\psi) + 4\sin\psi\}\cos\omega_o t \tag{11.52}$$

$$+ \frac{N'V_{dc}}{\pi}\sum_{n=2}^{\infty}\left\{\frac{4}{2n-1}\sin([2n-1]\psi) - M\left\{\frac{\sin 2n\psi}{n} + \frac{\sin(2[n-1]\psi)}{[n-1]}\right\}\right\} \times \cos([2n-1]\omega_o t)$$

$$+ \frac{2V_{dc}}{\pi^2}\sum_{m=1}^{\infty}\frac{1}{m}\sum_{n=-\infty}^{\infty} C_{mn}\cos(m\omega_c t + [2n+1]\omega_o t)$$

where

$$C_{mn} = J_{2n+1}(mN'\pi M)\cos([(N'-1)m + n]\pi)\left(\pi - 2\psi - \frac{\sin(2[2n+1]\psi)}{[2n+1]}\right)$$
$$+ \sum_{\substack{k=1 \\ (k \neq -n) \\ (k \neq n+1)}}^{\infty} J_{2k-1}(mN'\pi M)\cos([(N'-1)m + k]\pi) \times \left[\frac{\sin(2[k+n]\psi)}{[k+n]} + \frac{\sin(2[k-n-1]\psi)}{[k-n-1]}\right] \tag{11.53}$$

and ψ and N' are the same as have already been defined in Eq. (11.51) for multilevel PD modulation. Once more, the reduction of this solution to the three- and five-level solutions for multilevel APOD PWM presented earlier as Eqs. (11.41) and (11.46) is left as an exercise for the reader.

(a)

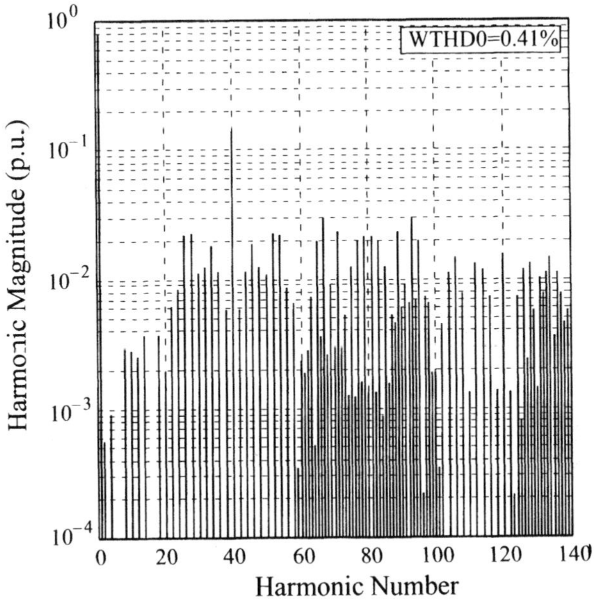

(b)

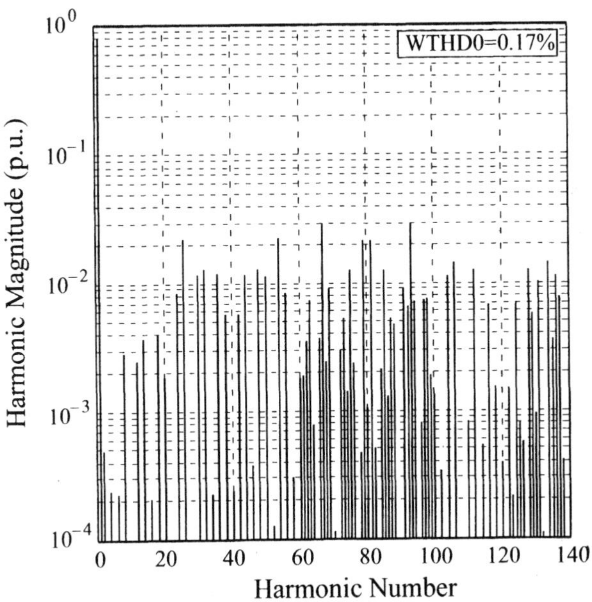

Figure 11.31 Theoretical harmonic spectra for seven-level three-phase inverter modulated by *PD naturally sampled* PWM: (a) phase leg a and (b) *l–l* output switched voltage waveforms, $M = 0.8$, $f_c/f_o = 40$.

11.9 Equivalent PD PWM for Cascaded Inverters

The results of the previous sections have shown that minimum *l–l* WTHD is realized for modulation control of a diode-clamped multilevel inverter by using phase disposition (PD) PWM. This strategy places harmonic energy into the first carrier component in each phase voltage and relies upon the cancellation of this component when the *l–l* voltages are created to achieve an improved WTHD compared to other modulation strategies.

In Section 11.7.2, it was identified that the conventional modulation of a cascaded inverter discussed in Section 11.1, phase-shifted cascaded PWM (PSCPWM), is equivalent to phase opposition disposition (APOD) PWM of diode-clamped inverters. Since APOD PWM is less effective than PD PWM, it is likely therefore that PSCPWM is not the most effective modulation strategy for a cascaded inverter, and that a strategy that places harmonic energy into a common mode carrier harmonic in each inverter phase leg would be preferable. Based on the principles of modulation of a single-phase inverter presented in Chapter 4, this would require that the carrier for each phase leg of each cascaded H-bridge is phase inverted by 180°, so that the carriers do not cancel between the two phase legs. Two possibilities can be considered:

- Single-phase two-level modulation of each H-bridge, which inverts both the fundamental reference and the carrier waveform between the two H-bridge phase legs. However, this strategy is clearly suboptimal, since it generates harmonics around the first carrier group and does not achieve the harmonic cancellation up to the second carrier group that makes PSCPWM so effective.

- Discontinuous single-phase modulation, where only one phase leg of each H-bridge is modulated at a time. This strategy also does not eliminate first carrier group harmonics, but since each phase leg only switches for one half of the fundamental, the carrier frequency can be doubled for the same number of overall switching transitions. Hence this is a more attractive approach to achieve equivalent PD PWM of a cascaded inverter. Note, however, that it is important to invert the carrier between phase legs within each cascaded H-bridge, in contrast to the approach to discontinuous modulation that was presented in Section 4.5.2 for single-phase inverters (where the objective was to eliminate the carrier harmonic between the phase legs).

Figure 11.32 shows the reference and carrier waveform arrangements required to achieve this form of modulation for three-level and five-level systems. Note that the individual phase legs of each full-bridge inverter no longer switch at every carrier cycle, much the same way as a diode-clamped inverter operates under PD modulation. It should be recalled again, however, that the carrier frequency must be doubled to maintain the same overall number of switch transitions compared to PWSPWM (this is a normal adjustment procedure for discontinuous modulation).

One limitation of PWM operation for diode-clamped inverters of five-levels or more, and this discontinuous modulation strategy for two or more cascaded inverters in a phase leg, is the unbalanced switching load that occurs between the various switching devices. For a diode-clamped inverter, the outer switches in each phase leg operate less often over a fundamental cycle. For a cascaded inverter, the outer H-bridge has a significant dead-time period where neither phase leg switches, as shown in Figure 11.32. While this unbalanced switching is generally unavoidable for a diode-clamped inverter, it can be easily compensated for with a cascaded inverter by swapping the reference waveform segments between the various H-bridge phase legs over multiple fundamental cycles. This is illustrated in Figure 11.33 for a five-level cascaded inverter, where the reference waveforms are swapped between the bridge phase-legs over two fundamental cycles to balance the number of switching transitions for each phase leg. Note again the inverted carriers for the even and odd phase legs within each cascaded bridge that are required to retain the first carrier harmonic in the overall phase leg so that it can cancel in the *l–l* voltage.

Figure 11.34 shows experimental voltage and filtered current waveforms for a five-level cascaded inverter operating with equivalent discontinuous modulation to PD PWM of a five-level diode-clamped inverter. Figure 11.35 shows the experimental phase leg voltage and *l–l* voltage spectra for this arrangement. It should be noted in particular how the use of a discontinuous modulation strategy for the cascaded inverter has retained the desired carrier harmonic in the phase leg switched voltage, as shown in Figure 11.35(a). As anticipated, this harmonic has been completely cancelled in the *l–l* voltage, as shown in Figure 11.35(b). (NOTE: While the experimental results shown are for regular sampled PWM, the difference compared to natural sampled PWM is only a slight magnitude variation between the sideband harmonics in each carrier group, and can be neglected.)

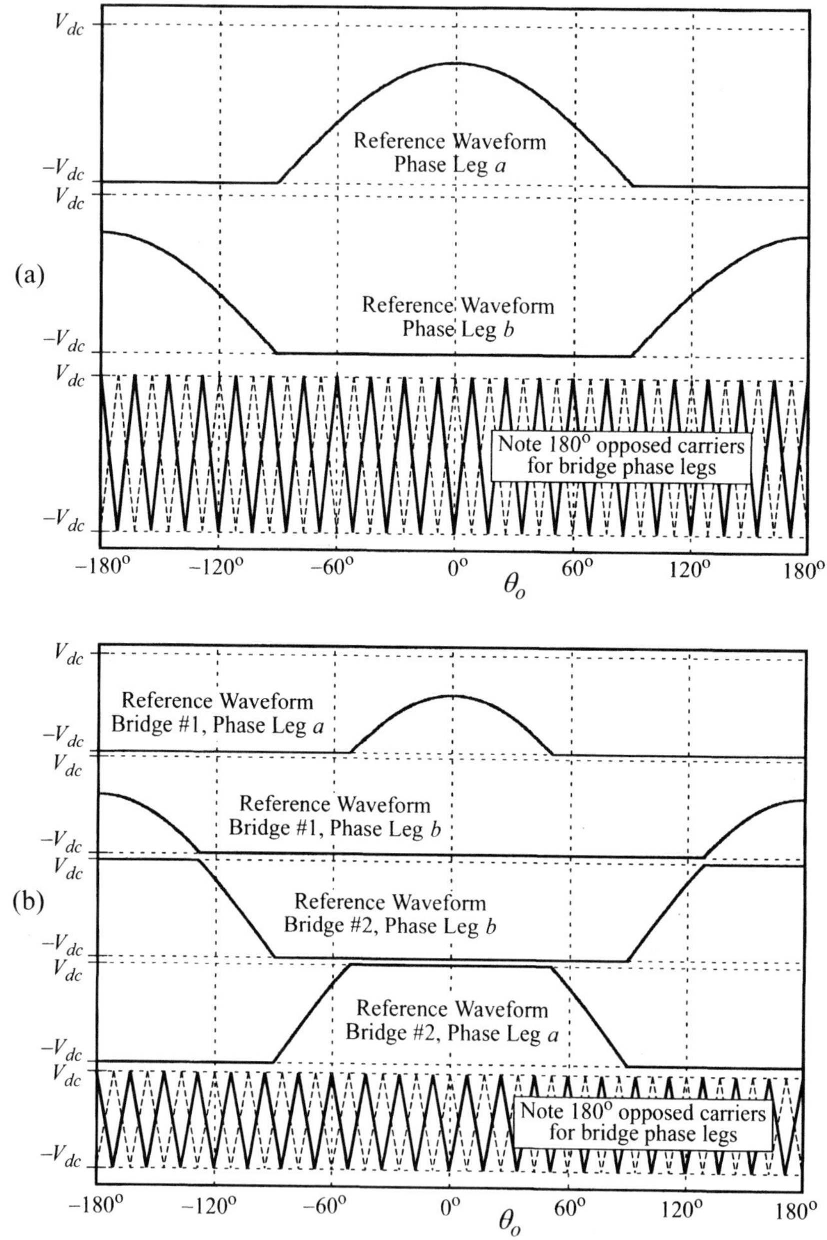

Figure 11.32 Reference and carrier waveforms required for equivalent PD modulation of (a) a three-level (one bridge per phase leg) and (b) a five-level (two bridges per phase leg) cascaded inverter.

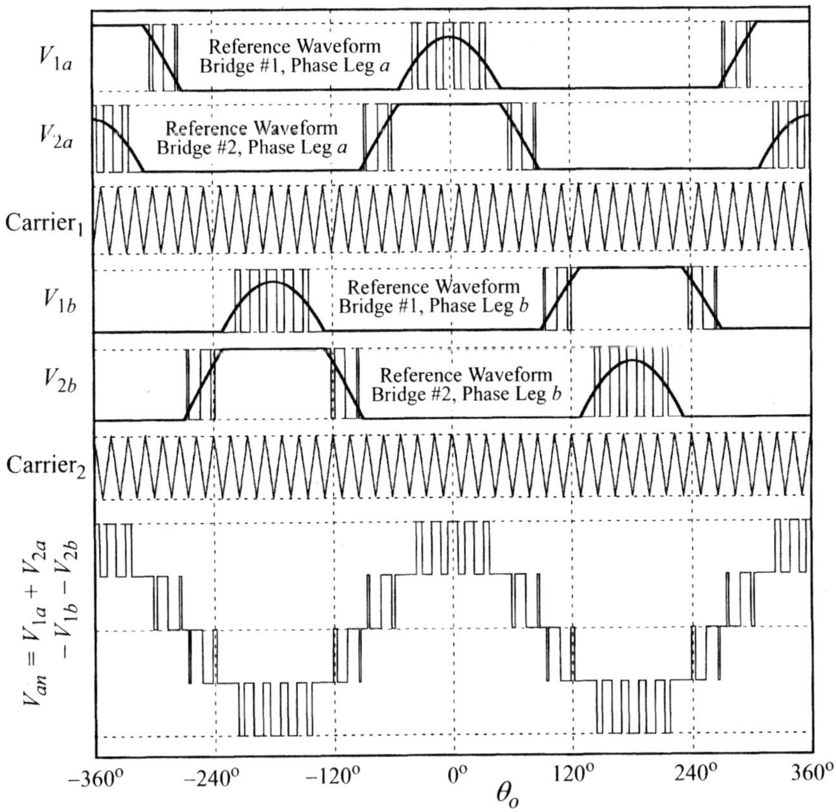

Figure 11.33 Reference and carrier waveforms required for equivalent PD modulation of a five-level cascaded inverter with cyclic rotation of switching duty between H-bridge phase legs.

11.10 Hybrid Multilevel Inverter

A further multilevel inverter topology called the hybrid inverter [10,11] has been derived from the cascaded inverter structure. The hybrid inverter utilizes cascaded series inverters with different internal DC bus voltages and with different switching devices for the different DC levels. The simplest configuration comprises two series-connected single-phase inverters per phase, with their DC bus voltages sized in the ratio 2:1. The devices of the higher voltage inverter are operated to switch only at fundamental frequency (block modulation) while the lower voltage inverter is controlled by PWM to achieve an overall sinusoidal output voltage. Suitable devices for the high-voltage inverter

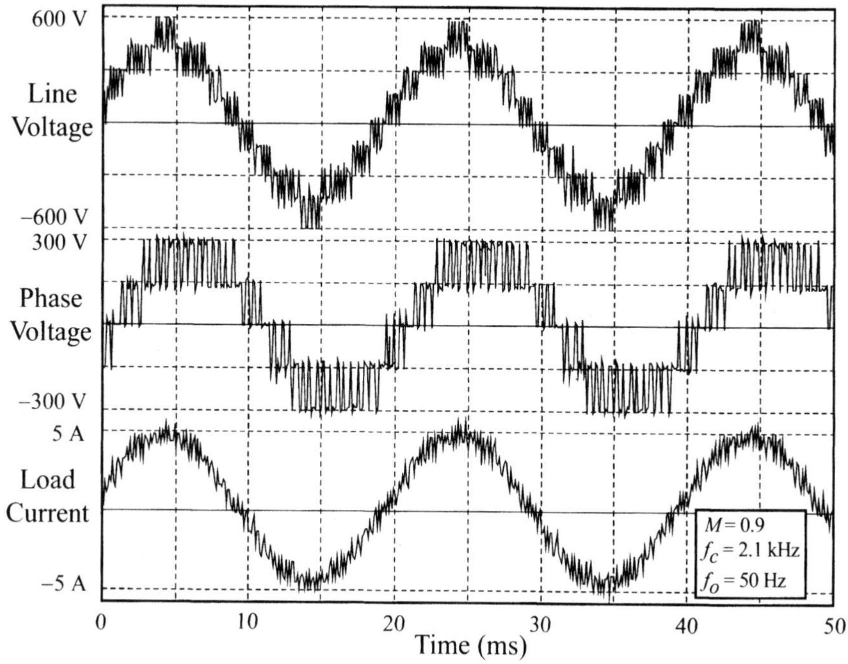

Figure 11.34 Experimental switched voltages and filtered load current for five-level cascaded three-phase inverter modulated by *equivalent PD regularly sampled PWM*, $M = 0.9$, $f_c/f_o = 42$ [6].

are relatively slow turn-off switches such as GTOs or IGCTs while the switch for the low-voltage inverter is typically an IGBT. The schematic for a hybrid inverter with two cascaded inverters per leg is shown in Figure 11.36, while the switched state of the inverters to achieve the various possible levels of phase leg output voltage are summarized in Table 11.6. It is important to observe that seven-levels are produced by this type of cascade arrangement rather than the usual five-levels produced by a normal cascaded inverter with $N = 2$. This is an important advantage in terms of reduced circuit complexity to achieve the same switched output quality.

Figure 11.37 shows the reference and switched voltage waveforms of the hybrid multilevel inverter. For each phase leg, the high-voltage (HV) inverter is modulated with the required fundamental reference sine wave with maximum magnitude $3V_{dc}$ and switches high if the reference exceeds $+V_{dc}$, low if the reference is less than $-V_{dc}$ or otherwise produces 0 V. The low-voltage (LV) inverter is controlled by subtracting the HV inverter output from the over-

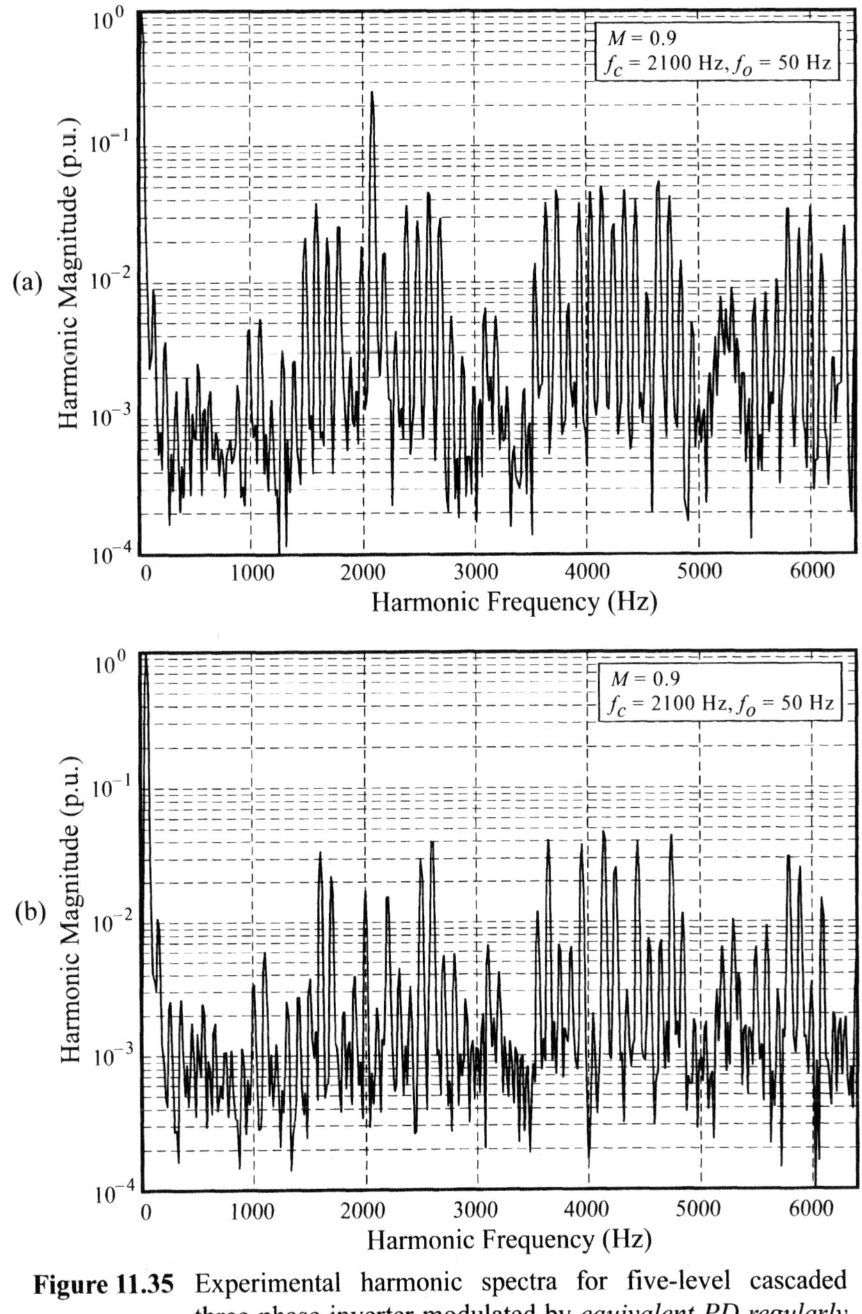

Figure 11.35 Experimental harmonic spectra for five-level cascaded three-phase inverter modulated by *equivalent PD regularly sampled* PWM: (a) phase leg *a* and (b) *l–l* output switched voltage waveforms, $M = 0.9$, $f_c/f_o = 42$ [6].

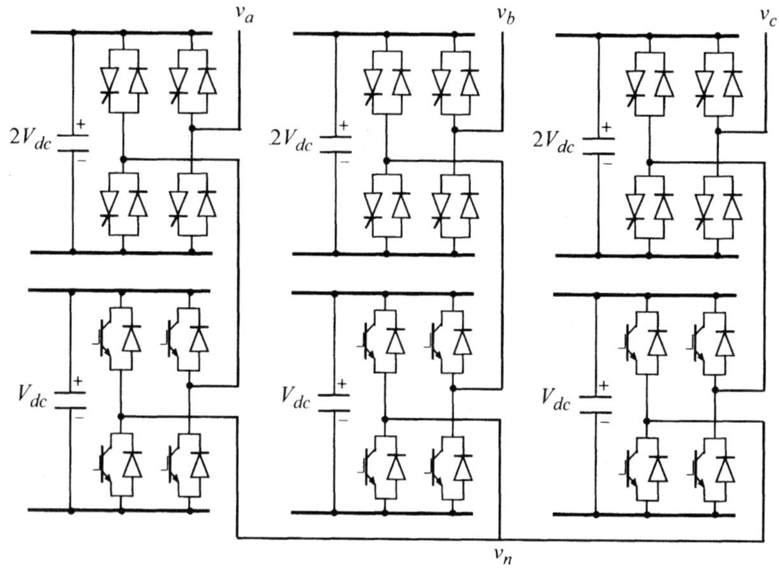

Figure 11.36 Schematic of the hybrid multilevel three-phase inverter comprising two series-connected cascade inverters with unequal DC bus voltage levels.

all reference waveform. In this manner the LV reference contains information about the harmonic content of the HV stage, and is used to modulate the LV inverter using conventional three-level PWM, as shown in Figure 11.37(b). The result is a waveform that contains the low-frequency harmonic components of the HV stage voltage as well as conventional high-frequency PWM sidebands. When the two switched waveforms are combined together, the result is a seven-level switched waveform as shown in Figure 11.37(c).

Table 11.6 Hybrid Inverter Modulation Process, $a \leftrightarrow b$ Denotes Switching Between LV Inverter Levels a and b

Output switches between	HV Inverter (IGCT)	LV Inverter (IGBT)
$-3V_{dc}$ and $-2V_{dc}$	$-2V_{dc}$	$0 \leftrightarrow -V_{dc}$
$-2V_{dc}$ and V_{dc}	$-2V_{dc}$	$0 \leftrightarrow -V_{dc}$
$-V_{dc}$ and 0	0	$0 \leftrightarrow -V_{dc}$
0 and V_{dc}	0	$0 \leftrightarrow V_{dc}$
V_{dc} and $2V_{dc}$	$2V_{dc}$	$0 \leftrightarrow V_{dc}$
$2V_{dc}$ and $3V_{dc}$	$2V_{dc}$	$0 \leftrightarrow V_{dc}$

(a)
(b)
(c)

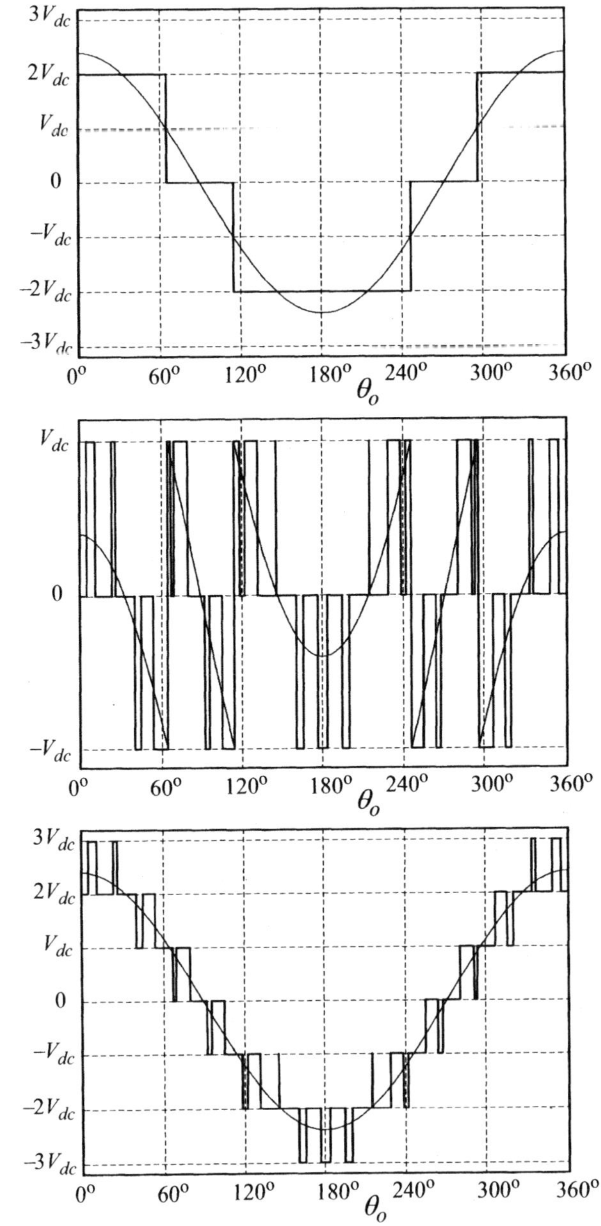

Figure 11.37 Hybrid inverter modulation waveforms: (a) high-voltage (HV) stage reference and switched output waveform, (b) low-voltage (LV) stage reference and switched output waveform, and (c) hybrid inverter switched output waveform and reference command.

Figure 11.38(a) shows more detail of the reference and carrier waveforms for each phase leg of the LV inverter. This inverter is modulated in exactly the same way as for a single-phase inverter operating under continuous three-level modulation as discussed in Chapter 4, except that the reference waveform is substantially more complex. In particular, the reference waveform is inverted between the phase legs while they both use the same carrier, to ensure maximum harmonic cancellation.

The harmonic performance of this modulation strategy can now be found by using a two-stage process where the spectra for the HV stage and the LV stage are determined separately and then added together.

The harmonic content of the HV stage can be found using conventional one-dimensional Fourier analysis. This approach yields the result

$$v_{HV}(t) = \frac{8V_{dc}}{\pi}\sum_{n=0}^{\infty}\frac{1}{[2n+1]}\sin\left\{[2n+1]\cos^{-1}\left(\frac{1}{3M}\right)\right\}\cos([2n+1]\omega_o t) \tag{11.54}$$

The spectral profile of the LV stage can then be found using the double Fourier series method of analysis. As before, the harmonic coefficients of Eq. (11.17) are found by evaluating the double integral of $f(x,y)$ over one cell in the (x,y) plane. Figure 11.38(b) shows the contour plot of $f(x,y)$ for one phase leg of the LV PWM stage of the hybrid inverter, where it should be noted that, similar to the approach presented in Chapter 3, $f(x,y)$ switches between 0 and $2V_{dc}$ to simplify the evaluation of Eq. (11.18), the double Fourier series integral. The integral can then be solved, with considerable effort, to achieve the expression

$$\begin{aligned}v_{LV}(t) = 3MV_{dc}\cos\omega_o t - \frac{8V_{dc}}{\pi}\sum_{n=0}^{\infty}\frac{1}{[2n+1]}\sin\left\{[2n+1]\cos^{-1}\left(\frac{1}{3M}\right)\right\}\\ \times\cos([2n+1]\omega_o t)\\ + \frac{2V_{dc}}{\pi}\sum_{m=1}^{\infty}\sum_{\substack{n=-\infty\\(n\neq 0)}}^{\infty}\frac{1}{m}J_{2n+1}(3m\pi M)\cos([m+n]\pi)\\ \times\cos(2m\omega_c t+[2n+1]\omega_o t)\end{aligned} \tag{11.55}$$

Summing Eqs. (11.54) and (11.55) gives the phase leg voltage with respect to the DC bus midpoint of

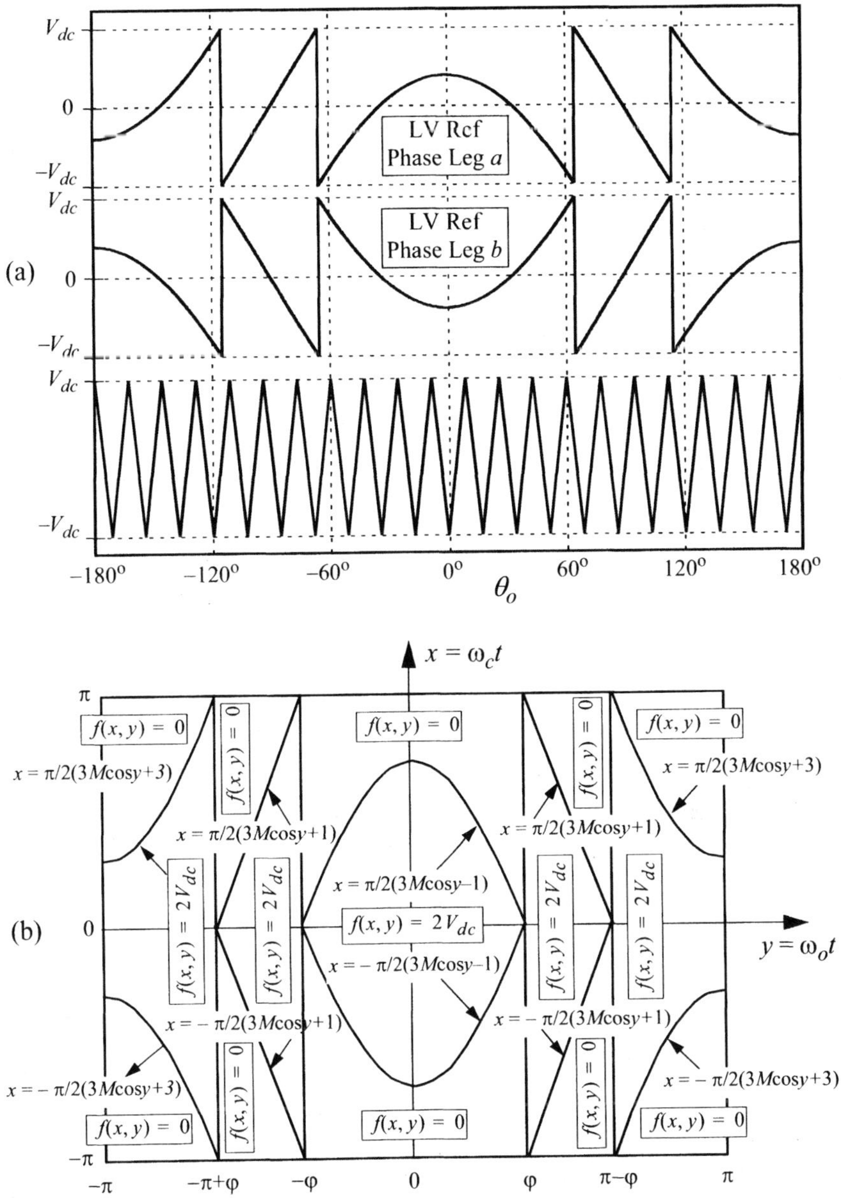

Figure 11.38 (a) Phase leg reference and carrier waveforms and (b) contour plot of $f(x, y)$, for LV stage of seven-level hybrid inverter, $\varphi = \cos^{-1}\left(\frac{1}{3M}\right)$.

$$v_{az}(t) = 3MV_{dc}\cos\omega_o t + \frac{2V_{dc}}{\pi}\sum_{m=1}^{\infty}\sum_{\substack{n=-\infty \\ (n\neq 0)}}^{\infty}\frac{1}{m}J_{2n+1}(3m\pi M)\cos([m+n]\pi) \times \cos(2m\omega_c t + [2n+1]\omega_o t) \tag{11.56}$$

and a line-to-line voltage of

$$v_{az}(t) = 3\sqrt{3}MV_{dc}\cos\left(\omega_o t + \frac{\pi}{6}\right) \tag{11.57}$$

$$+\frac{4V_{dc}}{\pi}\sum_{m=1}^{\infty}\sum_{\substack{n=-\infty \\ (n\neq 0)}}^{\infty}\frac{1}{m}J_{2n+1}(3m\pi M)\cos([m+n]\pi)\sin\left([2n+1]\frac{\pi}{3}\right) \times \cos(2m\omega_c t + [2n+1][\omega_o t - \pi/3] + \pi/2)$$

In comparison, using Eqs. (11.52) and (11.53), the phase leg solution for a seven-level diode-clamped inverter modulated by APOD PWM, *not* overmodulated, is

$$v_{az}(t) = 3MV_{dc}\cos\omega_o t + \frac{2V_{dc}}{\pi}\sum_{m=1}^{\infty}\frac{1}{m}\sum_{n=-\infty}^{\infty}J_{2n+1}(3m\pi M)\cos n\pi \times \cos(m\omega_c t + [2n+1]\omega_o t) \tag{11.58}$$

Clearly, this equation is almost identical to Eq. (11.56). In fact, the only significant difference is that for the diode-clamped inverter the first set of sideband harmonics are centered about the carrier frequency, while for the hybrid inverter these harmonics appear about the second multiple of the carrier. However, once the carrier frequencies are adjusted so that each scheme produces the same number of overall switch transitions per fundamental cycle, the spectra become the same except for the sign of the sidebands $[\cos([m+n]\pi)$ versus $\cos n\pi]$. This difference results from different placement of the active states within the carrier interval for the two modulation strategies. Theoretically it could influence sideband interactions as they sum between carrier groups, but in practice the effect is negligible for pulse ratios greater than 10.

Figure 11.39 shows the theoretical phase leg and *l–l* output voltage spectra for a seven-level hybrid inverter modulated by continuous (APOD equivalent) PWM. Figure 11.40 shows matching experimental results. The close match between the theoretical and experimental results confirms the validity of the previous analysis. However, it should be noted that there is a slight magnitude difference between theory and experiment for some of the sideband compo-

(a)

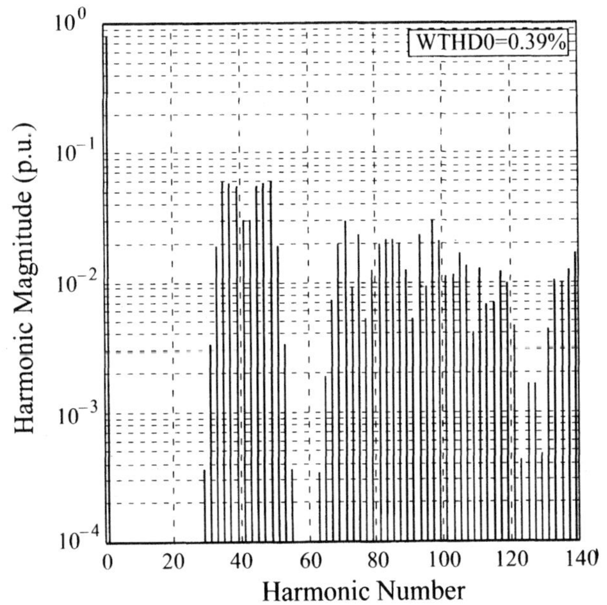

(b)

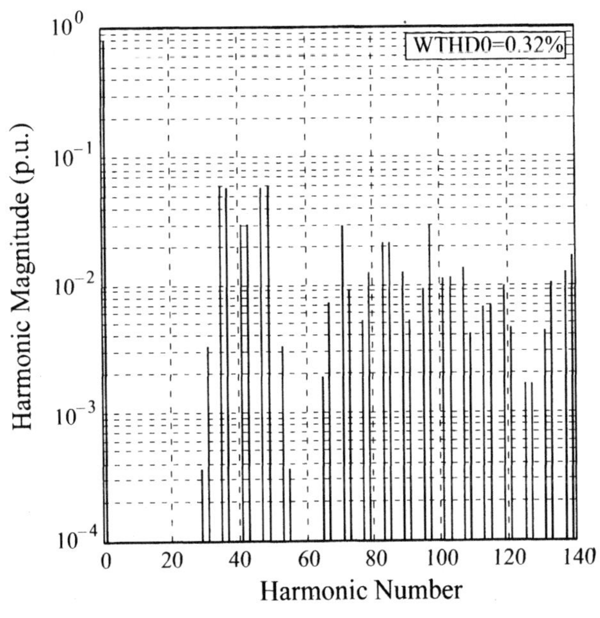

Figure 11.39 Theoretical harmonic spectra for seven-level hybrid three-phase inverter modulated by *continuous naturally sampled* PWM (APOD equivalent): (a) phase leg a and (b) l–l output switched voltage waveforms, $M = 0.8$, $f_c/f_o = 21$.

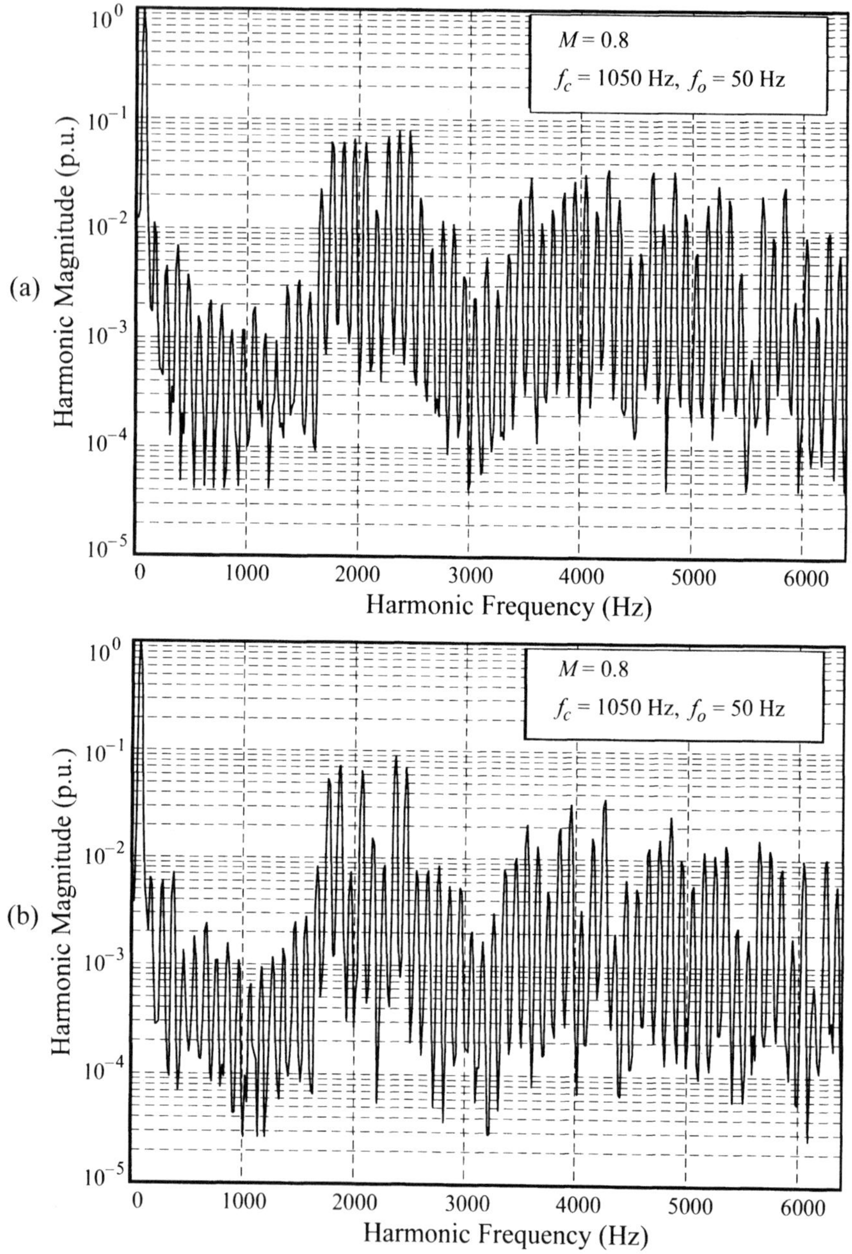

Figure 11.40 Experimental harmonic spectra for seven-level hybrid three-phase inverter modulated by *continuous asymmetrical regular sampled* PWM (APOD equivalent): (a) phase leg a and (b) l–l output switched voltage waveforms, $M = 0.8$, $f_c/f_o = 21$ [11].

nents, particularly the first sideband harmonics in the first carrier group. This is primarily because the experimental system was implemented using regular sampling, and it has been shown in previous chapters that regular sampling introduces some sideband magnitude skew compared to natural sampling for all PWM strategies. In most practical applications, the difference is minor and can be ignored.

11.11 Equivalent PD PWM for a Hybrid Inverter

Since the harmonically superior PD method of PWM has been shown to be realizable for a conventional cascade inverter, the question then becomes how to implement this strategy for the hybrid inverter. The key to achieving this objective comes again from analysis of the spectrum of the phase voltage for a seven-level diode-clamped inverter operated under PD modulation, Figure 11.31(a), where the most significant harmonic in the phase leg voltage is the first carrier component. It has been identified already that this placement of harmonic energy into a carrier component which cancels between phase legs is the key to the superior *l–l* performance of PD modulation. Thus to achieve the equivalent of PD modulation for a hybrid inverter, a carrier component must again be retained in the phase leg voltage spectrum.

For the cascaded inverter, PD equivalent PWM was achieved by using a discontinuous modulation strategy, with the switching frequency doubled to maintain the same overall number of switching transitions as for continuous modulation. Applying these concepts to the hybrid inverter suggests that the LV stage reference waveform must be similarly modified to operate under discontinuous modulation. Figure 11.41(a) shows the reference and carrier waveform arrangements used for the previous results. Figure 11.41(b) shows the modifications needed to achieve discontinuous three-level modulation. Note how the two reference waveforms are now arranged so that whenever a positive bridge output is required, the first phase leg is modulated and the second phase leg is held to the negative rail and whenever a negative bridge output is required while the first phase leg is held to the negative rail and the second phase leg is modulated. Also, the two phase legs now use carriers which are 180° inverted from each other to retain the carrier harmonic in the overall bridge output voltage.

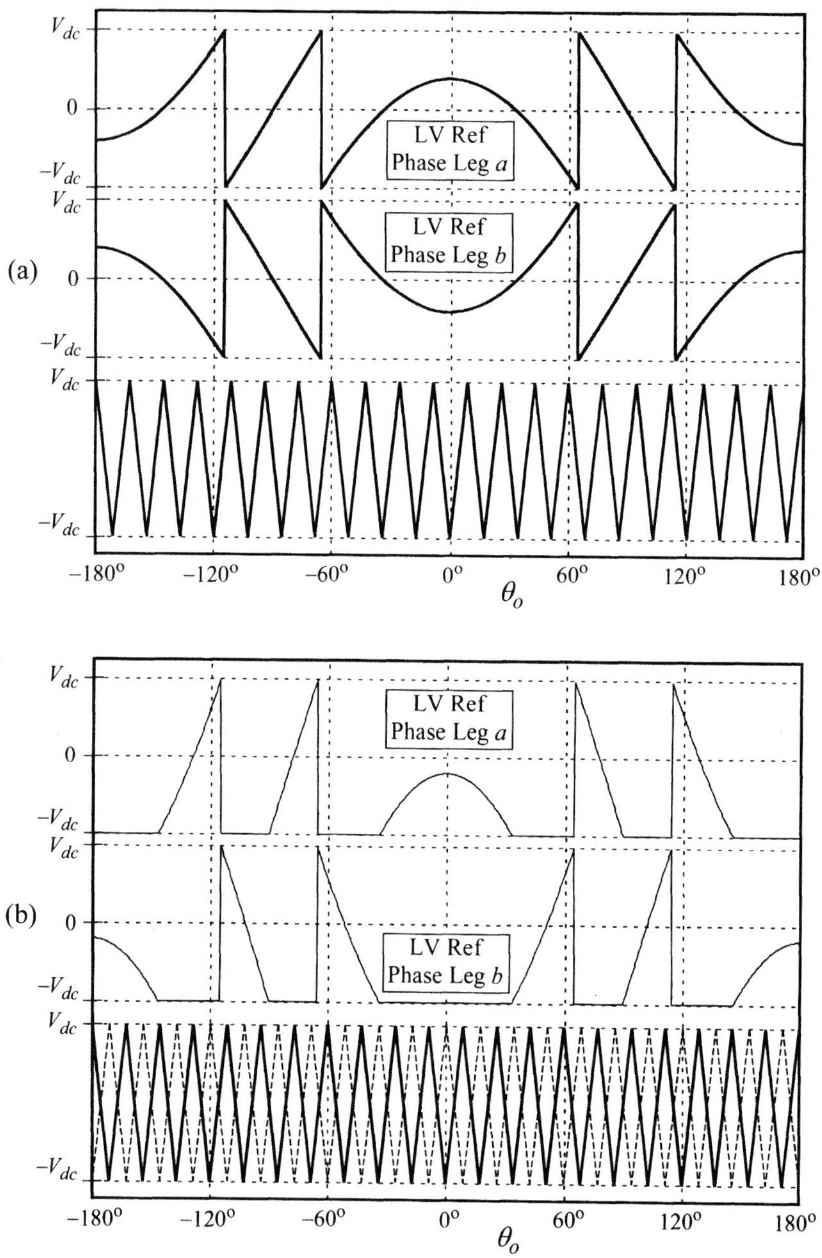

Figure 11.41 Phase leg reference and carrier waveforms for (a) continuous and (b) discontinuous modulation of the LV stage of a seven-level hybrid inverter.

Figure 11.42 shows the phase leg and line voltage spectra of the hybrid inverter under this form of discontinuous modulation obtained using a time-domain computer simulation since no analytical solution is available for this strategy. Figure 11.43 shows experimental results for the same conditions, which again match very closely except for slight differences in the sideband harmonics caused by regular sampling compared to natural sampling.

Figure 11.44 shows the experimental switched waveforms for discontinuous PWM of a hybrid inverter. An important feature of this strategy is that it does not require coordinated switching across all three-phase legs of the hybrid inverter, except for the use of a common carrier reference waveform. However, close examination of the switching waveforms reveals a small anomaly — a spurious pulse in the switched phase voltage each time the HV stage makes a switch transition (i.e., four times per cycle). These pulses occur because when the HV stage makes a transition, a corresponding LV transition must also occur at the same instant to avoid a transient voltage error. In a practical system, particularly with dissimilar devices for the HV and LV inverters, the two transitions will typically not occur simultaneously. Note that this is a drawback of the hybrid inverter topology in principle and is unrelated to the PWM process.

11.12 Third-Harmonic Injection for Multilevel Inverters

In a similar manner as was employed with two-level inverters, it is clearly possible to inject a third-harmonic component into the reference waveforms for each phase leg of a multilevel inverter. This would obviously achieve a 15% increase in modulation index before overmodulation nonlinearities occur, simply because of the reduced height of the three-phase reference envelope that is achieved by third-harmonic injection. However, the effect on the sideband harmonics and the WTHD is not so straightforward.

Figures 11.45, 11.46, and 11.47 show the phase leg and *l–l* voltage spectra for a five-level diode-clamped multilevel inverter modulated with PD, APOD, and POD PWM, respectively, all with third-harmonic injection into the fundamental references and a modulation index of $M = 0.8$. Note that these results were calculated using a time-domain switched simulation, since analytical solutions are very complex to evaluate for multilevel inverters with third-harmonic injected references.

(a)

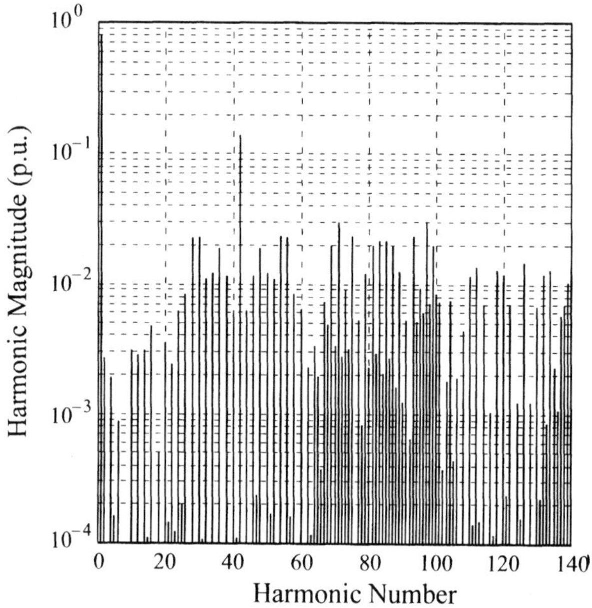

(b)

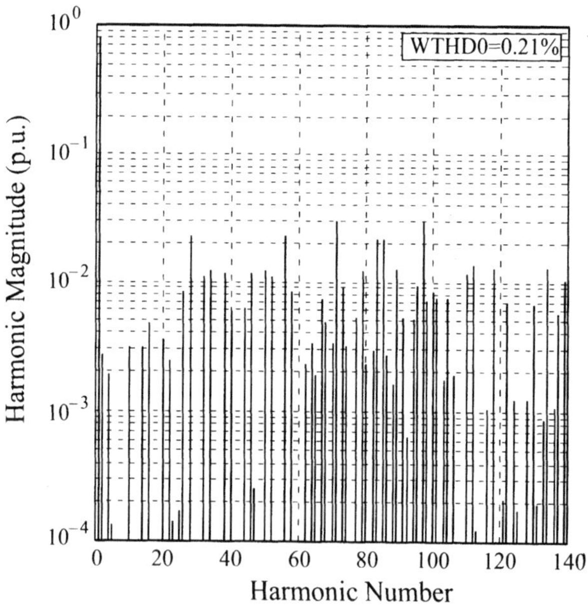

Figure 11.42 Theoretical harmonic spectra for seven-level hybrid three-phase inverter modulated by *discontinuous naturally sampled* PWM (PD equivalent): (a) phase leg a and (b) l–l output switched voltage waveforms, $M = 0.8$, $f_c/f_o = 42$.

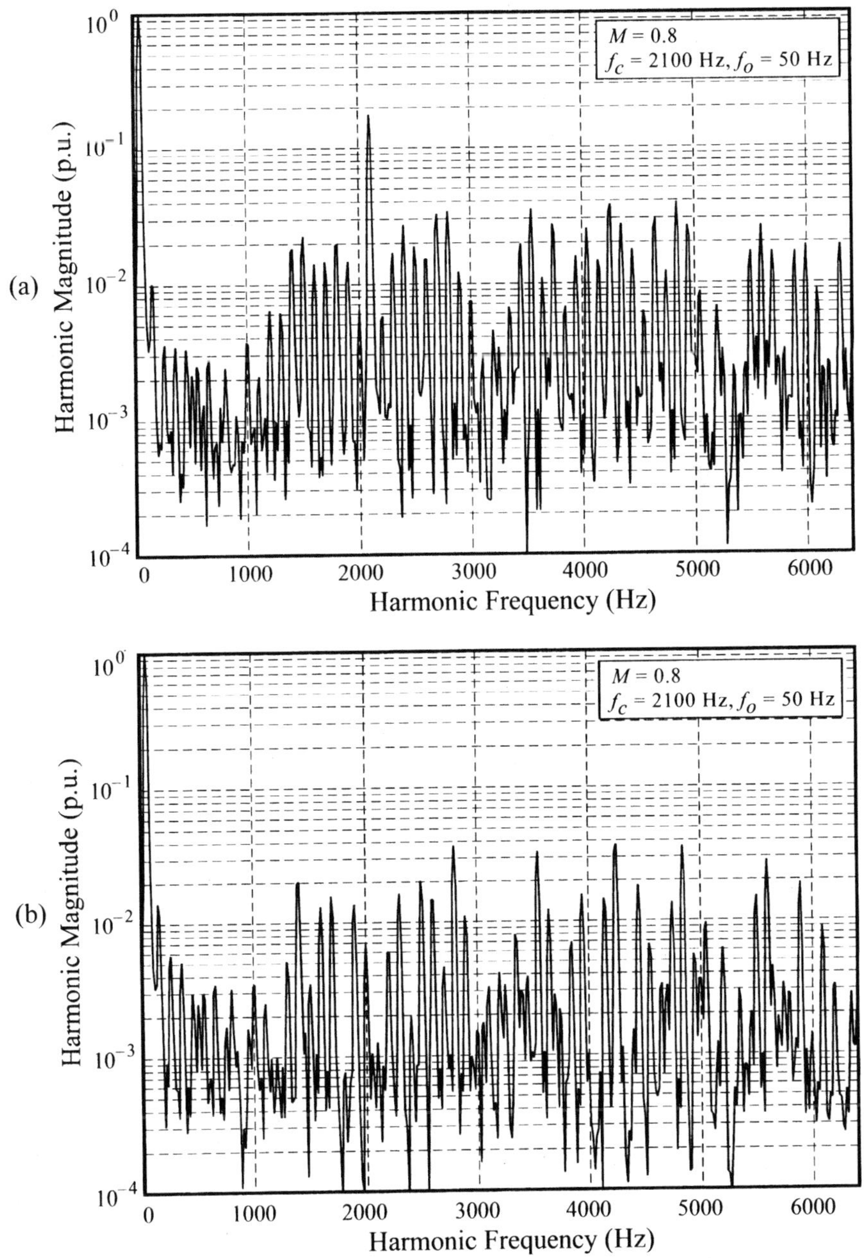

Figure 11.43 Experimental harmonic spectra for seven-level hybrid three-phase inverter modulated by *discontinuous naturally sampled* PWM (PD equivalent): (a) phase leg a and (b) l–l output switched voltage waveforms, $M = 0.8$, $f_c/f_o = 42$ [11].

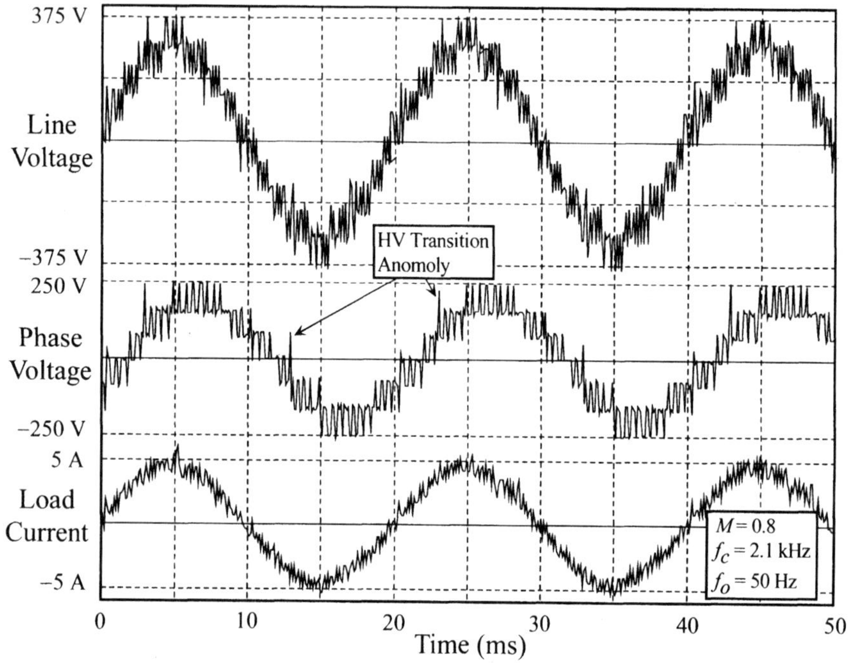

Figure 11.44 Experimental switched voltages and filtered load current for seven-level hybrid three-phase inverter modulated by *discontinuous* (*equivalent PD*) *regularly sampled PWM*, $M = 0.8$, $f_c/f_o = 42$ [11].

These figures can be directly compared against Figures 11.24, 11.27, and 11.30, being the phase leg and *l–l* voltage spectra for five-level PD, APOD, and POD PWM, respectively, *without* a third-harmonic component in the reference. From these results, it can be seen that the effect of a third-harmonic reference component has again been to transfer harmonic energy between sidebands, but not necessarily in an advantageous way. For example, for APOD modulation (Figure 11.27 versus Figure 11.46), the effect has been to concentrate energy into the first sideband harmonics in the first carrier group, and this has increased the WTHD with third-harmonic injection. For PD and POD PWM, the reduction in WTHD with third-harmonic injection is minimal. [Note also that the effect of third-harmonic injection on the performance of a cascaded inverter is identical to that of either APOD or PD PWM (for discontinuous or continuous PWM of the cascaded inverter, respectively), since it has been shown previously that these modulation strategies are identical.]

(a)

(b)

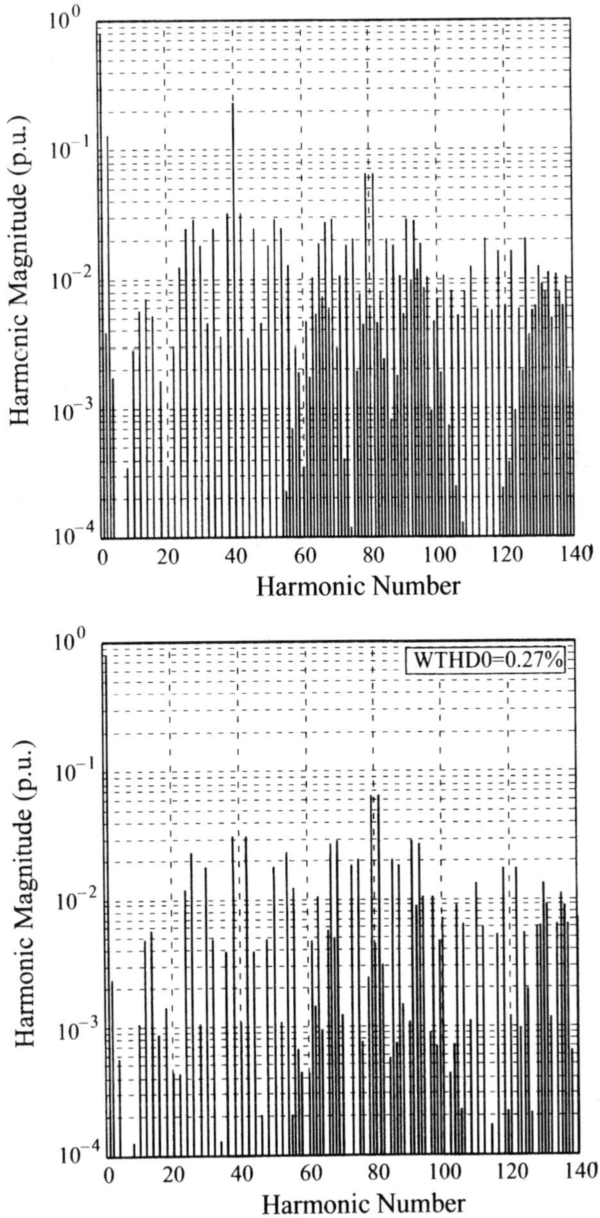

Figure 11.45 Theoretical harmonic spectra for five-level three-phase inverter modulated by *PD naturally sampled* PWM *with third-harmonic injection*: (a) phase leg a and (b) l–l output switched voltage waveforms, $M = 0.8$, $f_c/f_o = 40$.

(a)

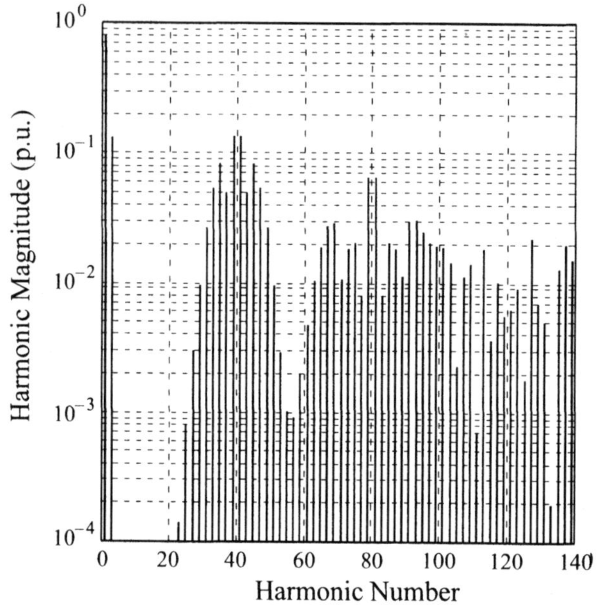

(b)

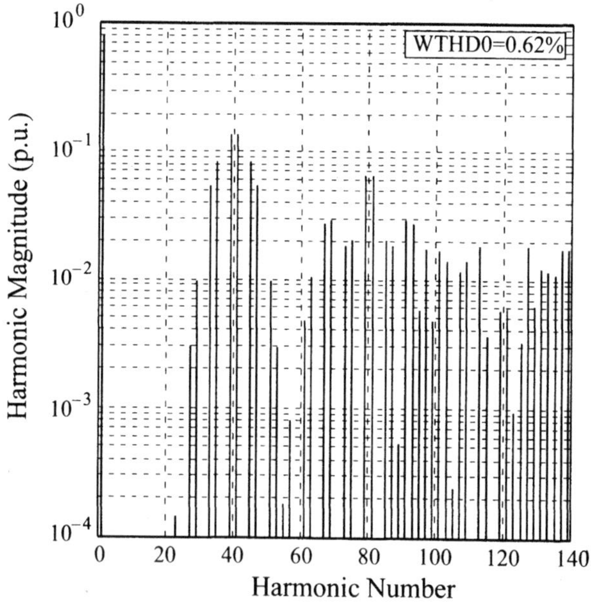

Figure 11.46 Theoretical harmonic spectra for five-level three-phase inverter modulated by *APOD naturally sampled* PWM *with third-harmonic injection*: (a) phase leg a and (b) l–l output switched voltage waveforms, $M = 0.8$, $f_c/f_o = 40$.

(a)

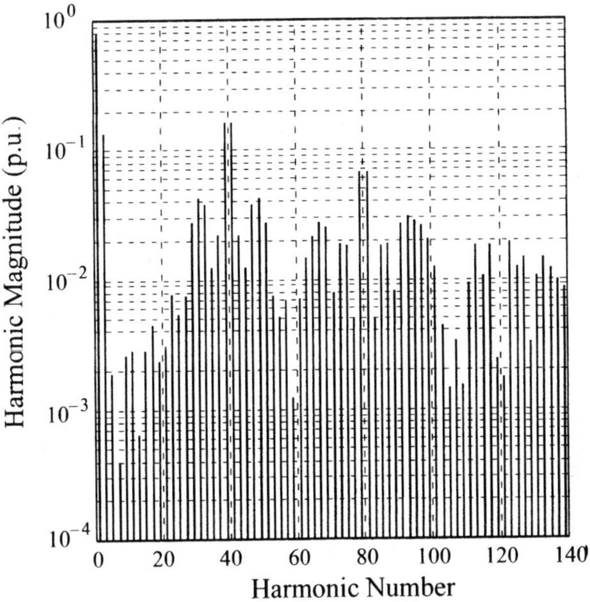

(b)

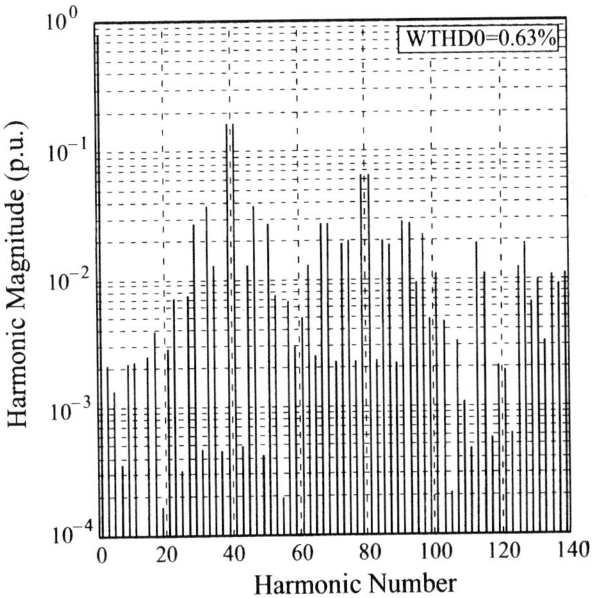

Figure 11.47 Theoretical harmonic spectra for five-level three-phase inverter modulated by *POD naturally sampled* PWM *with third-harmonic injection*: (a) phase leg a and (b) l–l output switched voltage waveforms, $M = 0.8$, $f_c/f_o = 40$.

While these results are only for a particular carrier/fundamental ratio and modulation index, they indicate that third-harmonic injection offers minimal harmonic advantage for PWM of multilevel inverters, and hence only has value to increase the available linear modulation region if this is required. Analysis over a wider range of operating conditions confirms this expectation. Furthermore, for operation at lower carrier/fundamental ratios, there can be a net loss in harmonic performance with third-harmonic injection because of the increased spread in significant harmonics caused by this reference. Significant harmonic sideband intrusion into the baseband region then becomes a real possibility and must be considered in a practical system.

Figure 11.48 illustrates this effect for PD PWM with a reduced carrier ratio, where the WTHD of the modulation strategy with a third-harmonic injected reference is becoming significantly higher than without a third-harmonic component in the reference. Also, the low-order sideband intrusion into the baseband region has become more significant with the third-harmonic reference alternative.

11.13 Operation of a Multilevel Inverter with a Variable Modulation Index

The analyses of multilevel inverter PWM presented so far all assume a relatively high modulation index, sufficient for the reference waveforms to link across all carrier waveforms. At lower modulation indices for hybrid and for carrier disposition modulation of higher level diode-clamped inverters, this will not be the case.

As the fundamental reference waveforms cease to cross the outer carrier waveforms, the modulation process reverts to that of an inverter with two less voltage levels. For example, for a diode-clamped inverter, when the modulation index falls below 50% for a five-level system, or 66% for a seven-level system, the modulation process will revert to that of a three-level inverter for the five-level system, and a five-level inverter for the seven-level system. It should be noted also that the seven-level inverter modulation process will have a further transition to that of a three-level inverter when the modulation index falls below 33%. For a hybrid inverter, once the modulation index falls below 33%, the PWM process will revert to that of a cascaded inverter with a single H-bridge in each phase leg. The harmonic performance of PWM of higher

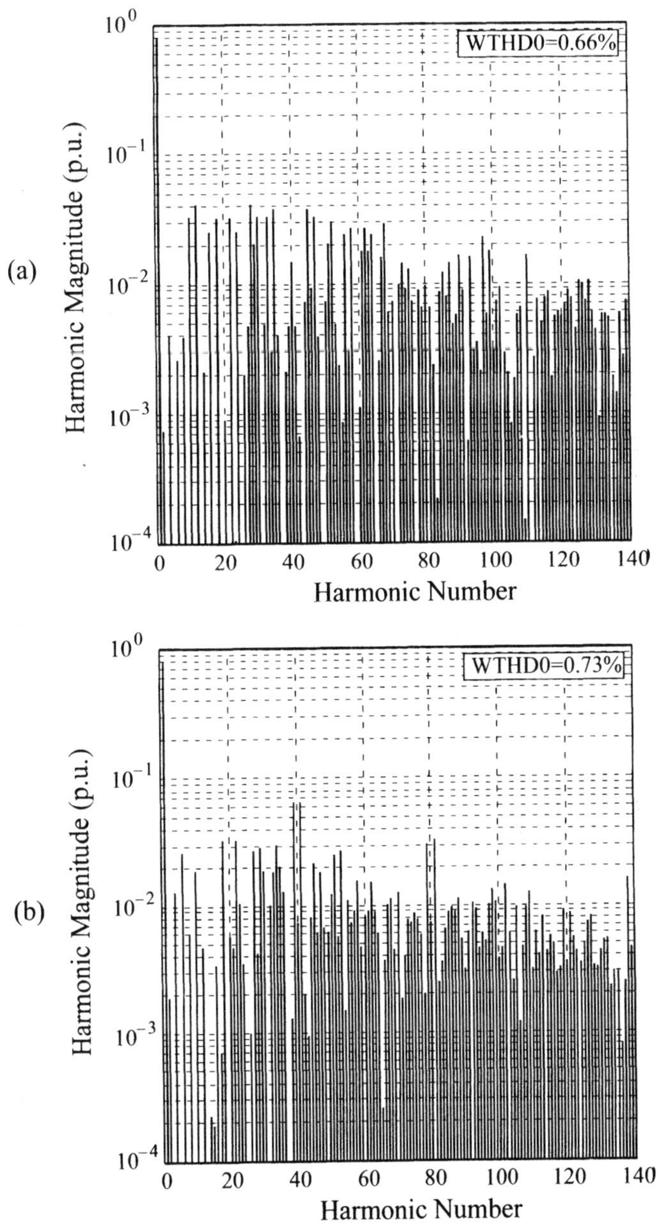

Figure 11.48 Theoretical *l–l* harmonic spectra for five-level hybrid three-phase inverter modulated by *PD naturally sampled* PWM: (a) without and (b) with *third-harmonic injection*, $M = 0.8$, $f_c/f_o = 20$.

level inverters under these low modulation conditions must then be determined from the performance of the appropriate lower level inverter, suitably scaled to suit the revised fundamental magnitude.

Figure 11.49 shows the variation in WTHD0 for three- and five-level diode-clamped inverters operating under a variety of PWM alternatives. It is particularly interesting to note the double peak characteristic of the five-level APOD and PD PWM strategies. The peak at the higher modulation index occurs during five-level switching, while the peak at the lower modulation index occurs during three-level switching. In fact, the five-level WTHD characteristic during the region $0 < M < 0.5$ is identical to the three-level WTHD characteristic, simply compressed and scaled appropriately.

Figure 11.49 also shows the WTHD0 characteristics for PWM of multilevel inverters where the fundamental references have a third-harmonic injected component. Note that the modulation index where the reference waveforms cross into the outer carriers increases by the factor $2/\sqrt{3}$ under these conditions. From this figure, it can be seen that third-harmonic injection compromises the performance of APOD PWM, improves the harmonic performance of five-level POD PWM, and does not make a significant difference to the performance of PD PWM. Under all conditions, PD modulation is clearly superior to other carrier disposition modulation strategies.

Separate results have not been included for the performance of a cascaded inverter under varying modulation indices, since from previous results presented in this chapter it has been shown that a cascaded inverter has a performance equivalent to APOD PWM under continuous modulation, and PD PWM under discontinuous modulation.

11.14 Summary

The various multilevel inverter circuits such as cascade, diode-clamped, hybrid, and capacitor-clamped topologies create another dimension of complexity to inverter modulation by introducing the potential to use multiple DC voltage levels. In this chapter, it has been shown that the harmonic performance obtained from these inverters can be determined by applying the basic principles of PWM laid out in Chapters 3, 4, and 5. It has also been shown that the phase disposition modulation principle produces the lowest harmonic distortion for the diode-clamped inverter by placing significant harmonic energy

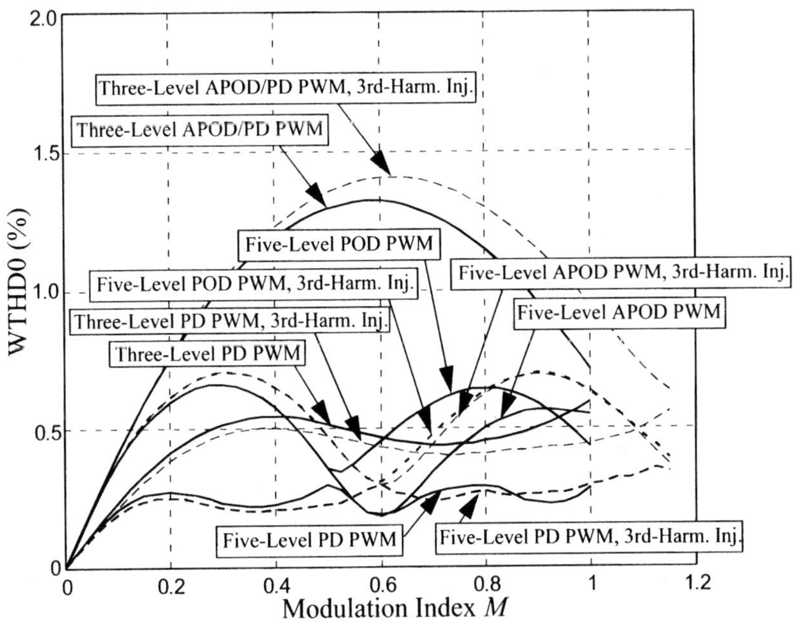

Figure 11.49 WTHD0 of *l–l* switched output voltage of multilevel three-phase inverters for various modulation alternatives and level numbers, $f_c/f_o = 20$.

into a common mode first carrier component, which cancels between phase legs of the inverter. This leads to the realization that a similar improvement in modulation performance for cascaded and hybrid multilevel inverter can be achieved by using discontinuous PWM, which also places significant harmonic energy into a common mode carrier component. Thus the modulation principles for diode-clamped, cascaded, and hybrid inverters are shown to share a common origin and can be linked together.

References

[1] A. Nabae, I. Takahashi, and H. Akagi, "A new neutral-point-clamped PWM inverter," *IEEE Trans. on Industry Applications*, vol. IA–17, no. 5, Sept./Oct. 1981, pp. 518–523.

[2] P. Bhagwat and V.R. Stefanovic, "Generalized structure of a multilevel PWM inverter," *IEEE Trans. on Industry Applications*, vol. 19, no. 6, Nov./Dec. 1983, pp. 1057–1069.

[3] M. Marchesoni, M. Mazzucchelli, and S. Tenconi, "Non-conventional power converter for plasma stabilization," in *Conf. Rec. IEEE Power Electronics Specialists Conf. (PESC)*, Kyoto, 1988, pp. 122–129.

[4] G. Carrara, D. Casini, S. Gardella, and R. Salutari, "Optimal PWM for the control of multilevel voltage source inverter," in *Conf. Rec. European Power Electronics Conf. (EPE)*, Brighton, 1993, pp. 255–259.

[5] B.P. McGrath, "Topologically independent modulation of multilevel inverters," Ph.D. Thesis, Monash University, Australia, 2002.

[6] B.P. McGrath and D.G. Holmes, "A comparison of multicarrier PWM strategies for cascaded and neutral point clamped multilevel inverters," in *Conf Rec. IEEE Power Electronics Specialists Conf. (PESC)*, Galway, 2000, pp. 674–679.

[7] B.P. McGrath and D.G. Holmes, "Opportunities for harmonic cancellation with carrier-based PWM for two-level and multilevel cascaded inverters," *IEEE Trans. on Industry Applications*, vol. 37, No. 2, Mar./April 2001, pp. 574–582.

[8] V.G. Agelidis and M. Calais, "Application specific harmonic performance evaluation of multicarrier PWM techniques," in *Conf. Rec. Power Electronics Specialists Conference*, *(PESC)*, Fukuoka,1998, pp. 172–178.

[9] B.P. McGrath and D.G. Holmes, "An analytical technique for the determination of spectral components of multilevel carrier based PWM methods," *IEEE Trans. on Industrial Electronics*, vol. 49, no. 4, Aug. 2002, pp. 724–738.

[10] M.D. Manjrekar, R. Lund, P. Steimer, and T.A. Lipo, "Hybrid multilevel power conversion system: A competitive solution for high power applications," *IEEE Trans. On Industry Applications*, vol. 36, no. 3, May/June 2000, pp. 834–841.

[11] B.P. McGrath, M. Manjrekar, D.G. Holmes, and T.A. Lipo, "An improved modulation strategy for the hybrid multilevel inverter," in *Conf. Rec. IEEE Industry Applications Society Annual Mtg.*, Rome, 2000, pp. 2086–2093.

12

Space Vector PWM for Multilevel Converters

In earlier chapters of this book, it has been demonstrated for two-level inverters that sine plus third-harmonic injected reference-triangle PWM is essentially identical to space vector PWM and that the major difference between the two methods is the treatment of the zero space vector. In this chapter it is shown how the space vector approach to pulse width modulation can be extended to include multilevel converters. Comparisons between the two modulation approaches then indicate that once more the two methods effectively produce the same result.

12.1 Optimized Space Vector Sequences

It should be evident that the space vector method can be readily extended to incorporate multilevel inverters. However, the large number of states offered by multilevel converters can impose massive computational overhead if not carefully optimized. Optimized space vector sequences were first introduced in [1] and later refined in [2] and [3]. Figure 12.1 shows space vector diagrams for three- and five-level diode-clamped converters where each digit of the space vector identifier represents the voltage level to which the *a*, *b*, and *c* phase legs are, respectively, switched. It is important to note that in both cases some of the switched states are redundant and create the same space vectors. For a three-phase inverter the minimum number of switched transitions in one switching cycle under continuous modulation is three (i.e., one per phase leg), so that the converter cycles through four switched states in each switching period in much the same manner as the conventional two-level SVM. At least the first and last of these must be a redundant space vector state if only the three nearest space vectors are to be used (e.g., 101 – 201 – 211 – 212). Depending upon where in the *d–q* space the reference vector is located, there are two alternatives for this sequence, viz:

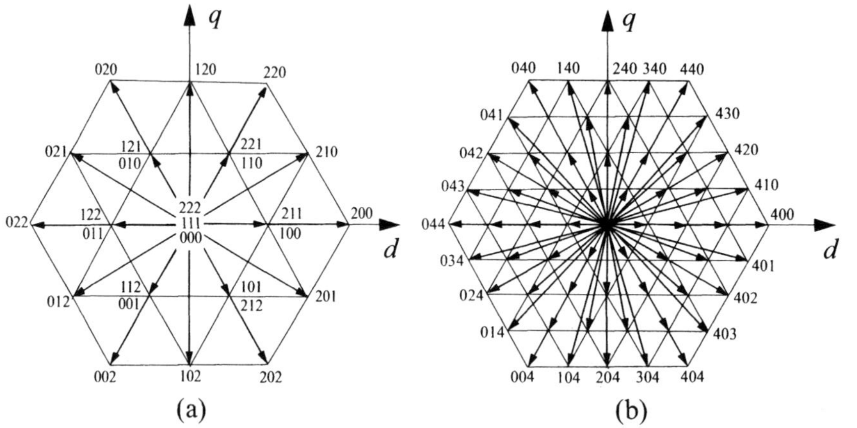

Figure 12.1 Space vector states for (a) three-level and (b) five-level diode-clamped converter.

1. Select two vectors of even redundancy and one vector of odd redundancy (e.g., 211/100, 221/110, and 210), or
2. Select one vector of even redundancy and two vectors of odd redundancy (e.g., 211/100, 200, and 210).

For two-level inverters it can be observed that only alternative 2 is possible since only the origin provides redundant vectors (000/111).

Figure 12.2 illustrates a portion of the five-level space vector plot. summarizes all possible sequences for this portion which achieves the required minimum of three space vector transitions in a switching period. It should be recalled from two-level SVM theory that these sequences should be alternately reversed over consecutive switching periods, and it is assumed that this is done again here. For triangles (b) and (d) there is only one possible sequence. For triangles (a) and (c) the most suitable sequence can be identified from the possible alternatives by ensuring that no extra switching transitions occur when moving between triangles. For example, sequence (c):(i) should be used when moving from triangle (b) to (c) since it begins with the same state as the sequence in (b), while sequence (c):(ii) should be used when moving from triangle (c) to (d) since it begins with the same state as the sequence in (d). Within triangle (c) sequences (c):(i) and (c):(ii) must be swapped at some point, and this is most conveniently done when the duty cycle for the space vector {431/320} exceeds that of {421/310} [4].

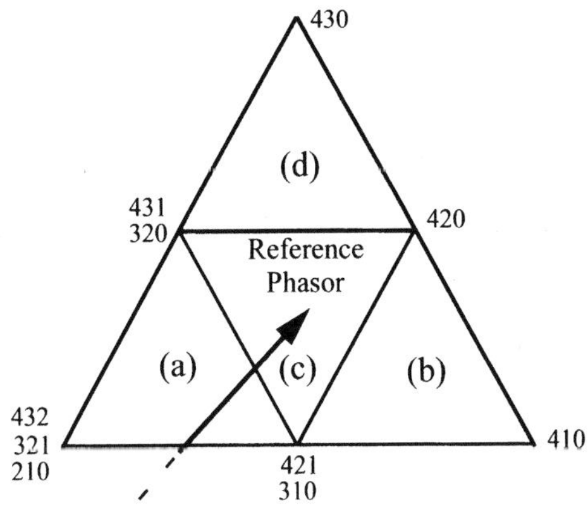

Figure 12.2 Portion of five-level space vector diagram in which triangles (a) and (c) have two vectors of even redundancy and one of odd redundancy and (b) and (d) have one vector of even redundancy and two of odd redundancy.

Applying this principle to triangle (a) means that sequences (a):(i) and (a):(ii) cannot be used because they will introduce extra switching transitions when moving into triangle (c). This is important if the reference vector lies near the boundary of triangles (a) and (c) because it will cross the linear boundary between triangles (a) and (c) twice in a fundamental cycle, and there are many type (c) triangles in the space vector plot. Hence, only sequences (a):(iii) and (a):(iv) can be used, and these are identical to sequences (c):(i) and (c):(ii)

Table 12.1 Possible Sequences in Five-Level Space Vector Subset of Figure 12.1

Triangle	Sequence
(a)	(i) {432 to 431 to 421 to 321}
	(ii) {210 to 310 to 320 to 321}
	(iii) {421 to 321 to 320 to 310}
	(iv) {431 to 421 to 321 to 320}
(b)	{421 to 420 to 410 to 310}
(c)	(i) {421 to 420 to 320 to 310}
	(ii) {431 to 421 to 420 to 320}
(d)	{431 to 430 to 420 to 320}

Note: Reverse sequences to those shown are available but are not explicitly shown.

except that state {420} has been replaced by state {321}. As a result only state {321} of the triple redundant vector {432/321/210} is usable.

Similar analyses for vectors with even redundancies greater than three reveals that only two states can ever be used to achieve minimal switching, and for vectors with odd redundancies greater than two, only one state is usable. Note also that all useful sequences begin and end with an even redundant space vector state.

While the above analysis is from the perspective of a diode-clamped topology, the only significant difference with the cascade topologies is that there is a greater variety of redundancies. A similar analysis for cascaded-type inverters leads to a similar restriction of states and sequences which can be used to achieve the minimum number of switching transitions in a fundamental cycle.

12.2 Modulator for Selecting Switching States

The task of the modulator is to determine the position and duration that the switches should assume during each duty cycle in order to synthesize the reference voltage vector. However, since the number of possible switching states of a multilevel converter increases with the cube of the number of converter levels, the number of possible vectors increases as

$$N_{\text{vectors}} = 1 + 6\sum_{i=1}^{L-1} i \tag{12.1}$$

where L is the number of levels.

For example, while a simple conventional two-level inverter has only $2^3 = 8$ switching states and $1 + 6 = 7$ possible vectors, a three-level inverter produces $3^3 = 27$ states and $1 + 18 = 19$ vectors. Furthermore, when the number of levels increases to 5, the number of states increases to $5^3 = 125$ and the number of vectors increases to 61. Clearly the computational effort needed to establish the correct switch conditions rapidly overwhelms even the fastest signal processor and a random search, for example, over the 91 possible triangles of a five-level inverter, becomes an impossible task. It obviously becomes necessary to develop a more effective approach that can rapidly determine the three nearest space vectors without requiring a blind search process.

12.3 Decomposition Method

Since 3 possible switching states exist for each of the three phases of a three-level inverter, there are 27 possible states for this inverter topology. The complexity of selecting the proper states for a given command voltage vector can be considerably simplified by viewing the three-level inverter space vector hexagon as consisting of a number of two-level hexagons, as shown in Figure 12.3 [5]. The origin of each of these hexagons are located on one of the apexes of the three-level inverter inner hexagon. Thus each of the two-level hexagons are shifted by $V_{dc}/3$ with respect to the origin of the three-level hexagon.

To proceed with the switching state determination, first one of these six hexagons is selected, based on the location of the target voltage vector. Second, the original reference voltage vector is decremented by the voltage vector that locates the origin of the selected two-level hexagon. This then allows the determination of the switching sequence and the calculation of the voltage vector duration to be done in the same manner as for a conventional two-level inverter using either space vector or sine-triangle modulation concepts.

The choice of two-level hexagon should be such that the reference voltage vector smoothly progresses from one hexagon to the next as it rotates. This requirement is satisfied by subdividing the vector space into six regions as shown in Figure 12.4, and selecting the two-level hexagon from Figure 12.3 whose number corresponds to the region number. Selection of the two-level hexagon in this way also facilitates a smooth progression from the inner (origin-centered) hexagon of the three-level inverter to the outer two-level hexagons, for reduced magnitude reference space vectors.

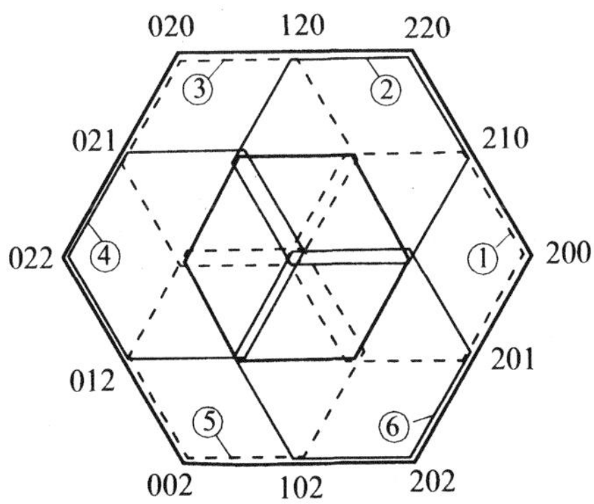

Figure 12.3 Space vector hexagon of a three-level inverter viewed as being composed of six two-level hexagons.

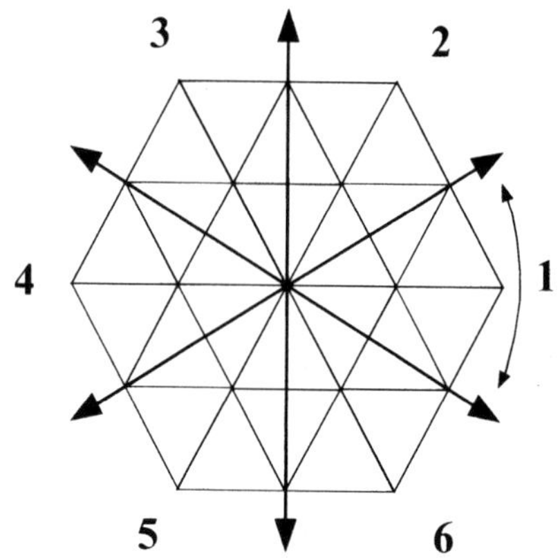

Figure 12.4 Segregation of three-level hexagon into six sectors $s = 1, \ldots, 6$. Hexagon s of Figure 12.3 is then used to generate the space vectors that make up the commanded voltage vector.

Once the sector is identified, the origin of the reference voltage vector is changed to the origin of the two-level hexagon associated with the sector. This is achieved by subtracting a vector locating the center of the selected hexagon from the original reference vector. The corrections on a phase leg basis are summarized in Table 12.2, and produce revised reference voltages $v_{as}^{*\prime}$, $v_{bs}^{*\prime}$, $v_{cs}^{*\prime}$ from each of the original reference voltages v_{as}^*, v_{bs}^*, v_{cs}^*. (A similar correction can be readily developed for stationary frame *d/q* reference voltages without difficulty for a space vector PWM implementation.)

The particular switching sequence of the three-level inverter is now determined by the vertices defining the selected hexagon. For example, in the case shown in Figure 12.5, the minimum transition switching sequence is 211–210–110–100. If the {211/100} redundant vector is taken to be the origin reference, this switching sequence becomes exactly the same as that of a conventional two-level space vector modulator (i.e., 111–110–010–000). Therefore any two-level modulation strategy can now be used, provided that the correct switches are selected from the four switches of each phase leg to achieve the required inverter switched state. This selection can be realized by simple logic circuitry or equivalent logic statements in a microcontroller for each hexagon.

The concept of partitioning the space vector hexagon of a higher level inverter into multiple hexagons of a smaller level inverters can be extended to other multilevel inverters. For example, consider the case of four-level inverter

Table 12.2 Two-level Hexagon Offset Adjustment for Reference Voltage Vectors in Sector s

Sector	Revised Reference Voltage $v_{as}^{*\prime}$	Revised Reference Voltage $v_{bs}^{*\prime}$	Revised Reference Voltage $v_{cs}^{*\prime}$
1	$v_{as}^* - V_{dc}/3$	$v_{bs}^* + V_{dc}/6$	$v_{cs}^* + V_{dc}/6$
2	$v_{as}^* - V_{dc}/6$	$v_{bs}^* - V_{dc}/6$	$v_{cs}^* + V_{dc}/3$
3	$v_{as}^* + V_{dc}/6$	$v_{bs}^* - V_{dc}/3$	$v_{cs}^* + V_{dc}/6$
4	$v_{as}^* + V_{dc}/3$	$v_{bs}^* - V_{dc}/6$	$v_{cs}^* - V_{dc}/6$
5	$v_{as}^* + V_{dc}/6$	$v_{bs}^* + V_{dc}/6$	$v_{cs}^* - V_{dc}/3$
6	$v_{as}^* - V_{dc}/6$	$v_{bs}^* + V_{dc}/3$	$v_{cs}^* - V_{dc}/6$

whose space vector diagram is shown in Figure 12.6. In this case each phase leg generates four voltage states and therefore there are a total of 64 switching states. However, if the four-level space vector diagram is decomposed into six three-level diagrams, three-level SVPWM algorithms can be used to obtain switching states and their durations. Then the switching of each of these three-level inverters can be determined by considering that they are composed of six two-level inverter hexagons as has just been considered. In this way, simple logic trees can be built up for any level inverter to allow the proper switching selections can be rapidly determined for any reference voltage phasor.

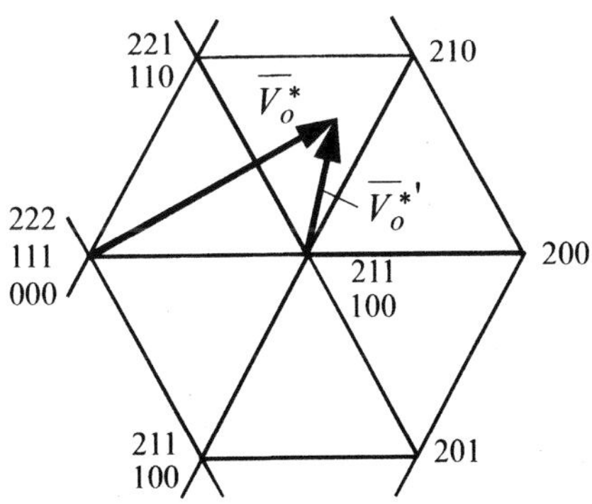

Figure 12.5 Example showing change of reference origin for a voltage reference vector $\overline{V}_o^*$ located in Sector 1.

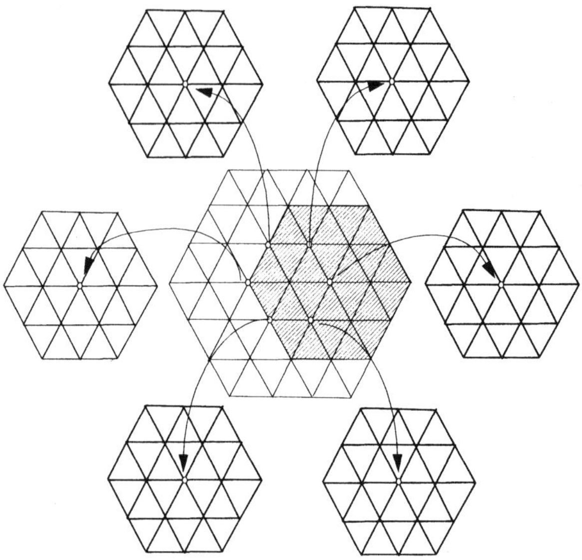

Figure 12.6 Application of hexagon decomposition to a four-level inverter [5].

12.4 Hexagonal Coordinate System

Much of the computational overhead for the determination of the triangle containing the target vector is introduced by representing each of the vector states as points in an orthogonal coordinate system. The coordinate system needed to completely define the three line voltages is, in reality, two dimensional since the fact that

$$v_{ab} + v_{bc} + v_{ca} = 0 \tag{12.2}$$

forms a constraint which places all of the line voltage vector states on a plane.

Since the possible states of both phase voltage and line voltage vectors naturally form equilateral triangles it has been shown for three-levels [1] and L-levels [2] that the computational complexity is markedly reduced by adopting a nonorthogonal system in which the basic vectors are directed along two non-colinear directions of the six possible line voltage vector directions. For example, one such system can be formed by first considering a three-level inverter, whereupon

$$\{\boldsymbol{u}, \boldsymbol{v}, \boldsymbol{z}\} = \frac{1}{2V_{dc}}\{\boldsymbol{V}_{1,0,-1}, \boldsymbol{V}_{0,1,-1}, \boldsymbol{V}_{1,1,1}\} = \left\{\begin{bmatrix}1\\0\\-1\end{bmatrix}, \begin{bmatrix}0\\1\\-1\end{bmatrix}, \begin{bmatrix}1\\1\\1\end{bmatrix}\right\} \tag{12.3}$$

The first vector $\bar{u}$ is colinear with vectors directed along the phase a axes as shown in Figure 12.7 while the second vector is colinear with vectors directed along phase c. The third vector corresponds to one of the two zero state vectors which form, in effect, a zero sequence component. Vectors along the b axis as well as the remaining vectors can be determined as components along the $\bar{u}$ and $\bar{v}$ vector directions.

It is useful to consider a transformation to another three-dimensional coordinate system, in which case

$$\{\boldsymbol{g}, \boldsymbol{h}, \boldsymbol{z}\} = \left\{\begin{bmatrix}1\\0\\0\end{bmatrix}, \begin{bmatrix}0\\1\\0\end{bmatrix}, \begin{bmatrix}0\\0\\1\end{bmatrix}\right\} = \boldsymbol{T}\{\boldsymbol{u}, \boldsymbol{v}, \boldsymbol{z}\} \tag{12.4}$$

It is not difficult to determine that the needed transformation matrix is

$$\boldsymbol{T} = \{\boldsymbol{u}, \boldsymbol{v}, \boldsymbol{z}\}^{-1} = \frac{1}{3}\begin{bmatrix}2 & -1 & -1\\-1 & 2 & -1\\1 & 1 & 1\end{bmatrix} \tag{12.5}$$

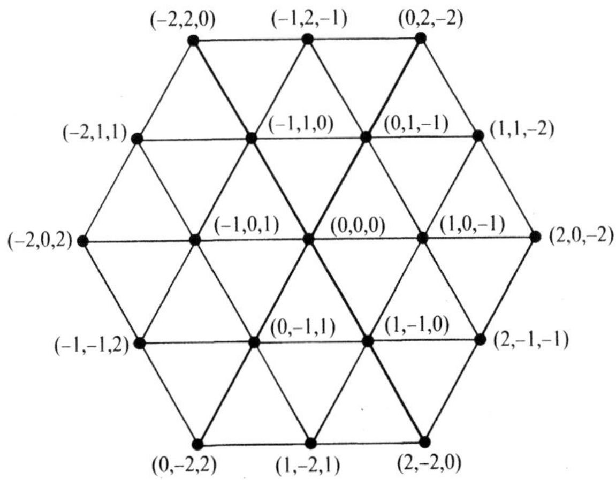

Figure 12.7 Line voltage vector states of a three-level inverter using a,b,c coordinates $\bar{u}$ and $\bar{v}$.

For practical purposes the zero sequence component is superfluous if the transformation of variables is used on the line voltages since this component will always be zero. The transformation necessary to simply transform from $\bar{u}, \bar{v}$ to $\bar{g}, \bar{h}$ coordinates is therefore

$$\boldsymbol{T} = \frac{\sqrt{2}}{3}\begin{bmatrix} 2 & -1 & -1 \\ -1 & 2 & 1 \end{bmatrix} \tag{12.6}$$

Figure 12.8 shows the two-dimensional representation of the space vectors and reference vector which is normalized to the link voltage V_{dc} for the case of a three-level inverter. In contrast to Figure 12.1(a), the vectors from the origin to any vertex can now be represented by only two rather than three integers.

The fact that all switching vectors have integer coordinates is useful in determining which triangle is occupied by a target voltage vector. For example, consider a target vector located within the (1,1), (1,0), (0,1) triangle as shown in Figure 12.8. The vector location can be localized to within a parallelogram by successively rounding the target vector g and h components up or down as follows:

$$\boldsymbol{V}_{ul} = \begin{bmatrix} \lceil V_{ref,\,g} \rceil \\ \lfloor V_{ref,\,h} \rfloor \end{bmatrix} = \begin{bmatrix} 1 \\ 0 \end{bmatrix} \tag{12.7}$$

$$\boldsymbol{V}_{lu} = \begin{bmatrix} \lfloor V_{ref,\,g} \rfloor \\ \lceil V_{ref,\,h} \rceil \end{bmatrix} = \begin{bmatrix} 0 \\ 1 \end{bmatrix} \tag{12.8}$$

$$\boldsymbol{V}_{uu} = \begin{bmatrix} \lceil V_{ref,\,g} \rceil \\ \lceil V_{ref,\,h} \rceil \end{bmatrix} = \begin{bmatrix} 1 \\ 1 \end{bmatrix} \tag{12.9}$$

$$\boldsymbol{V}_{ll} = \begin{bmatrix} \lfloor V_{ref,\,g} \rfloor \\ \lfloor V_{ref,\,h} \rfloor \end{bmatrix} = \begin{bmatrix} 0 \\ 0 \end{bmatrix} \tag{12.10}$$

where $\lceil x \rceil$ and $\lfloor x \rfloor$ denote rounding to the next upper and lower integer value of x, respectively. This operation places the target vector in either the (1,0), (0,1), (1,1) triangle or the (1,0), (0,1), (0,0) triangle of Figure 12.8. It can be seen that $\boldsymbol{V}_{ul}$ and $\boldsymbol{V}_{lu}$ will be two of the vertices of the triangle locating the target vector regardless of within which of the two triangles the vector resides.

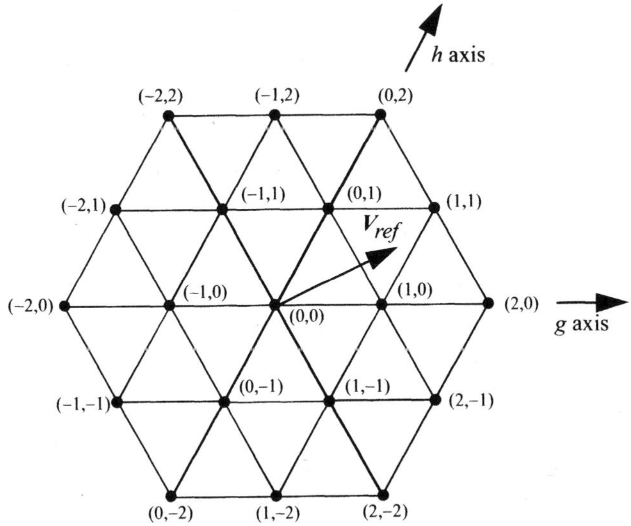

Figure 12.8 Line voltage vector states of a three-level converter in a hexagonal g,h coordinate system.

The third vertex can now be determined by noting that if V_g and V_h denote g and h components of any vector located on the diagonal, from $\boldsymbol{V}_{ul}$ and $\boldsymbol{V}_{lu}$

$$V_{ref,g} + V_{ref,h} = V_{ul,g} + V_{ul,h} = V_{lu,g} + V_{lu,h} \tag{12.11}$$

The identity of the third vertex can therefore be determined by evaluating the sign of the expression

$$V_{ref,g} + V_{ref,h} - (V_{ul,g} + V_{ul,h}) \tag{12.12}$$

If the value is positive, then $\boldsymbol{V}_{uu}$ locates the third vertex; if it is negative, then $\boldsymbol{V}_{ll}$ locates the third point.

Once the nearest three vectors have been identified, their corresponding duty cycles can be found by solving

$$\boldsymbol{V}_{ref} = d_{ul}\boldsymbol{V}_{ul} + d_{lu}\boldsymbol{V}_{lu} + d_3\boldsymbol{V}_3 \tag{12.13}$$

where $\boldsymbol{V}_3$ equals either $\boldsymbol{V}_{ll}$ or $\boldsymbol{V}_{uu}$ and d_3 equals its corresponding duty cycle. The solution is again subject to the constraint

$$d_{ul} + d_{lu} + d_3 = 0 \tag{12.14}$$

Since all switching state vectors have integer coordinates, the solutions for two of the vertices are effectively the fractional portions of $\boldsymbol{V}_{ref}$, that is,

$$\begin{aligned} d_{ul} &= V_{ref,g} - V_{ll,g} \\ d_{lu} &= V_{ref,h} - V_{ll,h} \\ d_{ll} &= 1 - d_{ul} - d_{lu} \end{aligned} \tag{12.15}$$

if $\boldsymbol{V}_3 = \boldsymbol{V}_{ll}$, or

$$\begin{aligned} d_{ul} &= -(V_{ref,h} - V_{uu,h}) \\ d_{lu} &= -(V_{ref,g} - V_{uu,g}) \\ d_{uu} &= 1 - d_{ul} - d_{lu} \end{aligned} \tag{12.16}$$

if $\boldsymbol{V}_3 = \boldsymbol{V}_{uu}$.

As an example, assume that it is necessary to realize a phase voltage of $0.5V_{dc}$ at the time instant where the phase position of the vector is 15° electrical with respect to the phase a axis reference. In this case the amplitude of the phase voltage target vector is 0.866 at an angle of 45°. (Clearly this step could be omitted if the phase voltage were commanded directly). The corresponding line voltages are

$$\boldsymbol{V}_{l-l} = \begin{bmatrix} V_{ab} \\ V_{bc} \\ V_{ca} \end{bmatrix} = V_{dc}\begin{bmatrix} 0.612 \\ 0.224 \\ -0.837 \end{bmatrix} \tag{12.17}$$

In terms of g and h components,

$$\boldsymbol{V}_{l-l,gh} = \boldsymbol{T}\boldsymbol{V}_{l-l} = V_{dc}\begin{bmatrix} 0.612 \\ 0.224 \end{bmatrix} \tag{12.18}$$

Hence

$$\boldsymbol{V}_{ul} = \begin{bmatrix} 1 \\ 0 \end{bmatrix} \quad \boldsymbol{V}_{uu} = \begin{bmatrix} 1 \\ 1 \end{bmatrix} \qquad \boldsymbol{V}_{ll} = \begin{bmatrix} 0 \\ 0 \end{bmatrix} \quad \boldsymbol{V}_{lu} = \begin{bmatrix} 0 \\ 1 \end{bmatrix} \tag{12.19}$$

The exact location of the vector is determined by evaluating

$$V_g + V_h - (V_{ul,g} + V_{ul,h}) = 0.612 + 0.224 - 1 = -0.163 < 0 \tag{12.20}$$

Thus, the vector is located in the triangle with vertices (0,0), (0,1), and (1,0), and the duty cycles for the three vectors defining the triangle are

$$\begin{aligned} d_{ul} &= V_{ref,g} - V_{ll,g} = 0.612 \\ d_{lu} &= V_{ref,h} - V_{ll,h} = 0.224 \\ d_{ll} &= 1 - d_{ul} - d_{lu} = 0.163 \end{aligned} \tag{12.21}$$

When the amplitude of the target vector increases from $0.5V_{dc}$ to $0.7V_{dc}$ then

$$V_{ref,g} + V_{ref,h} - (V_{ul,g} + V_{ul,h}) = 0.857 + 0.314 - 1 = 0.171 > 0 \tag{12.22}$$

in which case the triangle containing $\overline{V}_{ref}$ has vertices (0,1), (1,0), (1,1) and

$$\begin{aligned} d_{ul} &= -(V_{ref,h} - V_{uu,h}) = 0.686 \\ d_{lu} &= -(V_{ref,g} - V_{uu,g}) = 0.143 \\ d_{uu} &= 1 - d_{ul} - d_{lu} = 0.171 \end{aligned} \tag{12.23}$$

A variation of this method has also been reported that is said to provide a somewhat improved computational efficiency [3].

12.5 Optimal Space Vector Position within a Switching Period

Once the optimum space vector switching sequence for continuous modulation has been identified, it must be placed in each switching period to optimize the harmonic profile of the waveform. This means splitting the duty cycle of the first vector in the sequence, which is redundant with the last vector, across the first and last switching states. A technique reported by Fukuda [6] for two-level inverters can be used to resolve this issue.

Figure 12.9(a) shows the trajectory of the time integral of the converter output voltage (i.e., the flux) in the d–q space. The ideal trajectory is circular for a sinusoidal output voltage, but the switched nature of the converter operation restricts the real trajectory to a quasi-circular path as shown. Figure 12.9(b) shows the flux trajectories for one switching cycle. The switching sequence uses space vectors $\overline{V}_1$, $\overline{V}_2$, and $\overline{V}_3$ with respective duty cycles d_1, d_2, and d_3, and the space vector time durations are given by

$$\begin{aligned} T_{V_1(1)} &= kd_1\Delta T/2 \qquad & T_{V_2} &= d_2\Delta T/2 \\ T_{V_1(2)} &= (1-k)d_1\Delta T/2 \qquad & T_{V_3} &= d_3\Delta T/2 \end{aligned} \tag{12.24}$$

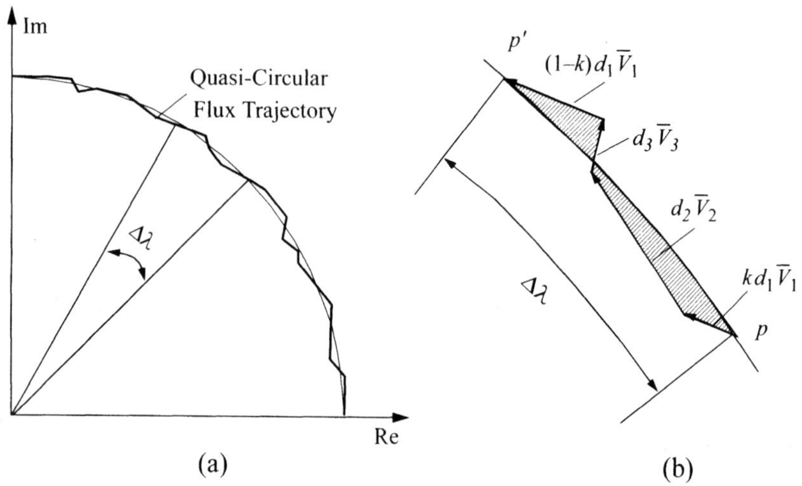

Figure 12.9 (a) Locus of the flux trajectory and (b) flux trajectory over one switching cycle showing the approximation of circular arc. Area between the space vector flux paths and the ideal flux trajectory shows the region to be minimized.

where $\Delta T/2$ denotes the switching period (half carrier interval) and k is an arbitrary factor that splits space vector $\overline{V}_1$ into two parts so as to occupy both the first and last state of the switching sequence,

The deviation between the switched quasi-circular trajectory and the ideal circular path at each switching cycle provides a measure of the harmonic distortion of the modulating process. This distortion can be evaluated by integrating the area between the two trajectories over each switching period. The issue then becomes determining a means to find the value of k that minimizes this area to achieve the optimum modulation strategy.

Figure 12.10 shows the optimum values for k for successive switching periods over a complete fundamental cycle for a three-, five- and seven-level inverter. These values were found from simulation by varying k within each switching period to minimize the flux error area. Clearly the optimal value of k varies insignificantly about the value of 0.5 so that for optimal harmonic performance the start and end redundant vector periods can be made equal, at least for any reasonably high pulse ratio.

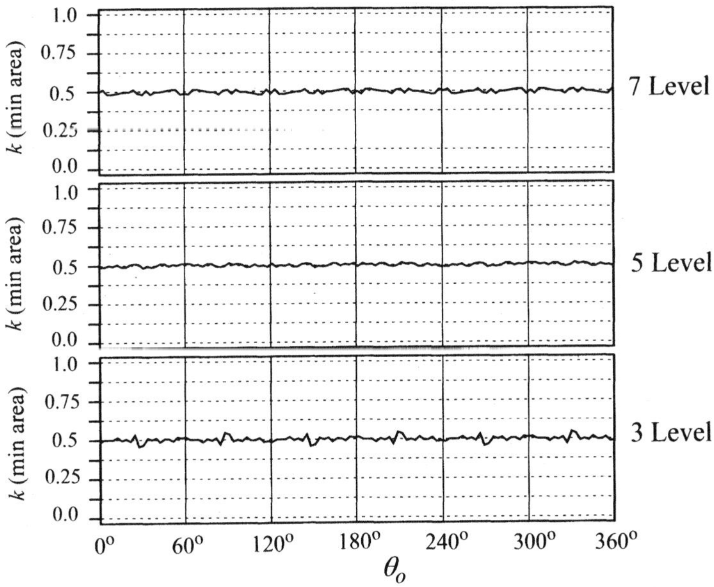

Figure 12.10 Value of k which yields minimum volt–second area for a pulse ratio of 120, modulation index $M = 0.8$.

12.6 Comparison of Space Vector PWM to Carrier-Based PWM

It has been shown in Section 6.3 that for two-level inverters the addition of a common offset voltage to the three phase references of

$$v_{off} = \frac{\max(v_{az}^*, v_{bz}^*, v_{cz}^*) + \min(v_{az}^*, v_{bz}^*, v_{cz}^*)}{2} \tag{12.25}$$

will center the active space vectors during the switching period of a carrier-based PWM algorithm and hence match the references of space vector modulation. Figure 12.11 shows the converter switching states which will result when this offset is used for a five-level system operating under phase disposition (PD) PWM. For the expanded switching condition shown, the sequence consists of two vectors of even redundancy ($\overline{V}_1$ and $\overline{V}_3$) and one vector of odd redundancy ($\overline{V}_2$). Examination of this sequence shows that the durations of the two states of $\overline{V}_1$ are not equal but rather center the vector of odd redundancy in the middle of the half carrier period. Hence, this offset does not achieve the optimum centering of the two middle vectors of the switching sequence.

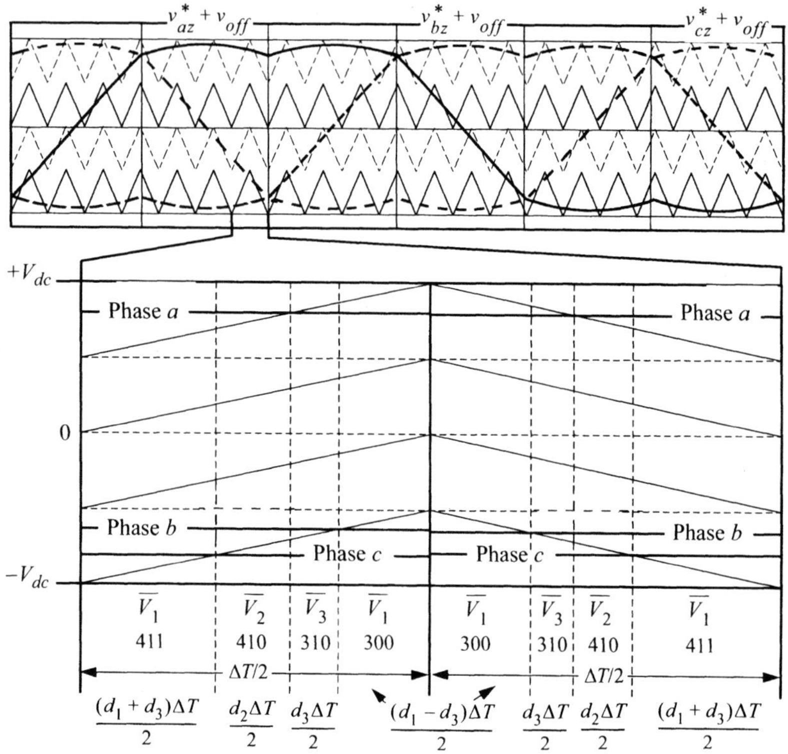

Figure 12.11 Space vector positions resulting from use of a two-level SVM common mode offset with a five-level inverter. The effect is to center the space vector with odd redundancy ($\overline{V}_2$) rather than making equal split dwell times for space vector $\overline{V}_1$.

The limitation of Eq. (12.25) is that it assumes that the first and last switching transition in each switching period is determined by the comparison of the absolute maximum and minimum reference values against the carrier. This is not necessarily the case for a multilevel inverter. For example, Figure 12.11 shows that the last switching transition in the first half carrier period is caused by the comparison of the V_b reference against the carrier, and this is the middle reference value. Hence, to proceed it is necessary to identify which of references v_{az}^*, v_{bz}^*, or v_{cz}^* is responsible for the first and last switching transitions in each half carrier period. This can be done by using a modulus offset function to vertically shift the two-level optimized reference voltages so that their carrier intersections lie within a common carrier band, to create modified references of

$$v_{kz}^{*\,'} = (v_k^* + v_{off} + V_{dc})\ \mathrm{mod}\left(\frac{2V_{dc}}{N-1}\right) \qquad k = a, b, c \tag{12.26}$$

It should be noted that a DC offset must also be added to the reference waveforms to avoid the modulus function becoming a negative number. An additional common mode voltage which correctly positions the first and last switching transitions in each switching period can then be determined using a similar max/min offset expression as for the two-level case, of

$$v'_{off} = \frac{V_{dc}}{N-1} - \frac{\max(v_{az}^{*\,'}, v_{bz}^{*\,'}, v_{cz}^{*\,'}) + \min(v_{az}^{*\,'}, v_{bz}^{*\,'}, v_{cz}^{*\,'})}{2} \tag{12.27}$$

The final reference waveforms are generated by adding the offset voltages described by Eq. (12.27) to the original sinusoidal phase voltage commands. Figure 12.12 shows the offset waveform and final v_{az} phase voltage reference command for three-, five- and seven-level systems [4].

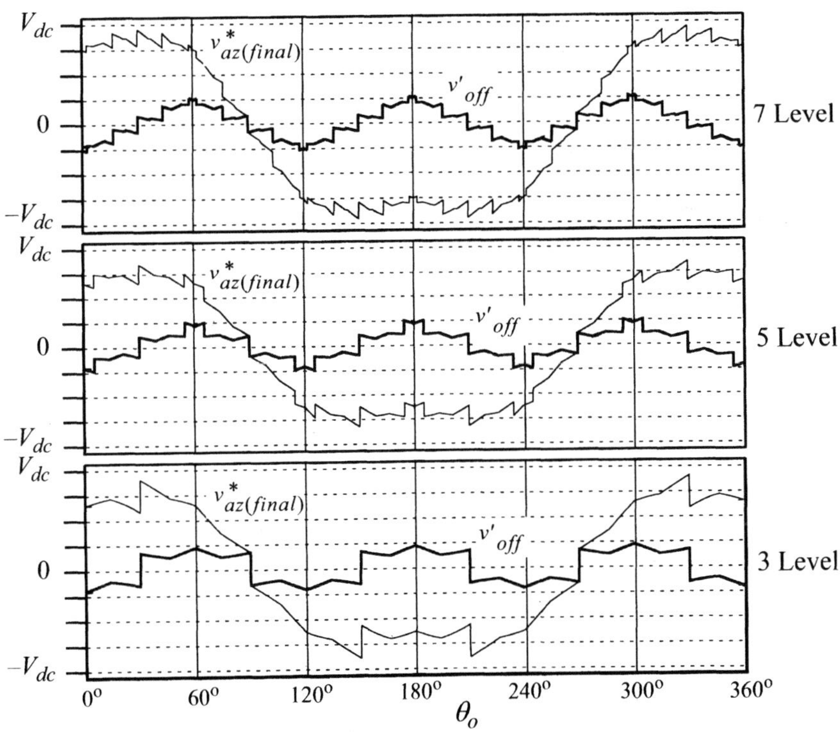

Figure 12.12 Multilevel inverter centered space vector modulation reference and offset waveforms [4].

12.7 Discontinuous Modulation in Multilevel Inverters

A further possibility with space vector positioning is to move to discontinuous SVM by eliminating either the first or last state in each switching sequence. While this approach does not achieve a minimum flux error, the increase in switching frequency of 3/2, which is possible because of the reduced number of switching transitions required, can yield harmonic benefits for certain modulation ranges.

The easiest approach for discontinuous modulation of a multilevel inverter is to simply lock particular phase legs to the upper or lower voltage rails for fractions of the fundamental cycle. For multilevel inverters this may require additional switching transitions which reduce the possible increase in switching frequency from the theoretical maximum, but there are still useful harmonic gains to be achieved. While the increased number of voltage levels does create a greater variety of possible modes of discontinuity, it has been found that all have similar harmonic profiles so that only the simplest alternative is considered here.

Figure 12.13 shows the zero sequence offset waveforms which should be added to the phase reference voltages to achieve 60° and 30° discontinuous switching for a carrier-based modulation system. They can be described mathematically for 60° DPWM1 as

$$v_{off} = \max[\mathrm{abs}(v_{az}^*), \mathrm{abs}(v_{bz}^*), \mathrm{abs}(v_{cz}^*)] \tag{12.28}$$

$$v_{zs,\,\mathrm{DPWM1}} = -\mathrm{sgn}(v_{off}) + v_{off} \tag{12.29}$$

and for 30° DPWM3 as

$$v_{off} = \mathrm{mid}[\mathrm{abs}(v_{az}^*), \mathrm{abs}(v_{bz}^*), \mathrm{abs}(v_{cz}^*)] \tag{12.30}$$

$$v_{zs,\,\mathrm{DPWM3}} = -\mathrm{sgn}(v_{off}) + v_{off} \tag{12.31}$$

In practice the required switching can be implemented by demodulating the phase voltages implicitly developed in equation form by a discontinuous implementation of the multilevel system controller.

Figure 12.14(a) shows experimental waveforms for centered PD modulation of a three-level cascaded inverter with appropriate zero sequence offsets added to the phase references to make the start and end redundant space vector periods equal. Figure 12.14(b) shows the experimental spectra for this system and Figure 12.14(c) shows the simulated harmonic spectra for offset centered

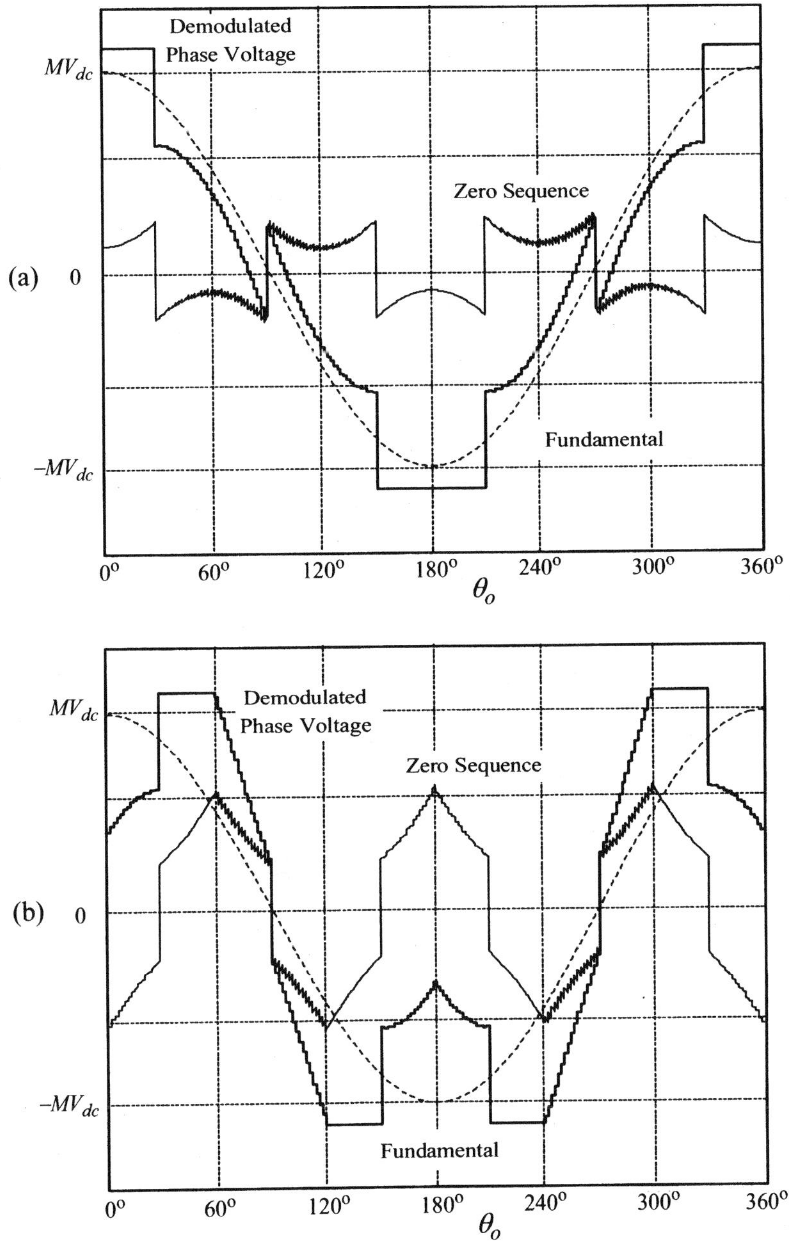

Figure 12.13 SVM demodulated phase voltages for: (a) three-level 60° DPWM1 discontinuous PWM and (b) three-level 30° DPWM3 discontinuous PWM.

SVM operating under equivalent modulation conditions. The extremely close match of the simulation and experimental results confirms the equivalence of SVM and PD carrier modulation when an appropriate zero sequence offset is added to the reference waveforms. Figure 12.15 shows similar results for a seven-level PD modulated hybrid converter compared to a seven-level centered SVM. Again the close match between the simulated and experimental spectra confirm the equivalence between SVM and PD carrier modulation for this converter topology when appropriate zero sequence offsets are added to the carrier. Figure 12.16 compares three-level cascaded inverter PD modulation under 30° discontinuous operation with zero sequence offsets against DPWM3 discontinuous three-level SVM simulation results. Once more a close match is in evidence. Finally, it is interesting to note the presence of low-order frequency distortion in these experimental results because of commutation between adjacent sectors. This signature is a typical result of the practical implementation of discontinuous modulation strategies.

12.8 Summary

This chapter has applied space vector modulation concepts to multilevel inverters. Strategies have been presented to select the optimum space vector sequences from the multitude of switching alternatives that are possible for multilevel inverters, and it has been shown how the results can be applied to any multilevel inverter topology. Then, using flux trajectory concepts, it has been identified that the optimum switching arrangement for a multilevel inverter centers the middle space vectors of a switching sequence within each "half-carrier" period. This leads to the development of zero sequence or common mode offset waveforms for multilevel inverters to achieve identical results using carrier-based modulation processes. Finally, it has been shown how discontinuous modulation can be achieved for multilevel inverters in the same way as for two-level inverters using space vector concepts.

Earlier in this book it was established that carrier and space vector modulation methods for two-level inverters create exactly the same phase leg switching sequences when appropriate zero sequence offsets are added to the reference waveforms for carrier modulation. In this chapter a similar equivalence has been demonstrated for PD carrier modulation and space vector modulation for diode-clamped, N-level cascaded, and for hybrid cascaded inverters.

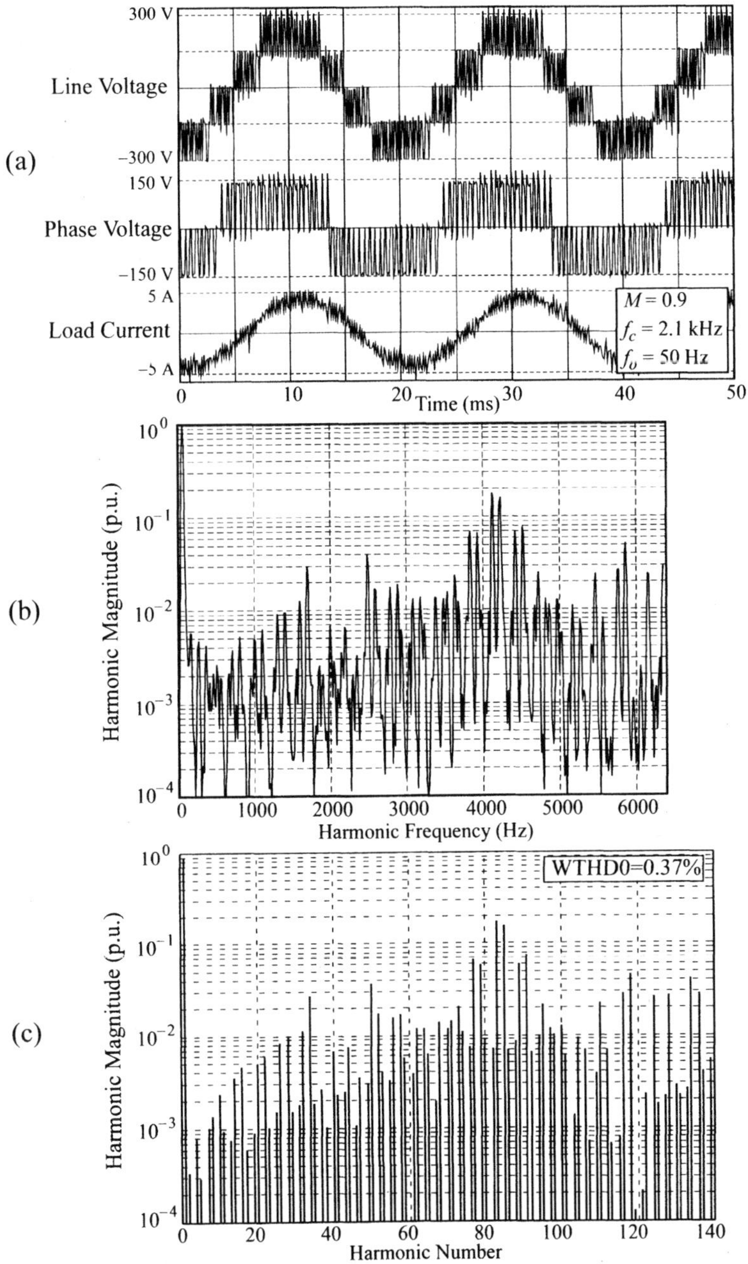

Figure 12.14 Three-level cascaded inverter with centered PD modulation: (a) Experimental switched voltages and filtered load current, (b) experimental l–l harmonic spectra, and (c) simulated l–l harmonic spectra for centered SVM, $M = 0.9$, $f_c/f_o = 42$ [4].

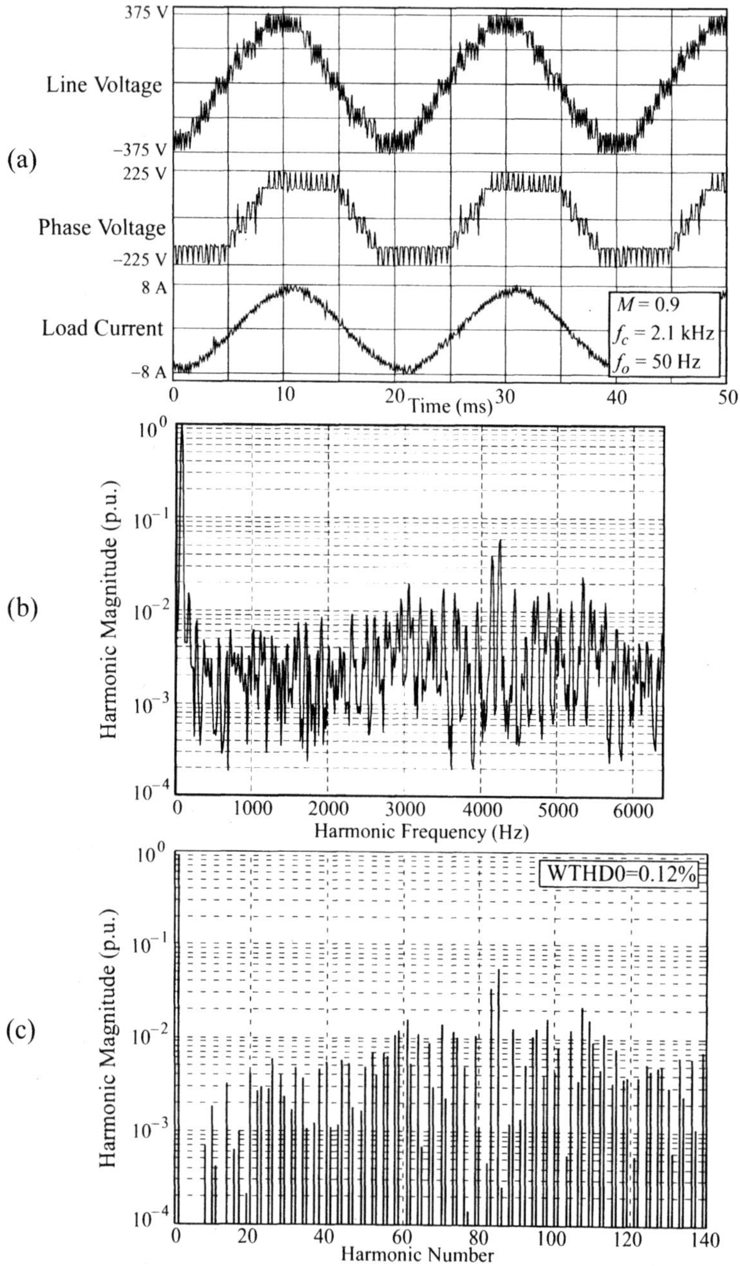

Figure 12.15 Seven-level cascaded inverter with centered PD modulation: (a) Experimental switched voltages and filtered load current, (b) experimental *l–l* harmonic spectra, and (c) simulated *l–l* harmonic spectra for centered SVM, $M = 0.9$, $f_c/f_o = 42$ [4].

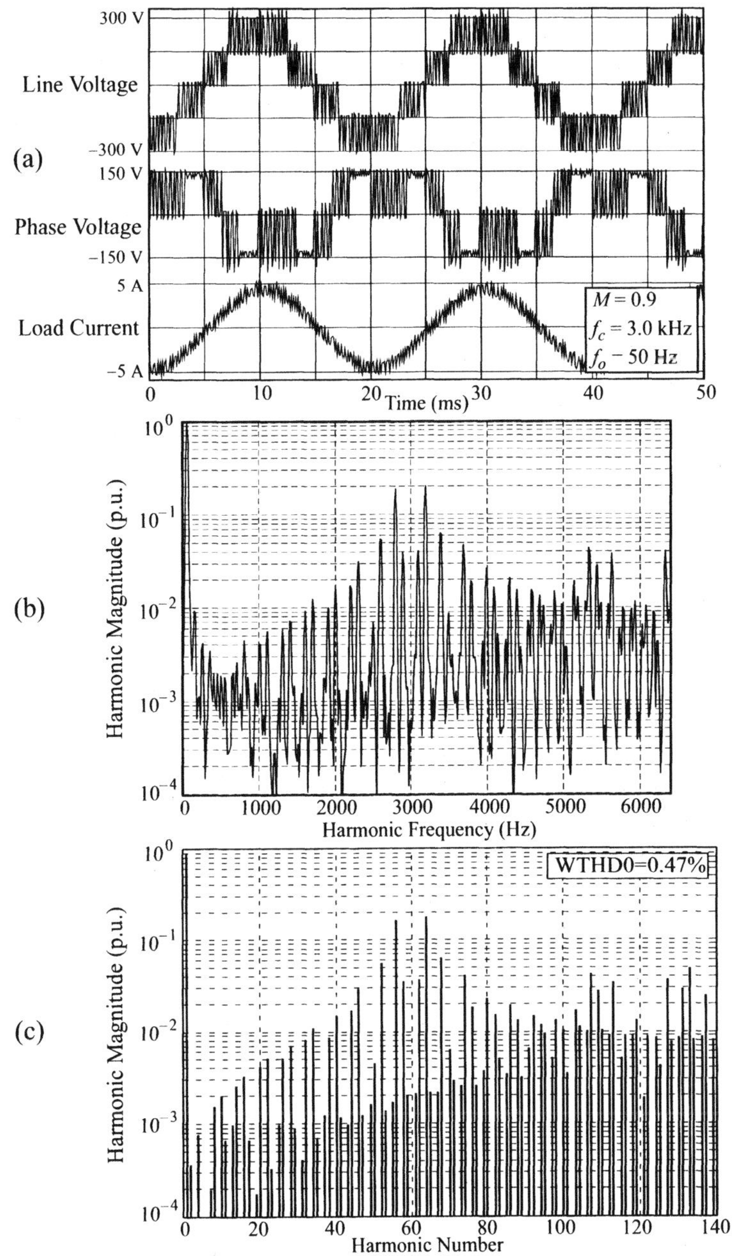

Figure 12.16 Three-level cascaded inverter with 30° DPWM3 discontinuous SVM: (a) Experimental switched voltages and filtered load current, (b) experimental *l–l* harmonic spectra, and (c) simulated *l–l* harmonic spectra, $M = 0.9$, $f_c/f_o = 42$ [4].

References

[1] R. Joetten and C. Kehl, "A fast space-vector control for a three-level voltage source inverter," in *Conf. Rec. European Power Electronics Conf.* (*EPE*), Florence, 1991, pp. 2:070–2:075.

[2] N. Celanovic and D. Boroyevich, "A fast space vector modulation algorithm for multilevel three-phase converters," *IEEE Trans. on Industry Applications,* vol. 37, no. 2, 2001, pp. 637–641.

[3] D. Peng, F.C. Lee, and D. Boroyevich, "A novel SVM algorithm for multilevel three-phase converters," in *Conf. Rec. IEEE 33rd Annual Power Electronics Specialists Conference*, Cairns, 2002, vol. 2, pp. 509–513.

[4] B.P. McGrath, D.G. Holmes, and T.A. Lipo, "Optimized space vector switching sequences for multilevel inverters," in *Conf. Rec. IEEE Applied Power Electronics Conf. and Exposition* (*APEC*), Vancouver, 2001, pp. 1123–1129.

[5] J. Seo, C. Choi and D. Hyun, "A new simplified space–vector PWM method for three–level inverters," *IEEE Trans. on Power Electronics*, vol. 16, July 2001, pp. 545–550.

[6] S. Fukuda and Y. Iwaji, "A single-chip microprocessor-based PWM technique for sinusoidal inverters," in *Conf. Rec. IEEE Industry Applications Society Annual Mtg.*, Pittsburgh, 1988, pp. 921–926.

13

Implementation of a Modulation Controller

This text has concentrated on the theoretical processes of pulse width modulation, exploring various alternative ways of controlling the switched state of power electronic converter phase legs and comparing their relative advantages and disadvantages. The understanding that comes from this analysis is important, since the first step in building a power electronic conversion system is to know in principle how the power electronic switches should be controlled to achieve a desired performance outcome. However, it is then equally important to build a physical system that will achieve this desired switching performance as closely as is possible within the constraints of the switching device characteristics. In other words, there is limited value in knowing the theoretical harmonic performance of an asymmetrical regular sampled modulation scheme if the hardware implementation switches incorrectly or (even worse) makes switching errors because noise or other interference is injected into the control electronics.

This chapter provides an outline of the structure and major component sections of power electronic conversion systems and their associated controllers. The material presented is not intended to be definitive — this would be impossible given the rapid advances in technology that have occurred over the last few years, and the material would be out of date almost before it was published anyway. Instead, the approach taken is to discuss the major functional areas of these systems, considering the practical issues and decisions that must be considered irrespective of the particular final solution adopted, and illustrating these issues by presenting examples taken from the systems that were used to generate the experimental results presented in this text. It is interesting to reflect that despite technological advances in electronics, the same issues associated with the physical implementation of these types of systems arise again and again, and the same style of technical solutions continue to be used.

13.1 Overview of a Power Electronic Conversion System

The operation of most pulse width modulated power electronic conversion systems can be considered at a number of levels. First, there is the primary control objective — what is the prime purpose of the conversion system? The outcome of this process will usually be a command to produce an AC voltage of particular magnitude and frequency to supply to the load. Second, the PWM switching algorithm must be selected, with modulation strategy and switching frequency determined to suit the converter and the application. This algorithm in turn must be implemented on an appropriate hardware/software platform, which produces logic level outputs as commands to turn particular inverter switches ON and OFF. And lastly, these outputs are translated by hardware into actual gate control signals which control the primary electronic switches in the converter system power stage. In addition, there are a number of secondary processes that must be managed and supported by an inverter controller if the overall system is to work properly and effectively.

There are many different types of power electronic conversion systems, and each has its own primary control objective. For example, the primary role of a simple AC induction motor variable speed drive system is to produce a variable voltage, variable frequency three-phase AC voltage, where the frequency is set to make the motor operate at the speed required by the load with a matching voltage based (most likely) on a constant V/f characteristic. Secondary objectives are often then added to these systems, such as controlled speed ramp rates, frequency jumps to avoid resonances, and speed feedback control for higher performance applications. However, irrespective of how complex and sophisticated the higher level control algorithm becomes, its end result will be to command a modulation system to supply a particular AC voltage and frequency to the motor at a particular point in time.

More sophisticated motor drive systems use field-orientated control strategies with inner current loop regulation to achieve improved motor control, flux control, slip control, or one of the many alternatives that have been developed over the years. But again, irrespective of the approach taken, the end result from the primary control system will most often be a commanded AC voltage of particular magnitude and frequency that must be matched by the modulation system through its hardware implementation and (eventually) the physical converter system.

Other types of systems may have different primary objectives. An uninterruptible power supply is typically required to produce an output voltage with a fixed magnitude and frequency to supply a widely varying load; a grid-connected rectifier must draw a sinusoidal AC current from the supply with a magnitude and phase angle appropriate to meet the real power demands of its load; and an active filter system will typically be required to inject low-order harmonic currents of varying magnitude, frequency, and phase, depending on the filter objective. Once again, for all of these systems, the eventual outcome of the control algorithm will be to command some type of modulation system to generate a particular AC voltage and frequency output from the converter system at any particular point in time.

It is noted in passing that these concepts are not completely appropriate for inverter systems that are controlled by hysteresis regulation. Hysteresis control strategies directly command the switches of the power electronic converter in response to a measured current error, and there is often no clear separation between a higher level control strategy and a lower level modulation control process. However, since modulation using hysteresis concepts is outside the scope of this book, this issue will not be pursued further.

The choice of a particular modulation strategy depends on a number of issues such as power levels, selection of switching devices, particular load requirements, and the capability of the hardware controller that is to be used. Many of these issues have been explored in previous chapters of this book and need not be discussed further here. The focus in this chapter is to consider the physical implementation of the switching processes used by the modulation system and the higher level control algorithms.

13.2 Elements of a PWM Converter System

Figure 13.1 shows the general structure of a shunt injection active filter system, comprising a three-phase VSI coupled into an AC grid system through series inductances and an optional coupling transformer. Figure 13.2 presents a more detailed view of this system, showing the internal hardware arrangements of the physical implementation. Figure 13.3 shows details of an alternative converter system, being a photovoltaic (PV) generation system supplying power back into an AC grid system. While these two systems obviously have quite a different functionality, it is equally clear that they have a very similar

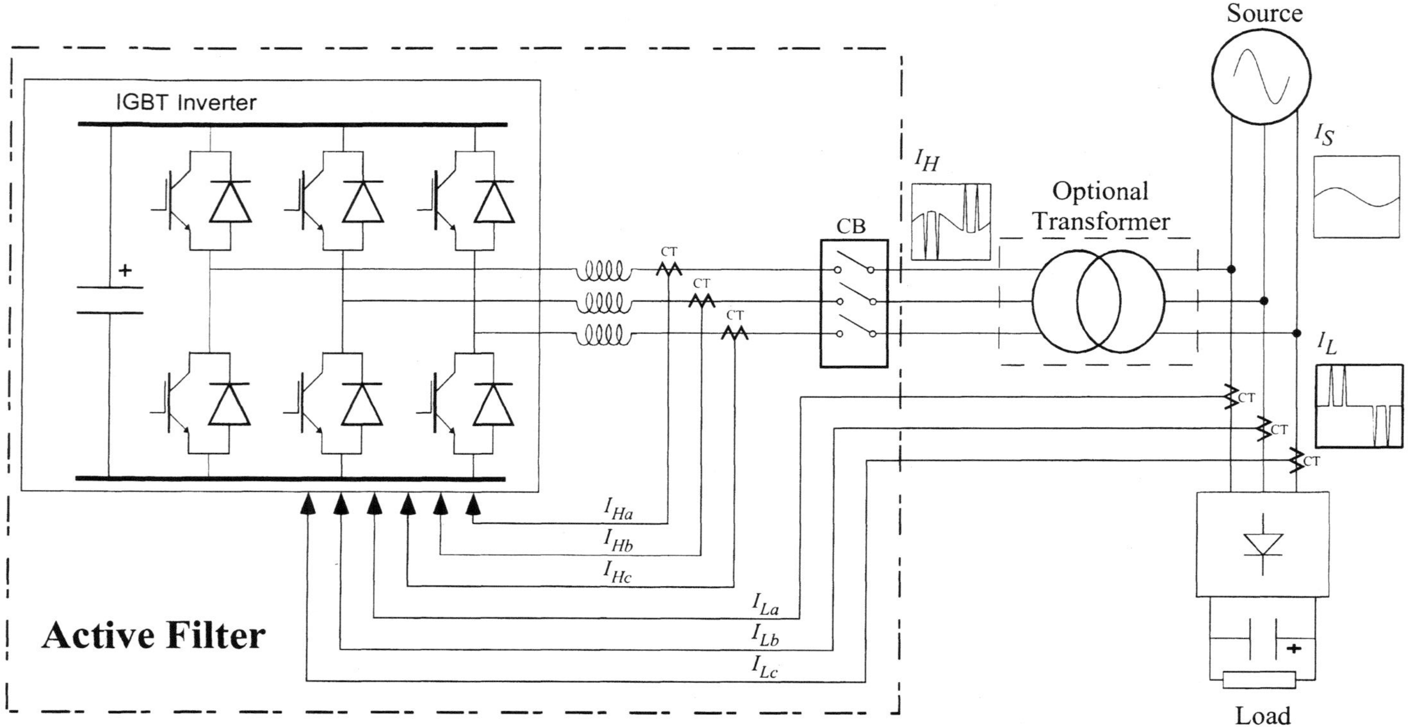

Figure 13.1 Block diagram of an active filter converter system.

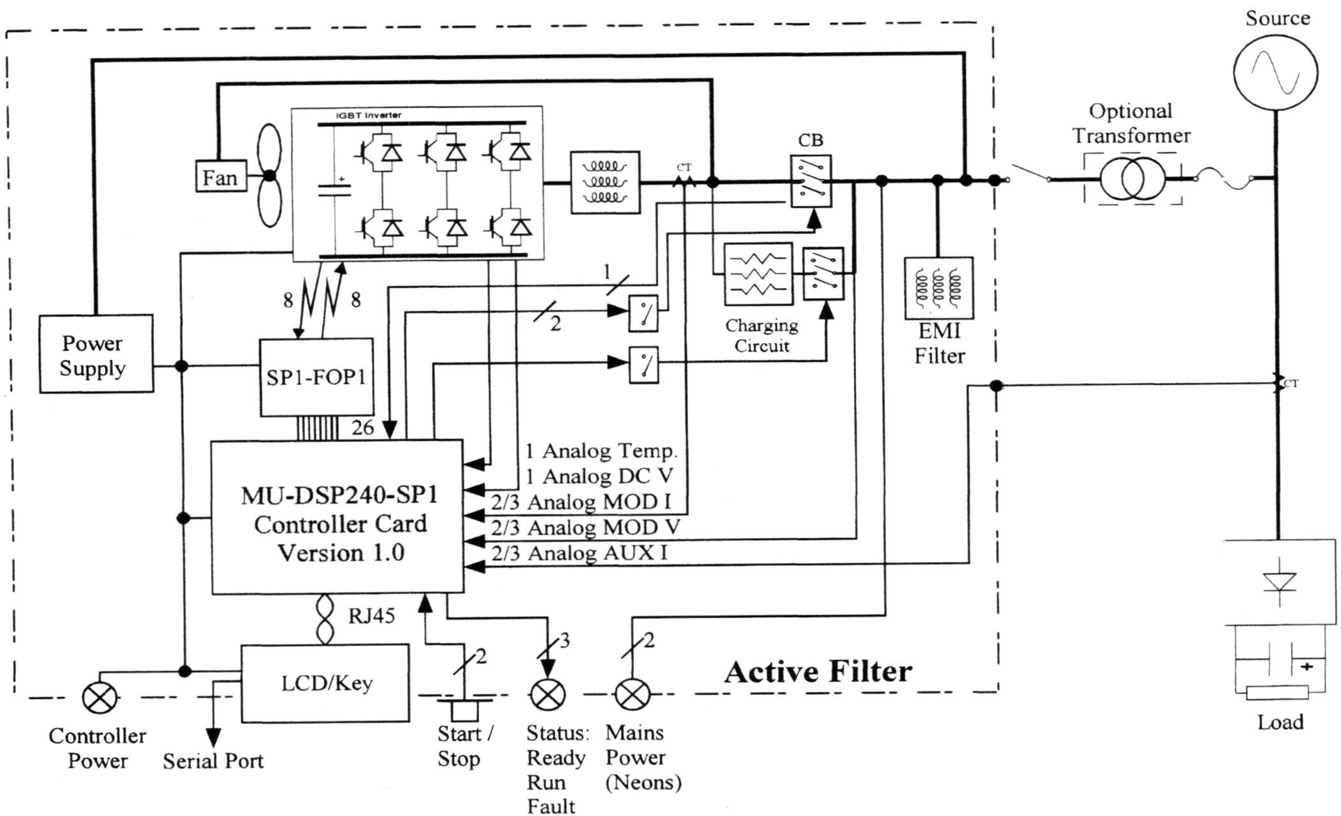

Figure 13.2 Detailed arrangement of active filter converter system.

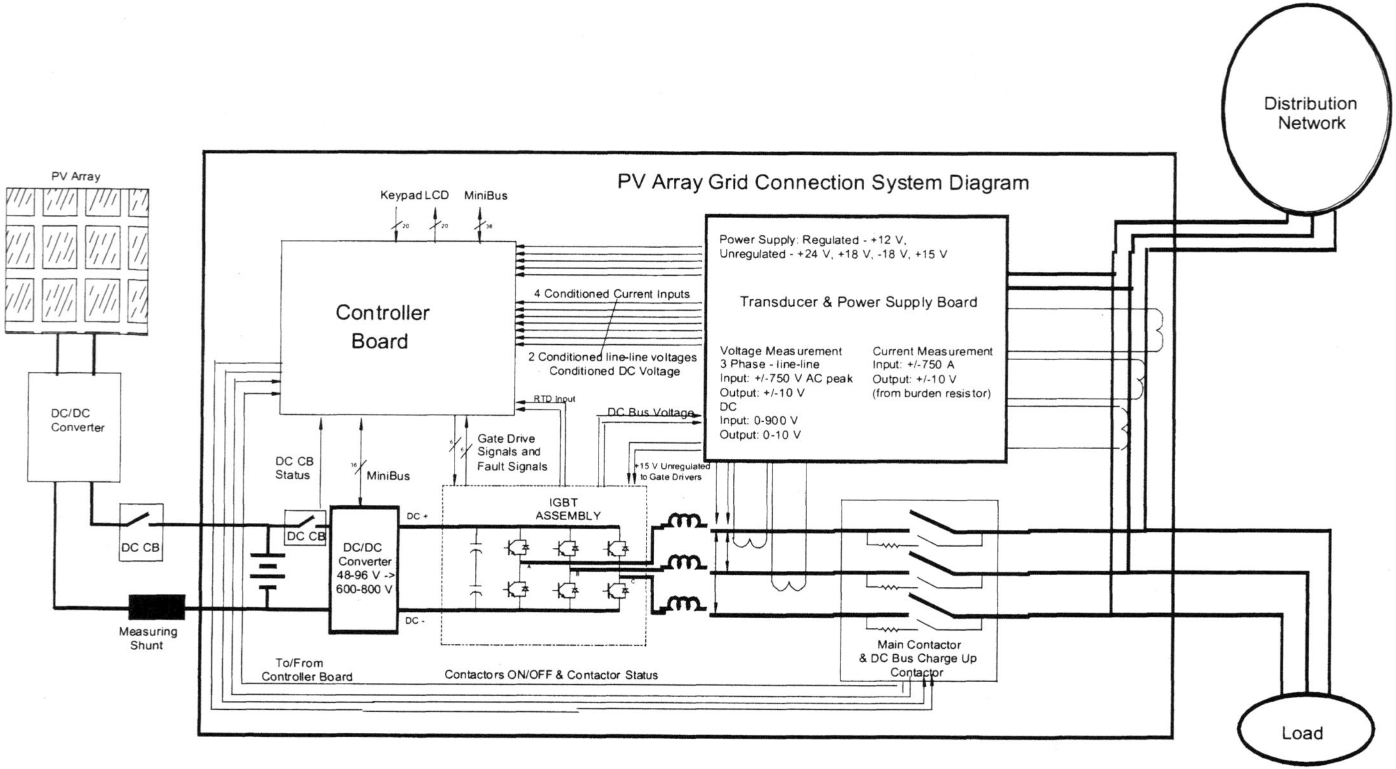

Figure 13.3 Detailed arrangement of grid-connected PV generation system.

physical structure both for the power stage and for the controller hardware. This is often the case with power electronic conversion systems — the same building blocks are used in different combinations to make up different types of systems. It is also illuminating to observe that as the level of detail increases, the relative size on the drawings of the main converter section reduces accordingly.

Figure 13.4 shows the power stage and controller arrangement for a dual converter series/shunt active filter compensation system. While this system has quite a different function compared to the previous two examples, once again it uses the same building blocks of VSIs and associated DSP-based controllers.

From these examples, a number of common building blocks can be identified as being part of most power electronic conversion systems, viz:

- *VSI Power Stage(s).* The center of any conversion system is the arrangement of the main semiconductor switches to provide the primary inverter structure.
- *Gate Driver Interface.* This circuitry translates the logical requirement to change the state of an inverter switch, into the physical voltage/current gate signals that are required to control the actual semiconductor switching device.
- *Auxiliary Logic Power Supply.* An essential part of all conversion system implementations is the auxiliary logic power supply, which powers the control system for the inverter.
- *Input and Output (I/O) Conditioning Circuitry.* All power electronic control systems require certain input and output signals to operate. These signals must be scaled and conditioned from their external voltage and current levels to match the electronic control system I/O levels.
- *PWM Controller Implementation.* This is the hardware/software structure that executes the control algorithm to convert a reference command into a series of logical signals that define the state of the main power inverter switches. The algorithm can be open or closed loop.
- *Ancillary Functions.* Most converters have a number of ancillary functions that must be provided by the controller implementation. Examples of these functions include remote serial port communications, speed and position encoder support, and provision of power supplies to energise external field equipment.

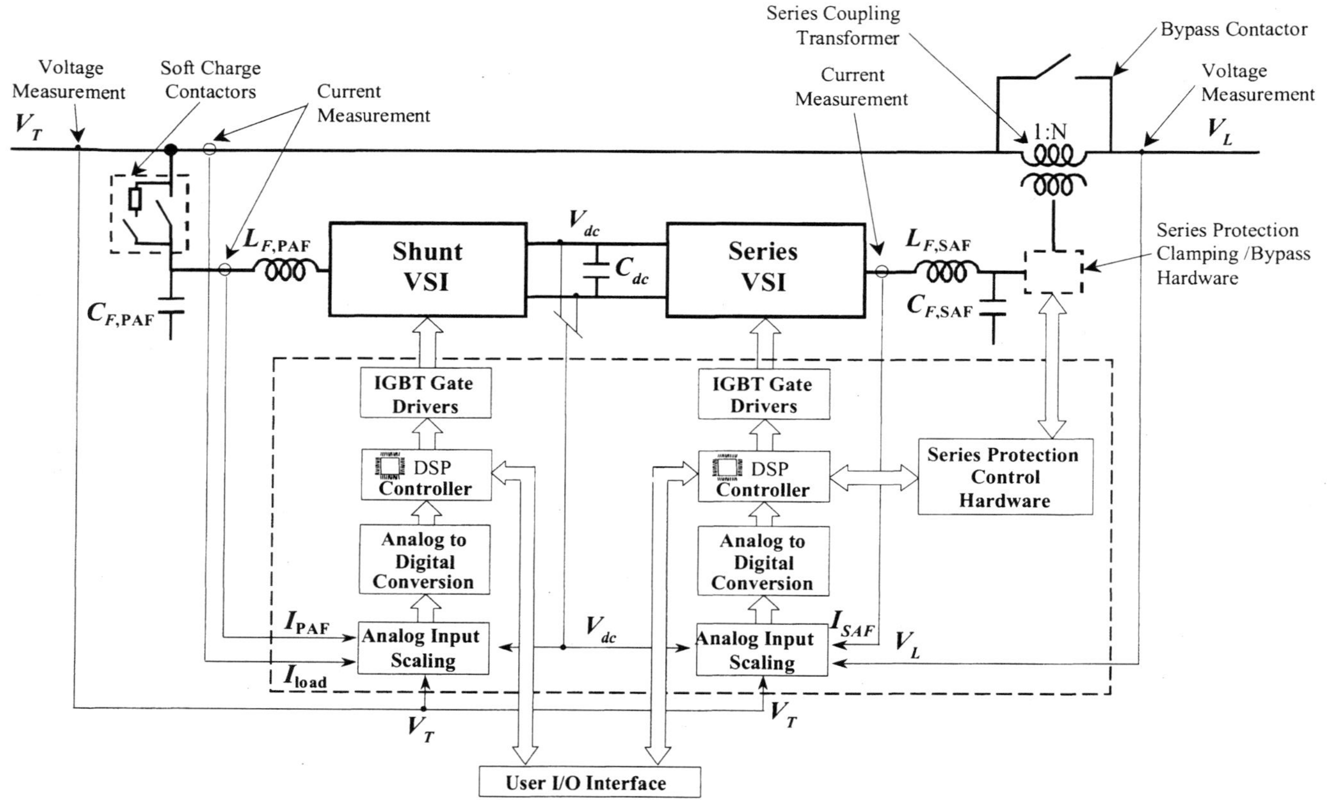

Figure 13.4 General arrangement of series/shunt active filter converter system.

13.2.1 VSI Power Conversion Stage

The basic arrangement of the semiconductor switches for a VSI was established in Chapter 1 and consists of two or three-phase legs (for a single-phase or three-phase inverter as appropriate) connected in parallel across a high-voltage DC bus. Each phase leg is made up of two power devices connected in series, with diodes paralleled across each device to conduct the reverse current. These diodes may be intrinsic, as is the case for field-effect transistor (FET) devices, packaged with the main switch, as is done for IGBT and similar devices, or sometimes even separately mounted. Figures 1.5, 1.6 and 1.8 in Chapter 1 illustrate different variations of the conventional VSI topology.

At the practical level, the construction of the power stage of a VSI is usually straightforward but does require some care with the physical layout. The switching devices must be mounted on an appropriate heatsink to suit the package, and this heatsink must be designed to dissipate the losses caused by both the conduction and the switching processes. Most device manufacturers supply comprehensive application notes to assist in the calculation of these losses, and the reader is referred to these for further information.

One of the major challenges with designing the power stage of an inverter is to minimize the electromagnetic interference (EMI) that can be produced by the high *dv/dt* and *di/dt* transitions that occur during each switch transition. Figure 13.5 illustrates this issue, by showing the path of the load current during a switching transition of phase leg a as switch S_1 turns off and the diode of switch S_2 picks up the load current. The important circuit parameters that control this current transfer are L_{d1} and L_{d2}, the parasitic inductances between the DC bus and the upper and lower devices, and L_c, the parasitic inductance between the DC bus and the main DC bus capacitor C_{dc}. Note that the source and load inductances have been assumed (as is usual) to be large enough that their currents remain unchanged during the entire switching transition period.

The switching transition commences when S_1 is turned off and ceases to conduct. As S_1 turns off, its forward voltage increases to $2V_{dc}$, at which point the diode of S_2 becomes forward biased and attempts to take over conducting the load current. However, this current transition requires that the load current ceases to flow through the upper parasitic inductance L_{d1} and starts flowing through the lower parasitic inductance L_{d2}, and this will not happen instantly. Hence the forward voltage will continue to rise until the current transition occurs, creating overvoltage stress across S_1 and an increased EMI hazard.

(a)

(b)

(c)

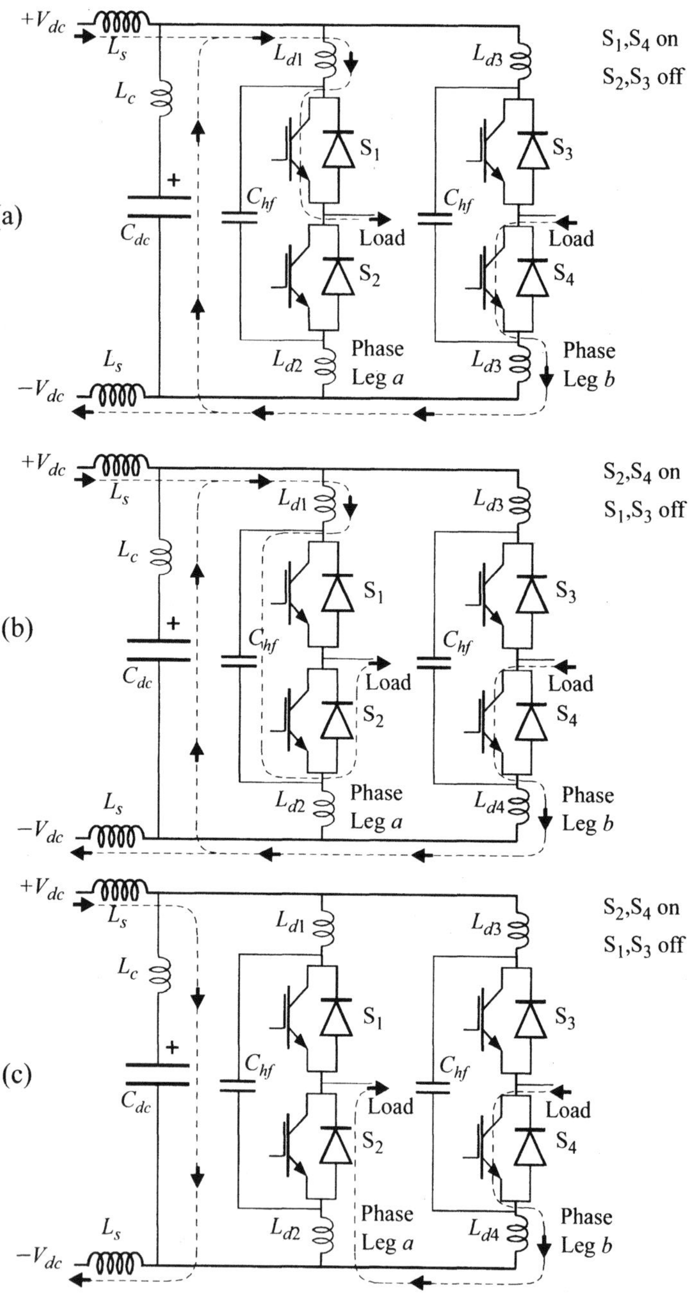

Figure 13.5 Switching sequence for single-phase VSI (a) before switching, (b) during a phase transition from $+V_{dc}$ to $-V_{dc}$, and (c) switching complete.

The usual solution to minimize this problem is to install high-frequency capacitors C_{hf} across the phase leg switches. These capacitors provide a short-term path for the transitioning load current as shown in Figure 13.5(b), and this minimizes the voltage overshoot.

A further issue can be caused by the parasitic inductance of the main DC bus capacitor, L_C. It can be seen from Figures 13.5(a) and 13.5(c) that the current through the DC bus capacitor reverses as phase leg *a* changes state. This current reversal also cannot happen too quickly because of L_c and hence will add to the potential for voltage overshoot as phase leg *a* changes state. The high-frequency capacitors C_{hf} shown in Figure 13.5 mitigate this effect as well. It is important, however, to design the physical arrangement of the inverter phase legs to minimize the parasitic inductances between the switching devices and the high-frequency capacitors. Otherwise their benefit is reduced. The usual practice is to mount the high-frequency capacitors as close to the switching device as is physically possible, and to use particular types of capacitors (such as film capacitors) that have a small internal inductance.

It should be noted that for higher power inverters, other parasitic inductances and capacitances associated with each phase leg circuit may need to be considered as well. Sometimes, for example, snubbing circuits must be installed directly across each phase leg output to adequately control the *dv/dt* and *di/dt* transitions that can occur. Consideration of these issues is beyond the scope of this book and will not be pursued. However, the interested reader is referred to the wealth of references relating to snubbing of power inverter circuits that have been published over the last few decades.

13.2.2 Gate Driver Interface

The gate driver circuitry converts the logical command to change the state of an inverter semiconductor switch, into the particular voltages and currents that are required to make the switch change state. To turn on IGBTs and FETs, an initial burst of current must be injected into the gate (up to several amps for larger power devices), and then a holding voltage is required to keep the device ON. For turn off, a similar burst of opposite polarity current is required to discharge the gate capacitance, and then a zero or (preferably) negative holding voltage is required to keep the device OFF. It is also important that the gate drive circuit that provides this supply presents a low output impedance to the switch gate, to ensure that Miller capacitance feedback during high *dv/dt* conditions across the switch does not cause accidental switch state transitions [1].

There are two main issues to be considered with gate driver systems. First, the circuitry for each switching device needs a local power supply, to supply the gate switching current. For the low-side devices, while it is possible to use the same power supply as the main controller, it is usually preferable to provide a separate isolated power supply to avoid cross-coupling interference back into the main controller system. For higher power systems, it is also usually better to provide separate individual power supplies for each low-side gate driver. This avoids problems caused by circulating currents flowing through a common source connection between the switching devices. For the high-side driver, two alternatives are common — a bootstrap circuit that creates a high-side driver supply from the low-side driver supply [2], or separate isolated power supplies for each individual high-side switch in the power stage. Figure 13.6 shows these alternative arrangements. For lower power systems, it is also possible to directly couple the gate drive signal through a pulse transformer, and this can sometimes avoid the need to provide a separate power supply for the gate drive circuitry.

Second, the logic signal that defines the state of the inverter switch must be coupled into the gate driver circuit. For the low-side switch, this signal can either be directly coupled or coupled through a isolation system. Almost invariably, it is preferable to couple through an isolation system to avoid inter-

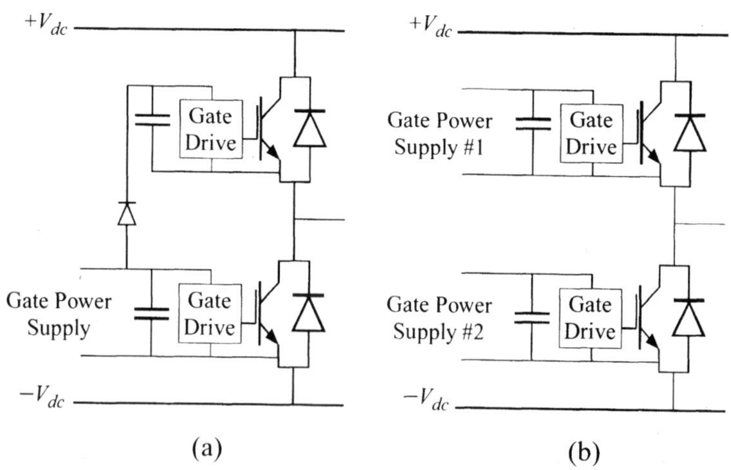

Figure 13.6 Alternative high side gate driver power supplies: (a) high side supply bootstrapped from low side and (b) separate isolated power supply for each switch.

ference from the high *di/dt* and *dv/dt* conditions that are associated with a switching transition propagating back into the main controller system and causing misbehavior of the controller logic. For the high-side switch, some form of isolation (or at least decoupling) is essential, since the source reference for this switch toggles approximately $2V_{dc}$ as the switch state changes, and this level of voltage change would swamp the gate drive voltage if it is not protected. The common strategies to achieve this isolation are optical interfacing using high-speed opto-couplers [3], fiber-optic transmission, or magnetic coupling using small pulse transformers [4].

Irrespective of the approach used, it is most important that the high-side gate drive circuitry is isolated from the main controller logic circuits across a wide frequency spectrum, because of the extremely high *dv/dt* that is impressed on the source reference voltage of this device as it changes state. Hence both the gate voltage supply and the control signal isolation must have a very large high-frequency common mode voltage rejection ratio, which in turn implies a very small coupling capacitance across any optical couplers or transformers that are used. Typical ratings should be in excess of 5000 V/μs for optical couplers, and interwinding capacitances of only a few picofarads for coupling transformers, to avoid interference problems.

13.2.3 Controller Power Supply

All inverter control systems require auxiliary logic supplies to energize the digital control logic, the analog interface circuitry, and the gate driver circuits and to provide a field power supply for external transducers, switches, and the like. These supplies can be derived from a simple AC transformer and rectifier/regulator if an AC supply is available or more commonly from a switched mode power supply (SMPS) that is designed to operate directly from the main DC bus of the inverter. A SMPS also has the advantage of being able to generate multiple isolated supplies for very little additional cost.

The typical power supply requirements for an inverter controller would be a +5 V supply for the digital logic, ±15 V for the analog circuitry, +15 V for a gate driver supply, and +24 V to provide an external-use field power supply. One of the most common topologies used is a flyback converter system, which has the particular advantage that it can be easily arranged to produce multiple output voltages with a minimum of additional components.

All of these supplies should be isolated from each other to avoid ground loops and unexpected interference problems. In addition, the external field supply may need output protection so that a fault in the external circuits does not cause the main SMPS supply to shut down. It is usually unacceptable for the main inverter controller to cease to function because of a fault in the external field wiring.

Finally, the gate supply usually needs special consideration, since up to six separate supplies can be required to power a fully isolated gate driver system for a standard VSI, and they must all have low coupling capacitances back to the main logic system to avoid EMI problems because of *dv/dt*. Often, a more effective solution is to include small separate SMPSs for each gate drive circuit, driven from one common supply generated by the main SMPS.

13.2.4 I/O Conditioning Circuitry

The analog and digital inputs and outputs associated with an inverter controller have a wide variety of voltage and current ranges. Analog inputs can span from up to 1000 V to measure a DC bus voltage or the peak of an AC input voltage and down to 100 mV to measure the voltage across a current shunt. Analog outputs usually span a lesser range but can still be required to supply many tens of milliamps at voltages in excess of 20 V peak. Digital inputs can range from the clean contact of a relay (typically using the field voltage supply as a pull up) to a high-voltage AC status input voltage; digital outputs can take the form of anything from a 5 V semiconductor logic level output, to energizing an AC mains-driven coil of a main contactor.

All of these signals require conditioning to convert between the low-power, low-voltage levels of the converter controller inputs and outputs and the physical incoming and outgoing field signals. Such circuitry is usually straightforward in principle, but it must be carefully designed and constructed to work correctly in the harsh electrical environment of a power electronic converter. Good grounding practices, low-pass filtering and bypassing, and careful consideration of component ratings are essential to create robust I/O circuitry that reliably interfaces to the central controller logic.

Modern converter controllers often also include a serial communication interface to "talk" to other control systems. It is important that this interface is appropriately isolated, to protect system operators and other equipment, and to avoid interference problems occurring between the communicating systems.

13.2.5 PWM Controller

The PWM controller translates a commanded reference value into a set of logic pulse trains that define the status of each phase leg of the inverter, using either a closed- or open-loop control strategy. The controller can be analog or digital circuitry, software or hardware, and each implementation variation has particular advantages and disadvantages. However, in general the most common arrangement is a digital controller based around a microcontroller or a digital signal processor (DSP), since this gives the greatest flexibility and potential to implement the more sophisticated PWM algorithms discussed in this book. In addition, the controller usually incorporates a number of additional functions that are required to make a complete inverter system, such as inverter protection functions, operator interface, field transducer control signals, and so on.

Figure 13.7 shows the general arrangement of the features that are often included in a typical inverter controller. Figure 13.8 shows a more detailed arrangement describing the Low Power Inverter (LPI) controller that was used to generate many of the experimental results presented in this book. This controller is based around a TMS320F240 DSP [5] that has been specifically designed to control a three-phase VSI. The controller includes all associated circuitry on the one printed circuit board and is able to directly measure incoming electrical quantities, perform all calculations required to determine the switching instants of each phase leg, and output a pulse train that defines the switching instants of the phase legs.

Specifically, the LPI controller includes

- DSP chip and associated memory and logic glue chips required for the processor to execute programs in a stand-alone mode.
- Eight isolated gate drivers based around Hewlett-Packard's HCPL-316J gate driver integrated circuit (IC). This IC integrates a high-speed opto coupler with a gate driver circuit to provide an isolated single-chip interface between the logical command to control an inverter switch, and the gate terminals of the IGBTs. Each gate drive subsystem is powered by a separate power supply fed through a high-frequency transformer and rectifier.
- Ten analog input circuits to measure AC input voltages and currents, DC bus voltage, temperature, and two user inputs. These circuits provide all conditioning necessary to reduce high input voltages to a 0–5 V

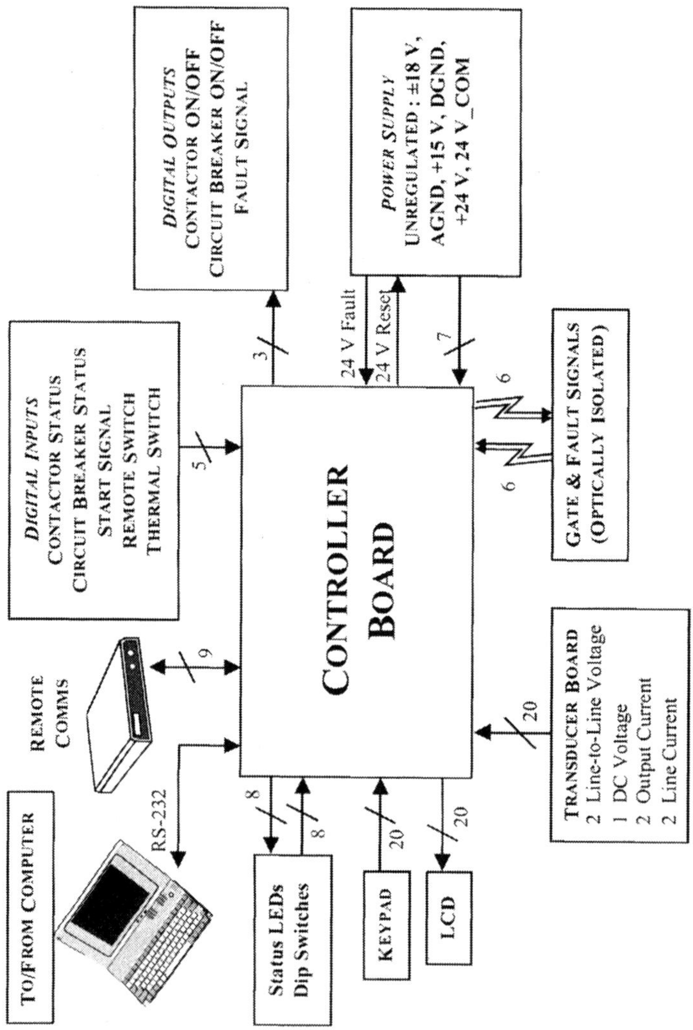

Figure 13.7 General arrangement of VSI controller board.

range, or to directly connect to Hall effect DC current transducers or AC current transformers, so that the signals can then be fed directly into the analog-to-digital converters that are incorporated into the DSP.

- Three output and two input fully isolated digital circuits, that can be used to control relays, measure the status of switches, etc.
- Fully isolated RS-232/RS-485 multidrop interface to allow communication with an operator terminal or higher level control system.

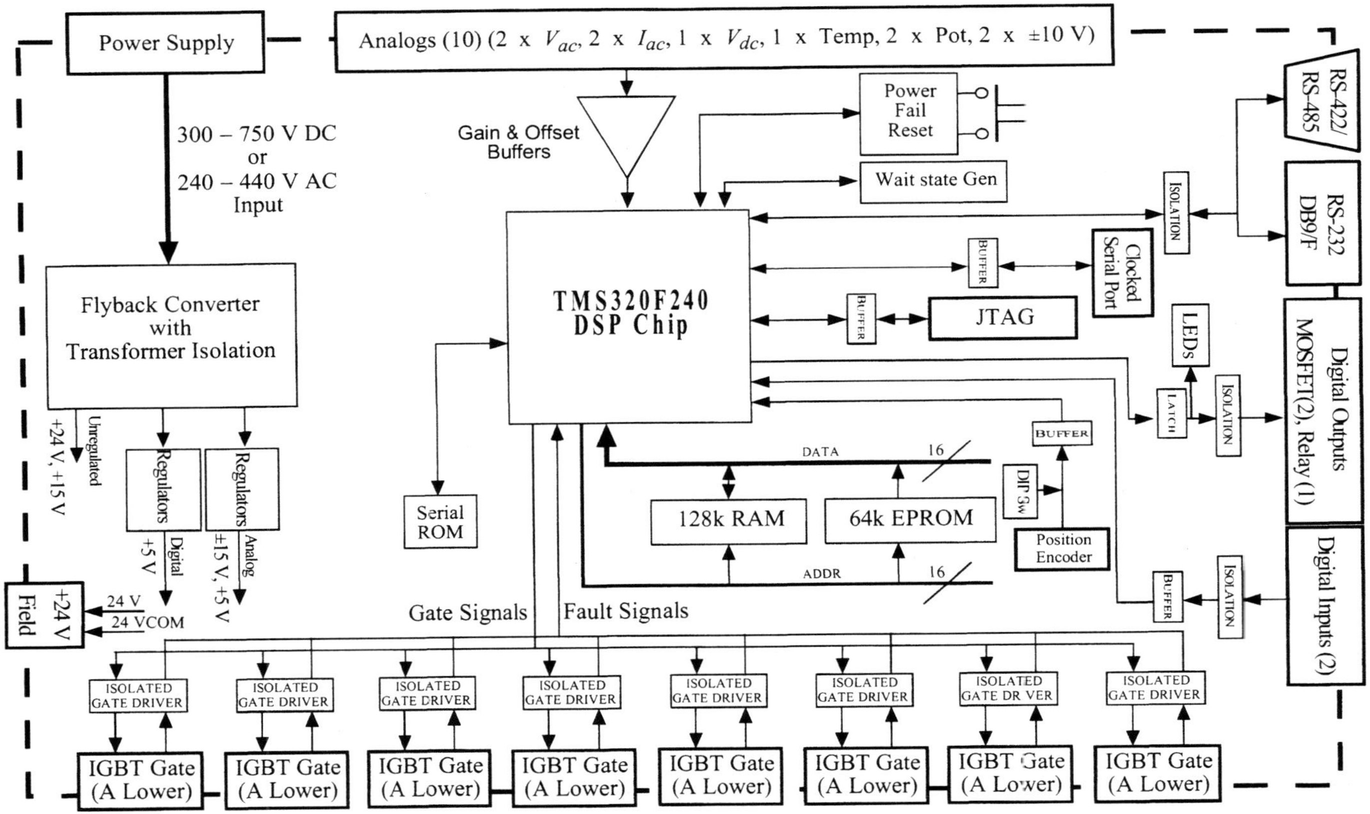

Figure 13.8 Detailed arrangement of LPI controller board for a VSI [6].

- High-frequency switched mode power supply to provide all required logic and analog supplies directly from the DC bus of the main inverter.
- Ancillary functions such as a position encoder interface, current limited +24 V field supply, display light emitting diodes (LEDs), option selection switches, and a serial read-only memory to store fixed controller parameters.

These types of facilities are typical of what would be expected in a modern VSI controller, although, of course, other systems will have detail variations depending on their specific requirements.

13.3 Hardware Implementation of the PWM Process

The role of a PWM controller is to generate a train of logic pulses that indicate the desired state of the phase leg(s) of an inverter. These pulses then feed through the gate driver interface to control the switches of the inverter. The pulses can be generated either by hardware or software, and both approaches have advantages and disadvantages.

13.3.1 Analog versus Digital Implementation

The most straightforward type of analog PWM controller implementation simply uses a linear comparator to compare a sinusoidal fundamental reference waveform against a triangular carrier, with the comparator output directly defining the inverter phase leg status. Each phase leg of an inverter requires a separate comparator system. However, while it is simple, this approach has a number of disadvantages. First, it can be difficult to accurately generate the sinusoidal reference waveforms and the triangular carrier using just analog circuitry. This can be a particular problem for a three-phase inverter system, where three sinusoidal reference waveforms are required with varying magnitude and frequency that are accurately scaled and phase shifted 120° from each other. Second, it is challenging to implement the more complex PWM strategies, such as discontinuous modulation. These strategies use segments of sinusoidal reference waveforms that again can be difficult to generate using analog circuitry. Next, the issues of overmodulation, shutdown, startup, and other nonlinear regions of operation need to be considered. Once again designing analog circuits to operate in these regions can be difficult. And finally, analog circuits have repeatability and long-term reliability issues to consider because of component tolerance and aging variations. Nevertheless, for a simple PWM controller system, analog circuitry does have its place, although in most cases it is usual to still incorporate a low-cost digital microcontroller into the system to provide background control and support functions.

Not unexpectedly, for these reasons most modern inverter controllers now prefer to use fully digital systems to implement the PWM algorithms. Digital controllers use a digital clock system to determine when an inverter phase leg should switch, and they offer greatly improved reliability and repeatability. They are also more flexible and can handle exceptional operating conditions much more readily than analog systems. However, digital controllers can have switching frequency limitations, depending on the type of digital system implementation. Also, digital controllers almost invariably must implement a *sampled* modulation strategy, since they operate by incrementing or decrementing a digital counting system to a preset time-out interval, and this time-out interval must be set by the controlling microprocessor some time before the half carrier interval within which the actual switch transition occurs.

Early approaches for digital PWM controllers used off-the-shelf microcomputers, with the modulation process implemented in software [5]. The approach used was to set up digital timer(s) to create software interrupt(s) when an inverter switch state change was required, with the interrupt software service program then generating the logic status signals that define the new switching state of the inverter. While many variations of this strategy have been developed, the approach in principle has substantial performance limitations since it requires a software response to the timer interrupt to output the changed status of the inverter switches. Inevitably this causes delays and uncertainty as to when the inverter state change actually occurs, because of both interrupt latency and the normal software overheads of interrupt processing. In addition, it becomes more and more difficult to precisely define the switching performance of this type of system when multiple phase legs are controlled, the switching transition points of the phase legs cross over in time as the fundamental output voltages vary, and/or the timer interrupt delays begin to run into each other within the microcontroller. These issues also become more intractable at higher switching frequencies, as the relatively fixed interrupt response delay of a microcomputer of several microseconds becomes a more significant fraction of the PWM carrier period.

To overcome these limitations, and as levels of digital logic integration have increased, the more modern approach is to implement the PWM timing process in a dedicated digital logic subsystem, with the inverter switch status directly defined by outputs from this subsystem. The digital logic can either be integrated with the microcontroller to create a single-chip solution or constructed in a separate gate array logic chip that is then interfaced to the microcomputer. In either case, the system operates by calculating the PWM timing intervals in software and loading these periods into the dedicated hardware timers during the preceding (half) carrier interval. The timers then count down to determine the specific switching point during the (half) carrier interval without further software intervention.

13.3.2 Digital Timer Logic Structures

The first imperative for any digital logic PWM implementation is to decide the level of accuracy that is required. Early work [7] has determined that an 8-bit timer system can provide sufficient accuracy for most purposes, since such systems can switch to at least 0.5% resolution (1 in 256) over the (half) carrier interval. For a converter switching frequency of say 10 kHz (which translates to a half carrier period of 50 μs), an 8-bit timer will require a clock frequency of about 5 MHz (which is typical for older lower performance microprocessors) to fully use the available counter range. However, with modern microcontrollers operating at clock speeds of up to 40 MHz, at least a 10-bit timer is now becoming commonplace, since at this level of resolution it is virtually guaranteed that switching accuracy will not be a significant concern for any type of modulation strategy that may be implemented.

Once the resolution level has been decided, the PWM logic subsystem can be developed. Two approaches in principle are presented below to illustrate how digital logic can be used to implement the PWM process. However, they are by no means definitive, and it is beyond the scope of this book to attempt a more complete presentation of the various alternatives that are possible.

One approach is to allocate individual timers to each phase leg of the inverter, and to count down these timers from a common system clock. Each timer is set to its time-out value at the start of each (half) carrier period, and triggers a state change for its phase leg when it reaches zero. For sawtooth modulation, the phase legs are switched high at the start of each carrier interval (there is no concept of a "half" carrier interval for sawtooth modulation), and switch low during the timing period as each timer times out. For double-sided triangular carrier PWM, some additional logic is required, since the countdown value in the second half carrier interval is $\Delta T/2$ minus the countdown value of the first half carrier period. Either the value loaded into the preset register of the timer must be varied to suit the particular half carrier interval that is in progress or perhaps two timer registers could be used, each loaded with the appropriate timer value for their particular half carrier cycle. Neither alternative is particularly satisfactory and usually also creates some extra software overhead. Furthermore, additional logic is required (typically a flip flop and some "glue" logic) to toggle the polarity of the phase leg transition at each timer count-to-zero event. Other complications with separate timer systems include the need to keep the timers precisely synchronized, initialization issues, and the need for additional logic to lock the timer outputs so that the phase legs can be clamped to the upper or the lower DC rails for extended periods, if discontinuous modulation strategies are to be used.

An alternative approach, now more commonly used, is to link a single timer system to a set of comparison registers that trigger an event when the timer count matches the comparison register contents. The comparison registers are "double buffered" with shadow registers that can be loaded at any time, and the contents of the shadow registers are transferred to the main comparison registers at the start of each half carrier interval. Each comparison register controls one phase leg, usually with some associated output logic so that the comparison event can force the phase leg status HIGH, LOW, TOGGLED, or OFF (both switches OFF). Sawtooth or triangular carrier modulation can be easily implemented with this system, either by loading the comparison registers with new values every half carrier interval as before, or more easily by arranging the timer to count "up" and then "down" over successive half carrier intervals, and toggling the logical output from the comparison register as the timer count passes the compare register value "upwards" and "downwards". This approach also has the advantage that the timer compare values calculated by the PWM algorithm no longer need to be subtracted from $\Delta T/2$ for the second half carrier interval, which can be a substantial advantage at higher switching frequencies. Figure 13.9 shows the arrangement of the event manager (EV) subsystem for the TMS320F240 DSP manufactured by Texas Instruments, which uses this type of structure. The TMS320F240 has a particularly flexible arrangement, since it provides three general-purpose 16-bit timers and twelve compare registers, with differing levels of logic on the outputs of each of the compare registers to allow alternative responses to a timer comparison event.

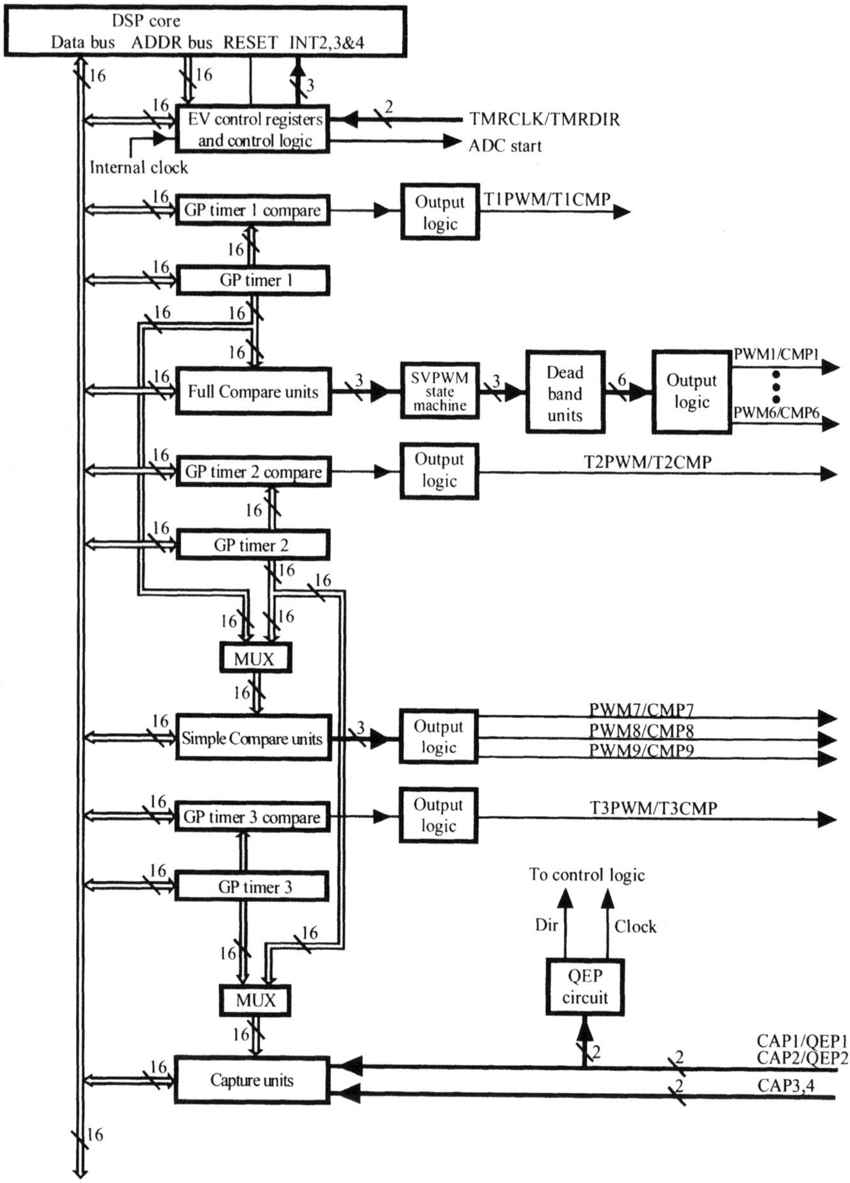

Figure 13.9 TMS320F240 event manager (EV) block diagram [8].

Another variation of this common timer system is to use a continuous 16-bit timer which rolls over every 2^{16} counts [9]. This system provides eight 16-bit comparison registers which are loaded from the controlling software into a content-addressable memory (CAM) file. All eight CAM file entries are compared against the continuous timer every eight processor clock cycles, and if a match occurs the output event determined by the 8-bit command byte associated with the CAM file entry is triggered. Programmable events include setting or toggling a high-speed output, initiating an analog input conversion (ADC) cycle or simply generating a software interrupt. Figure 13.10 shows the general structure of this system. The arrangement has the advantage that the CAM file registers can be loaded at any time (provided, of course, they are loaded before the switching time has passed!). Hence there is potential to reduce the computational delay by loading the registers as close to their event trigger time as possible. However, the approach also has the cost that the PWM calculation software has to determine each phase leg switching time as an absolute time, not a relative time within a carrier interval, and this introduces additional complexity since a rolling time offset must be added to the phase leg switching time calculated from the PWM algorithm.

A further complication with a digital modulation system is how to manage dead time — the delay that is required between turning OFF one switch of a phase leg and turning ON the other switch, to avoid the potentially damaging situation of *shoot through*, which occurs if both switches are momentarily

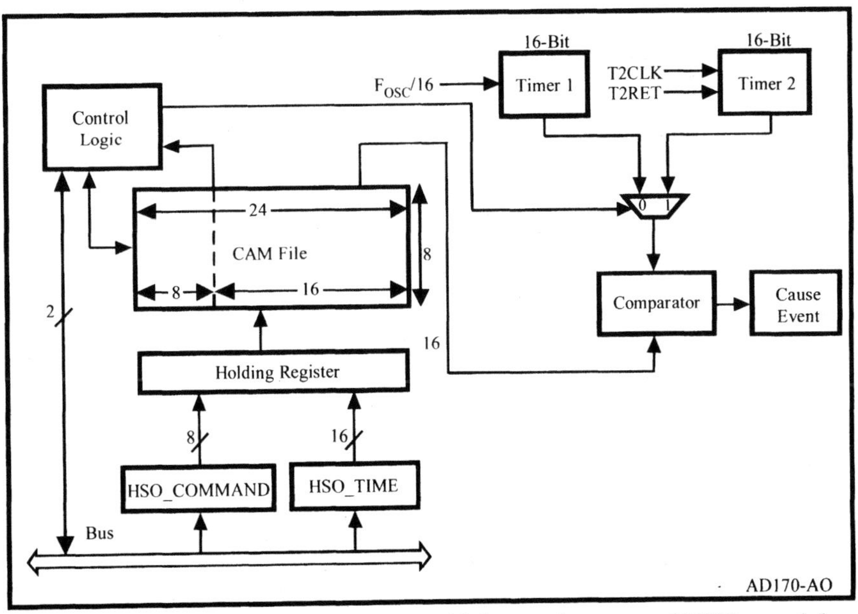

Figure 13.10 Intel 8XC196KC/KD high-speed output (HSO) module block diagram [9].

turned on together and hence short circuit the DC bus. One approach to manage this hazard is for the controlling software to add delays to the computed switching intervals to guarantee that the outgoing switch of a phase leg turns off before the incoming switch turns on. However, this adds to the software burden and requires that the two switches of each phase leg are controlled by separate counter comparison events. The alternative (and preferable) approach is to include additional timer logic after the main timer comparison system, to stagger the switching of the outgoing and incoming devices of each phase leg to ensure they cannot conduct together even momentarily. This approach can be seen on the outputs of the three "full compare" circuits in Figure 13.9.

The TMS320F240 DSP timer system also includes one additional variation to particularly support the implementation of space vector modulation. It can be recalled from Chapter 6 that SV modulation only requires the calculation of two time intervals, one for each of the two active space vectors. When this strategy is to be implemented with the TMS320F240 DSP, the full compare registers of the EV subsystem are used to time-out the active space vector periods. Additional logic is then inserted after each compare register output to map the compare event to the particular set of converter device status conditions that produce the required space vector. The supporting DSP software is required to update these mapping patterns as the sextant of the modulation process changes.

References [10] to [12] discuss other variations of these types of systems which also have integrated timers and logic systems with high-performance microcontrollers to create single-chip PWM solutions. At the time of writing this book, the authors are unaware of any single-chip product that provides a floating-point DSP alternative to support the controlling software. Hence, if a floating-point processor is required with hardware timer PWM support, this support would currently need to be implemented in a separate programmable gate array logic IC. This does not change the principles of the PWM logic described above, but it would require the user to program the logic themselves, with all the attendant concerns of checking and verification that this involves. Obviously, of course, the approach of using a separate discrete logic implementation does provide the advantage of being able to match the PWM logic specifically to suit the needs of an individual converter system, and this may be an overriding reason to use this approach. However, most floating-point DSPs currently available also do not incorporate an integrated ADC system, and hence this subsystem would have to be separately included into the digital controller circuitry. This would make a "single major chip" controller design become a "three major chip" controller design, with a substantial rise in cost as a consequence. However, such issues will not be discussed further since they are design implementation considerations and outside the scope of this book.

13.4 PWM Software Implementation

Implementing a PWM algorithm in a digital processing system requires time-critical software that can calculate values to be loaded into the digital PWM timers every (half) carrier cycle. Until recently, it has been almost mandatory that this software was written in the machine code (assembly language) of the processor to be used, to achieve a program that could execute sufficiently quickly to maintain reasonable switching frequencies. Even then, many implementations could only revise the timer variables once every full carrier cycle because of processor speed limitations, which of course corresponds to *symmetrical regular sampled* PWM. (The limitations of this PWM approach have been discussed in detail in the earlier chapters of this book and it is clear that *asymmetrical regular sampled* PWM is to be preferred.)

With the advent of fast digital signal processors, and also the development of optimized versions of the C programming language, the limitation of writing PWM programs in machine code have now been removed. Modern compilers can produce an optimized compiled version of a C program that will execute with minimal overhead compared to a hand-crafted machine code equivalent, and with much less programming effort. C language implementations for microcontrollers also include library functions to directly access the hardware peripherals, and most can include in-line assembly code to optimize those procedures that really need to be tuned for speed. Also, the much faster execution speed of DSPs means that timer updates every half carrier cycle are now quite feasible for switching frequencies beyond even 20 kHz (this implies the timer variables must be updated every 25 μs).

It is not intended here to attempt to present specific "smart" strategies to program really fast PWM software systems, particularly since the ways to optimize software are very processor specific and the field is always advancing. However, there are general principles of a software PWM system that will usually work well, and these are what will now be considered.

13.4.1 Background Software

Most time-critical software systems are designed with a concept of *foreground* and *background* software. Foreground software is triggered by external interrupt events (timers, serial communications, switch transitions, etc.) and overrides the normal background program execution to execute the time-critical system response. Smaller computer systems, appropriate for applications such as a PWM controller, will typically execute one background program, with several interrupt-driven programs able to execute over the top of this program when their triggering event occurs. More complex systems may support multiple background programs, with their execution scheduled by a central *real-time operating system* (RTOS), but this is not important in the context of the PWM software system and need not be considered further here.

The role of the background software of the PWM system is to determine the essential parameters of the PWM process. For example, for an open-loop constant V/f AC motor drive, the essential parameters are the target voltage magnitude and frequency that the inverter is required to create. For a closed-loop variable speed system, the target objective is the plant speed, and within that the demanded current if the system uses an inner current regulation system within the speed loop. The essential parameters required from this system is a demanded voltage phasor at any particular point in time. The execution of these procedures is the role of the background software.

Often, some parts of the background software may need to be executed within an interrupt routine. For example, for a closed-loop drive system, it is important that the proportional and integral feedback gain calculations are done at precise regular intervals to maintain stability, and this can only be achieved properly within a timer-interrupted software service routine. For convenience, and if there is sufficient processor speed capacity, these calculations can be executed within the PWM timer update routine at the converter switching frequency. Alternatively, they may be executed in fragments over several PWM cycles, or in various other ways at different levels of software priority and using other timers. In the context of this discussion, these details are unimportant — all of these variations are still considered as background software execution. All that is required is an eventual outcome that defines the operation of the PWM process sufficiently for the main *foreground* interrupt software to calculate timer values for the hardware digital timers at every half carrier interval and to load them into the comparison registers at $\Delta T/2$ intervals.

13.4.2 Calculation of the PWM Timing Intervals

With a regular sampled PWM system, it is critical that the reference waveform be sampled and held constant over each half carrier cycle. For a digital PWM implementation, this means that the digital timer values must only be updated at the triangular carrier waveform extremes (while this waveform is only "virtual" in a digital PWM system, it is still useful to retain the concept of a triangular carrier waveform in terms of sampling issues). This can be achieved by triggering a software interrupt from the PWM timer system as the main counter reaches its upper limit and zero (assuming an up/down counter implementation as discussed in Section 13.3.2). The interrupt routine then has the following half carrier interval to calculate the next timer values and to load them into the timer shadow registers ready for the next half carrier cycle after that. This does mean of course that the software is always running a half carrier delay behind the time where the calculated timer values are put into effect, so that a sampling delay adjustment of $3\Delta T/4$ is now required rather than $\Delta T/4$ as discussed in Section 3.6. Usually, this is not particularly problematic except for possible stability concerns in very high-speed closed-loop systems (such considerations are beyond the scope of this text), and an appropriate phase advance can simply be added to the reference waveform.

Each PWM timer calculation essentially involves determining a fractional timing period using a reference sinusoid of the form given in Eq. (3.11), with the sinusoid angular argument frozen at the carrier extreme (sampling) instant plus an appropriate phase advance correction to allow for sampling delay. The typical calculation process would be

$$\text{Timer}_{i,\,k+1} = \left(\frac{\Delta T}{2}\right) \bullet M\cos(\theta_o(k) + \delta_{advance}) + \frac{\Delta T}{4} \qquad (13.1)$$

where $\theta_o(k)$ is the phasor angle of the target sinusoid at the kth carrier extreme occurrence and $\delta_{advance}$ is defined by

$$\delta_{advance} = \frac{3\Delta T}{4\,T_o} \quad \text{where } T_o = \frac{1}{f_o} \qquad (13.2)$$

The calculation of the sinusoid can be readily achieved using a software look-up table, with perhaps 1024 points providing sufficient resolution for most applications. The variables $\theta_o(k)$ and $\delta_{advance}$ should be directly calculated as scaled quantities so that they can be used immediately as index pointers to this look-up table, and the use of a binary multiple table size allows a quick and easy wrap around calculation using a simple logical AND function. Note also that it is not necessary to store multiple sinusoidal look-up tables for multiple phase leg calculations — it is more efficient to store say two cycles of sinusoids in the table (2048 entries), and to add $2\pi/3$ offsets to $\delta_{advance}$ (scaled appropriately to become a direct index pointer) to calculate the other phase leg angular indexes. Third-harmonic, space vector, and discontinuous modulation strategies can also be easily implemented using this approach by varying the contents of the look-up table to reflect the particular phase leg reference waveform that is required (as presented in Chapters 5 and 6).

Once the sinusoid has been determined, the actual timer value can be calculated by multiplying it by $\Delta T/2 \bullet M$ and adding $\Delta T/4$ to make it always positive. This requires a separate multiply and addition process for each phase leg, but all modern microcontrollers, and particular modern DSPs, have high-speed multiply functions included in their instruction set, so the time penalty is minimal. Note also that the modulation index M can be prescaled by $\Delta T/2$ as it is varied by the background software, to reduce interrupt software overhead.

In summary, each PWM calculation cycle involves one table look-up, one multiply and one addition for each phase leg to determine the value to be loaded into its timer comparison register, once the target phasor angle and modulation index are known. In practice there is usually a little more software overhead required to check for saturation (number values less than 0 or exceeding $\Delta T/2$), and to correctly manage issues such as dead time, pulse dropping correction, etc., but this overhead is typically small. As an example, the overall PWM calculation time for a three-phase inverter using a TMS320F240 DSP is substantially less than 20 μs, including all overheads, and with almost all of the interrupt software routine programmed in C. This is typical of the performance that can be achieved using a modern DSP controller.

One final issue does need a little more discussion — the calculation of the phasor reference angle $\theta_o(k)$. In general with a digital PWM system, the ratios between the fundamental frequency, and switching frequency and the sinusoidal look-up table resolution will not be integer. The fundamental frequency can be an arbitrary value, the switching frequency will be a fixed submultiple of the microcontroller clock, and the sinusoid table size will be governed by memory constraints.

One obvious means to maintain the phasor angle of the target reference voltage is to add an increment to $\theta_o(k)$ at every PWM calculation cycle, so that

$$\theta_o(k) = \theta_o(k-1) + \Delta\theta_o \tag{13.3}$$

where

$$\Delta\theta_o = \text{Table length} \bullet \frac{\Delta T}{2} f_o = \text{Table length} \bullet \frac{\Delta T}{2} \frac{1}{T_o} \tag{13.4}$$

and "Table length" is the length of one fundamental cycle in the sinusoidal look-up table. A further refinement is to include $\delta_{advance}$ into the initial value of $\theta_o(0)$, so that it does not have to be added at each PWM calculation cycle.

The limitation with this approach is that for fixed word length systems, the resolution of Eq. (13.4) is inadequate, and hence the output frequency resolution will be too coarse to be useful. The straightforward solution to this problem is to calculate $\theta_o(k)$ at a much higher level of accuracy, say using a 16-bit variable instead of the 10 bits that are required to look up values in a 1024-element sinusoid table. The variable $\theta_o(k)$ can then be right shifted down 6 bits just before it is used as a table index pointer (with appropriate rounding compensation). This approach adds minimal additional software overhead.

With this level of resolution, each increment of $\theta_o(k)$ represents an angular shift of $360/65536 = 0.005493^\circ$. For a switching frequency of say 5 kHz (10 kHz half carrier update rate), this would mean a minimum fundamental output frequency resolution of $(10000 \times 0.005493)/360 = 0.153$ Hz, which is sufficient for many applications. If this is inadequate, an even finer frequency resolution can be achieved by updating $\theta_o(k)$ using dithered angular increments of $\theta_o(k)$ over several half carrier PWM intervals, so that the overall angular shift every n interrupt cycles equates to the required angular shift to achieve the desired fundamental frequency (note that the n cycles need not constitute a complete fundamental period). The dithered angular increments can be stored in another cyclic look-up table which is accessed sequentially as the PWM timer interrupts proceed, to minimize software overhead. The dither table would be updated by the background software whenever a fundamental frequency change is commanded.

13.5 Summary

This chapter has presented an overview of the approaches that can be used to build a PWM power electronic conversion system in practice. It has reviewed the physical hardware that is required and has considered the ways in which the modulation process can be implemented in either hardware (analog) or software (digital). The material presented in this chapter is intended to be a "snapshot" of the state of the art at the time of writing, and no doubt newer technologies will soon emerge that will offer better ways to implement the concepts presented. However, it is anticipated that at least the same functional objectives will need to be achieved, irrespective of the technology that is used.

References

[1] Microchip, "1.5A Dual High-Speed Power MOSFET Drivers TSC4426/TSC4427/TSC4428," Data Sheet No. TSC4426/7/8-8, 1996.

[2] International Rectifier, "IR2110/IR2113 High Side and Low Side Drive," Data Sheet No. PD60147–L.

[3] Hewlett-Packard, "2.0A Gate Drive Optocoupler with Integrated (V_{CE}) Desaturation Detection and Fault Status Feedback," HCPL–316J Technical Data Sheet, 1999.

[4] IXYS Corporation, "ISOSMART Half-Bridge Driver Chipset," Data Sheet IXBD4410, 2000.

[5] S.R. Bowes and M.J. Mount, "Microprocessor control of PWM inverters," *IEE Proceedings* (*London*), vol. 128, Pt. B, no. 6., Nov. 1981, pp. 293–305.

[6] Creative Power Technologies Pty. Ltd. (www.creativepower.com.au), "MU–DSP240–LPI Inverter Controller Card," Technical Manual, Rev2.3, Mar. 2001.

[7] D.A. Grant, M. Stevens, and J.A. Houldsworth, "The effect of word length on the harmonic content of microprocessor-based PWM waveform generators," *IEEE Trans. on Industry Applications,* vol. IA–21, no. 1, Jan./Feb. 1985, pp. 218–225.

[8] Texas Instruments Corp., "TMS320F40 Digital Signal Processor," Data Sheet, 1996 and rev2002.

[9] Intel Corp., "8XC196KC/8XC196KD User's Manual," Order Number: 272238–001, 1992.

[10] Analog Devices Inc., "ADSP–2106x SHARC™ DSP Microcomputer Family," Data Shee, 2000.

[11] Motorola, "DSP56F801 16–bit Digital Signal Processor," Data Sheet No. DSP56F801/D, Rev7.0 1/2002.

[12] Infimum Technologies, "C164CI/SI C164CI/SL 16–bit Single Chip Microcontroller," Data Sheet rev2.0, May 2001.

14 Continuing Developments in Modulation

This text has primarily focused on the development of PWM algorithms for the purpose of reducing losses in inductive, primarily motor-type, loads. The concept of weighted total harmonic distortion has been developed as a figure of merit to assess losses in these types of loads for various modulation algorithms. However, the loss issue is only one of a number of important concerns associated with inverters driving motor loads. Other concerns involve, for example, the degree of torque "smoothness" or "ripple torque" which can also be quantified by a suitable performance factor [1]. The PWM-induced current ripple produces rotating magneto-motive force (MMF) waves rotating in the forward and backward directions which interact with the main, fundamental component MMF wave to produce ripple torques. Fortunately, these torques are generally not of concern with modern PWM algorithms except during overmodulation in which the odd non-triplen multiples of the fundamental component begin to appear. The reader is referred to the extensive bibliography at the end of this book for a treatment of this issue.

The subject of converter modulation continues to evolve in many directions. For example, as a result of increasing concern for source side waveform purity, PWM has more recently been applied to control of the input side converter (rectifier) which has special requirements. Also, practical effects such as delays in the turn-on of semiconductor switches have received recent attention. This effect can result in baseband harmonics of the fundamental inverter output component, which must be eliminated (or at least minimized) to avoid introducing unwanted side effects such as torque ripple at low speed. Certain other types of modulation have also appeared which cannot be strictly categorized by their waveform. This chapter concludes this book with a selection of such ongoing practical issues.

14.1 Random Pulse Width Modulation

Generally, overmodulation begins to occur at frequencies of 50 Hz or more, so that the fifth, seventh, etc. harmonics at this point exist at frequencies which do not present a danger of producing a mechanical resonance. However, this same current ripple, in turn, produces a ripple in the rotating flux wave in the machine which exerts forces in the radial direction, proportional to the square of the flux density, causing the frame to resonate at the sideband harmonic frequencies produced by the PWM algorithm. When one of the spectral lines of the PWM algorithm coincides with the resonant frequency of one of the motor structural parts, cooling fins, fan blades, housings, etc., the resulting noise can become very objectionable. It is desirable then to spread out the frequency spectrum of the harmonic voltage in such a manner as to eliminate or minimize the presence of specific spectral lines associated with mechanical resonance.

While any PWM scheme with acoustic noise reduction must produce a spectrum without energy concentration at discrete frequencies, the switching frequency should be limited to a spectral range where substantial additional load harmonic losses do not occur. In addition, the linear transfer function associated with the fundamental input/output characteristic of the inverter should be retained. Not unexpectedly, either sine–triangle modulation or space vector modulation can be modified to achieve random modulation.

One approach to spread the PWM spectral energy is to randomly vary the period of the triangular carrier, as shown in Figure 14.1. However, in order to maintain the linearity of the modulation process, it is still necessary for the carrier wave to have straight line segments, i.e., it must remain triangular within each half carrier period. Similarly, as in the case of fixed-frequency modulation, it is assumed that the instantaneous carrier frequency f_c is always much larger than the command frequency, f_o. Hence the PWM duty cycle remains proportional to the value of the reference over any one period of the carrier so that linearity is maintained. The instantaneous value of the carrier frequency between any two time segments can then be written as

$$f_c = \frac{1}{t_{k+1} - t_k} = f_{c0} + \Delta f_c \tag{14.1}$$

where the change in the frequency of any given segment is given by

$$\Delta f_c = (f_{c,max} - f_{c0})n(t) \tag{14.2}$$

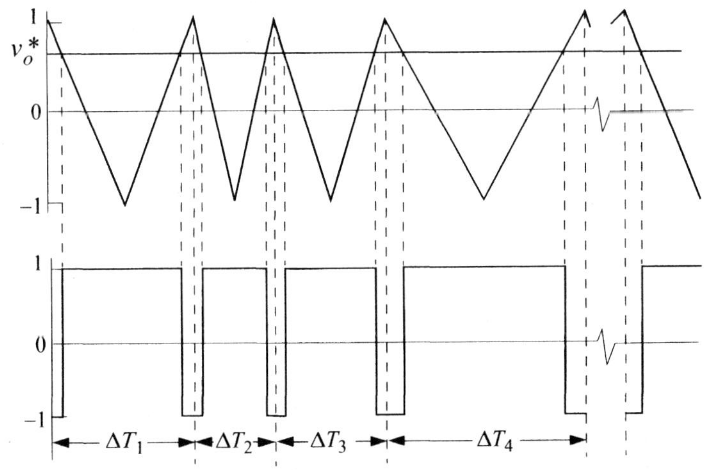

Figure 14.1 Generation of random PWM waveform using trianglar carrier-wave of random periodicity.

where f_{c0} is the average value of the period and $n(t)$ is a random variable between -1 and $+1$. The maximum frequency $f_{c,\,max}$ is set by switching limitations of the PWM inverter [2].

Figure 14.2 shows one method for implementing a random carrier PWM regulator [3]. In this approach the band-limited random noise is generated using a look-up table whose contents have been generated offline. A table which contains 1024 elements with an 8-bit value is used to determine the deviation from the center frequency. The width of the band is selected to be ± 1000 Hz, around a nominal switching frequency of 2500 Hz. The elements are randomly mixed and stored in an EPROM. A linear weight table of the random modulation scheme is shown as an example since it is easy to generate and all values are uniformly weighted. The weight function is shown in Figure 14.3(a) and the randomly modulated look-up table is shown in Figure 14.3(b). The frequency data is first converted to analog data and then is scaled to represent the change in the carrier slope by

$$\Delta m = 4\Delta f_c \tag{14.3}$$

The corresponding voltage spectrum is shown in Figure 14.4. A relatively smooth spectrum without the presence of discrete spectral lines is in evidence. However, some peaking at the sideband of twice the nominal PWM switching frequency remains (2 x 2500 = 5000 Hz). In addition peaking also occurs near

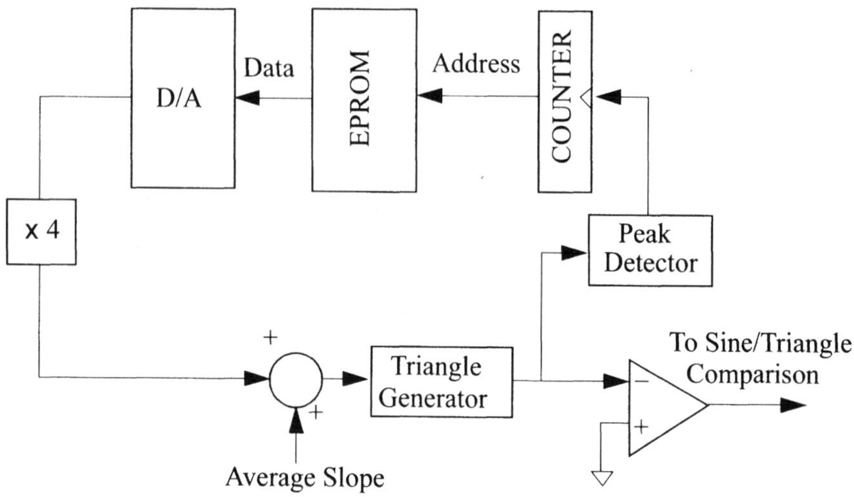

Figure 14.2 Block diagram illustrating one possible implementation of random carrier where noise is generated digitally and stored in a look-up table.

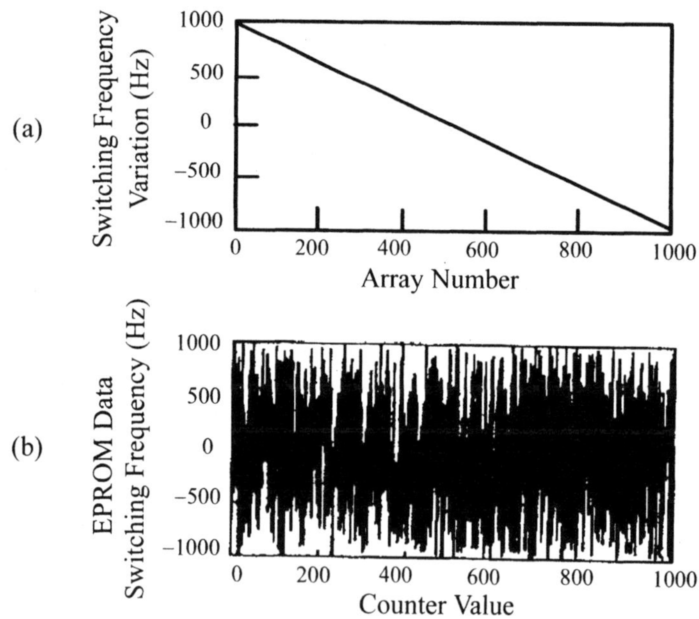

Figure 14.3 (a) Linear distribution of frequency values over 1024 points and (b) random elements in the EPROM table. (Courtesy of F. Blaabjerg [4].)

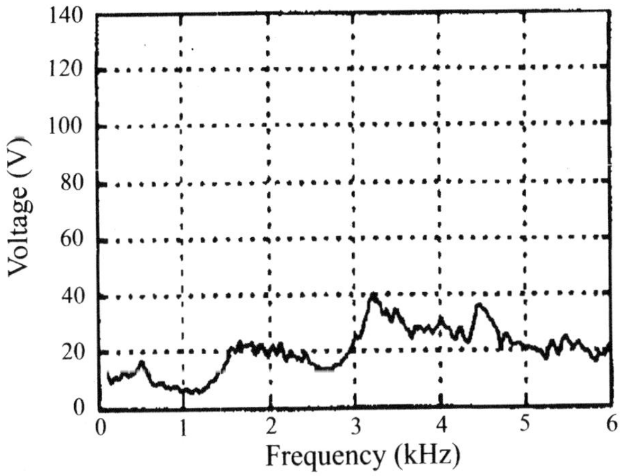

Figure 14.4 Voltage spectra corresponding to linear distribution of PWM frequencies in Figure 14.3. The center frequency is 2.5 kHz and the frequency band is ±1000 kHz. (Courtesy of F. Blaabjerg [4].)

frequencies corresponding to the nominal frequency plus/minus the maximum random switching frequency (1500 and 3500 Hz). It has been shown [4] that these (relatively modest) spectral peaks can be subdued by distributing half of the points as a linear distribution and half in a hyperbolic distribution (i.e., linear distribution with respect to the carrier wave period rather than frequency) as shown in Figure 14.5(a). In this case the distribution is well smoothed, shown in Figure 14.5(b).

Another distribution one might expect to be beneficial is a Gaussian distribution where the lower switching frequencies are somewhat reduced from the linear case and the higher switching frequencies are increased. This should serve to reduce the frequency-dependent load losses. However, the increased emphasis of the higher frequencies simply accentuates the spectral peaking near the maximum switching frequency, as shown in Figures 14.5(c) and 14.5(d).

While it is possible to control the distribution of the voltage, it should be mentioned that when supplying an electrical machine, one or more mechanical resonances can be triggered which would then dominate the noise spectra. These mechanical resonances are readily found by using the linear/hyperbolic distribution of frequencies which excite the machine with a nearly uniform

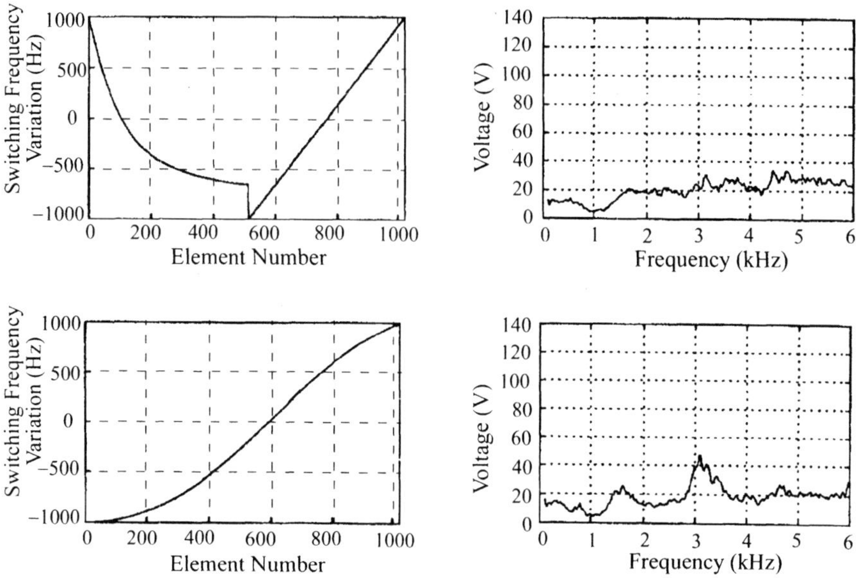

Figure 14.5 Two different look-up table distributions: (a),(b) linear/hyperbolic distribution and (c),(d) Gaussian distribution. (Courtesy of F. Blaabjerg [4].)

spectrum from $f_{c0} - \Delta f_c$ to $f_{c0} + \Delta f_c$. By omitting frequency spans around the resonant frequencies, the acoustical noise peaks can be minimized. The intelligent frequency spans can be realized by first generating a general random loop-up table and then excluding the unwanted frequencies in the EPROM table. Another method is to omit the unwanted spans when calculating the random look-up table, thus omitting the unwanted numbers which lead to resonances [4].

14.2 PWM Rectifier with Voltage Unbalance

It should be evident that all of the theory developed up to this point is equally valid for rectification as well as for inversion, since the power flow through the basic three-leg converter is bidirectional. However, in the vast majority of cases the inverter load is *balanced*, that is, the impedances seen by the three inverter phases are identical. Such a balanced load is typically a three-phase induction, synchronous, or permanent magnet motor. On the other hand, for a rectifier the input AC source voltages are frequently unbalanced, particularly in a weak grid system. Unevenly distributed single-phase loads or nonsymmetrical transformer windings as well as faults in the network can typically lead to an unbalanced supply. This possibility of unbalanced source voltages further complicates the issue of converter modulation.

Consider a set of general three-phase voltages of unequal amplitude and arbitrary phase. Assuming that the zero sequence voltage is zero, one can write the input as the three-element vector

$$\boldsymbol{v}_i = \begin{bmatrix} v_{ag} \\ v_{bg} \\ v_{cg} \end{bmatrix} \tag{14.4}$$

where

$$v_{ag} = V_{ag}\cos(\omega_e t + \phi_a) \tag{14.5}$$

$$v_{bg} = V_{bg}\cos(\omega_e t + \phi_b) \tag{14.6}$$

$$v_{cg} = V_{cg}\cos(\omega_e t + \phi_c) \tag{14.7}$$

and g denotes the ground point of the supply. From Euler's equation the voltage of phase ag can be written alternatively as the sum of two complex exponentials, as

$$v_{ag} = \frac{1}{2}\left[V_{ag}e^{j(\omega_e t + \phi_a)} + V_{ag}e^{-j(\omega_e t + \phi_a)}\right] \tag{14.8}$$

Similar equations are readily written for the other two phases.

If the phase of the exponential term is combined with the voltage amplitude, then one can define phasor magnitudes as

$$\bar{V}_{ag} = V_{ag}e^{j\phi_a} \tag{14.9}$$

and

$$\bar{V}_{ag}^{\dagger} = V_{ag}e^{-j\phi_a} \tag{14.10}$$

where † denotes the conjugate of a complex number. The equation for v_{ag} can now be written as

$$v_{ag} = \frac{1}{2}\bar{V}_{ag}e^{j\omega_e t} + \frac{1}{2}\bar{V}_{ag}^{\dagger}e^{-j\omega_e t} \tag{14.11}$$

Similar expressions can be written for v_{bg} and v_{cg}.

Substituting these voltage expressions into the defining equation for the grid input voltage space vector, Eq. (1.57), gives

$$\begin{aligned}\bar{v}_i &= \frac{2}{3}(v_{ag} + \bar{a}v_{bg} + \bar{a}^2 v_{cg})\\ &= \frac{1}{3}(\bar{V}_{ag} + \bar{a}\bar{V}_{bg} + \bar{a}^2\bar{V}_{cg})e^{j\omega_e t} + \frac{1}{3}(\bar{V}_{ag} + \bar{a}\bar{V}_{bg} + \bar{a}^2\bar{V}_{cg})^{\dagger} e^{-j\omega_e t}\\ &= \frac{1}{3}(\bar{V}_{ag} + \bar{a}\bar{V}_{bg} + \bar{a}^2\bar{V}_{cg})e^{j\omega_e t} + \frac{1}{3}(\bar{V}_{ag}^{\dagger} + \bar{a}\bar{V}_{bg}^{\dagger} + \bar{a}^2\bar{V}_{cg}^{\dagger})e^{-j\omega_e t}\end{aligned} \tag{14.12}$$

The two quantities in the parentheses form the positive and negative sequence voltage components, from symmetrical component theory. The positive sequence component is

$$v_{ap} = \frac{1}{3}\mathrm{Re}\left[(\bar{V}_{ag} + \bar{a}\bar{V}_{bg} + \bar{a}^2\bar{V}_{cg})e^{j\omega_e t}\right] \tag{14.13}$$

Substituting from Eq. (14.9) (and equivalent expressions for V_{bg} and V_{cg}) this expression can be written as

$$v_{ap} = \frac{1}{3}\mathrm{Re}\left[\left(V_{ag}e^{j\phi_a} + V_{bg}e^{j\left[\phi_b + \frac{2\pi}{3}\right]} + V_{cg}e^{j\left[\phi_c + \frac{4\pi}{3}\right]}\right)e^{j\omega_e t}\right] \tag{14.14}$$

Taking the real part of the bracketed expression, Eq. (14.14) becomes

$$v_{ap} = \frac{1}{3}\left[\begin{aligned}V_{ag}\cos(\omega_e t + \phi_a) + V_{bg}\cos\left(\omega_e t + \phi_b + \frac{2\pi}{3}\right)\\ + V_{cg}\cos\left(\omega_e t + \phi_c + \frac{4\pi}{3}\right)\end{aligned}\right] \tag{14.15}$$

Similarly,

$$v_{bp} = \frac{1}{3}\left[\begin{aligned}V_{ag}\cos\left(\omega_e t + \phi_a + \frac{4\pi}{3}\right) + V_{bg}\cos(\omega_e t + \phi_b)\\ + V_{cg}\cos\left(\omega_e t + \phi_c + \frac{2\pi}{3}\right)\end{aligned}\right] \tag{14.16}$$

and

$$v_{cp} = \frac{1}{3}\left[\begin{aligned}V_{ag}\cos\left(\omega_e t + \phi_a + \frac{2\pi}{3}\right) + V_{bg}\cos\left(\omega_e t + \phi_b + \frac{2\pi}{3}\right)\\ + V_{cg}\cos(\omega_e t + \phi_c)\end{aligned}\right] \tag{14.17}$$

When written as a vector, Eqs. (14.15) to (14.17) become

$$\boldsymbol{v}_{ip} = \begin{bmatrix} v_{ap} \\ v_{bp} \\ v_{cp} \end{bmatrix} = \frac{1}{3}\begin{bmatrix} V_{ag}\cos(\omega_e t + \phi_a) + V_{bg}\cos\left(\omega_e t + \phi_b + \frac{2\pi}{3}\right) \\ \qquad + V_{cg}\cos\left(\omega_e t + \phi_c + \frac{4\pi}{3}\right) \\ V_{ag}\cos\left(\omega_e t + \phi_a + \frac{4\pi}{3}\right) + V_{bg}\cos(\omega_e t + \phi_b) \\ \qquad + V_{cg}\cos\left(\omega_e t + \phi_c + \frac{2\pi}{3}\right) \\ V_{ag}\cos\left(\omega_e t + \phi_a + \frac{2\pi}{3}\right) + V_{bg}\cos\left(\omega_e t + \phi_b + \frac{4\pi}{3}\right) \\ \qquad + V_{cg}\cos(\omega_e t + \phi_c) \end{bmatrix} \tag{14.18}$$

Note that the voltage $v_{bp}(\omega_e t)$ can be considered as simply as $v_{ap}(\omega_e t - 2\pi/3)$, and similarly $v_{cp}(\omega_e t) = v_{ap}(\omega_e t - 4\pi/3)$. Thus the three positive voltage components form a balanced three-phase set.

In a similar manner the voltage vector representing the negative sequence components is

$$\boldsymbol{v}_{in} = \begin{bmatrix} v_{an} \\ v_{bn} \\ v_{cn} \end{bmatrix} = \frac{1}{3}\begin{bmatrix} V_{ag}\cos(\omega_e t + \phi_a) + V_{bg}\cos\left(\omega_e t + \phi_b + \frac{4\pi}{3}\right) \\ \qquad + V_{cg}\cos\left(\omega_e t + \phi_c + \frac{2\pi}{3}\right) \\ V_{ag}\cos\left(\omega_e t + \phi_a + \frac{2\pi}{3}\right) + V_{bg}\cos(\omega_e t + \phi_b) \\ \qquad + V_{cg}\cos\left(\omega_e t + \phi_c + \frac{4\pi}{3}\right) \\ V_{ag}\cos\left(\omega_e t + \phi_a + \frac{4\pi}{3}\right) + V_{bg}\cos\left(\omega_e t + \phi_b + \frac{2\pi}{3}\right) \\ \qquad + V_{cg}\cos(\omega_e t + \phi_c) \end{bmatrix} \tag{14.19}$$

The three negative sequence voltages again form a balanced three-phase set but with reverse rotation. Thus, the unbalanced input voltage vector in Eq. (14.4) can always be decomposed into two balanced components as

$$\boldsymbol{v}_i = \boldsymbol{v}_{ip} + \boldsymbol{v}_{in} \tag{14.20}$$

From Eq. (1.13) it was shown that the DC link current is related to the AC input currents by

$$I_{dc} = \frac{1}{2}(i_a m_a + i_b m_b + i_c m_c) \tag{14.21}$$

where m_a, m_b, and m_c represent the modulation switching functions. When the currents are balanced and sinusoidal as a result of suitable switching functions, then the average value of the link current I_{dc} is a constant. In a similar manner, it can be supposed that when the AC voltage components are balanced and sinusoidal, then the link DC voltage V_{dc} would be constant [5]. In effect,

$$V_{dc} = \frac{1}{2}(v_{ag} m_a + v_{bg} m_b + v_{cg} m_c) \tag{14.22}$$

where v_{ag}, v_{bg}, v_{cg} are now considered as voltages defined at each converter phase leg with respect to ground reference. In vector form,

$$2V_{dc} = \boldsymbol{m}_i^T \boldsymbol{v}_i \tag{14.23}$$

where T denotes the transpose of the vector $\boldsymbol{m}_i$.

Since Eq. (14.20) demonstrates that any unbalanced voltage is comprised of two balanced sequence components, a suitable modulation vector can be assumed for each sequence that will result in a constant DC link voltage. Hence Eq. (14.23) becomes

$$2V_{dc} = (\boldsymbol{m}_{ip} + \boldsymbol{m}_{in})^T (\boldsymbol{v}_{ip} + \boldsymbol{v}_{in}) \tag{14.24}$$

Expanding Eq. (14.24) gives

$$2V_{dc} = \boldsymbol{m}_{ip}^T \boldsymbol{v}_{ip} + \boldsymbol{m}_{in}^T \boldsymbol{v}_{in} + \boldsymbol{m}_{ip}^T \boldsymbol{v}_{in} + \boldsymbol{m}_{in}^T \boldsymbol{v}_{ip} \tag{14.25}$$

Assuming proper modulation, the first two right-hand terms produce the desired constant DC link voltage while the second two terms can be shown to produce second harmonics at the line frequency f_e. The proper forms for the modulation vectors are not difficult to devise. If the three positive sequence voltages are, for example, defined from Eq. (14.18) by the set of sine waves

$$v_{ap} = V_{mp}\cos(\omega_e t + \phi_p) \tag{14.26}$$

$$v_{bp} = V_{mp}\cos\left(\omega_e t + \phi_p + \frac{2\pi}{3}\right) \tag{14.27}$$

$$v_{cp} = V_{mp}\cos\left(\omega_e t + \phi_p + \frac{4\pi}{3}\right) \tag{14.28}$$

then the switching functions of the same form, namely

$$m_{ap} = \cos(\omega_e t + \phi_p + \phi_{mp}) \tag{14.29}$$

$$m_{bp} = \cos(\omega_e t + \phi_p + 2\pi/3 + \phi_{mp}) \tag{14.30}$$

$$m_{cp} = \cos(\omega_e t + \phi_p + 4\pi/3 + \phi_{mp}) \tag{14.31}$$

will produce the constant result

$$\boldsymbol{m}_{ip}^T \boldsymbol{v}_{ip} = \frac{3}{2} V_{mp} \cos(\phi_{mp}) \tag{14.32}$$

Similarly, choosing

$$\boldsymbol{m}_{in}(\omega_e t) = -\frac{\boldsymbol{v}_{in}(\omega_e t + \phi_{mn})}{V_{mp}} \tag{14.33}$$

results in the DC voltage

$$\boldsymbol{m}_{in}^T \boldsymbol{v}_{in} = -\frac{3}{2} \frac{V_{mn}^2}{V_{mp}} \cos(\phi_{mn}) \tag{14.34}$$

and the last two terms in Eq. (14.25) become

$$\boldsymbol{m}_{ip}^T \boldsymbol{v}_{in} + \boldsymbol{m}_{in}^T \boldsymbol{v}_{ip} = \frac{\boldsymbol{v}_{ip}^T(\omega_e t + \phi_{mp}) \boldsymbol{v}_{in}}{V_{mp}} - \frac{\boldsymbol{v}_{in}^T(\omega_e t + \phi_{mn}) \boldsymbol{v}_{ip}}{V_{mp}} \tag{14.35}$$

However, if one sets

$$\phi_{mp} = \phi_{mn} \tag{14.36}$$

and notes that

$$\boldsymbol{v}_{ip}^T \boldsymbol{v}_{in} = \boldsymbol{v}_{in}^T \boldsymbol{v}_{ip} \tag{14.37}$$

the unwanted terms can be made to cancel. The output voltage in Eq. (14.25) then becomes

$$\begin{aligned} 2V_{dc} &= \frac{\boldsymbol{v}_{ip}^T \boldsymbol{v}_{ip} - \boldsymbol{v}_{in}^T \boldsymbol{v}_{in}}{V_{mp}} \cos\phi_{mp} \\ &= \frac{\boldsymbol{v}_{ip}^2 - \boldsymbol{v}_{in}^2}{V_{mp}} \cos\phi_{mp} \\ &= \frac{3}{2}\left(V_{mp} - \frac{V_{mn}^2}{V_{mp}}\right) \cos\phi_{mp} \end{aligned} \tag{14.38}$$

Therefore the modulation function needed to produce a constant DC link voltage for any unbalanced condition is

$$\boldsymbol{m}_i = \frac{\boldsymbol{v}_{ip} - \boldsymbol{v}_{in}}{V_{mp}} \tag{14.39}$$

subject to the constraint of Eq. (14.36).

Upon substituting Eqs. (14.18) and (14.19) into (14.39), the proper modulation function expressed as an explicit function of time becomes

$$\boldsymbol{m}_i(\omega_e t) = \frac{1}{3V_{mp}}\begin{bmatrix} V_{bg}\cos\left(\omega_e t + \phi_b + \frac{\pi}{2}\right) + V_{cg}\cos\left(\omega_e t + \phi_c - \frac{\pi}{2}\right) \\ V_{cg}\cos\left(\omega_e t + \phi_c + \frac{\pi}{2}\right) + V_{ag}\cos\left(\omega_e t + \phi_a - \frac{\pi}{2}\right) \\ V_{ag}\cos\left(\omega_e t + \phi_a + \frac{\pi}{2}\right) + V_{bg}\cos\left(\omega_e t + \phi_b - \frac{\pi}{2}\right) \end{bmatrix} \tag{14.40}$$

Assuming half-wave symmetry, Eq. (14.40) can also be written as

$$\boldsymbol{m}_i(\omega_e t) = \frac{1}{3V_{mp}}\begin{bmatrix} V_{cg}\cos\left(\omega_e t + \phi_c - \frac{\pi}{2}\right) - V_{bg}\cos\left(\omega_e t + \phi_b - \frac{\pi}{2}\right) \\ V_{ag}\cos\left(\omega_e t + \phi_c - \frac{\pi}{2}\right) - V_{cg}\cos\left(\omega_e t + \phi_c - \frac{\pi}{2}\right) \\ V_{bg}\cos\left(\omega_e t + \phi_a - \frac{\pi}{2}\right) - V_{ag}\cos\left(\omega_e t + \phi_b - \frac{\pi}{2}\right) \end{bmatrix} \tag{14.41}$$

It can be noted that Eq. (14.41) involves only the amplitude of the positive sequence voltage which can be readily measured with a filter. Since the amplitude of the modulation signal can be adjusted via regulation of the DC link voltage, it is generally not necessary to measure the positive sequence voltage at all. The three voltages simply represent three line-to-line voltages. Equation (14.41) can be used to develop the control block diagram of Figure 14.6.

In general, it is possible to keep ϕ_{mp} at zero to achieve unity power factor operation at the AC terminals of the bridge for the positive sequence if the rectifier is a buck-type converter. However, if the converter is the more common boost type, the angle $\phi_{mp} = \phi_{mn}$ must be increased to achieve voltage boost. The angles ϕ_{mp} and ϕ_{mn} now correspond to the phase shifts on the input side of the boost inductor.

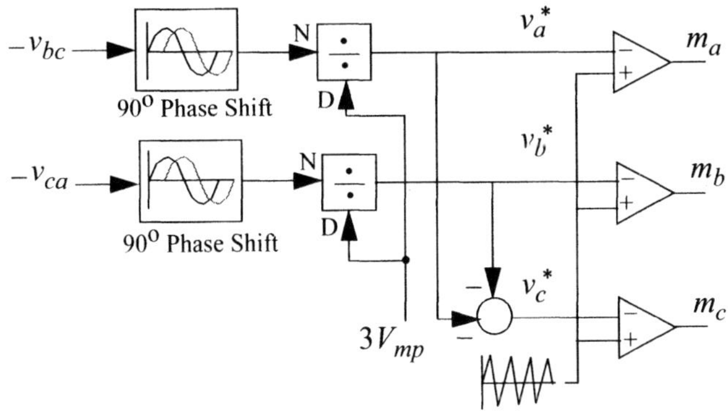

Figure 14.6 Modulator with unbalance correction, $\phi_{mp} = \phi_{mn}$ assumed.

The maximum DC link voltage that can be achieved decreases as the unbalance increases. It is useful to define an unbalance factor

$$U = \frac{V_{mn}}{V_{mp}} \tag{14.42}$$

in which case Eq. (14.38) becomes

$$2V_{dc} = \frac{3}{2}V_{mp}(1 - U^2) \tag{14.43}$$

The maximum DC link voltage plotted as a function of the unbalance factor is shown in Figure 14.7 [5].

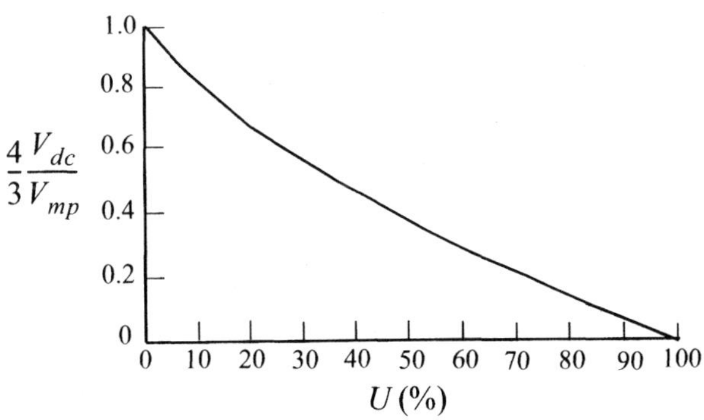

Figure 14.7 Normalized DC link voltage V_{dc} as a function of unbalance factor in percent.

14.3 Common Mode Elimination

This book has primarily been devoted to the calculation of harmonics resulting from the pulse width modulation processes. Solutions for these harmonics have generally been expressed in terms of the line-to-line voltages. These solutions form what is termed the *differential mode* since they express the relative voltage difference between two points of the load. However, voltages inevitably also build up across the entire load with respect to Earth ground. This component of voltage, neglected up to this point, is termed the *common mode* voltage (V_{com}). When the load is a motor, the common mode voltage enables voltages to build up through electrostatic coupling between the rotor and stator frame, resulting in current flow through the motor bearings when this shaft voltage exceeds the dielectric capability of the bearing grease. It has been determined that such shaft currents may cause premature motor bearing failures. In addition, the common mode voltage causes current through the stator winding insulation to the grounded frame, resulting in false tripping of ground current relays and contributing to deterioration of the ground insulation.

The common mode produced by a multilevel inverter can be nearly eliminated by selecting only the specific PWM states that result in no common mode voltage [6]. It is useful to consider, for example, the case of a three-level inverter as shown in Figure 14.8. One can recall that the numeric 2 denotes that a particular phase is connected to V_{dc}, 0 denotes connected to $-V_{dc}$, and 1 denotes connected to the link midpoint or 0 volts. The common mode represents the average value of the three load-phase voltages with respect to Earth ground. While the Earth ground is not explicitly expressed in the vector states,

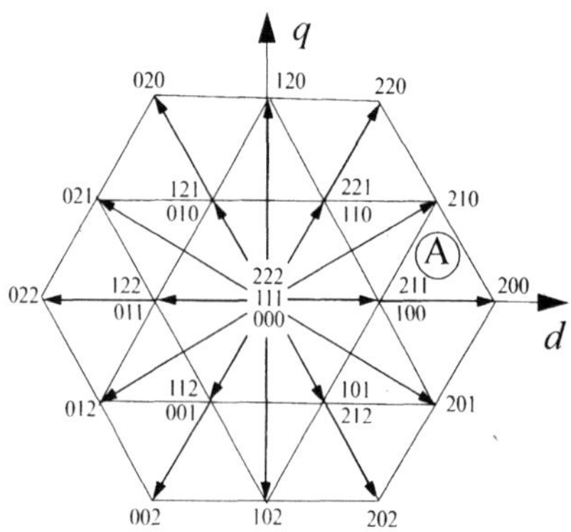

Figure 14.8 (a) All possible voltage vectors that can be realized with a three-level inverter.

the voltage of the bus center point with respect to Earth ground is constant, and can be defined as V_{zg}, where zg denotes the voltage at the DC bus midpoint z with respect to Earth ground g. For example, if it is assumed that a voltage vector is to be synthesized within triangle A of Figure 14.8, then for state 200

$$V_{com} = \frac{1}{3}[(V_{dc} + V_{zg}) + (-V_{dc} + V_{zg}) + (-V_{dc} + V_{zg})]$$

$$= -\frac{V_{dc}}{3} + V_{zg} \tag{14.44}$$

Similarly for state 210

$$V_{\text{com}} = V_{zg} \tag{14.45}$$

for state 211

$$V_{\text{com}} = \frac{V_{dc}}{3} + V_{zg} \tag{14.46}$$

and for state 100

$$V_{\text{com}} = -\frac{2V_{dc}}{3} + V_{zg} \tag{14.47}$$

Hence, in the process of synthesizing a voltage within triangle A using the space vector algorithm, the common mode voltage takes on four distinct values resulting in a high frequency being impressed with respect to ground during the modulation process.

It is clear that a constant common mode voltage can be obtained if only states having the same common mode voltage are selected. Five constant common mode voltage levels can be identified, corresponding to $(2V_{dc}/3 + V_{zg})$, $(V_{dc}/3 + V_{zg})$, (V_{zg}), $(-V_{dc}/3 + V_{zg})$, and $(-2V_{dc}/3 + V_{zg})$. The first and last of these common mode voltages, however, do not produce vector states on the outer hexagon, preventing the synthesis of phase voltages greater than $V_{dc}/3$. The three remaining possibilities are shown in Figure 14.9. It is apparent that the states producing a common mode voltage of V_{zg} will achieve the widest region for synthesis of a voltage vector. Also, this value of common mode voltage allows the presence of one zero state (i.e., 111) which is important for achieving smaller values of output voltage. The usable vector states produce a hexagon which is the same as the two-level inverter except phase shifted by 30°. It is clear that the modulation process for this reduced set of space vectors becomes essentially identical to that of a conventional two-level inverter except for this phase shift.

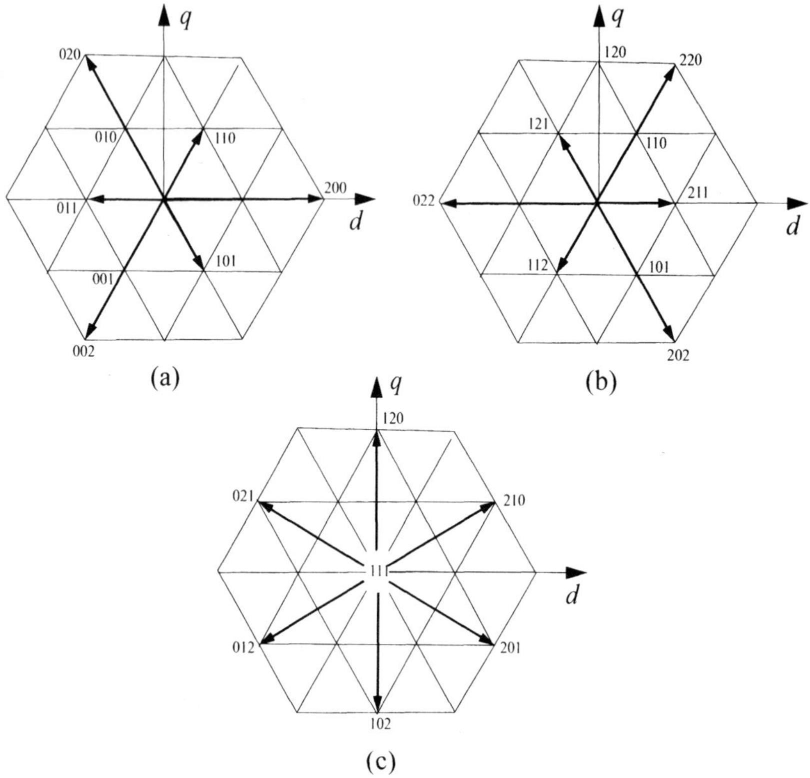

Figure 14.9 Vector states producing a common mode voltage of (a) $-V_{dc}/3+V_{zg}$, (b) $V_{dc}/3+V_{zg}$, and (c) V_{zg}.

The adoption of a PWM scheme based upon the vectors of Figure 14.9 unfortunately results in a reduced reachable region for sinusoidal modulation as shown in Figure 14.10 [7]. The maximum value of phase voltage that can be reached before overmodulation begins is $\sqrt{3}/2V_{dc}$ rather than V_{dc} as occurs when all states are used. In addition, since none of the states of the inner hexagon are used, the WTHD clearly deteriorates to that of a two-level inverter.

The process of eliminating states can be readily extended to inverters with an odd number of levels. Figure 14.11 shows the vector states which again produce a constant common mode voltage of V_{zg} for a five-level inverter, where it can be seen that the five-level inverter reduces to the equivalent of a three-level inverter. Once more the achievable maximum voltage available for continuous non-overmodulated PWM is reduced by $\sqrt{3}/2$.

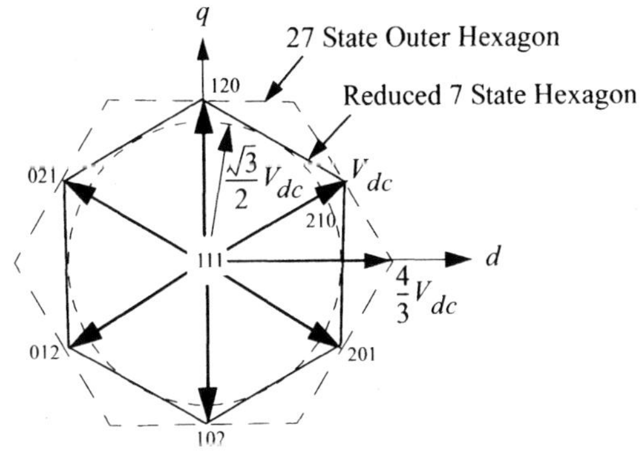

Figure 14.10 Reduced zero common-mode seven-state hexagon for three-level inverter, and achievable modulation region for balanced sinusoidal three-phase voltages.

Inverters with an even number of levels can also be modulated in such a manner as to achieve a constant common mode voltage. However, the results are less satisfactory than for odd levels. In general, switching algorithms can be chosen which will produce a common mode voltage of $(kV_{dc}/3 + V_{zg})$, $k = 0, \ldots, 9$. Many of these possibilities are trivial, since, for example, when $k = 0$ or 9, only vectors at the origin are possible, while values of $k = 1$, 2, 7, and 8 will only produce vectors in the inner hexagons which prevents the possibility of applying the full DC link voltage to the load. Figure 14.12 shows the

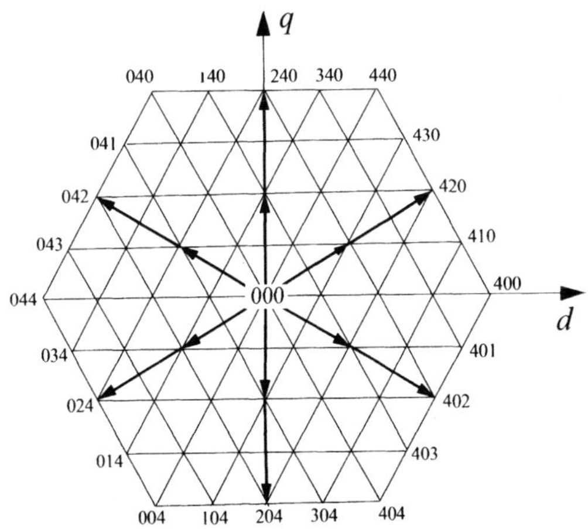

Figure 14.11 Permissible vector states for a five-level inverter that achieve a constant common mode voltage.

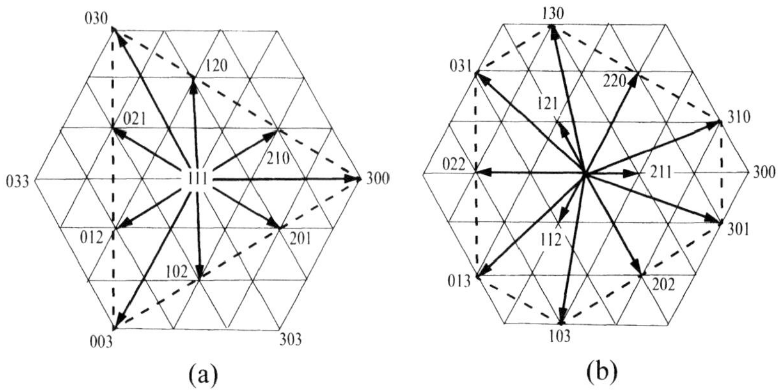

Figure 14.12 Permissible vector states for four-level inverter achieving a constant common mode voltage of (a) $(V_{dc} + V_{zg})$ and (b) $(4/3\,V_{dc} + V_{zg})$.

vectors for $k = 3$ and 4 which will provide common mode voltages of $(V_{dc} + V_{zg})$ and $(4/3\,V_{dc} + V_{zg})$, respectively. While a number of vectors can be identified for each case, the maximum output voltage for continuous sinusoidal modulation is severely restricted. It is readily determined that for $k = 3$ the maximum voltage is $3/2V_{dc}$ while for $k = 4$, the maximum increases to $2V_{dc}$. However, this is still poorer in a per unit sense than for the three-level case. Values of $k = 5$ and 6 are simply vector rotations of $k = 3$ and 4 having the same maximum voltage limitations.

The incompatibility of an inverter with an even number of levels supplying a load with an odd number of levels (three phase) points to an essential problem with control of common mode voltage, particularly with respect to the simple two-level inverter. Since most loads are three phase, common mode voltages inherently will always exist when a two-level inverter is used. It has been shown that the problem can be eliminated if the inverter levels and load phases are both odd. Another possibility, when using a two-level inverter, would be to implement a load with an even number of phases. This option is rarely possible for reasons of economy. However, it has been shown that an extra phase can be synthesized which serves the purpose of a fourth phase for sustaining the PWM switching harmonics but does not require a fundamental load component [8]. Thus the switches for this additional phase can be much smaller in current rating. The circuit concept is shown in Figure 14.13.

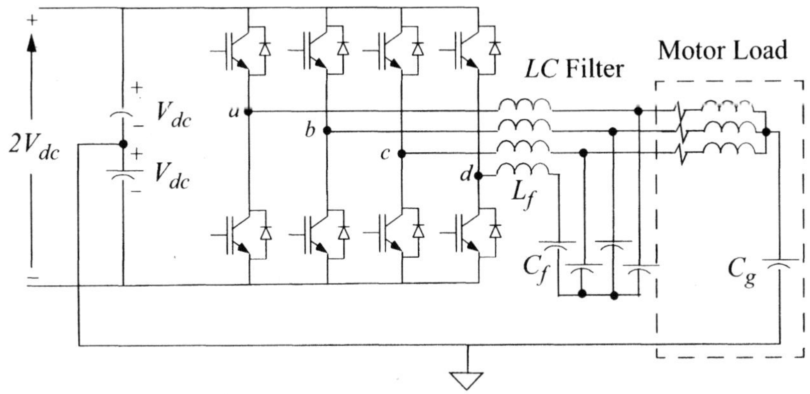

Figure 14.13 Four-leg inverter with second-order *LC* filter and motor load.

14.4 Four Phase Leg Inverter Modulation

The inverter topology shown in Figure 14.13 can also be used with a fully rated fourth phase leg for applications such as an unbalanced three-phase star connected load, where a physical neutral point connection is required. In such applications, the fourth phase leg *d* should be modulated to follow the common mode average of the other three phase legs, with a reference waveform definition given by

$$v_d^* = (v_a^* + v_b^* + v_c^*)/3 \tag{14.48}$$

Clearly, where the main phase leg reference waveforms are simple sinusoids, as defined by Eqs. (5.1), (5.2), and (5.3), phase leg *d* should be switched to achieve a zero average, which means a fixed 50% duty cycle. For modulation using more complex phase reference waveforms, such as the third harmonic injected references defined by Eqs. (5.19), (5.20), and (5.21), the reference waveform for phase leg *d* will be defined by

$$v_d^* = M_3 \cos 3\omega_o t = -\frac{M}{6}\cos 3\omega_o t \tag{14.49}$$

Space vector and the various discontinuous PWM algorithms discussed in Chapter 6 will each have their own average common mode voltage definitions, which can be determined by using Eq. (14.48) with the various phase leg reference waveform definitions given in Table 6.2 for SVM, and Table 6.5 through to Table 6.10 for discontinuous modulation.

Figure 14.14 shows the *l–n* output phase spectra (*a*/*b*/*c* phase to *d* phase) for the cases of simple sinusoidal and third-harmonic injected primary phase leg references. In both cases, the spectra are almost exactly the same as for the primary phase legs alone [Figures 5.4(a) and 5.7(a), respectively], except for a slight reduction in the carrier harmonic, and the elimination of the baseband third harmonic for the case of third-harmonic injection. This is only to be expected since the carrier harmonics of a simple square wave do not have the same magnitude as the carrier harmonics of a sinusoidally modulated system, while the baseband third harmonic should be eliminated from the *l–n* voltage, since it was injected as a common mode component in the first place.

Since there are now 16 inverter switch combinations, the concept of a space vector also needs to be extended to suit the four phase leg inverter. However, because the *l–l* output combinations between the primary three phase legs must remain unchanged, the only option is to split each active space vector into two alternatives — one with the fourth phase leg switched up and the other with the fourth phase leg switched down. Figure 14.15 shows the creation of these space vectors in the first sextant region, where it can be seen how $\overline{SV_2}$ has been split into the components $\overline{SV_{2L}}$ and $\overline{SV_{2U}}$ to reflect the change of state of the fourth phase leg during the $\overline{SV_2}$ interval. Of course, only the largest of the two active space vectors will split into two at any particular point in time, since the transition of the fourth phase leg will always occur during the longest active space vector interval because of the centering effect of Eq. (14.48).

The four phase leg inverter can be represented in a three-dimensional vector space by applying an extension of Eq. (1.37), called a Quad transformation [9], which when scaled for consistency with this text, gives

$$\begin{bmatrix} f_q^s \\ f_d^s \\ f_0 \\ f_z \end{bmatrix} = \sqrt{\frac{2}{3}} \begin{bmatrix} 1 & -\frac{1}{2} & -\frac{1}{2} & 0 \\ 0 & -\frac{\sqrt{3}}{2} & \frac{\sqrt{3}}{2} & 0 \\ \frac{1}{2\sqrt{2}} & \frac{1}{2\sqrt{2}} & \frac{1}{2\sqrt{2}} & -\frac{3}{2\sqrt{2}} \\ \frac{\sqrt{3}}{2\sqrt{2}} & \frac{\sqrt{3}}{2\sqrt{2}} & \frac{\sqrt{3}}{2\sqrt{2}} & -\frac{3\sqrt{3}}{2\sqrt{2}} \end{bmatrix} \begin{bmatrix} f_a \\ f_b \\ f_c \\ f_d \end{bmatrix} \tag{14.50}$$

where f_z is a fill-in variable that can be discarded when three-phase modulation constraints are imposed (as f_0 was discarded for two-level modulation).

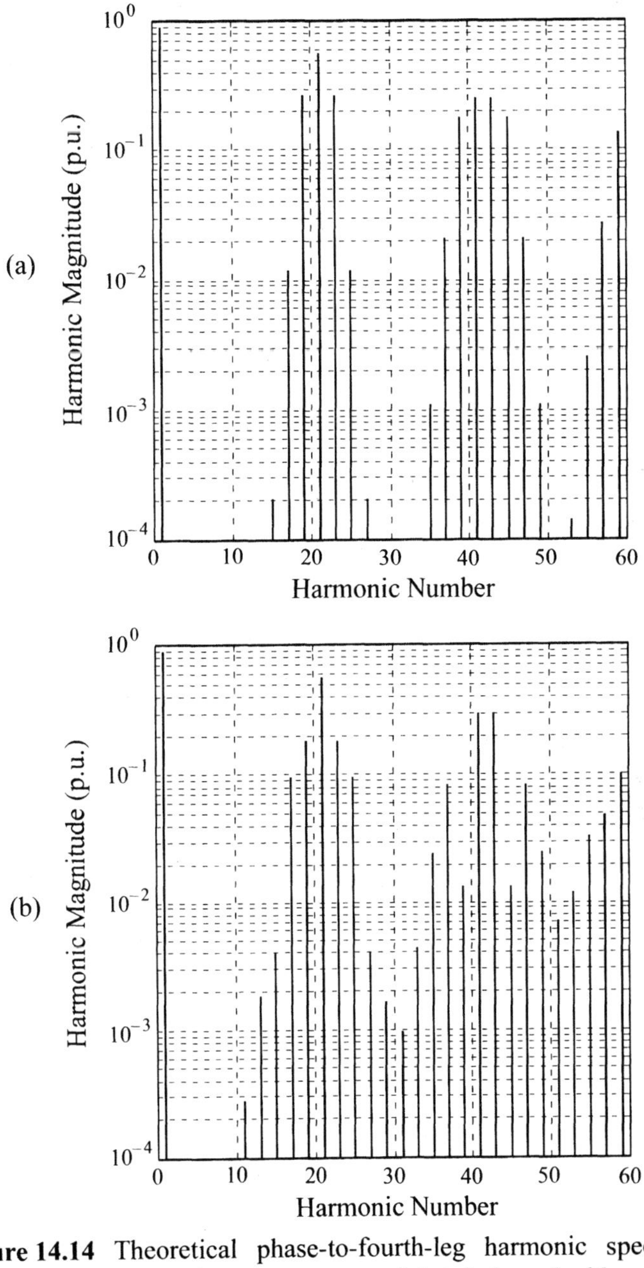

Figure 14.14 Theoretical phase-to-fourth-leg harmonic spectra for four phase leg inverter modulated by *double-edge naturally sampled* PWM: (a) simple sinusoidal references and (b) sinusoidal references with one-sixth third-harmonic injection, $M = 0.9$, $f_c/f_o = 21$.

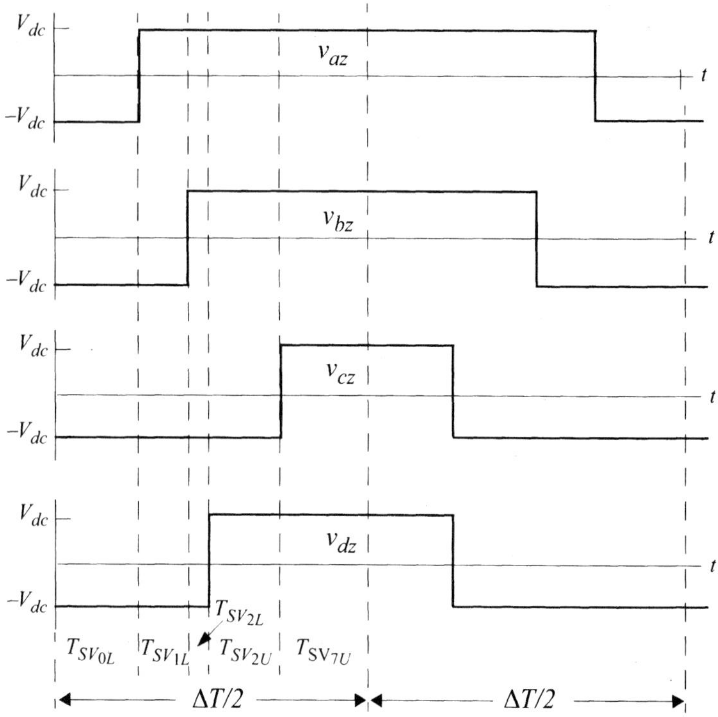

Figure 14.15 Pulse pattern of space vector modulation for four phase leg inverter in the first sextant, $0 \leq \theta_o \leq \pi/3$ with centered active space vectors.

Applying this transformation to the 16 possible switch combinations of the four phase leg inverter leads to the space vector diagram shown in Figure 14.16. This figure also shows the projection of the three-dimensional space vectors back onto the two-dimensional plane, where they match the conventional *l–l* space vectors that have been considered previously. The space vector modulation process of a four phase leg inverter can therefore be considered to be the same process of following a reference phasor that is rotating in a two-dimensional plane along the unit circle shown in Figure 14.16 (or, of course, a reduced diameter circle for reduced modulation indexes), with the alternative switching between the upper and lower space vector options arranged to either maintain a neutral point, or to address another criteria such as reduced common mode voltage (as mentioned in Section 14.3 earlier).

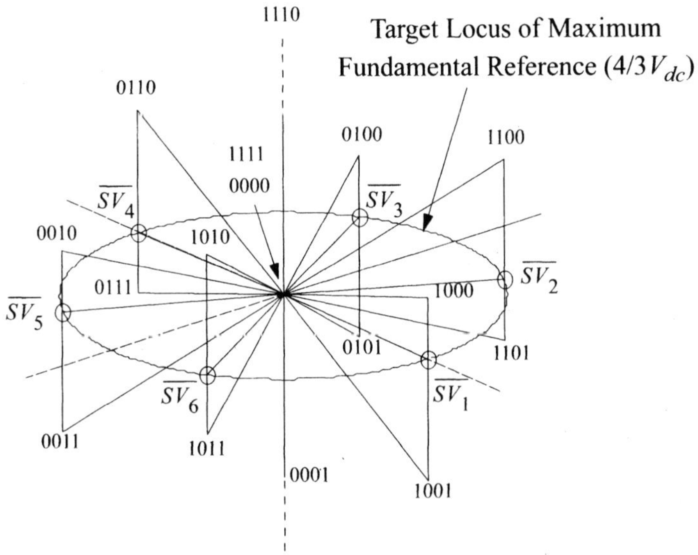

Figure 14.16 Three-dimensional space vector representation of the switching states of a four phase leg inverter.

14.5 Effect of Minimum Pulse Width

Harmonic analysis of three-phase inverter PWM has thus far been evaluated in terms of ideal waveforms. In reality, however, the timing of the output voltage waveforms is not the same as the ideal. The fundamental reason is that the pulse widths of the drive signals provided to the two switching devices in each phase leg are not simple complementary logic. The waveforms must be modified by including a short period of time between turn off of one of the two devices and turn on of the complementary device to prevent both devices conducting simultaneously and thereby short circuiting the DC rail. This disastrous condition is generally called *shoot-through*, and the intentional delay introduced into the switching is referred to as *dead time* or *dwell time*, T_d. The dead time provides a safety zone which allows for device-dependent effects such as transistor storage time to be mitigated.

When the inverter PWM voltage approaches its maximum value, the width of the pulses become increasingly narrow. When any pulse width is reduced to the dead time, it cannot be decreased further, and the minimum pulse width becomes t_d to prevent a shoot-through. This process is termed *lockout*. Further increases in voltages are achieved only by a sudden, discrete elimination of this pulse, a process termed *dropout*. In general, modern fast switching IGBT devices require dead times that do not produce a significant volt–second error during the lockout–dropout process as the inverter heads toward square-wave operation during overmodulation. However, as the power rating of the switches increases, slower IGBT devices or GTOs must be adopted, so that the switching frequency drops progressively and the dead time requirements increase. The switching frequency for multi-megawatt inverters are rarely greater than a few hundred hertz. The pulse dropout effect becomes increasingly important, especially for multilevel inverters where pulse dropouts occur during transitions between levels as well as during transition to square-wave operation [10]. Pulse lockout also prevents the common mode voltage algorithms described in Section 14.2 from achieving a perfectly constant common mode voltage even when common mode elimination algorithms are employed [6].

Consider, for example, a sector of a three-level inverter shown in Figure 14.17. In order to balance the neutral point DC voltage, both middle voltage vectors (V_{100}, V_{211} and V_{110}, V_{221}) must be selected during the time interval ΔT to ensure that current flows both into and out of the link voltage midpoint, thereby providing a means to equalize the two link capacitor voltages. Consider first operation within triangle A of Figure 14.18 which serves to illustrate the conventional PWM sequence. In this case all of the middle voltage states are visited over one half switching period $\Delta T/2$. The voltages are visited in reverse fashion over the next half period. However, when the target voltage vector is near the right-hand side of this triangle, it can be seen that the duration

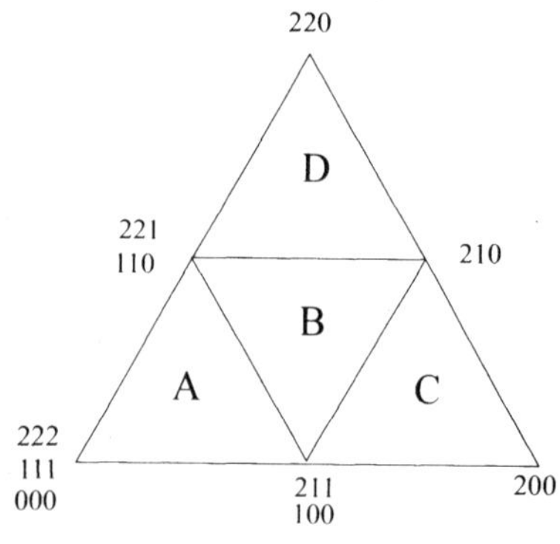

Figure 14.17 One sector of the hexagon for three-level inverter.

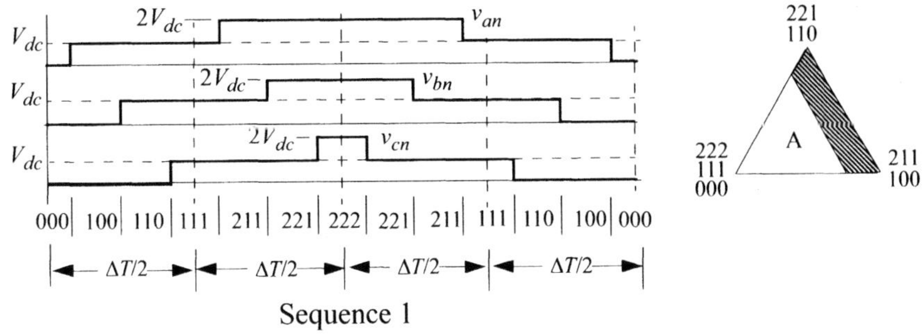

Figure 14.18 First sequence to obtain a voltage vector in triangle A.

of the zero voltage vectors becomes increasingly small. The minimum pulse width limitation cannot be satisfied in phases *a* and *c* along the right-hand edge so that this portion of the triangle cannot be reached with this PWM algorithm.

Two additional switching patterns, shown in Figure 14.19, can be devised which will achieve operation in the shaded region of the triangle of Figure 14.18. It can be noted that state 211 becomes abitrarily small along the left-hand side of triangle A while state 221 becomes small along the bottom border. Both patterns will produce voltages in the right-hand side region of the triangle. Hence, a combination of the three switching patterns will realize voltage

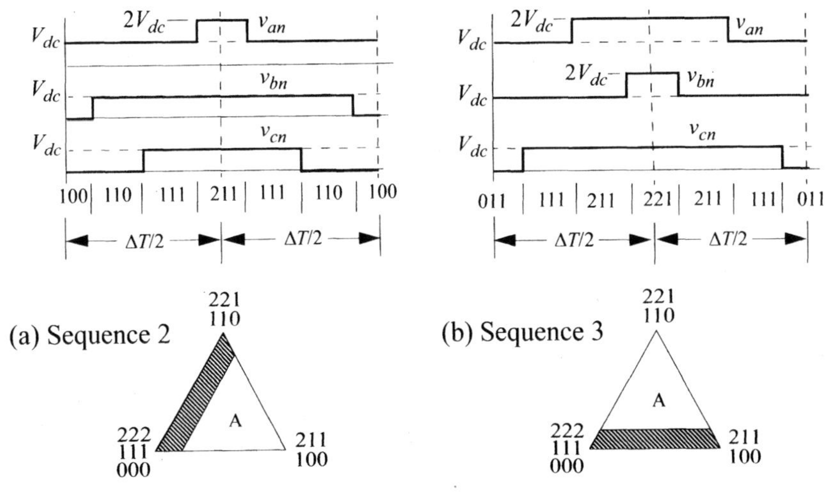

Figure 14.19 (a) Second and (b) third sequences to obtain a voltage vector in triangle A.

vectors at any point within triangle A. It should be noted that only one pair of the two pairs of redundant states are used in each of the two additional cases. As a result only certain switches are active while others do not switch at all. However, this difficulty balances out over a complete cycle of the fundamental.

In a similar manner sequences of voltage vectors can be devised for triangles B, C, and D. Figure 14.20 shows two possible sequences for triangle B. Unreachable regions again occur for the two sequences. However, again by proper selection of the sequence a voltage vector anywhere in the entire voltage triangle can be realized except for a small region near vertex 210. Sequence voltages for triangles C and D are illustrated in Figure 14.21. In this case only one sequence is possible for each of these triangles. Regions of inaccessibility again appear at the outer edge of each triangle corresponding to the region where the transition is occurring from pulse width modulation to square wave operation. This condition is, of course, identical to the case of a simple two-level inverter. The problem can again be alleviated by selective elimination of pulses in the same manner as outlined in Chapter 9. By proper choice of sequences, the minimum pulse limitation can be avoided except for the transition to square wave operation as shown in Figure 14.22. It should be clear that this issue extends to any number of levels of a diode-clamped type multilevel inverter.

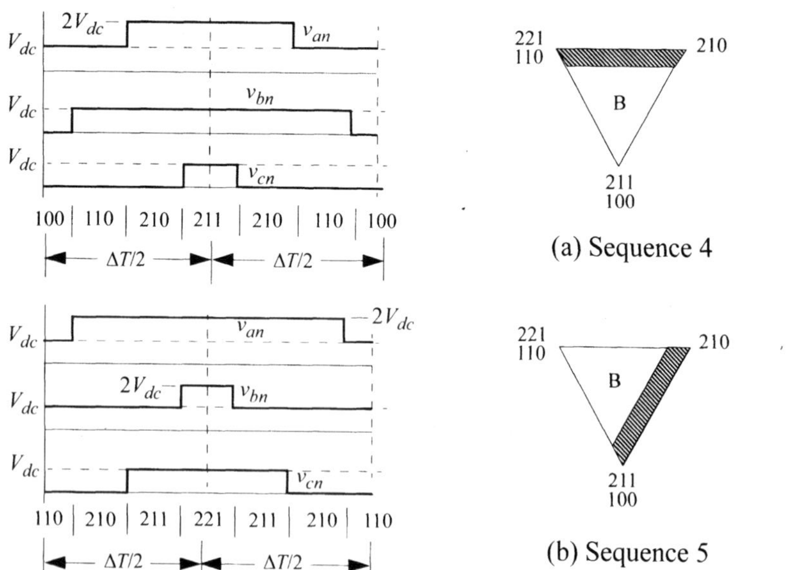

Figure 14.20 Two feasible sequences of voltages for triangle B.

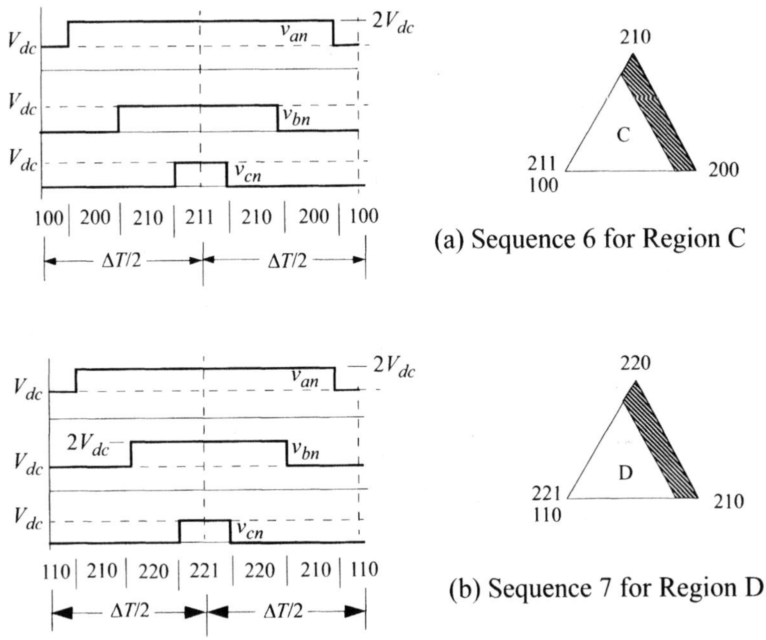

(a) Sequence 6 for Region C

(b) Sequence 7 for Region D

Figure 14.21 Sequences for triangles C and D showing unreachable regions.

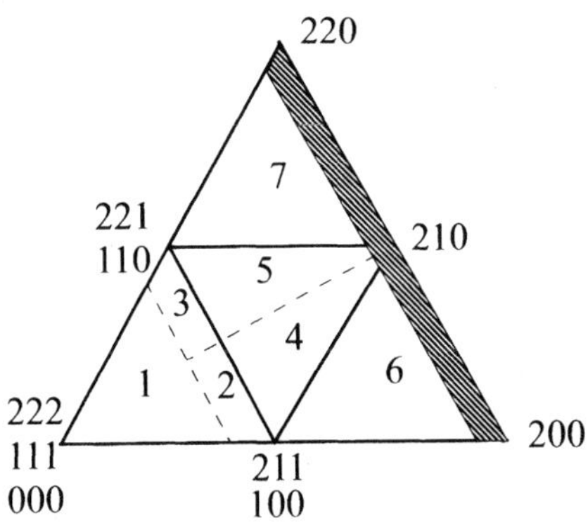

Figure 14.22 Regions of operation in one sector of three-level hexagon showing sequence selection due to minimum pulse width limitations. Remaining unreachable region is shaded.

14.6 PWM Dead-Time Compensation

In addition to invoking minimum pulse width limitations, dead time has another detrimental effect which occurs at every switching event. This effect can be examined by considering one leg of a three-phase PWM inverter (Figure 14.23). Four possible commutation sequences exist as shown in Figure 14.24 [11]. Assuming first that the output current is positive and transistor T_1 transitions from ON to OFF while T_2 transitions from OFF to ON after a slight delay. During the dead zone D_2 conducts and D_1 blocks the flow of current to the positive rail. This condition results in the correct voltage being applied to the output terminals. Second, with the output current still positive, let T_1 transition from OFF to ON while T_2 transitions from ON to OFF. During the dead zone D_2 again conducts and D_1 block current flow to the positive rail. It can be noted that this condition results in a *loss* of voltage since the output remains clamped to the negative rail during the dead time.

Two additional conditions exist when the output current is negative. Assuming first that T_1 transitions from OFF to ON and T_2 from ON to OFF, then D_1 conducts and D_2 blocks current flow. Again the correct voltage is applied to the output. Finally, assume that T_1 changes from ON to OFF and T_2 transitions slightly later from OFF to ON. Again D_1 conducts and D_2 blocks current. However, in this case the output is clamped to the positive bus whereas a transition to the negative bus is desired. Hence, this condition results in a *gain* in voltage at the motor terminals. When a complete cycle is considered the resulting waveforms are as shown in Figure 14.25, where the carrier ratio is only six to more readily illustrate the dead time effect [12]. One should note that the error voltage v_{err} is in phase with the current and therefore has the

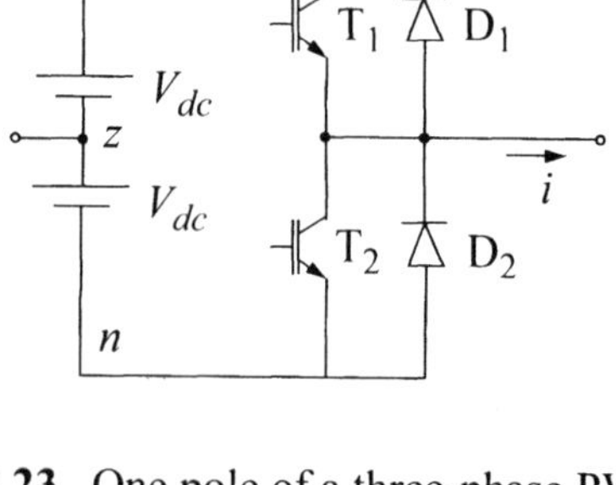

Figure 14.23 One pole of a three-phase PWM inverter.

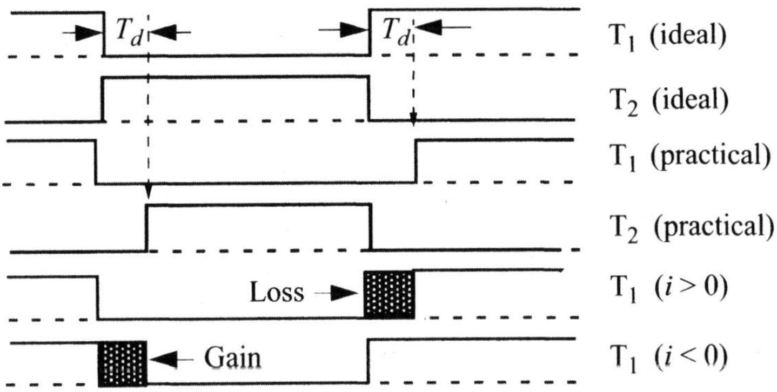

Figure 14.24 Ideal and practical pulse patterns corresponding to devices T1 and T2 for both positive and negative output current.

effect of adding a resistance in series with the load. When the load consists of any type of AC machine it has been shown that stability is adversely affected by the presence of series resistance and can actually produce a system instability resulting in unforced speed and torque oscillations [13,14]. Also, it is apparent that the voltage is essentially rectangular since the width of all error pulses are essentially the same. Therefore, since these pulses must result in an equal and opposite voltage across the load, low-frequency odd-harmonic components will now appear across the load. Since the average deviation for each pulse Δe is measured by

$$\Delta e = 2T_d V_{dc} \tag{14.51}$$

the average voltage deviation over a half cycle of the inverter output is given by

$$\begin{aligned}\Delta V &= \frac{N}{2}\frac{\Delta e}{\Delta T/2} \\ &= \frac{2NT_d V_{dc}}{\Delta T}\end{aligned} \tag{14.52}$$

where N is the number of switchings per cycle and ΔT is one cycle period.

Figure 14.26 illustrates how the average voltage deviation affects the inverter output voltage [15]. The voltage v^* represents the ideal fundamental component of voltage if there were no dead-time effect. Assuming that the inverter feeds an inductive load the current waveform lags v^* by the phase

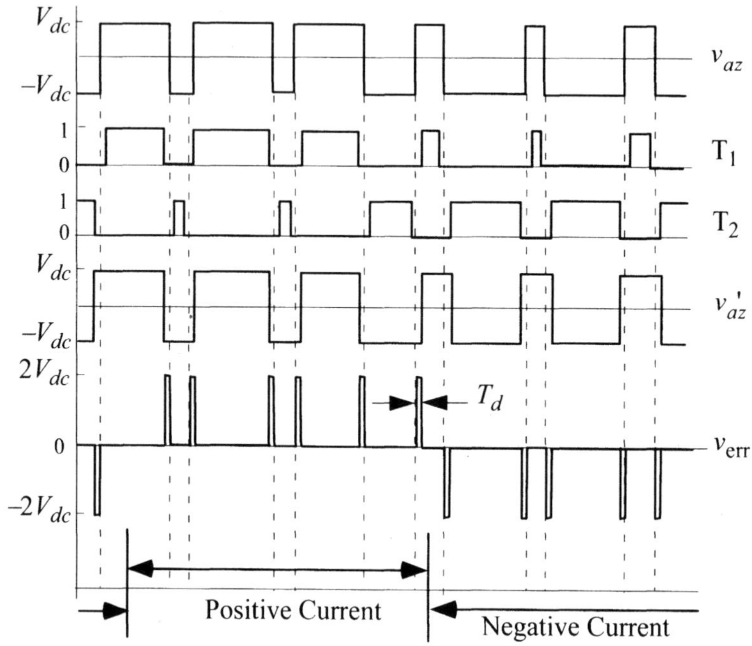

Figure 14.25 Effect of dead time on inverter voltage waveform, v_{az} = ideal switched output voltage with respect to DC link midpoint, T_1 and T_1 are the logic states of the phase leg switches after deadtime compensation, v_{az}' = actual switched output voltage and v_{err} = voltage error ($v_{az} - v_{az}'$).

angle ϕ'. Since the dead time increases (decreases) the output voltage for the negative (positive) half cycle of current, the average voltage deviation can be represented as a square wave Δv that has a magnitude of ΔV. If Δv_1 denotes the fundamental component of Δv, the RMS value of Δv_1 is therefore

$$\Delta V_1 = \frac{2\sqrt{2}}{\pi}\Delta V \tag{14.53}$$

The essential effect of dead time is the superposition of Δv on the ideal reference voltage v^* as illustrated in exaggerated form as the dashed line of Figure 14.26. Hence the effective fundamental component, including dead time v_1, becomes the sum of v^* and Δv_1 and is plotted as the heavy solid curve. If the harmonic components of the current are neglected, the phase displacement ϕ between v_1 and i corresponds to the fundamental power factor angle of the load. It is apparent that the resultant fundamental output voltage differs from the reference wave in both magnitude and phase.

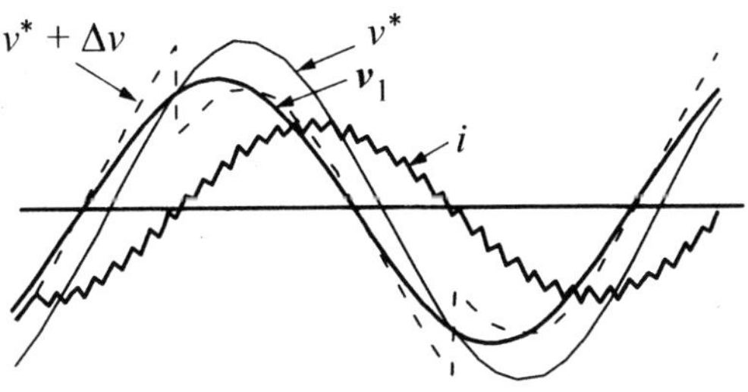

Figure 14.26 Effect of dead time on fundamental output voltage.

The relationships illustrated in Figure 14.26 are described in phasor form in Figure 14.27 [15]. From trigonometry it is readily established that

$$(V_1^*)^2 = \Delta V_1^2 + V_1^2 - 2(\Delta V_1)V_1\cos(\pi - \phi) \tag{14.54}$$

Solving for the actual output voltage V_1 as a per unit of the desired voltage,

$$\frac{V_1}{V_1^*} = -\frac{\Delta V_1}{V_1^*} + \sqrt{1 - \frac{\Delta V_1}{V_1^*}\sin^2\phi} \tag{14.55}$$

Since ΔV_1 is constant for a fixed ratio $N/\Delta T$, T_d, and V_{dc}, the effect of the voltage error becomes more pronounced as the reference voltage amplitude decreases toward zero.

Dividing Eq. (14.53) by the RMS value of the command signal V_1^* and then noting that $\sqrt{2}V_1^*/V_{dc} = M$, results in

$$\frac{\Delta V_1}{V_{dc}} = \frac{8}{\pi}\frac{NT_d}{\Delta T}\frac{1}{M} \tag{14.56}$$

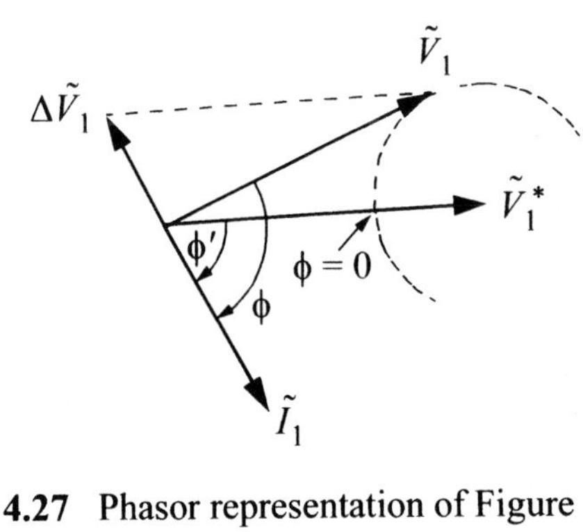

Figure 14.27 Phasor representation of Figure 14.26.

Equation (14.56) can be inserted into Eq. (14.55) to obtain an expression for the per unit output voltage as a function of modulation index. A plot illustrating the effect of dead time as a function of modulation index M and power factor is shown in Figure 14.28 for the case where $\Delta V_1/V_{dc} = 0.05$ $(NT_d/\Delta T = 0.02)$. As the power factor increases the voltage drop becomes smaller. Also, it can be noted that the output voltage decreases to zero when the voltage drop due to dead time is equal to the commanded voltage.

The waveforms of Figure 14.29 show how compensation for dead time can be achieved [12]. Figures 14.29(a) and 14.29(b) show the ideal switching waveforms which are produced for devices T1 and T2 using conventional dead-time implementation. These two logic signals are OR'd and then delayed by a time equal to the dead time T_d in Figures 14.29(c) and 14.29(d). The four signals that are required for compensation are given in Figures 14.29(e) through 14.29(h). Adjusted values of T_1 and T_2 are produced which take on one of two values depending upon whether the current is positive or negative. These signals can be derived by OR and AND logic operations on the ideal signals. The modified logic signals result in the ideal voltage waveform at the output being delayed, however, by a time equal to the dead time. This delay time can be compensated by a suitable lead circuit operating on the voltage com-

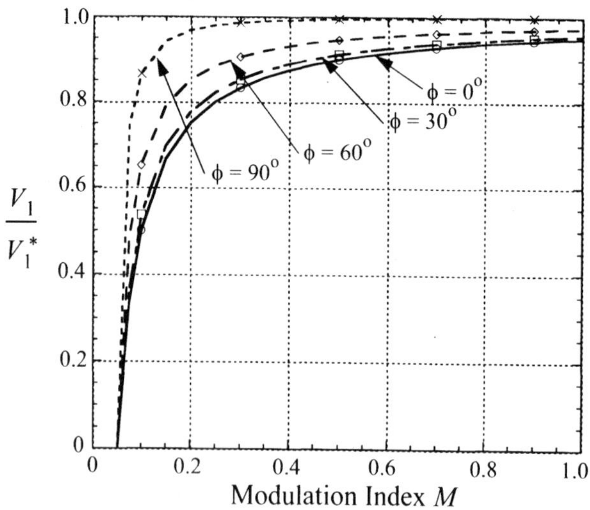

Figure 14.28 Variation of normalized output voltage with modulation index M assuming a 5% voltage drop at rated voltage due to dead time.

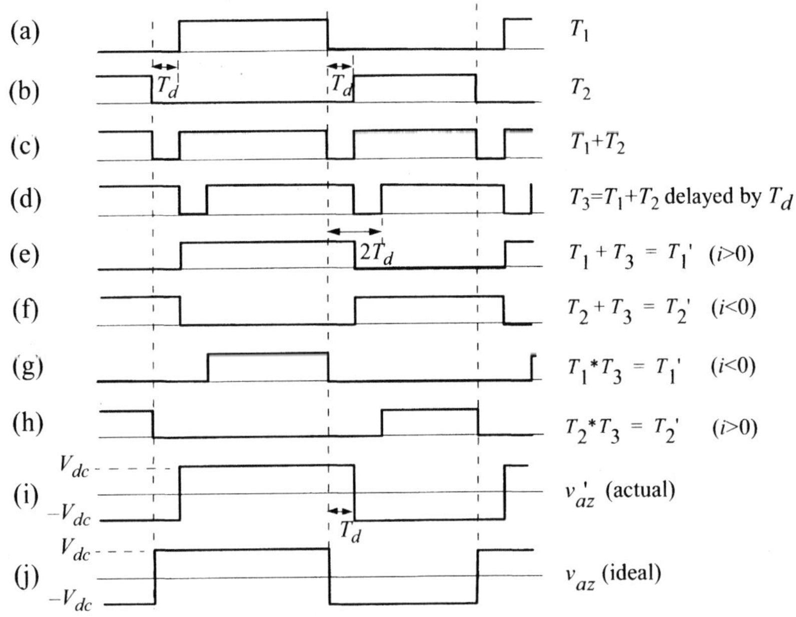

Figure 14.29 Generation of switching waveforms for dead time compensation, where T_1, T_2 = uncompensated gate logic signals, T_d = dead time delay, i = output current, T_1', T_2' = modified gate logic signals, v_{az} (ideal) = target switched phase leg voltage, v'_{az} (actual) = resultant switched phase leg voltage

mand signal. This phase advance procedure is a modification of a technique which has already been discussed in Chapter 3, Figure 3.15. A block diagram of the control logic to implement the scheme is illustrated in Figure 14.30. The comparator and switching block can be readily implemented with operational amplifiers. Measurement of the current simply requires determination of the current polarity which can be accomplished inexpensively. Purely digital implementations of dead-time compensation are also available in the literature [11].

Thus far it has been assumed that the dead time is a known quantity which is programmed into the inverter switching logic. It is important to emphasize that in practice this is not the entire case since turn-off delays occur as a result of recombination within the transistor device. Two additional delay times must be estimated and compensated; the storage time delay t_s and the voltage rise

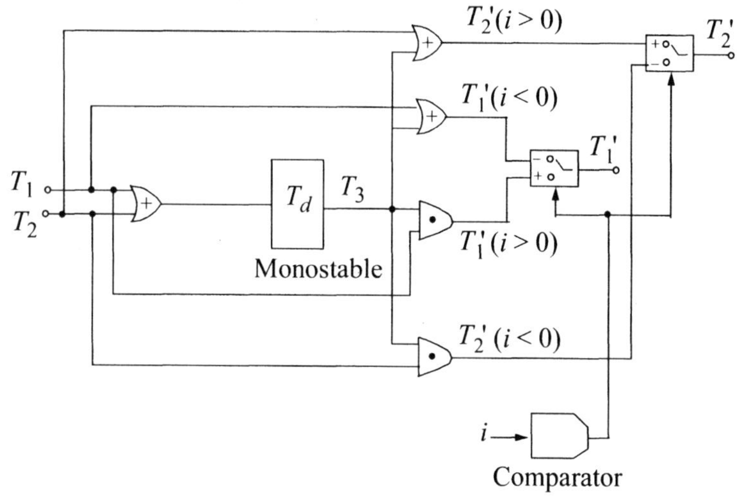

Figure 14.30 Logic circuit for implementing current-dependent logic signals for dead time compensation.

time t_{rv} [16]. The storage time delay is essentially a pure time delay similar to the dead time but is a function of the instantaneous charge present in the device as well as being temperature dependent. A similar but less severe device-dependent delay occurs during the turn-on of the oncoming device. Hence, compensation for the delay time can never be exact without elaborate instrumentation.

While the dead time effect is satisfactorily compensated for all significant values of current, the state of the comparator in Figure 14.30 becomes indeterminate when the current reverses through zero as a result of noise as well as current ripple due to the pulse width modulation process. The result is that the current remains at or near zero for several PWM cycles. This effect is termed *current clamping* [17]. An example of the current clamping effect is shown in Figure 14.31. Both feedback and feedforward techniques have appeared in the literature to combat this effect [18,19]. Both methods require an accurate measurement and regulation of current. However, the subject of closed-loop regulation is outside the scope of this text which has focused exclusively on open-loop pulse width modulation algorithms. Consequently, this issue forms a suitable point of closure for this text. The subject of closed-loop regulation methods will be taken up at length in a future textbook.

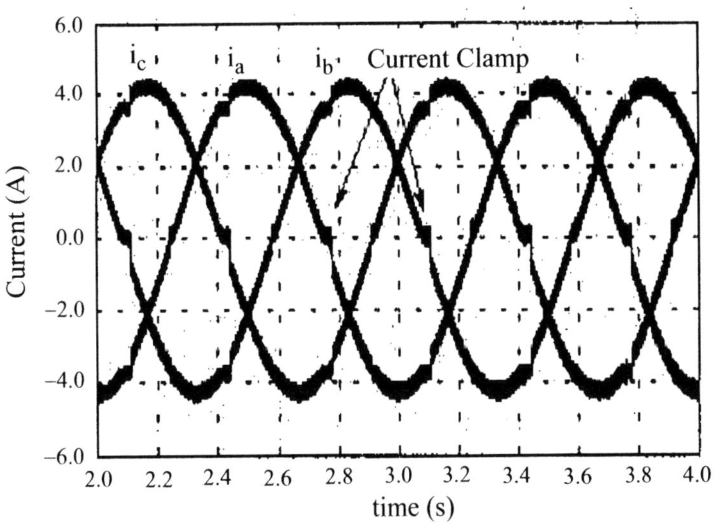

Figure 14.31 Illustration of the current clamping effect due to dead time delay. (Simulation courtesy of R.E. Betz [19].)

14.7 Summary

The subject of pulse width modulation has been an ongoing research activity worldwide for nearly 40 years and the intense interest in this subject shows no sign of abating. Only several of the significant recent developments in the area have been considered in this chapter and the reader is referred to the extensive bibliography for additional information. Surely the development of increasingly powerful microcontrollers and digital signal processors combined with lower cost transistor switches will propel continued activity in the field for the forseeable future. Attention has recently focused on active PWM rectifiers, matrix converters, electromagnetic interference (EMI) control, and various advanced types of current regulation, which are only a few of the interesting directions available for future researchers. This text was written to clearly establish the present status of work in this fascinating field. The future is up to the reader.

References

[1] S. Fukuda and K. Suzuki, "Using harmonic distortion determining factor for harmonic evaluation of carrier-based PWM methods," in *Conf Rec. IEEE Industry Applications Society Annual Mtg*, New Orleans, 1997, pp. 1534–1541.

[2] A.M. Trzynadlowski, S. Legowski, and R.L. Kirlin, "Random pulse width modulation technique for voltage controlled power inverters," in *Conf. Rec. IEEE Industrial Applications Society Annual Meeting,* 1987, pp. 863–868.

[3] T.G. Habetler and D.M. Divan, "Acoustic noise reduction in sinusoidal PWM drives using a randomly modulated carrier," *IEEE Trans. on Power Electronics*, vol. 6, no. 3, July 1991, pp. 356–363.

[4] J.K. Pedersen, F. Blaabjerg, and P.S. Fredericksen, "Reduction of acoustical noise emission in AC machines by intelligent distributed random modulation," in *Conf. Rec. Fifth European Conference on Power Electronics and Applications (EPE)*, Brighton, 1993, vol. 4, pp. 369–375.

[5] D. Vincenti and J. Jin, "A three-phase regulated PWM rectifier with on-line feedforward input unbalance correction," *IEEE Trans. on Industrial Electronics,* vol. 41, no. 5, Oct. 1994, pp. 526–532.

[6] H. Zhang, A. von Jouanne, S. Dai, A.K. Wallace, and F. Wang, "Multilevel inverter modulation schemes to eliminate common-mode voltages," *IEEE Trans. on Industry Applications,* vol. 36, no. 6, Nov./Dec. 2000, pp. 1645–1653 (corrections in vol. 37, no. 1, p3).

[7] P.C. Loh, D.G. Holmes, Y. Fukuta, and T.A. Lipo, "Reduced common mode carrier-based modulation strategies for cascaded multilevel inverters," in *Conf. Rec. Industry Applications Society Annual Conf.*, Pittsburgh, 2002, vol. 3, pp. 2002–2009.

[8] A. Julian, G. Oriti, and T.A. Lipo, "Elimination of common mode voltage in three phase sinusoidal power converters," *IEEE Trans. on Power Electronics*, vol. 15, no. 5, Sept. 1999, pp. 982–989.

[9] M. Ryan, R.D. Lorenz, and R.W. De Doncker, "Modeling of multileg sine-wave inverters: A geometric approach," *IEEE Trans. on Industrial Electronics*, vol. 46, no. 6, Dec. 1999, pp. 1183–1191.

[10] M. Koyama, T. Fujii, R. Uchida, and T. Kawabata, "Space voltage vector based new PWM method for large capacity three-level GTO inverter," in *Conf. Rec. Int. Conf. on Industrial Electronics, Control and Instrumentation (IECON)*, 1992, pp. 271–276.

[11] D. Leggate and R.J. Kerkman, "Pulse-based dead-time compensator for PWM voltage inverters," *IEEE Trans. on Industrial Electronics*, vol. 44, no. 2, April 1997, pp. 191–197.

[12] R.C. Dodson, P.D. Evans, H.T. Yazdi, and S.C. Harley, "Compensating for dead time degradation of PWM inverter waveforms," *IEE Proceedings (London)*, vol. 137, Pt. B, No. 2, Mar. 1990, pp. 73–81.

[13] R.H. Nelson, T.A. Lipo, and P.C. Krause, "Stability analysis of a symmetrical induction machine," *IEEE Trans. on Power Apparatus and Systems*, vol. PAS–88, no. 11, Nov. 1969, pp. 1710–1717.

[14] P.C. Krause and T.A. Lipo, "Analysis and simplified representation of a rectifier-inverter induction motor drive," *IEEE Trans. on Power Apparatus and Systems,* vol. PAS–88, no. 5, May 1969, pp. 588–596.

[15] S.G. Jeong and M.-H. Park, "The analysis and compensation of dead-time effects in PWM inverters," *IEEE Trans. on Industrial Electronics*, vol. 38, no. 2, April 1991, pp. 108–114.

[16] A.R. Munoz and T.A. Lipo, "On-line dead-time compensation technique for open-loop PWM-VSI drives," *IEEE Trans. on Power Electronics*, vol. 14, no. 4, July 1999, pp. 683–689.

[17] Y. Murai, A. Riyanto, H. Nakamura, and K. Matsui, "PWM strategy for high frequency carrier inverters eliminating current-clamps during switching dead-time," in *Conf. Rec. IEEE Industrial Applications Society Annual Meeting*, Houston, 1992, pp. 317–322.

[18] J.W. Choi, S.I. Yong, and S.-K. Sul, "Inverter output voltage synthesis using novel dead time compensation," in *Conf. Rec. IEEE Applied Power Electronics Conf. Record (APEC)*, Orlando, 1994, pp. 100–106.

[19] T. Summers and R.E. Betz, "Dead-time issues in predictive current control," in *Conf. Rec. IEEE Industry Applications Society Annual Meeting*, Pittsburgh, 2002, pp. 2086–2093.

Appendix 1

Fourier Series Representation of a Double Variable Controlled Waveform

The principle of Fourier decomposition is that any regular time-varying waveform $f(t)$ can be expressed as an infinite series of sinusoidal harmonics, viz:

$$f(t) = \frac{a_0}{2} + \sum_{m=1}^{\infty} (a_m \cos m\omega t + b_m \sin m\omega t) \tag{A1.1}$$

where

$$a_m = \frac{1}{\pi} \int_{-\pi}^{\pi} f(t) \cos m\omega t \; d(\omega t) \qquad m = 0, 1, \ldots, \infty \tag{A1.2}$$

$$b_m = \frac{1}{\pi} \int_{-\pi}^{\pi} f(t) \sin m\omega t \; d(\omega t) \qquad m = 1, 2, \ldots, \infty \tag{A1.3}$$

For a waveform $f(x, y)$ which varies as a function of two time variables

$$x(t) = \omega_c t + \theta_c \tag{A1.4}$$

$$y(t) = \omega_o t + \theta_o \tag{A1.5}$$

its Fourier series at a particular value of $y = y_1$ can be written as

$$f(x, y_1) = \frac{a_0(y_1)}{2} + \sum_{m=1}^{\infty} [a_m(y_1) \cos mx + b_m(y_1) \sin mx] \tag{A1.6}$$

where

$$a_m(y_1) = \frac{1}{\pi} \int_{-\pi}^{\pi} f(x, y_1) \cos mx \; dx \qquad m = 0, 1, \ldots, \infty \tag{A1.7}$$

$$b_m(y_1) = \frac{1}{\pi} \int_{-\pi}^{\pi} f(x, y_1) \sin mx \; dx \qquad m = 1, 2, \ldots, \infty \tag{A1.8}$$

The coefficients $a_m(y_1)$ and $b_m(y_1)$ are clearly just particular values of two functions $a_m(y)$ and $b_m(y)$ which vary cyclically over the entire range of y. Since they are cyclic, these functions can also be expressed as Fourier series:

$$a_m(y) = \frac{c_{m0}}{2} + \sum_{n=1}^{\infty} [c_{mn} \cos ny + d_{mn} \sin ny] \quad m = 0, 1, \ldots, \infty \quad \text{(A1.9)}$$

$$b_m(y) = \frac{e_{m0}}{2} + \sum_{n=1}^{\infty} [e_{mn} \cos ny + f_{mn} \sin ny] \quad m = 1, 2, \ldots, \infty \quad \text{(A1.10)}$$

where

$$c_{mn} = \frac{1}{\pi} \int_{-\pi}^{\pi} a_m(y) \cos ny \, dy = \frac{1}{\pi^2} \int_{-\pi}^{\pi} \int_{-\pi}^{\pi} f(x, y) \cos mx \cos ny \, dx \, dy$$

$$m = 0, 1, \ldots, \infty \quad n = 0, 1, \ldots, \infty \quad \text{(A1.11)}$$

$$d_{mn} = \frac{1}{\pi} \int_{-\pi}^{\pi} a_m(y) \sin ny \, dy = \frac{1}{\pi^2} \int_{-\pi}^{\pi} \int_{-\pi}^{\pi} f(x, y) \cos mx \sin ny \, dx \, dy$$

$$m = 0, 1, \ldots, \infty \quad n = 1, 2, \ldots, \infty \quad \text{(A1.12)}$$

$$e_{mn} = \frac{1}{\pi} \int_{-\pi}^{\pi} b_m(y) \cos ny \, dy = \frac{1}{\pi^2} \int_{-\pi}^{\pi} \int_{-\pi}^{\pi} f(x, y) \sin mx \cos ny \, dx \, dy$$

$$m = 1, 2, \ldots, \infty \quad n = 0, 1, \ldots, \infty \quad \text{(A1.13)}$$

$$f_{mn} = \frac{1}{\pi} \int_{-\pi}^{\pi} b_m(y) \sin ny \, dy = \frac{1}{\pi^2} \int_{-\pi}^{\pi} \int_{-\pi}^{\pi} f(x, y) \sin mx \sin ny \, dx \, dy$$

$$m = 1, 2, \ldots, \infty \quad n = 1, 2, \ldots, \infty \quad \text{(A1.14)}$$

Examining these equations, it is seen that the solution obtained is general for any time t since the x and y values are periodic.

Equations (A1.11) to (A1.14) expand using trigonometric identities to

$$c_{mn} = \frac{1}{2\pi^2}\int_{-\pi}^{\pi}\int_{-\pi}^{\pi} f(x,y)\cos(mx+ny)\,dx\,dy + \frac{1}{2\pi^2}\int_{-\pi}^{\pi}\int_{-\pi}^{\pi} f(x,y)\cos(mx-ny)\,dx\,dy$$

$$m = 0, 1, \ldots, \infty \qquad n = 0, 1, \ldots, \infty \qquad \text{(A1.15)}$$

$$d_{mn} = \frac{1}{2\pi^2}\int_{-\pi}^{\pi}\int_{-\pi}^{\pi} f(x,y)\sin(mx+ny)\,dx\,dy - \frac{1}{2\pi^2}\int_{-\pi}^{\pi}\int_{-\pi}^{\pi} f(x,y)\sin(mx-ny)\,dx\,dy$$

$$m = 0, 1, \ldots, \infty \qquad n = 1, 2, \ldots, \infty \qquad \text{(A1.16)}$$

$$e_{mn} = \frac{1}{2\pi^2}\int_{-\pi}^{\pi}\int_{-\pi}^{\pi} f(x,y)\sin(mx+ny)\,dx\,dy + \frac{1}{2\pi^2}\int_{-\pi}^{\pi}\int_{-\pi}^{\pi} f(x,y)\sin(mx-ny)\,dx\,dy$$

$$m = 1, 2, \ldots, \infty \qquad n = 0, 1, \ldots, \infty \qquad \text{(A1.17)}$$

$$f_{mn} = \frac{1}{2\pi^2}\int_{-\pi}^{\pi}\int_{-\pi}^{\pi} f(x,y)\cos(mx-ny)\,dx\,dy - \frac{1}{2\pi^2}\int_{-\pi}^{\pi}\int_{-\pi}^{\pi} f(x,y)\cos(mx+ny)\,dx\,dy$$

$$m = 1, 2, \ldots, \infty \qquad n = 1, 2, \ldots, \infty \qquad \text{(A1.18)}$$

The general Fourier component form for $f(x,y)$ can now be written as

$$f(x,y) = \frac{c_{00}}{4} + \frac{1}{2}\sum_{n=1}^{\infty}[c_{0n}\cos ny + d_{0n}\sin ny] \qquad \text{(A1.19)}$$

$$+ \frac{1}{2}\sum_{m=1}^{\infty}[c_{m0}\cos mx + e_{m0}\sin mx]$$

$$+ \sum_{m=1}^{\infty}\sum_{n=1}^{\infty}\left[\begin{array}{l}(c_{mn}\cos ny + d_{mn}\sin ny)\cos mx \\ \qquad + (e_{mn}\cos ny + f_{mn}\sin ny)\sin mx\end{array}\right]$$

Again by the use of trigonometric identities, Eq. (A1.19) can be rearranged to

$$f(x, y) = \frac{c_{00}}{4} + \frac{1}{2}\sum_{n=1}^{\infty}[c_{0n}\cos ny + d_{0n}\sin ny] \tag{A1.20}$$

$$+\frac{1}{2}\sum_{m=1}^{\infty}[c_{m0}\cos mx + e_{m0}\sin mx]$$

$$+\sum_{m=1}^{\infty}\sum_{n=1}^{\infty}\begin{bmatrix}\frac{1}{2}[c_{mn}\cos(mx+ny) + c_{mn}\cos(mx-ny)] \\ +\frac{1}{2}[d_{mn}\sin(mx+ny) - d_{mn}\sin(mx-ny)] \\ +\frac{1}{2}[e_{mn}\sin(mx+ny) + e_{mn}\sin(mx-ny)] \\ +\frac{1}{2}[f_{mn}\cos(mx-ny) - f_{mn}\cos(mx+ny)]\end{bmatrix}$$

which can be combined to form

$$f(x, y) = \frac{c_{00}}{4} + \frac{1}{2}\sum_{n=1}^{\infty}[c_{0n}\cos ny + d_{0n}\sin ny] \tag{A1.21}$$

$$+\frac{1}{2}\sum_{m=1}^{\infty}[c_{m0}\cos mx + e_{m0}\sin mx]$$

$$+\frac{1}{2}\sum_{m=1}^{\infty}\sum_{n=1}^{\infty}\begin{bmatrix}(c_{mn}-f_{mn})\cos(mx+ny) + (e_{mn}+d_{mn})\sin(mx+ny) \\ +(c_{mn}+f_{mn})\cos(mx-ny) + (e_{mn}-d_{mn})\sin(mx-ny)\end{bmatrix}$$

From Eqs. (A1.15) to (A1.18) the coefficients of the last part of Eq. (A1.21) for $m = 1, 2, \ldots, \infty$, $n = 1, 2, \ldots, \infty$ are

$$\frac{1}{2}(c_{mn}-f_{mn}) = \frac{1}{2\pi^2}\int_{-\pi}^{\pi}\int_{-\pi}^{\pi} f(x, y)\cos(mx+ny)\,dx\,dy \tag{A1.22}$$

$$\frac{1}{2}(c_{mn}+f_{mn}) = \frac{1}{2\pi^2}\int_{-\pi}^{\pi}\int_{-\pi}^{\pi} f(x, y)\cos(mx-ny)\,dx\,dy \tag{A1.23}$$

$$\frac{1}{2}(e_{mn}+d_{mn}) = \frac{1}{2\pi^2}\int_{-\pi}^{\pi}\int_{-\pi}^{\pi} f(x,y)\sin(mx+ny)\,dx\,dy \tag{A1.24}$$

$$\frac{1}{2}(e_{mn}-d_{mn}) = \frac{1}{2\pi^2}\int_{-\pi}^{\pi}\int_{-\pi}^{\pi} f(x,y)\sin(mx-ny)\,dx\,dy \tag{A1.25}$$

Finally, it can be seen from Eq. (A1.12) and Eq. (A1.14) that for any given value of n, $d_{mn} = -d_{m(-n)}$ and $f_{mn} = -f_{m(-n)}$ so that the $(mx - ny)$ terms in Eq. (A1.21) can alternately be obtained by summing the $(mx + ny)$ terms over negative n.

Hence the complete solution for $f(x, y)$ becomes

$$\begin{aligned} f(x,y) = \frac{c_{00}}{4} &+ \frac{1}{2}\sum_{n=1}^{\infty}[c_{0n}\cos ny + d_{0n}\sin ny] \\ &+ \frac{1}{2}\sum_{m=1}^{\infty}[c_{m0}\cos mx + e_{m0}\sin mx] \\ &+ \frac{1}{2}\sum_{m=1}^{\infty}\sum_{\substack{n=-\infty \\ (n\neq 0)}}^{\infty}\left[\begin{matrix}(c_{mn}-f_{mn})\cos(mx+ny) \\ \qquad +(d_{mn}+e_{mn})\sin(mx+ny)\end{matrix}\right] \end{aligned} \tag{A1.26}$$

This result can be cast into complex form by multiplying Eq. (A1.24) by j and adding to Eq. (A1.22), to define the resultant coefficients as

$$\bar{C}_{mn} = A_{mn} + jB_{mn} = \frac{(c_{mn}-f_{mn})}{2} + j\frac{(d_{mn}+e_{mn})}{2} \tag{A1.27}$$

where

$$\begin{aligned} A_{mn}+jB_{mn} &= \frac{1}{2\pi^2}\int_{-\pi}^{\pi}\int_{-\pi}^{\pi} f(x,y)[\cos(mx+ny)+j\sin(mx+ny)]\,dx\,dy \\ &= \frac{1}{2\pi^2}\int_{-\pi}^{\pi}\int_{-\pi}^{\pi} f(x,y)e^{j(mx+ny)}\,dx\,dy \end{aligned} \tag{A1.28}$$

Equation (A1.26) can now be expressed in terms of these alternative coefficients as

$$f(x,y)=\frac{A_{00}}{2}+\sum_{n=1}^{\infty}[A_{0n}\cos ny+B_{0n}\sin ny]+\sum_{m=1}^{\infty}[A_{m0}\cos mx+B_{m0}\sin mx] \tag{A1.29}$$

$$+\sum_{m=1}^{\infty}\sum_{\substack{n=-\infty\\(n\neq 0)}}^{\infty}[A_{mn}\cos(mx+ny)+B_{mn}\sin(mx+ny)]$$

Finally, Eq. (A1.29) can be expressed in terms of time by substituting from Eq. (A1.4) and Eq. (A1.5), to become

$$f(t)=\frac{A_{00}}{2}+\sum_{n=1}^{\infty}[A_{0n}\cos(n[\omega_o t+\theta_o])+B_{0n}\sin(n[\omega_o t+\theta_o])] \tag{A1.30}$$

$$+\sum_{m=1}^{\infty}[A_{m0}\cos(m[\omega_c t+\theta_c])+B_{m0}\sin(m[\omega_c t+\theta_c])]$$

$$+\sum_{m=1}^{\infty}\sum_{\substack{n=-\infty\\(n\neq 0)}}^{\infty}\left[\begin{aligned}&A_{mn}\cos(m[\omega_c t+\theta_c]+n[\omega_o t+\theta_o])\\&\qquad+B_{mn}\sin(m[\omega_c t+\theta_c]+n[\omega_o t+\theta_o])\end{aligned}\right]$$

Equation (A1.30) is the form commonly used in the literature [1]. The first line defines the DC component, the fundamental component (defined when $n = 1$) and baseband harmonic components which are integer multiples of the fundamental component. The second line defines carrier harmonic components as multiples of the modulation frequency. The third line defines sideband harmonic components displaced either side of the main carrier harmonics by integer multiples of the fundamental frequency component.

Note that in the general form of Eq. (A1.30), the angular relationship between the baseband harmonics including the fundamental, and the carrier and sideband harmonics, is arbitrary, by virtue of the phase offset angles θ_c and θ_o defined in Eqs. (A1.4) and (A1.5).

Reference

[1] S.R. Bowes, "New sinusoidal pulse-width modulated inverter," *IEE Proceedings (London)*, vol. 122, no. 11, Nov. 1975, pp. 1279–1285.

Appendix 2

Jacobi–Anger and Bessel Function Relationships

A2.1 Jacobi–Anger Expansions

A common expression that appears in the development of the analytical Fourier solution of a PWM waveform under various modulation strategies is $e^{\pm j\xi\cos\theta}$, i.e., a sinusoid which has as its argument another sinusoid. In the development of the analytical solutions for PWM, this expression must be multiplied by the term $\cos n\theta$ and integrated over some fraction of 2π, and this is facilitated by expanding the $e^{\pm j\xi\cos\theta}$ expression into a Bessel series form using the Jacobi–Anger expansion [1] of

$$e^{\pm j\xi\cos\theta} = J_0(\xi) + 2\sum_{k=1}^{\infty} j^{\pm k} J_k(\xi)\cos k\theta \tag{A2.1}$$

Equation (A2.1) can be restated from its complex form as

$$\cos(\xi\cos\theta) = J_0(\xi) + 2\sum_{k=1}^{\infty} \cos k\frac{\pi}{2} J_k(\xi)\cos k\theta$$

$$= J_0(\xi) + 2\sum_{k=1}^{\infty} \cos k\pi J_{2k}(\xi)\cos 2k\theta \tag{A2.2}$$

and

$$\sin(\xi\cos\theta) = 2\sum_{k=1}^{\infty} \sin k\frac{\pi}{2} J_k(\xi)\cos k\theta$$

$$= 2\sum_{k=0}^{\infty} \cos k\pi J_{2k+1}(\xi)\cos([2k+1]\theta) \tag{A2.3}$$

Additionally, using the substitution of $\theta = \theta' - \pi/2$, Eqs. (A2.2) and (A2.3) lead to the following relationships of

$$\cos(\xi \sin\theta') = J_0(\xi) + 2\sum_{k=1}^{\infty} J_{2k}(\xi)\cos(2k\theta') \tag{A2.4}$$

and

$$\sin(\xi \sin\theta') = 2\sum_{k=0}^{\infty} J_{2k+1}(\xi)\sin([2k+1]\theta') \tag{A2.5}$$

A further useful result is the expansion of the expression $\cos(\delta + \xi\sin\theta)$, which proceeds as follows:

$$\cos(\delta + \xi\sin\theta) = \cos\delta\cos(\xi\sin\theta) - \sin\delta\sin(\xi\sin\theta) \tag{A2.6}$$

Substituting from Eqs. (A2.4) and (A2.5) into Eq. (A2.6) gives

$$\begin{aligned}
\cos(\delta + \xi\sin\theta) = {} & \cos\delta\left[J_0(\xi) + 2\sum_{k=1}^{\infty} J_{2k}(\xi)\cos(2k\theta)\right] \\
& - \sin\delta\left[2\sum_{k=0}^{\infty} J_{2k+1}(\xi)\sin([2k+1]\theta)\right] \\
= {} & J_0(\xi)\cos\delta + \sum_{k=1}^{\infty} J_{2k}(\xi)\{\cos(\delta + 2k\theta) + \cos(\delta - 2k\theta)\} \qquad \text{(A2.7)} \\
& + \sum_{k=0}^{\infty} J_{2k+1}(\xi)\{\cos(\delta + [2k+1]\theta) - \cos(\delta - [2k+1]\theta)\}
\end{aligned}$$

Grouping like sinusoids and after some manipulation and simplification gives the final form of

$$\cos(\delta + \xi\sin\theta) = \sum_{k=-\infty}^{\infty} J_k(\xi)\cos(\delta + k\theta) \tag{A2.8}$$

A2.2 Bessel Function Integral Relationships

The product of $e^{j\xi\cos\theta}$ and $\cos(n\theta)$ integrated over a 2π interval is

$$\int_{-\pi}^{\pi} e^{\pm j\xi\cos\theta}\cos n\theta\, d\theta = \int_{-\pi}^{\pi}\left[J_0(\xi) + 2\sum_{k=1}^{\infty} j^{\pm k} J_k(\xi)\cos k\theta\right]\cos n\theta\, d\theta \qquad \text{(A2.9)}$$

which becomes

$$\int_{-\pi}^{\pi} e^{\pm j\xi\cos\theta}\cos n\theta\, d\theta = \int_{-\pi}^{\pi} J_0(\xi)\cos n\theta\, d\theta + \int_{-\pi}^{\pi} 2\sum_{k=1}^{\infty} j^{\pm k} J_k(\xi)\cos k\theta\cos n\theta\, d\theta \qquad \text{(A2.10)}$$

The only term of the right-hand side of Eq. (A2.10) which does not integrate to zero over 2π is when $k = n$, so that Eq. (A2.10) becomes

$$\int_{-\pi}^{\pi} e^{\pm j\xi\cos\theta}\cos n\theta\, d\theta = \int_{-\pi}^{\pi} 2j^{\pm n} J_n(\xi)\cos n\theta\, d\theta \qquad \text{(A2.11)}$$

which simplifies to

$$\int_{-\pi}^{\pi} e^{\pm j\xi\cos\theta}\cos n\theta\, d\theta = 2\pi j^{\pm n} J_n(\xi) \qquad \text{(A2.12)}$$

Using a similar development, $e^{\pm j\xi\cos\theta}$ multiplied by $\sin n\theta$ and integrated over 2π gives

$$\int_{-\pi}^{\pi} e^{\pm j\xi\cos\theta}\sin n\theta\, d\theta = 0 \qquad \text{(A2.13)}$$

Hence from Eq. (A2.12), for any positive value of n and positive exponent of the natural logarithm,

$$\int_{-\pi}^{\pi} e^{j\xi\cos\theta} e^{jn\theta}\, d\theta = 2\pi j^n J_n(\xi) \qquad \text{(A2.14)}$$

In particular, when $n = 0$,

$$\int_{-\pi}^{\pi} e^{j\xi\cos\theta}\, d\theta = 2\pi J_0(\xi) \qquad \text{(A2.15)}$$

An additional useful identity can be developed for n having a negative value. Looking again at Eq. (A2.10), the only nonzero term on the right-hand side of the integration over 2π will occur when $k = |n|$, irrespective of the sign of n. Hence from Eq. (A2.12), for n negative

$$\int_{-\pi}^{\pi} e^{j\xi\cos\theta}\cos(-n\theta)d\theta = 2\pi j^{|n|}J_{|n|}(\xi) = 2\pi j^{-n}J_{-n}(\xi) \tag{A2.16}$$

From Eq. (A2.16), the following identity can be stated:

$$J_{-n}(\xi) = (-1)^n J_n(\xi) \tag{A2.17}$$

Another important identity can be derived for the case where the exponent of the natural logarithm is negative, that is, taking the case of a negative exponent of the natural logarithm from Eq. (A2.12)

$$\int_{-\pi}^{\pi} e^{-j\xi\cos\theta}\cos n\theta\, d\theta = 2\pi j^{-n}J_n(\xi) \tag{A2.18}$$

Equation (A2.18) can be alternatively written as

$$\int_{-\pi}^{\pi} e^{-j\xi\cos\theta}\cos n\theta\, d\theta = 2\pi j^{n}J_n(-\xi) = 2\pi j^{-n}J_n(\xi) \tag{A2.19}$$

from which the following identity can be stated

$$J_n(-\xi) = (-1)^n J_n(\xi) \tag{A2.20}$$

In particular when $n = 0$,

$$J_0(-\xi) = J_0(\xi) \tag{A2.21}$$

Figure A2.1 shows the shape of the first seven Bessel functions for arguments up to 10. Note that only the first Bessel function $J_0(\xi)$ has a nonzero value for an argument of zero.

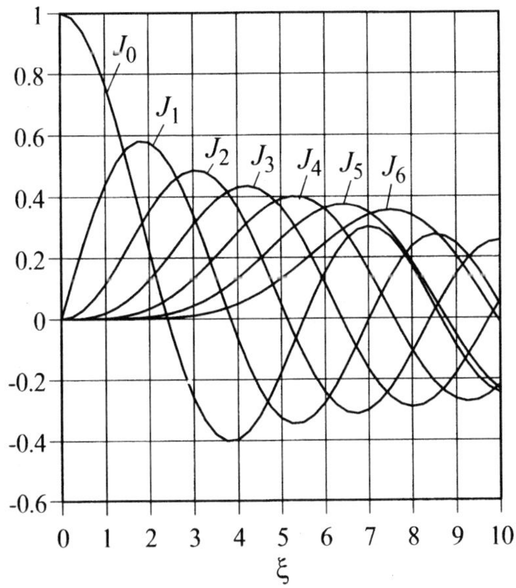

Figure A2.1 Bessel functions $J_n(\xi)$ for $n = 0, 1, \ldots, 6$.

Reference

[1] G.N. Watson, *Theory of Bessel functions*, Cambridge University Press, Cambridge, England, 1944.

Appendix 3

Three-Phase and Half-Cycle Symmetry Relationships

Assume that the phase (and consequently line) voltages of a three-phase system satisfy half-cycle and three-phase symmetry but are otherwise arbitrary steady state functions of time. These voltages have the form [1]

$$v_a = v(\theta) \tag{A3.1}$$

$$v_b = v\left(\theta - \frac{2\pi}{3}\right) \tag{A3.2}$$

$$v_c = v\left(\theta + \frac{2\pi}{3}\right) \tag{A3.3}$$

where $\theta = \omega t$ and the angular frequency is constant. By half-cycle symmetry

$$v_a(\theta + \pi) = -v_a(\theta) \tag{A3.4}$$

$$v_b(\theta + \pi) = -v_b(\theta) \tag{A3.5}$$

$$v_c(\theta + \pi) = -v_c(\theta) \tag{A3.6}$$

and by three-phase symmetry

$$v_b(\theta + \pi) = v_a\left(\theta + \frac{\pi}{3}\right) \tag{A3.7}$$

$$v_c(\theta + \pi) = v_b\left(\theta + \frac{\pi}{3}\right) \tag{A3.8}$$

$$v_a(\theta + \pi) = v_c\left(\theta + \frac{\pi}{3}\right) \tag{A3.9}$$

Hence, if the solution for a balanced three-phase system is known over the interval $0 \le \theta \le \pi/3$, by combining Eqs. (A3.4) to (A3.9) the solution for the next interval, $\pi/3 \le \theta + \pi/3 \le 2\pi/3$, can be determined as

$$v_a\left(\theta + \frac{\pi}{3}\right) = -v_b(\theta) \tag{A3.10}$$

$$v_b\left(\theta + \frac{\pi}{3}\right) = -v_c(\theta) \tag{A3.11}$$

$$v_c\left(\theta + \frac{\pi}{3}\right) = -v_a(\theta) \tag{A3.12}$$

where $v_a(\theta)$, $v_b(\theta)$, and $v_c(\theta)$ are known from Eqs. (A3.1) to (A3.3).

The solution for the remaining four 60° intervals of a complete fundamental cycle are readily obtained by progressively increasing the arguments of Eqs. (A3.10) to (A3.12).

Reference

[1] T.A. Lipo, "The analysis of induction motors with voltage control by symmetrically triggered thyristors," *IEEE Trans. on Power Apparatus and Systems,* vol. PAS–90, no. 2, Mar./April 1971, pp. 515–525.

Appendix 4

Overmodulation of a Single-Phase Leg

A4.1 Naturally Sampled Double-Edge PWM

Figure A4.1 shows the unit cell for naturally sampled PWM under overmodulation conditions.

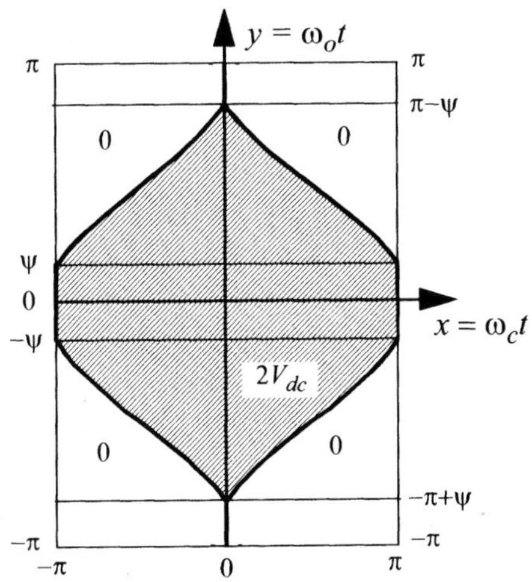

Figure A4.1 Unit cell for double-edge naturally sampled two-level PWM for an overmodulation condition, $\psi = \cos^{-1}(1/M)$.

The integral defining the harmonic coefficients over this unit cell is

$$C_{mn} = \frac{V_{dc}}{\pi^2}\left[\int_{-\pi+\psi}^{-\psi}\int_{-\frac{\pi}{2}1+M\cos y}^{\frac{\pi}{2}1+M\cos y} e^{jmx+ny}\,dx\,dy + \int_{-\psi}^{\psi}\int_{-\pi}^{\pi} e^{jmx+ny}\,dx\,dy + \int_{\psi}^{\pi-\psi}\int_{-\frac{\pi}{2}(1+M\cos y)}^{\frac{\pi}{2}(1+M\cos y)} e^{j(mx+ny)}\,dx\,dy\right] \qquad (A4.1)$$

where M is the modulation index and $\psi = \cos^{-1}(1/M)$.

A4.1.1 Evaluation of Double Fourier Integral for Overmodulated Naturally Sampled PWM

For $m = n = 0$, this integral is clearly real ($B_{00} = 0$) and can be written as

$$C_{00} = \frac{V_{dc}}{\pi^2}\left[\int_{-\pi+\psi}^{-\psi}\int_{-\frac{\pi}{2}(1+M\cos y)}^{\frac{\pi}{2}(1+M\cos y)} dx\,dy + \int_{-\psi}^{\psi}\int_{-\pi}^{\pi} dx\,dy + \int_{\psi}^{\pi-\psi}\int_{-\frac{\pi}{2}(1+M\cos y)}^{\frac{\pi}{2}(1+M\cos y)} dx\,dy\right] \tag{A4.2}$$

which can be integrated to form

$$C_{00} = \frac{V_{dc}}{\pi^2}\left[\int_{-\pi+\psi}^{-\psi}\pi(1+M\cos y)dy + \int_{-\psi}^{\psi}2\pi\,dy + \int_{\psi}^{\pi-\psi}\pi(1+M\cos y)dy\right] \tag{A4.3}$$

Further integration then results in

$$C_{00} = \frac{V_{dc}}{\pi^2}\left[\begin{aligned}&\left[\pi^2 - 2\pi\psi - \pi M\sin\psi - \pi M\sin(\psi-\pi)\right] + 4\pi\psi\\ &\qquad + [\pi^2 - 2\pi\psi + \pi M\sin(\pi-\psi) - \pi M\sin\psi]\end{aligned}\right] \tag{A4.4}$$

Hence

$$C_{00} = A_{00} + jB_{00} = 2V_{dc} \tag{A4.5}$$

For $m = 0, n > 0$, Eq. (A4.1) simplifies to

$$C_{0n} = \frac{V_{dc}}{\pi^2}\left[\begin{aligned}&\int_{-\pi+\psi}^{-\psi}\int_{-\frac{\pi}{2}(1+M\cos y)}^{\frac{\pi}{2}(1+M\cos y)} e^{jny}\,dx\,dy + \int_{-\psi}^{\psi}\int_{-\pi}^{\pi} e^{jny}\,dx\,dy\\ &\qquad + \int_{\psi}^{\pi-\psi}\int_{-\frac{\pi}{2}(1+M\cos y)}^{\frac{\pi}{2}(1+M\cos y)} e^{jny}\,dx\,dy\end{aligned}\right] \tag{A4.6}$$

Eliminating the inner variable x leads to

$$C_{0n} = \frac{V_{dc}}{\pi^2}\left[\begin{array}{r} \displaystyle\int_{-\pi+\psi}^{-\psi} \pi(1+M\cos y)e^{jny}\,dy + \int_{-\psi}^{\psi} 2\pi e^{jny}\,dy \\ \displaystyle + \int_{\psi}^{\pi-\psi} \pi(1+M\cos y)e^{jny}\,dy \end{array}\right] \tag{A4.7}$$

and partially evaluating the outer integral expression results in

$$C_{0n} = \frac{V_{dc}}{\pi}\left[\begin{array}{r} \displaystyle\frac{4}{n}\sin n\psi + \frac{1}{jn}[e^{-jn\psi} - e^{jn\psi}e^{-jn\pi}] + \int_{-\pi+\psi}^{-\psi} M\cos y e^{jny}\,dy \\ \displaystyle + \frac{1}{jn}[e^{jn\pi}e^{-jn\psi} - e^{jn\psi}] + \int_{\psi}^{\pi-\psi} M\cos y e^{jny}\,dy \end{array}\right] \tag{A4.8}$$

Substituting $y' = y + \pi$ into the first integral expression of Eq. (A4.8) gives

$$C_{0n} = \frac{V_{dc}}{\pi}\left[\begin{array}{r} \displaystyle\frac{4}{n}\sin n\psi + \frac{1}{jn}[e^{-jn\psi} - e^{jn\psi}e^{-jn\pi}] - \int_{\psi}^{\pi-\psi} M\cos y' e^{jny'}e^{jn\pi}\,dy' \\ \displaystyle + \frac{1}{jn}[e^{jn\pi}e^{-jn\psi} - e^{jn\psi}] + \int_{\psi}^{\pi-\psi} M\cos y\, e^{jny}\,dy \end{array}\right] \tag{A4.9}$$

For <u>n even</u>, Eq. (A4.9) can be reduced to

$$C_{0n}\Big|_{n\text{ even}} = \frac{V_{dc}}{\pi}\left[\begin{array}{l} \displaystyle\frac{4}{n}\sin n\psi - \frac{2}{n}\sin n\psi - \frac{2}{n}\sin n\psi \\ \displaystyle - \int_{\psi}^{\pi-\psi} M\cos y e^{jny}\,dy + \int_{\psi}^{\pi-\psi} M\cos y e^{jny}\,dy \end{array}\right] = 0 \tag{A4.10}$$

which means there are no even harmonic baseband harmonic components even under overmodulation conditions.

For <u>n odd</u>, Eq. (A4.9) can be reduced to

$$C_{0n}\Big|_{n\text{ odd}} = \frac{V_{dc}}{\pi}\left\{\frac{4}{n}\sin n\psi + \frac{2}{jn}\cos n\psi - \frac{2}{jn}\cos n\psi + 2\int_{\psi}^{\pi-\psi} M\cos y e^{jny}\,dy\right\}$$

$$= \frac{V_{dc}}{\pi}\left[\frac{4}{n}\sin n\psi + M\int_{\psi}^{\pi-\psi}(e^{j[n+1]y} + e^{j[n-1]y})\,dy\right] \qquad \text{(A4.11)}$$

For <u>$n = 1$</u>, Eq. (A4.11) can be further reduced to

$$C_{01} = \frac{V_{dc}}{\pi}\left[4\sin\psi + M\int_{\psi}^{\pi-\psi}(e^{j2y} + 1)dy\right]$$

$$= \frac{V_{dc}}{\pi}[4\sin\psi + M(\pi - 2\psi - \sin 2\psi)] \qquad \text{(A4.12)}$$

For <u>$n > 1$</u>, Eq. (A4.11) can be reduced to (since n must always be odd)

$$C_{0(2n-1)} = \frac{V_{dc}}{\pi}\left\{\frac{4\sin([2n-1]\psi)}{[2n-1]} + M\int_{\psi}^{\pi-\psi}(e^{j2ny} + e^{j[2n-2]y})\,dy\right\}$$

$$= \frac{V_{dc}}{\pi}\left\{\frac{4\sin([2n-1]\psi)}{[2n-1]} - M\left\{\frac{\sin 2n\psi}{n} + \frac{\sin(2[n-1]\psi)}{[n-1]}\right\}\right\} \qquad \text{(A4.13)}$$

For <u>$m > 0, n = 0$</u>, Eq. (A4.1) can be simplified to

$$C_{m0} = \frac{V_{dc}}{\pi^2}\left[\int_{-\pi+\psi}^{-\psi}\int_{-\frac{\pi}{2}(1+M\cos y)}^{\frac{\pi}{2}(1+M\cos y)} e^{jmx}\,dx\,dy + \int_{\psi}^{\pi-\psi}\int_{-\frac{\pi}{2}(1+M\cos y)}^{\frac{\pi}{2}(1+M\cos y)} e^{jmx}\,dx\,dy + \int_{-\psi}^{\psi}\int_{-\pi}^{\pi} e^{jmx}\,dx\,dy\right]$$

(A4.14)

This expression can be integrated to form the result

$$C_{m0} = \frac{V_{dc}}{jm\pi^2}\left[\begin{array}{l}\displaystyle\int_{-\pi+\psi}^{-\psi}\left(e^{jm\frac{\pi}{2}}e^{jm\frac{\pi}{2}M\cos y} - e^{-jm\frac{\pi}{2}}e^{-jm\frac{\pi}{2}M\cos y}\right)dy \\ \displaystyle + \int_{\psi}^{\pi-\psi}\left(e^{jm\frac{\pi}{2}}e^{jm\frac{\pi}{2}M\cos y} - e^{-jm\frac{\pi}{2}}e^{-jm\frac{\pi}{2}M\cos y}\right)dy \\ \displaystyle + \int_{-\psi}^{\psi}(e^{jm\pi} - e^{-jm\pi})\,dy\end{array}\right] \tag{A4.15}$$

The last integral term of Eq. (A4.15) is obviously zero for all m.

Substituting $y' = y + \pi$ into the first integral then gives

$$C_{m0} = \frac{V_{dc}}{jm\pi^2}\left[\begin{array}{l}\displaystyle\int_{\psi}^{\pi-\psi}\left(e^{jm\frac{\pi}{2}}e^{-jm\frac{\pi}{2}M\cos y'} - e^{-jm\frac{\pi}{2}}e^{jm\frac{\pi}{2}M\cos y'}\right)dy' \\ \displaystyle + \int_{\psi}^{\pi-\psi}\left(e^{jm\frac{\pi}{2}}e^{jm\frac{\pi}{2}M\cos y} - e^{-jm\frac{\pi}{2}}e^{-jm\frac{\pi}{2}M\cos y}\right)dy\end{array}\right] \tag{A4.16}$$

Since y and y' are simply variables of integration, Eq. (A4.16) simplifies to

$$C_{m0} = \frac{V_{dc}}{jm\pi^2}\int_{\psi}^{\pi-\psi}\left[\begin{array}{l}e^{jm\frac{\pi}{2}}\left(e^{jm\frac{\pi}{2}M\cos y} + e^{-jm\frac{\pi}{2}M\cos y'}\right) \\ \qquad - e^{-jm\frac{\pi}{2}}\left(e^{jm\frac{\pi}{2}M\cos y} + e^{-jm\frac{\pi}{2}M\cos y'}\right)\end{array}\right]dy$$

$$= \frac{2V_{dc}}{jm\pi^2}\int_{\psi}^{\pi-\psi}\cos\left(m\frac{\pi}{2}M\cos y\right)\left(e^{jm\frac{\pi}{2}} - e^{-jm\frac{\pi}{2}}\right)dy$$

$$= \frac{4V_{dc}}{m\pi^2}\sin m\frac{\pi}{2}\int_{\psi}^{\pi-\psi}\cos\left(m\frac{\pi}{2}M\cos y\right)dy \tag{A4.17}$$

To solve Eq. (A4.17) the Jacobi–Anger expression, Eq. (A2.2), can be used to form

$$C_{m0} = \frac{4V_{dc}}{m\pi^2}\sin m\frac{\pi}{2}\int_{\psi}^{\pi-\psi}\left(J_0\left(m\frac{\pi}{2}M\right) + 2\sum_{k=1}^{\infty}J_{2k}\left(m\frac{\pi}{2}M\right)\cos k\pi\cos 2ky\right)dy$$

$$= \frac{4V_{dc}}{m\pi^2}\sin m\frac{\pi}{2}\left[J_0\left(m\frac{\pi}{2}M\right)(\pi-2\psi) + 2\sum_{k=1}^{\infty}J_{2k}\left(m\frac{\pi}{2}M\right)\cos k\pi\int_{\psi}^{\pi-\psi}\cos 2ky\,dy\right]$$

$$= \frac{4V_{dc}}{m\pi^2}\sin m\frac{\pi}{2}\left[J_0\left(m\frac{\pi}{2}M\right)(\pi-2\psi) - 2\sum_{k=1}^{\infty}\frac{1}{k}J_{2k}\left(m\frac{\pi}{2}M\right)\cos k\pi\sin 2k\psi\right] \tag{A4.18}$$

For $m > 0,\ n \neq 0$, the inner integral of Eq. (A4.1) can be evaluated to give

$$C_{mn} = \frac{V_{dc}}{jm\pi^2}\left[\int_{-\pi+\psi}^{-\psi}e^{jny}\left[e^{jm\frac{\pi}{2}}e^{jm\frac{\pi}{2}M\cos y} - e^{-jm\frac{\pi}{2}}e^{-jm\frac{\pi}{2}M\cos y}\right]dy + \int_{\psi}^{\pi-\psi}e^{jny}\left[e^{jm\frac{\pi}{2}}e^{jm\frac{\pi}{2}M\cos y} - e^{-jm\frac{\pi}{2}}e^{-jm\frac{\pi}{2}M\cos y}\right]dy + \int_{-\psi}^{\psi}e^{jny}(e^{jm\pi} - e^{-jm\pi})\,dy\right] \tag{A4.19}$$

Again the last integral term in Eq. (A4.19) is zero.

Substituting $y' = y + \pi$ once more into the first integral term then gives

$$C_{mn} = \frac{V_{dc}}{jm\pi^2}\left[\begin{array}{l}\displaystyle\int_{\psi}^{\pi-\psi} e^{jn\pi} e^{jny'}\left[e^{jm\frac{\pi}{2}} e^{-jm\frac{\pi}{2}M\cos y'} - e^{-jm\frac{\pi}{2}} e^{jm\frac{\pi}{2}M\cos y'}\right]dy' \\ \displaystyle + \int_{\psi}^{\pi-\psi} e^{jny}\left[e^{jm\frac{\pi}{2}} e^{jm\frac{\pi}{2}M\cos y} - e^{-jm\frac{\pi}{2}} e^{-jm\frac{\pi}{2}M\cos y}\right]dy\end{array}\right]$$

$$= \frac{V_{dc}}{jm\pi^2}\int_{\psi}^{\pi-\psi} e^{jny}\left[\begin{array}{l}\left[e^{jn\pi} e^{jm\frac{\pi}{2}} - e^{-jm\frac{\pi}{2}}\right] e^{-jm\frac{\pi}{2}M\cos y} \\ \qquad + \left[e^{jm\frac{\pi}{2}} - e^{jn\pi} e^{-jm\frac{\pi}{2}}\right] e^{jm\frac{\pi}{2}M\cos y}\end{array}\right]dy \quad \text{(A4.20)}$$

Substituting the Jacobi–Anger relationship, Eq. (A2.1), into this integral gives

$$C_{mn} = \frac{V_{dc}}{jm\pi^2}\int_{\psi}^{\pi-\psi} e^{jny}\left[\begin{array}{l}\left[e^{jn\pi} e^{jm\frac{\pi}{2}} - e^{-jm\frac{\pi}{2}}\right]\left[\begin{array}{l}J_0\left(m\frac{\pi}{2}M\right) \\ \displaystyle + 2\sum_{k=1}^{\infty} J_k\left(m\frac{\pi}{2}M\right) e^{-jk\frac{\pi}{2}} \cos ky\end{array}\right] \\ + \left[e^{jm\frac{\pi}{2}} - e^{jn\pi} e^{-jm\frac{\pi}{2}}\right]\left[\begin{array}{l}J_0\left(m\frac{\pi}{2}M\right) \\ \displaystyle + 2\sum_{k=1}^{\infty} J_k\left(m\frac{\pi}{2}M\right) e^{jk\frac{\pi}{2}} \cos ky\end{array}\right]\end{array}\right]dy$$

$$= \frac{V_{dc}}{jm\pi^2}\int_{\psi}^{\pi-\psi} e^{jny}\left[\begin{array}{l} J_0\left(m\frac{\pi}{2}M\right)\left[e^{jn\pi}e^{jm\frac{\pi}{2}} - e^{-jm\frac{\pi}{2}} + e^{jm\frac{\pi}{2}} - e^{jn\pi}e^{-jm\frac{\pi}{2}}\right] \\ + 2\sum_{k=1}^{\infty} J_k\left(m\frac{\pi}{2}M\right)\left[\begin{array}{l}\left[e^{jn\pi}e^{jm\frac{\pi}{2}} - e^{-jm\frac{\pi}{2}}\right]e^{-jk\frac{\pi}{2}} \\ + \left[e^{jm\frac{\pi}{2}} - e^{jn\pi}e^{-jm\frac{\pi}{2}}\right]e^{jk\frac{\pi}{2}}\end{array}\right]\cos ky \end{array}\right] dy$$

$$= \frac{2V_{dc}}{m\pi^2}\int_{\psi}^{\pi-\psi} e^{jny}\left[\begin{array}{l} J_0\left(m\frac{\pi}{2}M\right)\sin m\frac{\pi}{2}(1+\cos n\pi) \\ + 2\sum_{k=1}^{\infty}\left[\begin{array}{r}\left[\sin\left([m+k]\frac{\pi}{2}\right) + \cos n\pi\sin\left([m-k]\frac{\pi}{2}\right)\right] \\ \times J_k\left(m\frac{\pi}{2}M\right)\cos ky\end{array}\right]\end{array}\right] dy \tag{A4.21}$$

With some manipulation, it can be shown that

$$\cos n\pi\sin\left([m-k]\frac{\pi}{2}\right) = \cos([n+k]\pi)\sin\left([m+k]\frac{\pi}{2}\right) \tag{A4.22}$$

Substituting this result back into Eq. (A4.21) gives

$$C_{mn} = \frac{2V_{dc}}{m\pi^2}\int_{\psi}^{\pi-\psi} e^{jny}\left[\begin{array}{l} J_0\left(m\frac{\pi}{2}M\right)\sin m\frac{\pi}{2}(1+\cos n\pi) \\ + 2\sum_{k=1}^{\infty}\left[\begin{array}{l} J_k\left(m\frac{\pi}{2}M\right)\sin\left([m+k]\frac{\pi}{2}\right) \\ \quad \times\{1+\cos([n+k]\pi)\}\cos ky\end{array}\right]\end{array}\right] dy$$

$$= \frac{2V_{dc}}{m\pi^2} \int_{\psi}^{\pi-\psi} \left[\begin{array}{l} J_0\left(m\frac{\pi}{2}M\right) \sin m\frac{\pi}{2}(1 + \cos n\pi) e^{jny} \\ + \sum_{k=1}^{\infty} \left[\begin{array}{l} J_k\left(m\frac{\pi}{2}M\right) \sin\left([m+k]\frac{\pi}{2}\right) \\ \quad \times \{1 + \cos([n+k]\pi)\} \{e^{j[n+k]y} + e^{j[n-k]y}\} \end{array} \right] \end{array} \right] dy \tag{A4.23}$$

Integrating Eq. (A4.23) gives

$$C_{mn} = \frac{2V_{dc}}{m\pi^2} \left[\begin{array}{l} \frac{1}{jn} J_0\left(m\frac{\pi}{2}M\right) \sin m\frac{\pi}{2}(1 + \cos n\pi)(e^{jn[\pi-\psi]} - e^{jn\psi}) \\ + \sum_{\substack{k=1 \\ k \neq |n|}}^{\infty} \left[\begin{array}{l} J_k\left(m\frac{\pi}{2}M\right) \sin\left([m+k]\frac{\pi}{2}\right) \{1 + \cos([n+k]\pi)\} \\ \times \left[\frac{e^{j[n+k][\pi-\psi]} - e^{j[n+k]\psi}}{j[n+k]} + \frac{e^{j[n-k][\pi-\psi]} - e^{j[n-k]\psi}}{j[n-k]} \right] \end{array} \right] \\ + 2J_n\left(m\frac{\pi}{2}M\right) \sin\left([m+n]\frac{\pi}{2}\right) \left[\frac{e^{j2n[\pi-\psi]} - e^{j2n\psi}}{j2n} + \pi - 2\psi \right] \Bigg|_{k=n} \\ + 2J_{-n}\left(m\frac{\pi}{2}M\right) \sin\left([m-n]\frac{\pi}{2}\right) \left[\frac{e^{j2n[\pi-\psi]} - e^{j2n\psi}}{j2n} + \pi - 2\psi \right] \Bigg|_{k=-n} \end{array} \right]$$

$$= \frac{4V_{dc}}{m\pi^2} \left[\begin{array}{l} -\frac{1}{n} J_0\left(m\frac{\pi}{2}M\right) \sin m\frac{\pi}{2}(1 + \cos n\pi) \sin n\psi \\ + J_n\left(m\frac{\pi}{2}M\right) \sin\left([m+n]\frac{\pi}{2}\right) \left[\pi - 2\psi - \frac{1}{n} \sin 2n\psi \right] \Bigg|_{k=n} \\ + J_n\left(m\frac{\pi}{2}M\right) \cos n\pi \sin\left([m-n]\frac{\pi}{2}\right) \left[\pi - 2\psi - \frac{1}{n} \sin 2n\psi \right] \Bigg|_{k=-n} \\ - \sum_{\substack{k=1 \\ k \neq |n|}}^{\infty} \left[\begin{array}{l} J_k\left(m\frac{\pi}{2}M\right) \sin\left([m+k]\frac{\pi}{2}\right) \{1 + \cos([n+k]\pi)\} \\ \qquad \times \left[\frac{\sin([n+k]\psi)}{[n+k]} + \frac{\sin([n-k]\psi)}{[n-k]} \right] \end{array} \right] \end{array} \right] \tag{A4.24}$$

Using Eq. (A4.22), $\cos n\pi \sin([m-n]\pi/2)$ reduces to $\sin([m+n]\pi/2)$, so that the final form of Eq. (A4.24) becomes

$$C_{mn} = \frac{4V_{dc}}{m\pi^2}\left[\begin{array}{l} -\frac{1}{n}J_0\left(m\frac{\pi}{2}M\right)\sin m\frac{\pi}{2}(1+\cos n\pi)\sin n\psi \\ +J_n\left(m\frac{\pi}{2}M\right)\sin\left([m+n]\frac{\pi}{2}\right)\left[\pi-2\psi-\frac{1}{n}\sin 2n\psi\right] \\ -\sum_{\substack{k=1 \\ k\neq|n|}}^{\infty}\left[\begin{array}{l} J_k\left(m\frac{\pi}{2}M\right)\sin\left([m+k]\frac{\pi}{2}\right)\{1+\cos([n+k]\pi)\} \\ \qquad\times\left[\frac{\sin([n+k]\psi)}{[n+k]}+\frac{\sin([n-k]\psi)}{[n-k]}\right]\end{array}\right]\end{array}\right] \tag{A4.25}$$

A4.1.2 Harmonic Solution for Overmodulated Single-Phase Leg under Naturally Sampled PWM

Assembling Eqs. (A4.5), (A4.12), (A4.13), (A4.18), and (A4.25), the full solution for naturally sampled PWM of a single-phase leg under overmodulation, using a similar approach as was used to develop Eq. (3.10), is

$$v_{az}(t) = V_{dc} + \frac{V_{dc}}{\pi}[4\sin\psi + M(\pi-2\psi-\sin 2\psi)]\cos\omega_o t \tag{A4.26}$$

$$+\frac{V_{dc}}{\pi}\sum_{n=2}^{\infty}\left[\begin{array}{l}\frac{4}{[2n-1]}\sin([2n-1]\psi) \\ \qquad -M\left[\frac{\sin 2n\psi}{n}+\frac{\sin(2[n-1]\psi)}{[n-1]}\right]\end{array}\right]\cos([2n-1]\omega_o t)$$

$$+\frac{4V_{dc}}{\pi^2}\sum_{m=1}^{\infty}\frac{1}{m}\sin m\frac{\pi}{2}\left[\begin{array}{l} J_0\left(m\frac{\pi}{2}M\right)(\pi-2\psi) \\ \qquad -2\sum_{k=1}^{\infty}\frac{1}{k}J_{2k}\left(m\frac{\pi}{2}M\right)\cos k\pi\sin 2k\psi\end{array}\right]\cos m\omega_c t$$

$$+\sum_{m=1}^{\infty}\sum_{\substack{n=-\infty \\ (n\neq 0)}}^{\infty} C_{mn}\cos(m\omega_c t + n\omega_o t)$$

where C_{mn} is given by

$$C_{mn} = \frac{4V_{dc}}{m\pi^2}\left[\begin{array}{l} -\frac{1}{n}J_0\left(m\frac{\pi}{2}M\right)\sin m\frac{\pi}{2}(1+\cos n\pi)\sin n\psi \\ +J_n\left(m\frac{\pi}{2}M\right)\sin\left([m+n]\frac{\pi}{2}\right)\left[\pi - 2\psi - \frac{1}{n}\sin 2n\psi\right] \\ -\sum_{\substack{k=1\\k\neq|n|}}^{\infty}\left[\begin{array}{l} J_k\left(m\frac{\pi}{2}M\right)\sin\left([m+k]\frac{\pi}{2}\right)\{1+\cos([n+k]\pi)\} \\ \qquad \times\left[\frac{\sin([n+k]\psi)}{[n+k]}+\frac{\sin([n-k]\psi)}{[n-k]}\right]\end{array}\right]\end{array}\right] \quad (4.27)$$

and ψ is defined by

$$\psi = \left\{\begin{array}{ll} \cos^{-1}\left(\frac{1}{M}\right) & \text{for } M>1 \\ 0 & \text{for } M\leq 1 \end{array}\right\} \quad (A4.28)$$

A4.1.3 Linear Modulation Solution Obtained from Overmodulation Solution

In the linear modulation region, ψ becomes zero, and Eq. (A4.26) reduces to

$$v_{az}(t) = V_{dc} + V_{dc}M\cos\omega_o t + \frac{4V_{dc}}{\pi}\sum_{m=1}^{\infty}\frac{1}{m}J_0\left(m\frac{\pi}{2}M\right)\sin m\frac{\pi}{2}\cos m\omega_c t \quad (A4.29)$$

$$+\frac{4V_{dc}}{\pi}\sum_{m=1}^{\infty}\sum_{\substack{n=-\infty\\(n\neq 0)}}^{\infty}\frac{1}{m}J_n\left(m\frac{\pi}{2}M\right)\sin\left([m+n]\frac{\pi}{2}\right)\cos(m\omega_c t + n\omega_o t)$$

This result is the same as Eq. (3.39) obtained in Chapter 3 when $\theta_c = \theta_o = 0$.

A4.1.4 Square-Wave Solution Obtained from Overmodulation Solution

When the commanded modulation index increases without bounds, all of the solution terms containing Bessel functions approach zero (see Fig. A2.1).

Hence only the first two rows of Eq. (A4.26) remain, representing the DC, fundamental component, and baseband harmonics. Recall that by definition

$$M = \frac{1}{\cos\psi} \tag{A4.30}$$

Substituting this relationship into Eq. (A4.26) gives

$$v_{az}(t) = V_{dc} + \frac{4V_{dc}}{\pi}\left[\sin\psi + \frac{(\pi - 2\psi - \sin 2\psi)}{4\cos\psi}\right]\cos\omega_o t \tag{A4.31}$$

$$+ \frac{V_{dc}}{\pi}\sum_{n=2}^{\infty}\left[\begin{array}{l} \frac{4}{[2n-1]}\sin([2n-1]\psi) \\ -\left[\frac{\sin 2n\psi}{n\cos\psi} + \frac{\sin(2[n-1]\psi)}{[n-1]\cos\psi}\right]\end{array}\right]\cos([2n-1]\omega_o t)$$

As M increases toward infinity, $\psi \to \pi/2$ whereupon

$$v_{az}(t) = V_{dc} + \frac{4V_{dc}}{\pi}\left[1 + \frac{0}{0}\right]\cos\omega_o t \tag{A4.32}$$

$$+ \frac{V_{dc}}{\pi}\sum_{n=2}^{\infty}\left[\begin{array}{l} \frac{4}{[2n-1]}\sin\left([2n-1]\frac{\pi}{2}\right) \\ -\left[\frac{0}{0} + \frac{0}{0}\right]\end{array}\right]\cos([2n-1]\omega_o t)$$

L'Hospital's rule can be used to resolve the $\frac{0}{0}$ terms, i.e., for the first harmonic

$$\lim_{\psi \to \frac{\pi}{2}} \frac{[\pi - 2\psi - \sin 2\psi]}{4\cos\psi} = \lim_{\psi \to \frac{\pi}{2}} \frac{-2 - 2\cos 2\psi}{-4\sin\psi} = 0 \tag{A4.33}$$

Therefore, when M approaches infinity, the magnitude of the fundamental harmonic component of the phase leg output becomes

$$V_1 = \frac{4}{\pi}V_{dc} \tag{A4.34}$$

which is the same result as for a square wave. In a similar manner it can be readily shown that the baseband harmonics become

$$V_n = -\frac{4}{\pi}V_{dc}\left[\frac{(-1)^n}{2n-1}\right] \qquad n = 2,3,\ldots \tag{A4.35}$$

which is again the same result as for a square wave.

A4.2 Symmetric Regular Sampled Double-Edge PWM

Using the same strategies as were developed in Chapter 3 for linear modulation, the above solution process can now be modified to take into account the effects of regular sampling. This is done by changing the limits of the inner integral of Eq. (A4.1) to account for the reference waveform sampling in each carrier interval. The staircase variable that defines these limits is y', which is frozen at $x = 0$ in each carrier interval. For the pth carrier cycle,

$$y' = \frac{\omega_o}{\omega_c} 2\pi p \tag{A4.36}$$

or terms of the continuous variables x and y,

$$y' = y - \frac{\omega_o}{\omega_c}(x - 2\pi p) \tag{A4.37}$$

Applying this relationship to the double integral defined by Eq. A4.1 gives

$$C_{mn} = \frac{V_{dc}}{\pi^2}\left[\begin{array}{l} \displaystyle\int_{-\pi+\psi}^{-\psi} \int_{-\frac{\pi}{2}(1 + M\cos y')}^{\frac{\pi}{2}(1 + M\cos y')} e^{j(mx + ny)} dx\, dy + \int_{-\psi}^{\psi} \int_{-\pi}^{\pi} e^{j(mx + ny)} dx\, dy \\ \displaystyle\qquad + \int_{\psi}^{\pi-\psi} \int_{-\frac{\pi}{2}(1 + M\cos y')}^{\frac{\pi}{2}(1 + M\cos y')} e^{j(mx + ny)} dx\, dy \end{array}\right] \tag{A4.38}$$

Substituting from Eq. (A4.37) and changing the integral variable to y' gives

$$C_{mn} = \frac{V_{dc}}{\pi^2}\left[\begin{array}{l} \displaystyle\int_{-\pi+\psi}^{-\psi} \int_{-\frac{\pi}{2}(1 + M\cos y')}^{\frac{\pi}{2}(1 + M\cos y')} e^{j(qx + ny')} dx\, dy' + \int_{-\psi}^{\psi} \int_{-\pi}^{\pi} e^{j(qx + ny')} dx\, dy' \\ \displaystyle\qquad + \int_{\psi}^{\pi-\psi} \int_{-\frac{\pi}{2}(1 + M\cos y')}^{\frac{\pi}{2}(1 + M\cos y')} e^{j(qx + ny')} dx\, dy' \end{array}\right] \tag{A4.39}$$

where $q = m + n(\omega_o/\omega_c)$.

It can be noted that this result is almost identical to the expression for naturally sampled PWM, Eq. (A4.1), except that the m index has been replaced by the q index. Equation (A4.39) can now be solved for various values of m and n.

A4.2.1 Evaluation of Double Fourier Integral for Overmodulated Symmetric Regular Sampled PWM

For $m = n = 0$, Eq. (A4.39) can be reduced to

$$C_{00} = \frac{V_{dc}}{\pi^2}\left[\int_{-\pi+\psi}^{-\psi}\int_{-\frac{\pi}{2}(1+M\cos y)}^{\frac{\pi}{2}(1+M\cos y)} dx\,dy + \int_{-\psi}^{\psi}\int_{-\pi}^{\pi} dx\,dy + \int_{\psi}^{\pi-\psi}\int_{-\frac{\pi}{2}(1+M\cos y)}^{\frac{\pi}{2}(1+M\cos y)} dx\,dy\right] \tag{A4.40}$$

which is clearly the same as Eq. (A4.2) in Section A4.1.1. Hence once more

$$C_{00} = 2V_{dc} \tag{A4.41}$$

For $m > 0, n = 0$, Eq. (A4.39) can be simplified to

$$C_{m0} = \frac{V_{dc}}{\pi^2}\left[\int_{-\pi+\psi}^{-\psi}\int_{-\frac{\pi}{2}(1+M\cos y)}^{\frac{\pi}{2}(1+M\cos y)} e^{jmx}\,dx\,dy + \int_{\psi}^{\pi-\psi}\int_{-\frac{\pi}{2}(1+M\cos y)}^{\frac{\pi}{2}(1+M\cos y)} e^{jmx}\,dx\,dy + \int_{-\psi}^{\psi}\int_{-\pi}^{\pi} e^{jmx}\,dx\,dy\right] \tag{A4.42}$$

This is clearly the same as Eq. (A4.14) so that the solution of Eq. (A4.18) can be used again.

For $m > 0, n \neq 0$, the inner integral of Eq. (A4.39) can be evaluated to give

$$C_{mn} = \frac{V_{dc}}{jq\pi^2}\left[\begin{array}{l} \displaystyle\int_{-\pi+\psi}^{-\psi} e^{jny}\left[e^{jq\frac{\pi}{2}}e^{jq\frac{\pi}{2}M\cos y} - e^{-jq\frac{\pi}{2}}e^{-jq\frac{\pi}{2}M\cos y}\right]dy \\ \displaystyle + \int_{\psi}^{\pi-\psi} e^{jny}\left[e^{jq\frac{\pi}{2}}e^{jq\frac{\pi}{2}M\cos y} - e^{-jq\frac{\pi}{2}}e^{-jq\frac{\pi}{2}M\cos y}\right]dy \\ \displaystyle + \int_{-\psi}^{\psi} e^{jny}(e^{jq\pi} - e^{-jq\pi})\,dy \end{array}\right] \tag{A4.43}$$

Since this is much the same form as Eq. (A4.19), the same solution process as led to Eq. (A4.25) can be used. The only difference is that the last integral term is no longer zero, and must be evaluated. With this extension, and after some development, Eq. (A4.43) becomes finally

$$C_{mn} = \frac{4V_{dc}}{q\pi^2}\left[\begin{array}{l} \displaystyle\frac{1}{n}\left[\sin(q\pi) - J_0\left(q\frac{\pi}{2}M\right)\sin q\frac{\pi}{2}(1+\cos n\pi)\right]\sin n\psi \\ \displaystyle + J_n\left(q\frac{\pi}{2}M\right)\sin\left([q+n]\frac{\pi}{2}\right)\left[\pi - 2\psi - \frac{1}{n}\sin 2n\psi\right] \\ \displaystyle - \sum_{\substack{k=1 \\ k\neq|n|}}^{\infty}\left[\begin{array}{l} J_k\left(q\frac{\pi}{2}M\right)\sin\left([q+k]\frac{\pi}{2}\right)\{1+\cos([n+k]\pi)\} \\ \displaystyle \quad\times\left[\frac{\sin([n+k]\psi)}{[n+k]} + \frac{\sin([n-k]\psi)}{[n-k]}\right] \end{array}\right] \end{array}\right] \tag{A4.44}$$

For $m = 0, n > 0$, q reduces to $n(\omega_o/\omega_c)$, and Eq. (A4.44) is still valid.

A4.2.2 Harmonic Solution for Overmodulated Single-Phase Leg under Symmetric Regular Sampled PWM

Assembling Eqs. (A4.41), (A4.42), and (A4.44), the full solution for symmetric regular sampled PWM of a single-phase leg under overmodulation is

$$v_{az}(t) = V_{dc} + \frac{4V_{dc}}{\pi^2}\sum_{n=1}^{\infty} C_{0n}\cos n\omega_o t \tag{A4.45}$$

$$+ \frac{4V_{dc}}{\pi^2}\sum_{m=1}^{\infty}\frac{1}{m}\sin m\frac{\pi}{2}\left[\begin{matrix} J_0\left(m\frac{\pi}{2}M\right)(\pi - 2\psi) \\ -2\sum\limits_{k=1}^{\infty}\frac{1}{k}J_{2k}\left(m\frac{\pi}{2}M\right)\cos k\pi\sin 2k\psi \end{matrix}\right]\cos m\omega_c t$$

$$+ \sum_{m=1}^{\infty}\sum_{\substack{n=-\infty \\ (n\neq 0)}}^{\infty} C_{mn}\cos(m\omega_c t + n\omega_o t)$$

where C_{0n} is given by

$$C_{0n} = \frac{4V_{dc}}{\left[n\frac{\omega_o}{\omega_c}\right]^2\pi}\left[\begin{matrix} \frac{1}{n}\left[\sin\left(n\frac{\omega_o}{\omega_c}\pi\right) - J_0\left(n\frac{\omega_o}{\omega_c}\frac{\pi}{2}M\right)\sin\left(n\frac{\omega_o}{\omega_c}\frac{\pi}{2}\right)(1+\cos n\pi)\right]\sin n\psi \\ + J_n\left(n\frac{\omega_o}{\omega_c}\frac{\pi}{2}M\right)\sin\left(\left[n\frac{\omega_o}{\omega_c}+n\right]\frac{\pi}{2}\right)\left[\pi - 2\psi - \frac{1}{n}\sin 2n\psi\right] \\ -\sum\limits_{\substack{k=1 \\ k\neq|n|}}^{\infty}\left[\begin{matrix} J_k\left(n\frac{\omega_o}{\omega_c}\frac{\pi}{2}M\right)\sin\left(\left[n\frac{\omega_o}{\omega_c}+k\right]\frac{\pi}{2}\right) \\ \times\{1+\cos([n+k]\pi)\} \\ \times\left[\frac{\sin([n+k]\psi)}{[n+k]} + \frac{\sin([n-k]\psi)}{[n-k]}\right] \end{matrix}\right] \end{matrix}\right] \tag{A4.46}$$

and C_{mn} is given by

$$C_{mn} = \frac{4V_{dc}}{q\pi^2}\left[\begin{array}{l} \frac{1}{n}\left[\sin(q\pi) - J_0\left(q\frac{\pi}{2}M\right)\sin q\frac{\pi}{2}(1+\cos n\pi)\right]\sin n\psi \\ \quad + J_n\left(q\frac{\pi}{2}M\right)\sin\left([q+n]\frac{\pi}{2}\right)\left[\pi - 2\psi - \frac{1}{n}\sin 2n\psi\right] \\ \quad - \sum_{\substack{k=1\\k\neq|n|}}^{\infty}\left[\begin{array}{l} J_k\left(q\frac{\pi}{2}M\right)\sin\left([q+k]\frac{\pi}{2}\right)\{1+\cos([n+k]\pi)\} \\ \qquad \times\left[\frac{\sin([n+k]\psi)}{[n+k]} + \frac{\sin([n-k]\psi)}{[n-k]}\right]\end{array}\right]\end{array}\right] \tag{A4.47}$$

with $q = m + n(\omega_o/\omega_c)$ and ψ defined by Eq. (A4.28).

A4.2.3 Linear Modulation Solution Obtained from Overmodulation Solution

Once again in the linear region $\psi = 0$, and Eq. (A4.45) can be reduced to

$$\begin{aligned} v_{az}(t) = V_{dc} &+ \frac{4V_{dc}}{\pi}\sum_{n=1}^{\infty}\frac{1}{\left[n\frac{\omega_o}{\omega_c}\right]}J_n\left(n\frac{\omega_o}{\omega_c}\frac{\pi}{2}M\right)\sin\left(\left[n\frac{\omega_o}{\omega_c}+n\right]\frac{\pi}{2}\right)\cos n\omega_o t \\ &+ \frac{4V_{dc}}{\pi}\sum_{m=1}^{\infty}\frac{1}{m}\sin m\frac{\pi}{2}J_0\left(m\frac{\pi}{2}M\right)\cos m\omega_c t \\ &+ \frac{4V_{dc}}{\pi}\sum_{m=1}^{\infty}\sum_{\substack{n=-\infty\\(n\neq 0)}}^{\infty}\frac{1}{q}J_n\left(q\frac{\pi}{2}M\right)\sin\left([q+n]\frac{\pi}{2}\right)\cos(m\omega_c t + n\omega_o t) \end{aligned} \tag{A4.48}$$

This result is the same as Eq. (3.78) obtained in Chapter 3. Note also that this solution produces baseband even harmonics as a consequence of the symmetric regular sampling process. Once again, this reflects the poorer harmonic performance of symmetric regular sampled PWM compared to asymmetric regular sampled PWM.

A4.3 Asymmetric Regular Sampled Double-Edge PWM

For asymmetrical regular sampling, the limits of the inner integral of Eq. (A4.1) must now be changed to account for two reference waveform sampling points in each carrier interval. The staircase variables that define these limits are y_r' for the switched pulse rising edge and y_f' for the falling edge. These variables are frozen at $x = -\pi/2$ and $x = \pi/2$, respectively, in each carrier interval. For the pth carrier cycle,

$$y_r' = \frac{\omega_o}{\omega_c}\left(2\pi p - \frac{\pi}{2}\right) \qquad y_f' = \frac{\omega_o}{\omega_c}\left(2\pi p + \frac{\pi}{2}\right) \tag{A4.49}$$

or terms of the continuous variables x and y,

$$y_r' = y - \frac{\omega_o}{\omega_c}\left(x - 2\pi p + \frac{\pi}{2}\right) \qquad y_f' = y - \frac{\omega_o}{\omega_c}\left(x - 2\pi p - \frac{\pi}{2}\right) \tag{A4.50}$$

These variables can now be used to redefine Eq. (A4.1), as

$$C_{mn} = \frac{V_{dc}}{\pi^2}\left[\begin{aligned} &\int_{-\pi+\psi}^{-\psi}\left[\int_{-\frac{\pi}{2}(1+M\cos y_r')}^{0} e^{j(mx+ny)}dx\,dy + \int_{0}^{\frac{\pi}{2}(1+M\cos y_f')} e^{j(mx+ny)}dx\,dy\right] \\ &+ \int_{-\psi}^{\psi}\left[\int_{-\pi}^{0} e^{j(mx+ny)}dx\,dy + \int_{0}^{\pi} e^{j(mx+ny)}dx\,dy\right] \\ &+ \int_{\psi}^{\pi-\psi}\left[\int_{-\frac{\pi}{2}(1+M\cos y_r')}^{0} e^{j(mx+ny)}dx\,dy + \int_{0}^{\frac{\pi}{2}(1+M\cos y_f')} e^{j(mx+ny)}dx\,dy\right] \end{aligned}\right] \tag{A4.51}$$

where it can be seen that each integration section has been split into two halves, once for the start of the switched pulse and one for the end of the switched pulse.

Utilizing Eq. (A4.50), the integration variables in Eq. (A4.51) can now be changed to give

$$C_{mn} = \frac{V_{dc}/2}{\pi^2}\left[\begin{array}{l} \displaystyle\int_{-\pi+\psi}^{-\psi}\left[\begin{array}{l}\displaystyle\int_{-\frac{\pi}{2}(1+M\cos y_r')}^{0} e^{j\left(qx+n\left[y_r'+\frac{\omega_o}{\omega_c}\frac{\pi}{2}\right]\right)}\,dx\,dy_r' \\ \displaystyle+\int_{0}^{\frac{\pi}{2}(1+M\cos y_f')} e^{j\left(qx+n\left[y_f'-\frac{\omega_o}{\omega_c}\frac{\pi}{2}\right]\right)}\,dx\,dy_f'\end{array}\right] \\ \displaystyle+\int_{-\psi}^{\psi}\left[\int_{-\pi}^{0} e^{j\left(qx+n\left[y_r'+\frac{\omega_o}{\omega_c}\frac{\pi}{2}\right]\right)}\,dx\,dy_r' + \int_{0}^{\pi} e^{j\left(qx+n\left[y_f'-\frac{\omega_o}{\omega_c}\frac{\pi}{2}\right]\right)}\,dx\,dy_f'\right] \\ \displaystyle+\int_{\psi}^{\pi-\psi}\left[\begin{array}{l}\displaystyle\int_{-\frac{\pi}{2}(1+M\cos y_r')}^{0} e^{j\left(qx+n\left[y_r'+\frac{\omega_o}{\omega_c}\frac{\pi}{2}\right]\right)}\,dx\,dy_r' \\ \displaystyle+\int_{0}^{\frac{\pi}{2}(1+M\cos y_f')} e^{j\left(qx+n\left[y_f'-\frac{\omega_o}{\omega_c}\frac{\pi}{2}\right]\right)}\,dx\,dy_f'\end{array}\right]\end{array}\right] \tag{A4.52}$$

where again $q = m + n(\omega_o/\omega_c)$. Equation (A4.52) can now be solved for various values of m and n.

A4.3.1 Evaluation of Double Fourier Integral for Overmodulated Asymmetric Regular Sampled PWM

For $m = n = 0$, Eq. (A4.52) can be reduced to

$$C_{00} = \frac{V_{dc}}{\pi^2}\left[\begin{aligned}&\int_{-\pi+\psi}^{-\psi}\left(\int_{-\frac{\pi}{2}(1+M\cos y_r')}^{0} dx\,dy_r' + \int_{0}^{\frac{\pi}{2}(1+M\cos y_f')} dx\,dy_f'\right)\\ &+\int_{-\psi}^{\psi}\left(\int_{-\pi}^{0} dx\,dy_r' + \int_{0}^{\pi} dx\,dy_f'\right)\\ &+\int_{\psi}^{\pi-\psi}\left(\int_{-\frac{\pi}{2}(1+M\cos y_r')}^{0} dx\,dy_r' + \int_{0}^{\frac{\pi}{2}(1+M\cos y_f')} dx\,dy_f'\right)\end{aligned}\right] \tag{A4.53}$$

which is clearly the same as Eq. (A4.2) and therefore has a solution of

$$C_{00} = A_{00} + jB_{00} = 2V_{dc} \tag{A4.54}$$

For $m > 0, n = 0$, Eq. (A4.52) can be simplified to

$$C_{mn} = \frac{V_{dc}}{\pi^2}\left[\begin{aligned}&\int_{-\pi+\psi}^{-\psi}\left(\int_{-\frac{\pi}{2}(1+M\cos y_r')}^{0} e^{jmx} dx\,dy_r' + \int_{0}^{\frac{\pi}{2}(1+M\cos y_f')} e^{jmx} dx\,dy_f'\right)\\ &+\int_{-\psi}^{\psi}\left(\int_{-\pi}^{0} e^{jmx} dx\,dy_r' + \int_{0}^{\pi} e^{jmx} dx\,dy_f'\right)\\ &+\int_{\psi}^{\pi-\psi}\left(\int_{-\frac{\pi}{2}(1+M\cos y_r')}^{0} e^{jmx} dx\,dy_r' + \int_{0}^{\frac{\pi}{2}(1+M\cos y_f')} e^{jmx} dx\,dy_f'\right)\end{aligned}\right] \tag{A4.55}$$

which is the same as Eq. (A4.14) since $y_r' = y_f' = y$, and so has a solution of

$$C_{mn} = \frac{4V_{dc}}{m\pi^2}\sin m\frac{\pi}{2}\left\{J_0\left(m\frac{\pi}{2}M\right)(\pi - 2\psi) - 2\sum_{k=1}^{\infty}\frac{1}{k}J_{2k}\left(m\frac{\pi}{2}M\right)\cos k\pi \sin 2k\psi\right\} \tag{A4.56}$$

For $m > 0, n \neq 0$, the inner integral of Eq. (A4.52) can be evaluated to give

$$C_{mn} = \frac{V_{dc}}{jq\pi^2}\left[\begin{array}{l} \displaystyle\int_{-\pi+\psi}^{-\psi}\left[\begin{array}{l} e^{jn\left[y_r' + \frac{\omega_o}{\omega_c}\frac{\pi}{2}\right]}\left[1 - e^{-jq\frac{\pi}{2}(1 + M\cos y_r')}\right]dy_r' \\ \qquad + e^{jn\left[y_f' - \frac{\omega_o}{\omega_c}\frac{\pi}{2}\right]}\left[e^{jq\frac{\pi}{2}(1 + M\cos y_f')} - 1\right]dy_f' \end{array}\right] \\ + \displaystyle\int_{-\psi}^{\psi}\left[e^{jn\left[y_r' + \frac{\omega_o}{\omega_c}\frac{\pi}{2}\right]}[1 - e^{-jq\pi}]dy_r' + e^{jn\left[y_f' - \frac{\omega_o}{\omega_c}\frac{\pi}{2}\right]}[e^{jq\pi} - 1]dy_f'\right] \\ + \displaystyle\int_{\psi}^{\pi-\psi}\left[\begin{array}{l} e^{jn\left[y_r' + \frac{\omega_o}{\omega_c}\frac{\pi}{2}\right]}\left[1 - e^{-jq\frac{\pi}{2}(1 + M\cos y_r')}\right]dy_r' \\ \qquad + e^{jn\left[y_f' - \frac{\omega_o}{\omega_c}\frac{\pi}{2}\right]}\left[e^{jq\frac{\pi}{2}(1 + M\cos y_f')} - 1\right]dy_f' \end{array}\right] \end{array}\right] \tag{A4.57}$$

Replacing y_r' and y_f' with a common integration variable y, and rearranging, yields

$$C_{mn} = \frac{V_{dc}}{jq\pi^2}\left[\begin{array}{l} \displaystyle\int_{-\pi+\psi}^{-\psi} e^{jny}\left[\begin{array}{l} e^{jn\frac{\omega_o}{\omega_c}\frac{\pi}{2}} - e^{-jn\frac{\omega_o}{\omega_c}\frac{\pi}{2}} \\ + e^{-jn\frac{\omega_o}{\omega_c}\frac{\pi}{2}} e^{jq\frac{\pi}{2}(1 + M\cos y)} - e^{jn\frac{\omega_o}{\omega_c}\frac{\pi}{2}} e^{-jq\frac{\pi}{2}(1 + M\cos y)} \end{array}\right] dy \\ + \displaystyle\int_{\psi}^{\pi-\psi} e^{jny}\left[\begin{array}{l} e^{jn\frac{\omega_o}{\omega_c}\frac{\pi}{2}} - e^{-jn\frac{\omega_o}{\omega_c}\frac{\pi}{2}} \\ + e^{-jn\frac{\omega_o}{\omega_c}\frac{\pi}{2}} e^{jq\frac{\pi}{2}(1 + M\cos y)} - e^{jn\frac{\omega_o}{\omega_c}\frac{\pi}{2}} e^{-jq\frac{\pi}{2}(1 + M\cos y)} \end{array}\right] dy \\ + \displaystyle\int_{-\psi}^{\psi} e^{jny}\left[e^{jn\frac{\omega_o}{\omega_c}\frac{\pi}{2}} - e^{-jn\frac{\omega_o}{\omega_c}\frac{\pi}{2}} + e^{-jn\frac{\omega_o}{\omega_c}\frac{\pi}{2}} e^{jq\pi} - e^{jn\frac{\omega_o}{\omega_c}\frac{\pi}{2}} e^{-jq\pi}\right] dy \end{array}\right]$$

$$
= \frac{V_{dc}}{jq\pi^2}\left[\begin{aligned}
&\int_{-\pi+\psi}^{-\psi} e^{jny}\left[\begin{aligned}&2j\sin\left(n\frac{\omega_o}{\omega_c}\frac{\pi}{2}\right)\\ &+e^{j\left[q-n\frac{\omega_o}{\omega_c}\right]\frac{\pi}{2}}e^{jq\frac{\pi}{2}M\cos y}-e^{-j\left[q-n\frac{\omega_o}{\omega_c}\right]\frac{\pi}{2}}e^{-jq\frac{\pi}{2}M\cos y}\end{aligned}\right]dy\\
&+\int_{\psi}^{\pi-\psi} e^{jny}\left[\begin{aligned}&2j\sin\left(n\frac{\omega_o}{\omega_c}\frac{\pi}{2}\right)\\ &+e^{j\left[q-n\frac{\omega_o}{\omega_c}\right]\frac{\pi}{2}}e^{jq\frac{\pi}{2}M\cos y}-e^{-j\left[q-n\frac{\omega_o}{\omega_c}\right]\frac{\pi}{2}}e^{-jq\frac{\pi}{2}M\cos y}\end{aligned}\right]dy\\
&+\int_{-\psi}^{\psi} e^{jny}\left[2j\sin\left(n\frac{\omega_o}{\omega_c}\frac{\pi}{2}\right)+2j\sin\left(q\pi-n\frac{\omega_o}{\omega_c}\frac{\pi}{2}\right)\right]dy
\end{aligned}\right]
\tag{A4.58}
$$

which finally produces

$$
C_{mn} = \frac{V_{dc}}{jq\pi^2}\left[\begin{aligned}
&\int_{-\pi+\psi}^{-\psi} e^{jny}\left[e^{jm\frac{\pi}{2}}e^{jq\frac{\pi}{2}M\cos y}-e^{-jm\frac{\pi}{2}}e^{-jq\frac{\pi}{2}M\cos y}\right]dy\\
&+\int_{\psi}^{\pi-\psi} e^{jny}\left[e^{jm\frac{\pi}{2}}e^{jq\frac{\pi}{2}M\cos y}-e^{-jm\frac{\pi}{2}}e^{-jq\frac{\pi}{2}M\cos y}\right]dy\\
&+\int_{-\pi+\psi}^{-\psi} 2je^{jny}\sin\left(n\frac{\omega_o}{\omega_c}\frac{\pi}{2}\right)dy+\int_{\psi}^{\pi-\psi} 2je^{jny}\sin\left(n\frac{\omega_o}{\omega_c}\frac{\pi}{2}\right)dy\\
&+\int_{-\psi}^{\psi} e^{jny}\left[2j\sin\left(n\frac{\omega_o}{\omega_c}\frac{\pi}{2}\right)+2j\sin\left(m\pi+n\frac{\omega_o}{\omega_c}\frac{\pi}{2}\right)\right]dy
\end{aligned}\right]
\tag{A4.59}
$$

The top two integral terms of Eq. (A4.59) can be recognized as the same form as the top two terms of Eq. (A4.19), except that m is replaced with n in the $e^{jq\frac{\pi}{2}M\cos y}$ and $e^{-jq\frac{\pi}{2}M\cos y}$ terms. Hence the solution of Eq. (A4.25) can be applied to this part of the integration with a simple variable substitution.

The last two integral terms of Eq. (A4.59) can be readily evaluated to

$$\frac{2}{n}\sin\left(n\frac{\omega_o}{\omega_c}\frac{\pi}{2}\right)\left[e^{-jn\psi}-e^{-jn[\pi-\psi]}+e^{jn[\pi-\psi]}-e^{jn\psi}+e^{jn\psi}-e^{-jn\psi}+(e^{jn\psi}-e^{-jn\psi})\cos m\pi\right]$$

$$= 4j\sin\left(n\frac{\omega_o}{\omega_c}\frac{\pi}{2}\right)(\cos m\pi-\cos n\pi)\sin n\psi \qquad \text{(A4.60)}$$

Gathering up these terms gives a complete solution for C_{mn} of

$$C_{mn}=\frac{4V_{dc}}{q\pi^2}\left[\begin{array}{l}\dfrac{1}{n}\left[\begin{array}{l}\sin\left(n\dfrac{\omega_o}{\omega_c}\dfrac{\pi}{2}\right)(\cos m\pi-\cos n\pi)\\ \qquad -J_0\left(q\dfrac{\pi}{2}M\right)\sin m\dfrac{\pi}{2}(1+\cos n\pi)\end{array}\right]\sin n\psi\\ +J_n\left(q\dfrac{\pi}{2}M\right)\sin\left([m+n]\dfrac{\pi}{2}\right)\left[\pi-2\psi-\dfrac{1}{n}\sin 2n\psi\right]\\ -\displaystyle\sum_{\substack{k=1\\k\neq|n|}}^{\infty}\left[\begin{array}{l}J_k\left(q\dfrac{\pi}{2}M\right)\sin\left([m+k]\dfrac{\pi}{2}\right)\{1+\cos([n+k]\pi)\}\\ \qquad\times\left[\dfrac{\sin([n+k]\psi)}{[n+k]}+\dfrac{\sin([n-k]\psi)}{[n-k]}\right]\end{array}\right]\end{array}\right] \qquad \text{(A4.61)}$$

For $m = 0, n > 0$, q reduces to $n(\omega_o/\omega_c)$, and Eq. (A4.61) is still valid.

A4.3.2 Harmonic Solution for Overmodulated Single-Phase Leg under Asymmetric Regular Sampled PWM

Assembling Eqs. (A4.54), (A4.56), and (A4.61), the full solution for asymmetric regular sampled PWM of a single-phase leg under overmodulation is

$$
\begin{aligned}
v_{az}(t) &= V_{dc} + \sum_{n=1}^{\infty} C_{0n}\cos n\omega_o t \qquad \text{(A4.62)}\\
&+ \frac{4V_{dc}}{\pi^2}\sum_{m=1}^{\infty}\frac{1}{m}\sin m\frac{\pi}{2}\left[\begin{array}{l} J_0\left(m\frac{\pi}{2}M\right)(\pi-2\psi) \\ \quad -2\sum_{k=1}^{\infty}\frac{1}{k}J_{2k}\left(m\frac{\pi}{2}M\right)\cos k\pi\sin 2k\psi \end{array}\right]\cos m\omega_c t\\
&+ \sum_{m=1}^{\infty}\sum_{\substack{n=-\infty\\(n\neq 0)}}^{\infty} C_{mn}\cos(m\omega_c t + n\omega_o t)
\end{aligned}
$$

where C_{0n} is given by

$$
C_{0n} = \frac{4V_{dc}}{\left[n\frac{\omega_o}{\omega_c}\right]\pi^2}\left[\begin{array}{l}
\frac{1}{n}\sin\left(n\frac{\omega_o}{\omega_c}\frac{\pi}{2}\right)(1-\cos n\pi)\sin n\psi \\
\qquad + J_n\left(n\frac{\omega_o}{\omega_c}\frac{\pi}{2}M\right)\sin n\frac{\pi}{2}\left[\pi - 2\psi - \frac{1}{n}\sin 2n\psi\right] \\
- \sum_{\substack{k=1\\k\neq|n|}}^{\infty}\left[\begin{array}{l} J_k\left(n\frac{\omega_o}{\omega_c}\frac{\pi}{2}M\right)\sin k\frac{\pi}{2}\{1+\cos([n+k]\pi)\} \\ \qquad \times\left[\frac{\sin([n+k]\psi)}{[n+k]} + \frac{\sin([n-k]\psi)}{[n-k]}\right]\end{array}\right]
\end{array}\right]
$$

(A4.63)

and C_{mn} is given by

$$C_{mn} = \frac{4V_{dc}}{q\pi^2}\left[\begin{array}{l} \frac{1}{n}\left[\begin{array}{l}\sin\left(n\frac{\omega_o}{\omega_c}\frac{\pi}{2}\right)(\cos m\pi - \cos n\pi) \\ \qquad - J_0\left(q\frac{\pi}{2}M\right)\sin m\frac{\pi}{2}(1+\cos n\pi)\end{array}\right]\sin n\psi \\ + J_n\left(q\frac{\pi}{2}M\right)\sin\left([m+n]\frac{\pi}{2}\right)\left[\pi - 2\psi - \frac{1}{n}\sin 2n\psi\right] \\ - \sum_{\substack{k=1 \\ k \neq |n|}}^{\infty}\left[\begin{array}{l} J_k\left(q\frac{\pi}{2}M\right)\sin\left([m+k]\frac{\pi}{2}\right)\{1+\cos([n+k]\pi)\} \\ \qquad \times\left[\frac{\sin([n+k]\psi)}{[n+k]} + \frac{\sin([n-k]\psi)}{[n-k]}\right]\end{array}\right]\end{array}\right] \tag{A4.64}$$

with $q = m + n(\omega_o/\omega_c)$ and ψ defined by Eq. (A4.28).

A4.3.3 Linear Modulation Solution Obtained from Overmodulation Solution

Once more $\psi = 0$ when the phase leg is modulated in the linear region and Eq. (A4.62) can be reduced to

$$\begin{aligned} v_{az}(t) = V_{dc} &+ \frac{4V_{dc}}{\pi}\sum_{n=1}^{\infty}\frac{1}{\left[n\frac{\omega_o}{\omega_c}\right]}J_n\left(n\frac{\omega_o}{\omega_c}\frac{\pi}{2}M\right)\sin n\frac{\pi}{2}\cos n\omega_o t \\ &+ \frac{4V_{dc}}{\pi}\sum_{m=1}^{\infty}\frac{1}{m}J_0\left(m\frac{\pi}{2}M\right)\sin m\frac{\pi}{2}\cos m\omega_c t \\ &+ \frac{4V_{dc}}{\pi}\sum_{m=1}^{\infty}\sum_{\substack{n=-\infty \\ (n\neq 0)}}^{\infty}\frac{1}{q}J_n\left(q\frac{\pi}{2}M\right)\sin\left([m+n]\frac{\pi}{2}\right)\cos(m\omega_c t + n\omega_o t) \end{aligned} \tag{A4.65}$$

This result is the same as Eq. (3.98) obtained in Chapter 3. Note also that once again there are no even baseband harmonic components in this solution, just as for the development in Chapter 3.

Appendix 5

Numeric Integration of a Double Fourier Series Representation of a Switched Waveform

A5.1 Formulation of the Double Fourier Integral

Appendix 1 has developed the double Fourier integral expression for a PWM switched waveform, $f(t) = f(x, y)$, as

$$f(x,y) = \frac{A_{00}}{2} + \sum_{n=1}^{\infty}[A_{0n}\cos ny + B_{0n}\sin ny] + \sum_{m=1}^{\infty}[A_{m0}\cos mx + B_{m0}\sin mx] \tag{A5.1}$$

$$+ \sum_{m=1}^{\infty}\sum_{\substack{n=-\infty \\ (n\neq 0)}}^{\infty}[A_{mn}\cos(mx+ny) + B_{mn}\sin(mx+ny)]$$

where

$$A_{mn} + jB_{mn} = \frac{1}{2\pi^2}\int_{-\pi}^{\pi}\int_{-\pi}^{\pi} f(x,y)e^{j(mx+ny)}\,dx\,dy \tag{A5.2}$$

and

$$x(t) = \omega_c t + \theta_c \tag{A5.3}$$

$$y(t) = \omega_o t + \theta_o \tag{A5.4}$$

With this representation, for all PWM algorithms the process of determining the magnitudes of the various harmonics that make up the switched waveform becomes the process of evaluating the double Fourier integral, Eq. (A5.2), as the switched output changes state. For two-level inverters, the switched output can only take the two values of 0 and $2V_{dc}$. For multilevel inverters, the switched output can vary in discrete steps from 0 to $V_{dc(\text{total})}$. (It should be noted that $2V_{dc}$ is the usual way of defining the overall DC bus voltage for a two-level inverter, while V_{dc} is often used to define the overall sum of the series DC voltages for a multilevel inverter.)

Furthermore, the switched output only changes state at defined times within each carrier interval as determined by the intersection between the (usually triangular) carrier waveform and the fundamental reference waveform. (It should be noted also that the reference waveform is not necessarily sinusoidal.) Assuming that only one rising and one falling switched transition occur in each carrier interval (the usual arrangement), the instant of these transitions can be defined as x_r and x_f, respectively. Hence Eq. (A5.2) can be alternatively expressed as

$$A_{mn} + jB_{mn} = \frac{1}{2\pi^2}\int_{-\pi}^{\pi}\left(\int_{-\pi}^{x_r} f_l e^{j(mx+ny)}dx + \int_{x_r}^{x_f} f_u e^{j(mx+ny)}dx + \int_{x_f}^{\pi} f_l e^{j(mx+ny)}dx\right)dy \tag{A5.5}$$

where f_l is the value of $f(x,y)$ before the rising and after the falling switched edges, and f_u is the value of $f(x,y)$ between these edges. Note that for two-level modulation, $f_l = 0$ and $f_u = 2V_{dc}$ and Eq. (A5.5) reduces to

$$A_{mn} + jB_{mn} = \frac{V_{dc}}{\pi^2}\int_{-\pi}^{\pi}\int_{x_r}^{x_f} e^{j(mx+ny)}dx\,dy \tag{A5.6}$$

For simplicity of explanation, only this form will be considered further in this appendix.

More complex reference waveforms such as those used in space vector modulation are made up from segments of sinusoids, and hence the outer integral of Eq. (A5.6) cannot be evaluated as a single continuous integral. However, it can be restated as a summation of integral segments, i.e.,

$$A_{mn} + jB_{mn} = \frac{V_{dc}}{\pi^2}\sum_{i=1}^{j}\int_{y_s(i)}^{y_e(i)}\int_{x_r(i)}^{x_f(i)} e^{j(mx+ny)}dx\,dy \tag{A5.7}$$

where j is the number of segments in the reference waveform and $y_s(i)$ and $y_e(i)$ are the start and finish angles of segment i. For space vector modulation, $j = 6$; for 120° discontinuous modulation, $j = 3$ (of which one segment always has a magnitude of zero and can be discarded), and so on.

It should be noted also that for naturally sampled modulation, $x_r(i) = -x_f(i)$ under all conditions.

The further complication of regular sampled modulation is managed by changing the reference waveform variable y into a sampled staircase waveform as presented in Chapter 3, using the substitutions of

$$y' = y - \frac{\omega_o}{\omega_c}x \tag{A5.8}$$

for symmetrical regular sampled modulation, and

$$y_r' = y - \frac{\omega_o}{\omega_c}\left(x + \frac{\pi}{2}\right) \qquad y_f' = y - \frac{\omega_o}{\omega_c}\left(x - \frac{\pi}{2}\right) \tag{A5.9}$$

for asymmetrical regular sampled modulation.

Under these conditions, Eq. (A5.7) becomes

$$A_{mn} + jB_{mn} = \frac{V_{dc}}{\pi^2}\sum_{i=1}^{j}\int_{y_s(i)}^{y_e(i)}\int_{x_r(i)}^{x_f(i)} e^{j\left(\left[m + n\frac{\omega_o}{\omega_c}\right]x + ny'\right)}\,dx\,dy' \tag{A5.10}$$

for double-edge symmetrical regular sampled modulation, and

$$A_{mn} + jB_{mn} = \frac{V_{dc}}{\pi^2}\sum_{i=1}^{j}\left[\int_{y_s(i)}^{y_e(i)}\int_{x_r(i)}^{0} e^{j\left(mx + n\left[y_r' + \frac{\omega_o}{\omega_c}x + \frac{\omega_o}{\omega_c}\frac{\pi}{2}\right]\right)}\,dx\,dy_r' + \int_{y_s(i)}^{y_e(i)}\int_{0}^{x_f(i)} e^{j\left(mx + n\left[y_f' + \frac{\omega_o}{\omega_c}x - \frac{\omega_o}{\omega_c}\frac{\pi}{2}\right]\right)}\,dx\,dy_f'\right] \tag{A5.11}$$

for double-edge asymmetrical regular sampled modulation.

The substitution of $q = m + n(\omega_o/\omega_c)$ allows Eqs. (A5.10) and (A5.11) to be simplified to

$$A_{mn} + jB_{mn} = \frac{V_{dc}}{\pi^2}\sum_{i=1}^{j}\int_{y_s(i)}^{y_e(i)}\int_{x_r(i)}^{x_f(i)} e^{j(qx + ny')}\,dx\,dy' \tag{A5.12}$$

for double-edge symmetrical regular sampled modulation, and

$$A_{mn} + jB_{mn} = \frac{V_{dc}}{\pi^2} \sum_{i=1}^{j} \left[\int_{y_s(i)}^{y_e(i)} \int_{x_r(i)}^{0} e^{j\left(qx + ny_r' + n\frac{\omega_o \pi}{\omega_c 2}\right)} dx\, dy_r' + \int_{y_s(i)}^{y_e(i)} \int_{0}^{x_f(i)} e^{j\left(qx + ny_f' - n\frac{\omega_o \pi}{\omega_c 2}\right)} dx\, dy_f' \right] \quad \text{(A5.13)}$$

for double-edge asymmetrical regular sampled modulation.

A5.2 Analytical Solution of the Inner Integral

For $\underline{m = 0}$, the inner integral of Eq. (A5.7) can be evaluated to

$$A_{0n} + jB_{0n} = \frac{V_{dc}}{\pi^2} \sum_{i=1}^{j} \int_{y_s(i)}^{y_e(i)} e^{jny} \{x_f(i) - x_r(i)\} dy \quad \text{(A5.14)}$$

for naturally sampled modulation only. This expression can in turn usually be readily evaluated since the inner integral limits $x_r(i)$ and $x_f(i)$ almost always contain sinusoidal expressions with y as the argument, and these expressions combine with the e^{jny} term using trigonometric expansions into easily integrable single-term sinusoidal expressions.

For $\underline{m \neq 0}$, and for sampled modulation strategies in any case, the inner integral of Eqs. (A5.7), (A5.12), or (A5.13) as appropriate is readily evaluated to give

$$A_{mn} + jB_{mn} = \frac{V_{dc}}{jm\pi^2} \sum_{i=1}^{j} \int_{y_s(i)}^{y_e(i)} e^{jny} (e^{jmx_f(i)} - e^{jmx_r(i)})\, dy \quad \text{(A5.15)}$$

for naturally sampled modulation,

$$A_{mn} + jB_{mn} = \frac{V_{dc}}{\pi^2} \sum_{i=1}^{j} \int_{y_s(i)}^{y_e(i)} e^{jny} (e^{jqx_f(i)} - e^{jqx_r(i)})\, dy \quad \text{(A5.16)}$$

for symmetrical regular sampled modulation, and

$$A_{mn} + jB_{mn} = \frac{V_{dc}}{\pi^2} \sum_{i=1}^{j} \left[\int_{y_s(i)}^{y_e(i)} e^{jny'_r} e^{jn\frac{\omega_o}{\omega_c}\frac{\pi}{2}} (1 - e^{jqx_r(i)})\, dy'_r + \int_{y_s(i)}^{y_e(i)} e^{jny'_f} e^{-jn\frac{\omega_o}{\omega_c}\frac{\pi}{2}} (e^{jqx_f(i)} - 1)\, dy'_f \right] \tag{A5.17}$$

for asymmetrical regular sampled modulation.

Equation (A5.17) can be simplified by combining the two integral expressions to use a common integration variable y, so that it becomes

$$A_{mn} + jB_{mn} = \frac{V_{dc}}{\pi^2} \sum_{i=1}^{j} \int_{y_s(i)}^{y_e(i)} e^{jny} \left[e^{jn\frac{\omega_o}{\omega_c}\frac{\pi}{2}} (1 - e^{jqx_r(i)}) + e^{-jn\frac{\omega_o}{\omega_c}\frac{\pi}{2}} (e^{jqx_f(i)} - 1) \right] dy$$

$$= \frac{V_{dc}}{\pi^2} \sum_{i=1}^{j} \int_{y_s(i)}^{y_e(i)} e^{jny} \left[\left(e^{jn\frac{\omega_o}{\omega_c}\frac{\pi}{2}} - e^{-jn\frac{\omega_o}{\omega_c}\frac{\pi}{2}} \right) + e^{-jn\frac{\omega_o}{\omega_c}\frac{\pi}{2}} e^{jqx_f(i)} - e^{jn\frac{\omega_o}{\omega_c}\frac{\pi}{2}} e^{jqx_r(i)} \right] dy$$

$$= \frac{V_{dc}}{\pi^2} \sum_{i=1}^{j} \int_{y_s(i)}^{y_e(i)} e^{jny} \left[2j \sin n \frac{\omega_o}{\omega_c}\frac{\pi}{2} + e^{-jn\frac{\omega_o}{\omega_c}\frac{\pi}{2}} e^{jqx_f(i)} - e^{jn\frac{\omega_o}{\omega_c}\frac{\pi}{2}} e^{jqx_r(i)} \right] dy \tag{A5.18}$$

Since Eq. (A5.18) must always be evaluated over a complete fundamental cycle, the first term of this expression sums to

$$\int_{-\pi}^{\pi} 2j \sin n \frac{\omega_o}{\omega_c}\frac{\pi}{2} e^{jny}\, dy = 0 \tag{A5.19}$$

and hence can be ignored. Equation (A5.18) therefore reduces to

$$A_{mn} + jB_{mn} = \frac{V_{dc}}{\pi^2} \sum_{i=1}^{j} \int_{y_s(i)}^{y_e(i)} e^{jny} \left[e^{-jn\frac{\omega_o}{\omega_c}\frac{\pi}{2}} e^{jqx_f(i)} - e^{jn\frac{\omega_o}{\omega_c}\frac{\pi}{2}} e^{jqx_r(i)} \right] dy \tag{A5.20}$$

The difficulty in proceeding from this point in general is that since the inner integral limits $x_r(i)$ and $x_f(i)$ almost always contain sinusoidal expressions with y as the argument, they create $e^{j\lambda\cos y}$ terms in the outer integrals of Eqs. (A5.15), (A5.16), and (A5.20). This makes these equations complex and difficult to analytically solve. The alternative approach presented here is to solve them numerically only, and to simply evaluate them for particular values of m and n.

A5.3 Numeric Integration of the Outer Integral

Substitution of the appropriate inner integral limits into Eqs. (A5.15), (A5.16), and (A5.20) creates smoothly varying expressions in y, which can readily be evaluated using numeric integration. Furthermore, since these expressions only include the carrier frequency as a constant, they can be evaluated using relatively coarse time steps over the fundamental period. Hence the computational requirements to determine any harmonic component are essentially fixed irrespective of the carrier-to-fundamental ratio. And as an added bonus, there is no requirement to use an integer pulse ratio between the carrier and the fundamental component.

The approach is illustrated here for the case of asymmetrical regular sampled PWM with one-sixth third-harmonic injection. For this case, the inner integration limits become

$$x_r = -\frac{\pi}{2}\left(1 + M\cos y - \frac{M}{6}\cos 3y\right) \tag{A5.21}$$

$$x_f = \frac{\pi}{2}\left(1 + M\cos y - \frac{M}{6}\cos 3y\right) \tag{A5.22}$$

Note that for this example, only one outer integral summation term is required, since the reference waveform definition of $M\cos y - M\cos 3y/6$ is continuous.

Substitution of these limits into Eq. (A5.20) gives

$$A_{mn} + jB_{mn} = \frac{V_{dc}}{\pi^2}\int_{-\pi}^{\pi} e^{jny}\left[\begin{matrix} e^{-jn\frac{\omega_o}{\omega_c}\frac{\pi}{2}} e^{jq\frac{\pi}{2}\left\{1 + M\cos y - \frac{M}{6}\cos 3y\right\}} \\ - e^{jn\frac{\omega_o}{\omega_c}\frac{\pi}{2}} e^{-jq\frac{\pi}{2}\left\{1 + M\cos y - \frac{M}{6}\cos 3y\right\}} \end{matrix}\right] dy \tag{A5.23}$$

which simplifies to

$$A_{mn} + jB_{mn} = \frac{V_{dc}}{\pi^2} \int_{-\pi}^{\pi} e^{jny} \left[e^{jm\frac{\pi}{2}} e^{jq\frac{\pi}{2}\left\{M\cos y - \frac{M}{6}\cos 3y\right\}} - e^{-jm\frac{\pi}{2}} e^{-jq\frac{\pi}{2}\left\{M\cos y - \frac{M}{6}\cos 3y\right\}} \right] dy \qquad \text{(A5.24)}$$

Equation (A5.24) is readily evaluated numerically over the range -π to π to determine the co-efficients of harmonics defined by any values of m and n.

Whilst it is not a closed form solution, numerical integration of the outer Fourier integral provides an easy and reliable way to determine the magnitude and phase of harmonics for any type of PWM strategy. And in particular for this text, numerical integration has been used extensively to verify the various closed form solutions that have been presented.

Bibliography

[1] L. Abraham and R. Blümel, "Optimization of three phase pulse pattern by variable zero sequence component," in *Conf. Rec. European Power Electronics Conf. (EPE)*, Florence, 1991, pp. 3–272 to 3–277.

[2] L.A. Aga and L. Norum, "Compact control for a PWM-VSI fed asynchronous motor drive," in *Conf. Rec. of Int. Conf. on Industrial Electronics, Control and Instrumentation (IECON)*, Maui, 1993, pp. 1129–1133.

[3] V.G. Agelidis and D. Vincenti, "Optimum non-deterministic pulse-width modulation for three phase inverters," in *Conf. Rec. of Int. Conf. on Industrial Electronics, Control and Instrumentation (IECON)*, Maui, 1993, pp. 1234–1239.

[4] V.G. Agelidis and D. Vincenti, "Non-deterministic AM-PWM strategy for three-phase VSI," in *Conf. Rec. of Int. Conf. on Industrial Electronics, Control and Instrumentation (IECON)*, Bologna, 1994, pp. 73–78.

[5] V.G. Agelidis and M. Calais, "Application specific harmonic performance evaluation of multicarrier PWM techniques," in *Conf. Rec. Power Electronics Specialists Conference, (PESC)*, Fukuoka,1998, pp. 172–178.

[6] T.-H. Ai, J.-F. Chen, and T.-J. Liang, "A random switching method for HPWM full-bridge inverters," *IEEE Trans. on Industrial Electronics*, vol. 49, no. 3, June 2002, pp. 595–597.

[7] M. Alakula, "Vector controlled ac/ac converters with a minimum of energy storage," in *Conf. Rec. Industry Applications Society Annual Mtg.,* Denver, 1994, pp. 1130–1134.

[8] A. Alesina and M. Venturini, "Solid-state power conversion: a fourier analysis approach to generalized transformer synthesis," *IEEE Trans. on Circuits and Systems,* vol. CAS–28, no. 4, April 1981, pp. 319–330.

[9] A. Alesina and M. Venturini, "Intrinsic amplitude limits and optimum design of 9–switch direct PWM ac-ac converters," in *Conf. Rec. IEEE Power Electronics Specialists Conf. (PESC)*, Kyoto, 1988, pp. 1284–1291.

[10] A. Alesina and M. Venturini, "Analysis and design of optimum-amplitude nine-switch direct ac-ac converters," *IEEE Trans. Power Electronics,* vol. 4, no. 1, Jan. 1989, pp. 101–112.

[11] D.R. Alexander and S.M. Williams, "An optimal PWM algorithm implementation in a high performance 125 kVA inverter," in *Conf. Rec. Applied Power Electronics Conf. and Exposition, (APEC)*, San Diego, 1993, pp. 771–771.

[12] P.L. Alger, G. Angst, and E.J. Davies, "Stray-load losses in polyphase induction machines," *AIEE Trans. Power Applications & Systems*, vol. 78, pt. III–A, June 1949, pp. 349–357.

[13] K.A. Almarri, J.C. Balda, and K. Carr, "Optimized selection of the random PWM switching frequency in a limited pool," in *Conf. Rec. Applied Power Electronics Conf. and Exp.(APEC)*, New Orleans, 2000, pp. 569–573.

[14] K.A. Almarri, J.C. Balda, and K. Carr, "An improved RPWM method based on a novel multi-level congruential random-number generator," in *Conf. Rec. Power Electronics Specialists Conf. (PESC)*, Galway, 2000, pp. 218–223.

[15] A. Alolah, L.N. Hulley, and W. Shepherd, "A three-phase neutral point clamped inverter for motor cotnrol," in *Conf. Rec. IEE Conf. on Power Electronics and Variable Speed Drives*, Birmingham, 1987, pp. 144–148.

[16] O. Alonso-Sadaba, J. Lopez-Taberna, I. Munoz-Morales, and P. Sanchis-Gurpide, "Voltage harmonics generator by 3 level converters using PWM natural sampling," in *Conf. Rec. IEEE Power Electronics Specialists Conf. (PESC)*., Vancouver, 2001, pp. 1561–1565

[17] R.N.C. Alves, "Analysis and implementation of the theory of PWM for use in three phase inverters," Ph.D. Thesis (in Portuguese), 1998.

[18] R.N.C. Alves, E.R.C. da Silva, A.M.N. Lima, and C.B. Jacobina, "Reduced-loss PWM strategy for three-phase voltage-type inverters," in *Conf. Rec. Brazilian Conf. on Power Electronics (COBEP)*, Brazil, 1999, pp. 181–186.

[19] R. Alves, R. Orizondo, G. Pesse, and R. Rojas, "ATP/ EMTP simulation of a controlled three-phase rectifier with PWM for power factor improvement and harmonic reduction," in *Conf. Rec. Int. Conf. on Electrical Machines (ICEM)*, Brugge, 2002, in CD ROM.

[20] G. Amler, "A PWM current-source inverter for high quality drives," *European Power Electronics Journal*, vol. 1, no. 1, July 1991, pp. 21–32.

[21] A. Ammous, Y. Ounajjar, M. Ayedi, K.Al Haddad, and F. Sallami, "PWM-switch modeling of converters including semiconductor device non-linearities," in *Conf. Rec. 7th International Conf. on Modeling and Simulation of Electric Machines, Converters and Systems (ElectIMACS)*,2002, in CD ROM.

[22] E.C. Andresen, S. Gediga, and H.J. Schwartz, "A continuous PWM-square wave transition method for voltage vector control of induction motors," in *Conf. Rec. European Conf. on Power Electronics and Applications (EPE)*, Florence, 1991, pp. 2/156–2/160.

[23] M. Andrews and J.T. Boys, "Improvements in estimating the spectra of random PWM waveforms," *Electronics Letters*, Oct. 1993, vol. 29, No. 21, pp. 1822–23.

[24] C. Andrieuz and M. Lajoie-Mazenc, "Analysis of different current control systems for inverter-fed synchronous machine," in *Conf Rec. European Power Electronics Conf. (EPE)*, Brussels, 1985 pp. 2.159–2.165.

[25] V.G. Angelidis and M. Calais, "Application specific harmonic performance evaluation multicarrier PWM techniques," in *Conf. Rec. IEEE Power Electronics Specialists Conf. (PESC)*, Kyoto, 1988, pp. 172–178.

[26] V.G. Angelicas, P.D. Ziogas, and G. Joos, ""Dead-band" PWM switching patterns," *IEEE Trans. on Power Electronics,* vol. 11, no. 4, July 1996, pp. 522–531.

[27] C. Attaianese and G. Tomasso, "Optimized modulation for PWM rectifiers," in *Conf. Rec. 2002 Power Engineering Society Winter Meeting*, vol. 2, pp. 1264–1269.

[28] C. Bacigalupo, M. Baragli, M. Carpita, L. Puglisi, and G. Striano, "Modulation techniques applied to high power 3 level GTO's converter for asynchronous motor drives," in *Conf. Rec. European Power Electronics Conf.* (*EPE*), Trondheim, 1997, pp. 1.58–1.64.

[29] A.R. Bakhshai, G. Joos, P.R.Jain, and H. Jin, "Incorporating the overmodulation range in space vector patter generators using a classification algorithm," *IEEE Trans on Power Electronics*, vol. 15, no. 1, Jan. 2000, pp. 83–91.

[30] T.H. Barton, "Pulse-width modulation waveform – the bessel approximation," in *Conf. Rec. IEEE Industry Applications Society Annual Mtg.,* Toronto, 1978, pp. 1125–1130.

[31] A. Baskys and V. Gobis, "Generation of 3–phase voltage using space vector modulation method," in *Conf. Record of 9th European Conf. on Power Electronics and Applications* (*EPE*), Graz, 2001, in CD ROM.

[32] F. Bauer and H.D. Heining, "Quick response space vector control for a high power three-level-inverter drive system," in *Conf. Rec. European Conference on Power Electronics and Applications*, Aachen, 1989, pp. 417–421.

[33] M.M. Bech, J.K. Pedersen, F. Blaabjerg, and A.M. Trzynadlowski, "A methodology for true comparison of analytical and measured frequency domain spectra in random PWM converters," *IEEE Trans. on Power Electronics*, vol. 14, no. 3, May 1999, pp. 578–586.

[34] M. Bech, F. Blaabjerg, and J. K. Pedersen "Random modulation techniques with fixed switching frequency for three-phase power converters," *IEEE Trans. on Power Electronics*, vol. 15, no. 4, pp.753–761, 2000.

[35] G. Beinhold, F.R. Jakob, and M. Nahrstaedt, "A new range of medium voltage multilevel inverter drives with floating capacitor technology," in *Conf. Rec.European Power Electronics Conf.* (*EPE*), Graz, 2001, in CD–ROM.

[36] A. Bellini and G. Figalli, "On the selection of the commutating instants for induction motor drives," *IEEE Trans. on Industry Applications*, vol. IA–15, no. 5, Sept./Oct 1979, pp. 501–506.

[37] A. Bellini, G. Figalli, and G. Ulivi, "A three-phase modulation technique suitable to supply induction motors," in *Conf. Rec. of International Power Electronics Conf.*, Tokyo, 1983, pp. 396–406.

[38] A. Ben Abdeighani, C.A. Martins, X. Roboam, and T.A. Maynard, "Use of extra degrees of freedom in multilevel drives," *IEEE Trans. on Industrial Electronics*, vol. 49, no. 5, Oct. 2002, pp. 965–977.

[39] L. Ben-Brahim,."The analysis and compensation of dead-time effect in three-phase PWM inverters," in *Conf. Rec. of Int. Conf. on Industrial Electronics, Control and Instrumentation* (*IECON*), Aachen, pp. 792–797, 1998.

[40] A.R. Bendre, I.T. Wallace, J. Nord, and G. Venkataramanan, "A current source PWM inverter with actively commutated SCRs," in *Conf. Rec. IEEE Power Electronics Specialists Conf.* (*PESC*), Vancouver, 2001, vol. 3, pp. 1571–1576.

[41] W.R. Bennett, "New results in the calculation of modulation products," *The Bell System Technical Journal*, vol. 12, April 1933, pp. 228–243.

[42] P. Bhagwat and V.R. Stefanovic, "Generalized structure of a multilevel PWM inverter," *IEEE Trans. on Industry Applications*, vol. 19, no. 6, Nov./Dec. 1983, pp. 1057–1069.

[43] S. Bhattacharya, D.G. Holmes, and D.M. Divan, "Optimizing three phase current regulators for low inductance loads," in *Conf. Rec. IEEE Industry Applications Society Annual Mtg.*, Orlando, 1995, pp. 2357–2366.

[44] P. P. Biringer and I. Nagy, "General analysis of voltage sourced inverters with tolerance band control," in *Conf. Rec. Electrical Drives & Power Electronics Conf.,* Kosice, 1992, pp. 213-218.

[45] F. Blaabjerg and J.K. Pedersen, "Digital quasi-random modulated SFAVM PWM in an AC drive system," *IEEE Trans. on Industrial Electronics*, vol. 41, Oct. 1994, pp. 518–525.

[46] F. Blaabjerg, J.K. Pedersen, and P. Thoegersen, "Improved modulation techniques for PWM-VSI drive," *IEEE Trans. on Industrial Electronics*, vol. 44, no. 1, Feb. 1997, pp. 87–95.

[47] H.S. Black, *Modulation Theory*, Van Nostrand, New York, 1953.

[48] V. Blasko, "Analysis of a hybrid PWM based on modified space-vector and triangle-comparison methods," *IEEE Trans. on Industry Applications*, vol. 33, May/June 1997, pp. 525–530.

[49] V. Blasko, M.M. Bech, F. Blaabjerg, and J.K. Pedersen, "A new hybrid random pulse width modulator for industrial drives," in *Conf. Rec. Applied Power Electronics Conf. and Exp.* (*APEC*), New Orleans, 2000, pp. 932–938.

[50] A. Boglietti, G. Griva, M. Pastorelli, F. Profumo, and T. Adam, "A comparison of different modulation techniques," in *Conf. Rec. Electrical Drives & Power Electronics Conf.*, Kosice, 1992, pp. 525–531.

[51] S. Bolognani, R. Conton, and M. Zigliotto, "Experimental analysis of the EMI reduction in PWM inverters using random space vector modulation," in *Conf. Rec. Int. Symp. on Industrial Electronic*s, Warsaw, 1996, pp. 482–487.

[52] R. Bonert and R.S. Wu, "Improved three phase pulsewidth modulation for overmodulation," *IEEE Trans. on Industry Applications,* vol. IA–20, no. 5, Sept./Oct. 1985, pp. 1224–1228.

[53] M.A. Boost and P.D. Ziogas, "State-of-the-art carrier PWM techniques: a critical evaluation," *IEEE Trans. on Industry Applications,* vol. 24, no. 2, Mar., 1988, pp. 271–290.

[54] J. Bordanau, M. Cosan, D. Borojevic, H. Mao, and F.C. Lee, "State-space model for the comprehensive dynamic analysis of three-level voltage-source inverters," in *Conf. Rec. IEEE Power Electronics Specialists Conf.* (*PESC*), St. Louis, 1997, pp. 942–948.

[55] B.K. Bose and H.A. Sutherland, "A high-performance pulsewidth modulator for an inverter-fed drive system using a microcomputer," *IEEE Trans. on Industry Applications*, vol. IA–10, no. 2, Mar./April 1983, pp. 235–243.

[56] J.M. Bourgeois, "PWM technique for acoustic noise reduction in power apparatus," in *Conf. Rec. of Applied Power Electronics Conf.* (*APEC*), San Diego, 1993, pp. 141–145.

[57] S.R. Bowes, "New sinusoidal pulse-width modulated inverter," *IEE Proceedings* (*London*), vol. 122, no. 11, Nov. 1975, pp. 1279–1285.

[58] S.R. Bowes, "Developments in PWM switching strategies for microprocessor-controlled inverter drives," in *Conf. Rec. IEEE Industry Applications Society Annual Mtg.*, Denver, 1987, pp. 323–329.

[59] S.R. Bowes, "Regular-sampled harmonic elimination/minimisation PWM technique, in *Conf. Rec. IEEE Applied Power Electronics Conf.* (*APEC*), Los Angeles, 1990, pp. 532–540.

[60] S.R. Bowes, "Novel real-time harmonic minimized PWM control for drives and static power converters," *IEEE Trans. on Power Electronics*, vol. 9, no. 3, May 1994, pp. 256–262.

[61] S. Bowes and B.M. Bird, "Novel approach to the analysis and synthesis of modulation processes in power converters," *IEE Proceedings* (*London*), vol. 122, no. 5, May 1975, pp. 507–513.

[62] S.R. Bowes and R. Bullough, "PWM switching strategies for current-fed inverter drives," *IEE Proceedings* (*London*), vol. 131 Pt. B, no. 5., Sept. 1984, pp. 195–202.

[63] S.R. Bowes and R.I. Bollough, "Optimal PWM microprocessor-controlled current-source inverter drive," *IEE Proc. B – Electric Power Applications,* vol. 135, no. 2, Mar. 1988, pp. 59–75.

[64] S.R.Bowes and P.R. Clark, "Transputer based optimal PWM control of inverter drives," in *Conf. Rec. of the IEEE Industry Applications Society Annual Meeting,* Pittsburgh, 1988, vol. 1, pp. 314–321.

[65] S.R. Bowes and P.R. Clark, "Transputer based harmonic-elimination PWM control of inverter drives," in *Conf. Rec. IEEE Industry Applications Society Annual Meeting,* San Diego, 1989, vol. 1, pp. 744–752.

[66] S.R. Bowes and P.R. Clark, "Simple microprocessor implementation of new regular-sampled harmonic elimination PWM techniques," in *Conf. Rec. IEEE Industry Applicatons Society Annual Meeting*, Seattle, 1990, vol. 1, pp. 341–347.

[67] S.R. Bowes and P.R. Clark, "Transputer based optimal PWM control of inverter drives," in *IEEE Trans. on Industry Applications,* vol. 28, no. 1, Jan./Feb.1992, pp. 81–88.

[68] S.R. Bowes and P.R. Clark, "Simple microprocessor implementatioin of new regular-sampled harmonic elimination PWM techniques," *IEEE Trans. on Industry Applications*, vol. 28, no. 1, Jan./Feb. 1992, pp. 89–95.

[69] S.R. Bowes and P.R. Clark, "Regular-sampled harmonic-elimination PWM control of invertrer drives," *IEEE Trans. on Power Electronics*, vol. 10, no. 5, Sept. 1995, pp. 521–531.

[70] S.R. Bowes and T. Davies, "Microprocessor-based development system for PWM variable-speed drives," *IEE Proceedings* (*London*), vol. 132, Pt. B, no. 1, Jan. 1985, pp. 18–45.

[71] S.R. Bowes and S. Grewal, "Simplified harmonic elimination PWM control strategy," *Electronics Letters*, vol. 34, no. 4, 19 Feb. 1998, pp. 325–326.

[72] S.R. Bowes and S. Grewal, "Modulation strategy for single phase PWM inverters," *Electronics Letters*, vol. 34, no. 5, 5th Mar. 1998, pp. 420–422.

[73] S.R. Bowes and S. Grewal, "A novel harmonic eliminatioin PWM strategy," in *Conf. Rec. IEE Int. Conf. on Power Electronics and Variable Speed Drives*, London, 1998, pp. 426–432.

[74] S.R. Bowes and S. Grewal, "Novel harmonic elimination PWM control strategies for three-phase PWM inverters using space vector techniques," *IEE Proc.*(*London*) *-Electr. Power Appl.*, vol. 146, no. 5, Sept. 1999, pp. 495–514.

[75] S.R. Bowes, S. Grewal, "Novel space-vector-based harmonic elimination inverter control," *IEEE Trans on Industry Applications*, vol. 36, no. 2, Mar./April, 2000, pp. 549–557.

[76] S.R. Bowes, S. Grewal, and D. Holliday, "Single-phase harmonic elimination PWM," *Electronics Letters*, vol. 36, no. 2, 20 Jan, 2000, pp. 106–108.

[77] S.R Bowes, S. Grewal, and D. Holiday, "Simplified ultrasonic regular-sampled PWM technique," *Electronics Letters*, vol. 36, no. 9, 27th April, 2000, pp. 854–855.

[78] S.R. Bowes, S. Grewal, and D. Holliday, "Single-phase three-level regular-sampled selective harmonic elimination PWM," *IEE Proceedings – Electric Power Applications*, vol. 148, No. 2, Mar. 2001.

[79] S.R. Bowes, D. Holliday, and S. Grewal, "High frequency PWM technique for two and three level single-phase inverters," *IEE Proceedings – Electric Power Applications*, vol. 147, no. 3, May 2001.

[80] S.R. Bowes and Y.-S. Lai, "Investigation into optimising high switching frequency regular sampled PWM control for drives and static power converters," *IEEE Proc. – Electric Power Applicatons*, vol. 143, no. 4, July 1996.

[81] S.R. Bowes and Y.-S. Lai, The relationship between space-vector modulation and regular-sampled PWM," *IEEE Trans. on Industrial Electronics,* vol. 44, no. 5, Oct. 1997, pp. 670–679.

[82] S.R. Bowes and A. Midoun, "Suboptimal switching strategies for microprocessor-controlled PWM inverter drives," *IEE Proceedings* (*London*), vol. 132, Pt. B, no. 3., May 1985, pp. 133–148.

[83] S.R. Bowes and A. Midoun, "New PWM switching strategy for microprocessor controlled inverter drives," *IEE Proceedings* (*London*), vol. 133, Pt. B, no. 4, July 1986, pp. 237–254.

[84] S.R. Bowes and A. Midoun, "Microprocessor implementation of new optimal PWM switching strategy," *IEE Proc. B – Electric Power Applications*, vol. 135, no.5 Sept. 1988, pp. 269–280.

[85] S.R. Bowes and M.J. Mount, "Microprocessor control of PWM inverters," *IEE Proceedings* (*London*), vol. 128, Pt. B, no. 6., Nov. 1981, pp. 293–305.

[86] J.T. Boys and B.E. Walton, "A loss minimised sinusoidal PWM inverter," *IEE Proceedings* (*London*), vol. 132, Pt. B, no. 5, Sept. 1985, pp. 260–268.

[87] J.T. Boys and P.G. Handley, "Harmonic analysis of space vector modulated PWM waveforms," *IEE Proceedings* (*London*), Pt. B, vol. 137, no. 4, July 1990, pp. 197–204.

[88] J.T. Boys and P.G. Handley, "Spread spectrum switching: low noise modulation technique for PWM inverter drives," *IEE Proceedings* (*London*), Pt. B., vol. 139, no. 3, May 1992, pp. 252–260.

[89] J.T. Boys and M. Andrews, "Random PWM inverter drive systems: theory and practice," in *Conf. Rec. Proceedings of the Int. Conf. on Industrial Electronics, Control and Instrumentation* (*IECON*), Maui, 1993, pp. 695–700.

[90] J.T. Boys, "Theoretical spectra for narrow-band random PWM waveforms," *IEE Proceedings* (*London*), Pt. B, vol. 140, no. 6, Nov. 1993, pp. 393–400.

[91] E.O. Brigham, *The Fast Fourier Transform*, Prentice-Hall, Englewood Cliffs, NJ, 1974.

[92] T. Bruckner and D.G. Holmes, "Optimal pulse width modulation for three-level inverters," in *Conf. Rec. IEEE Power Electronics Specialists Conf.* (*PESC*), Acapulco, 2003, pp. 165–170.

[93] G.S. Buja and G. Indri, "Improvement of pulse width modulation techniques," *Archiv für Elektrotechnik*, vol. 57, 1975, pp. 281–289.

[94] G.S. Buja and G.B. Indri, "Optimal pulse width modulation for feeding AC motors," *IEEE Trans. on Industry Applications*, vol. IA–13, no. 1, Jan./Feb. 1977, pp. 38–44.

[95] G.S. Buja, "Optimum output waveforms in PWM inverters," *IEEE Trans. on Industry Applications*, vol. IA–16, no. 6, Nov./Dec. 1980, pp. 830–836.

[96] M. Calais, V.G. Agelidis, L.J. Borle, and M.S. Dymond, "A transformerless five level cascaded inverter based single phase photovoltaic system," in *Conf. Rec. IEEE Power Electronics Specialists Conf.*, (*PESC*), Galway, 2000, pp. 1173–1178.

[97] M. Calais, V.G. Agelidis, and L. Borle, "Analysis of multicarrier PWM methods for single-phase five level inverter," in *Conf. Rec. IEEE Power Electronics Specialists Conf.* (*PESC*), Vancouver, 2001, in CD ROM.

[98] V.M. Cardenas G., S. Horta M., and R. Echavarria, "Elimination of dead time effects in three phase inverters," in *Conf. Rec. IEEE Intl. Power Electronics Congress (CIEP)*, 1996, pp. 258–262.

[99] R.F. Carnargo, F. Botteron, H.L. Hey, H.A. Grundling, J.R. Pinheiro and H. Pinheiro, "New limiting algorithms for space vector modulated three-phase four-leg voltage source inverter," in *Conf. Rec. IEEE Power Electronics Specialists Conf. (PESC)*, Cairns, 2002, pp. 232–237.

[100] M. Carpita and S. Tenconi, "A novel multilevel structure for voltage source inverter," in *Conf. Rec. European Power Electronics Conference,* Florence, 1991, pp. 96–95.

[101] G. Carrara, D. Casini, S. Gardella, and R. Salutari, "Optimal PWM for the control of multilevel voltage source inverter," in *Conf. Rec. European Power Electronics Conf. (EPE)*, Brighton, 1993, pp. 255–259.

[102] G. Carrara, S. Gardella, M. Marchesoni, R. Salutari, and G. Sciutto, "A new multilevel PWM method: A theoretical analysis," *IEEE Trans. on Power Electronics*, vol. 7, no. 3, July 1992, pp. 497–505.

[103] D. Casadei, G. Serra, A. Tani, and P. Nielsen, "Performance of SVM controlled matrix converter with input and output unbalanced conditions," in *Conf. Rec. IEEE European Power Electronics Conf. (EPE)*, Sevilla, 1995, pp. 2.628–2.633.

[104] J.B. Casteel and R.G. Hoft, "Optimum PWM waveforms of a microprocessor controlled inverter," in *Conf. Rec. IEEE Power Electronics Specialists Conf. (PESC)*, Syracuse NY, 1978, pp. 243–250.

[105] N. Celanovic and D. Boroyevich, "A fast space vector modulation algorithm for multilevel three-phase converters," *IEEE Trans. on Industry Applications,* vol. 37, no. 2, 2001, pp. 637–641.

[106] N. Celanovic and D. Boroyevich, "Comprehensive study of neutral-point voltage balancing problem in three-level neutral-point-clamped voltage source PWM inverters," *IEEE Trans. on Power Electronics,* vol. 15, 2000, pp. 242–249.

[107] E. Cengelci, P. Enjeti, C. Singh, F. Blaabjerg, and J.K. Pederson, "New medium voltage PWM inverter topologies for adjustable speed ac motor drive system," in *Conf. Rec. Thirteenth Annual Applied Power Electronics Conference and Exposition (APEC)*, Anaheim, 1998, vol. 2, pp. 565–571.

[108] E. Cengelci, S.U. Sulistijo, B.O. Woom, P. Enjeti, R. Teodorescu, and F. Blaabjerg, "A new medium voltage PWM inverter topology for adjustable speed drive," in *Conf. Rec. IEEE Industrial Applications Society Annual Meeting,* St. Louis, 1998, pp. 1416–1423.

[109] B.J. Chalmers and B.R. Indri, "Optimal motor losses due to nonsinusoidal supply waveforms," *IEE Proc. (London)*, vol 115, no. 12, Dec. 1968, pp. 1777–1782.

[110] Y. Chen, B. Mwinyiwiwa, Z. Wolanski, and B.-T. Ooi, "Regulating and equalizing DC capacitance voltages in multilevel STATCOM," *IEEE Trans. on Power Delivery,* vol. 12, 1997, pp. 901–907.

[111] T. P. Chen, Y. S. Lai, and C. H. Liu, "A new space vector modulation technique for inverter control," in *Conf. Rec.. IEEE Power Electronics Specialists Conf. (PESC)*, Charleston, 1999, pp.777–782.

[112] T.P. Chen and C.H. Liu, "Load power angle tracking harmonic-elimination discontinuous PWM strategy for switching loss minimization," in *Conf. Rec. IEEE Power Electronics Specialists Conf. (PESC)*, Vancouver, 2001, vol. 1, pp. 96–101.

[113] Y. Chen and B.-T. Ooi, "STATCOM based on multimodules of multilevel converters under multiple regulation feedback control," *IEEE Trans. Power Electronics*, vol. 14, Sept. 1999, pp. 959–956.

[114] K.M. Cho, W.S. Oh, and C.G. In, "A new switching strategy for PWM power converters," in *Conf. Rec. IEEE Power Electronics Specialists Conf. (PESC)*, Cairns, 2002, vol. 1, pp. 221–225.

[115] N.S. Choi, J.G. Cho, and G.H. Cho, "A general circuit topology of multilevel inverter," in *Conf. Rec. IEEE Power Electronics Specialists Conf. (PESC)*, Cambridge MA, 1991, pp. 96–103.

[116] J.W. Choi, S.I. Yong, and S.-K. Sul, "Inverter output voltage synthesis using novel dead time compensation," in *Conf. Rec. IEEE Applied Power Electronics Conf. Record (APEC)*, Orlando, 1994, pp. 100–106.

[117] J.-W. Choi and S.-K Sul, "New dead time compensation eliminating zero current clamping in voltage-fed PWM inverter," in *Conf. Rec. Industry Applications Society Annual Meeting*, Denver, vol. 2, 1994, pp. 977–984.

[118] J.-W. Choi and S.-K. Sul, "Inverter output voltage synthesis using novel dead time compensation," *IEEE Trans. on Power Electronics*, vol. 11, no. 2, Mar. 1996, pp. 221–227.

[119] A.H. Chowdury, A. Mansoor, M.A.Choudhury, and M.A. Rahman, "On-line improved inverter waveform by variable step delta modulation," in *Conf. Rec. IEEE Power Electronics Specialists Conf. (PESC)*, Taipei, 1994, pp. 143–148.

[120] C.F. Christiansen, M.I. Valla, and C.H. Rivetta, "A synchronization technique for static delta-modulated PWM inverters," *IEEE Trans. on Industrial Electronics*, vol. 35, no. 4, Nov. 1988, pp. 502–507.

[121] D.W. Chung, J.S. Kim, and S.K. Sul, "Unified voltage modulation technique for real-time three-phase power conversion," *IEEE Trans. on Industry Applications*, vol. 34, Mar./April 1998, pp. 374–380.

[122] D.-W. Chung and S.-K. Sul, "Minimum-loss strategy for three-phase PWM rectifier," *IEEE Trans. on Industrial Electronics*, vol. 46, no. 3, June 1999, pp. 517–526.

[123] L. Clotea, O. Clotea, and I. Matlac, "A novel microcontroller-based space-vector PWM strategy implementation," in *Conf. Rec. Optimization of Electrical*

and Electronic Equipment, (OPTIM),, Brazov Romania, 16–17 May, 2002, in CD ROM.

[124] B. Cook, A. Cantoni, and R. Evans, "A microprocessor based, 3–phase pulse width modulator," in *Conf. Rec. IEEE Industrial Applications Society Int. Semiconductor Power Converter Conf.*, Orlando, 1982, pp. 375–384.

[125] K. Corzine, S.D. Sudhoff, and C.A. Whitcomb, "Performance characteristics of a cascaded two-level converter," *IEEE Trans. on Energy Conversion*, vol. 14, Sept. 1999, pp. 433–439.

[126] K. Corzine and J.R. Baker, "Reduced-parts-count multilevel rectifiers," *IEEE Trans. on Industrial Electronics*, vol. 49, no. 4, Aug. 2002, pp. 766–774.

[127] K. Corzine and J.R. Baker, "Multilevel voltage-source duty-cycle modulation: analysis and implementation," *IEEE Trans. on Industrial Electronics*, vol. 49, no. 5, Oct. 2002, pp. 1009–1016.

[128] K. Corzine and Y.L. Familiant, "A new cascaded multi-level H-Bridge drive," *IEEE Trans. on Power Electronics*, vol. 17, Jan. 2002, pp. 125–131.

[129] G.A. Covic and J.T. Boys, "Noise quieting with random PWM AC drives," *IEE Proc (London) – Electr. Power Appl.*, vol. 145, no. 1, Jan. 1988, pp. 1–10.

[130] D. Daum, "Unterdrückung von oberschwingungen durch puls-breitensteuerung," *ETZ–A*, 1972, pp. 310–317.

[131] P. Degobert and J.P. Hautier, "Low voltage control of a PWM current source inverter fed AC drive in electric vehicle traction applications," in *Conf. Rec. Int. Conf. on Electrical Machines (ICEM)*, Brugge, 2002, in CD ROM.

[132] D.A. Deib and H.W. Hill, "The advantages of harmonic-distortion minimization over traditional harmonic-elimination techniques," in *Conf. Rec. IEEE Power. Electronics Specialists Conf. (PESC)*, Seattle, 1993, pp. 1048–1054.

[133] P. Delarue, A. Bouscayrol, and E. Semail, "Generic control method of multileg voltage-source-converters for fast practical implementation," *IEEE Trans. on Power Electronics*, vol. 18, no. 2, Mar. 2003, pp. 517–526.

[134] A. Dell'Aquila, R. Forososa, E. Montaruli, and P. Zanchetta, "Novel multilevel PWM inverter implementation," in *Conf. Rec. of Int. Conf. on Industrial Electronics, Control and Instrumentation (IECON)*, New Orleans, 1997, pp. 710–715.

[135] M. Depenbrock, "Pulsewidth control of a 3–phase inverter with non-sinusoidal phase voltages," in *Conf Rec. IEEE Industry Applications Society Int. Semiconductor Power Converter Conf.*, 1997, pp. 399–403.

[136] S.B. Dewan and J.B. Forsythe, "Harmonic analysis of a synchronized pulse width modulated three phase inverter," *Trans. on IEEE Industry Applications*, vol. 14–74, Jan./Feb. 1974, pp. 117–122.

[137] A. Diaz and E. G. Strangas, "A novel wide range pulse width overmodulation method," in *Conf. Rec. IEEE Applied Power Electronics Conf. (APEC)*, New Orleans, 2000, pp. 556–561.

[138] J. W. Dixon, S. M. Tepper, and L. T. Moran, "Analysis and evaluation of different modulation techniques for active power filters," in *Conf. Rec. IEEE Applied Power Electronics Conf. (APEC)*, Orlando, 1994, pp. 894-900.

[139] R.C. Dodson and P.D. Evans, "Application of dead time compensation to variable speed drives," in *Conf. Rec. 4th Int. Conf. on Electrical Machines and Drives*, London, 1989, pp. 369–373.

[140] R.C. Dodson, P.D. Evans, H.T. Yazdi, and S.C. Harley, "Compensating for dead time degradation of PWM inverter waveforms," *IEE Proceedings (London)*, vol. 137, Pt. B, No. 2, Mar. 1990, pp. 73–81.

[141] A. Donzel and G. Bornard, "New control law for capacitor voltage balance in multilevel inverter with switching rate control (CVC)," in *Conf. Rec.IEEE Industry Applications Society Annual Mtg.*, Rome, 2000, pp. 2037–2044.

[142] J. Doval, A. Nogueiras, L. Eguizabal, and C.M. Penalver, "Random modulation in inverters feeding induction motors," in *Conf. Rec. of Power Electronics Specialists Conf. (PESC)*, St. Louis, 1997, pp. 601–605.

[143] J. Doval-Gandoy, A. Iglesias, C. Castro, and C.M. Penalver "Three-alternatives for implementing space vector modulation with the DSP TMS320F240," in *Conf. Rec. of Int. Conf. on Industrial Electronics, Control and Instrumentation (IECON)*, San Jose, 1999, pp. 336–341.

[144] W.G. Dunford and J.D. Van Wyk, "Harmonic imbalance in asynchronous PWM schemes," in *Conf. Rec. IEEE Power. Electronics Specialists Conf. (PESC)*, Cambridge MA, 1991, pp. 397–401.

[145] H.B. Dwight, *Table of Integrals and Other Mathematical Data*, 3rd ed., MacMillan, New York, 1957.

[146] M.A. Dzieniakowski and M.P. Kazmierkowski, "Microprocessor based novel current regulator for VSI PWM inverters," in *Conf. Rec. IEEE Power Electronics Specialists Conf. (PESC)*, Toledo Spain, 1992, pp. 459–464.

[147] J. Eibel and R. Joetten, "Control of a 3-level-switching inverter feeding," *ETG-Facber.*, vol. 11, 1982.

[148] P.N. Enjeti and J.F. Lindsay, "Solving nonlinear equations of harmonic elimination PWM in power control," *IEE Electronics Letters*, vol. 23, no.12, June 1987, pp. 656–657.

[149] P.N. Enjeti, P.D. Ziogas, and J.F. Lindsay, "Programmed PWM techniques to eliminate harmonics – a critical evaluation," in *Conf. Rec. IEEE Industry Applications Society Annual Mtg.*, Pittsburgh, 1988, pp. 418–430.

[150] P.N. Enjeti, P.D. Ziogas, and J.F. Lindsay, "A current source PWM inverter with instantaneous current control capability," in *Conf. Rec. IEEE Industry Applications Society Annual Mtg.*, Pittsburgh, 1988, pp. 927–933.

[151] P.N. Enjeti, P.D. Ziogas, J.F. Lindsay, and M.H. Rashid, "A new PWM speed control system for high-performance ac motor drives," *IEEE Trans. on Industrial Electronics*, vol. 37, no. 2, April 1990, pp. 143–151.

[152] P.N. Enjeti and W. Shireen, "A new technique to reject DC-link voltage ripple from inverters operating programmed PWM waveforms," *IEEE Trans. Power Electronics*, Jan. 1992, pp. 171–180.

[153] P.N. Enjeti and R. Jakkli, "Optimal power control strategies for neutral point clamped (NPC) inverter topology," *IEEE Trans. on Industry Applications*, vol. 28, no. 3, May/June 1992, pp. 558–566.

[154] W.L. Erdman, R. Hudson, J. Yang, and R.G. Hoft, "A 7.5 kW ultrasonic inverter drive employing MOS-controlled thyristors," *IEEE Trans. Industry Applications*, vol. 26, July/Aug. 1990, pp. 756–768.

[155] M.F. Escalante, J.-C. Vannier, and A. Arzande, "Flying capacitor multilevel inverters and DTC motor drive applications," *IEEE Trans. on Industrial Electronics*, vol. 49, no. 4, Aug. 2002, pp. 809–815.

[156] J.R. Espinoza and G. Joos, "Current source converter on-line pattern generator switching frequency minimization," *IEEE Trans. on Industrial Electronics*, vol. 44, 1997, pp. 198–206.

[157] P.D. Evans and P.R. Close, "Harmonic distortion in PWM inverter output waveforms," *IEE Proc. (London)*, Pt. B, vol. 134, no. 4, 1987, pp. 224–232.

[158] C. Feng and V.G. Agelidis, "PWM control methods for five-level flying capacitor multilevel converters," in *Conf. Rec. 7th International Conf. on Modeling and Simulation of Electric Machines, Converters and Systems (ElectIMACS)*, Aug. 2002, in CD ROM.

[159] C. Feng and V.G. Agelidis, "On the comparison of fundamental and high frequency carrier-based PWM techniques for multilevel NPC inverters," in *Conf. Rec. IEEE Power Electronics Specialists Conf. (PESC)*, Cairns, 2002, pp. 520–525.

[160] M. Fracchia, T. Ghiara, M. Marchesoni, and M. Mazzucchelli, "Optimized modulation techniques for the generalized n-level converter," in *Conf. Rec. IEEE Power Electronics Specialists Conf. (PESC)*, Toledo Spain, 1992, pp. 1205–1213.

[161] G. Franzo, M. Mazzucchelli, L. Puglisi, and G. Sciutto, "Analysis of PWM techniques using uniform sampling in variable-speed electrical drives with large speed range," in *Conf. Rec. IEEE Industry Applications Annual Mtg.*, 1984, pp. 568–575.

[162] S. Fukuda and N. Takada, "PWM current source rectifier with sinusoidal line current," in *Conf. Rec. IEEE Industry Applications Society Annual Mtg.*, Atlanta, 1987, pp. 679–684.

[163] S. Fukuda and H. Hasegawa, "Current source rectifier/inverter system with sinusoidal currents," in *Conf. Rec. IEEE Industry Applications Society Annual Mtg.*, Pittsburgh, 1988, pp. 909–914.

[164] S. Fukuda and Y. Iwaji, "A single-chip microprocessor-based PWM technique for sinusoidal inverters," in *Conf. Rec. IEEE Industry Applications Society Annual Mtg.*, Pittsburgh, 1988, pp. 921–926.

[165] S. Fukuda, H. Hasegawa, and Y. Iwaji, "PWM technique for inverter with sinusoidal output current," *IEEE Trans. on Power Electronics*, vol. 5, no. 1, Jan. 1990, pp. 54–61.

[166] S. Fukuda, K. Suzuki, and Y. Iwaji, "Harmonic evaluation of an NPC PWM inverter employing the harmonic distortion determining factor," in *Conf. Rec. IEEE Industrial Applications Society Annual Meeting*, Orlando, 1995, pp. 2417–2421.

[167] S. Fukuda and K. Suzuki, "Using harmonic distortion determining factor for harmonic evaluation of carrier-based PWM methods," in *Conf Rec. IEEE Industry Applications Society Annual Mtg.,* New Orleans, 1997, pp. 1534–1541.

[168] S. Fukuda and K. Suzuki, "Harmonic evaluation of two-level carrier-based PWM methods," in *Conf. Rec. European Power Electronics Conf. (EPE)*, Trondheim, 1997, pp. 2.331–2.336.

[169] S. Fukuda, "LQ control of sinusoidal current PWM rectifiers," *IEE Proc. (London) – Electr. Power Appl.*, vol. 144, no. 2, Mar. 1997, pp. 95–100.

[170] S. Funabiki and Y. Sawada, "A computative decision of pulse width in three-phase PWM inverter," in *Conf. Rec. IEEE Industry Applications Society Annual Mtg.*, Pittsburgh, 1988, pp. 694–699.

[171] G. Gateau, M. Fadel, P. Maussion, R. Bensaid, and T.A. Meynard, "Multicell converters: active control and observation of flying-capacitor voltages," *IEEE Trans. on Industrial Electronics*, vol. 49, no. 5, Oct. 2002, pp. 998–1008.

[172] H. Gheraia, E. M. Berkouk, and G. Maness, "High voltage two PWM rectifiers - seven- level NPC voltage source inverter cascade. Application to the induction machine drive," in *Conf. Rec. Int. Conf. on Electrical Machines (ICEM),* 2002, in CD ROM.

[173] J. Seun-Gi, K. Ban-Sup, K. Kyung-Seo, and P. Min-Ho, "The analysis and compensation of dead time effects," in *Conf. Rec. Int. Conf. on Industrial Electronics, Control and Instrumentation (IECON)*, Singapore, 1988, pp. 667–671.

[174] K.P. Gokhale, A. Kawamura, and R.G. Hoft, "Dead beat microprocessor control of PWM inverter for sinusoidal output waveform synthesis," *IEEE Trans. on Industry Applications*, vol. IA–23, no. 5, Sep./Oct. 1987, pp. 901–909.

[175] G.A. Goodarzi and R.G. Hoft, "GTO inverter optimal PWM waveforms," in *Conf. Rec. IEEE Industry Applications Society Annual Mtg.*, Atlanta, 1987, pp. 312–316.

[176] D.A. Grant and R. Seidner, "Ratio changing in pulse-width-modulated inverters," *IEE Proc. (London)*, Part B, vol. 128, no. 5, 1981, pp. 243–248.

[177] D.A. Grant, J.A. Houldsworth, and K. Lower, "A new high-quality PWM ac drive," in *Conf. Rec. IEEE Industry Applications Society Annual Mtg.*, 1982, pp. 530–535.

[178] D.A. Grant and J.A. Houldsworth, "PWM AC motor drive employing ultrasonic carrier," in *Conf. Rec. IEE Conf. Power Electronics and Variable Speed Drives*, London, 1984, pp. 2345–240.

[179] D.A. Grant, M. Stevens, and J.A. Houldsworth, "The effect of word length on the harmonic content of microprocessor-based PWM waveform generators," *IEEE Trans. on Industry Applications,* vol. IA–21, no. 1, Jan./Feb. 1985, pp. 218–225.

[180] R.M. Green and J.T. Boys, "PWM sequence selection and optimization: a novel approach," *IEEE Trans. on Industry Applications*, vol. IA–18, no. 2, Mar./April 1982, pp. 146–151.

[181] T.C. Green, J.C. Salmon, and B.W. Williams, "A novel three-phase waveform generator for inverter drives," in *Conf. Rec. European Conf. on Power Electronics and Applications, (EPE),* Grenoble, 1987, pp. 191–196.

[182] T.C. Green, J.C. Salmon, and B.W. Williams, "Investigation of delta modulation spectra and of sub-harmonic elimination techniques," in *Conf. Rec. Power Electronics Specialists Conf. (PESC)*, Kyoto, 1988, pp. 290–297.

[183] T.C. Green and B.W. Williams, "Spectra of delta sigma modulated inverters: An analytical treatment," *IEEE Trans. on Power Electronics*, vol. 7, no. 4, Oct. 1992, pp. 644–654.

[184] G. Griva, T.G. Habetler, F. Profumo, and M. Pastorelli, "Performance evaluation of a direct torque controlled drive in the continuous PWM-square wave transition region," *IEEE Trans on Power Electronics*, vol. 10, no. 4, July 1995, pp. 464–471.

[185] H. Grotstollen, "Line Voltage Modulation – A new possibility of PWM for three phase inverters," in *Conf. Rec. IEEE Industry Applications Society Annual Mtg.*, Toronto, 1993, pp. 567–574.

[186] Y. Guijun and L. Norum, "Low cost digital controller for PWM converter," in *Conf. Rec. IFAC Conf on Low Cost Automation*, Milan, 1989, pp. 91–97.

[187] T.G. Habetler and D.M. Divan, "Performance characterization of a new discrete pulse modulated current regulator," in *Conf. Rec. IEEE Industry Applications Society Annual Mtg.*, Pittsburgh, 1988, pp. 395–405.

[188] T.G. Habetler and D.M. Divan, "Acoustic noise reduction in sinusoidal PWM drives using a randomly modulated carrier," *IEEE Trans. on Power Electronics*, vol. 6, no. 3, July 1991, pp. 356–363.

[189] T.G. Habetler, "A space vector-based rectifier regulator for ac/dc/ac converters," *IEEE Trans. on Power Electronics*, vol. 8, no. 1, Jan. 1993, pp. 30–36.

[190] S. Halász, "Voltage spectrum of pulse-width-modulated inverters". *Periodica Pol. El. Eng*, Vol. 25. No. 2. 1981.

[191] S. Halász., "Optimal control of pwm inverters for a given number of commutations," in *Conf. Rec. of International Conference on Electrical Machines (ICEM'82)*, Budapest, 1982.

[192] S. Halász and K. Polgár, "The problem of elimination of harmonics ac inverter drives," in *Conf. Rec. International Conference on Electrical Machines (ICEM'84)*, Lausanne, 1984.

[193] S. Halász and T. Frank, "AC inverter drives with optimal PWM strategies," *in Conf. Rec. of Conf. on Applied Motion Control*, 1985. Minneapolis, USA.

[194] S. Halász, "Selection of reasonable commutation frequency of inverters for ac drives," in *Conf. Rec. International Conference on Electrical Machines (ICEM'86)*, Münich, 1986.

[195] S. Halász, "Analysis of pulsewidth modulation techniques for induction motor drive" *IEEE Int. Symposium on Ind. Electronics*, Budapest, 1993, pp. 300–204.

[196] S. Halász, "Comparison of Sinusoidal PWM Methods," *Periodica Pol. El.Eng.* Vol. 37., No.4. 1993. pp.273–290.

[197] S. Halász, G. Csonka, and A. A. Hassan. "Sinusoidal PWM techniques with additional of the zero-sequence harmonics," in *Conf. Rec. Int. Conf. on Industrial Electronics, Control and Instrumentation (IECON)*, Bologna, 1994, pp. 85–90.

[198] S. Halász, G. Csonka, A.A.M. Hassan, and B.T. Huu, "Unipolar PWM in three level inverter drive.," in *Conf. Rec. of IEEE Conf. on Ind. Appl.* in *Power Systems, Comp. Science and Telecomm.* (Melecon'96), Bari Italy, 1996,.

[199] S. Halász, G. Csonka, and A.A.M. Hassan, "Generalized harmonic loss curves of ac motor fed from two and three level inverters," in *Conf. Rec. of the Int. Conf. on Ind. Electronics Control and Instrumentation*, Taipei, 1996, pp. 957–962.

[200] S. Halasz, A.A.M. Hassan, and B.T. Hu, "Optimal control of three-level PWM inverters," *IEEE Trans. on Industrial Electronics*, Vol. 44, No. 1, Feb. 1997, pp. 96–106.

[201] S. Halász and B.T. Huu, "Two phase PWM technique for inverter fed ac drives," in *Conf. Rec. Industrial Electronics Conf. (IECON)*, New Orleans, 1997, pp. 688–693.

[202] S. Halász and B.T. Huu, "Generalized harmonic loss-factor as a novel important quality index of PWM techniques," *PCC – Nagaoka,* Nagaoka Japan, 1997. pp. 787–792.

[203] S. Halász and A.Zaharov, "Voltage spectra of two-phase PWM techniques in inverter fed ac drives," in *Proceedings of IEEE International Symposium on Industrial Electronics*, July 7–10. 1998. Pretoria, South Africa, pp. 202–207.

[204] S. Halász and A.Zaharov, "Voltage spectra of two-phase PWM technique with 120° cycle," in *Proceedings of 8th International Power Electronics and Motion Control Conference (PEMC)*, Prague,1998.

[205] S. Halasz and A. Zaharov, "Analysis of two-phase PWM technique in inverter ac drives," in *Conf. Rec. of 8th European Conference on Power Electronics and Applications (EPE)*, Lausanne, 1999, in CD ROM.

[206] S. Halász and A.Zaharov, "Various realization methods of two-phase PWMs in inverter fed ac drives," in *Conf. Rec. of International Conference on Electrical Drives and Power Electronics*, High Tatras Slovakia, 1999, pp. 215–220.

[207] S. Halász, Bin. T. Huu, and A. Zakharov, "Two-phase modulation technique for three-level inverter-fed ac drives," *IEEE Transactions on Industrial Electronics*, Dec. 2000, Volume 47, No 6, pp. 1200–1212.

[208] S. Halász and S. Pravadalioglu, "Inverter ac drives with two-phase and three-phase PWM techniques.," in *Conf. Rec. 8th European Conference* in *Power Electronics and Applications* (*EPE*), Graz Austria, 2001, in CD ROM.

[209] T. Halkosaari, K. Kuusela, and H. Tuusa, "Effect of non-idealities on the performance of the 3–phase current source PWM converter," in *Conf. Rec. IEEE Power Electronics Specialists Conf.* (*PESC*), Vancouver, 2001, pp. 654–659.

[210] F. Hamma, T.A. Meynard, F. Tourkhani, and P. Viarouge, "Characteristics and design of multilevel choppers," in *Conf. Rec. IEEE Power Electronics Specialists Conf.* (*PESC*), Atlanta, 1995, pp. 1208–1214.

[211] P.W. Hammond, "A new approach to enhance power quality for medium voltage AC drives," *IEEE Trans. on Industry Applications*, vol. 33, no. 1, Jan/Feb. 1997, pp. 202–208.

[212] P.W. Hammond, "Enhancing the reliability of modular medium-voltage drives," *IEEE Trans. on Industrial Electronics*, vol. 49, no. 5, Oct. 2002, pp. 948–954.

[213] P.G. Handley and J.T. Boys, "Space vector modulation: an engineering review," in *Conf. Rec. IEEE Power Electronics and Variable-Speed Drives Conf.*, 1991, pp. 87–91.

[214] A. Haras, "Space vector modulation in orthogonal and natural frames including the overmodulation range," in *Conf. Rec. European Power Electronics Conf.* (*EPE*), Trondheim, 1997, pp. 2.337–2.342.

[215] M. Hashad and J. Iwaszkiewicz, "A novel orthogonal-vectors-based topology of multilevel inverters," *IEEE Trans. on Industrial Electronics*, vol. 49, no. 4, Aug. 2002, pp. 868–874.

[216] A.M. Hava, "Carrier based PWM voltage source inverter in the overmodulation range," Ph.D. Thesis, University of Wisconsin, 1998.

[217] A. M. Hava, S. K. Sul, R. J. Kerkman, and T.A. Lipo, "Dynamic overmodulation characteristic of triangle intersection PWM methods," in *Conf. Rec. IEEE Industry Applications Soociety Annual Mtg.*, New Orleans, pp.1520–1527, 1997.

[218] A.M. Hava, R.J. Kerkman, and T.A. Lipo, "A high performance generalized discontinuous PWM algorithm," *IEEE Trans. on Industry Applications*, vol. 33, no. 5, Sept./Oct 1997, pp. 1059–1071.

[219] A.M. Hava, R.J. Kerkman, and T.A. Lipo, "Carrier-Based PWM-VSI overmodulation strategies: analysis, comparison and design," *IEEE Trans. on Power Electronics*, vol. 13, no. 4, July 1998, pp. 674–689.

[220] A.M. Hava, T. Lipo, and R.J. Kerkman, "Simple analytical and graphical tools for carrier based PWM methods," *IEEE Trans. on Power Electronics*, vol. 14, no. 1, Jan. 1999, pp. 49–61.

[221] K. Heintze et. al., "Pulsewidth modulating static inverters for the speed control of induction motors," *Siemens-Z.*, vol. 45, no. 3, 1971, pp. 154–161.

[222] S.J. Henriksen, R.E. Betz, and B.J. Cook, "Digital hardware implementation of a current controller for IM variable-speed drives," *IEEE Trans. on Industry Applications*, vol. 35, no. 5, Sept./Oct. 1999, pp. 1021–1029.

[223] D. Hintze and D. Schröder, "PWM Current Source Inverter with On Line Optimized Pulse Pattern Generation for Voltage and Current Control," in *Conf. Rec. CICEM 91*, Wuhan China, 1991, pp. 189–192.

[224] F. Herty, "Methods reducing distortions on output current and voltage waveforms in PWM voltage source inverters," in *Conf. Rec. 9th European Conf. on Power Electronics and Applications* (*EPE*), Graz, 2001, in CD ROM.

[225] W.A. Hill and C.D. Harbourt, "Performance of medium voltage multi-level inverters," in *Conf. Rec. of Industry Applications Society Annual Mtg.*, Phoenix, 1999, pp. 1186–1192.

[226] C. Hochgraf, R. Lasseter, D. Divan, and T.A. Lipo, "Comparison of multilevel inverters for static VAR compensation," in *Conf. Rec. IEEE Industry Applications Society Annual Mtg.*, Denver, 1994, pp. 921–928.

[227] D.G. Holmes, "The general relationship between regular-sampled-pulse-width-modulation and space vector modulation for hard switched converters," in *Conf. Rec. IEEE Industry Applications Society Annual Mtg.*, Houston, 1992, pp. 1002–1010.

[228] D.G. Holmes, "A unified modulation algorithm for voltage and current source inverters based on ac-ac matrix converter theory," *IEEE Trans. on Industry Applications,* vol. 28, no. 1, Feb. 1992, pp. 31–40.

[229] D.G. Holmes, "The significance of zero space vector placement for carrier-based PWM schemes," *IEEE Trans. on Industry Applications,* vol. 32, no. 5, Oct. 1996, pp. 1122–1129.

[230] D.G. Holmes, "A generalised approach to the modulation and control of hard switched converters," Ph.D. Thesis, Monash University, Australia, 1997.

[231] D.G. Holmes, "A general analytical method for determining the theoretical harmonic components of carrier based PWM strategies," in *Conf. Rec. IEEE Industry Applications Society Annual Mtg.,* St. Louis, 1998, pp. 1207–1214.

[232] D.G. Holmes and D.A. Martin, "Implementation of a direct digital predictive current controller for single and three-phase voltage source inverters," Internal Report, Department of Electrical and Computer Systems Engineering, Monash University, 1999.

[233] D.G. Holmes and B.P. McGrath, "Opportunities for harmonic cancellation with carrier based PWM for two-level and multi-level cascaded inverters," in *Conf. Rec. IEEE Industry Applications Society Annual Mtg.*, Phoenix, 1999, pp. 781–788.

[234] J. Holtz, "Pulsewidth modulation in motion control," in *Conf. Rec. Singapore International Conference on Intelligent Control and Instrumentation* (*SICICI '92*), Volume: 1, 17–21 Feb 1992, pp. 115–120.

[235] J. Holtz, "Pulsewidth modulation – a survey," *IEEE Trans. on Industrial Electronics,* vol. 39, no. 5, Dec. 1992, pp. 410–420.

[236] J. Holtz, "Pulsewidth modulation for electronic power conversion," *Proc. IEEE,* vol. 82, no. 8, Aug. 1994, pp. 1194-1214.

[237] J. Holtz, "On the performance of optimal pulsewidth modulation techniques," *European Power Electronics Journal,* 1993, pp. 17–26.

[238] J. Holtz and B. Beyer, "Off-line optimized synchronous pulsewidth modulation with on-line control during transients," *European Power Electronics Journal.,* 1991, pp. 183–190.

[239] J. Holtz and B. Beyer, "Optimal synchronous pulsewidth modulation with a trajectory tracking scheme for high dynamic performance," in *Conf. Record Applied Power Electronics Conf. (APEC),* Boston, 1992, pp. 147–154.

[240] J. Holtz and B. Beyer, "The trajectory tracking approach – a new method for minimum distortion PWM in dynamic high-power drives," in *Conf. Rec. IEEE Industry Applications Society Annual Mtg.*, Houston, 1992, pp. 331–338.

[241] J. Holtz and B. Beyer, "Optimal pulsewidth modulation for ac servos and low-cost industrial drives," *IEEE Trans. on Industry Applications,* vol. 30, no. 4, July/Aug. 1994, pp. 1039–1047.

[242] J. Holtz and E. Bube, "Field-orientated asynchronous pulse-width modulation for high-performance ac machine drives operating at low switching frequency," *IEEE Trans. on Industry Applications*, vol. 27, no. 3, May/June 1991, pp. 574–581.

[243] J. Holtz, A.M. Khambadkone, and W. Lotzkat, "On continuous control of PWM inverters in the overmodulation range including the six-step mode," *IEEE Trans. on Power Electronics*, vol. 8, no. 4, Oct. 1993, pp. 546–553.

[244] J. Holtz, P. Lammert, and W. Lotzkat, "High-speed drive system with ultrasonic mosfet-PWM-inverter and single-chip-microprocessor control," *IEEE Trans. on Industry Applications, vol.* IA–23, no. 6, Nov./Dec. 1987, pp. 1010–1015.

[245] J. Holtz, W. Lotzkat, and K.H. Werner, "A high power multi transistor-inverter uninterruptible power supply system," in *Conf. Rec. IEEE Power Electronics Specialists Conf. (PESC)*, Blacksburg, 1986, pp. 311–320.

[246] J. Holtz and S.F. Salama, "Megawatt GTO-inverter with three-level PWM control and regenerative snubber circuits," in *Conf. Rec. IEEE Power Electronics Specialists Conference (PESC)*, Kyoto, 1988, pp. 214–217.

[247] J. Holtz and L. Springob, "Reduced harmonics PWM controlled line-side converter for electric drives," *IEEE Trans. on Industry Applications*, vol. 29, no. 4, Nov. 1993, pp. 814–819.

[248] J. Holtz and S. Stadtfeld, "A predictive controller for the stator current vector of ac machines fed from a switched voltage source," in *Conf. Rec. IPEC Conf.,* Tokyo, 1983, pp. 1665–1675.

[249] J. Holtz and S. Stadtfeld, "A PWM inverter drive system with on-line optimized pulse patterns," in *Conf. Rec. European Power Electronics Conf.* (*EPE*), Brussels, 1985, pp. 3.21–3.25.

[250] J. Holtz, S. Stadtfeld, and P. Lammert, "An economic very high power PWM inverter for induction motor drives," in *Conf. Rec. European Conference on Power Electronics and Applications* (*EPE*), Brussels, 1985, pp. 3.75–3.80.

[251] J. Holtz, S. Stadtfeld, and H.P. Wurm, "A novel PWM technique minimizing the peak inverter current at steady state and transient operation," *Electrische Bahnen*, vol. 81, Feb. 1983, pp. 55–61.

[252] J. Holtz and H.P. Wurm, "A new type of voltage fed inverter for the megawatt range," *Electrische Bahnen,* vol. 80, no. 7, 1982, pp. 214–221.

[253] M. Hombu, S. Ueda, and A. Ueda, "A current source GTO inverter with sinusoidal inputs and outputs," in *Conf. Rec. IEEE Industry Applications Society Annual Mtg.*, Toronto, 1985, pp. 1033–1039.

[254] J.A. Houldsworth and D.A. Grant, "The use of harmonic distortion to increase the output voltage of a three-phase PWM inverter," *IEEE Trans. on Industry Applications,* vol. 20, no. 5, Sept., 1984, pp. 1224–1228.

[255] S.Y.R. Hui, S. Sathiakumar, and K.K. Sung, "Novel random PWM schemes with weighted switches decision," in *Conf. Rec. Int. Conf. on Power Electronics and Variable Speed Drives*, 23–25 Sept. 1996, pp. 348–353.

[256] B. Huo, A.M. Trzynadlowski, I. Panahi, A. Mohammed, and Z. Yu, "Novel random pulse width modulator with constant sampling frequency based on the TMS320F240 DSP controller,"*Conf. Rec. Int. Conf. on Industrial Electronics, Control and Instrumentation* (*IECON*), San Jose, 1999, pp. 342–347.

[257] B.T. Huu, S. Halász, and G. Csonka, "Generalized harmonic losses of inverter fed AC motors," *Proceedings of the 7th International Power Electronics and Motion Control Conference*, Budapest, 1996, pp. 3/121–3/125.

[258] D.-S. Hyun, "A novel PWM scheme for a three-level voltage source inverter," in *Conf. Rec. Industry Applications Society Annual Meeting*, Denver, 1994, pp. 1151–1157.

[259] T. Ishida, K. Matsuse, K. Sugita, L. Huang, and K. Sasagawa, "DC voltage control strategy for a five-level converter," *IEEE Trans. on Power Electronics,* vol. 15, 2000, pp. 508–515.

[260] T. Ishida, K. Matsuse, K. Sasagawa, and L. Huang, "Fundamental characteristics of a five-level double converter for induction motor drive," in *Conf. Rec. IEEE Industry Applications Society Annual Meeting*, Rome, 2000, pp. 2189–2196.

[261] M. Iwahori and K. Kousaka, "Three-phase current source GTO rectifier adopting new PWM control techniques," in *Conf. Rec. IEEE Industry Applications Society Annual Mtg.*, San Diego, 1989, pp. 855–860.

[262] C.B. Jacobina, A.M.N. Lima, E.R.C. da Silva, and R.L. deA. Ribeireo, "Current control for a random PWM voltage source inverter," in *Conf. Rec. IEEE Power Electronics Specialists Conf.* (*PESC*), St. Louis, 1997, vol. 2, 1440–1446.

[263] S.P. Jackson, "Multiple pulse modulation in static inverters reduces selected output harmonics and provides smooth adjustment of fundamentals," *IEEE Trans. on Industry and General Applications*, vol. IGA–6, no. 4, July/Aug. 1970, pp. 357–360.

[264] F. Jenni and D. Wuest, "The optimization parameters of space vector modulation," in *Conf. Rec. of the European Power Electronics Conf. (EPE)*, Brighton, 1993, pp. 376–381.

[265] F. Jenni and D. Wuest, *Steuerverfaahren fur selbstgefuhrte Stromrichter,* (Control Techniques for Self-Operated Static Inverters), B.G. Teubner, Stuttgart, 1995.

[266] S.G. Jeong and M.-H. Park, "The analysis and compensation of dead-time effects in PWM inverters," *IEEE Trans. on Industrial Electronics*, vol. 38, no. 2, April 1991, pp. 108–114.

[267] Q. Jiang, D.G. Holmes, and D.B. Giesner, "A method of linearising optimal PWM switching strategies to enable their computation on-line in real-time," in *Conf. Rec. IEEE Industry Applications Society Annual Mtg.*, Dearborn, 1991, pp. 819–825.

[268] Q. Jiang and T.A. Lipo, "Switching angle and DC link voltage optimization for multilevel cascade inverters," *Electric Machines and Power Systems,* vol. 28, July 2000, pp. 605–612.

[269] R. Joetten and C. Kehl, "A fast space-vector control for a three-level voltage source inverter," in *Conf. Rec. European Power Electronics Conf. (EPE)*, Florence, 1991, pp. 2:070–2:075.

[270] G. Joos, P.D. Ziogas, and B. Vincenti, "A three phase model reference adaptive PWM technique with improved features," in *Conf. Rec. IEEE Industry Applications Society Annual Mtg.*, Seattle, 1990, pp. 997–1004.

[271] G. Joos, G. Moschopoulos, and P.D. Ziogas, "A high performance current source inverter," in *Conf. Rec. IEEE Power Electronics Specialists Conf. (PESC)*, Cambridge MA, 1991, pp. 123–130.

[272] G. Joos, X. Huang and B.T. Ooi, "Direct-coupled multilevel cascaded series VAR compensators," *IEEE Trans. on Industry Applications*, vol. 34, no. 5, Sept./Oct. 1998, pp. 1156–1163.

[273] A. Julian, G. Oriti, and T.A. Lipo, "Elimination of common mode voltage in three phase sinusoidal power converters," *IEEE Trans. on Power Electronics*, vol. 15, no. 5, Sept. 1999, pp. 982–989.

[274] B. Kaku, I Miyashita, and S. Sone, "Switching loss minimized space vector PWM method for IGBT three-level inverter," *IEE Proc. (London) – Electrical Power Appl.*, vol. 144, May 1997, pp. 182–190.

[275] T. Kalker, A. Ackva, and U. Jansen, "Novel digital controller for induction machines considering the inverter switching times and a fluctuating dc-link voltage," in *Conf. Rec. European Power Electronics Conf. (EPE),* Florence, 1991, pp. 2.058-2.063.

[276] B.J. Kang and C.M. Liaw, "Random hysteresis PWM inverter with robust spectrum shaping," *IEEE Trans. on Aerospace and Electronic Systems*, vol. 37, no. 2, April 2001, pp. 619–629.

[277] C.A. Karybakas, "A DC to AC inverter based on the delta modulation principle," *Int J. Electronics*, vol. 49, no. 1, Jan. 1980, pp. 59–66.

[278] T. Kato, "Precise PWM waveform analysis of inverter for selected harmonic elimination," in *Conf. Rec. Industry Applications Society Industry Applications Society Annual Mtg.*, Denver, 1986, pp. 611–616.

[279] T. Kato and K. Miyao, "Modified hysteresis control with minor loops for single-phase full-bride inverters," in *Conf. Rec. IEEE Industry Applications Society Annual Mtg.*, Pittsburgh, 1988, pp. 689–693.

[280] T. Kawabata, T. Miyashita, and Y. Yamamoto, "Dead beat control of three phase inverter," *IEEE Trans. on Power Electronics*, vol. 5, no. 1, Jan. 1990, pp. 21–28.

[281] T. Kawabata, K. Kawaguchi, and S. Godo, "Novel modulation method of single-phase bridge inverter for improved utilization of power devices," in *Conf. Rec. IEEE Power Electronics Specialists Conf. (PESC)*, Cairns, 2002, pp. 226–231.

[282] Y. Kawabata, M. Nasu, T. Nomoto, E.C. Ejiogu, and T. Kawabata, "High-efficiency and low acoustic noise drive system using open-winding AC motor and two space-vector-modulated inverters," *IEEE Trans. on Industrial Electronics*, vol. 49, no. 4, Aug. 2002, pp. 783–789.

[283] A. Kawamura, T. Haneyoshi, and R. G. Hoft, "Deadbeat controlled PWM inverter with parameter estimation using only voltage sensor," in *Conf. Rec. IEEE Power Electronics Specialists Conf. (PESC)*, Vancouver, 1986, pp. 576-583.

[284] M.P. Kazmierkowski and W. Sulkowski, "A novel vector control scheme for transistor PWM inverter-fed induction motor drive," *IEEE Trans. on Industrial Electronics*, vol. 38, no. 1, Feb. 1991, pp. 41–47.

[285] M.P. Kazmierkowski, M.A. Dzieniakowski, and W. Sulkowski, "Novel space vector based current controllers for PWM-inverters," *IEEE Trans. on Power Electronics*, vol. 6, no. 1, Jan. 1991, pp. 158–165.

[286] C. Keller, R. Jakob, and S. Salama, "Topology and balance control of medium voltage multilevel drives," in *Conf. Rec. European Power Electronics Conf. (EPE)*, Graz, 2001, in CD–ROM.

[287] R.J. Kerkman, D. Leggate, B.J. Seibel, and T.M. Rowan, "Operation of PWM voltage source inverters in the overmodulation region," *IEEE Trans on Industrial Electronics,* vol. 41, no. 1, Feb. 1996, pp. 132–141.

[288] A. Khambadkone and J. Holtz, "Low switching frequency and high dynamic pulsewidth modulation based on field-orientation for high power inverter drive," *IEEE Trans. on Power Electronics*, vol. 6, no. 3, May/June 1991, pp. 574–581.

[289] A. Khambadkone and J. Holtz, "Low switching frequency high-power inverter drive based on field-oriented pulsewidth modulation," in *Conf. Rec. European Conf. on Power Electronics and Appl. (EPE)*, Florence, 1991, pp. 4/672–677.

[290] Y. Kim and M. Ehsani, "An algebraic algorithm for microcomputer-based (direct) inverter pulse width modulation," *IEEE Trans. on Industry Applications*, vol. 23, no. 4, July/Aug. 1987, pp. 654–660.

[291] Y. Kim and M. Ehsani, "New modulation methods for force-commutated direct frequency changers," in *Conf. Rec. IEEE Industry Applications Society Annual Mtg.*, San Diego, 1989, pp. 798–809.

[292] Y. Kim and M. Ehsani, "Control of force-commutated direct frequency changers," in *Conf. Rec. IEEE Industry Applications Society Annual Mtg.*, Seattle, 1990, pp. 1163–1170.

[293] Y.-S. Kim, B.-S. Seo, and D.-S. Hyun, "A novel structure of multi-level high voltage source inverter," in *Conf. Rec. European Power Electronics Conf. (EPE)*, Sevilla,1995, pp. 132–137.

[294] J.-S. Kim, J.-W. Choi, and S.-K. Sul, "Analysis and compensation of voltage distortion by zero current clamping in voltage-fed PWM inverter," in *Conf. Rec. Int. Power Electronics Conf,* Yokohama, 1995, pp. 265–270.

[295] J.-S. Kim, "A novel voltage modulation technique of the space vector PWM," *Trans. Inst. Elect. Eng. Japan*, vol. 116–D, no. 8, 1996, pp. 820–825.

[296] R.Y. Kim, Y.-H. Lee, and D.-S. Hyun, "New link potential control scheme for four-level inverter with passive rectifier," in *Conf. Rec. IEEE Applied Power Electronics Conf.*, Dallas, 1999, pp. 528–534.

[297] H.-J. Kim, H.-D. Lee, and S.-K. Sul, "A new PWM strategy for common mode voltage reduction in neutral-point clamped inverter-fed AC motor drives," in *Conf. Rec. IEEE Industrial Applications Society Annual Mtg.*, Rome, 2000, in CD ROM.

[298] J. Kimball and P.T. Krein, "Real-time optimization of dead time for motor control inverters," in *Conf. Rec. IEEE Power Electronics Specialists Conf. (PESC)*, St. Louis, 1997, vol. 1, pp. 597–600.

[299] R.L. Kirlin, S. Kwok, S. Legowski, and A.M. Trzynadlowski, "Power spectra of a PWM inverter with randomized pulse pattern," *IEEE Trans. on Power Electronics*, vol. 9, no. 5, Sept. 1994, pp. 463–472.

[300] R.L. Kirlin, S. Legowski, and A.M. Trzynadlowski, "An optimal approach to random pulse width modulation in PWM inverters," in *Conf. Rec. of Power Electronics Specialists Conf. (PESC)*, Atlanta, 1995, pp. 313–318.

[301] R.L. Kirlin, M. Bech, and A.M. Trzynadlowski, "Power spectral density analysis of randomly switched pulse modulation for DC/AC converters," in *Conf. Rec. of Workshop on Statistical Signal and Array Processing*, Pocono Manor PA, 2000, pp. 373–377.

[302] R.L. Kirlin, M.M. Bech, A.M. Trzynadlowski, and B. Huo, "Power and power spectral density in PWM inverters with randomized switching frequency," in

Conf. Rec. IEEE Power Electronics Specialists Conf. (PESC), Vancouver, 2001, vol. 1, pp. 188–192.

[303] R.L. Kirlin, M.M. Bech, and A.M. Trzynadlowski, "Analysis of power and power spectral density in PWM inverters with randomized switching frequency," *IEEE Trans. on Industrial Electronics*, vol. 49, no. 2, April 2002, pp. 486–499.

[304] M.C. Klabunde, Y. Zhao, and T.A. Lipo, "Current Control of a 3–Level Rectifier/Inverter Drive System," in *Conf. Rec. IEEE Industry Applications Society Annual Mtg.*, Denver, 1994, pp. 859–866.

[305] J. Klima, "Analytical analysis of space- vector PWM four- switch voltage source inverter - feeding an induction motor drive," in *Conf. Rec. Int. Conf. on Electrical Machines (ICEM)*, 2002, in CD ROM.

[306] G.B. Kliman and A.B. Plunkett, "Development of a modulation strategy for a PWM inverter drive," *IEEE Trans. on Industry Applications*, vol. IA–15, no. 1, Jan./Feb. 1979, pp. 702–709.

[307] R.D. Klug, "Effects and correction of switching dead-times in 3–phase PWM inverter drives," *European Conf. on Power Electronics and Applications (EPE)*, Aachen, 1989, pp. 1261–1266.

[308] H. Kohlmeier, O. Niermeyer, and D. Schröder, "High Dynamic Four-Quadrant AC-Motor Drive with improved Power-Faktor and On-Line optimized Pulse Pattern with PROMC,"*IEEE Transactions on Industry Applications*, Nov./Dec. 1987 vol. IA–23, No. 6, pp. 1001–1009.

[309] H.A. Kojori, H.R. Karshenas, and S.B. Dewan, "Generalized switching strategies for current source inverter/converters," in *Conf. Rec. Power Electronics Specialists Conf. (PESC)*, Taipei, 1994, pp. 135–142.

[310] J.W. Kolar, U. Drofenik, and F.C. Zach, "Space vector based analysis of the variation and control of the neutral point potential of hysteresis current controlled three phase/switch/level PWM rectifier systems," in *Conf. Rec. Int. Conf. on Power Electronics and Drive Systems, (PEDS)*, 1995.

[311] J.W. Kolar, H. Ertl, and F.C. Zach, "Calculation of the passive and active component stress of three phase PWM converter system with high pulse rate," in *Conf. Rec. European Power Electronics (EPE)*, Aachen, 1989, pp. 1303–1311.

[312] J.W. Kolar, H. Ertl, and F.C. Zach, "Influence of the modulation method on the conduction and switching losses of a PWM converter system," *IEEE Trans. Ind. Applications*, vol. 27, no. 6., Nov./Dec. 1991, pp. 1063–1075.

[313] J. Koulischer, P. Mathys, and G. Bingen, "Computer-aided choice of digital PWM strategies," in *Conf. Rec. IEEE Industry Applications Society Annual Mtg.*, Denver, 1986, pp. 49–57.

[314] M. Koyama, T. Fujii, R. Uchida, and T. Kawabata, "Space voltage vector based new PWM method for large capacity three-level GTO inverter," in *Conf. Rec. Int. Conf. on Industrial Electronics, Control and Instrumentation (IECON)*, 1992, pp. 271–276.

[315] P.C. Krause and T.A. Lipo, "Analysis and simplified representation of a rectifier-inverter induction motor drive," *IEEE Trans. on Power Apparatus and Systems,* vol. PAS–88, no. 5, May 1969, pp. 588–596.

[316] P.C. Krause and T.A. Lipo, "Analysis and simplified representation of rectifier-inverter reluctance-synchronous motor drives," *IEEE Trans. on Power Apparatus and Systems,* vol. PAS–88, no. 6, June 1969, pp. 962–970.

[317] J.-S. Lai and F.Z. Peng, "Multilevel converters – a new breed of power converters," *IEEE Trans. on Industry Applications,* vol. 32, 1996, pp. 509–517.

[318] Y. S. Lai and S. R. Bowes, "A universal space vector modulation strategy based on regular-sampled pulse width modulation," in *Conf. Rec. Int. Conf. on Industrial Electronics, Control and Instrumentation* (*IECON*), Taipei, 1996, pp. 120–126.

[319] R.S. Lai and K.D.T. Ngo, "A PWM method for reduction of switching loss in a PWM inverter," *IEEE Trans. on Power Electronics*, vol. 10, May 1995, pp. 326–332.

[320] Y.-S. Lai, "Random switching techniques for inverter control," *Electronics Letters*, vol. 33, no. 9, 24 April 1997, pp. 747–749.

[321] Y.-S. Lai, "Random space vector modulation technique for high switching frequency inverter control," *Electronics Letters*, vol. 33, no. 17, 14 Aug. 1997, pp. 1425–1426.

[322] Y.-S. Lai, "New random inverter control technique for common mode voltage mitigation of motor drives," *IEEE Proc.* (*London*) – *Electr. Power Appl.*, vol. 146, no. 3, May 1999, pp. 289–296.

[323] Y.-S. Lai, Y.-T. Chang, and J. Wang, "Design and implementation of induction motor drives using random switching technique," in *Conf. Rec. Power Engineering Society Summer Meeting*, Seattle, 2000, pp. 2487–2492.

[324] Y.-S. Lai and S.R. Bowes, "A universal space vector modulation strategy based on regular-sampled pulse-width modulation," in *Conf. Rec. Int. Conf. on Industrial Electronics, Control and Instrumentation* (*IECON*), Taipei, 1996, pp. 120–126.

[325] H. Largiader, "Design aspects of induction motors for traction applications with supply through static frequency changers," *Brown Boveri Review*, vol. 57, April 1970, pp. 152–167.

[326] G. Ledwich, "Current source inverter modulation," *IEEE Trans. on Power Electronics,* vol. 6, no. 4, Oct. 1991, pp. 618–623.

[327] D.C. Lee and G.-M. Lee, "A novel overmodulation technique for space-vector PWM inverters," *IEEE Trans. on Power Electronics*, vol. 13, no. 6, Nov. 1998, pp. 1144–1151.

[328] D.H. Lee, S.R. Lee, and F.C. Lee, "An analysis of the midpoint balance for neutral-point clamped three-level VSI," in *Conf. Rec. IEEE Power Electronics Specialists Conf.* (*PESC*), Fukuoka, 1998, pp. 193–199.

[329] Y.-H. Lee, B.-S. Suh, and D.-S. Hyun, "A novel PWM scheme for a three level voltage source inverter with GTO thyristors," *IEEE Trans. on Industry Applications,* vol. 32, no. 2, Mar./April 1996, pp. 260–268.

[330] Y.-H. Lee, R.-Y. Kim, and D.-S. Hyun, "A novel SVPWM strategy considering DC-link balancing for a multi-level voltage source inverter," in *Conf. Rec. IEEE Applied Power Electronics Conf. (APEC)*, Dallas, 1999, pp. 509–514.

[331] Y.-H. Lee, B.-S. Suh, C.-H. Choi, and D.-S. Hyun, "New neutral point current control for a 3–level converter/inverter pair system," in *Conf. Rec. IEEE Industry Applications Society Annual Mtg.,*, Phoenix, 1999, pp. 1528–1534.

[332] Y.-H. Lee, D.-H. Kim, and D-S. Hyun, "Carrier based SVPWM method for multi-level system with reduced HDF," in *Conf. Rec. IEEE Industry Applications Society Annual Mtg.*, Rome, 2000, pp. 1996–2002.

[333] S.G. Lee, D.W. Kang, Y.H. Lee, and D.S. Hyun, "The carrier based PWM method for voltage balance of flying capacitor multilevel inverter," in *Conf. Rec. Power Electronics Specialists Conf. (PESC)*, Vancouver, 2001, vol. 1, pp. 126–131.

[334] D. Leggate and R.J. Kerkman, "Pulse-based dead-time compensator for PWM voltage inverters," *IEEE Trans. on Industrial Electronics*, vol. 44, no. 2, April 1997, pp. 191–197.

[335] S. Legowski and A.M. Trzynadlowski, "Incremental method of pulse width modulation for three-phase inverters," *Int. J. Electronics*, vol. 64, no. 5, 1988, pp. 803–814.

[336] S. Legowski, J. Bei, and A.M. Trzynadlowski, "Analysis and implementation of a PWM technique based on voltage space vectors," in *Conf. Rec. of Applied Power Electronics Conference (APEC)*, Boston, 1992, pp. 586–593.

[337] W. Leonhard, *Control of Electrical Drives*, 2nd Edition, Springer, Berlin, 1996.

[338] L. Li, D. Czarkowski, Y. Liu, and P. Pillay, "Multilevel space vector PWM technique based on phase-shift harmonic suppression," in *Conf. Rec. IEEE Applied Power Electronics Conf. (APEC)*, New Orleans, 2000, pp. 535–541.

[339] L. Li, D. Czarkowski, Y. Liu, and P. Pillay, "Multilevel selective harmonic elimination PWM technique in series-connected voltage inverters," *IEEE Trans. on Industry Applications,* vol. 36, no. 1, Jan./Feb. 2000, pp. 160–170.

[340] S. Li and L. Xu, "A DSP peripheral design for three-level inverter space vector PWM modulations," in *Conf. Rec. IEEE Power Electronics Specialists Conf. (PESC)*, Acapulco, 2003, pp. 189–195.

[341] Y. Liang and C.O. Nwankpa, "A new type of STATCOM based on cascading voltage source inverters with phase-shifted unipolar SPWM," in *Conf. Rec. Industry Applications Society Annual Mtg.*, Pittsburgh, 1988, pp. 978–985.

[342] Y. Liang and C.O. Nwankpa, "Power-line conditioner based on flying-capacitor multilevel voltage-source converter with phase-shift SPWM," *IEEE Trans. on Industry Applications,* vol. 36, 2000, pp. 965–971.

[343] T.-J. Liang, J.-F. Chen, and J.-L. Shyu, "Novel multi-random PWM technique for inverter design," in *Conf. Rec. of Int. Conf. on Power Electronics and Drives Systems*, Hong Kong, 27–29 July 1999, pp. 942–946.

[344] C.M. Liaw and Y.M. Lin, "Random slope PWM inverter using existing system background noise: analysis, design and implementation," *IEE Proceedings - Electric Power Applic.*, vol. 147, no. 1, Jan. 2000, pp. 45–54.

[345] C.M. Liaw, Y.M. Lin, C.H. Wu, and K.I. Hwu, "Analysis, design and implementation of a random frequency PWM inverter," *IEEE Trans. on Power Electronics*, vol. 15, no. 5, Sept. 2000, pp. 843–854.

[346] B.-R. Lin and J.-H. Lu, "Implementation of nondeterministic PWM for inverter drives," in *Conf. Rec. Int. Symposium on Industrial Electronics* (*ISIE*), Bled Slovenia, 12–16 July 1999, pp. 813–818.

[347] B.-R. Lin, "High power factor AC/DC/AC converter with random PWM," *IEEE Trans. on Aerospace and Electronic Systems*, vol. 35, no. 3, July 1999, pp. 935–943.

[348] B.-R. Lin, "Implementation of nondeterministic pulsewidth modulation inverter drives," *IEEE Trans. on Aerospace and Electronic systems*, vol. 36, no. 2, April 2000, pp. 482–490.

[349] J.-L. Lin, "A new approach of dead-time compensation for PWM voltage inverters," *IEEE Trans. on Circuits and Systems*, vol. 49, no. 4, April 2002, 476–483.

[350] T.A. Lipo, "The analysis of induction motors with voltage control by symmetrically triggered thyristors," *IEEE Trans. on Power Apparatus and Systems,* vol. PAS–90, no. 2, Mar./April 1971, pp. 515–525.

[351] T.A. Lipo, "A Cartesian vector approach to reference frame theory of AC machines," in *Conf. Rec. International Conf. on Electrical Machine* (*ICEM*), Lausanne, 1994, pp. 239–242.

[352] T.A. Lipo, *Introduction to AC Machine Design*, Vol. 1, University of Wisconsin Power Electronics Research Center, University of Wisconsin, Madison, 1996.

[353] H.L. Liu, N.S. Choi, and G.H. Cho, "Space vector PWM for three-level inverter with DC-link voltage balancing," in *Conf. Rec. Int. Conf. on Industrial Electronics, Control and Instrumentation* (*IECON*), Kobe, 1991, pp. 197–203.

[354] H.L. Liu and G.H. Cho, "Three-level space vector PWM in low index modulation region avoiding narrow pulse problem," *IEEE Trans. on Power Electronics*, vol. 9, no. 5, Sept. 1994, pp. 481–486.

[355] H.L. Liu, G.H. Cho, and S.S. Park, "Optimal PWM design for high power three-level inverter through comparative studies," *IEEE Trans. on Power Electronics*, vol. 10, no. 1, Jan. 1995, pp. 38–47.

[356] P.C. Loh, D.G. Holmes, Y. Fukuta, and T.A. Lipo, "Reduced common mode carrier-based modulation strategies for cascaded multilevel inverters," in *Conf. Rec. Industry Applications Society Annual Conf.*, Pittsburgh, 2002, vol. 3, pp. 2002–2009.

[357] P.C. Loh, D.G. Holmes, and T.A. Lipo, "Synchronisation of distributed PWM cascaded multilevel inverters with minimal harmonic distortion and common mode voltage," in *Conf. Rec. IEEE Power Electronics Specialists Conf. (PESC)*, Acapulco, 2003, pp. 177–182.

[358] R. Lund, M. Manjrekar, P. Steimer, and T.A. Lipo, "Control strategies for a hybrid seven level inverter," in *Conf. Rec. European Power Electronics Conf. (EPE)*, Lausanne, 1999, in CD ROM.

[359] X. Ma, "High-performance PWM frequency changers," *IEEE Trans. on Industry Applications*, vol. IA–22, no. 2, Mar./April, 1986, pp. 267–280.

[360] M. Malinowski, "Adaptive modulator for three-phase PWM rectifier/inverter," in *Conf. Rec. of the 2000 EPE–PEMC Conf.*, Kosice, pp. 1.35–1.41.

[361] L. Malesani and P. Tenti, "Three-phase ac/dc PWM converter with sinusoidal ac currents and minimum filter requirements," *IEEE Trans. on Industry Applications*, vol. IA–23, no. 1, Jan./Feb., 1987, pp. 71–77.

[362] M. Manjrekar and G. Venkataramanan, "Advanced topologies and modulation strategies for multilevel inverters," in *Conf. Rec. IEEE Power Electronics Specialists Conf. (PESC)*, Baveno, 1996, pp. 1013–1018.

[363] M. Manjrekar, "Topologies, analysis, controls and generalization in H–bridge multilevel power conversion," Ph.D. dissertation, University of Wisconsin, 1999.

[364] M.D. Manjrekar, R. Lund, P. Steimer, and T.A. Lipo, "Hybrid multilevel power conversion system: A competitive solution for high power applications," *IEEE Trans. On Industry Applications*, vol. 36, no. 3, May/June 2000, pp. 834–841.

[365] M. Manjrekar and T.A. Lipo, "A hybrid multilevel inverter topology for drive applications," in *Conf. Rec. IEEE Applied Power Electronics Conf. and Exposition*, New Orleans, 2000, pp. 525–529.

[366] M. Marchesoni, M. Mazzucchelli, and S. Tenconi, "Non-conventional power converter for plasma stabilization," in *Conf. Rec. IEEE Power Electronics Specialists Conf. (PESC)*, Kyoto, 1988, pp. 122–129.

[367] M. Marchesoni, M. Mazzucchelli, and P. Tenca, "About the DC-link capacitors voltage balance in multi-point clamped converters," in *Conf. Rec. Int. Conf. on Industrial Electronics, Control and Instrumentation (IECON)*, Aachen, 1998, pp. 548–553.

[368] M. Marchesoni, M. Mazzucchelli, F. Robinson, and P. Tenca, "Analysis of DC-link capacitor voltage balance in AC-DC-AC diode-clamped multilevel converters," in *Conf. Record European Power Electronics Conf. (EPE)*, Lausanne, 1999, in CD ROM.

[369] M. Marchesoni, M. Mazzucchelli, and P. Tenca, "An optimal controller for voltage balance and power losses reduction in MPC AC/DC/AC converters," in *Conf. Rec. IEEE Power Electronics Specialists Conf. (PESC)*, Galway, 2000, pp. 662–667.

[370] M. Marchesoni and P. Tenca, "Theoretical and practical limits in multilevel MPC inverters with passive front ends," in *Conf. Record European Power Electronics Conf. (EPE)*, Graz, 2001, in CD ROM, 12 pp.

[371] M. Marchesoni and P. Tenca, "Diode-clamped multilevel converters: A practicable way to balance DC-link voltages," *IEEE Trans. on Industrial Electronics*, vol. 49, no. 4, Aug. 2002, pp. 752–765.

[372] A.M. Massoud, S.J. Finney, and B.W. Williams, "Control techniques for multilevel voltage source inverters," in *Conf. Rec. IEEE Power Electronics Specialists Conf. (PESC)*, Acapulco, 2003, pp. 171–176.

[373] K. Matsuse, T. Katahata, and K. Sugita, "A vector control system for induction motor using a five-level inverter with DC choppers," in *Conf. Rec. IEEE Power Electronics Specialists Conference, (PESC)*, Fukuoka, 1998, pp. 984–989.

[374] B.P. McGrath, "Modulation and control of pulse width modulated multilevel converters," Internal Report, Power Electronics Group, Monash University Australia, 1999.

[375] B.P. McGrath, "Topologically independent modulation of multilevel inverters," Ph.D. Thesis, Monash University, Australia, 2002

[376] B.P. McGrath and D.G. Holmes, "A comparison of multicarrier PWM strategies for cascaded and neutral point clamped multilevel inverters," in *Conf Rec. IEEE Power Electronics Specialists Conf. (PESC)*, Galway, 2000, pp. 674–679.

[377] B.P. McGrath, M. Manjrekar, D.G. Holmes, and T.A. Lipo, "An improved modulation strategy for the hybrid multilevel inverter," in *Conf. Rec. IEEE Industry Applications Society Annual Mtg.*, Rome, 2000, pp. 2086–2093.

[378] B.P. McGrath, D.G. Holmes, and T.A. Lipo, "Optimized space vector switching sequences for multilevel inverters," in *Conf. Rec. IEEE Applied Power Electronics Conf. and Exposition (APEC)*, Vancouver, 2001, pp. 1123–1129.

[379] B.P. McGrath and D.G. Holmes, "Opportunities for harmonic cancellation with carrier-based PWM for two-level and multilevel cascaded inverters," *IEEE Trans. on Industry Applications*, vol. 37, No. 2, Mar./April 2001, pp. 574–582.

[380] B.P. McGrath and D.G. Holmes, "An analytical technique for the determination of spectral components of multilevel carrier based PWM methods," *IEEE Trans. on Industrial Electronics*, vol. 49, no. 4, Aug. 2002, pp. 724–738.

[381] B.P. McGrath and D.G. Holmes, "Multicarrier PWM strategies for multilevel inverters," *IEEE Trans. on Industrial Electronics*, vol. 49, no. 4, Aug. 2002., pp. 858–867.

[382] B.P. McGrath and D.G. Holmes, "Sinusoidal PWM of multilevel inverters in the overmodulation region," in *Conf. Rec. IEEE Power Electronics Specialists Conf. (PESC)*, Cairns, 2002, vol. 2, pp. 485–490.

[383] M.A. Mendes, Z.M. Peixoto, P.F. Seixas, and P. Donoso-Garcia, "A space vector PWM method for three-level flying capacitor inverters," in *Conf. Rec. IEEE Power Electronics Specialists Conf. (PESC)*, Vancouver, 2001, pp. 182–187.

[384] R.W. Menzies, and T. Seger, "Constrained optimal pulsewidth modulation," in *Conf. Rec. of Int. Conf. on Electrical Machines* (*ICEM*), 1980, pp. 1198–1205.

[385] T.A. Meynard and H. Foch, "Multi-level conversion: high voltage choppers and voltage-source inverters," in *Conf. Rec. IEEE Power Electronics Specialists Conf.* (*PESC*), Toledo Spain, 1992, pp. 397–403.

[386] T.A. Meynard, M. Fadel, and N. Aouda, "Modeling of multilevel converters," *IEEE Trans. on Industrial Electronics,* vol. 44, no. 3, June 1997, pp. 356–364.

[387] T.A. Meynard, H. Foch, P. Thomas, J. Courault, R. Jakob, and M. Nahrstaedt, "Multicell converters: basic concepts and industry applications," *IEEE Trans. on Industrial Electronics*, vol. 49, no. 5, Oct. 2002, pp. 955–964.

[388] T.A. Meynard, H. Foch, F. Forest, C. Turpin, F. Richardeau, L. Delmas, G. Gateau, and E. Lefeuvre, "Multicell converters: derived topologies," *IEEE Trans. on Industrial Electronics*, vol. 49, no. 5, Oct. 2002, pp. 978–987.

[389] N. Mohan, T. Undeland, and W.P. Robbins, *Power Electronics, Converters, Applications and Design*, J. Wiley & Sons, New York, 1989.

[390] B. Mokrytzki, "Pulse width modulated inverters for ac motor drives," *IEEE Trans. on Industry and General Applications*, vol. IGA–3, no. 6, Nov./Dec. 1967, pp. 493–503.

[391] S.K. Mondal, B.K. Bose, V. Oleschuk, and J.O.P. Pinto, "Space vector pulse width modulation of three-level inverter extending operation into overmodulation region," *IEEE Trans. on Power Electronics*, vol. 18, no. 2, Mar. 2003, pp. 604–611.

[392] H. du Toit Mouton, "Natural balancing of three-level neutral-point-clamped PWM inverters," *IEEE Trans. on Industrial Electronics*, vol. 49, no. 5, Oct. 2002, pp. 1017–1025.

[393] J.F. Moynihan, M.G. Egan, and J.M.D. Murphy, "Theoretical spectra of space-vector-modulated waveforms," *IEE Proc.* (*London*) *Electr. Power Applications*, vol. 145, no. 1, Jan. 1998, pp. 17–24.

[394] A.R. Munoz and T.A. Lipo, "On-line dead-time compensation technique for open-loop PWM-VSI drives," *IEEE Trans. on Power Electronics*, vol. 14, no. 4, July 1999, pp. 683–689.

[395] Y. Murai, Y. Gohshi, K. Matsui, and I. Hosono, "High-frequency split zero-vector PWM with harmonic reduction for induction motor drives," *IEEE Trans. on Industry Applications*, vol. 28, no. 1, Jan./Feb. 1992, pp. 105–112.

[396] Y. Murai, K. Ohashi, and I. Hosono, "New PWM method for fully digitized inverters," in *Conf. Rec. IEEE Industry Applications Society Annual Mtg.*, Toronto, 1985, pp. 448–453.

[397] Y. Murai, A. Riyanto, H. Nakamura, and K. Matsui, "PWM strategy for high frequency carrier inverters eliminating current-clamps during switching dead-time," in *Conf. Rec. IEEE Industrial Applications Society Annual Meeting*, Houston, 1992, pp. 317–322.

[398] Y. Murai, Y. Tanizawa, and M. Yoshida, "Three-phase current-waveform-detection on PWM inverters from DC link current-steps," in *Conf. Rec. IEE* (*Japan*) *Int. Power Electronics Conf.* (*IPEC*), Yokohama, 1995, pp. 271–275.

[399] Y. Murai, T. Watanabe, and H. Iwasaki, "Waveform distortion and correction circuit for PWM inverters with switching lag-times," *IEEE Trans. on Industry Applications,* vol. 23, no. 5, Sept./Oct., 1987, pp. 881–886.

[400] J.M.D. Murphy and M.G. Egan, "An analysis of induction motor performance with optimum PWM waveforms," in *Conf Rec. Int. Conf. on Electrical Machines* (*ICEM*), Athens, 1980, pp. 642–656.

[401] J.M.D. Murphy and M.G. Egan, "A comparison of PWM strategies for inverter-fed induction motors," *IEEE Trans. on Industry Applications*, vol. IA–19, no. 3, May/June 1983, pp. 363–369.

[402] M. Mutoh, K. Sakai, and A. Ueda, "Stabilizing methods at high frequency for an induction motor drive driven by a PWM inverter," in *Conf. Rec. European Power Electronics Conf.* (*EPE*), Florence, 1991, pp. 2/352–358.

[403] B. Mwinyiwiwa, Z. Wolanski, Y. Chen, and B.-T. Ooi, "Multi-modular multi-level converters with input/output linearity," *IEEE Trans. on Industry Applications*, vol. 33, no. 5, Sept./Oct. 1997, pp. 1214–1219.

[404] B. Mwinyiwiwa, Z. Wolanski, and B.-T. Ooi, "Microprocessor implemented SPWM for multiconverters with phase-shifted triangle carriers," *IEEE Trans. on Industry Applications*, vol. 34, no. 3, May/June 1998, pp. 495–500.

[405] S.-H. Na, Y.-G. Jung, Y.-C. Lim, and S.-H. Yang, "Reduction of audible switching noise in induction motor drives using random position space vector PWM," *IEE Proceedings – Electric Power Applications* (*London*), vol. 149, no. 3, May 2002, pp. 195–200.

[406] A. Nabae, I. Takahashi, and H. Akagi, "A new neutral-point-clamped PWM inverter," *IEEE Trans. on Industry Applications*, vol. IA–17, no. 5, Sept./Oct. 1981, pp. 518–523.

[407] H. Nagase, Y. Matsuda, K. Ohnishi, H. Ninomiya, and T. Koike, "High Performance Induction Motor Drive System Using a PWM Inverter," *IEEE Trans. on Industry Applications*, vol. IA–20, no. 6, Nov./ Dec. 1984, pp. 1482–1489.

[408] G. Narayanan and V.T. Ranganathan, "Synchronised PWM strategies based on space vector approach. Part 1: Principles of waveform generation," *IEE Proc.* (*London*) *– Electr. Power Appl.*, vol. 146, no. 3, May 1999, pp. 267–275.

[409] G Narayanan and V.T. Ranganathan, "Synchronised PWM strategies based on space vector approach Part 2: Performance assessment and application to V/f drives," *IEE Proc. (London) – Electr. Power Appl.*, vol. 146, no. 3, May 1999, pp. 276–281.

[410] R.H. Nelson, T.A. Lipo, and P.C. Krause, "Stability analysis of a symmetrical induction machine," *IEEE Trans. on Power Apparatus and Systems*, vol. PAS–88, no. 11, Nov. 1969, pp. 1710–1717.

[411] C. Newton and M. Sumner, "Neutral point control for multi-level inverters: theory, design and operational limitations," in *Conf. Rec. IEEE Industry Applications Society Annual Mtg.*, New Orleans, 1997, pp. 1336–1343.

[412] C. Newton and M. Sumner, "Multi-level convertors a real solution to medium/high voltage drives," *Power Engineering Journal*, vol. 12, 1998, pp. 21–26.

[413] Y. Nishida, S. Nakamura, M. Ishikawa, and K. Yasui, "Practical Evaluation of a three-phase PWM buck-mode PFC rectifier," in *Conf. Rec. European Power Electronics-Power Electronics and Motion Control Conference (EPE-PEMC)*, 2002, in CD ROM.

[414] S. Nonaka and Y. Neba, "A PWM current source type converter-inverter system for bi-directional power flow," in *Conf. Rec. IEEE Industry Applications Society Annual Mtg.*, Pittsburgh, 1988, pp. 296–301.

[415] L. Norum, "Microprocessor applications in power electronics," Ph.D. Thesis, The University of Trondheim, Norway, 1986.

[416] L. Norum, "A dual-processor computer for advanced embedded control applications in power electronic systems," *Power Electronics Forum*, University of Minnesota, Oct. 1993, pp. 1–7.

[417] D.W. Novotny and T.A. Lipo, *Vector Control and Dynamics of AC Drives*, Oxford University Press, London, 1996.

[418] C.D.M. Oates, "Spectral assessment of inverter waveforms," in *Conf. Rec. 3rd Int. Conf. on Power Electronics and Variable Speed Drives*, London, July 1988, pp. 79–82.

[419] O. Ogasawara, H. Akagi, and A. Nabae, "A novel PWM scheme of voltage source inverters based on space vector theory," in *Conf. Rec. European Power Electronics Conf. (EPE)*, Aachen, 1989, pp. 1197–1202.

[420] S. Ogasawara and H. Akagi, "Analysis of variation of neutral point potential in neutral-point-clamped voltage source PWM inverters," in *Conf. Rec. IEEE Industry Applications Society Annual Mtg.*, Toronto, 1993, pp. 965–970.

[421] K. Oguchi, H. Hama, and T. Kubota, "48/72–step voltage double three level converters coupled with line-side reactors," in *Conf. Rec. Int. Conf. on Industrial Electronics, Control and Instrumentation (IECON)*, Aachen, 1998, pp. 602–606.

[422] W.S. Oh, Y.T. Kim, and H.J. Kim, "Dead time compensation of current controlled inverter using space vector modulation method," in *Proceedings of 1995 International Conference on Power Electronics and Drive Systems*, vol. 1, pp. 374–378.

[423] T. Ohnishi and H. Okitsu, "A novel PWM technique for three phase inverter/converter," in *Conf. Rec. Int. Power Electronics Conf.*, Tokyo, 1983, pp. 384–395.

[424] T. Ohnuki, O. Miyashita, T. Haneyoshi, and E. Ohtsuji, "High power factor PWM rectifiers with an analog pulsewidth prediction controller," *IEEE Trans. on Power Electronics*, vol. 11, no. 3, May 1996, pp. 460–465.

[425] O. Ojo, P. Kshirsagar, and O. Osaloni, "Concise modulation strategies for 4–leg voltage source inverter," in *Conf. Rec. IEEE Power Electronics Specialists Conf. (PESC)*, Cairns, 2002, vol. 1, 238–243.

[426] O. Ojo and P. Kshirsagar, "A new carrier based discontinuous PWM modulatioin methodology for four-leg voltage source inverters," in *Conf. Rec. IEEE Power Electronics Specialists Conf. (PESC)*, Acapulco, 2003, pp. 201–207.

[427] V. Oleschuk and B. K. Bose, "Quasi sliding strategy of modulation for modified vector PWM for frequency controlled 3–phase voltage source inverters.," in *Conf. Rec. Int. Conf. on Industrial Electronics, Control and Instrumentation (IECON)*., Aachen, 1998, pp.537–541.

[428] V. Oleschuk and F. Blaabjerg, "Quasi- linear algorithms of synchronous PWM for speed controlled induction motor drives," in *Conf. Rec. Int. Conf. on Electrical Machines (ICEM)*, 2002, in CD ROM.

[429] V. Oleschuk and F. Blaabjerg, "Direct synchronized PWM techniques with linear control functions for adjustable speed drives," in *Conf. Rec. IEEE Applied Power Electronics Conf. (APEC)*, Dallas, 2002, pp. 76–82.

[430] V. Oleschuk, B.K. Bose, and Z. Chen, "Synchronized overmodulation techniques for neutral-clamped inverters," in *Conf. Rec. IEEE Power Electronics Specialists Conference (PESC)*, Acapulco, 2003, pp. 41–46.

[431] A.C. Oliveira, A.M.N. Lima, and C..B. Jacobina, "Varying the switching frequency to compensate the dead-time in pulse width modulated voltage source inverters," in *Conf. Rec. IEEE Power Electronics Specialists Conf. (PESC)*, Cairns, 2002, pp. 244–249.

[432] B. Ooi and M. Kazerani, "Application of dyadic matrix converter theory in conceptual design of dual field vector and displacement factor controls," in *Conf. Rec. IEEE Industry Applications Society Annual Mtg.*, Denver, 1994, pp. 903–910.

[433] R.H. Osman, "A medium voltage drive utilizing series-cell multilevel topology for outstanding power quality," in *Conf. Rec. IEEE Industry Applications Society Annual Mtg.*, Phoenix, 1999, pp. 2662–2669.

[434] R.C. Panaiescu and N. Mohan, "A simple space-vector PWM algorithm for VSI-fed AC motor drives," in *Conf. Rec. IEEE Applied Power Electronics Conf. (APEC)*, Dallas, 2002, pp. 72–75.

[435] Z. Papazacharopoulos, A. Kladas, and S. Manias, "Investigation of the switching frequency harmonics impact on PWM induction motor drive efficiency," in *Conf. Rec. IEEE Power Electronics Specialists Conf. (PESC)*, Vancouver, 2001, pp. 1203–1208.

[436] H.S. Patel and R.G. Hoft, "Generalized techniques of harmonic elimination and voltage control in thyristor inverters: Part I — harmonic elimination," *IEEE Trans. on Industry Applications*, vol. IA–9, no. 3, May/June 1973, pp. 310–317.

[437] H.S. Patel and R.G. Hoft, "Generalized techniques of harmonic elimination and voltage control in thyristor inverters: Part II — voltage control techniques,"

IEEE Trans. on Industry Applications, vol. IA–10, no. 5, Sept./Oct 1974, pp. 666–673.

[438] S.C. Peak and A.B. Plunkett, "Transistorized PWM inverter induction motor drive system," in *Conf. Rec. IEEE Industry Applications Society Annual Mtg.*, 1982, pp. 892–898.

[439] J.K. Pedersen, F. Blaabjerg, and P.S. Fredericksen, "Reduction of acoustical noise emission in AC machines by intelligent distributed random modulation," in *Conf. Rec. Fifth European Conference on Power Electronics and Applications* (*EPE*), Brighton, 1993, vol. 4, pp. 369–375.

[440] J.K. Pedersen and F. Blaabjerg, "Blanking-time and on-state compensation in a PWM-VSI inverter with one current sensor only," in *Conf. Rec. Symposium on Power Electronics, Electrical Drives, Advanced Electrical Motors (SPEEDAM)*, Taormina, 1994.

[441] F.Z. Peng, "A generalized multilevel inverter topology with self voltage balancing," in *Conf. Rec. IEEE Industry Applications Society Annual Mtg.*, Rome, 2000, pp. 2024–2031.

[442] F.Z. Peng, J.S. Lai, J.W. McKeever, and J. Van Coevering, "Multilevel voltage-source converter system with balanced DC voltages," in *Conf. Rec. IEEE Power Electronics Specialists Conf.* (*PESC*), Atlanta, 1995, pp. 1144–1150.

[443] F.Z. Peng, J.S. Lai, J.W. McKeever, and J. Van Coevering, "A multilevel voltage-source inverter with separate DC sources for static VAR generation," *IEEE Trans. on Industry Applications*, vol. 32, no. 5, Sept./Oct. 1996, pp. 1130–1138.

[444] F.Z. Peng, "A generalized multilevel inverter topology with self voltage balancing," *IEEE Trans. on Industry Applications*, vol. 37, Mar./Apr. 2001, pp. 611–618.

[445] D. Peng, F.C. Lee, and D. Boroyevich, "A novel SVM algorithm for multilevel three-phase converters," in *Conf. Rec. IEEE 33rd Annual Power Electronics Specialists Conference*, Cairns, 2002, pp. 509–513.

[446] G. Pfaff, A. Weschta, and A. Wick, "Design and experimental results of a brushless ac servo-drive," *IEEE Trans. on Industry Applications,* vol. IA–20, no. 4, July/Aug. 1984, pp. 814–821.

[447] J.O. Pinto, B.K. Bose, L.E. Borges, and M.P. Kazmierkowski, "A neural network based space vector PWM controller for voltage-fed inverter induction motor drive," in *Conf. Rec. IEEE Power Electronics Specialists Conf.* (*PESC*), Charleston, 1999, pp. 2614–2662.

[448] A.B. Plunkett and T.A. Lipo, "Source impedance effects in the control of inverter-induction motor drives," *World Electrotechnical Congress*, Moscow, USSR, June 21–25, 1977.

[449] A. Pollmann, "Comparison of PWM modulation techniques," in *Conf. Rec. Microelectronics in Power Electronics and Controlled Electrical Drives*, 1982.

[450] A. Pollmann, "A digital pulsewidth modulator employing advanced modulation techniques," *IEEE Trans. on Industry Applications*, vol. IA–19, no. 3, May/June 1983, pp. 409–414.

[451] A. Pollmann, "Software pulsewidth modulation mp control of ac drives," *IEEE Trans. on Industrial Applications*, vol. 22, no. 4, 1986, pp. 691–696.

[452] J. Pontt, J. Rodriguez, R. Huerta, and J. Pavez, "Mitigation of non-eliminated harmonics of SHEPEM three-level multipulse three-phase active front end converters with low switching frequency for meeting standard IEEE-519-92," in *Conf. Rec. IEEE Power Electronics Specialists Conf. (PESC)*, Acapulco, 2003, pp. 531–536.

[453] V.H. Prasad, D. Boroyevich, and R. Zhang, "Analysis and comparison of space vector modulation schemes for a four-leg voltage source inverter," in *Conf. Rec. IEEE Applied Power Electronics Conf. and Exposition (APEC)*, Atlanta, 1997, pp. 864–871.

[454] S. Rahmani, K. Al-Haddad, and F. Fnaiech, "A new PWM control technique applied to single-phase shunt active power filter," in *Conf. Rec. ElectrIMACS*, 2002, in CD ROM.

[455] J. Richardson and O. Kukrer, "Implementation of a PWM regular sampling strategy for ac drives," *IEEE Trans. on Power Electronics.*, vol. 6, no. 4, Oct. 1989, pp. 649–655.

[456] S. Richter, "On the proper application of synchronous space vector modulation methods to high power induction motor drives," in *Conf. Rec. Symposium on Power Electronics, Electrical Drives and Advanced Electrical Motors (SPEEDAM)*, 1994, pp. 103–108.

[457] J. Rodriguez, L. Moran, A. Gonzalez, and C. Silva, "High voltage multilevel converter with regeneration capability," in *Conf. Rec. Power Electronics Specialists Conference (PESC)*, Charleston, 1999, pp. 1077–1082.

[458] J. Rodriguez, L. Moran, C. Silva, and P. Correa, "A high performance vector control of a 11–level inverter, in *Conf. Rec. 3rd Int. Power Electronics And Motion Control Conf.*, Beijing, 2000, pp. 1116–1121.

[459] J. Rodriguez, J. Pontt, G. Alzamora, N. Becker, O. Einenkel, J.L. Cornet, and A. Weinstein, "Novel 20 MW downhill conveyor system using three-level converters," in *Conf. Rec. IEEE Industrial Applications Society Annual Meeting*, Chicago, 2001, pp. 1396–1403.

[460] J. Rodriguez, P. Correa, and L. Moran, "A vector control technique for medium voltage multilevel inverters," in *Conf. Rec. IEEE Applied Power Electronics Conf.*, Anaheim, 2001, pp. 173–178.

[461] J. Rodriguez, J-S. Lai, and F.Z. Peng, "Multilevel inverters: a survey of topologies, controls, and applications, *IEEE Trans. on Industrial Electronics*, vol. 49, no. 4, Aug. 2002, pp. 724–738.

[462] J. Rodriguez, L. Moran, P. Correa, and C. Silva, "A vector control technique for medium-voltage multilevel inverters," *IEEE Trans. on Industrial Electronics*, vol. 49, no. 4, Aug. 2002, pp. 882–888.

[463] J. Rodriguez, J. Pontt, E. Silva, J. Espinoza, and M. Perez, "Topologies for regenerative cascaded multilevel inverters," in *Conf. Rec. IEEE Power Electronics Specialists Conf. (PESC)*, Acapulco, 2003, pp. 519–524.

[464] R. Rojas, T. Ohnishi, and T. Suzuki, "Neutral-point-clamped inverter with improved voltage waveform and control range," in *Conf. Rec. Int. Conf. on Industrial Electronics, Control and Instrumentation* (*IECON*), Maui, 1993, pp. 1240–1245.

[465] R. Rojas, T. Ohnishi, and T. Suzuki, "An improved voltage vector control method for neutral-point clamped inverters," IEEE Trans. Power Electronics, vol. 10, Nov. 1995, pp. 666–672.

[466] R. Rojas, T. Ohnishi, and T. Suzuki, "Simple structure multilevel inverter useful for ac motor drives," *Electric Machines and Power Systems*, vol. 235, 1997, pp. 327–340.

[467] R. Rojas and T. Ohnishi, "A four-level inverter PWM technique with balanced capacigtorvoltages," in *Conf. Rec. Brazilian Conf. on Power Electronics* (*COBEP*), 1997, pp. 97–102.

[468] R.D.F. Rossi, B.R. Menezes, and S.R. Silva, "Vector control of voltage fed three phase inverters: variable switching regions," in *Conf. Rec. IEEE Power Electronics Specialists Conf.* (*PESC*), Taipei, 1994, pp. 219–224.

[469] A. Rufer and M. Veenstra, "Asymmetric multilevel converter for high resolution voltage phasor generation," in *Conf. Rec. European Power Electronics Conf.* (*EPE*), Lausanne, 1999, in CD ROM.

[470] M. Ryan, R.D. Lorenz, and R.W. De Doncker, "Modeling of multileg sine-wave inverters: A geometric approach," *IEEE Trans. on Industrial Electronics*, vol. 46, no. 6, Dec. 1999, pp. 1183–1191.

[471] M. Ryan, R.W. De Doncker, and R.D. Lorenz, "Decoupled control of a 4–leg inverter via a new 4x4 transformation matrix," in *Conf. Rec. IEEE Power Electronics Specialists Conf.* (*PESC*), Charleston, 1999, pp. 187–192.

[472] M. Saeedifard, A.R. Bakhshai, and P. Jain, "Introducing the low switching frequency space vector modulated multi-modular three-level converters for high power applications," in *Conf. Rec. IEEE Power Electronics Specialists Conf.* (*PESC*), Acapulco, 2003, pp. 1639–1644.

[473] N.P. Schibli, T. Nguyen, and A.C. Rufer, "A three-phase multilevel converter for high-power induction motors," *IEEE Trans. on Power Electronics*, vol. 13, no. 5, Sept./Oct. 1998, pp. 987–985.

[474] A. Schonung and H. Stemmler, "Static frequency changers with subharmonic control in conjunction with reversible variable speed AC drives," *Brown Boveri Review*, 1964, pp. 555–577.

[475] R.H. Sepe and J.H. Lang, "Inverter non linearities and discrete-time vector current control," *IEEE Trans. on Industry Applications*, vol. 30, no. 1, Jan./Feb. 1994, pp. 62–70.

[476] P.F. Seixas, M.A. Severo Mendes, P. Donoso Garcia, and A.M.N. Lima, "A space vector PWM method for three-level voltage source inverters," in *Conf. Rec. Brazilian Conf. on Power Electronics* (*COBEP*), Brazil, 1999, pp. 187–191.

[477] J. Seo, C. Choi and D. Hyun, "A new simplified space–vector PWM method for three–level inverters," *IEEE Trans. on Power Electronics*, vol. 16, July 2001, pp. 545–550.

[478] J.-K. Seok and S.-K. Sul, "A new overmodulation strategy for induction motor drive using space vector PWM," in *Conf. Rec. Applied Power Electronics Conf. (APEC)*, Dallas, vol. 1, 1995, pp. 211–216.

[479] J.-K. Seok, J.-S. Kim, and S.-K. Sul, "Overmodulation strategy for high-performance torque control," *IEEE Trans. on Power Electronics*, vol. 13, no. 4, July 1998, pp. 784–792.

[480] J. Seung-Gi, L. Bang-Sup, K. Kyung-Seo, and P. Min-Ho, "The analysis and compensation of dead time effects," in *Conf. Rec. Int. Conf. on Industrial Electronics, Control and Instrumentation* (*IECON*), Singapore, 1988, pp. 667–671.

[481] M.A. Severo Mendes, Z.M. Assis Peixoto, and P. Donoso-Garcia, "A space vector PWM method for three-level flying-capacitor inverters," in *Conf. Rec. IEEE Power Electronics Specialists Conf. (PESC)*, Vancouver, 2001, vol. 1, pp. 182–187.

[482] M.A. Severo, P.F. Seixas, Z.M. Assis Peixoto, A.M. Lima, and B. G. de Andrade, "Improving the space vector PWM for the three-level flying-capacitor inverter," in *Conf. Rec. ElectrIMACS*, 2002, in CD ROM.

[483] S. Shekahawat and J. Dhyanchand, "Microprocessor-based sinusoidally sampled PWM technique for neutral-point clamped inverter," in *Conf. Rec. Power Electronics Specialists Conf. (PESC)*, Kyoto, 1988, pp. 50–58.

[484] N.P. Schibli, T. Nguyen, and A.C Rufer, "A three-phase multilevel converter for high-power induction motors," *IEEE Trans. on Power Electronics*, vol. 13, Sept. 1998, pp. 978–986.

[485] Y. Shrivastava, S.Y. Hui, and S. Sathiakumar, "Noise analysis of DC-AC random PWM schemes," *IEEE Trans. on Power Electronics*, vol. 14, no. 4, July 1999, pp. 761–770.

[486] Y. Shrivastava and S.Y.Hui, "Analysis of random PWM switching methods for three-level inverters," *IEEE Trans. on Power Electronics*, vol. 14, no. 6, Nov. 1999, pp. 1156–1163.

[487] Y. Shrivastava and H.S.H. Chung, "Comparison of RPWM and PWM space vector switching schemes for 3–level power inverters," in *Conf. Rec. 2001 IEEE Power Electronics Specialists Conf. (PESC)*, Vancouver, 2001, pp. 138–145.

[488] J.L. Shyu, T.J. Liang and J.F. Chen, "Digitally-controlled PWM inverter modulated by multi-random technique with fixed switching frequency," *IEE Proceedings – Electric Power Applications*, vol. 148, no. 1, Jan. 2001, pp. 62–68.

[489] G. Sinha and T.A. Lipo, "A four level rectifier-inverter drive system for drive applications," in *Conf. Rec. Industry Applications Society Annual Mtg.*, San Diego, 1996, pp. 980–987.

[490] G. Sinha and T.A. Lipo, "A new modulation strategy for improved DC bus utilization in hard and soft switched multilevel inverters,"in *Conf. Rec. of Int. Conf.*

on Industrial Electronics, Control and Instrumentation (IECON), New Orleans, 1997, pp. 670–675.

[491] G. Sinha and T.A. Lipo, "A four-level inverter based drive with a passive front end," *IEEE Trans. on Power Electronics*, vol. 15, no. 2, Mar. 2000, pp. 285–294.

[492] S. Sirisukprasert, J.-S. Lai nd T.-H. Liu, "Optimum harmonic reduction with a wide range of modulation indexes," *IEEE Trans. on Industrial Electronics*, vol. 49, no. 4, Aug. 2002, pp. 875–881.

[493] D. Soto and T.C. Green, "A comparison of high-power converter topologies for the implementation of FACTS controllers," *IEEE Trans. on Industrial Electronics*, vol. 49, no. 5, Oct. 2002, pp. 1072–1080.

[494] A.M. Stankovic, G.C Verghese, and D.J. Perreault, "Analysis and synthesis of randomized modulation schemes for power converters," *IEEE Trans. on Power Electronics*, vol. 10, Nov. 1995, pp. 680–693.

[495] A.V. Stankovic and T.A. Lipo, "A novel control method for input output harmonic elimination of the PWM boost type rectifier under unbalanced operating conditions," *IEEE Trans. on Power Electronics*, vol. 16, no. 5, Sept. 2001, pp. 603–611.

[496] R. Steele, *Delta Modulation Systems*, Pentech Press, London, 1975.

[497] V.R. Stefanovic and S.N. Vukosavic, "Space-vector PWM voltage control with optimized switching strategy," in *Conf. Rec. IEEE Industry Applications Society Annual Mtg.*, Houston, 1992, pp. 1025–1032.

[498] J.K. Steinke, "Control of a neutral-point-clamped PWM inverter for high power ac traction drives," in *Conf. Rec. Int. Conf. on Power Electronics and Variable Speed Drives*, London, 1988, pp. 214–217.

[499] J.K. Steinke, "PWM control of a three-level-inverter – principles and practical experience," in *Conf Rec. IEE Fourth Int. Conf. on Power Electronics and Variable Speed Drives*, London, 1990, pp. 98–103.

[500] J.K. Steinke, "Switching frequency optimal PWM control of a three level inverter," *IEEE Trans. on Power Electronics*, vol. 7, no. 3, July 1992, pp. 189–204.

[501] T. Sukegawa, K. Kamiyama, T. Matsui, and T. Okuyama, "Fully digital vector-controlled PWM VSI-fed AC drives with an inverter dead-time compensation strategy," in *Conf Rec. IEEE Industry Application Society Annual Meeting*, Pittsburgh, 1988, pp. 463–469.

[502] B.-S. Suh and D.-S. Hyun, "A new N-level high voltage inversion system," *IEEE Trans. on Industrial Electronics*, vol. 44, no. 1, Feb. 1997, pp. 107–115.

[503] B.-S. Suh, G. Sinha, M.D. Manjrekar, and T.A. Lipo, "Multilevel power conversion - An overview of topologies and modulations strategies," in *Conf. Rec. of the 6th International Conference on Optimization of Electrical and Electronic Equipment (OPTIM)*, Brazov Rumania, vol. 2, 1998, pp. 11–24.

[504] J.-H. Suh, C.-H. Choi, and D.-S. Hyun, "A new simplified space-vector PWM method for three-level inverters," in *Conf. Rec. IEEE Applied Power Electronics Conf. (APEC)*, Dallas, 1999, pp. 515–520.

[505] B.-S. Suh, C.-H. Choi, D.-W. Kang, D.-S. Hyun, and Y.-H. Lee, "An improved carrierwave-based SVPWM method using phase voltage redundancies for generalized cascaded multilevel inverter topology," in *Conf. Rec. 2000 Applied Power Electronics Conference and Exposition (APEC)*, New Orleans, 2000, vol. 1, pp. 542–548.

[506] Y.-S. Suh, P.G. Albano, V. Tijeras, and T.A. Lipo, "Unity power factor control of a three-pole converter under single-phase input," in *Conf. Rec. IEEE Power Electronics Specialists Conf. (PESC)*, Cairns, 2002, pp. 1425–1430.

[507] Y.-S. Suh, V. Tijeras, and T.A. Lipo, "A control method in dq synchronous frame for PWM boost rectifier under generalized unbalanced operating conditions," in *Conf. Rec. IEEE Power Electronics Specialists Conf. (PESC)*,Cairns, 2002, pp. 1425–1430.

[508] Y.-S. Suh, V. Tijeras, and T.A. Lipo, "A nonlinear control of the instantaneous power in dq synchronous frame for PWM AC/DC converter under generalized unbalanced operating conditions," in *Conf. Rec. Industrial Applications Society Annual Meeting*, Pittsburgh, 2002, pp. 1189–1196.

[509] T. Summers and R.E. Betz, "Dead-time issues in predictive current control," in *Conf. Rec. IEEE Industry Applications Society Annual Meeting*, Pittsburgh, 2002, pp. 2086–2093.

[510] J. Sun, S. Beineke, and H. Grostollen, "Optimal PWM based on real-time solution of harmonic elimination equations," *IEEE Trans. on Power Electronics*, vol. 11, no. 4, July 1996, pp. 612–621

[511] I. Takahashi and H. Mochikawa, "A new control of PWM inverter waveform for minimum loss operation of an induction motor drive," *IEEE Trans. on Industry Applications*, vol. IA–21, no. 4, May/June 1985, pp. 580–587.

[512] I. Takahashi and H. Mochikawa, "Optimum waveforms on an inverter for decreasing acoustic noise of an induction motor," in *IEEE Industrial Applications Society Annual Meeting Conf. Rec.* 1985, pp. 641–646.

[513] I. Takahashi and T. Noguchi, "A new quick-response and high-efficiency control strategy of an induction motor," *IEEE Trans. on Industry Applications*, vol. IA–22, no. 5, Sept./Oct. 1986, pp. 820–827.

[514] I. Takahashi and K. Iwaya, "100 kHz,10 KW switching type power amplifier using multilevel inverter," in *Conf. Rec. of IEEE:Power Electronics and Drives Systems Conf. (PEDS)*, 2001, vol. 1, pp. 286–291.

[515] I. Takahashi, S. Sekiguchi, and S. Miyari, "Control of PWM inverter output wave with reduced low-order harmonics," *Electrical Engineering* in *Japan*, vol. 97, no. 3, 1977, pp. 57–63.

[516] S. Tamai, "3–level GTO converter-inverter pair system for large capacity induction motor drive," in *Conf. Rec. European Power Electronics Conference, (EPE)*, Brighton, 1993, pp. 45–50.

[517] K. Taniguchi, S. Kaku, and H. Irie, "A three-phase sinusoidal PWM inverter," in *Conf. Rec. IEEE Industry Applications Society Annual Mtg.*, Toronto, 1985, pp. 1269–1273.

[518] K. Taniguchi, Y. Ogino, and H. Irie, "PWM technique for power mosfet inverter," *IEEE Trans. on Power Electronics,* vol. 3, no. 3, May/June 1988, pp. 328–334.

[519] S. Tenconi, M. Carpita, C. Bacigalupo, and R. Cali, "Multilevel voltage source converters for medium voltage adjustable speed drives," in *Conf. Rec. IEEE Int. Symposium on Industrial Electronics,* Athens, 1995, pp. 91–98.

[520] P. Tenti, L. Malesani, and L. Rossetto, "Optimum control of PWM multi converter systems," in *Conf. Rec. IEEE Industry Applications Society Annual Mtg.*, Pittsburgh, 1988, pp. 888–894.

[521] P. Tenti, A. Zuccato, L. Rossetto, and M. Bortolotto, "Optimum digital control of PWM rectifiers," in *Conf. Rec. Int. Conf. on Industrial Electronics, Control and Instrumentation* (*IECON*), Bologna, 1994, pp. 382-387.

[522] R. Teodorescu, F. Blaabjerg, J.K. Pedersen, E. Cengelci, S. Sulisijo, B. Woo, and P.N. Enjeti, "Multilevel converters – a survey," in *Conf. Rec. of European Power Electronics Conf.* (*EPE*) Lausanne, 1999, in CD–ROM

[523] R. Teodorescu, F. Blaabjerg, J.K. Pedersen, E. Cengelci, and P.N. Enjeti, "Multilevel inverter by cascading industrial VSI," *IEEE Trans. on Industrial Electronics*, vol. 49, no. 4, Aug. 2002, pp. 832–838.

[524] L.M. Tolbert and T.G. Habetler, "Novel multilevel inverter carrier-based PWM methods," *IEEE Trans. on Industry Applications*, vol. 35, no. 5, Sept./Oct. 1999, pp. 1098–1107.

[525] L.M. Tolbert, F.Z. Peng, and T.G. Habetler, "Multilevel PWM methods at low modulation indices," in *Conf. Rec. IEEE Applied Power Electronics Conf. and Exposition* (*APEC*), Dallas, 1999, pp. 1032–1038.

[526] L.M. Tolbert, F.Z. Peng, and T.G. Habetler, "Multilevel converters for large electric drives," *IEEE Trans. on Industry Applications*, vol. 35, no. 1, Jan./Feb. 1999, pp. 36–43.

[527] V.G. Torok, "Near-optimum on-line modulation of PWM inverters," in *Conf. Rec. IFAC Control* in *Power Electronics and Electrical Drives*, Lausanne, 1983, pp. 247–254.

[528] A.M. Trzynadlowski, "Non-sinusoidal modulating functions for three-phase inverters," in *Conf. Rec. Power Electronics Specialists Conf.* (*PESC*), Kyoto, 1988, pp. 477–484.

[529] A.M. Trzynadlowski, "An overview of modern PWM techniques for three-phase, voltage-controlled, voltage-source inverters," in *Conf. Rec. IEEE International Symposium on Industrial Electronics* (*ISIE'96*), Warsaw, 1996, pp. 25–39.

[530] A.M. Trzynadlowski, M.M. Bech, F. Blaabjerg, and J.J. Pedersen, "An integral space-vector PWM technique for DSP-controlled voltage-source inverters,"

IEEE Trans. on Industry Applications, vol. 35, no. 5, Sept./Oct. 1999, pp. 1091–1097.

[531] A.M. Trzynadlowski, M.M. Bech, F. Blaabjerg, J.K. Pedersen, R.L. Kirlin, and M. Zigliotto, "Optimization of switching frequencies in the limited-pool random space vector PWM strategy for inverter-fed drives," *IEEE Trans. on Power Electronics*, vol. 16, no. 6, Nov. 2002, pp. 852–857.

[532] A.M. Trzynadlowski and S. Legowski, "Minimum-loss vector PWM strategy for three phase inverters," *IEEE Trans. on Power Electronics,* vol. 9, no. 1, Jan. 1994, pp. 26–34.

[533] A.M. Trzynadlowski, R.L. Kirlin, and S.F. Legowski, "Space vector PWM technique with minimum switching losses and a variable pulse rate," *IEEE Trans. on Industrial Electronics*, vol. 44, no. 2, April 1997, 173–181.

[534] A.M. Trzynadlowski, S. Legowski, and R.L. Kirlin, "Random pulse width modulation technique for voltage controlled power inverters," in *Conf. Rec. IEEE Industrial Applications Society Annual Meeting,* Atlanta, 1987, pp. 863–868.

[535] A.M. Trzynadlowski, M. Zigliotto, S. Bolognani, and M.M. Bech, "Reduction of the electromagnetic interference conducted in the mains in inverter-fed AC drives using random pulse width modulation," in *Conf. Rec. Ind. Applic. Conf.*, St. Louis, 12–15 Oct. 1998, pp. 739–744.

[536] K.K. Tse, H.S-H. Chung, S.Y. Huo, and H.C. So, "Analysis and spectral characteristics of a spread-spectrum technique for conducted EMI suppression," *IEEE Trans. on Power Electronics*, vol. 15, no. 2, Mar. 2000, pp. 399–410.

[537] F.G. Turnbull, "Selected harmonic reduction in static DC-AC inverters," *IEEE Trans. Communication and Electronics*, vol. 83, July 1964, pp. 374–378.

[538] C. Turpin, P. Baudesson, F. Richardeau, F. Forest, and T.A. Meynard, "Fault management of multicell converters," *IEEE Trans. on Industrial Electronics*, vol. 49, no. 5, Oct. 2002, pp. 978–987.

[539] H.W. Van der Broeck, "Analysis of the harmonics in voltage fed converter drives caused by PWM schemes with discontinuous switching operation," in *Conf. Rec. European Power Electronics Conf. (EPE)*, Florence, 1991, pp. 3:261–3:266.

[540] H.W. Van der Broeck, H. Skudelny, and G. Stanke, "Analysis and realization of a pulse width modulator based on voltage space vectors," *IEEE Trans. on Industry Applications.*, vol. 24, no. 1, Jan./Feb., 1988, pp. 142–150.

[541] H.W. Van der Broeck and H.C. Skudelny, "Analytical analysis of the harmonic effects of a PWM ac drive," *IEEE Trans. on Power Electronics,* vol. 3, no. 2, Mar./April 1988, pp. 216–223.

[542] M. Veenstra and A. Rufer, "PWM-control of multi-level voltage-source inverters," in *Conf. Rec. IEEE Power Electronics Specialists Conf. (PESC)*, Galway, 2000, pp. 1387–1393.

[543] B. Velaerts, P. Mathys, E. Tatakis, and G. Bingen, "A novel approach to the generation and optimization of three-level PWM waveforms," in *Conf. Rec. Power Electronics Specialists Conf. (PESC)*, Kyoto, 1988, pp. 1255–1262.

[544] A. Veltman, P.P.J. van den Bosch, and R.J.A. Gorter, "On-line optimal switching patterns for 2–level and 3–level inverters using the fish method," in *Conf. Rec. IEEE Power Electronics Specialists Conf. (PESC)*, Seattle, 1993, pp. 1061–1067.

[545] G. Venkataramanan and A. Bendre, "Reciprocity-transposition-based sinusoidal pulsewidth modulation for diode-clamped multilevel converters," *IEEE Trans. on Industrial Electronics*, vol. 49, no. 5, Oct. 2002, pp. 1035–1047.

[546] M. Venturini and A. Alesina, "The generalized transformer: a new bi-directional sinusoidal waveform frequency converter with continuously adjustable input power factor.," in *Conf. Rec. IEEE Industry Applications Society Annual Mtg.*, 1980, pp. 242–252.

[547] D. Vincenti, P.D. Ziogas, and R.V. Patel, "An on-line pc-based converter PWM modulator," in *Conf. Rec. Industry Applications Society Annual Mtg.*, Pittsburgh, 1988, pp. 895–902.

[548] D. Vincenti and J. Jin, "A three-phase regulated PWM rectifier with on-line feedforward input unbalance correction," *IEEE Trans. on Industrial Electronics,* vol. 41, no. 5, Oct. 1994, pp. 526–532.

[549] A. von Jouanne, S. Dai, and H. Zang, "A multilevel inverter approach providing DC-link balancing, ride through enhancement and common-mode voltage elimination," *IEEE Trans. on Industrial Electronics*, vol. 49, no. 4, Aug. 2002, pp. 739–745.

[550] A.M. Walczyna, "On reduction of harmonic reactive distortions and subharmonics of drives with VSI-fed induction motors controlled by direct torque and flux control methods," in *Conf. Rec. IEEE Power Electronics Specialists Conf. (PESC)*, Atlanta, 1995, pp. 408–414.

[551] G.R. Walker, "Modulation and control of multilevel converters," Ph.D. Thesis, The University of Queensland, Australia, July 1998.

[552] G.R. Walker and G. Ledwich, "Bandwidth considerations for multilevel converters," *IEEE Trans. on Power Electronics*, vol. 14, no. 1, Jan. 1999, pp. 74–81.

[553] F. R. Walsh, J. F. Moynihan, P. J. Roche, M. G. Egan, and J. M. D. Murphy, "Analysis and influence of modulation scheme on the sizing of the input filter in a PWM rectifier system., in *Conf. Rec. of the European Power Electronics Conf. (EPE)*, Trondheim, 1997, pp. 2.929–2.933.

[554] F. Wang, "Sine-triangle vs. space vector modulation for three-level PWM voltage source inverters," in *Conf. Rec. IEEE Industry Applications Society Annual Mtg.*, Rome, 2000, pp. 2482–2488.

[555] Y. Wang and H. Grotstellen, "Control strategies for the discontinuous current mode of AC drives with PWM inverters," in *Conf. Rec. European Conf. on Power Electronics and Applications (EPE)*, Florence, 1991, pp. 3/217–222.

[556] G.N. Watson, *Theory of Bessel Functions*, Cambridge University Press, Cambridge England, 1944.

[557] P. Wood, *Switching Power Converters*, Van Nostrand Reinhold, New York, 1981.

[558] R. Wu, S.B. Dewan, and G.R. Slemon, "A PWM ac-to-ac converter with fixed switching frequency," *IEEE Trans. on Industry Applications*, vol. 26, no. 5, Sept./Oct. 1990, pp. 880–885.

[559] D. Wuest and F. Jenni, "Space vector based current control schemes for voltage source inverters," in *Conf. Rec. IEEE Power Electronics Specialists Conf. (PESC)*, Seattle, 1993, pp. 986–992.

[560] K. Yamamoto and K. Shinohara, "Analysis of zero-current-clamping phenomena during switching dead-time in PWM inverter," in *Conf. Rec Int. Power Electronics Conf.*, Yokohama, 1995, pp. 1519–1524.

[561] K. Yamamoto and K. Shinohara, "Comparison between space vector modulation and subharmonic methods for current harmonics of DSP-based permanent magnet AC servo motor drive system," *IEE–Electr. Power Appl.*, vol. 143. no.2, Mar. 1996, pp. 151–156.

[562] K. Yamamoto, K. Shinohara, and H. Ohga, "Effect of parasitic capacitance of power device on output voltage deviation during switching dead-time in voltage-fed PWM inverter," in *Conf. Rec. IEEE Power Conversion Conf.*, Yokohama, pp. 777–782.

[563] X. Yuan and I. Barbi, "Fundamentals of a new diode clamping multilevel inverter," *IEEE Trans. on Power Electronics,* vol. 15, no. 4, July 2000, pp. 711–718.

[564] X. Yuan, H. Stemmler, and I. Barbi, "Investigation on the clamping voltage self-balancing of the three-level capacitor clamping inverter," in *Conf. Rec. Power Electronics Specialists Conf. (PESC)*, Charleston, 1999, pp. 1059–1064.

[565] J. H. Youn and B. H. Kwon, "An effective software implementation of the Space-Vector Modulation," *IEEE Trans. on Ind. Electronics,* vol. 46, no 4,, Aug. 1999, pp. 866–868.

[566] F. Zach and H. Ertl, "Efficiency optimal control for ac drives with PWM inverters," *IEEE Trans. on Industry Applications,* vol. IA–21, no. 4, July/Aug. 1985, pp. 987–1000.

[567] F. Zach and F.A. Theil, "Pulse width modulated (PWM) inverters for efficiency optimal control of ac drives – Switching angles and efficiency/loss profiles," in *Conf. Rec. IFAC Symposium,* Lausanne, 1983, pp. 231–238.

[568] F. Zach, R. Martinez, S. Keplinger, and A. Selser, "Dynamically optimal switching patterns for PWM inverter drives (for minimization of torque and speed ripple)," *IEEE Trans. on Industry Applications*, vol. IA–21, no. 4, July/Aug. 1985, pp. 975–986.

[569] K. Zakrzewski, W. Kubiak, J. Szulakowski, and F. Sobczak, "Load evaluation for standard induction motors fed from a PWM converter – analytical method based on experiment," in *Conf. Rec. Int. Conf. on Electrical Machines (ICEM)*, 2002, in CD ROM.

[570] F. Zare and G. Ledwich, "Space vector modulation technique with reduced switching losses," in *Conf. Rec. European Power Electronics Conf. (EPE)*, Lausanne, 1999, in CD ROM.

[571] R. Zhang, D. Boroyevich, V. H. Prasad, H. Mao, F.C. Lee, and S. Dubovsky, "A three-phase inverter with a neutral leg with space vector modulation," in *Conf. Rec. IEEE Applied Power Electronics Conf. and Exposition (APEC)*, Atlanta, 1997, pp. 857–863.

[572] H. Zhang, A. von Jouanne, S. Dai, A.K. Wallace, and F. Wang, "Multilevel inverter modulation schemes to eliminate common-mode voltages," *IEEE Trans. on Industry Applications,* vol. 36, no. 6, Nov./Dec. 2000, pp. 1645–1653 (corrections in vol. 37, no. 1, p3).

[573] M. Zigliotto and A.M. Trzynadlowski, "Effective random space vector modulation for EMI reduction in low-cost PWM inverters," in *Conf. Rec. IEE Int. Conf. on Power Electronics and Variable Speed Drives*, London, 21–23 Sept. 1998, pp. 163–168.

[574] P.D. Ziogas, L. Moran, G. Joos, and D. Vincenti, "A refined scheme for voltage and current source converters," in *Conf. Rec. IEEE Industry Applications Society Annual Mtg.*, Seattle, 1990, pp. 977–983.

[575] D.N. Zmood and D.G. Holmes, "Generalised approach to the modulation of current source inverters," in *Conf. Rec. IEEE Power Electronics Specialist Conf. (PESC)*, Fukuoka, 1998, pp. 739–745.

[576] D.N. Zmood and D.G. Holmes, "Practical performance limitations for PWM strategies," in *Conf. Rec. IEEE Industry Applications Society Annual Mtg.*, Pittsburgh, 1988, pp. 1245–1252.

[577] D.N. Zmood, "A systematic development of improved linear regulators for sinusoidal power converters," Ph.D. Thesis, Monash University, Australia, 2002.

[578] J. Zubek, A. Abbondanti, and C.J. Nordby, "Pulse width modulated inverter motor drives with improved modulation," *Trans. on IEEE Industry Applications*, vol. IA–75, no. 6, Nov./Dec. 1975, pp. 695–703.

Index

A

AC resistance, *see* Induction motor
Alternative phase opposition disposition (APOD)
  see also Contour plot
  three-level
    integral limits, 482
    principles, 481–484
  five-level
    carrier, 470
    integral limits, 495
    principles, 492–497
Analytical harmonic solutions
  multilevel cascaded inverter
    phase-shifted cascaded modulation
      naturally sampled, 456
      overmodulated, naturally sampled, 465
      regularly sampled, asymmetrical, 458
      three-phase *l–l*, naturally sampled, 458
      three-phase *l–l*, regularly sampled, asymmetrical, 460
  multilevel diode-clamped inverter
    alternative phase opposition disposition modulation
      three-level, 484
      five-level, 494
      seven-level, 514
      higher-levels, 502
    phase disposition modulation
      three-level, 479
      three-level, overmodulated, 488
      five-level, 492
      higher-levels, 499–501
    phase opposition disposition modulation
      three-level, 484
      five-level, 498
  multilevel hybrid inverter, 512, 514
  one phase leg
    naturally sampled modulation
      double-edge, 118
      single-edge, 111
    overmodulation
      naturally sampled, 353–354
      regularly sampled, asymmetrical, 357–358
      regularly sampled, symmetrical, 356–357
Analytical harmonic solutions (*continued*)
    regularly sampled modulation
      double-edge, asymmetrical, 144
      double-edge, symmetrical, 137
      single-edge, 133
  single-phase inverter
    naturally sampled modulation
      double-edge, 159–160, 162
      double-edge, discontinuous, 192–193
      revised for multilevel cascaded inverter, 453
      single-edge, 162
    overmodulation, naturally sampled, 360–361
    regularly sampled modulation
      double-edge, asymmetrical, 163
      double-edge, discontinuous, both phase legs switched, 196
      double-edge, discontinuous, one phase leg switched, 203, 207
      double-edge, symmetrical, 163
      single-edge, 162
  three-phase inverter
    120° discontinuous modulation
      naturally sampled, 316, 318, 321
      regularly sampled, 322
    naturally sampled modulation
      sinusoidal reference, double-edge, 219–220
    regular sampled modulation
      sinusoidal reference, asymmetrical, 223
    regularly sampled modulation, third harmonic injection
      double-edge, asymmetrical, 238
      double-edge, symmetrical, 237
      single-edge, 237
    space vector modulation
      naturally sampled, 282
      regularly sampled, asymmetrical, 285
      regularly sampled, symmetrical, 284
Asymmetrical regular sampling, advantages compared to symmetrical sampling, 168

B

Bar factor, 83
Baseband harmonics, *definition*, 103

Bidirectional chopper, 4
Bootstrap gate power supply, 566
Break frequency ratio, induction motor resistance and inductance, 78
Break frequency, *definition*, 77

C

Capacitor voltage balance, multilevel inverters, 435, 438
Capacitor-clamped multilevel inverter structure, 49–51
Carrier
 three-level
  APOD/POD modulation, 482
  PD modulation, 472
 five-level
  APOD modulation, 494
  diode-clamped multilevel inverter, 470
  PD modulation, 491
  POD modulation, 497
 harmonics, *definition*, 104
 maximum slope, 367
 phase-shifted, multilevel cascaded inverter, 455
 sawtooth, 106
 triangular, 114
Carrier/fundamental frequency ratio, 106
 integer versus non-integer, 148–149
Cascaded voltage source multilevel inverter
 see Phase-shifted cascaded modulation (PSCPWM)
 structure, 51–54, 438–439
Chopper, 3–4
Clamped waveform sections
 discontinuous modulation, three-phase inverter, 306
Common mode elimination, *see* Pulse width modulation strategies, common mode elimination
Compensated modulation, *see* Modulation, compensated
Contour plot
 minimum-loss modulation, effective weighted total harmonic distortion, 428, 429
 multilevel hybrid inverter, 513
 multilevel inverter
  three-level
   APOD/PD modulation, 482
   PD modulation, 472
   PD overmodulation, 486
  five-level
   APOD modulation, 494
   PD modulation, 491
Contour plot (*continued*)
   POD modulation, 498
 optimum modulation, weighted total harmonic distortion, 418, 421
Converter
 controller details, 569–572
 elements of, 557–561
Current clamping, dead time, 619
Current error
 optimized space vector modulation, two-level inverter, 393, 394
Current source inverter, 21–23
 autosequentially commutated, 23
 control logic signals, 340
 equivalent zero states
  sawtooth carrier, 342
  triangular carrier, 341
 switching continuity constraints, 21

D

DC link voltage, *definition*, 6
DC link voltage, variation with AC voltage unbalance, 597
Dead time
 definition, 607
 compensation, 612–618
 phasor representation, 615
 voltage deviation, 613, 616
Deep bar effect, 73
Diode, inverse parallel, 5
Diode-clamped multilevel inverter
 see Alternative phase opposition disposition (APOD)
 see Phase disposition (PD)
 see Phase opposition disposition (POD)
 structure, 42–49, 434–437
Direct modulation
 definition, 96
 evaluation of, 146
Discontinuous modulation
 multilevel inverter, 548–550
 single-phase inverter
  both phase legs switched, 186–198
  one phase leg switched, 200–207
 three-phase inverter
  120° switching, 299–301
  30° switching, 302–306
  60° switching, 302–306
  clamped waveform sections, 306
  phase leg references
   120° switching, 309, 313, 314
   30° switching, 312–314
   60° switching, 310, 311, 313–315
  switched waveforms, 301, 302, 304, 305

Double-edge modulation
naturally sampled, 115
regularly sampled, 126
*d–q–*0 stationary reference frame, *see* Stationary reference frame
DSP, event manager (EV) system, 576
Dwell time, 607

E

Effective modulation index
carrier based overmodulation
60° discontinuous reference, 371
sinusoidal reference, 365
space vector reference, 369
versus modulation index
60° discontinuous reference, 372
sinusoidal reference, 366
space vector overmodulation, 381
space vector reference, 370
Effective weighted total harmonic distortion
minimum-loss modulation, use in, 427
Electromagnetic interference (EMI), 563
Equalization of voltage and current stress, *see* Switching loss, equalization

F

Figures of merit
square-wave switching, 72
weighted total harmonic distortion, 93
Flux trajectory, space vector, 544
Flying-capacitor inverter, *see* Capacitor-clamped multilevel inverter
Fortesque zero sequence component, 29
Four phase leg inverter, *see* Pulse width modulation strategies, four phase leg inverter
Fourier coefficients
multilevel inverter, block switching, 440
Fourier transform, application to pulse width modulation, 102

G

Gain, during overmodulation, 365, 371
inverse, 373
versus modulation index, 366, 370, 372
Gate power supply, bootstrap, 566
Gaussian distribution, random carrier period, 590

H

Half-bridge inverter, topology, 6
Harmonic cancellation, phase-shifted cascaded modulation, 456, 457, 459, 461, 462, 464
Harmonic components of switched waveform, 102
Harmonic distortion factor
see also Weighted total harmonic distortion
current, 61–64
one phase leg, 173, 174
single-phase inverter, three-level modulation, 174, 177
three-phase inverter, 247, 248
comparison, 328
discontinuous modulation strategies, 327
space vector modulation, 294
summary, 327
variation with modulation depth, 248
voltage, 57–61
weighted total harmonic distortion, relationship with, 250
Harmonic elimination
generalized output waveform, 397
multilevel inverter
equal voltage levels, 440–441
optimum switching angles
equal voltage levels, 441
unequal voltage levels, 447
solution results, individual, 442
unequal voltage levels, 443–446
principles, 396–407
two-level inverter
elimination of 5th and 7th harmonics, solution, 402
solution results, individual, 404, 406, 409–413
solution results, summary, 408
use in overmodulation region, 414
weighted total harmonic distortion, revised, 404–405
Harmonic losses, calculation
one phase leg, 169–173
single-phase inverter, 174–177
three-phase inverter
discontinuous modulation, 326–330
sine-triangle modulation, 241–248
space vector modulation, 291–294
Harmonic solutions, *see* Analytical harmonic solutions
Harmonic spectra
current source inverter, experimental, 344, 346–348
four phase leg inverter, 605
multilevel cascaded inverter

Harmonic spectra (*continued*)
centered PD modulation, experimental, 551, 552
discontinuous modulation, experimental, 553
equivalent PD modulation
five-level, experimental, 509
phase-shifted modulation
experimental, 464
naturally sampled, 457, 461
overmodulated, 468
regularly sampled, asymmetrical, 459, 462
multilevel diode-clamped inverter
three-level
APOD/POD modulation, 485
PD modulation, 480
PD overmodulation, 490
five-level
APOD modulation, 496
APOD modulation with third-harmonic injection, 524
PD modulation, 493
PD modulation with third-harmonic injection, 523
POD modulation, 500
POD modulation with third-harmonic injection, 525
seven-level
PD modulation, 503
multilevel hybrid inverter
APOD modulation, 515
APOD modulation, experimental, 516
PD equivalent modulation
experimental, 521
theoretical, 520
PD modulation, comparison with third-harmonic injection, 527
one phase leg
direct modulation, 148
naturally sampled modulation
double edge, 119
single-edge, 112
overmodulation
naturally sampled, 355
regularly sampled, asymmetrical, 359
regularly sampled, symmetrical, 359
regularly sampled modulation
double-edge, asymmetrical, 145
double-edge, symmetrical, 138
single-edge, 134
single-phase inverter
naturally sampled modulation
double-edge, 161

Harmonic spectra (*continued*)
double-edge, discontinuous, both phase legs switched, 194
double-edge, discontinuous, one phase leg switched, 205, 208
single-edge, 164
overmodulation, naturally sampled, 362
regularly sampled modulation
double-edge, asymmetrical, 167
double-edge, discontinuous, asymmetrical, 199
double-edge, discontinuous, symmetrical, 197
double-edge, symmetrical, 166
double-edge, two- and three- level switching, comparison, 212
single-edge, 165
three-phase inverter
comparison
sinusoidal reference, 224
space vector modulation, 273
third-harmonic injection, 239
discontinuous modulation
120° switching, 323
60° switching, 324
alternative zero vector position, 334
comparison, 325
experimental, 255, 256
naturally sampled modulation
sinusoidal reference, 221
third-harmonic injection, 235
overmodulation, naturally sampled, 362
space vector modulation
naturally sampled, 286
regularly sampled, asymmetrical, 288
regularly sampled, symmetrical, 287
H-bridge single-phase inverter, *see* Single-phase full-bridge inverter
Hexagon space vector locus, two-level voltage source inverter, 33
Hexagonal coordinate system, space vector, 538–543
Hybrid voltage source inverter
structure, 54–55, 510
voltage switching levels, 510
Hyperbolic distribution, random carrier period, 590

I

IGCT and IGBT, use in hybrid inverter, 54
Inductance
induction motor, 63
power stage, parasitic, 565

Induction motor
AC resistance and inductance
nonrectangular rotor bar, 78
rectangular rotor bar, 76, 77
bar factor, 83
equivalent inductance, 63
frequency dependence
rotor leakage inductance, 84
rotor parameters, frequency regions, 84
rotor resistance, 82
rectangular rotor bar, analysis, 73–78
rotor referred equivalent circuit, 81
Integral limits
multilevel diode-clamped inverter
three-level APOD modulation, 482
three-level POD modulation, 482
five-level APOD modulation, 495
five-level PD modulation, 489
five-level POD modulation, 497
three-level PD modulation, 472
one phase leg
naturally sampled modulation
double-edge, 115
single-edge, 107
overmodulation, 352
regularly sampled modulation
double-edge, asymmetrical, 139
double-edge, symmetrical, 135
single-edge, 131
single-phase inverter
naturally sampled modulation
double-edge, discontinuous, both phase legs switched, 187
double-edge, discontinuous, one phase legs switched, 201
regularly sampled modulation
discontinuous, both phase legs switched, 195
three-phase inverter
120^0 modulation, 315
space vector modulation
naturally sampled, 274
regular sampled, asymmetrical, 283
regular sampled, symmetrical, 282
Inverse Gain, *see* Gain, during overmodulation, inverse

L

Losses, switching, *see* Switching loss, equalization

M

Maximum modulation index
single-phase inverter, 157
three-phase inverter
sinusoidal reference, 218
space vector modulation, 262
third harmonic injection, 229
Minimum pulse width, effect of, 607–611
Minimum-loss modulation, 421–429
effective weighted total harmonic distortion, 427
variation with α, 428, 429
harmonic loss coefficient, 427
multilevel inverter, 447–448
solution results, individual, 447, 448
solution results, summary, 449
two-level inverter
solution results, composite, 429
Modulation gain, during overmodulation, *see* Gain, during overmodulation
Modulation index (depth), *definition*, 105
Modulation index, effective during overmodulation, *see* Effective modulation index
Modulation strategies, *see* Pulse width modulation strategies
Modulation, compensated, 373–375
Multilevel inverters, topologies, 42–55, 434–439

N

Naturally sampled pulse width modulation
definition, 96
current source inverter, 343
principles, 105–107
Negative sequence voltage, 592, 593
Neutral-point clamped inverter
topology, 38
Newton's method, application to harmonic elimination, 399–401
NPC, *see* Neutral-point clamped inverter

O

Objectives, pulse width modulation, 96
Offset, space vector modulation
multilevel inverter
reference voltage, 537
zero, 547
two-level inverter, zero, 271, 545
Optical isolation, power stage, 567
Optimized space vector modulation, *see* Space vector modulation, optimized, two-level

Optimum modulation, 416–421
solution results
composite, 425
individual, 416–424
switching angles for minimum weighted total harmonic distortion, 418
weighted total harmonic distortion, variation with α, 418, 421
Output reference voltage, *see* Reference voltage, target
Overmodulation
compensated phase-to-neutral voltages, 375
gain
60° reference, 371–372
sinusoidal reference, 364–367
space vector reference, 367–371
region, 350
space vector approach, 376–381
strategies for various levels of overmodulation, 381

P

Performance comparison
pulse width modulation strategies
one phase leg, 150
three-phase inverter, 290
Performance index, optimum modulation, 415
Phase disposition (PD)
see also Contour plot
three-level
principles, 469–479
five-level
carrier, 470
integral limits, 489
principles, 489–492
three-level
integral limits, 472
equivalent for cascaded inverters, 504–505
Phase opposition disposition (POD)
see also Contour plot
three-level
integral limits, 482
principles, 481–484
five-level
carrier, 470
integral limits, 497
principles, 497–499
Phase shift modulation, 17–19
harmonic components, 19
Phase-shifted cascaded modulation (PSCPWM), 453–467
Phase-to-neutral voltages, overmodulated, after compensation, 375
Phasor angle, calculation of, 583
Placement, 294
Positive sequence voltage, 592, 593
Power converter, structure, *see* Converter, elements of
Power electronic converters, families of, 2
Power supply, converter controller, 567
Pulse placement, best and worst case, 184
Pulse ratio, *see* Carrier/fundamental frequency ratio
Pulse sequence, *see* Sequence, pulse
Pulse width modulation
evaluation, 97–98
Fourier analysis, 99–102
fundamental concepts, 96–97
hardware implementation, 572–579
objectives, 96
software implementation, 579–583
three-phase inverter, *l–l* harmonic cancellation, 220
timer calculation, 581
timer logic structure, 576, 577
Pulse width modulation strategies
see also Space vector modulation
common mode elimination, 598–602
four phase leg inverter, 603–606
Quad transformation, 604
harmonic elimination in overmodulation region, use of, 414
harmonic elimination, *see* Harmonic elimination
hybrid multilevel inverter, 507–517
equivalent PD modulation, 517–519
minimum-loss modulation, *see* Minimum-loss modulation
multilevel inverter
alternative phase opposition disposition (APOD), *see* Alternative phase opposition disposition (APOD)
phase disposition (PD), *see* Phase disposition (PD)
phase opposition disposition (POD), *see* Phase opposition disposition (POD)
phase-shifted cascaded modulation (PSCPWM), 453–467
third-harmonic injection, 519–526
variable modulation index, 526–528
one phase leg
direct modulation, 146–148
duty cycle variation
sine-sawtooth, 120–123
sine-triangle, 123–125

Pulse width modulation strategies (*continued*)
naturally sampled modulation
sine-sawtooth, 105–111
sine-triangle, 114–118
performance comparison, 150
regularly sampled modulation
sine-sawtooth, 130–133
sine-triangle, asymmetrical, 139–144
sine-triangle, symmetrical, 134–137
optimum modulation, *see* Optimum modulation
random carrier period, 586–590
three-phase inverter
discontinuous switching, 299–311
optimized space vector, 384–395
sinusoidal reference, 216–225
space vector, 259–290
third harmonic injection, 226–240
Pulse width, minimum, *see* Minimum pulse width, effect of
PWM controller, compensated for overmodulation, 374
PWM, *see* Pulse width modulation, Pulse width modulation strategies

Q

Quad transformation, 604
Quasi-square-wave switching, *see* Six-step switching

R

Random pulse width modulation, *see* Pulse width modulation strategies, random carrier period
Reference voltage, target
dead time, effect of, 615
four phase leg inverter, 603
multilevel cascaded inverter, equivalent PD modulation, 506, 507
multilevel hybrid inverter, 513, 518
multilevel inverter
discontinuous modulation, 549
one phase leg, 105
overmodulation
60° discontinuous reference, 371
sinusoidal reference, 351
space vector reference, 368
phasor angle, calculation of, 583
single-phase inverter, 157
three-phase inverter
discontinuous modulation, 309–314
sinusoidal reference, 216, 218
space vector modulation, 269
third harmonic injection, 227
Regularly sampled pulse width modulation
definition, 96
principles, 125–129
sampling delay correction, 127, 129
Resolution, phasor angle, 583
Root-mean-square (RMS)
definition, 13, 57
expressed using harmonics, 58
expressed using THD, 59

S

Sampling delay correction
optimized space vector modulation, 393
practical implementation, 581
regularly sampled pulse width modulation, 127, 129
Sawtooth carrier, 106
Sequence, pulse
discontinuous modulation, 331
space vector modulation, 331
Sideband harmonics, *definition*, 104
Sideband modulation, 177–182
variation of weighted total harmonic distortion with frequency dither, 182
Single-edge modulation
naturally sampled, 107
regularly sampled, 126
Single-phase full-bridge inverter
three-level modulation, 157–160
topology, 7, 156
two-level modulation, 207–211
Six-step switching
fundamental voltage, 13
harmonic components, 12
voltage waveforms, 11
Skin depth, 76
Skin effect, 77
Space vector diagram
three-level inverter, 598
current source inverter, 338
diode-clamped multilevel inverter
three- and five-level, 532
four phase leg inverter, 607
hexagon decomposition, 535, 538
reduced common mode
three-level inverter, 601
four-level inverter, 602
five-level inverter, 601
voltage source inverter, two-level, 261
Space vector modulation
concept, 25
multilevel inverter
hexagonal co-ordinate system, 538–543
offset adjustment, reference voltage, 537

Space vector modulation (*continued*)
optimal position, space vector, 543–544
optimized sequences, 531–534
optimized, two-level
current error, 393, 394
optimal switching intervals, 390–392
optimal switching time, 389
principles, 384–395
sampling delay correction, 393
weighted total harmonic distortion, comparison with conventional, 395
sequences, optimized, 532
two-level inverter
active times, 261, 264
active times during overmodulation, 379
diagram, 31, 261
integral limits, 274
maximum modulation index, 262
naturally sampled, 270–272
overmodulation strategies, 381
principles, 259–267
references, phase leg, 269, 270
relationship to DC link voltage, 35, 263
relationship to phase variables, 34
switch combinations, 31, 260
vector sequence, 266
zero offset, 271
zero vector placement, 294–298
Spectral weighting factor, use in WTHD, 69
Square-wave switching
figures of merit, 72
see also Six-step switching, Total harmonic distortion, Weighted total harmonic distortion
Stationary reference frame
definition, 25
transformation to/from three-phase variables, 29
Step-down chopper, 3
Step-up chopper, 4
Subharmonics, 253–256
Switching
fundamental constraints, 2
power stage, 564
pulse placement, best and worst case, 184
representation as functions, 14
three-phase inverter
phase leg switching times, 222
space vector switching times, 261
Switching cells, 4–5, 21
Switching function, unbalanced AC voltages, 596
Switching instants, *see* Integral limits
Switching intervals
optimized space vector modulation, 390–392
Switching loss, equalization
block switching, 441–442, 445, 446
multilevel cascaded inverter, 505, 507
Switching waveforms
dead time
compensation for, 617
effect of, 613, 614
four phase leg inverter, 606
harmonic elimination, 397
multilevel cascaded inverter
centered PD modulation, experimental, 551, 552
discontinuous modulation, experimental, 553
equivalent PD modulation, experimental, 508
phase-shifted modulation experimental, 463
multilevel hybrid inverter, 511, 522
multilevel inverter
block switching
cascaded inverter, 440, 445, 446
general arrangement, 439
effect of minimum pulse width, 609, 610, 611
harmonic elimination
five-level switching, 443
seven-level switching, 444
one phase leg
overmodulation, 350
random carrier period, 587
single-phase inverter
discontinuous modulation, both phase legs switched, 188
discontinuous modulation, one phase leg switched, 202
discontinuous modulation, one phase leg switched, carrier reversed, 206
three-level modulation, 158
two-level modulation, 209
three-phase inverter
120° discontinuous modulation, 301, 307
30° discontinuous modulation, 305
60° discontinuous modulation, 302, 304
naturally sampled modulation, 217, 218
space vector modulation, 266
Synchronous reference frame, 35–38

T

Target reference voltage, *see* Reference voltage, target
THD, *see* Total harmonic distortion
Third-harmonic injection
  alternative reference magnitudes, 230
  maximum modulation index, 229
  multilevel inverter modulation, 519–526
Three level modulation, *see* Single-phase full-bridge inverter, three-level modulation
Three-phase inverter
  carrier modulation
    phase-leg switching times, 222
    sinusoidal reference, 216–225
    third harmonic injection, 226–240
  space vector modulation, 259–272
  topology, 216
TMS320F240 EV system, 576
Total harmonic distortion
  current, *definition*, 62
  expressed using harmonics, 59
  expressed using RMS voltage, 59
  performance indicator, use as, 67
  phase shift voltage control, 60, 62
  six-step (three-phase), 64
  square-wave (single-phase), 60
  three-level inverter, square-wave switching, 66
  voltage, *definition*, 58
Trailing-edge modulation, *see* Single-edge modulation
Triangular carrier, 114
Triplen carrier ratios, 251–253
Triplen harmonics, *l–l* cancellation, 12
Two-level modulation, single-phase inverter, *see* Single-phase full-bridge inverter, two-level modulation
Two-phase inverter, topology, 8

U

Unbalanced AC voltages, rectifier, 590–597
Unit cell
  general, 100
  two-level modulation
    double-edge, 114
    overmodulation, 352
    single-edge, 106
Up-down chopper, 4

V

Vector transformation, to synchronous rotating frame, 36
Voltage control
  double bridge, 19
  multilevel inverter, block switching, 436–437, 439
  phase shift modulation, *see* Phase shift modulation
  Volts/Hertz criterion, 17
Voltage reference point, *definition*
  negative of DC link, 9
  neutral of star-connected load, 9
  positive of DC link, 9
  zero of DC link, 9
Voltage source inverter
  four phase leg bridge, 603
  single-phase bridge, topology, 156
  switching continuity constraints, 14
  three-level inverter, 38–42
  three-phase bridge, topology, 9, 216
  three-phase double bridge, 20
Voltage stiff inverter, *see* Voltage source inverter
Volts/Hz criterion, 17
Volt–second average, 96
VSC, *see* Multilevel inverters

W

Weighted total harmonic distortion
  definitions, 63, 85, 89, 92
  see also Harmonic distortion factor
  closed form, 70
  harmonic distortion factor, relationship with, 250
  harmonic elimination, revised for, 404–405
  induction motor, effect of frequency dependent parameters, 85, 86, 89
  normalized against V_{dc}, 92
  optimized space vector modulation, comparison with conventional, 395
  performance indicator, use as, 69
  phase shift voltage control, effect of, 63
  pulse width modulation of
    multilevel inverter, variation with modulation index, 529
    one phase leg, variation with modulation strategy, 150
      pulse ratio, 113
    single-phase inverter, variation with modulation strategy, 169
      sideband modulation, 182
    three-phase inverter, variation with pulse ratio, 251
  square-wave switching
    single-phase inverter, 63

Weighted total harmonic distortion (*continued*)
three-level inverter, 72
three-phase current source inverter, 66
three-phase voltage source inverter (six-step), 65
WTHD
definition, 63
see also Weighted total harmonic distortion
WTHD0, *definition*, 92
WTHD01, *definition*, 92
WTHD02, *definition*, 92
WTHD1, *definition*, 89
WTHD2, *definition*, 85

Z

Zero component, d–q plane, 25
Zero offset, space vector modulation
multilevel inverter, 547, 548
two-level, 271
Zero space vector, placement, 294–298
Zero states, current source inverter, 341, 342